AF816869

Cover Page Designed by:

Ms. Gowri Vemuri

First Edition

May, 2019

Author:

MATH-KNOTS

Index

Preface

The word "Science"brings to mind many different pictures: a fat textbook, bunch of equations, space shuttle, medicines and microscopes etc., simply discovering and learning about the nature around us. There are many branches of science and physics the most basic area of sciences, which deals with energy, matter, space, time and their interactions. The word physics is derived from the Latin word *physica*, which means "natural thing."

Physics as a fundamental science helps to uncover the laws of nature. The language of expression is mathematics. During ancient times, humans studied Stars and celestial bodies that lead to the development of disciplines like astronomy and mathematics. Based on size and speed, Physics as a subject is categorized several categories, which you can find from the following picture.

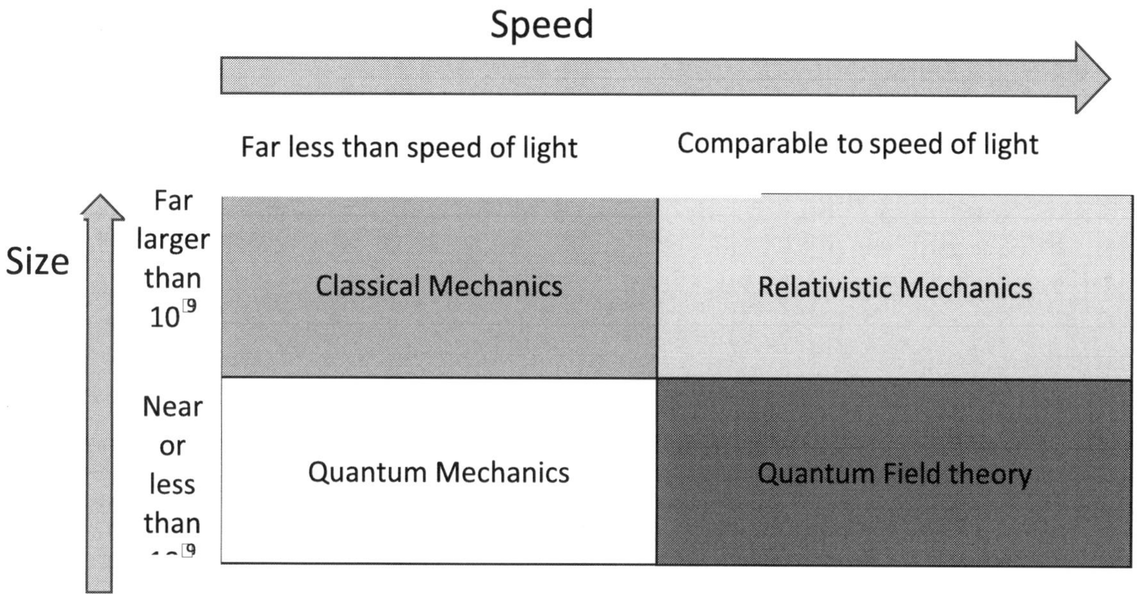

There are many ways to learn physics, but often students face difficulty in solving a problem by the concept they have learnt in the class. Our book is designed to help students in this area. Each chapter in the book, provides a summary of a concept, along with many practice problems.

Chapter - 1

(PART - 1)

ELECTRO MAGNETIC INDUCTION

ELECTRO MAGNETIC INDUCTION

SYNOPSIS

- The phenomenon in which electric current is induced by varying magnetic fields is called electromagnetic induction.

- **Magnetic Flux (ϕ) :** The number of magnetic lines of force passing normally through given area is called magnetic flux.

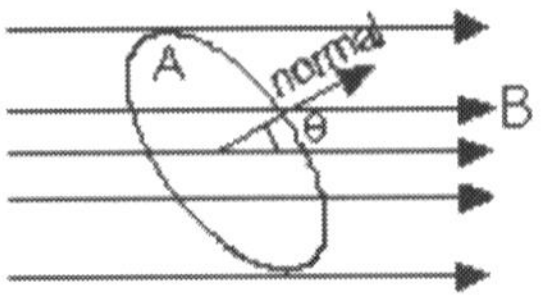

When a surface of area A is placed in a uniform magnetic field of induction B, such that the unit vector along the normal ($\hat{n}$) makes an angle ' θ ' with direction of magnetic field then the flux passing through it is given by $\phi = \overline{B}.\overline{A} = BA\cos\theta$

- If magnetic field is non uniform then $\phi = \int \overline{B}.\overline{ds}$
- The SI Unit of flux is weber (Wb).
  CGS unit of flux is maxwell (Mx)
  1 weber = 1 tesla - meter^2
  1 weber = 10^8 maxwell
  Dimensional formula of the magnetic flux is $ML^2T^{-2}A^{-1}$
  Magnetic flux is a scalar
- Magnetic flux can be positive, negative or zero depending upon the angle between area vector and field direction.

- When a cylinder is placed in a uniform magnetic field as shown in the below figure

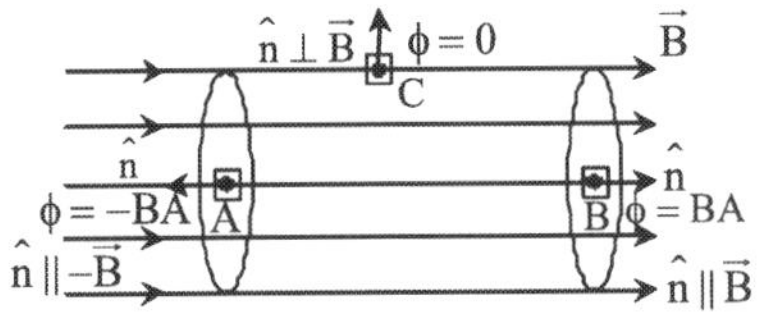

i) When the plane of the surface is parallel to the direction of the magnetic field (or) normal drawn to the surface is perpendicular to the magnetic field $\left(\hat{n} \perp \vec{B}\right)$ then magnetic flux linked with the surface is zero i.e., $\phi = 0\left[\because \theta = 90^0\right]$

ii) When the plane of the surface is perpendicular to magnetic field (or) normal drawn to the surface is parallel to the magnetic field $\left(\hat{n} \parallel \vec{B}\right)$, then magnetic flux linked with the surface is maximum.

i.e., $\phi_{max} = BA\left(\because \theta = 0^0\right)$

iii) When the flux entering the surface is opposite to the area vector ($\hat{n}$) then $\phi = -BA\left(\because \theta = 180^0\right)$

- The magnetic flux linked with a coil $\left(\phi = NBA\cos\theta\right)$ can be changed by

  a) Changining the no. of turns (N)

  b) Varying the magnetic field (B)

  c) Changing the area of the magnetic field bounded by the coil by moving the coil into or out of the magnetic field

  d) Changing the angle made by the coil with the direction of the field

- **The change of flux due to rotation of the coil:**

  When the coil is rotated from an angle of θ_1 to an angle of θ_2 (both are measured w.r.t normal) in a uniform magnetic field then the initial flux through the coil is

  $\phi_i = \text{NBA}\cos\theta_1$

  The final flux through the coil after rotation is

  $\phi_f = \text{NBA}\cos\theta_2$

  The change in the flux associated with the coil is

  $\Delta\phi = \phi_f - \phi_i$

  $\Delta\phi = \text{NBA}\left(\cos\theta_2 - \cos\theta_1\right)$

  if $\theta_1 = 0^0$ and $\theta_2 = 90^0$ then $\Delta\phi = -\text{NBA}$

  if $\theta_1 = 90^0$ and $\theta_2 = 180^0$ then $\Delta\phi = -\text{NBA}$

  if $\theta_1 = 0^0$ and $\theta_2 = 180^0$ then $\Delta\phi = -2\text{NBA}$

WE-1 : A rectangular loop of area 0.06 m² is placed in a uniform magnetic field of 0.3T with its plane (i) normal to the field (ii) inclined 30⁰ to the field (iii) parallel to the field. Find the flux linked with the coil in each case.

Sol: $\phi = \text{NBA}\cos\theta$

i) $\phi = 1\times0.06\times0.3\times\cos0^0 = 0.018$ weber

ii) $\phi = 1\times0.06\times0.3\times\cos60^0 = 0.009$ weber

iii) $\phi = 1\times0.06\times0.3\times\cos90^0 = 0$

WE-2 : At a certain location in the northern hemisphere, the earth's magnetic field has magnitude of $42\mu T$ and points downwards at 53⁰ to the vertical. Calculate the flux through a horizontal surface of area $2.5m^2.\left[\sin53^0 = 0.8\right]$

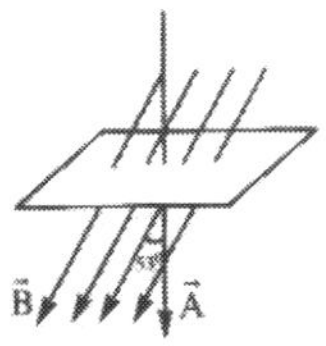

Sol: $\phi_B = \text{BA}\cos\theta = 42\times10^{-6}\times2.5\times\cos53^0 = 63\mu\text{Wb}$

FARADAY'S LAWS OF ELECTRO MAGNETIC INDUCTION

First Law :- Whenever the magnetic flux linked with an electric circuit (coil) changes, an emf is induced in the circuit (coil). The induced emf exists as long as the change in magnetic flux continues.

Second Law :- The induced emf produced in the coil is equal to the negative rate of change of magnetic flux linked with it.

$$e = -\frac{d\phi}{dt}$$

where ϕ = flux through each turn

If the coil contains N turns, an emf appears in every turn all these emfs are to be added. Then, the induced emf is given by

$$e = -N.\frac{d\phi}{dt} = -\frac{d}{dt}(N\phi)$$

Where '$N\phi$' is total flux linked with the coil of N turns. (or)

$$e = -\frac{d}{dt}(N\phi) = -\frac{d}{dt}(\text{NBA}\cos\theta)$$

Negative sign is in accordance with Lenz's law. The above law is also called **Neumann's law.**

LENZ'S LAW AND CONSERVATION OF ENERGY :

"The direction of the induced emf is always such that it tends to produce a current which opposes the change in magnetic flux"

- Induced emf can exist whether the circuit is opened or closed. But induced current can exist only in the closed circuits.
- A metallic ring is held horizontally and a bar magnet is dropped through the ring with its length along the axis of the ring, as shown in figure.

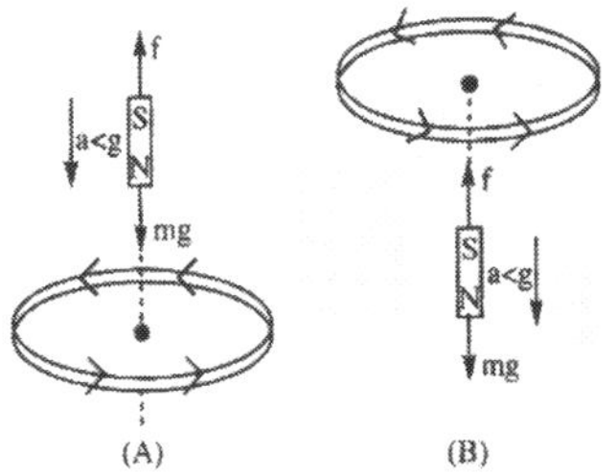

In both the cases net force on the magnet is $F_{net} = mg - f$

Hence net acceleration of the fall is $a_{net} = g - \frac{f}{m} \Rightarrow a_{net} < g$

where f=force exerted by the induced magnetic field of ring on the magnet.

- When the magnet is allowed to fall through an open ring (or) cut ring, then

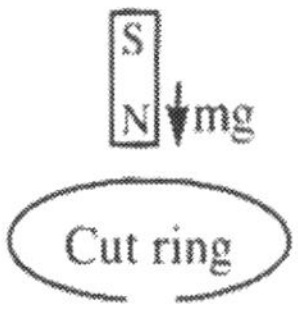

a) an emf is induced

b) No current is induced (since the ring is not closed) and hence no induced magnetic field.

c) No opposition to the motion of the magnet.

d) $F_{net} = mg$

e) $a_{net} = g$

Magnet falls with an acceleration = g

➢ When a magnet is allowed to fall through two identical metal coils at different temperatures then magnet falls slowly through the coil at low temperature as its resistance is less more induced current flows so more is the opposition.

➢ A magnet allowed to fall through a long cylindrical pipe then the acceleration of magnet is always less than 'g' and the acceleration continuously decreases due to induced currents. But the velocity increases until the magnet moves with acceleration. At a particular instant the acceleration becomes zero and the magnet moves downwards with uniform velocity, called terminal velocity.

➢ When north pole of a magnet is moved perpendicular to the plane of the coil as shown in figure, then

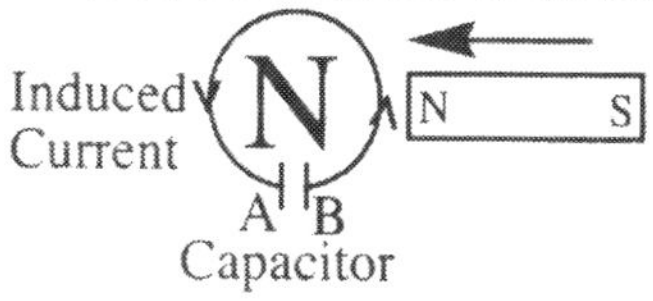

a) emf is induced

b) Induced current flows from B to A along the coil when A and B are connected through resistor.

c) Electrons flow from A to B along the coil

d) Hence plate A will become positively charged and plate B becomes negatively charged.

➢ When the two magnets are moved perpendicular to plane of coil as shown, then

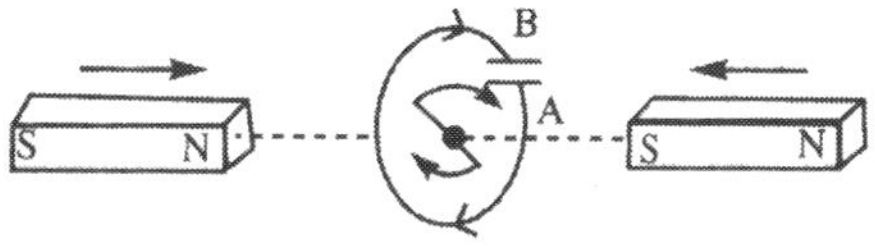

a) emf is induced

b) Induced current flows from A to B along the coil when A and B are connected through resistor.

c) Electrons flow from B to A along the coil

d) Hence plate A will become negatively charged and plate B becomes positively charged.

➢ The directions of iduced current in coil for different kinds of motion of magnets

a) 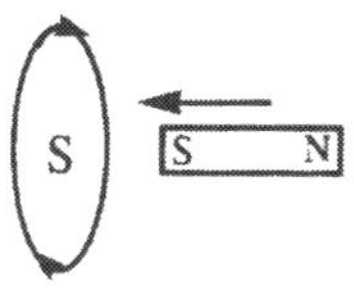

Clockwise induced current

b) 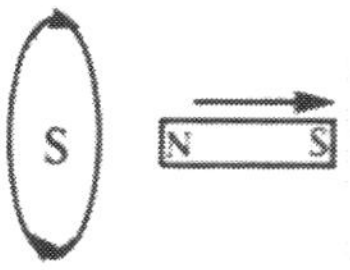

Clockwise induced current

c) 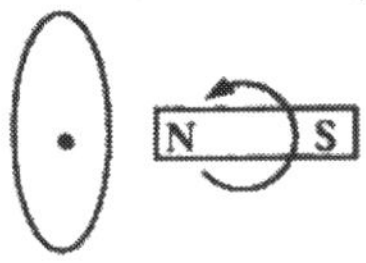 No induced current

(because there is no change of flux linked with the coil)

- When a current carrying conductor is placed beside a closed loop in its plane then the induced current direction for the following are

a) Current in conductor is constant.

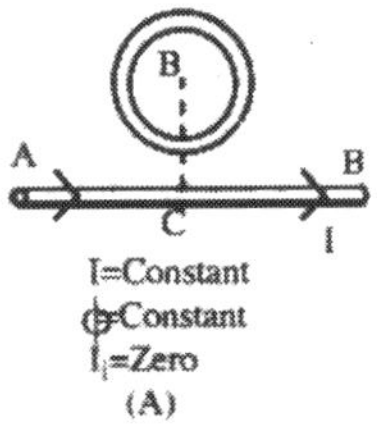

∴ No induced current

b) Current through the conductor increases as shown.

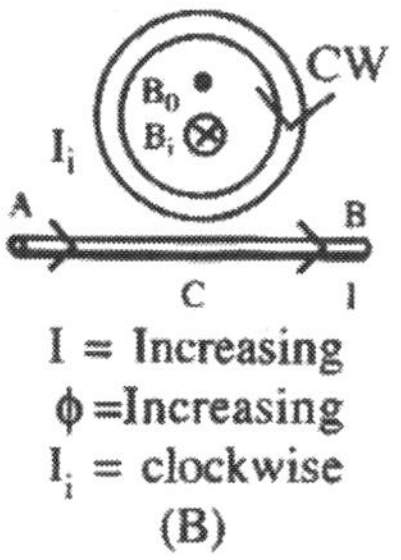

B_0 - original field
B_i - inducedfield

- In this case, the flux through the loop due to current carrying wire is out of the plane of the coil.
- As current is increasing, the outward flux through the coil also increases.
- Hence to oppose this, an inward flux is created by the clock wise induced current.

c) Current through the conductor decreases as shown.

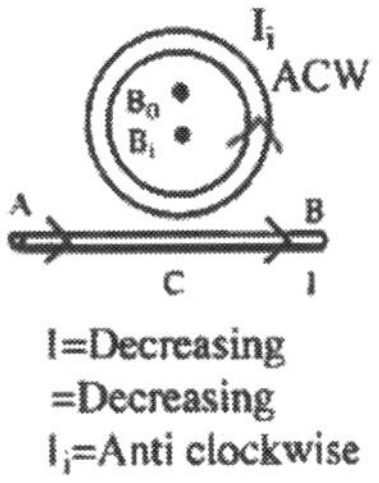

B_0 - original field
B_i - induced field

In this case, the flux through the loop due to current carrying wire is out of the plane of the coil.
As current is decreasing, the outward flux through the coil also decreases.
Hence to oppose this, an outward flux is created by the anti-clock wise induced current.

EXPRESSIONS FOR INDUCED EMF, INDUCED CURRENT AND INDUCED CHARGE :

- According to Faraday's second law and Lenz's law the induced emf is given by $e = -\frac{d\phi}{dt}$

If the coil has N turns then $e = -N\frac{d\phi}{dt}$

$$\Rightarrow e = -N\frac{(\phi_2 - \phi_1)}{dt}$$

- As $\phi = BAN\cos\theta$ and $e = -\frac{d\phi}{dt}$

  The emf is induced (or) change in flux is caused by changing B (or) A (or) N (or) θ

- If 'B' is changed then

  a) Average induced emf $e = -AN\cos\theta\frac{(B_2 - B_1)}{(t_2 - t_1)}$

  Here B_1 is magnetic field induction at an instant t_1

  B_2 is magnetic field induction at an instant t_2

  b) If the plane of the coil is perpendicular to magnetic field, then $\theta = 0^0 \Rightarrow \cos\theta = 1$

  then $e = -AN\frac{(B_2 - B_1)}{(t_2 - t_1)}$

  c) Instantaneous emf $e = -AN\cos\theta\frac{dB}{dt}$

- If 'A' is changed then

  a) Average induced emf $e = -BN\cos\theta\frac{(A_2 - A_1)}{(t_2 - t_1)}$

  b) If the plane of the coil is perpendicular to magnetic field, then $\theta = 0^0 \Rightarrow \cos\theta = 1$

  then $e = -BN\frac{(A_2 - A_1)}{(t_2 - t_1)}$

  c) Instantaneous emf $e = -BN\cos\theta\frac{dA}{dt}$

- If 'θ' is changed (i.e., if coil is rotated)

  a) Average induced emf

  $$e = -BAN\frac{(\cos\theta_2 - \cos\theta_1)}{(t_2 - t_1)}$$

  b) Instantaneous emf $e = -BAN\frac{d}{dt}(\cos\theta)$

  If the coil is rotated with constant angular velocity 'ω' then $\theta = \omega t$ and

  $$e = -BAN\frac{d}{dt}(\cos\omega t) = BAN\omega\sin\omega t$$

  $\therefore e = BAN\omega\sin\omega t$

c) $\omega t = 90^0$, if the plane is parallel to the magnetic field then induced emf is maximum. Then Peak emf.

$$e_0 = BAN\omega$$

$$\therefore e = e_0 \sin \omega t$$

This is the principle of AC generator.

INDUCED CURRENT

➢ If the magnetic flux in a coil of resistance R changes from ϕ_1 to ϕ_2 in a time 'dt', then a current 'i' is induced in the coil as $i = \frac{e}{R}$

$$i = \frac{N(\phi_2 - \phi_1)}{Rdt} \qquad \left(\because e = -N.\frac{d\phi}{dt}\right)$$

$\therefore$ Induced current is given by

Magnitude of current

$$i = \frac{\text{Induced emf}}{\text{Resistance in the circuit}} = \frac{N}{R}\left(\frac{d\phi}{dt}\right)$$

INDUCED CHARGE

➢ The amount of charge induced in a conductor is given as follows

We know, $I = \frac{e}{R}$

(or) $I = \frac{1}{R}\left(-\frac{d\phi}{dt}\right)$

$$\Rightarrow \frac{dq}{dt} = -\frac{1}{R}\frac{d\phi}{dt} \text{ (or) } dq = -\frac{1}{R}d\phi$$

$\therefore$ Induced charge, $q = -\frac{1}{R}\int_{\phi_i}^{\phi_f} d\phi$

$$q = -\frac{1}{R}\left[\phi_f - \phi_i\right] \text{ (or) } q = \frac{\phi_i - \phi_f}{R} \text{ (magnitude of charge)}$$

$\therefore$ In general, induced charge is given by

$$q = \frac{\text{change of magnetic flux}}{\text{resistance}}$$

For N turns, the induced charge is $q = \frac{N}{R}(d\phi)$

➢ Induced emf is independent of total resistance of the circuit but depends on time of change of flux.

➢ Induced current depends on both time of change of flux and resistance of circuit

- Induced charge is independent of time but depends on the resistance of circuit.
- When a magnet is moved towards a stationary coil(i) slowly and (ii) quickly, then
  a) induced charge is same in both cases b) induced emf is more in second case
  c) induced current is more in second case

WE-3: The magnetic flux through a coil perpendicular to its plane is varying according to the relation $\phi_B = \left(5t^3 + 4t^2 + 2t - 5\right)$ weber. Calculate the induced current through the coil at $t = 2$ second. The resistance of the coil is 5Ω.

Sol: $\phi = 5t^3 + 4t^2 + 2t - 5$

$$|e| = \frac{d\phi}{dt} = 15t^2 + 8t + 2$$

at $t = 2\,\text{sec}, e = 78V$

$$i \times 5 = 15 \times 4 + 8 \times 2 + 2 \Rightarrow i = 15.6A$$

WE-4: A circular coil of 500 turns of wire has an enclosed area of $0.1m^2$ per turn. It is kept perpendicular to a magnetic field of induction 0.2T and rotated by 180^0 about a diameter perpendicular to the field in 0.1s. How much charge will pass when the coil is connected to a galvanometer with a combined resistance of 50Ω.

Sol: $$q = \frac{\phi_i - \phi_f}{R} = \frac{NBA - (-NBA)}{R} = \frac{2NBA}{R}$$

$$q = \frac{2 \times 500 \times 0.2 \times 0.1}{50} = 0.4C.$$

WE-5: Some magnetic flux is changed from a coil of resistance 10Ω. As a result an induced current is developed in it, which varies with time as shown in figure. What is the magnitude of change in flux through the coil?

Sol: The induced charge is $q = \frac{\Delta\phi}{R}$

But, Area of i-t curve gives charge

$\therefore \Delta\phi = R \times$ Area of $i - t$ curve

$\Delta\phi = qR$

$$= \frac{1}{2} \times 4 \times 0.1 \times 10$$

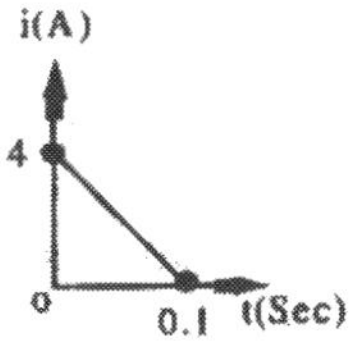

$\therefore \Delta\phi = 2wb$

WE-6: A long solenoid with 1.5 turns per cm has a small loop of area $2.0\,cm^2$ placed inside the solenoid normal to its axis. If the current in the solenoid changes steadily from 2.0 A to 4.0 A in 1.0s. The emf induced in the loop is

Sol: The magnetic field along the axis of solenoid is $B = \mu_0 ni$ where n is no. of turns per unit length.

flux through the smaller loop placed in solenoid is $= \phi B \times A$

Since current in solenoid is changing, emf induced in loop is $e = \frac{d\phi}{dt} = \frac{d}{dt}[\mu_0 niA]$

$$e = \mu_0 nA\left(\frac{di}{dt}\right)$$

$$= 4\pi\times10^{-7}\times1.5\times10^{2}\times2\times10^{-4}\times\left(\frac{4-2}{1-0}\right)$$

$$= 0.75\times10^{-6}V$$

WE-7: A sqaure loop of side 10cm and resistance $0.5\,\Omega$ is placed vertically in the east-west plane. A uniform magnetic field of 0.10T is set up across the plane in the north-east direction. The magnetic field is decreased to zero in 0.70s at a steady state. The magnitude of current in this time-interval is.

Sol: The initial magnetic flux is given by

$$\phi = BA\cos\theta$$

Given, B=0.10 T, area of square loop $= 10\times10 = 100cm^2$

$$= 10^{-2}m^2$$

$$\therefore \quad \phi = \frac{0.1\times10^{-2}}{\sqrt{2}}Wb$$

Final flux, $\phi_{min} = 0$

The change in flux is brought about in 0.70 s

The magnitude of the induced emf is

$$e = \frac{\Delta\phi}{\Delta t} = \frac{|\phi - 0|}{\Delta t} = \frac{10^{-3}}{\sqrt{2}\times0.7} = 1mV$$

The magnitude of current is

$$I = \frac{e}{R} = \frac{10^{-3}}{0.5} = 2mA$$

WE-8: A square loop ACDE of area $20cm^2$ and resistance 5Ω is rotated in a magnetic field B = 2T through 180^0

a) in 0.01 s and b) in 0.02 s.

Find the magnetiude of e,i and Δq in both the cases.

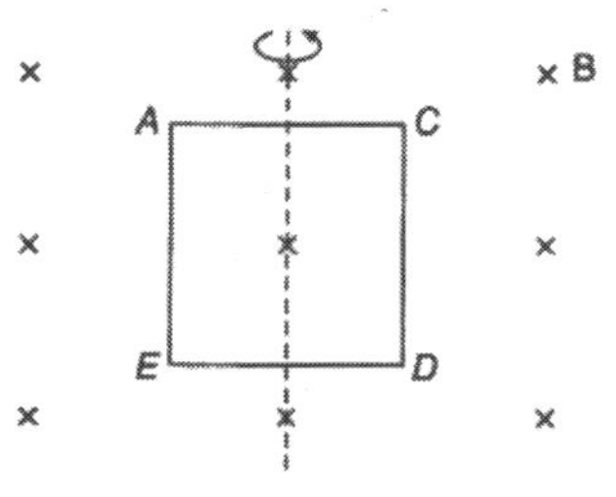

Solution : Let us take the area vector S perpendicular to plane of loop inwards. So intially dS parallel to B and when it is rotated by 180^0, S is anti parallel to B. Hence, initial flux passing through the loop,

$$\phi_i = BS\cos 0^0 = (2)(20\times10^{-4})(1) = 4\times10^{-3}\,Wb$$

Flux passing through the loop when it is rotated by 180^0, $\phi_f = BS\cos180^0$

$$= (2)(20\times10^{-4})(-1) = -4.0\times10^{-3}\,Wb$$

Therefore, change in flux,

$$\Delta\phi_B = \phi_f - \phi_i = -8\times10^{-3}\,Wb$$

(a) Given $\Delta t = 0.01\,s,\ R = 5\Omega$

$$\therefore\ |e| = \left|\frac{\Delta\phi_B}{\Delta t}\right| = \frac{8\times10^{-3}}{0.01} = 0.8V$$

or $i = \dfrac{|e|}{R} = \dfrac{0.8}{5} = 0.16A$

and $\Delta q = i\Delta t = 0.16\times0.01 = 1.6\times10^{-3}C$

b) $\Delta t = 0.02s$

$$\therefore\ |e| = \left|-\frac{\Delta\phi_B}{\Delta t}\right| = \frac{8\times10^{-3}}{0.02} = 0.4\text{V}$$

$$i = \frac{|e|}{R} = \frac{0.4}{5} = 0.08A$$

and $\Delta q = i\Delta t = (0.08)(0.02) = 16\times10^{-3}C$

MOTIONAL EMF

- The motional emf is the emf which results from relative motion between a conductor and the source of magnetic field.
- When a condutor of length l is moved with a velocity v perpendicular to its length in uniform magnetic field (B), which is perpendicular to both its length and as well as its velocity, the emf induced across its ends $e=Blv$
- If the rod moved making an angle θ with its length, then $e = Blv\sin\theta$
- In vector form $e = \vec{B}.(\vec{l}\times\vec{v})$ or $\vec{l}.(\vec{v}\times\vec{B})$
- among $\vec{B},\vec{l}$ and $\vec{v}$, if any two are parallel the emf induced across the conductor is zero

WE-9: A rectangular loop of length 'l' and breadth 'b' is placed at a distance of x from an infinitely long wire carrying current 'i' such that the direction of current is parallel to breadth. If the loop moves away from the current wire in a direction perpendicular to it with a velocity 'v', the magnitude of the e.m.f. in the loop is : (μ_0 = permeability of free space)

Electro Magnetic Induction

Sol:

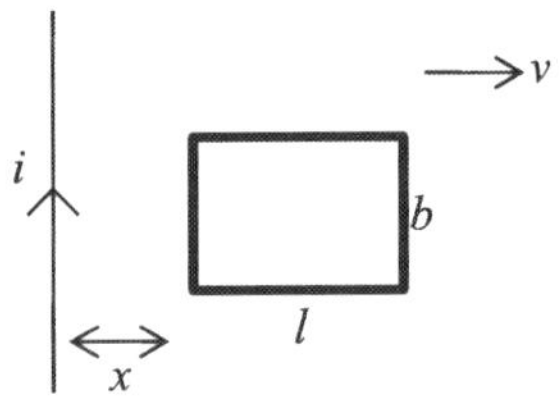

$emf = Blv$

$= Bbv$

$= (B_1 - B_2)bv$

$$= \left[\frac{\mu_0 i}{2\pi x} - \frac{\mu_0 i}{2\pi(x+l)}\right]bv = \frac{\mu_0 ibv}{2\pi}\left[\frac{1}{x} - \frac{1}{l+x}\right] = \frac{\mu_0 ilbv}{2\pi x(x+l)}$$

WE-10: A horizontal magnetic field B is produced across a narrow gap between the two square iron pole pieces. A closed square loop of side a, mass m and resistance R is allowed to fall with the top of the loop in the field. The loop attains a terminal velocity equal to :

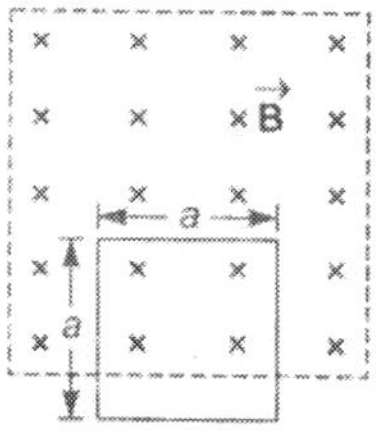

Sol: Induced emf in the loop, when it is falling with terminal velocity

$e = BVa$

$$i = \frac{e}{R} = \frac{Bva}{R}$$

Vertically upward force experienced by loop due to this

$F = Bia$

$$= B\left(\frac{Bva}{R}\right)a$$

$$= \frac{B^2va^2}{r}$$

When the loop attains terminal velocity 'v'

$$mg = \frac{B^2va^2}{R}$$

$$V = \frac{mgR}{B^2a^2}$$

WE-11: A conducting wire of mass m slides down two smooth conducting bars, set at an angle θ to the horizontal as shown in figure. The seperation between the bars is l. The system is located in the magnetic field B, perpendicular to the plane of the sliding wire and bars. The constant velocity of the wire is

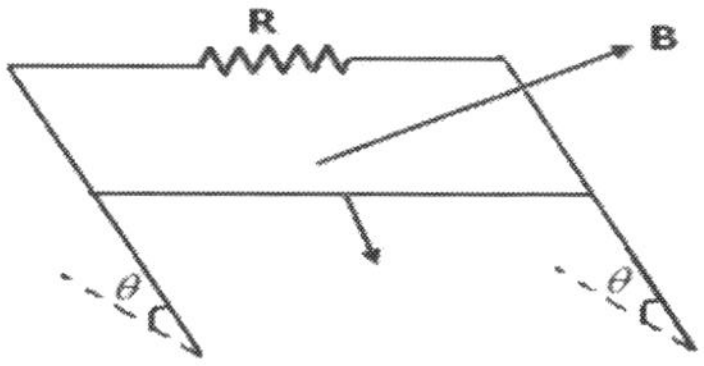

Sol: Along inclined plane the force acting downwards $= mg\sin\theta$ ....(1)

magnetic force acting upwards

$\Rightarrow F = Bil \qquad \Rightarrow B\left(\frac{Blv}{R}\right)l$

$= \frac{B^2l^2v}{R}$ ......(2)

From (1) and (2)

$\frac{B^2l^2v}{R} = mg\sin\theta$

$v = \frac{mgR\sin\theta}{B^2l^2}$

FLEMINGS'S RIGHT HAND RULE :

➢ Stretch the first three fingers of right hand such that they are mutually perpendicular to each other. If the fore finger represents the direction of magnetic field and the thumb represents the direction of the motion of the conductor, then the central finger indicates the direction of induced current

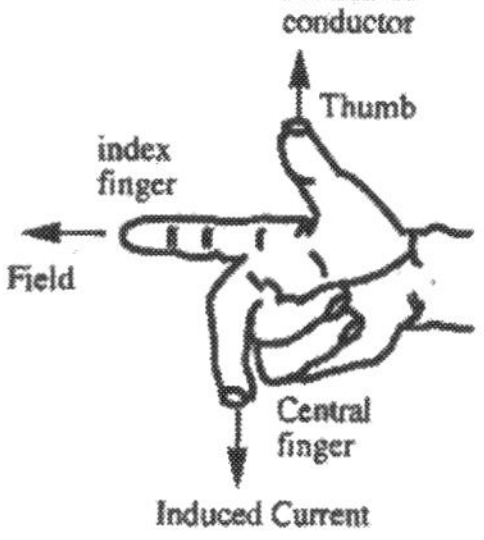

➢ A conductor of length 'l' measured from P to Q is moved with a speed of 'v' in a uniform magnetic field 'B' as shown in figure.

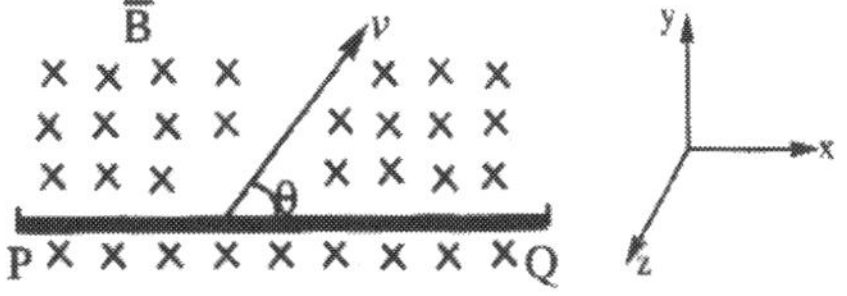

Here $\vec{B}=B(-\hat{k}), \vec{l}=l(\hat{i})$ and $\vec{v}=v\cos\theta\hat{i}+v\sin\theta\hat{j}$

Induced emf is $e=\vec{l}.\left(\vec{v}\times\vec{B}\right)=l(\hat{i}).\left(v\cos\theta\hat{i}+v\sin\theta\hat{j}\right)\times B(-\hat{k}) = -Blv\sin\theta$

The change in the flux in the time of 'Δt' is

$\therefore \Delta\phi = e\Delta t = -Blv\sin\theta\Delta t$

- A conductor of length 'l' is bent at its midpoint and is moved along its perpendicular bisector with a constant speed of 'v' in a uniform magnetic field of strength 'B' as shown in figure

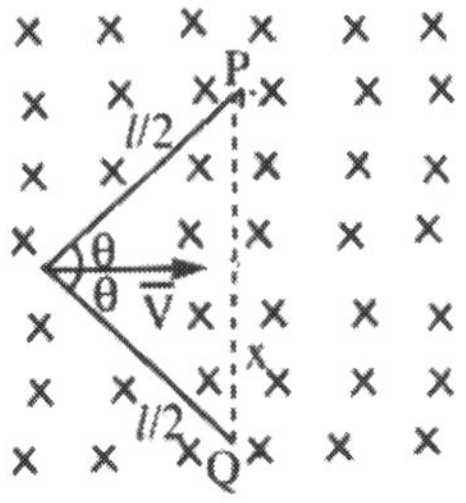

From the figure $\sin\theta = \dfrac{x}{l/2} \Rightarrow x = \dfrac{l}{2}\sin\theta$

Here $\vec{B}=B(-\hat{k}), \vec{v}=v\hat{i}$ and effective length of the conductor

$\vec{l}=2x(-\hat{j})=l\sin\theta(-\hat{j})$

Induced emf is $e=\vec{l}.\left(\vec{v}\times\vec{B}\right)=l\sin\theta(-\hat{j}).v\hat{i}\times B(-\hat{k})=-Blv\sin\theta$

The change in the flux associated in time internal of 'Δt' is

$\Delta\phi = e\Delta t = -Blv\sin\theta\Delta t$

Here the effective length between free ends of conductor is $l\sin\theta$.

- The emf induced across the ends of the conductor shown in the figure is

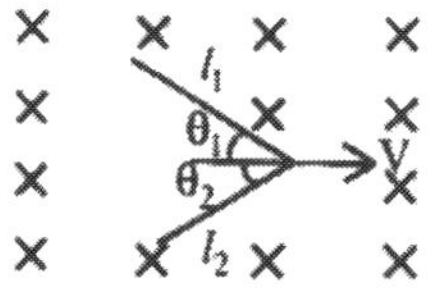

$e = BVl = BV\left(l_1\sin\theta_1 + l_2\sin\theta_2\right)$

i) If a conductor is moving vertically downwards with constant velocity v with its ends pointing east-west, it will cut the horizontal component of earth's field B_H as shown in figure (A) and hence the flux linked with the area generated by the motion of the conductor, and induced emf will be

$$\phi_H = B_H\left(ly\right) \text{ and } e = \frac{d\phi_H}{dt} = B_H v_y l$$

$$\left(\text{with } v_y = \frac{dy}{dt}\right)$$

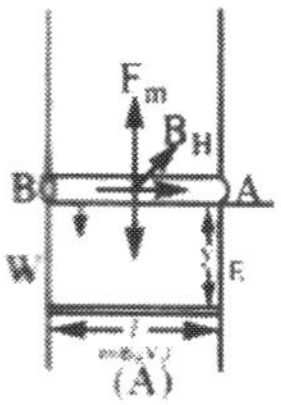

ii) However in case of vertical motion, if the ends of the conductor point north-south, both B_H and B_V will be parallel to the plane of area generated by the motion of the conductor as shown figure (B) and hence it doesn't cut the magnetic lines. so

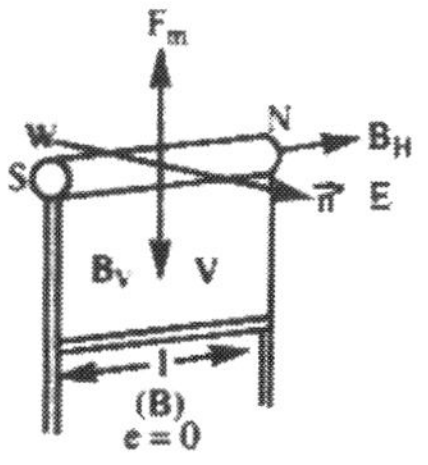

$$\phi = 0 \text{ and } e = \frac{d\phi}{dt} = 0$$

iii) If the wire is moving in a horizontal plane in any direction as shown in figure(C), it will cut flux of B_V(as B_H will always be parallel to area) and so

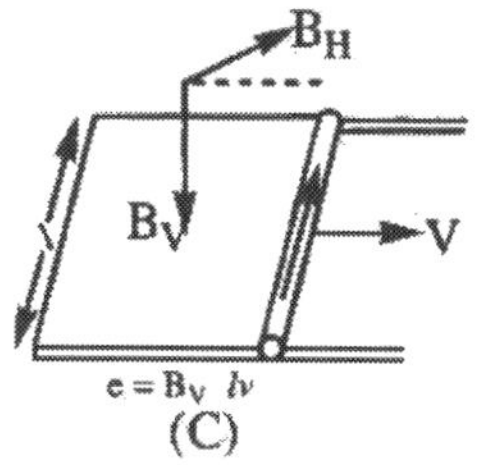

$$\phi_V = B_v ls \text{ and } e = \frac{d\phi_V}{dt} = B_V vl$$

$$\left[\text{with } v = \frac{ds}{dt}\right]$$

WE-12: A wire of length $2l$ is bent at mid point so that the angle between two halves is 60^0. If it moves as shown with a velocity v in a magnetic field B find the induced emf.

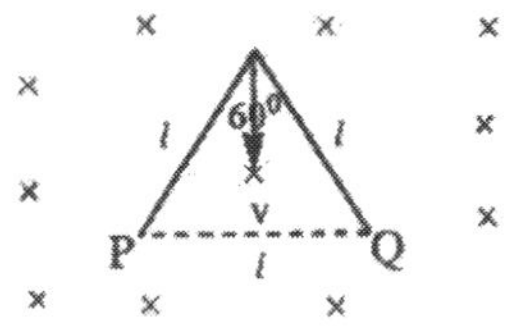

Sol: $e=Blv$. Here l = Effective length = PQ

WE-13 : A conductor of length 0.1m is moving with a velocity of 4m/s in a uniform magnetic field of 2T as shown in the figure. Find the emf induced?

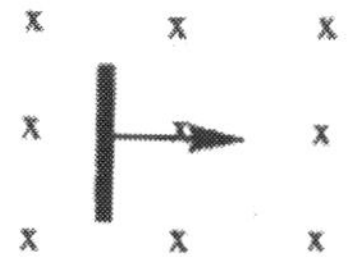

Sol: $e = Blv \sin 90^0 = (2)(0.1)(4) = 0.8$ Volt

WE-14: Figure shows a conducting rod PQ in contact with metal rails RP and SQ, which are 0.25m apart in a uniform magnetic field of flux density 0.4T acting perpendicular to the plane of the paper. Ends R and S are connected through a $5\,\Omega$ resistance. What is the emf when the rod moves to the right with a velocity of $5ms^{-1}$? What is the magnitude and direction of the current through the $5\,\Omega$ resistance? If the rod PQ moves to the left with the same speed, what will be the new current and its direction ?

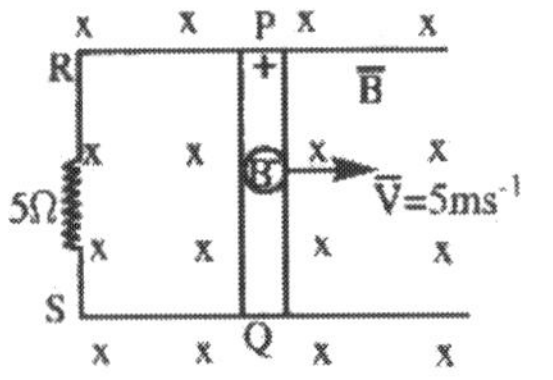

Sol: $|e| = Blv = 0.4 \times 0.25 \times 5 = 0.5V$

Current, $I = \frac{|e|}{R} = \frac{0.5V}{5\Omega} = 0.1A$

As the rod 'PQ' moves to right as shown, the free electrons in it experience a Lorentz force. According to F.L.H., the force is towards the end 'Q' of rod. $\therefore$ They move from P to Q, hence the end of the rod P becomes deficient of electrons $\Rightarrow V_P > V_Q$

Applying Flemming's right hand rule, the current in the rod shall flow from Q to P.

(b) : If the rod PQ moves to the left with the same speed, then the current of 0.1 A will flow in the rod PQ from P to Q

WE-15: A loop ABCD containing two resistors as shown in figure is placed in a uniform magnetic field B directed outwards to the plane of page. A sliding conductor EF of length l and of negligible resistance moves to the right with a uniform velocity v as shown in Fig. Determine the current in each branch.

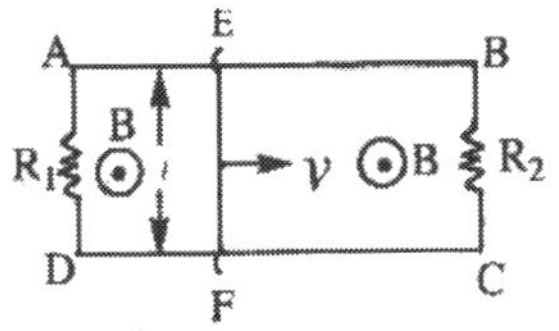

Sol: The magnetic field induction B, length l and the velocity v of the conductor EF are mutually perpendicular, hence the emf induced in it is e=Blv (with end F of the rod at higher potential)

$\therefore$ The effective electric circuit can be redrawn as shown in Fig.

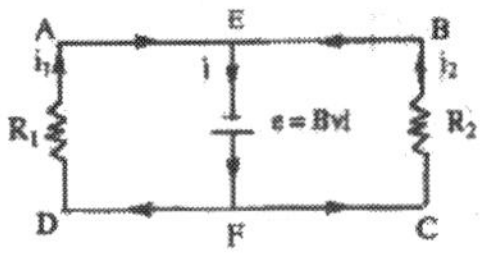

T he resistance R_1 and R_2 are in parallel, so the equivalent resistance R is given by $\frac{1}{R} = \frac{1}{R_1} + \frac{1}{R_2}$

From Ohms law, the total current is $i = \frac{e}{R}$

$$i = Blv\left(\frac{1}{R_1} + \frac{1}{R_2}\right)$$

Current in AD is $i_1 = \frac{BlV}{R_1}$; Current in BC is $i_2 = \frac{BlV}{R_2}$

WE-16: A rectangular loop with a slide wire of length l is kept in a uniform magnetic field as shown in the figure. The resistance of slider is R. Neglecting self inductance of the loop find the current in the connector during its motion with a velocity v.

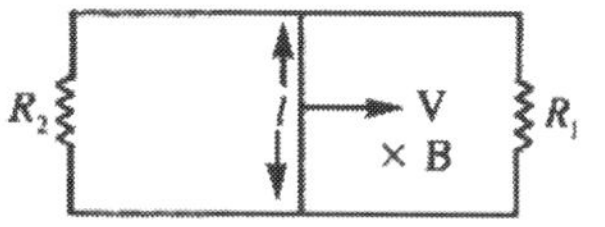

Sol: The equivalent circuit is

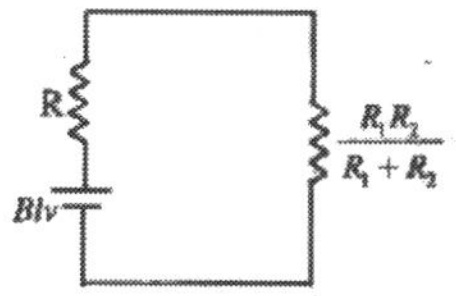

The equivalent resistance of the circuit is $R = \left(R + \frac{R_1 R_2}{R_1 + R_2}\right)$

Hence the current in the connector is $i = \frac{e}{R}$

$\therefore \quad i = \frac{Blv(R_1 + R_2)}{(RR_1 + RR_2 + R_1 R_2)}$

WE-17: A conducting rod PQ of length $L = 1.0$m is moving with a uniform speed v=2.0m/s in a uniform magnetic field B=4.0T directed into the paper. A capacitor of capacity $C = 10\mu F$ is connected as shown in the figure. Then what are the charges on the plates A and B of the capacitor.

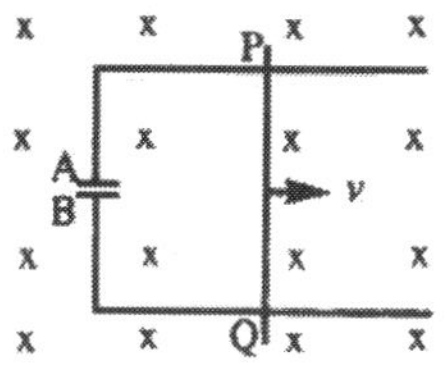

Sol: The motional emf is

$\therefore$ p.d across the capacitor $= Blv = 4 \times 1 \times 2 = 8V$

$q = CV = 10 \times 8 = 80\mu C$

A is +Ve w.r.t. B (from fleming right hand rule)

The charge on plate A is $q_A = 80\mu C$

The charge on plate B is $q_B = -80\mu C$

WE-18: Two parallel rails with negligible resistance are 10.0 cm apart. They are connected by a $5.0\,\Omega$ resistor. The circuit also contains two metal rods having resistances of $10.0\,\Omega$ and $15.0\,\Omega$ along the rails. The rods are pulled away from the resistor at constant speeds 4.00 m/s and 2.00 m/s respectively. A uniform magnetic field of magnitude 0.01T is applied perpendicular to the plane of the rails. Determine the current in the $5.0\,\Omega$ resistor.

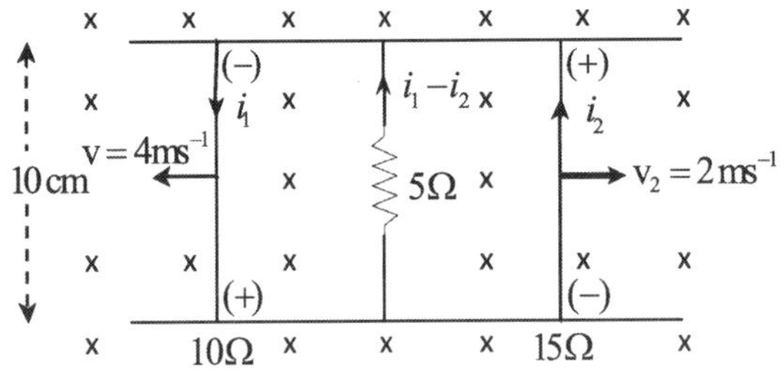

Sol: In the figure

$R = 5.0\Omega, r_1 = 10\Omega, r_2 = 15\Omega$,

$e_1 = Blv_1 = 0.01\times0.1\times4 = 4\times10^{-3}V$

$e_2 = Blv_2 = 0.01\times0.1\times2 = 2\times10^{-3}V$

Applying Kircoff's law to the left loop :

$10i_1 + 5(i_1 - i_2) = 4\times10^{-3}$

$\Rightarrow 15i_1 - 5i_2 = 4\times10^{-3} \qquad \rightarrow (1)$

Right loop : $15i_2 - 5(i_1 - i_2) = 2\times10^{-3}$

$\Rightarrow 20i_2 - 5i_1 = 2\times10^{-3} \qquad \rightarrow (2)$

Solving (1) and (2) gives :

$i_1 = \frac{18}{55}\times10^{-3}A$ and $i_2 = \frac{10}{55}\times10^{-3}A$

$\Rightarrow$ Current through $5\Omega = i_1 - i_2 = \frac{8}{55}\times10^{-3}A = \frac{8}{55}mA$

WE-19: A conducting rod *MN* moves with a speed *v* parallel to a long straight wire which carries a constant current *i*, as shown in Fig. The length of the rod is normal to the wire. Find the emf induced in the total length of the rod. State which end will be at a lower potential.

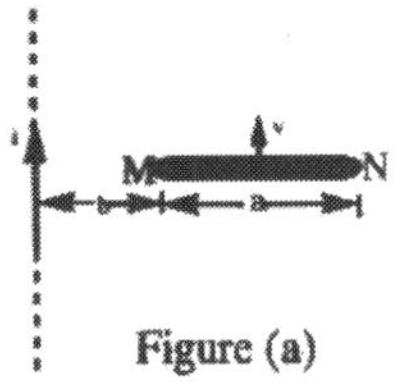

Figure (a)

Sol: The magnetic field induction due to current *i* is different at different sections of the rod, because they are at different distances from the wire.

Let us, first of all, subdivide the entire length of the conductor *MN* into elementary sections. Consider a section (shown shaded in the figure (b)) of thickness *dx* at a distance *x* from the wire. As all the three, *v*, *B* and (*dX*) are mutually normally to each other, so the emf induced in it is *de=Bvdx*. (from *N* to *M* by Fleming's right hand rule)

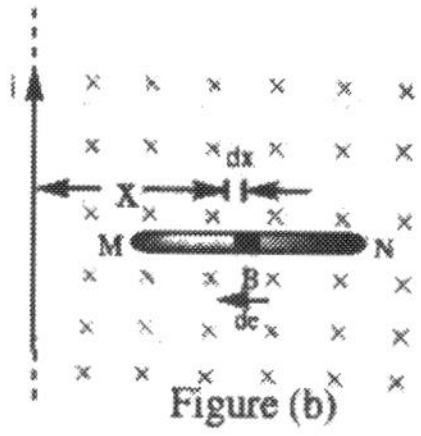

Figure (b)

For the rest of sections, the induced emf is in the same sense, (i.e., from N to M)

$\therefore$ Total emf induced in the conductor is $e = \int de = \int_b^{b+a} Bv\,dx$

Substituting for $B = \dfrac{\mu_0 i}{2\pi x}$, the above equation gets changed to

$$e = \int_b^{b+a} \frac{\mu_0 iv\,dx}{2\pi x}$$

$$e = \frac{\mu_0 iv}{2\pi}\left[\ln x\right]_b^{b+a} \text{ or, } e = \frac{\mu_0 iv}{2\pi}\ln\left(1 + a/b\right)$$

WE-20 : A square loop of side a is placed in the same plane as a long straight wire carrying a current i. The centre of the loop is at a distance r from the wire where $r >> a$. The loop is moved away from the wire with a constant velocity v. The induced e.m.f. in the loop is

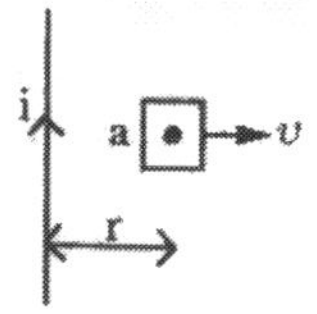

Sol: Magnetic field by the straight wire of current i at a distance r is $B = \dfrac{\mu_0 i}{2\pi r}$

flux associated with the loop is $\phi = BA = \dfrac{\mu_0 i}{2\pi r}a^2$

$$\therefore e = \frac{-d\phi}{dt} = \frac{-\mu_0}{2\pi}ia^2\frac{d}{dt}\left(\frac{1}{r}\right) = \frac{-\mu_0}{2\pi}ia^2\left(\frac{-1}{r^2}\right)\frac{dr}{dt}$$

Hence the induced emf in the loop is $e = \dfrac{\mu_0}{2\pi}i\dfrac{a^2}{r^2}v \quad \left(\because \dfrac{dr}{dt} = v\right)$

WE-21 : Two conducting rings of radii r and $2r$ move in apposite directions with velocities 2υ and υ respectively on a conducting surface S. There is a uniform magnetic field of magnitude B perpendicular to the plane of the rings. The potential difference between the highest points of the two rings is

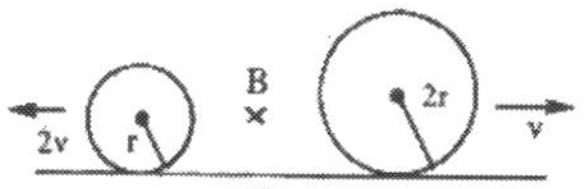

Sol: Replace the induced emfs in the rings by cells emfs

$e_1 = B2r(2v) = 4Brv$

$e_2 = B(4r)v = 4Brv$

The equivalent circuit is

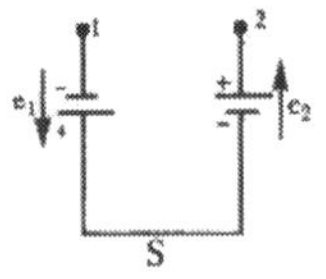

Hence the potential difference between the highest points of the two rings is $V_2 - V_1 = e_1 + e_2 = 8Brv$

WE-22 : A metallic square loop ABCD is moving in its own plane with velocity v in a uniform magnetic field perpendicular to its plane as shown in the figure. Find

a) In which sides of the loop electric field is induced.

b) Net emf induced in the loop

c) If one 'BC' is outside the field with remaining loop in the field and is being pulled out with a costant velocity then induced current in the loop.

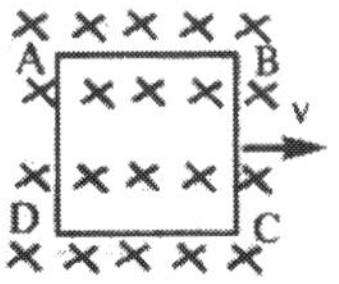

Sol: a) The metallic square loop moves in its own plane with velocity v.

A uniform magnetic field is imposed perpendicular to the plane of the square loop.

AD and BC are $\perp$ to the velocity as well as $\perp$ to field applied. Hence electric field is induced across the sides AD and BC only.

b) As there is no change of flux through the entire coil net emf induced in the coil is zero.

c) Induced current $i = \frac{e}{R}$

Where R is the resistance of the coil.

$\Rightarrow i = \frac{Blv}{R}$ (Only the side AD cuts the flux)

MOTIONAL EMF INDUCED IN A ROTATING BAR

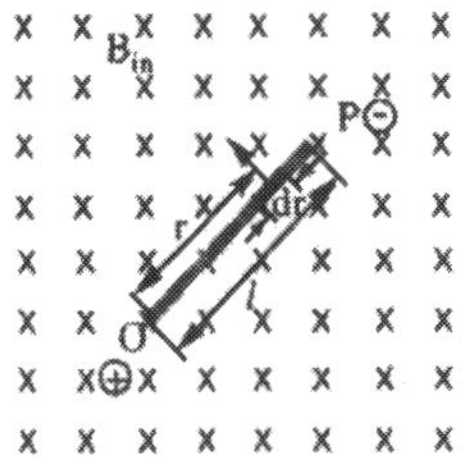

- If a rod of length l is rotated with a constant angular velocity 'ω' about an axis passing through its end (O) and perpendicular to its length and if a uniform magnetic field $\vec{B}$ is present perpendicular to it, then emf across its ends is given by $e = \frac{1}{2}Bl^2\omega$
- In the above case if the rod is rotated about an axis passing through its centre (O) and perpendicular to tis length them emf across its ends is zero

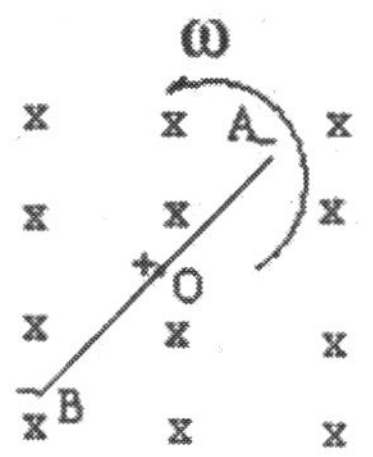

emf across OA is $e = +\frac{1}{8}Bl^2\omega$

emf across OB is $e = -\frac{1}{8}Bl^2\omega$

Net emf across AB is zero

end 'A' is -ve with respect to 'O'

end 'B' is -ve with respect to 'O'

- A spoked wheel of spoke length 'l' is rotated about its axis with an angular velocity 'ω' in a plane normal to uniform magnetic field B as shown.

The emf induced across the ends of each spoke is $e = \frac{1}{2}Bl^2\omega$, with axle (centre) at higher potential.

Since all the spokes are parallel between axle and rim, the emf induced between axle and rim is $e = \frac{1}{2}Bl^2\omega$.

It is independent of number of spokes.

WE-23 : A copper rod of length $2m$ is rotated with a speed of 10 rps, in a uniform magnetic field of 1 tesla about a pivot at one end. The magnetic field is perpendicular to the plane of rotation. Find the emf induced across its ends

Sol: $e = \frac{1}{2}B\omega l^2 = \frac{1}{2}B(2\pi n)l^2 = \pi Bnl^2$

$e = 3.14 \times 1 \times 10 \times 2 \times 2 = 125.6$ volt

WE-24: A wheel with 10 metallic spokes, each 0.50 m long, is rotated with a speed of 120 rev/minute in a plane normal to the earth's magnetic field at the place. If the magnitude of the field is 0.40 gauss, what is the induced emf between the axle and the rim of the wheel ?

Sol: Here each spoke of wheel act as a source of an induced emf (cell) and emf's of all spokes are parallel.

f=120 rev/min = 2 rev/second,

B=0.40 gauss $0.4 \times 10^{-4}T$,

Area swept, by each spoke per second, $A = \pi r^2 f$

Magnetic flux cut by each spoke per second, $\frac{d\phi_B}{dt} = BA = B\pi r^2 f$

Induced emf, $e = B\pi r^2 f$ (numerically)

$e = 0.4 \times 10^{-4} \times \frac{22}{7} \times 0.5 \times 0.5 \times 2$

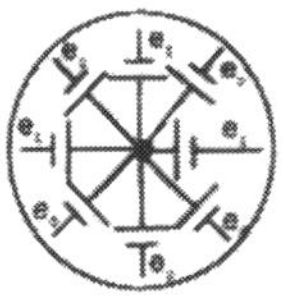

$e = 6.29 \times 10^{-5}$ volt

Induced emf in a wheel is independent of no. of spokes.

WE-25: A metal rod of resistance $20\,\Omega$ is fixed along a diameter of a conducting ring of radius 0.1m and lies on *x-y* plane. There is a magnetic field $\vec{B} = (50T)\hat{k}$. The ring rotates with an angular velocity $\omega = 20\,rad/s$ about its axis. An external resistance of $10\,\Omega$ is connected across the centre of the ring and rim. The current through external resistance is

Sol:

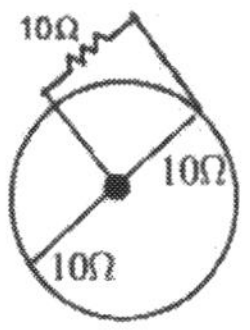

The equivalent circuit is

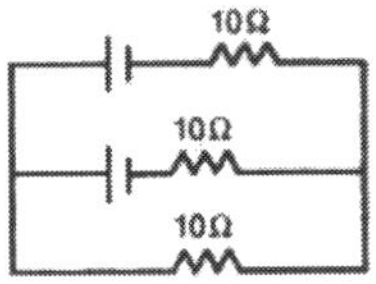

$e = \frac{1}{2}Bl^2\omega = \frac{1}{2} \times 50 \times 0.1 \times 0.1 \times 20$;

$\therefore e = 5V$

Hence the current through the external resistance is

$i = \frac{e}{R} \quad \therefore i = \frac{5}{15} = \frac{1}{3}A$

MOTIONAL EMF INDUCED IN A ROTATING DISC

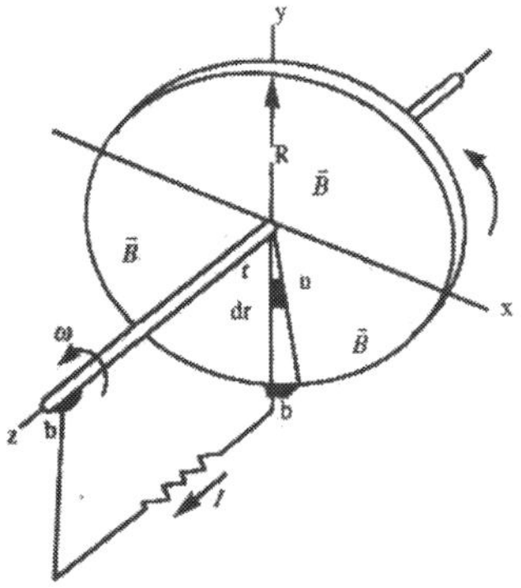

- A circular disc of radius 'R' is rotating with an angular velocity 'ω' about an axis passing through centre and plane of rotation is normal to an uniform magnetic field of induction B. It is equivalent to a spoked wheel with a large number of spokes each of length 'R' between centre and rim without any air gap. The emf induced between centre and rim is independent of number of spokes.

  So, the emf induced between centre and rim is $e = \frac{1}{2}Bl^2\omega = \frac{1}{2}BR^2\omega$

WE-26: A copper disc of radius 1m is rotated about its natural axis with an angular velocity 2 rad/sec in a uniform magnetic field of 5 telsa with its plane perpendicular to the field. Find the emf induced between the centre of the disc and its rim.

Sol: $e = \frac{1}{2}B\omega r^2; \quad e = \frac{1}{2}\times 5\times 2\times 1\times 1 = 5\,volt$

ENERGY CONSIDERATION

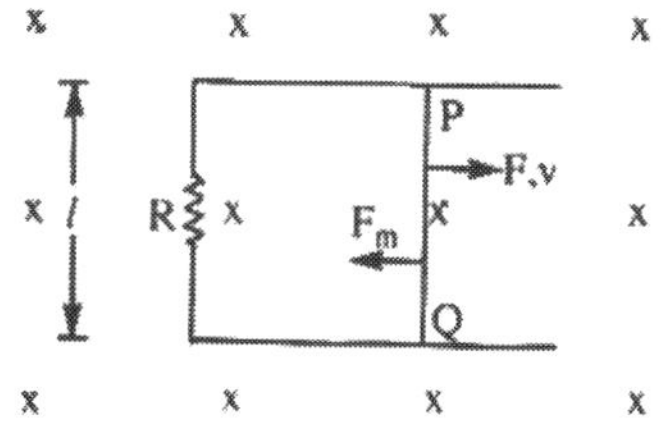

- A conductor PQ is moved with a constant velocity v on parallel sides of a U shaped conductor in a magnetic field as shown in figure. Let R be the resistance of the closed loop.

  The emf induced in the rod is $e=Blv$

  The current in the circuit is $i = \frac{e}{R} = \frac{Blv}{R}$

  As current flows in the conductor PQ from Q *to* P of the conductor. So, an equal and opposite force F has to be applied on the conductor to move the conductor with a constant velocity v.

  Thus, $F = F_m = \frac{B^2l^2v}{R}$

  The rate at which work is done by the applied force to move the rod is,

  $P_{\text{applied}} = Fv = \frac{B^2l^2v^2}{R}$

  The rate at which energy is dissipated in the circuit is,

  $P_{\text{dissipated}} = i^2R = \left(\frac{Bvl}{R}\right)^2 R = \frac{B^2l^2v^2}{R}$

  This is just equal to the rate at which work is done by the applied force.

WE-27: A 0.1 m long conductor carrying a current of 50 A is perpendicular to a magnetic field of 1.25 mT. The mechanical power to move the conductor with a speed of $1\,ms^{-1}$ is

Sol: Power $P=Fv$

$P=Bilv$

$l=0.1m$

$i=50$

$B = 1.25\times 10^{-3}$

$v = 1\,m/sec$

$\therefore p = Bilv$

$= 1.25 \times 10^{-3} \times 50 \times 0.1 \times 1$

$= 6.25 \times 10^{-3}$

$= 6.25\ mW$

WE-28: A short - circuited coil is placed in a time varying magnetic field. Electrical power is dissipated due to the current induced in the coil. If the number of turns were to be quadrupled and the radius of the wire is to be halved, then find the electrical power dissipated.

Sol: Current is induced in the short-circuited coil due to the imposed time - varying magnetic field.

Power $P = \frac{e^2}{R}$

Here $e = -\frac{d\phi}{dt}$ where $\phi = NBA$

and $R = \frac{\rho l}{\pi r^2}$ wher l and r are length and radius of the wire.

$$\therefore P = \frac{\pi r^2}{\rho l}\left[\frac{d}{dt} NBA\right]^2 \text{ or}$$

$$P = \frac{\pi r^2}{\rho l} N^2 A^2 \left(\frac{dB}{dt}\right)^2 \text{ or P=(constant)} \frac{N^2 r^2}{l}$$

when $r_2 = \frac{r_1}{2}$ then $t_2 = 4l_1$

$$\therefore \frac{P_2}{P_1} = \frac{(4N)^2}{N^2} \times \left(\frac{r}{2r}\right)^2 \times \left(\frac{l}{4l}\right)$$

$$\therefore \frac{P_2}{P_1} = \frac{16N^2 \times r^2 \times l}{N^2 \times 4r^2 \times 4l} \text{ or } \frac{P_2}{P_1} = \frac{1}{1}$$

$\therefore$ Power dissipated is the same.

WE-29 A pair of parallel horizontal conducting rails of negligible resistance, shorted at one end is fixed on a table. The distance between R can slide on the rails frictionlessly. The rod is tied to a massless string which passes over a pulley fixed to the edge of the table. A mass m, tied to the other end of the string, hangs vertically. A constant magnetic field B exists perpendicular to the table. If the system is released from rest, calculate :

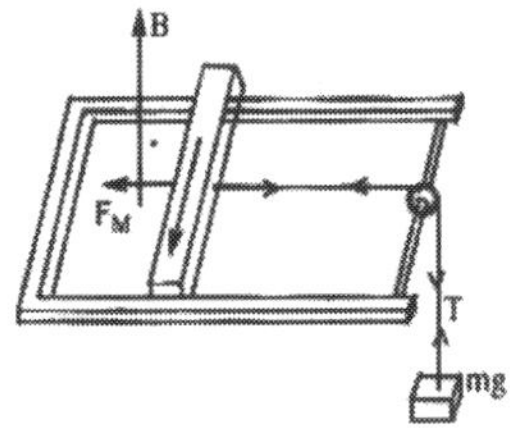

i) The terminal velocity achieved by the rod.

ii) The acceleration of the mass at the instant when the velocity of the rod, is half the terminal velocity.

Sol: i) the velocity of rod = V

Intensity of magnetic field = B

$\therefore$ emf induced in rod (e)=BLV

$\therefore$ current induced in rod $(i) = \dfrac{BlV}{R}$

Force on the rod $F = BiL = \dfrac{B^2VL^2}{R}$

Net force on the system = mg - T

mg - T = ma

but $T = F = \dfrac{B^2VL^2}{R}$

Hence, $mg - \dfrac{B^2VL^2}{R} = ma$

or $a = g - \dfrac{B^2VL^2}{mR}$ ....(i)

For rod to achieve terminal velocity $V_T, a = 0$

$\therefore 0 = g - \dfrac{B^2V_TL^2}{mR}$

or Terminal velocity $(V_T) = \dfrac{mgR}{B^2L^2}$ ....(ii)

ii) Acceleration of mass when $V = \dfrac{V_T}{2}$

or $V = \dfrac{mgR}{2B^2L^2}$. Put this value of V in (i)

$\therefore a = g - \dfrac{B^2L^2}{mR} \times \left(\dfrac{mgR}{2B^2L^2}\right)$ or $a = g - \dfrac{g}{2}$

or $a = \dfrac{g}{2}$ ....(iii)

WE-30: Two parallel vertical metallic bars XX^1 and YY^1, of negligible resistance and separated by a length 'l', are as shown in Fig. The ends of the bars are joined by resistance R_1 and R_2. A uniform magnetic field of induction B exists in space normal to the plane of the bars. A horizontal metallic rod PQ of mass m starts falling vertically, making contact with the bars. It is observed that in the steady state the powers dissipated in the resistance R_1 and R_2 and the terminal velocity attained by the rod PQ.

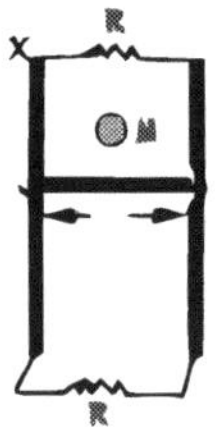

Sol: Let V_0 be the terminal velocity attained by the rod PQ (in the steady state). If i_1 and i_2 be the currents flowing through R_1 and R_2 in this state, then current flowing through the rod PQ is $i = i_1 + i_2$ (see the circuit diagram) as shown in Fig.

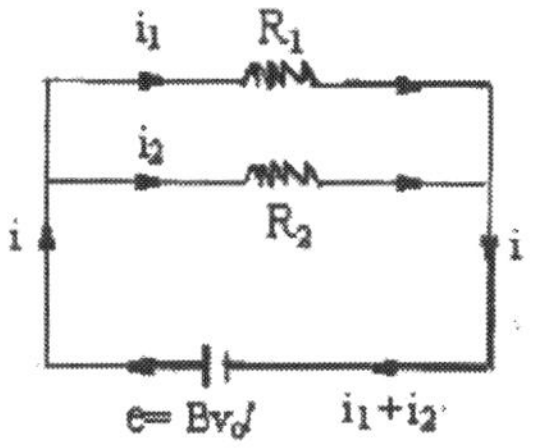

$\therefore$ Applying Kirchoff's loop rule, yields.

$i_1R_1 = BV_0l$ and $i_2R_2 = BV_0l$

$$\therefore i_1 + i_2 = BV_0l\left(\frac{1}{R_1} + \frac{1}{R_2}\right) \qquad ......(i)$$

Given that, $P_1 = i_1^2R_1 = \dfrac{B^2V_0^2l^2}{R_1}$ .....(ii)

and $P_2 = i_2^2R_2 = \dfrac{B^2V_0^2l^2}{R_2}$ .....(iii)

Also in the steady state, the acceleration of $PQ=0$

$\Rightarrow mg = B(i_1 + i_2)l$

(or) $mg = B^2l^2V_0\left(\dfrac{1}{R_1} + \dfrac{1}{R_2}\right) = P_1 + P_2$

[From equation (ii) and (iii)]

$\therefore$ The terminal velocity is $V_0 = \dfrac{P_1 + P_2}{mg}$

Substituting for V_0 in equation (ii),

$$P_1 = \frac{B^2l^2}{R_1}\left(\frac{P_1 + P_2}{mg}\right)^2 \Rightarrow R_1 = \left[\frac{Bl(P_1 + P_2)}{mg}\right]^2 \times \frac{1}{P_1}$$

Similarly from equation (iii)

$$R_2 = \left[\frac{Bl(P_1+P_2)}{mg}\right]^2 \times \frac{1}{P_2}$$

WE-31: The loop ABCD is moving with velocity 'v' towards right. The magnetic field is 4T. The loop is connected to a resistance of $8\,\Omega$. If steady current of 2A flows in the loop then value of 'v' if loop has a resistance of $4\,\Omega$, is : (Given AB=30cm, AD=30 cm)

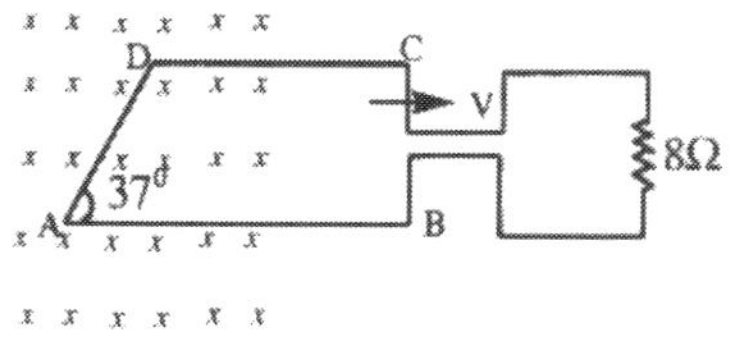

Sol: The induced emf in the loop is $e = Blv$

$$e = B(AD)\sin 37^0 v = 4\times 0.3\sin 37^0 v$$

Effective resistance of the circuit is $R = (4+8) = 12\Omega$

Hence $i = \frac{e}{R} = \frac{Blv}{R} \Rightarrow 2 = \frac{4\times 0.3\times \sin 37^0 v}{(4+8)}$

$$\therefore v = \frac{100}{3} m/s$$

WE-32: A square loop of side 12cm with its sides parallel to x and y-axes is moved with a velocity 8 cm/s along positive x-direction in an environment containing magnetic field along +ve z-direction. The field has a gradient of 10^{-3} tesla/em along -ve x-direction (increasing along -ve x-axis) and also decreases with time at the rate of 10^{-3} tesla/s. The emf induced in the loop is

Sol: The magnetic field in loop varies with position 'x' of loop and also with time simultaneously.

The rate of change of flux due to variation of 'B' with time is $\frac{d\phi}{dt} = A\times\frac{dB}{dt}$

The rate of change of flux due to variation B with position 'x' is

$$\frac{d\phi}{dt} = A\times\frac{dB}{dt} = A\frac{dB}{dx}\times\frac{dx}{dt} = A\frac{dB}{dx}\times v$$

Since both cause decrease in flux, the two effects will add up

$\therefore$ The net emf induced

$$e = \frac{d\phi}{dt} = A\frac{dB}{dt} + A\frac{dB}{dx}\times v = A\left[\frac{dB}{dt} + v.\frac{dB}{dx}\right]$$

$$= 144\times 10^{-4}\left[10^{-3} + 8\times 10^{-3}\right]$$

$$= 144\times 9\times 10^{-7} = 129.6\times 10^{-6} V$$

WE-33: A bar of mass m and length l moves on two frictionless parallel rails in the presence of a uniform magnetic field directed into the plane of the paper. The bar is given an initial velocity v_i to the right and released. Find the velocity of bar, induced emf across the bar and the current in the circuit as a function of time

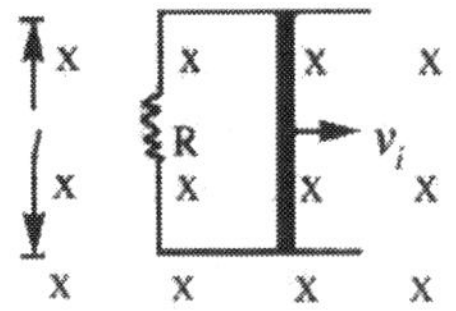

Sol: The induced current is in the counter clockwise direction and the magnetic force on the bar is given by $F_B = -ilB$. The negative sign indicates that the force is towards the left and retards motion.

F=ma

$$-ilB = m.\frac{dv}{dt}$$

Because the force depends on current and the current depends on the speed, the force is not constant and the acceleration of the bar is not constant. The induced current is given by $i = \frac{Blv}{R}$

$$-ilB = m.\frac{dv}{dt}$$

$$-\left(\frac{Blv}{R}\right)lB = m.\frac{dv}{dt} \Rightarrow \frac{dv}{v} = -\frac{B^2l^2}{mR}dt$$

$$\int_{v_1}^{v}\frac{dv}{v} = -\frac{B^2l^2}{mR}\int_0^t dt$$

$$\ln\left(\frac{v}{v_1}\right) = -\frac{B^2l^2}{mR}t = \frac{-t}{T}$$

where $T = \frac{mR}{B^2l^2} \Rightarrow v = v_i e^{\frac{t}{T}}$

The speed of the bar therefore decreases exponentially with time under the action of magnetic retarding force.

emf $= iR = Blv_i e^{\frac{t}{T}}$

current : $i = \frac{Blv}{R} = \frac{Bl}{R}v_1 e^{\frac{t}{T}}$

WE-34: The arm PQ of the rectangular conductor is moved from $x=0$, outwards in the uniform magnetic field which extends from $x=0$ to $x=b$ and is zero for $x>b$ as shown. Only the arm PQ possesses substantial resistance r. Consider the situation when the arm PQ is pulled outwards from $x=0$ to $x=2b$, and is then moved back to $x=0$ with constant speed v. Obtain expressions for the flux, the induced emf, the force necessary to pull the arm and the power dissipated as Joule heat. Sketch the variation of these quantities with distance.

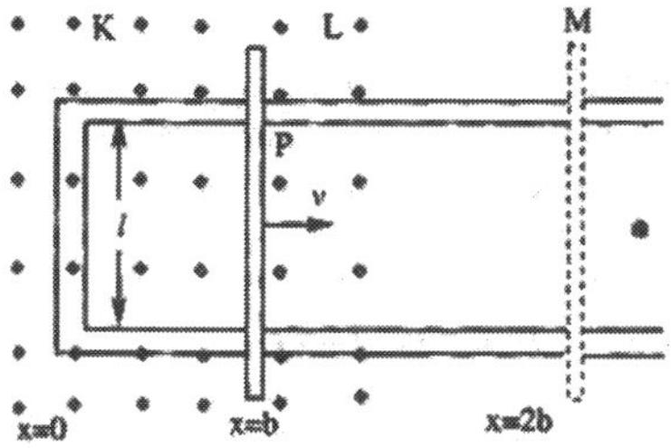

Sol: Let us first consider the forward motion from $x=0$ to $x=2b$. The flux ϕ_B linked with the circuit SPQR is

The induced emf is,

$$\varepsilon = -\frac{d\phi_B}{dt} = -Blv \qquad 0 \le x < b$$

$\varepsilon = 0 \quad b \le x < 2b$

When the induced emf is nonzero,

the current I is $I = \dfrac{Blv}{r}$ (in magnitude)

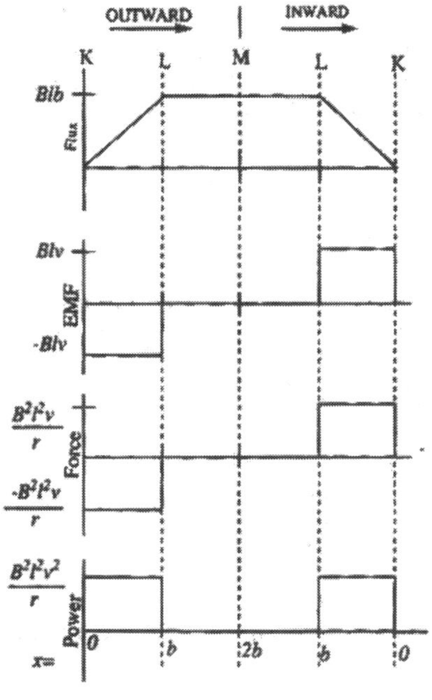

The force required to keep the arm PQ in constant motion is IlB. Its direction is to the left.

$$F = \frac{B^2 l^2 v}{r} \quad 0 \le x < b : F = 0 \quad b \le x < 2b$$

The Joule heating loss is

$$P_J = I^2 r = \frac{B^2 l^2 v^2}{r} \qquad 0 \le x < b$$

$$P_J = 0 \qquad b \le x < 2b$$

One obtains similar expressions for the inward motion from $x = 2b$ to $x = 0$. One can appreciate the whole process by examining the sketch of various quantities displayed in Fig

EDDY CURRENTS

- When bulk pieces of conductors are subjected to changing magnetic flux, induced currents are produced in them.
- The flow patterns of induced currents resemble the whirling eddies in water. This effect was discovered by Foucault and these currents are called eddy currents (or) Foucault currents.
- A copper plate is allowed to swing like a simple pendulum between the pole pieces of a strong magnet, its motion is damped and the plate comes to rest inthe magnetic field due to eddy currents in the plate.
- If rectangular slots are made in the copper plate area available to the flow of eddy currents is less. So, electromagnetic damping is reduced and the plate swings more freely.
- The eddy currents heat up the metallic cores and dissipate electrical energy inthe form of heat in the devices like transformers, electric motors and other such devices.
- The eddy currents are minimized by using. laminations of metal to make a metal core. The laminations are separated by an insulating material like lacquer.
- The plane of the laminations must be arranged parallel to the magnetic field, so that they cut across the eddy current paths reduces the strength of the eddy current.
- **Advantages :**

  Eddy currents are used in

  a) Magnetic braking in trains.

  b) Electromagnetic damping.

  c) Induction furnace.

  d) Electric power meters.

SELF INDUCTION :

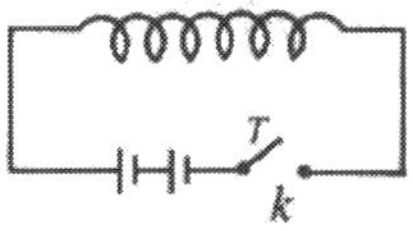

- If current flowing in a coil changes, the magnetic flux linked with the coil changes. Then emf induced in the coil is called self induced emf and the phenomenon is called self induction.
- If 'i' is the current flowing through the coil and 'ϕ' is magnetic flux linked with the coil, then

  $$\phi \propto i \quad \Rightarrow \phi = Li, \qquad \therefore L = \frac{\phi}{i}$$

  Here 'L' is called coefficient of self induction of the coil or self inductance of the coil.

  $$\phi \propto i \quad \Rightarrow \phi = Li, \qquad \therefore L = \frac{\phi}{i}$$

- Self induced e.m.f is given by

$$e = \frac{-d\phi}{dt} = -L\frac{di}{dt}$$

- Self inductance of a coil is magnetic flux linked with the coil when unit current flows through it (or) emf induced in the coil when current changes in it at the rate of 1 A/sec.
- S.I. Unit of self inductance : Henry.

  Other Units : weber / ampere, volt-second/ampere, $J/amp^2, Wb^2/J, volt sec^2 coul^{-1}$. Dimensional formula of L is $\left[ML^2T^{-2}I^{-2}\right]$
- A coil having high self inductance is called inductor.
- Self induction is also known as inertia of electricity as it opposes the growth or decay of the current in the circuit.
- Inductance may be viewed as electrical inertia. It is analogous to inertia in mechanics. It does not oppose the current, but is opposes the change in current.

SELF INDUCTANCE OF A FLAT CIRCULAR COIL :

Let us consider a circular coil of radius r and containing N-turns. Suppose it carries a current 'i'. The magnetic field at the centre due to this current $B = \frac{\mu_0 Ni}{2r}$

And total flux $= NBA = N\left(\frac{\mu_0 Ni}{2r}\right)\pi r^2 = \frac{\mu_0 \pi N^2 ri}{2}$

Now comparing with $N\phi_B = Li$ we get

$$L = \frac{\mu_0 \pi N^2 r}{2}$$

SELF INDUCTANCE OF A SOLENOID :

Consider a long solenoid of length l, area of cross section A and number of turns per unit length n and length is very large when compared with radius of cross section.

Let I be the current flowing through the solenoid. The magnetic field inside the long solenoid is uniform and is given by $B = \mu_0 nI$

Total number of turns in the solenoid of length l is $N=nl$.

Now, the magnetic flux linked with each turn of the solenoid $B \times A = \mu_0 nIA$

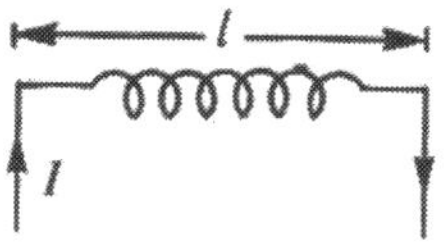

$\therefore$ Total magnetic flux linked with the whole solenoid, ϕ =magnetic flux with each turn $\times$ number of turns in the solenoid.

$\phi = \mu_0 nIA \times nl = \mu_0 n^2 IAl$ ....(1)

But $\phi = LI \Rightarrow LI = \mu_0 n^2 IAl$ from (1) & (2) $\therefore L = \mu_0 n^2 Al$

Since $n = \frac{N}{l}, L = \mu_0 \frac{N^2}{l} A$

- Self inductance of coil depends on
  i) Geometry of the coil
  i.e., a) Number of turns of the coil
  b) The length (l) of the solenoid,
  c) The area of cross-section (A) of the solenoid,

ii) Medium inside the coil (permeability)

iii) Nature of the material of the core of the solenoid.

- More is the permeability of the medium, more is the self inductance
- An inductor will have large inductance and low resistance.
- Resistor opposes the current, inductor opposes the change of current
- One can have resistance without inductance
- One cannot have inductance without resistance.
- An ideal inductor has inductance and no resistance.
- When the current in the coil either increases or decreases at a rate, then the coil can be imagined to be a cell of emf $e = L.\frac{di}{dt}$
- One can have self inductance without mutual inductance.
- One cannot have mutual inductance without self inductance.

The direction of induced emf for different states of current in a coil :

a) Steady current

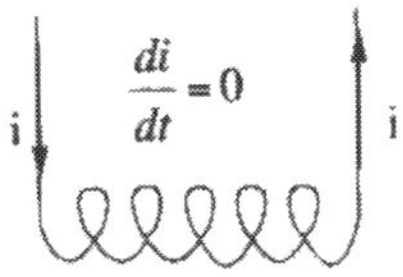

e = 0 no opposition

b) Make of circuit or increasing current

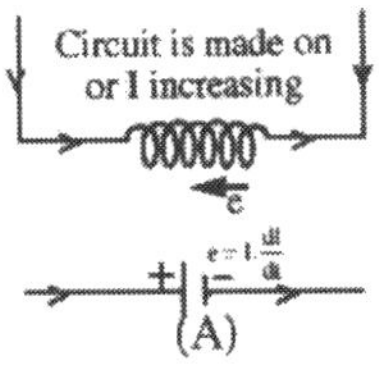

c) Breaking of circuit of decreasing of current

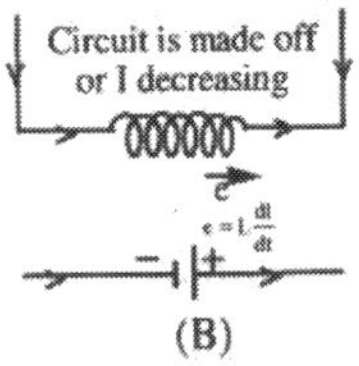

MUTUAL INDUCTION

- When current in one coil changes, magnetic flux linked with the second coil placed near by it also changes. The emf induced in secondary is called mutually induced emf and the phenomenon is called mutual induction.

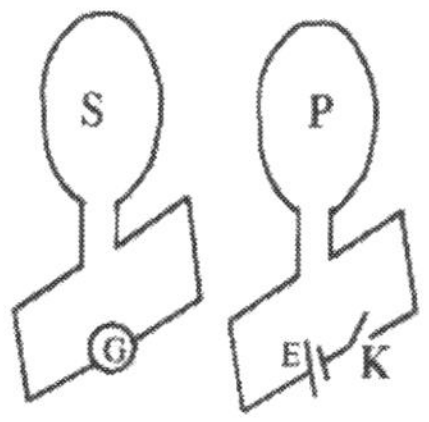

- If 'i_p' is current flowing in the primary coil, 'ϕ_S' is magnetic flux linked with secondary coil, then

  $\phi_S \propto i_p$

  $\Rightarrow \phi_S = Mi_p, \qquad \therefore M = \frac{\phi_S}{i_p}$

  Here 'M' is called coefficient of mutual induction or mutual inductance.
- Induced emf in secondary coil is

  $$e = \frac{-d\phi}{dt} = -M\left(\frac{di_p}{dt}\right) \text{ (or) } M = \frac{e}{-di_p / dt}$$

- Mutual indictance between two coils is equal to the magnetic flux linked in the secondary coil when unit current passes through the primary coil (or) emf induced in one coil when current in the other coil changes at the rate of 1 Amp/second.
- S.I. unit : Herry
- Dimensional formula of self inductance or mutual inductance is $ML^2T^{-2}A^{-2}$
- The value of mutual inductance depends on
  1) Distance between the two coils
  2) Number of turns of coils
  3) Geometrical shape of the coil
  4) Material of the core medium between the coils
  5) Orientation of the coils i.e., angle between the axes of the coils.

  If the axes the parallel, then M is maximum

  If the axes are perpendicular then M is minimum

MUTUAL INDUCTANCE OF TWO LONG COAXIAL SOLENOIDS :

- Consider two solenoids S_1 and S_2 such that the solenoid S_2 completely surrounds the solenoid S_1.

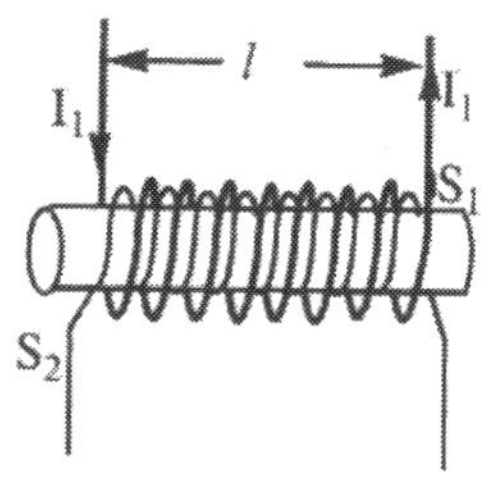

Let l be length of each solenoid (or length of primary coil) and of nearly same area of cross-section A. N_1 and N_2 are the total number of turns of solenoid S_1 and S_2 respectively.

$\therefore$ Number of turns per unit length of solenoid S_1 is, $n_1 = \frac{N_1}{l}$

Number of turns per unit length of solenoid S_2 is, $n_2 = \frac{N_2}{l}$

Magnetic field inside the solenoid S_1 is given by $B_1 = \mu_0 n_1 I_1 = \mu_0 \frac{N_1}{l} I_1$

$\therefore$ Magnetic flux linked with each turn of solenoid $S_2 = B_1 A = \mu_0 \frac{N_1}{l} I_1 A$

$\therefore$ Total magnetic flux linked with N_2 turns of the solenoid S_2 is

$$\phi_2 = N_2 (B_1 A) = \mu_0 \frac{N_1}{l} I_1 A \times N_2$$

$$\phi_2 = \frac{\mu_0 N_1 N_2 I_1 A}{l} \text{ .... (i)}$$

But $\phi_2 = M_{12} I_1$ .... (ii)

Where M_{12} is the mutual inductance when current varies in solenoid S_1 and makes magnetic flux linked with solenoid S_2,

from (i) and (ii) we get

$$M_{12} I_1 = \frac{\mu_0 N_1 N_2 I_1 A}{l}$$

$$\therefore \quad M_{12} = \frac{\mu_0 N_1 N_2 A}{l}$$

Similarly, $M_{21} = \frac{\mu_0 N_1 N_2 A}{l}$, where M_{21} is the mutual inductance when current varies in solenoid S_2 and makes magnetic flux linked with solenoid S_1.

It can be proved that $M_{12} = M_{21} = M$

The above equation is treated as a general result, if the two solenoids are wound on a magnetic substance of relative permeability μ_r, then the mutual inductance is given by

$$M = \frac{\mu_0 \mu_r N_1 N_2 A}{l} = \mu_0 \mu_r n_1 n_2 A l$$

WE-35: Two different coils have self inductance $L_1 = 8\text{mH}$ and $L_2 = 2\text{mH}$. The currents in both are increasing at the same constant rate. At a certain instant of time, the power given to the two coils is the same. At this moment the current, the induced voltage and energy stored in the first coil are i_1, V_1 and U_1 respectively. The corresponding values in the second coil are i_2, V_2 and U_2 respectively. Then the values of $\frac{i_1}{i_2}$, $\frac{V_1}{V_2}$ and $\frac{U_1}{U_2}$ are respectively

Sol: $\frac{i_1}{i_2} = \frac{L_2}{L_1} = \frac{2}{8} = \frac{1}{4}$

$\frac{v_1}{v_2} = \frac{L_1}{L_2} = \frac{8}{2} = 4$

$\frac{U_1}{U_2} = \frac{L_2}{L_1} = \frac{2}{8} = \frac{1}{4}$

WE-36: Two coaxial solenoids are made by winding thin insulated wire over a pipe of cross-sectional area $A = 10cm^2$ and length $= 20$cm. If one of the solenoids has 300 turns and the other 400 turns, their mutual inductance is $\left(\mu_0 = 4\pi \times 10^{-7} TmA^{-1}\right)$

Sol: $M = \frac{\mu_0 N_1 N_2 A}{L}$

$$M = \frac{4\pi \times 10^{-7} \times 3 \times 10^2 \times 4 \times 10^2 \times 10^{-3}}{2 \times 10^{-1}} = 2.4\pi \times 10^{-4} H$$

ENERGY STORED IN AN INDUCTOR

➢ Consider an ideal inductor of inductance 'L' connected with a battery. Let I be the current inthe circuit at any instant 't'

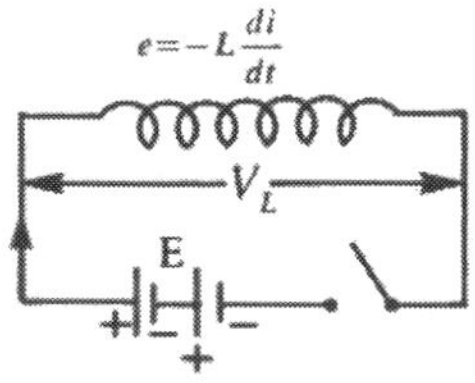

This induced emf is given by $e = -L\frac{dI}{dt}$

-ve sign shows that 'e' opposes the change of current I in the inductor.

To drive the current through the inductor against the induced emf 'e', the external voltage is applied. Here external voltage is emf of the battery = E

According to Kirchoff's voltage law, E+e=0

$E = -e$

$E = L\frac{dI}{dt}$

Let an infinitesimal charge dq be driven through the inductor in time dt. So, the rate of work done by the external voltage is given by

$$\frac{dW}{dt} = EI = L\frac{dI}{dt} \times I = LI\frac{dI}{dt}$$

The total work done in establishing a current through the inductor from 0 to I is given by

$$W = \int dW = \int_0^I LI\, dI$$

$$W = L\left(\frac{I^2}{2}\right) = \frac{1}{2}LI^2$$

$$\boxed{W = \frac{1}{2}LI^2}$$

The work done in maintaining the current through the inductor is stored as the potential energy (U) in its magnetic field. Hence energy stored inthe inductor is given by

$$\boxed{U = \frac{1}{2}LI^2}$$

- The equation $U = \frac{1}{2}LI^2$ is similar to the expression for kinetic energy $E = \frac{1}{2}mv^2$. It shows that L is analogues to mass 'm' and self inductance is called electrical inertia.
- The self inductance of a coil is numerically equal to twice the energy stored in it when unit current flows through it.
  i.e., When i=1A, L=2U
- Induced power $P = e \times i = Li\left(\frac{di}{dt}\right)$.
- In case of solenoid $L = \mu_0 n^2 Al$
- Magnetic energy stored per unit volume

  $$u_B = \frac{\frac{1}{2}Li^2}{Al} \Rightarrow u_B = \frac{1}{2}u_0 n^2 i^2$$

  Hence $u_B = \frac{B^2}{2\mu_0}$
- The magnetic energy stored per unit volume similar to electrostatic energy stored per unit volume in a parallel plate capacitor $u_B = \frac{1}{2}\varepsilon_0 E^2$

  In both cases the energy is proportional to the square of field stregth.

WE-37 : The self-inductance of a coil having 200 turns is 10 milli henry. Calculate the magnetic flux through the cross-section of the coil corresponding to current of 4 milliampere. Also determine the total flux linked with each turn.

Sol: Total magnetic flux linked with the coil,

$$N\phi = LI = 10^{-2} \times 4 \times 10^{-3} = 4 \times 10^{-5} Wb$$

$$\therefore \text{Flux per turn, } \phi = \frac{4 \times 10^{-5}}{200} = 2 \times 10^{-7} Wb$$

WE-38 : A coil of inductance 0.2 henry is connected to 600 volt battery. At what rate, will the current in the coil grow when circuit is completed ?

Sol: As the battery and inductor are in parallel, at any instant, emf of the battery and self emf in the inductor are equal

$$|e| = L\frac{dI}{dt} \text{ or } \frac{dI}{dt} = \frac{|e|}{L} = \frac{600V}{0.2H} = 3000A\ s^{-1}$$

WE-39 : An inductor of 5H inductance carries a steady current of 2A. How can a 50V self-induced emf be made to appear in the inductor ?

Sol: $L = 5H; |e| = 50V$

Let us produce the required emf by reducing current to zero

Now, $|e| = L\frac{dI}{dt}$ or $dt = \frac{LdI}{|e|} = \frac{5\times 2}{50}s$

$\frac{10}{50}s = \frac{1}{8}s = 0.2\,s$

So, the desired emf can be produced by reducing the given current to zero in 0.2 second

WE-40 : Two different coils have self-inductances $L_1 = 16\,mH$ and $L_2 = 12\,mH$. At a certain instant, the current in the two coils is increasing at the same rate of power supplied to the two coils is the same. Find the ratio of i) induced voltage ii) current iii) energy stored in the two coils at that instant.

Sol: i) $V_1 = L_1\frac{dI}{dt}; V_2 = L_2\frac{dI}{dt}$

$$\frac{V_1}{V_2} = \frac{L_1}{L_2} = \frac{16}{12} = \frac{4}{3}$$

ii) $P = V_1 I_1 = V_2 I_2 \Rightarrow \frac{I_1}{I_2} = \frac{V_1}{V_2} = \frac{3}{4}$

iii) $\frac{U_1}{U_2} = \frac{\frac{1}{2}L_1 I_1^2}{\frac{1}{2}L_2 I_2^2} = \left(\frac{L_1}{L_2}\right)\left(\frac{I_1}{I_2}\right)^2 = \frac{4}{3}\left(\frac{3}{4}\right)^2 = \frac{3}{4}$

WE-41 : The network shown is a part of the closed circuit in which the current is changing. At an instant, current in it is 5A. Potential difference between the points A and B if the current is

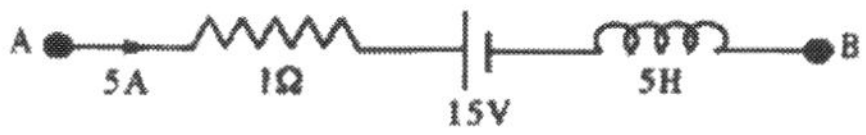

1) Increasing at 1A/sec

2) Decreasing at 1A/sec

Sol: 1) The coil can be imagined as a cell of emf

$$e = L\left(\frac{di}{dt}\right) = 5\times 1 = 5V$$

$\therefore$ Equivalent circuit is

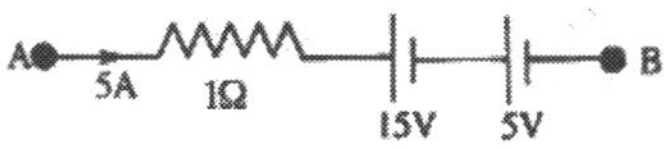

$V_A - 5(1) - 15 - 5 = V_B$

Hence $V_A - V_B = 5 + 15 + 5 = 25V$

2) The coil can be imagined as a cell of emf

$$e = L\left(\frac{di}{dt}\right) = 5 \times 1 = 5V$$

$\therefore$ Equivalent circuit is

$V_A = 5(1) - 15 + 5 = V_B$

Hence $V_A - V_B = 5 + 15 - 5 = 15V$

RELATION BETWEEN L_1, L_2 AND M :

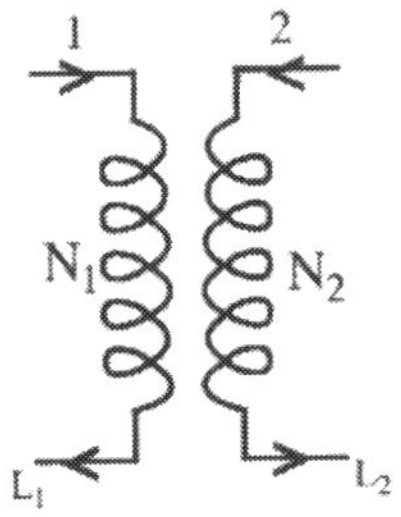

The flux linked with coil 1 is

$$N_1\phi_1 = L_1 i_1 \Rightarrow L_1 = \frac{N_1\phi_1}{i_1}$$

The flux linked with coil 2 is

$$N_2\phi_2 = L_2 i_2 \Rightarrow L_2 = \frac{N_2\phi_2}{i_2}$$

M on 1 because of 2 ; $M_{12} = \frac{N_1\phi_1}{i_2}$

M on 2 because of 1 ; $M_{21} = \frac{N_2\phi_2}{i_1}$

- If the flux in linkage is maximum, then $M_{12} = M_{21} = M$

$$M_{12} \times M_{21} = \frac{N_2\phi_2}{i_1} \times \frac{N_1\phi_1}{i_2}$$

$$M^2 = L_1 L_2$$

$$\therefore \qquad M = \sqrt{L_1 L_2}$$

This is the maximum mutual inductance when all the flux linked with one coil is also completely linked with the other.

In general, only a fraction of the total flux will be linked with the coil due to the flux leakage.

$\therefore \quad M = K\sqrt{L_1 L_2}$

Where K-coefficient of coupling $(K \le 1)$

For tight coupling (or) if the coils are closely wound, then K=1.

$\therefore \quad M_{max} = \sqrt{L_1 L_2}$

INDUCTORS IN SERIES :

If two coils of inductances L_1 and L_2 are connected in series then the potential divides.

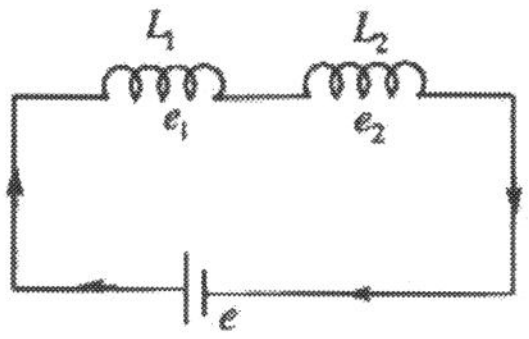

i.e., $e = e_1 + e_2$ (or) $L_S \dfrac{di}{dt} = L_1 \dfrac{di}{dt} + L_2 \dfrac{di}{dt}$

Since in series, $\dfrac{di}{dt}$ is same for all coils

$\therefore L_S = L_1 + L_2$

If n coils of inductances $L_1, L_2, L_3 ....L_n$ are connected in series then effective inductance of the arrangement,

$L = L_1 + L_2 + L_3 + ...+ L_n$

(when coils are far away)

INDUCTORS IN PARALLEL :

If two coils of inductances L_1 and L_2 are connected in parallel then the current divides.

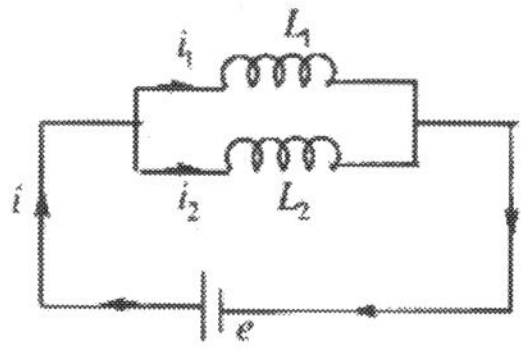

i.e., $i = i_1 + i_2$ (or) $\dfrac{di}{dt} = \dfrac{di_1}{dt} + \dfrac{di_2}{dt}$

$\Rightarrow \dfrac{e}{L_P} = \dfrac{e_1}{L_1} + \dfrac{e_2}{L_2}$

However in parallel as potential difference remains same i.e, $e = e_1 = e_2$, so

$\dfrac{1}{L_P} = \dfrac{1}{L_1} + \dfrac{1}{L_2}$ (or) $L_P = \dfrac{L_1 L_2}{(L_1 + L_2)}$

If n coil of inductances $L_1, L_2, L_3 .... L_n$ are connected in parallel then effetive inductance of the arrangement,

$$\frac{1}{L_P} = \frac{1}{L_1} + \frac{1}{L_2} + ... + \frac{1}{L_n}$$

(when coils are far away)

- Let two coils of inductances L_1 and L_2 are connected in series and M is their mutual inductance. The flux linked with one coil will be the sum of two fluxes which exist independently. When the flux in the two coils support each other

$$N_1\phi_1 = L_1 i_1 + M_{12} i_2$$

From Faraday's law, $e_1 = -L_1 \frac{di_1}{dt} - M_{12}\frac{di_2}{dt}$

Similarly $N_2\phi_2 = L_2 i_2 + M_{21} i_1$

$$e_2 = -L_2 \frac{di_2}{dt} - M_{21}\frac{di_1}{dt}$$

$$e = e_1 + e_2 = -L_1 \frac{di_1}{dt} - M_{12}\frac{di_2}{dt} - L_2\frac{di_2}{dt} - M_{21}\frac{di_1}{dt}$$

In series the current i and the change in current di is same $e = -\left(L_1 + M_{21} + L_2 + M_{12}\right)\frac{di}{dt}$

$$L = \left(L_1 + M_{21} + L_2 + M_{12}\right) = L_1 + L_2 + 2M$$

If the two coils oppose each other, then

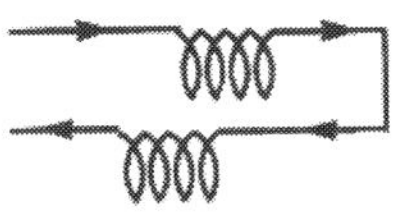

$$L = \left(L_1 - M\right) + \left(L_2 - M\right) = L_1 + L_2 - 2M$$

WE-42 : Calculate the mutual inductance between two coils when a current of 2A changes to 6A in 2 seconds and induces an emf of 20 mV in the secondary coil

Sol: $|e| = M\frac{dI}{dt}$

$$20\times10^{-3} = M\frac{(6-2)}{2} \text{ (or) } M = 10\text{mH}$$

WE-43 : If the coefficient of mutual induction of the primary and secondary coils of an induction coil is 6H and a current of 5A is cut off in 1/5000 second, calculate the emf induced in the secondary coil.

Sol: $|e| = M\frac{dI}{dt}$; $e = 6\times\frac{5}{1/5000}V = 15\times10^4 V$

WE-44 : A solenoid is of length 50 cm and has a radius of 2cm. It has 500 turns. Around its central section a coil of 50 turns is wound. Calculate the mutual inductance of the system.

Sol: $N_P = 500, N_S = 50$

$A = \pi \times 0.02 \times 0.02 m^2$

$\mu_0 = 4\pi \times 10^{-7} Hm^{-1}, 1 = 50 cm = 0.5 m$

Now, $M = \dfrac{\mu_0 N_P N_S A}{l}$

$= \dfrac{4\pi \times 10^{-7} \times 500 \times 50 \times \pi \times (0.02)^2}{0.5} H$

$= 789.8 \times 10^{-7} H = 78.98 \mu H$

WE-45 : A solenoidal coil has 50 turns per centimetre along its length and a cross-sectional area of $4 \times 10^{-4} m^2$. 200 turns of another wire is wound round the first solenoid co-axially. The two coils are electrically insulated from each other. Calculate the mutual inductance between the two coils.

Sol: $n_1 = 50$ turns per cm

$= 5000$ turns per metre

$n_2 l = 200, A = 4 \times 10^{-4} m^2$

$M = \mu_0 n_1 (n_2 l) A$

$= 4\pi \times 10^{-7} \times 5000 \times 200 \times 4 \times 10^{-4} H = 5.03 \times 10^{-4} H$

WE-46 : Two circular coils, one of smaller radius r_1 and the other of very large radius r_2 are placed co-axially with centres coinciding. Obtain the mutual inductance of the arrangement.

Sol: Suppose a current I_2 flows through the outer circular coil. The field at the centre of the coil is

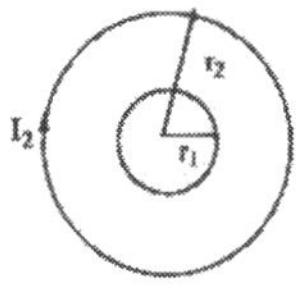

$B_2 = \dfrac{\mu_0 I_2}{2r_2}$

The second co-axially placed coil has very small radius. So B_2 may be considered constant over its cross-sectional area.

Now, $\phi_1 = \pi r_1^2 B_2 = \pi r_1^2 \left(\dfrac{\mu_0 I_2}{2r_2}\right)$

or $\phi_1 = \dfrac{\mu_0 \pi r_1^2}{2r_2} I_2$

Comparing with $\phi_1 = M_{12} I_2$, we get

$M_{12} = \dfrac{\mu_0 \pi r_1^2}{2r}$

Also, $M_{21} = M_{12} = \frac{\mu_0 \pi r_1^2}{2r_2} \Rightarrow M \propto \frac{r_1^2}{r_2}$

* It would have been difficult to calculate the flux through the bigger coil of the nouniform field due to the current in the smaller coil and hence the mutual inductance M_{12}. The equality $M_{12} = M_{21}$ is helpful. Note also that mutual inductance depends solely on the geometry.

WE-47: A small square loop of wire of side l is placed inside a large square loop of wire of side $L(>> l)$. The loops are coplanar and their centres coincide. What is the mutual inductance of the system?

Sol: Considering the large loop to be made up of four rod each of length L, the field at the centre, i.e., at a distance (L/2) from each rod, will be

$$B = 4\times\frac{\mu_0}{4\pi}\frac{I}{d}[\sin\alpha + \sin\beta]$$

i.e., $B = 4\times\frac{\mu_0}{4\pi}\frac{I}{(L/2)}\times 2\sin 45$

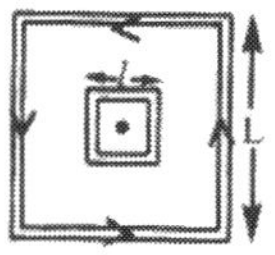

i.e., $B_1 = \frac{\mu_0}{4\pi}\frac{8\sqrt{2}}{L}I$

So the flux linked with smaller loop

$$\phi_2 = B_1 S_2 = \frac{\mu_0}{4\pi}\frac{8\sqrt{21}}{L}l^2$$

and hence, $M = \frac{\phi_2}{I} = 2\sqrt{2}\frac{\mu_0}{\pi}\frac{l^2}{L} \Rightarrow M \propto \frac{l^2}{L}$

WE-48: (a) A toroidal solenoid with an air core has an average radius of $0.15m$, area of cross section $12\times10^{-4}m^2$ and 1200 turns. Obtain the self inductance of the toroid. Ignore field variation across the cross section of the toroid. (b) A second coil of 300 turns is wound closely on the toroid above. If the current in the primary coil is increased from zero to 2.0 A in $0.05s$, obtain the induced emf in the secondary coil.

Sol: (a) $B = \mu_0 n_1 I = \frac{\mu_0 N_1 I}{l} = \frac{\mu_0 N_1 I}{2\pi r}$

Total magnetic flux, $\phi_B = N_1 BA = \frac{\mu_0 N_1^2 IA}{2\pi r}$

But $\phi_B = LI$ $\quad\therefore L = \frac{\mu_0 N_1^2 A}{2\pi r}$

$$L = \frac{4\pi\times10^{-7}\times1200\times1200\times12\times10^{-4}}{2\pi\times0.15}H$$

$= 2.3 \times 10^{-3} H = 2.3 mH$

(b) $|e| = \frac{d}{dt}(\phi_2)$, where ϕ_2 is the total magnetic flux linked with the second coil.

$$|e| = \frac{d}{dt}(N_2 BA) = \frac{d}{dt}\left[N_2 \frac{\mu_0 N_1 I}{2r} A\right]$$

or $$|e| = \frac{\mu_0 N_1 N_2 A}{2r}\frac{dI}{dt}$$

or $$|e| = \frac{4\pi \times 10^{-7} \times 1200 \times 300 \times 12 \times 10^{-4} \times 2}{2 \times 0.15 \times 0.05} V$$

=0.023 V

AC GENERATOR :

- An ac generator converts mechanical energy into electrical energy. The device used for the purpose is called ac generator.
- When the coil having N turns is rotated with a constant angular speed ω, the angle between the area vector A and the magnetic field vector B is at any instant *t* is $\theta = \omega t$ (assuming $\theta = 0^0$ at t=0). The flux linked with the coil at any instant *t* is $\phi_B = NBA\cos\theta = NBA\cos\omega t$

  From Faraday's law, the induced emf for the rotating coil of N turns is,

$$\varepsilon = -\frac{d\phi_B}{dt} = -\frac{d}{dt}(NBA\cos\omega t) = NBA\ \omega\sin\omega t$$

- The magnitude of induced emf is

  $\varepsilon = NBA\ \omega\sin\omega t = \varepsilon_0 \sin\omega t$

  where $e_0 = NBA\omega$ is the maximum value of the emf.

  ε_0 is called the amplitude or peak value of emf.

- The induced emf depends upon (i) strength of the magnetic field, (ii) area of the coil, (iii) speed of rotation, and (iv) the number of turns of the coil.

  If f be the frequency of rotation of coil, then $\varepsilon = \varepsilon_0 \sin 2\pi f t$

- A graph plotted between ε and ωt, is a sine curve as shown in Fig.

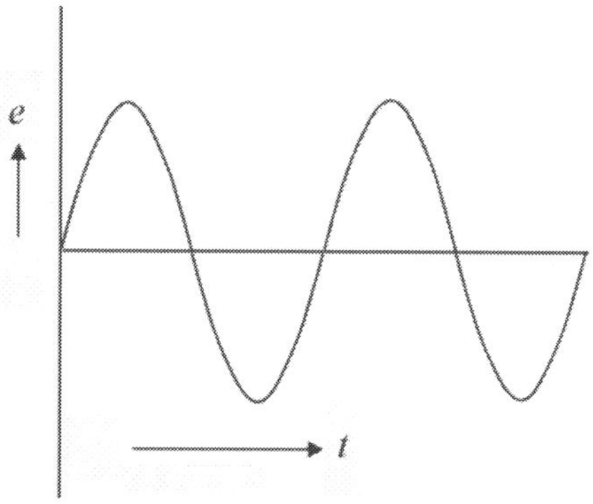

WE-49: A boy pedals a stationary bicycle at one revolution per second. The pedals are attached to 100 turns coil of are $0.1m^2$ and placed in a uniform magnetic field of 0.1T. What is the maximum voltage generated in the coil ?

Sol: $\varepsilon_0 = NBA\omega = NBA(2\pi f)(\because f = 1)$

$\varepsilon_0 = 100 \times 0.1 \times 0.1(2 \times 3.14 \times 1)V = 6.28V$

WE-50: A coil of 800 turns and 50 cm^2 area makes 10 rps about an axis in its own plane in a magnetic field of 100 gauss perpendicular to this axis. What is the instantaneous induced emf in the coil?

Sol: $A = 50cm^2 = 50 \times 10^{-4} m^2$

$n = 10\, rps,\ N = 800$

$B = 100\, gauss = 100 \times 10^{-4} T = 10^{-2} T$

Now, $\varepsilon = \varepsilon_0 \sin \omega t = NBA\omega \sin \omega t$

$= 800 \times 10^{-2} \times 50 \times 10^{-4} \times 2\pi \times 10 \sin(20\pi t)$

or $\varepsilon = 2.5 \sin(20\pi t)\, volt$

WE-51: A person peddles a stationary bicyle the pedals of the bicycle are attached to a 100 turn coil of area $0.10\ m^2$. The coil rotates at half a revolution per second and it is placed in a uniform magnetic field of 0.01 T perpendicular to the axis of rotation of the coil, What is the maximum voltage generated in the coil ?

Sol: Here $f = 0.5Hz$: $N = 100$, $A = 0.1m^2$ and $B = 0.01T$ from the equation

$\varepsilon = \varepsilon_0 \sin \omega t = NBA\omega \sin \omega t$ maximum emf $\varepsilon_0 = NBA\omega = NBA(2\pi f)$

$\varepsilon_0 = 100 \times 0.01 \times 0.1 \times 2 \times 3.14 \times 0.5 = 0.314V$

INDUCED ELECTRIC FIELDS :

When a conducting loop is placed in a varying magnetic field, a varying electric field produced in the loop, is called induced electric field.

An electric field is always generated by a changing magnetic field, even in free space where no charges are present.

Consider a conducting loop of radius R, situated in a uniform magnetic field $\overline{B}$ that is perpendicular to the plane of the loop as shown in the figure

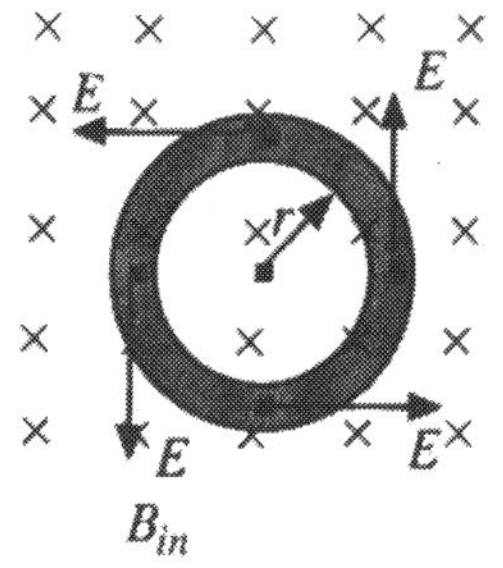

If the magnetic field changes with time, then an emf $e = \frac{-d\phi}{dt}$ is induced in the loop. The induced current thus produced implies the presence of an induced electric field E that must be tangential to the loop in order to provide an electric force on the charge around the loop.

The work done by the electric field on the loop in moving a test charge q once around the loop=qe. Because the magnitude of electric force onthe charge is qE, the work done by the electric field can also be expressed as $qE(2\pi r)$, where $2\pi r$ is the circumference of the loop. These two expressions for the work must be equal; therefore, we see that

$$qe = qE(2\pi r)$$

$$E = \frac{e}{2\pi r}$$

Using this result along with Faraday's law and the fact that $\phi_B = BA = B\pi r^2$ for a circular loop, the induced electric field can be expressed as

$$E = \frac{1}{2\pi r}\left(-\frac{d\phi_B}{dt}\right) = -\frac{1}{2\pi r}\frac{d}{dt}\left(B\pi r^2\right) = -\frac{r}{2}\frac{dB}{dt}$$

The emf for any closed path can be expressed as the line integral of $\vec{E}.\vec{dl}$ over that path. Hence, the general form of Faraday's law of induction is

$$e = \oint \vec{E}.\vec{dl} = \frac{d\phi_B}{dt}$$

It is important to recognize that the induced electric field E that appears in the equation is a non-convervative field that is generated by a changing magnetic field.

➢ Points to remember about induced electric field.

1) The induced electric field is produced only by changing magnetic field and not by charged particles.
2) One cannot define potentials w.r.t this induced field
3) The lines of induced electric field are closed curves and have no starting and terminating points.
4) As long as the magnetic field keeps on changing, the induced electric field will be present because this electric field is produced only by variable magnetic field.

WE-52: A uniform magnetic field of induction B is confined in a cylinderical region of radius R. If the field is increasing at a constant rate of $\frac{dB}{dt} = \alpha\, T/s$, then the intensity of the electric field induced at point P, distant r from the axis as shown in the figure is proportional to :

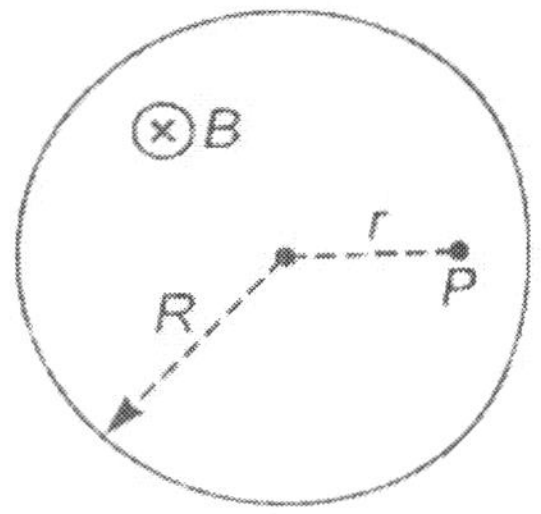

Sol: For r < R

$$e = \int E.ds$$

$$= \frac{d\phi_B}{dt}$$

$$E.2\pi r = -A\left(\frac{dB}{dt}\right)$$

$$E.2\pi r = -\pi r^2\left(\frac{dB}{dt}\right)$$

$$E = -\frac{r}{2}\left(\frac{dB}{dt}\right)$$

$$E = -\frac{r}{2}\alpha$$

$$\left(\overline{E}\right) = \frac{r}{2}\alpha$$

$E\alpha r$

WE-53: Magnetic flux linked with a stationary loop of resistance R varies with respect to time during the time period T as follows: $\phi = at\left(T-t\right)$ the amount of heat generated in the loop during that time (inductance of the coil is negligible) is

Sol: Give that $\phi = at\left(T-t\right)$

induced emf, $E = \frac{d\phi}{dt}$

$$= \frac{d}{dt}\left[at\left(T-t\right)\right]$$

$$= at\left(0-1\right) + a\left(T-1\right)$$

$$= a\left(T-2t\right)$$

So, induced emf is also a function of time Heat generated

$$H = \int_0^T \frac{E^2}{R}dt$$

$$= \frac{a^2}{R}\int_0^T \left(T-2t\right)dt$$

$$= \frac{a^2T^3}{3R}$$

WE-54: A closed loop of cross-sectional area 10^{-2}m^2 which has inductance L=10 mH and negligible resistance is placed in a time-varying magnetic field. Figure shows the variation of B with time for the interval 4 s. The field is perpendicular to the plane of the loop (given at $t=0, B=0, I=0$). The value of the maximum current induced in the loop is

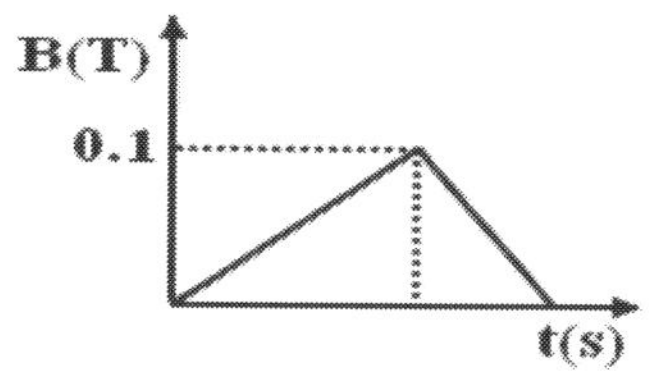

Sol: Induced emf(e) $= L\dfrac{di}{dt}$

$$\Rightarrow A\left(\frac{dB}{dt}\right) = L\frac{di}{dt} \quad \Rightarrow di = \frac{A}{L}\left(\frac{dB}{dt}\right) \times dt$$

$$\int_0^I di = \int_0^B \left(\frac{A}{L}\right) dB$$

$$I = \frac{A}{L}B$$

$$\Rightarrow I_{\max} = \frac{A}{L}B_{\max}$$

$$= \frac{10^{-2}}{10 \times 10^{-3}} \times 0.1 \quad = 0.1A = 100mA$$

WE-55: A magnetic field directed into the page changes with time according to the expression $B = \left(0.03t^2 + 1.4\right)T$, where t is in seconds. The field has a circular cross - section of radius R -2.5cm. What is the magnitude and direction of electric field at P, when t = 3.0s and r = 0.02 m.

Sol: $e = \oint E.dl = \dfrac{+d\phi}{dt}$

$$E(2\pi r) = A.\frac{dB}{dt} = \pi r^2 \times \frac{d}{dt}\left(0.03t^2 + 1.4\right)$$

$$E = \frac{\pi r^2}{2\pi r} \times (0.06t) = \frac{r}{2}(0.06t)$$

$$|E| = \frac{0.02}{2} \times 0.06 \times 3 = 18 \times 10^{-4} N/C$$

WE-56: The magnetic field at all points within the cylindrical region whose cross- section is indicated in the accompanying figure starts increasing at a constant rate 'α'. Find the magnitude of electric field as a function of r, the distance from the geometric centre of the region.

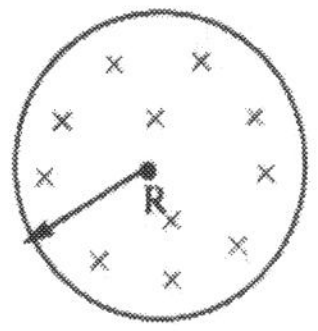

Sol: Case -1 : For r < R

Case -2 : r = R

$E.2\pi r=-A\frac{dB}{dt}$ $E.2\pi R=-\pi R^2\frac{dB}{dt}$

$E.2\pi r=-\pi r^2\frac{dB}{dt}$ $E=\frac{R}{2}\frac{dB}{dt}$

$E=-\frac{r}{2}\frac{dB}{dt}=-\frac{r}{2}\alpha$ $E=-\frac{R\alpha}{2}$

E α r

Case - 3 r > R

$E.2\pi r=-\pi R^2\frac{dB}{dt}$

$E=-\frac{R^2}{2r}\frac{dB}{dt}$

$E=-\frac{R^2}{2r}\alpha$

$E_{out}\alpha\frac{1}{r}$

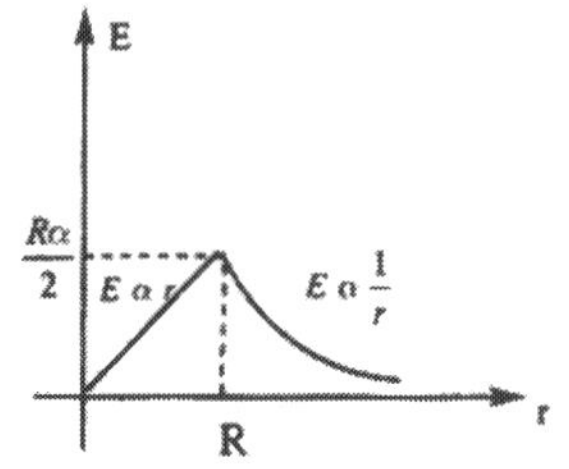

WE-57: A wire is bent in the form of a square of side 'a' in a varying magnetic field $\overline{B}=\alpha\ B_0 t\hat{k}$. If the resistance per unit length is λ, then find the following.

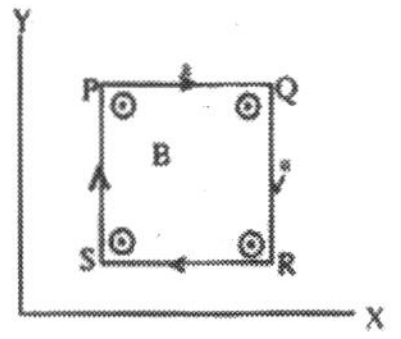

i) The direction of induced current
ii) The current in the loop
iii) Potential difference between P and Q

Sol: i) Direction of current is closewise.

ii) $|e|=\frac{d\phi}{dt}=\frac{d}{dt}(BA)=a^2(\alpha B_0.t)=a^2\alpha B$

Current : $i=\frac{e}{R}=\frac{a^2\alpha B_0}{4a\lambda}(\because R=4a\lambda)=\frac{a\alpha\ B_0}{4\lambda}$

iii) $v_P+\frac{e}{4}-i.a\lambda=V_Q$,

where 'e' is the total emf induced

or $V_P-V_Q=ia\lambda-\frac{e}{4}$

or $V_P-V_Q=\frac{e}{4a\lambda}.a\lambda-\frac{e}{4}$

$V_P-V_Q=\frac{e}{4}-\frac{e}{4}=0$

WE-58 : Shown in the figure is a circular loop of radius r connected to a resistance R. A variable magnetic field of induction $B = e^{-t}$ is established inside the coil. If the key(k) is closed, find the electric power developed ?

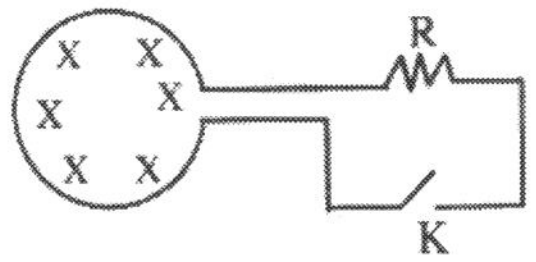

Sol: $E = \frac{-d\phi}{dt} = -A.\frac{dB}{dt}$

or $E = -\pi r^2 \frac{d}{dt}\left(e^{-t}\right)$

$\Rightarrow E = \pi r^2 \; e^{-t}$

$P = \frac{E^2}{R} = \frac{\pi^2 r^4 e^{-2t}}{R}$

at t = 0 ; $P = \frac{\pi^2 r^4}{R}$

D.C. CIRCUITS

Growth and decay of current in an inductor Resistor (L - R) circuit

I. Growth of current

Consider a circuit shown in the diagram

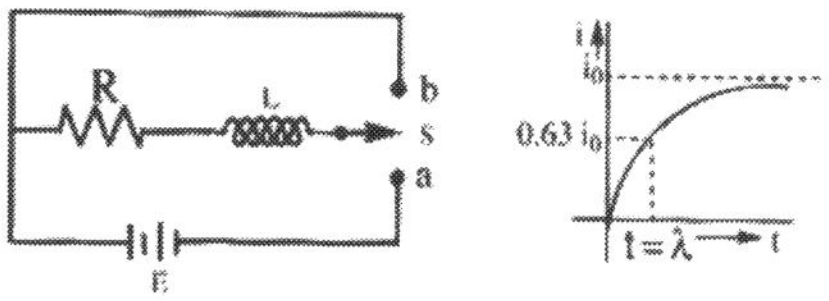

a) When a switch S is connected to 'a' , the current in the circuit beings to increase from zero to a maximum value 'i_0'. The Inductor opposes the growth of the current.

$$\therefore \qquad E - L\frac{di}{dt} = Ri$$

Where 'i' is the current in the circuit at any instant 't' and $i = i_0\left\{1 - e^{\frac{-t}{\lambda}}\right\}$

Where i_0 is the maximum current.

Here $\lambda = \frac{L}{R}$ called Inductive time constant

b) At $t = \lambda, i = i_0\left(1 - \frac{1}{e}\right) = 0.63 \; i_0$

c) Thus the inductive time constant of a circuit is defined as the time in which the current rises from zero to 63% of its final value.

d) Greater the value of 'λ' smaller will be the rate of growth of current.

e) Current reaches i_0 after infinite time.

f) When current attains maximum value, Inductor doesn't work.

$\therefore \quad i_0 = \frac{E}{R}$

II. Decay of Current

a) When circuit is disconnected from the battery and switch 's' is connected to point 'b', the current now beings to fall. But inductor opposes decay of current $\therefore -L\frac{di}{dt} = Ri$

Where i is the current at any instant and $i = i_0 e^{\frac{-t}{\lambda}}$

where $t = \lambda = \frac{L}{R}$

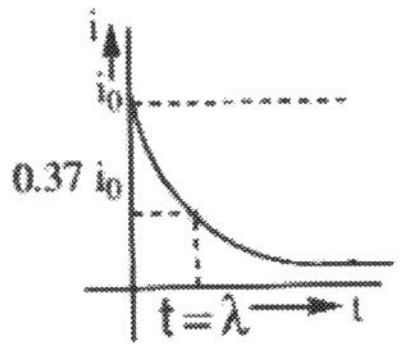

b) At $t = \lambda, i = \frac{i_0}{e} = 0.37 i_0$

c) The inductive time constant (λ) can also be defined as the time interval during which the current decays to 37% of the maximum current.

d) For small value of 'L', rate of decay of current will be large.

e) Current becomes zero after infinite time.

WE-59 : In the given circuit, current through the 5 mH inductor in steady state is

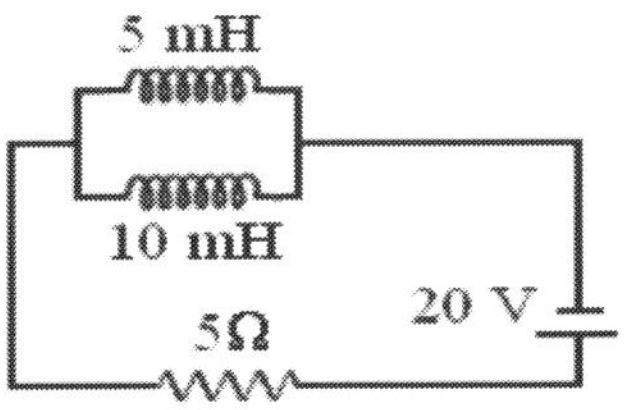

Sol: 5mH, 10mH are connected in parallel

$\therefore$ Equivalent inductance

$$L_{eq} = \frac{5 \times 10}{5 + 10} = \frac{50}{15}$$

$$= \frac{10}{3} mH$$

Current at steady state

$$I = \frac{20}{5} = 4A$$

As L_1 and L_2 are in parallel

$$I_1 = \left(\frac{L_2}{L_1 + L_2}\right) I$$

$$= \left(\frac{10}{10+5}\right) \times 4$$

$$= \frac{10}{15} \times 4$$

$$= \frac{8}{3} Amp$$

WE-60 : In the given circuit diagram, key K is switched on at $t = 0$. The ratio of current i through the cell at $t = 0$ to that at $t = \infty$ will be

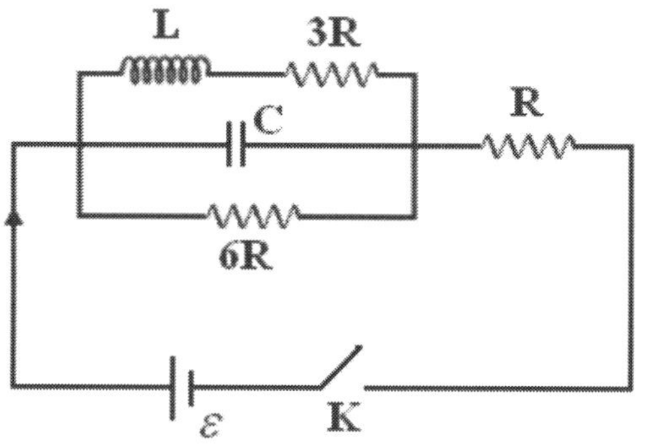

Sol: At t = 0, the branch containing L will offer infinite resistance while the branch containing the capacitor will be effectively a short circuit.

Hence, $(i)_{t=0} = \frac{E}{R}$

Similarly, at $t = \infty$, L will offer zero, resistance, where as 'c' will be an open circuit.

Hence, effective resistance $= R + \frac{6R \times 3R}{6R + 3R}$

$$(i)_{t=\infty} = \frac{e}{3R}$$

The required ratio

$$= \frac{e}{R} \times \frac{3R}{e} = 3 : 1$$

WE-61 : An inductor of inductance $L = 400\,mH$ and resistors of resistance $R_1 = 4\Omega$ and $R_2 = 2\Omega$ are connected to battery of emf 12V as shown in the figure. The internal resistance of the battery is negligible. The swich S is closed at $t = 0$. The potential drop across L as a function of time is

Sol: $I_1 = \frac{E}{R_1} = \frac{12}{2} = 6A$

$$E = L\frac{dl_2}{dt} + R_2 \times l_2$$

$$I_2 = I_0\left(1 - e^{-t/t_c}\right)$$

$$\Rightarrow I_0 = \frac{E}{R_2} = \frac{12}{2} = 6A$$

$$t_c = \frac{L}{R} = \frac{400\times10^{-3}}{2} = 0.2$$

$$I_2 = 6\left(1 - e^{-t/0.2}\right)$$

Potential drop across L

$$V_L = E - R_2I_2 = 12 - 2\times6\left(1 - e^{-bt}\right)$$

$$= 12e^{-5t}$$

WE-62 : An inductor of 3H is connected to a battery of emf 6V through a resistance of $100\,\Omega$. Calculate the time constant. What will be the maximum value of current in the circuit ?

Sol: Give that $L = 3H, E = 6V, R = 100\Omega$

Time constant $\tau_L = \frac{L}{R} = \frac{3}{100} = 0.03\,\text{sec}$

Maximum Current $I_0 = \frac{E}{R} = \frac{6}{100}\,\text{amp} = 0.06\,\text{amp}$

WE-63 : A cell of 1.5V is connected across an inductor of 2 mH in series with a $2\,\Omega$ resistor. What is the rate of growth of current immediately after the cell is switched on.

Sol: $E = L\frac{dI}{dt} + IR$, therefore, $\frac{dI}{dt} = \frac{E- = IR}{L}$

$$E = 1.5Volt, R = 2\Omega, L = 2mH = 2\times10^{-3}H$$

When the cell is switched on, I = 0

Hence $\frac{dI}{dt} = \frac{E}{L} = \frac{1.5}{2\times10^{-3}}As^{-1} = 750\,As^{-1}$

WE-64 : A coil having resistance $15\,\Omega$ and inductance 10H is connected across a 90 Volt dc supply. Determine the value of current after 2sec. What is the energy stored in the magnetic field at that instant.

Sol: Give that

$$R = 15\Omega, L = 10H, E = 90Volt$$

Peak value of current

$$I_0 = \frac{E}{R} = \frac{90}{15} A = 6A \text{ also, } \tau_L = \frac{L}{R} = \frac{10}{15} = 0.67\text{sec}$$

Now, $I = I_0\left(1 - e^{\frac{-Rt}{L}}\right)$, After 2sec,

$$I = 6\left[1 - e^{-2/0.67}\right] = 6\left[1 - 0.05\right] = 5.7A$$

Energy stored in the magnetic field

$$U = \frac{1}{2}LI^2 = \frac{1}{2} \times 10 \times (5.7)^2 J = 162.45J.$$

WE-65 : Calculate the back e.m.f of a 10H, 200Ω coil 100 ms after a 100V d.c supply is connected to it.

Sol: The value of current at 100ms after the switch is closed is

$$I - I_0\left[1 - e^{\frac{-t}{T_0}}\right], \text{ Here, } I_0 = \frac{100}{200} = 0.5\text{ amp;}$$

$$\tau_0 = \frac{L}{R} = \frac{10}{200} = 0.05\text{ sec; } t = 0.1\text{ sec}$$

$$I = 0.5\left(1 - e^{-0.1/0.05}\right) = 0.5\left(1 - e^{-2}\right) = 0.4325A$$

Now, $E = IR + L\frac{dI}{dt}$, or

$$100 = 0.4325 \times 200 + L\frac{dI}{dt}$$

Back e.m.f $= L\frac{dI}{dt} = 100 - 0.4325 \times 200 = 13.5V$

WE-66 : A coil of resistance 20Ω and inductance 0.5 henry is switched to dc 200 volt supply. Calculate the rate of increase of current:

a) At the instant of closing the switch and

b) After one time constant

c) Find the steady state current in the circuit.

Sol: a) This is the case of growth of current in an L - R circuit. Hence, current at time *t* is given by

$$i = i_0\left(l - e^{\frac{-t}{\tau_L}}\right)$$

Rate of increase of current, $\frac{di}{dt} = \frac{i_0}{\tau_L} e^{\frac{-t}{\tau_L}}$

At $t = 0 \frac{di}{dt} = \frac{i_0}{\tau_L} = \frac{E/R}{L/R} = \frac{E}{L}$

$$\frac{di}{dt}=\frac{200}{0.5}=400\,A/s$$

b) At $t=\tau_L$, $\frac{di}{dt}(400)e^{-1}=(0.37)(400)=148A/s$

c) The steady state current in the circuit, is $i_0=\frac{E}{R}=\frac{200}{20}10A$

Growth and decay of charge in a capacitor - Resistor (C - R) circuit

I. Growth of Charge : Consider a circuit shown in the diagram

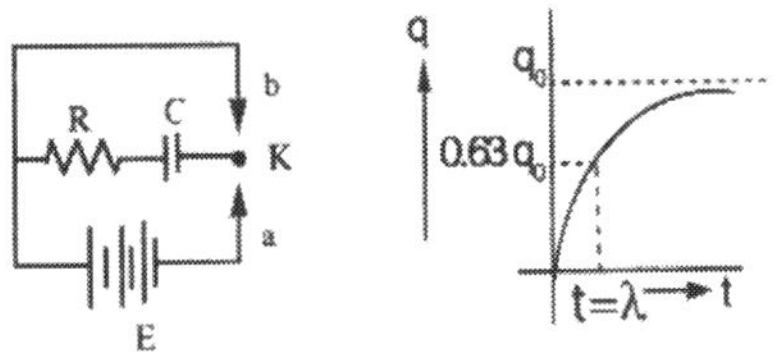

a) When the key's is connected to point 'a', the charging of capacitor takes place until the potential difference across the plates of the condenser becomes E.

b) But charge attained already on the plates opposes further introduction of charge

$$E-\frac{q}{c}=Ri\ \text{(or)}\ E-\frac{q}{c}=R\frac{dq}{dt}$$

Where 'q' is the instantaneous charge, i is the instaneous current in the circuit.

and $q=q_0\left(1-e^{\frac{-t}{\lambda}}\right)$

where q_0 is the maximum charge.

Where $\lambda=CR$, called **capacitive time constant**

c) When $t=\lambda$.

$$q=q_0\left(1-\frac{1}{e}\right)=0.63\,q_0$$

d) Thus the capacitive time constant is the time in which the charge on the plates of the capacitor becomes $0.63\,q_0$

e) Smaller the value of CR, more rapid is the growth of charge on the condenser.

f) Charge on the capacitor becomes maximum after infinite time and it is $q_0=EC$. Then current in the circuit becomes zero.

II. Decay of charge :

a) When the capacitor is fully charged the key is connected to point 'b'.

b) Charge slowly reduces to zero after infinite time.

$$\therefore -\frac{q}{c}=Ri\ \text{(or)}\ \frac{-q}{c}=R\frac{dq}{dt}\ \text{and}\ q=q_0e^{\frac{-t}{\lambda}}$$

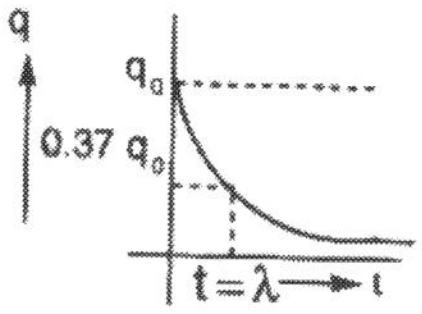

c) At $t=\lambda, q=\frac{q_0}{e}=0.37\ q_0$

d) Thus capacitive time constant can also be defined as the time interval in which the charge decreases to 37% of the maximum charge

e) Smaller the time constant, quicker is the discharge of the condenser.

Problem 1.59: In the circuit shown in figure switch S is closed at time $t=0$. Find the current through different wires and charge stored on the capacitor at any time t.

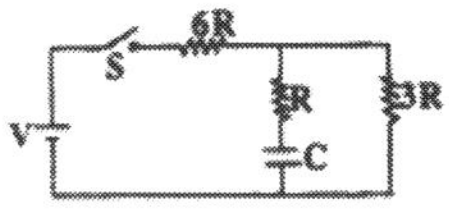

Sol: Calculation of equivalent time constant

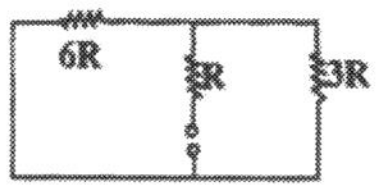

In the circuit shown in figure, after short circuiting the battery 3R and 6R are parallel, so their combined resistance is $\frac{(6R)(3R)}{6R+3R}=2R$. Now this 2R is in series with the remaining R.

Hence, $R_{net}=2R+R=3R$

$\tau_c=(R_{net})C=3RC$

Calculation of steady state charge q_0:

At $t=\infty$, capacitor is fully charged and no current flows through it.
P.D across capacitor = P.D across 3R

$$=\left(\frac{V}{9R}\right)(3R)=\frac{V}{3}, \qquad q_0=\frac{CV}{3}$$

Now, let charge on the capacitor at any time t be q and current through it is i_1. Then

$$q=q_0\left(1-e^{-t/\tau_c}\right) \text{ i.e., } q=q_0\left(1-e^{\frac{t}{3RC}}\right)$$

$$\text{and } i_1=\frac{dq}{dt}=\frac{q_0}{\tau_c}e^{-t/\tau_c}=\frac{q_0}{3RC}e^{\frac{1}{3RC}} \qquad \text{...(1)}$$

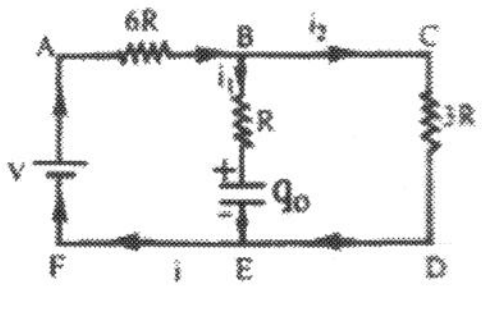

Applying Kirchhoff's second law in loop

ACDFA, we have $-6iR-3i_2R+V=0$

$$2i+i_2=\frac{V}{3R} \qquad ...(ii)$$

Applying Kirchoff's junction law at B, we have

$$i=i_1+i_2 \qquad ...(iii)$$

Solving Eqs. (i), (ii) and (iii), we have

$$i_2=\frac{V}{9R}-\frac{2}{3}i_1=\frac{V}{9R}-\frac{2q_0}{3t_c}e^{-t/t_c}$$

where $q_0=\frac{CV}{3}$

$$\text{i.e., } i_2=\frac{V}{9R}-\frac{2q_0}{3RC}e^{\frac{t}{3RC}}$$

$$i=\frac{V}{9R}+\frac{q_0}{3t_c}e^{-t/t_c}=\frac{V}{9R}+\frac{q_0}{3RC}e^{\frac{t}{3RC}}$$

WE-67 : A parallel - plate capacitor, filled with a dielectric of dielectric constant k, is charged to a potential V_0. It is now disconnected fromthe cell and the slab is removed. If it now discharges, with time constant τ, through a resistance, then find time after which the potential difference across it will be V_0?

Sol: When slab is removed, the potential difference across capacitor increases to kV_0

$$CV_0=kCV_0\,e^{-\frac{t}{\tau}} \text{ as } q_0=KCV_0$$

$$\frac{1}{k}=e^{-\frac{t}{\tau}}\Rightarrow k=e^{\frac{t}{\tau}}$$

$$\therefore ln=\frac{t}{\tau}\Rightarrow t=\tau lnk$$

WE-68 : $4\mu F$ capacitor and a resistance $2.5\,M\Omega$ are in series with 12V battery. Find the time after which the potential difference across the capacitor is 3 times the potential difference across the resistor. [Given $ln(2)$=0.693]

Sol: a) Charging current $i=\frac{V_0}{R}e^{\frac{t}{RC}}$

$\therefore$ Potential difference across R is

$$V_R-iR=V_0e^{-\frac{t}{RC}}$$

$\therefore$ Potential difference across 'C' is $V_C=V_0-V_R$

$$=V_0-V_0e^{-\frac{t}{RC}}=V_0\left(1-e^{-\frac{t}{RC}}\right)$$

Chapter - 2
(Part - 2)
ELECTRO MAGNETIC INDUCTION

but given $V_C = 3V_R$, we get

$1 - e^{-t/RC} = 3e^{-t/RC}$ or $4e^{-t/RC}$

$$e^{\frac{-t}{RC}} = 4 \Rightarrow \frac{t}{RC} = ln\ 4 \Rightarrow t = 2RC\ ln\ 2$$

$t = 2.5\times10^{6}\times4\times10^{-6}\times2\times0.693$

or $t = 13.86$ sec

WE-69 : In a circuit inductance L and capacitance C are connected as shown in figure and A_1 and A_2 are ammeters. When key k is pressed to complete the circuit, then just after closing key k, the reading of A_1 and A_2 will be :

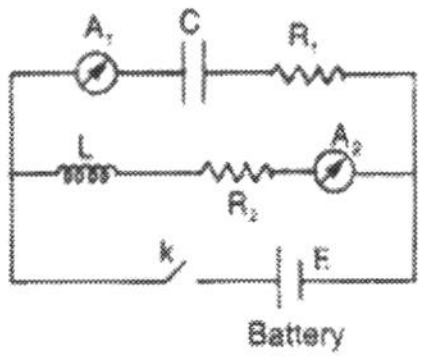

Sol: At $t = 0$ capacitor offers zero resistance and acts like a short circuit. While inductor offers infinite resistance and it acts like an open circuit. Therefore no current flow through inductor branch and maximum current flows through capacitor branch.

Hence reading of A_2 is zero and reading A_1 is given by $\frac{E}{R_1}$

<u>CONCEPTUAL</u>

FARADAY'S EXPERIMENT, INDUCED E.M.F & LENZ'S' LAW

1. **When ever the flux linked with a coil changes, then**
   1) current is always induced
   2) an emf and a current are always induced
   3) an emf is induced but a current is never induced
   4) an emf is always induced and a current is induced, when the coil is a closed one
2. **Whenever the magnet flux linked with a coil changes, then there is an induced emf in the circuit. This emf lasts**
   1) For a short time
   2) For a long time
   3) For ever
   4) So long as the change in the flux takes place
3. **A magnet is moved towards the coil (i) quickly (ii) slowly then the induced emf is**
   1) Larger in case (i)
   2) Smaller in case (i)
   3) Equal in both
   4) Larger or smaller depending upon the radius of the coil
4. **The laws of electromagnetic induction have been used in the construction of a**
   1) galvanometer
   2) voltmeter
   3) electric motor
   4) electric generator
5. **When a rate of change of current in a circuit is unity, the induced emf is equal to**
   1) Total flux linked with the coil
   2) induced charge
   3) Number of turns in the circle
   4) Coefficient of self induction
6. **A bar magnet is dropped along the axis of copper ring held horizontally. The acceleration of fall is**
   1) Equal to 'g' at the place
   2) Less than 'g'
   3) More than 'g'
   4) Depends upon diameter of the ring and length of the magnet

13. **An annular circular brass disk of inner radius 'r' and outer radius 'R' is rotating about an axis passing through its center and perpendicular to its plane with a uniform angular velocity 'ω' in a uniform magnetic filed of induction 'B' normal to the plane of the disk. The induced emf between the inner and outer edge of the annular disk is**

1) $\frac{B\omega(r^2+R^2)}{2}$ 2) $\frac{B\omega(R^2-r^2)}{2}$ 3) $\frac{B\omega(r-R)}{2}$ 4) $\frac{B\omega(r+R)}{2}$

14. **Consider the situation shown in the figure. If the current I in the long straight conducting wire XY is increased at a steady rate then the induced e.m.f.'s in loops A and B will be**

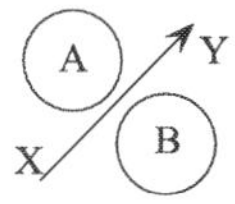

1) clockwise in A, anti clockwise in B 2) anti clockwise in A, clockwise in B
3) clockwise in both A and B 4) anti clockwise in both A and B

FLEMING'S RIGHT HAND RULE

15. **The direction of the induced e.m.f. is determined by**

1) Fleming's left hand rule 2) Fleming's right hand rule
3) Maxwell's right hand screw rule 4) Ampere's rule of swimming

16. **A wire moves with a velocity "v" through a magnetic field and experiences an induced charge separation as shown. Then the direction of the magnetic field is**

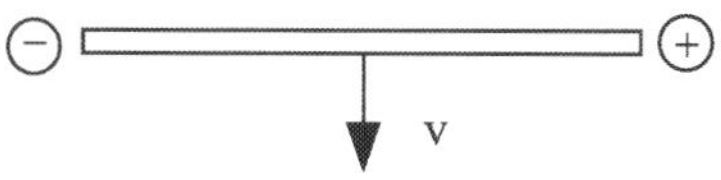

1) in to the page 2) out of the page
3) towards the bottom of the page 4) towards the top of the page

17. **An electric potential difference will be induced between the ends of the conductor shown in the figure, if the conductor moves in the direction shown by**

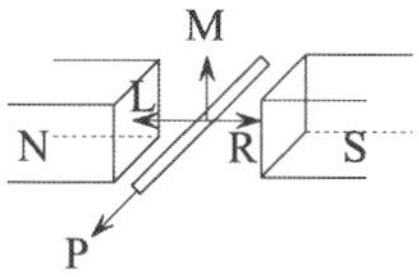

1) P 2) R 3) L 4) M

18. **A horizontal straight conductor when placed along south-north direction falls under gravity; there is**

1. an induced current form south-to-north direction
2. an induced current from north-to-south direction
3. no induced emf along the length of the conductor
4. an induced emf along the length of the conductor

19. **Two circular, similar, coaxial loops carry equal currents in the same direction. If the loops are brought nearer, what will happen?**

1. Current will increase in each loop 2. Current will decrease in each loop
3. Current will remain same in each loop
4. Current will increase in one and decrease in the other

20. A long conducting wire AH is moved over a conducting triangular wire CDE with a constant velocity v in a uniform magnetic field $\vec{B}$ directed into the plane of the paper. Resistance per unit length of each wire is ρ. Then

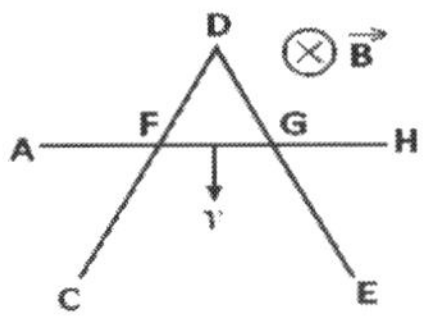

1. a constant clockwise induced current will flow in the closed loop
2. an increasing anticlockwise induced current will flow in the closed loop
3. a decreasing anticlockwise induced current will flow in the closed loop
4. a constant anticlockwise induced current will flow in the closed loop

21. A square coil *ACDE* with its plane vertical is released from rest in a horizontal uniform megnetic field $\vec{B}$ of length 2 L. The acceleration of the coil is

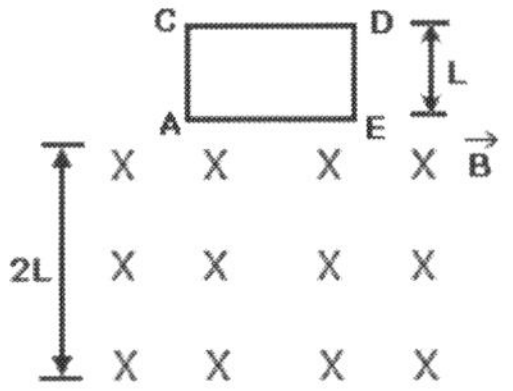

1. less than 'g' for all the time till the loop crosses the magnetic field completely
2. less than 'g' when it enters the field and greater than 'g' when it comes out of the field
3. 'g' all the time
4. less than 'g' when it enters and comes out of the field but equal to 'g' when it is within the field

22. A conducting wire frame is placed in a magnetic field which is directed into the plane of the paper. The magnetic field is increasing at a constant rate. The directions of induced currents in wires *AB* and *CD* are

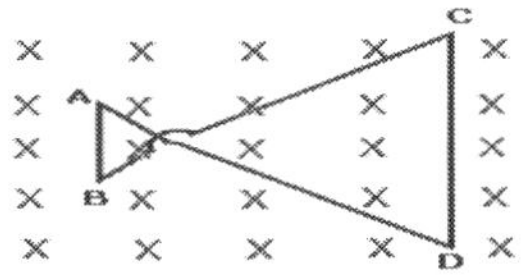

1. *B* to *A* and *D* to *C* 2. *A* to *B* and *C* to *D* 3. *A* to *B* and *D* to *C* 4. *B* to *A* and *C* to *D*

23. A rectangular loop of wire with dimensions shown in figure is coplanar with a long wire carrying current 'I'. The distance between the wire and the left side of the loop is r. The loop is pulled to the right as indicated. What are the directions of the induced current in the loop and the magnetic forces on the left and the right sides of the loop when the loop is pulled?

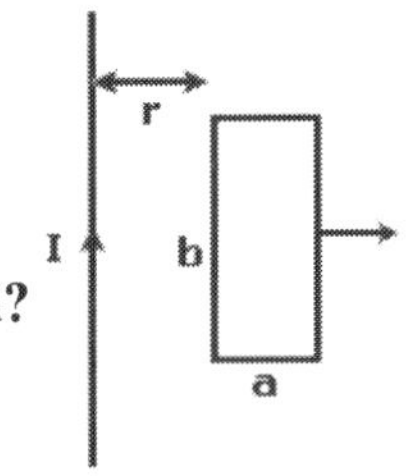

Induced Current	Force on Left side	Force on Right side
a. Counterclockwise	To the left	To the left
b. Counterclockwise	To the right	To the left
c. Clockwise	To the right	To the left
d. Clockwise	To the left	To the right

24. The four wire loops shown figure have vertical edge lengths of either $L, 2L$ or $3L$. They will move with the same speed into a region of uniform magnetic field $\vec{B}$ directed out of the page. Rank them according to the maximum magnitude of the induced emf greatest to least.

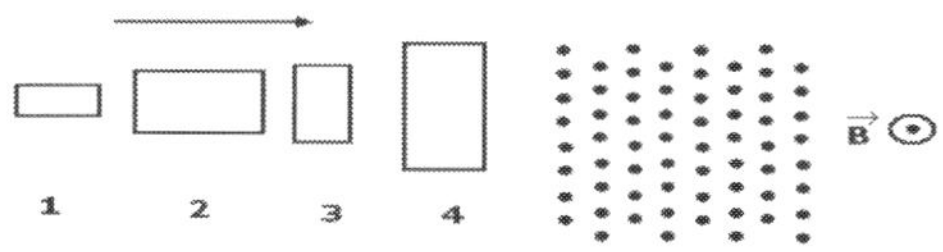

1. 1 and 2 tie, then 3 and 4 tie
2. 3 and 4 tie, then 1 and 2 tie
2. 4,2,3,1
4. 4 then, 2 and 3 tie and then 1

25. A rod lies across frictionless rails in a uniform magnetic field $\vec{B}$ as shown in figure. The rod moves to the right with speed V. In order to make the induced emf in the circuit to be zero, the magnitude of the magnetic field should

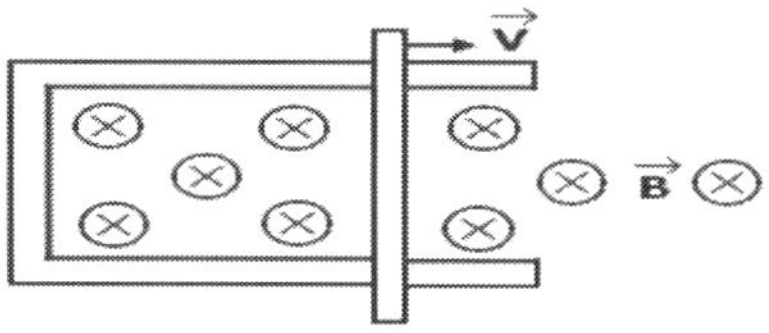

1. not change
2. increase linearly with time
3. decrease linearly with time
4. decrease nonlinearly with time

26. An electron moves on a straight line path YY' as shown in figure. A coil is kept on the right such that YY' is the plane of the coil. At the instant when the electron gets closest to the coil (neglect self-induction of the coil)

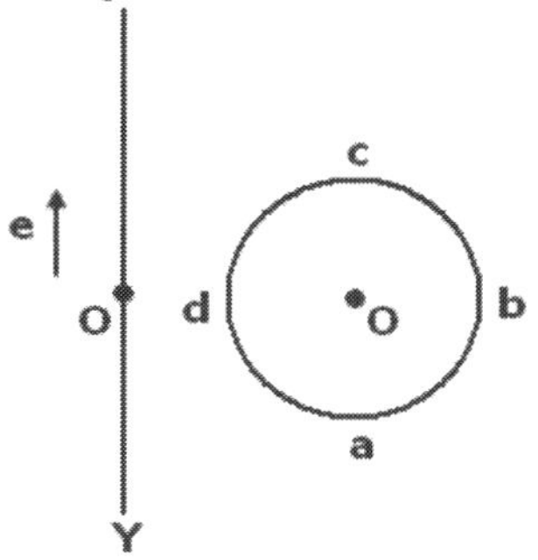

1. The current in the coil flows clockwise
2. The current in the coil flows anticlockwise
3. The current in the coil is zero
4. The current in the coil does not change the direction as the electron crosses point O

27. **In figure, there is a conducting ring having resistance R placed in the plane of paper in a uniform magnetic field B_0. If the ring is rotating in the plane of paper about an axis passing through point O and perpendicular to the plane of paper with constant angular speed ω in clockwise direction, then**

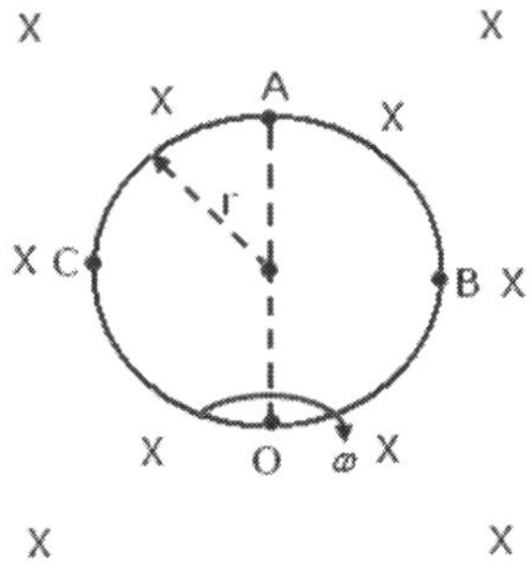

1. point O will be at higher potential than A
2. the potential of point B and C will be different
3. the current in the ring will be zero
4. the current in the ring will be $2B_0\omega r^2 / R$

28. **In the space shown a non-uniform magnetic field $\vec{B} = B_0(1+x)(-\hat{k})$ tesla is present. A closed loop of small resistance, placed in the xy plane is given velocity V_0. The force due to magnetic field on the loop is**

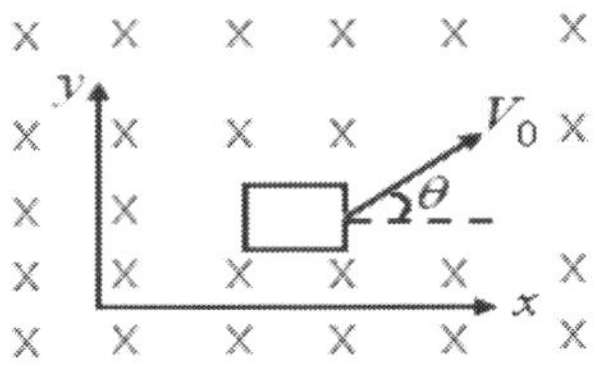

1. zero
2. Along $+x$ direction
3. along $-x$ direction
4. along $+y$ direction

29. **Two identical cycle wheels (geometrically) have different number of spokes connected from centre to rim. One is having 20 spokes and the other having only 10 (the rim and the spokes are resistanceless). One resistance of value R is connected between centre and rim. The current in R will be**
1. double in the first wheel than in the second wheel
2. four times in the first wheel than in the second wheel
3. will be double in the second wheel than that of the first wheel
4. will be equal in both these wheels

30. **AB and CD are fixed conducting smooth rails placed in a vertical palne and joined by a constant current source at its upper end. PQ is a conducting rod which is free to slide on the rails. A horizontal uniform magnetic field exists in space as shown in figure. If the rod PQ is released from rest then,**

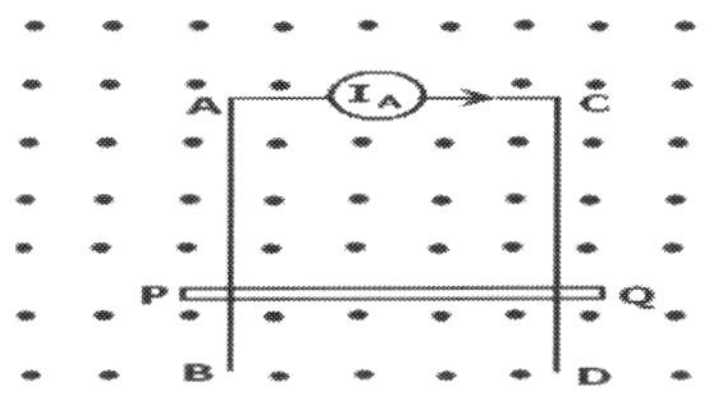

1. the rod PQ will move downward with constant acceleration

2. the rod PQ will move upward with constant acceleration
3. the rod will remain at rest
4. any of the above

31. Three identical coils A, B and C carrying currents are placed coaxially with their planes parallel to one another. A and C carry current as shown in figure B is kept fixed while A and C both are moved towards B with the same speed. Initially, B is equally seperated from A and C. The direction of the induced current in the coil B is

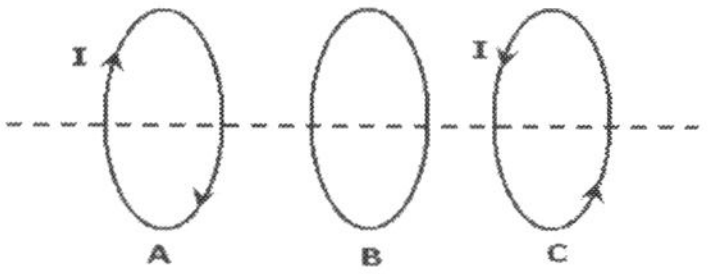

1. same as that in coil A
2. same as that in coil B
3. zero
4. none of these

32. Two identical conductors P and Q are placed on two frictionless rails R and S in a uniform magnetic field directed into the plane. If P is moved in the direction shown in figure with a constant speed, then rod Q

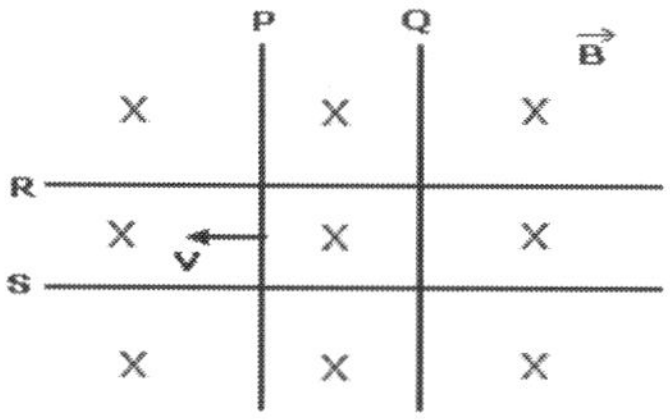

1. will be attracted towards P
2. will be repelled away from P
3. will remain stationary
4. may be repelled away or attracted towards P

Self Induction And Mutual Induction

33. An inductance stores energy in the

1) electric filed
2) magnetic field
3) resistance of the coil
4) electric and magnetic fields

34. If 'N' is the number of turns in a coil, the value of self inductance varies as

1) N^0
2) N
3) N^2
4) N^{-2}

35. A series combination of L and R is connected to a battery of emf E having negligible internal resistance. The final value of current depends upon

1) L and R only
2) E and L only
3) E and R only
4) L, R and E only

36. Two coils of inductances L_1, and L_2 are linked such that their mutual inductance is M

1) L_1+L_2
2) $\frac{1}{2}(L_1+L_2)$
3) $(L_1 \pm L_2)$
4) $\sqrt{L_1L_2}$

37. The coefficient of self inductance and the coefficient of mutual inductance have

1) same units but different dimensions
2) different units but same dimensions
3) different units and different dimensions
4) same units and same dimensions

38. The mutual inductance between a pair of coils each of 'N' turns is 'M'. If a current is 'I' in the first coil is bought to zero in a time t, then the average emf induced in the second coil is

1) MI/t
2) Mt/I
3) Mt/IN
4) It/MN

39. **A circuit contains two inductors of self-inductance L_1 and L_2 in series. If M is the mutual inductance then the effective inductance of the circuit shown will be**

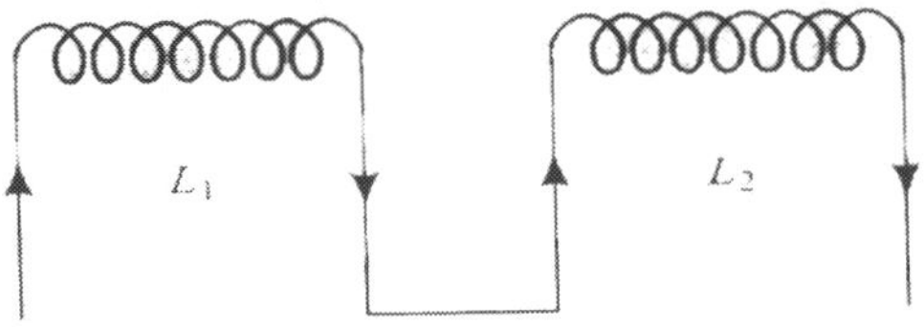

1. L_1+L_2 2. L_1+L_2-2M 3. L_1+L_2+M 4. L_1+L_2+2M

40. **In the circuit of figure, (1) and (2) are ammeters. Just after key K is pressed to complete the circuit, the reading is**

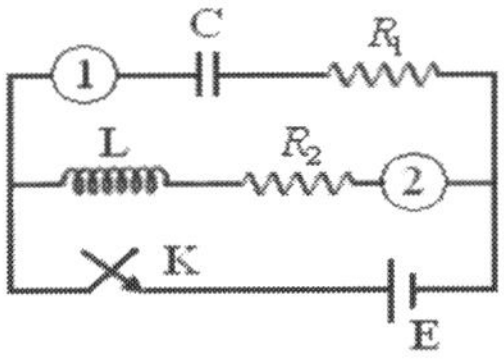

1. maximum in both (1) and (2)   2. zero in both (1) and (2)
3. zero in (1), minimum in (2)   4. maximum in (1), zero in (2)

41. **A pure inductor L, a capacitor C and a resistance R are connected across a battery of emf E and internal resistance r as shown in the figure. Switch S_W is closed at $t=0$, select the correct alternative(s).**

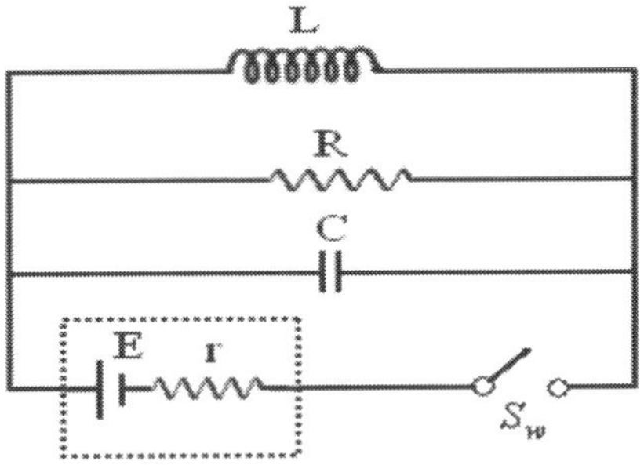

1. current through resistance R is zero all the time
2. current through resistance R is zero at $t=0$ and $t\rightarrow\infty$
3. maximum charge stored in the capacitor is CE
4. maximum enrgy stored in the inductor is equal to the maximum energy stored in the capacitor

42. **In the circuit shown in figure, a conducting wire HE is moved with a constant speed v towards left. The complete circuit is placed in a uniform magnetic field $\vec{B}$ perpendicular to the plane of the circuit inwards. The current in $HKDE$ is**

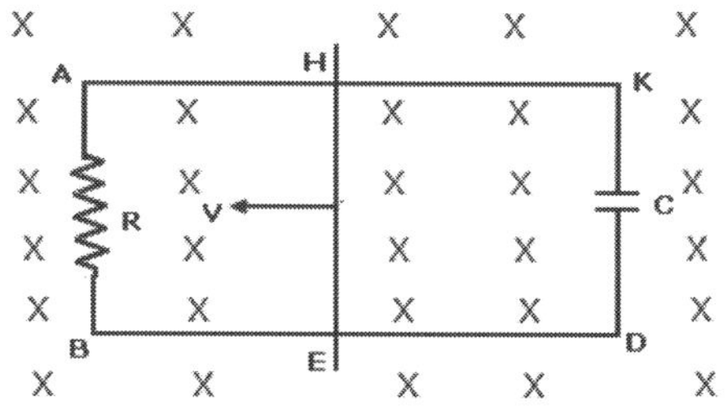

1. clockwise 2. anticlockwise 3. alternating 4. zero

43. In which of the following cases the emf is induced due to time varying magnetic field (induced field emf)?

Case I A magnet is moving along the axis of a conducting coil

Case II A loop having varying area (due to moving jumper) is placed in a magnetic field

Case III The resistance of the coil is changing, which is connected to an ideal battery. Case IV A current carrying wire is approaching a conducting ring.

1) I, II and III only 2) I, III and IV only 3) I, II and IV only 4) All the four

44. A closed conducting ring is placed in between two bar magnets as shown in the figure. The pole strength of M_1 **is double that of M_2. When the two bar magnets are at same distance from the centre of the ring, the bar magnet M_1 has given a velocity $2v$ while M_2 is given velocity v in the direction as shown in the figure.**

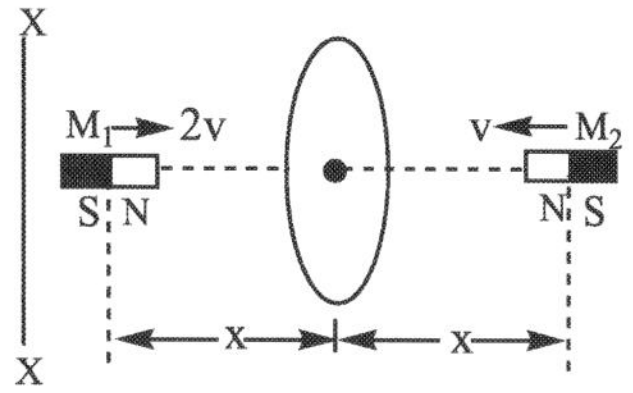

The direction of induced current in the ring as seen from XX from this moment to the moment till bar magnets collide is

1) always clockwise
2) always anticlockwise
3) first clockwise, and then anticlockwise
4) first anti-clockwise, and then clockwise

45. Two identical ciruclar loops of metal wire are lying on a table without touching each other. Loop A carries a current which increases with time. In response, the loop B

1) remains stationary
2) is attracted by the loop A
3) is repelled by the loop A
4) rotates about its CM, with CM fixed

46. A metallic square loop $ABCD$ is moving in its own plane with velocity v in a uniform magnetic field perpendicular to its plane as shown in the figure. Electric field is induced.

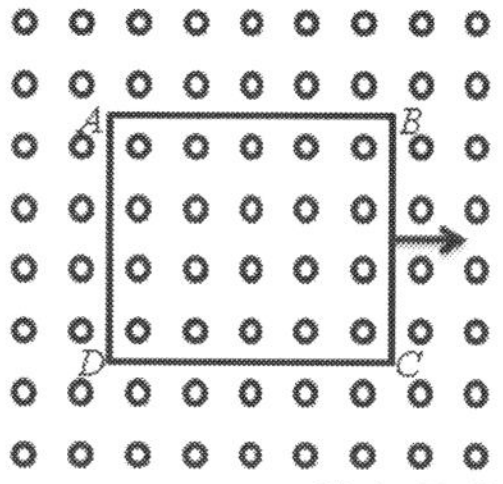

1) in AD, but not in BC
2) in BC, but not in AD
3) neither in AD nor in BC
4) in both AD and BC

47. **Two circular coils can be arranged in any of the three situations shown in the figure. Their mutual inducatnce will be**

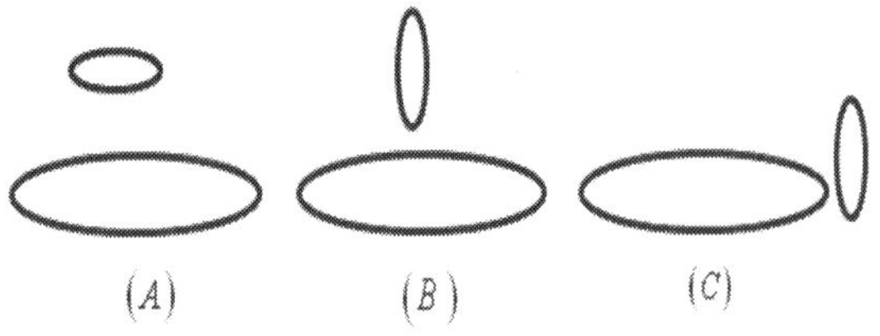

1) maximum in situation (A)
2) maximum in situation (B)
3) maximum in situation (C)
4) the same in all situations

48. **As shown in the figure, P and Q are two coaxil conducting loops separated by some distance. When the switch S is closed, a clockwise current I_p flows in P (as seen by E) and an induced current I_{Q_1} flows in Q. The switch remains closed for a long time. When S is opened, a current I_{Q_2} flows in Q. Then the direction I_{Q_1} and I_{Q_2} (as seen by E) are**

1) respectively clockwise and anticlockwise
2) both clockwise
3) both anticlockwise
4) respectively anticlockwise and clockwise

49. **The variation of induced emf (e) with time (t) in a coil if**

a short bar magnet is moved along axis of the coil shown with a constant velocity is best represented as

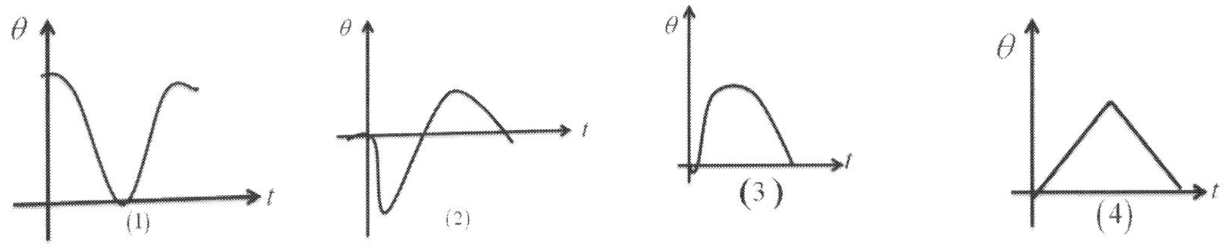

50. **An infinitely long cylinder is kept parallel to an uniform magnetic field B directed along positive z-axis. The direction of induced current as seen from the z-axis will be**

1) clockwise of the +ve z-axis
2) anticlockwise of the +ve z-axis
3) zero
4) along the magnetic field

51. **The figure shows certain wire segments joined together to form a coplanar loop. The loop is placed in a perpendicular magnetic field in the direction going into the plane of the figure. The magnitude of the field increases with time. I_1 and I_2 are the currents in the segments ab and cd. Then,**

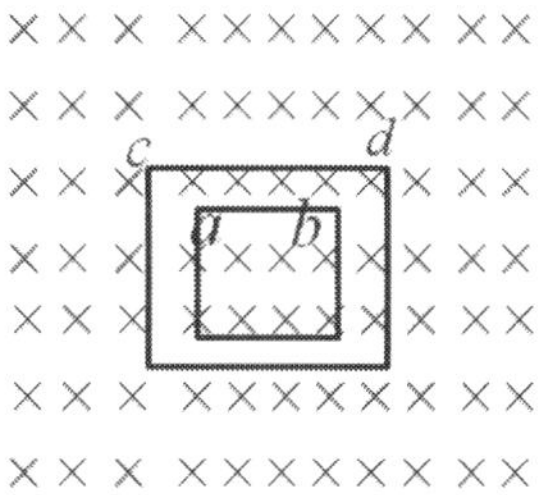

1) $I_1 > I_2$
2) $I_1 < I_2$
3) I_1 is in the direction ba and I_2 is in the direction cd
4) I_1 is in the direction ab and I_2 is in the direction dc

52. **A coil is suspended in a uniform magnetic field with the plane of the coil parallel to the magnetic lines of force. When a current is passed through the coil, it starts oscillating; it is very difficult to stop. But if an aluminium plate is placed near to the coil, it stops. This is due to**
1) development of air current when the plate is placed
2) induction of electrical charge on the plate
3) shileding of magnetic lines of force as aluminium is a paramagnetic material
4) electromagnetic induction in the aluminium plate giving rise to electromagnetic damping

53. **Which of the following units denotes the dimensions $\left[ML^2/Q^2\right]$, where Q denotes the electric charge?**
1) Wb/m^2
2) henry (H)
3) H/m^2
4) weber (Wb)

54. **A rod of length l rotates with a small but uniform angular velocity ω about its perpendicular bisector. A uniform magnetic field B exists parallel to the axis of rotation. The potential difference between the centre of the rod and an end is**
1) zero
2) $\frac{1}{8}\omega Bl^2$
3) $\frac{1}{2}\omega Bl^2$
4) $B\omega l^2$

55. **A rod of length l rotates with a uniform angular velocity ω about its perpendicular bisector. A uniform magnetic field B exists parallel to the axis of rotation. The potential difference between the two ends of the rod is**
1) zero
2) $\frac{1}{2}Bl\omega^2$
3) $Bl\omega^2$
4) $2Bl\omega^2$

56. **Consider the situation shown in figure . If the switch is closed and after some time it is opened again, the closed loop will show**

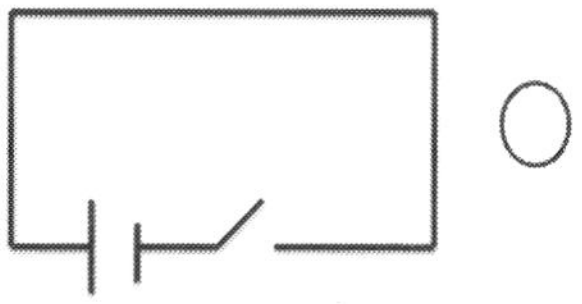

1) an anticlockwise current-pulse
2) a clockwise current-pulse
3) an anticlockwise current-pulse and then a clockwise current-pulse
4) a clockwise current-pulse and then an anticlockwise current-pulse

57. **A bar magnet is released from rest along the axis of a very long, vertical copper tube. After some time the magnet**

1) will stop in the tube
2) will move with almost contant speed
3) will move with an acceleration g
4) will oscillate

ASSERTION & REASON

Instructions for the Assertion & reason type questions:

A. Both Assertion and Reason are true and Reason is the correct explanation of Assertion

B. Both Assertion and Reason are true and Reason is not the correct explanation of Assertion

C. Assertion is true but reason is false

D. Assertion is false but reason is true.

58. **Assertion :** Magnetic flux is a vector qunatity

Reason: Value of magnetic flux can be positive, negative or zero

1) A 2) B 3) C 4) D

59. **Assertion :** Lenz'a law violates the principle of conservation of energy

Reason: Induced emf always oppose the change in magnetic flux responsible for its production

1) A 2) B 3) C 4) D

60. **Assertion:** When number of turns in a coil is doubled, coefficient of self-inductance of the coil becomes 4 times.

Reason: This is because $L \propto N^2$

1) A 2) B 3) C 4) D

61. **Assertion :** The induced emf and current will be same in two identical loops of copper and aluminium, when rotated with same speed in the same magnetic field.

Reason: Mutual induction does not depend on the orientation of the coils

1) A 2) B 3) C 4) D

62. **Assertion :** When two coils are wound on each other, the mutual induction between the coils is maximum.

Reason: Mutual induction does not depend on the orientation of the coils.

1) A 2) B 3) C 4) D

63. **Assertion (A):** Only a change in magnetic flux will maintain an induced current in a closed coil.

Reason (R): The presence of large magnetic flux through a coil maintains a current in the coil if the coil is continuous.

1) A 2) B 3) C 4) D

64. **Magnetic flux in a circular coil of resistance 10Ω changes with time as shown in figure. Symbol $\otimes$ indicates a direction perpendicular to paper inwards. Match the following:**

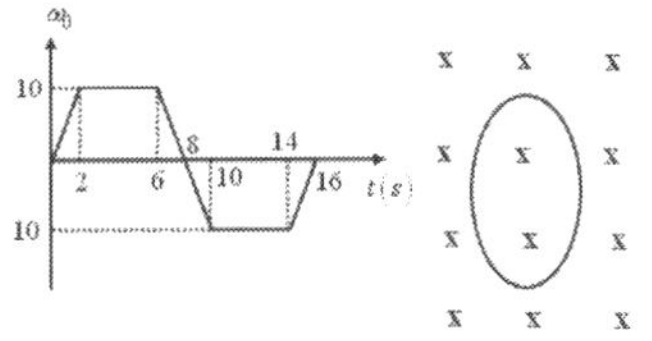

Table - 1

1) At 1s is induced current is

2) At 5s induced current is

3) At 9s induced current is

4) At 15s induced current is

Table-2

p) clockwise

q) anticlockwise

r) zero

s) 2A

t) None

1) a-q; b-r; c-p; d-q 2) a-p; b-r; c-q; d-p 3) a-r; b-p; c-q; d-q 4) a-p; b-r; c-s; d-q

65. **Three coils are placed infront of each other as shown. Currents in 1 and 2 are in same direction, while that in 3 is in opposite direction. Match the following table.**

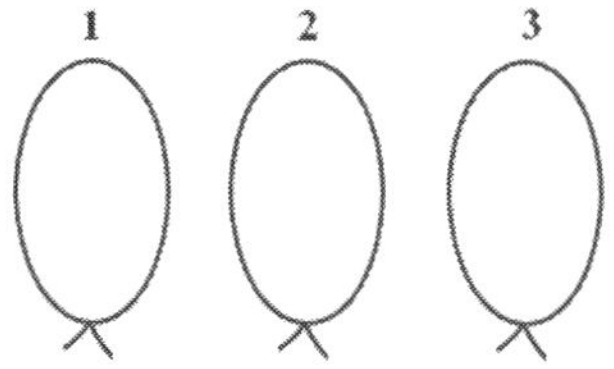

Table - 1	**Table - 2**
a) When current is increased	**p) Current in 1 will increase**
b) When current in 2 is increased	**q) Current in 2 will increase**
c) When current in 3 is increased	**r) Current in 3 will increase**
	s) None

1) a-r; b-r; c-p,q 2) a-p; b-p; c-q 3) a-q; b-q; c-r 4) a-r; b-q; c-p

KEY CONCEPTUAL

1) 4	2) 4	3) 1	4) 4	5) 4	6) 2	7) 3	8) 3	9) 2	10) 2
11) 1	12) 3	13) 2	14) 1	15) 2	16) 1	17) 4	18) 3	19) 2	20) 4
21) 4	22) 1	23) 4	24) 4	25) 4	26) 3	27) 3	28) 3	29) 4	30) 4
31) 3	32) 1	33) 2	34) 3	35) 3	36) 4	37) 4	38) 1	39) 4	40) 4
41) 2	42) 4	43) 2	44) 2	45) 3	46) 4	47) 1	48) 4	49) 2	50) 3
51) 4	52) 4	53) 2	54) 2	55)1	56) 4	57) 2	58) 4	59) 4	60) 1
61) 3	62) 3	63) 3	64) 1	65) 1					

LEVEL - I (CW)

MAGNETIC FLUX & FARADY'S LAWS, INDUCED EMF,CURRENT AND CHARGE

1. **A field of strength $5 \times 10^4/\pi$ ampere turns / meter acts at right angles to the coil of 50 turns of area 10^{-2} m^2. The coil is removed from the field in 0.1 second. Then the induced e.m.f in the coil is**

   1) 0.1 V 2) 0.2 V 3) 1.96 V 4) 0.98 V

2. **A coil has 1,000 turns and 500 cm^2 as its area. The plane of the coil is placed at right angles to a magnetic induction field of 2 X 10^{-5} web/m^2. The coil is rotated through 180° in 0.2 seconds. The average emf induced in the coil, in milli volts, is:**

   1) 5 2) 10 3) 15 4) 20

3. **A square loop of side 22cm is changed to a circle in time 0.4 sec with its plane normal to a magnetic field 0.2T. The emf induced is**

   1) $+6.6mv$ 2) $-6.6mv$ 3) $+13.2mv$ 4) $-13.2mv$

4. **A coil of 1200 turns and mean area of 500 cm^2 is held perpendicular to a uniform magnetic field of induction $4\times10^{-4}T$. The resistance of the coil is 20 ohms. When the coil is rotated through 180^0 in the magnetic field in 0.1 seconds the average electric current (in mA) induced is :**

   1) 12 2) 24 3) 36 4) 48

5. **A closed coil with a resistance R is placed in a magnetic field. The flux linked with the coil is ϕ. If the magnetic field is suddenly reversed in direction, the charge that flows through the coil will be**

1) $\phi/2R$ 2) ϕ/R 3) $2\phi/R$ 4) *zero*

6. **To measure the field 'B' between the poles of an electromagnet, a small test loop of area 1 cm², resistance 10 Ω and 20 turns is pulled out of it. A galvanometer shows that a total charge of $2\mu C$ passed through the loop. The value of 'B' is**

1) 0.001 T 2) 0.01 T 3) 0.1 T 4) 1.0 T

7. **A thin circular ring of area A is perpendicular to uniform magnetic field of induction B. A small cut is made in the ring and a galvanometer is connected across the ends such that the total resistance of circuit is R. When the ring is suddenly squeezed to zero area, the charge flowing through the galvanometer is**

1. $\frac{BR}{A}$ 2. $\frac{AB}{R}$ 3. ABR 4. $\frac{B^2A}{R^2}$

8. **A short- circuited coil is placed in a time varying magnetic field. Electrical power is dissipated due to the current induced in the coil. If the number of turns were to be quadrupled (four times) and wire radius halved, the electrical power dissipated would be**

1) halved 2) the same 3) doubled 4) quadrupled

MOTIONAL E.M.F

9. **An aeroplane with wing span 50 m is flying horizontally with a speed of 360 km/hr over a place where the vertical component of the earth's magnetic field is 2×10^{-4} Wb/m². The potential difference between the tips of the wings would be:**

1) 0.1V 2) 1.0V 3) 0.2V 4) 0.01V

10. **The horizontal component of the earth's magnetic field at a place is $3\times10^{-4}T$ and the dip is $\theta = \tan^{-1}(4/3)$. A metal rod of length 0.25m placed in the north-south position is moved at a constant speed of 10 cm/s towards the east. The e.m.f induced in the rod will be:**

1) zero 2) 1 mV 3) 5 mV 4) 10mV

11. **A metal bar of length 1m falls from rest under the action of gravity remaining horizontal with its ends in east-west direction. The induced e.m.f in it at the instant when it has fallen for 10s is ($B_H = 1.7 \times 10^{-5}$T and g = 10ms^{-2})**

1) 2.5mV 2) 3.2 mV 3) 1.7mV 4) 0.5mV

12. **A thin semicircular conducting ring of radius R is falling with its plane vertical in a horizontal magnetic induction B (figure). At the position MNQ the speed of the ring is V. The potential difference developed across the ring is :**

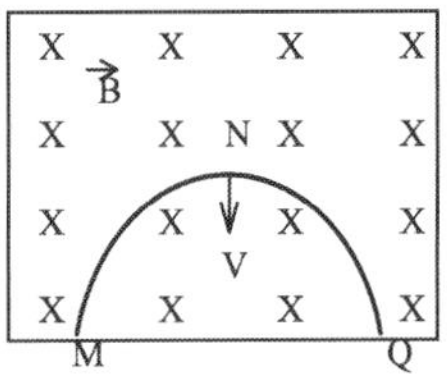

1) zero 2) $BV\pi R^2/V$ and M is at higher potential

3) πRBV and Q is at higher potential 4) 2RBV and Q is at higher potential

13. Two thick rods AB, CD are placed parallel to each other at a distance l. their ends are joined to a resistance R. A magnetic field of induction B is applied perpendicular to the plane containing the rods. If the rods are vertical, the terminal uniform velocity of the rod PQ of mass m is given by

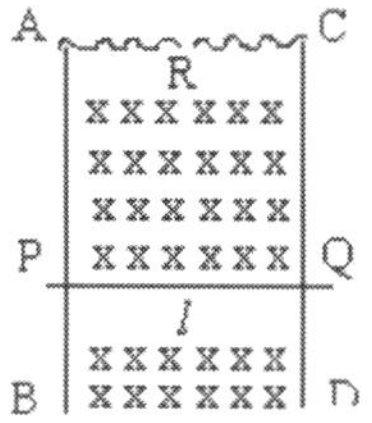

1) $\frac{mg.R}{B^2l^2}$ 2) $\frac{mg.R}{Bl}$ 3) $\frac{mg}{BlR}$ 4) $\frac{mgl}{BR}$

14. A conducting ring of radius 'r' is rolling without slipping with a constant angular velocity ω (figure). If the magnetic field strength is B and is directed into the page then the emf induced across PQ is

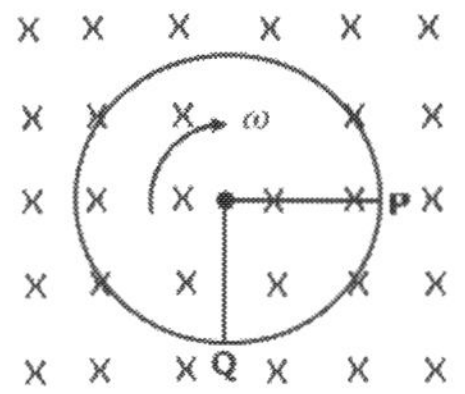

1. $B\omega r^2$ 2. $\frac{B\omega r^2}{2}$ 3. $4B\omega r^2$ 4. $\frac{\pi^2 r^2 B\omega}{8}$

15. A cycle wheel with 64 spokes is rotating with N rotations per second at right angles to horizontal component of magnetic field. The induced e.m.f. generated between its axle and rim is E. If the number of spokes is reduced to 32 then the value of induced e.m.f. will be

1) E 2) 2E 3) E/2 4) E/4

16. A uniform circular metal disc of radius R is rotating about a vertical axis passing through its center and perpendicular to its plane with constant frequency f. If B_H and B_V are horizontal and vertical components of the Earth's magnetic field respectively, then the induced e.m.f between its center and the rim is

1) $\pi B_V fR^2$ 2) $\pi B_H fR^2$ 3) $2\pi B_V fR^2$ 4) Zero

17. A copper disc of diameter 20 cm makes 1200 r.p.m. about its natural axis kept parallel to a uniform magnetic field of 10^{-2} T. The potential difference between the centre and edge of the disc is

1) 6.28×10^{-3} V 2) 62.8×10^{-3} V 3) 0.628×10^{-3} V 4) 0.628 V

18. A conducting square looop of side L and resistance R movesin its plane with a uniform velocity v perpendicular to one of tis sides. A magnetic induction B, constant in time and space, pointing perpendicular to and into the plane of the loop exists every where.

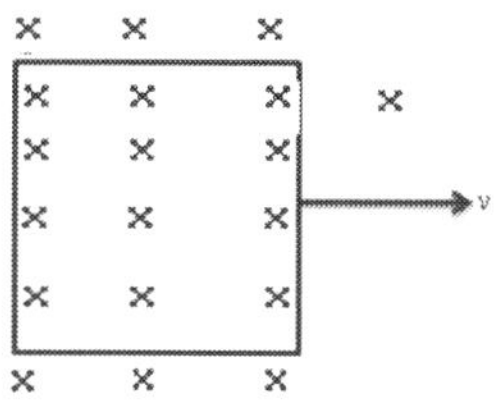

The current induced in the loop is

1) BLv / R clock wise
2) BLv / R anticlockwise
3) $2BLv / R$ anticlockwise
4) zero

19. A metal rod moves at a constant velocity in a direction perpendiuclar to its length. A constant uniform magnetic field exists in space in a direction perpendicular to the rod as well as its velocity. Select the correct statement (s) from the following.

1) The entire rod is at the same electric potential
2) There is an electric field in the rod
3) The electric potential is highest at the centre of the rod and decrease towards its ends
4) The electric potential is lowest at the centre of the rod and increases towards its ends

20. A thin flexible wire of length L is connected to two adjacent fixed points and carries a current I in the clockwise direction, as shown in the figure. When the system is put in a uniform magnetic field of strength B going into the plane of the paper, the wire takes the shape of a circle. The tension in the wire is

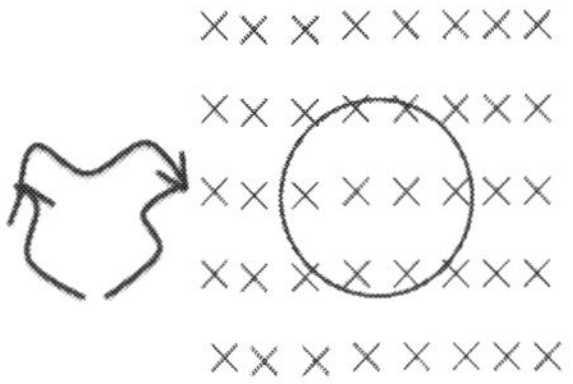

1) IBL
2) $\frac{IBL}{\pi}$
3) $\frac{IBL}{2\pi}$
4) $\frac{IBL}{4\pi}$

21. In an AC generator, a coil with N turns, all of the same area A and total resitance R, rotates with frequency ω in a magnetic field B. The maximum value of emf generated in the coil is

1) $NABR\omega$
2) NAB
3) $NABR$
4) $NAB\omega$

INDUCED ELECTRIC FIELDS

22. A flat circular coil having N turns (tightly wounD) is placed in a time varying magnetic field $B = B_0 \sin \omega t$. The outer radius of the coil is R. Determine the maximum value of the induced emf in the circuit.

1) $\pi R^2 NB_0\omega$
2) $3\pi R^2 NB_0\omega$
3) $\frac{\pi R^2 NB_0}{\omega}$
4) $\frac{\pi R^2 NB_0\omega}{3}$

23. A uniform but time-varying magnetic field $B(t)$ exists in a circular region of radius a and is directed into the plane of the paper as shown. The magnitude of the induced electirc field at point P at a distance r from the centre of the circular region

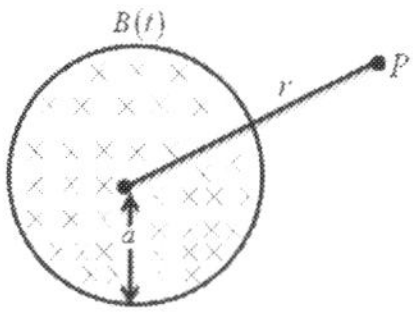

1) is zero 2) decreases as $1/r$ 3) increases as r 4) decreases $1/r^2$

SELF INDUCTION & MUTUAL INDUCTION

24. A coil has self inductance of 0.01H. The current through it is allowed to change at the rate of 1A in 10^{-2}s. The induced emf is

1) 1V 2) 2V 3) 3V 4) 4V

25. The average self-induced emf in a $25mH$ solenoid when the current in it falls from 0.2 A to 0 A in 0.01 second, is

1) 0.05 V 2) 0.5 V 3) 500 V 4) 50 V

26. Two inductors each of inductance L are joined in parallel. Their equivalent inductance is

1) zero 2) 2L 3) L/2 4) L

27. A coil of 100 turns when a current of 5A produced a magnetic flux of 1 μ Wb and each turn of the coil. The coefficient of self induction is

1) 10 μ H 2) 20 μ H 3) 30 μ H 4) 40 μ H

28. In an inductance coil the current increases from zero to 6 ampere in 0.3 second by which an induced e.m.f. of 60 volt is produced in it. The value of coefficient of self-induction of coil is

1) 1 henry 2) 1.5 henry 3) 2 henry 4) 3 henry

29. Two coils are at fixed locations. When coil 1 has no current and the current in coil 2 increases at the rate of 15.0 A/s, the emf in coil 1 is 25.0 mV, when coil 2 has no current and coil 1 has a current of 3.6 A, the flux linkage in coil 2 is

1. 16 mWb 2. 10 mWb 3. 4.00 mWb 4. 6.00 mWb

30. The current in a coil is changed from 5A to 10A in 10^{-2}s. An emf of 50mV is induced in coil near by it. The mutual inductance of two coils is

1) 100 μ H 2) 200 μ H 3) 300 μ H 4) 400 μ H

31. A small square loop of wire of side l is placed inside a large square loop of wire of side $L(L >> l)$. The loops are coplanar and their centres coincide. The mutual inductance of the system is proportional to

1) l/L 2) l^2/L 3) L/l 4) L^2/l

D.C TRANSIENT CIRCUITS

32. A coil of inductance 0.20 H is connected in series with a switch and a cell of emf 1.6 V. The total resistance of the circuit is 4Ω. What is the initial rate of growth of the current when the switch is closed?

1. $0.050As^{-1}$ 2. $0.40As^{-1}$ 3. $0.13As^{-1}$ 4. $8.0As^{-1}$

33. Two inductance coils made of different metal wires are having the same inductance. But their time constants are in the ratio 1 : 2. Then the ratio of their resistances is

1) 1 : 2 2) $1:\sqrt{2}$ 3) $\sqrt{2}:1$ 4) 2 : 1

34. **The time constant of an inductor is τ_1. When a pure resistor of $R\Omega$ is connected in series with it, the time constant is found to decrease to τ_2. The internal resistance of the inductor is**

1) $\frac{R\tau_2}{\tau_1-\tau_2}$ 2) $\frac{R\tau_1}{\tau_1-\tau_2}$ 3) $\frac{R(\tau_1-\tau_2)}{\tau_1}$ 4) $\frac{R(\tau_1-\tau_2)}{\tau_2}$

35. **A coil of inductane 8.4mH and resistance 6Ω is connected to a 12V battery. The current in the coil is 1A at approximately the time**

1) 500 s 2) 20 s 3) 35 ms 4) 1 ms

36. **An ideal coil of 10H is connected series with a resistance of 5Ω and a battery of 5V.2s after the connection is made, the current flowing (in ampere) in the circuit is**

1) $(I-e)$ 2) e 3) e^{-I} 4) $(I-e^{-1})$

KEY LEVEL - I

01) 1	02) 2	03) 2	04) 2	05) 3	06) 2	07) 2	08) 4	09) 2	10) 2
11) 3	12) 4	13) 1	14) 1	15) 1	16) 3	17) 2	18) 4	19) 2	20) 3
21) 4	22) 4	23) 2	24) 1	25) 2	26) 3	27) 2	28) 4	29) 4	30) 1
31) 2	32) 4	33) 4	34) 1	35) 4	36) 4				

HINTS LEVEL-I

1. $B=\mu_0 H=\frac{\mu_0\times 5\times 10^4}{\pi}$

$e=\frac{NBA}{time}=\frac{50\times 2\times 10^{-2}\times 10^{-2}}{0.1}=0.11$

2. $e=\frac{2NBA}{t}$

3. $e=-B\frac{\Delta A}{\Delta t}$

4. $E=\frac{2NBA}{t},\ i=\frac{E}{R}$

5. $q=\frac{d\phi}{R}$

6. $q=\frac{e}{R}dt=\frac{(B_1-B_2)NA}{R}$

$2\mu c=\frac{(B-0)20\times 10^{-4}}{10}=0.01T$

7. $Q=\frac{\Delta\phi}{R}=\frac{\phi_2-\phi_1}{R}=\frac{BA-0}{R}=\frac{BA}{R}=6\times 10^{-3}=6\text{mWb}$

8. Power $P = \frac{e^2}{R}$

Here, $e = induced\ emf = -\left(\frac{d\phi}{dt}\right) where\ \phi = NBA$

$e = -NA\left(\frac{dB}{dt}\right)$

Also, $R \propto \frac{1}{r^2}$

Where R= resistance , r= radius, l=length

$\therefore P \propto N^2 r^2$

$\therefore \frac{P_2}{P_1} = 4$

9. $e = B_H lv$

10. $e = B_V lv$ $\tan\theta = \frac{B_V}{B_H}$

11. $e = B_H lv$ where $v = gt$

12. Conceptual

13. $mg = Bil$ and $i = \frac{Blv_T}{R}$

14. $|\vec{E}| = \frac{Bl^2}{2}\omega, \quad l = \sqrt{2}r$

15 .E.M.F is independent of no of spokes here.

16. $e = \frac{1}{2}BR^2\omega$ where $\omega = 2\pi f$

17. $e = \frac{1}{2}BR^2\omega$ where $\omega = 2\pi f$

18. Net change in magnetic flux passing through the coil is zero.

$\therefore$ Current(of emf) induced in the loop is zero

19. A motional emf, $e = Blv$ is induced in the rod. Or we can say a potential difference is induced between the tow ends of the rod AB,with A at higher potential and B at lower potential. Due to this potential difference, there is an electric field in the rod.

20. $L = 2\pi R$

$\therefore R = \frac{L}{2\pi}$

$2T\sin(d\theta) = d\theta$

For small angles, $\sin(d\theta) = d\theta$

$\therefore 2T(d\theta) = I(dL)B\sin 90^0$

$= I(2R.d\theta).B$

$\therefore T = IRB = \frac{ILB}{2\pi}$

$\therefore$ Correct option is (c)

21. The emf generated would be maximum when flux(cutting) would be maximum i.e., angle betweena area vector of coil and magnetic field is 0^0. The emf generated is given by [as a function of time]

$e = NBA\omega\cos\omega t$

$\Rightarrow e_{max} = NAB\omega$

22. Conceptual

23. $\int \vec{E}.\vec{dl} = \left|\frac{d\phi}{dt}\right|$

$= S\left|\frac{dB}{dt}\right|$

or $E(2\pi r) = \pi a^2\left|\frac{dB}{dt}\right| for\ r \geq a$

$\therefore E = \frac{a^2}{2r}\left|\frac{dB}{dt}\right|$

$\therefore$ induced electric field $\propto \frac{1}{r}$

for $r \leq a$

$E(2\pi r) = \pi r^2\left|\frac{dB}{dt}\right|$

or $E = \frac{r}{2}\left|\frac{dB}{dt}\right| or\ E \propto r$

At $r = a,\ E = \frac{a}{2}\left|\frac{dB}{dt}\right|$

Therefore, variation of E with r (distance from centre) will be as follows

24. $e = L\frac{di}{dt}$

25. $e = L\frac{di}{dt}$

26. $L_p = \frac{L_1 L_2}{L_1 + L_2}$

27. $n\phi = L\,i$

28. $e = -L\left(\frac{di}{dt}\right)$

29. Coefficient of mutual inductance M is given by $|M| = \frac{e_1}{(di_2/dt)} = \frac{\phi_2}{i_1}$

$$\therefore \phi_2 = \frac{e_1 i_1}{(di_2/dt)} = \frac{(25.0\times10^{-3})(3.6)}{(15)}$$

30. $E = M\times\frac{di}{dt}$

31. Magnetic field produced by a current i in a large square loop at its centre

$B \propto \frac{i}{L}$ say $B = K\frac{i}{L}$

$\therefore$ Magnetic flux linked with smaller looop,

$\phi = B.S \quad \phi = \left(K\frac{i}{L}\right)(l^2)$

Therefore, the mutual inductance $M = \frac{\phi}{i} = K\frac{l^2}{L}$ or $M \propto \frac{l^2}{L}$

32. $V = RI + L\frac{dI}{dt}$, at $t = 0, I = 0$, thus we have $\frac{dI}{dt} = \frac{V}{L} = \frac{1.6}{0.2} = 8A/s$

33. $\frac{t_1}{t_2} = \frac{R_2}{R_1}$

34. $\tau_1 = \frac{L}{r}$; $\tau_2 = \frac{L}{R+r}$ Solve for r from the above equations

35. The current-time $(i-t)$ equation in $L-R$ circuit is given by [Growth of current in $L-R$ circuit]

$i = i_0\left(1-e^{-i/t_L}\right)------(i)$

Where $i_0 = \frac{V}{R} = \frac{12}{6} = 2A$

and $t_L = \frac{L}{R} = \frac{8.4\times10^{-3}}{6}$

and $i = 1A$

$t = ?$

Substituting these values in Eq. (i), we get

$t = 0.97\times10^{-3}s$ or $t = 0.97\,ms \quad t = 1ms$

36. $I_0 = \frac{E}{R} = 1A$

$\tau = \frac{L}{R} = \frac{10}{5} = 2s$

$\therefore \left(I = 1 - e^{-1}\right) A$

LEVEL - I (HW)

MAGNETIC FLUX AND MOTIONAL EMF

1. In a coil of area 10cm^2 and 10 turns with magnetic field directed perpendicular to the plane and is changing at the rate of 10^8 gauss/ second. The resistance of the coil is 20Ω. The current in the coil will be

1) 0.5A 2) 5A 3) 50A 4) 5 x 10^8A

2. A magnetic flux of 500 micro-webers passing through a 200 turns coil is reversed in 20×10^{-3} seconds. The average emf induced in the coil in volts, is :

1) 2.5 2) 5.0 3) 7.5 4) 10.0

3. A rectangular coil of 200 turns and area 100 cm^2 is kept perpendicular to a uniform magnetic field of induction 0.25 tesla. If the field is reversed in direction in 0.01 second, the average induced emf in the coil is

1) 10^6 V 2) 10^4 V 3) 10^2 V 4) zero

4. A coil having an area $2m^2$ is placed in a magnetic field which changes from 1Wb/m^2 to 4Wb/m^2 in an interval of 2 second.The average e.m.f. induced in the coil will be

1) 4V 2)3V 3)1.5V 4)2V

5. A flip coil consists of N turns of circular coils which lie in a uniform magnetic field. Plane of the coils is perpendicular to the magnetic field as shown in figure. The coil is connected to a current integrator which measures the total charge passing through it. The coil is turned through 180^0 about the diameter. The charge passing through the coil is

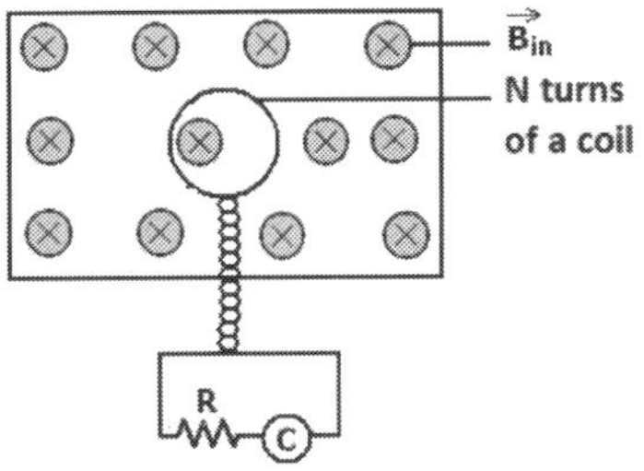

1. $\frac{NBA}{R}$ 2. $\frac{\sqrt{3}NBA}{2R}$ 3. $\frac{NBA}{\sqrt{2}R}$ 4. $\frac{2NBA}{R}$

6. A conductor AB of length l moves in xy plane with velocity $\vec{v} = v_0\left(\hat{i} - \hat{j}\right)$. A magnetic field $\vec{B} = B_0\left(\hat{i} + \hat{j}\right)$ exists in the region. The induced emf is

1. zero 2. $2B_0lv_0$ 3. B_0lv_0 4. $\sqrt{2}B_0lv_0$

SELF INDUCTION & MUTUAL INDUCTION

7. A coil has an inductance of 0.05H and 100 turns and 0.02A current is passed through it. Flux linked with coil is

1) 10^{-2} Wb 2) 10^{-3} Wb 3) 10^{-4} Wb 4) 10^{-5} Wb

8. A current of 2 A is increasing at the rate of 4 A/s through a coil of inductance 2 H. The energy stored in the inductor per unit time is

1) 2 W 2) 1 W 3) 16 W 4) 4 W

9. The current decays from 5 A to 2 A in 0.01s in a coil. The emf induced in a coil nearby it is 30V. The mutual inductance between the coils is

1) 1.0 H 2) 0.1 H 3) 0.001 H 4) 10 H

10. A varying current in a coil change from 10A to 0A in 0.5sec. If the average emf induced in the coil is 220V, the self inductance of the coil is

1) 5H 2) 6 H 3) 11H 4) 12 H

11. An air - cored solenoid is of length 0.3*m*, area of cross section $1.2\times10^{-3}m^{-2}$ and has 2500 urns. Around its central section, a coil of 350 turns is wound. The solenoid and the coil are electrically insulated from each other. Calculate the emf induced in the coil if the initial current of 3A in the solenoid is reversed in 0.25s.

1) 0.1056V 2) 1.056V 3) 10.56V 4) 0.01056 V

12. A solenoid of length 50cm with 20 turns per centimetre and area of cross-section $40cm^2$ completely surrounds another coaxial solenoid of the same length, area of cross-section $25cm^2$ with 25 turns per centimetre. Calculate the mutual inductance of the system.

1) 9.7 mH 2) 7.9 mH 3) 8.9 mH 4) 6.8 mH

L-R AND C-R CIRCUITS (D.C.)

13. A coil is connected to a battery of 12 V emf and negligible internal resistance. The current in the solenoid grows to 63% of its final steady state value in 0.3 s. If the final steady state current is 0.6 A, the inductance of the solenoid is

1) 0.6 H 2) 6.0 H 3) 0.015 H 4) 0.15 H

LEVEL - I (HW)- KEY

01) 2 02) 4 03) 3 04) 2 05) 4 06) 1 07) 4 08) 3 09) 2 10) 3
11) 1 12) 2 13) 2

LEVEL - I (HW) - HINTS

1. $i=\frac{1}{R}NA.\frac{dB}{dt}=\frac{1}{20}\times10\times10^{-3}\times10^{4}$

2. $e=N\frac{d\phi}{dt}$

3. $e=\frac{-NA\left(B_2-B_1\right)}{time}=\frac{-NA(-B-B)}{time}=\frac{2NAB}{time}$

4. $e=A\frac{dB}{dt}$

5. Initial flux through the coil, $\phi_{Bi} = +NBA$

Final flux through the coil, $\phi_{Bi} = +NBA$

When the coil is turned through 180^0 its flux reverses; the angle between magnetic field and area vector is reversed.

$\Delta\phi_B = \phi_{Bf} - \phi_{Bi} = -NBA - (NBA) = -2NBA$

$Q = \frac{\Delta\phi_B}{R} = \frac{2NBA}{R}$

6. $\vec{l}, \vec{v}$ and $\vec{B}$ are coplanar.

7. $n\phi = Li$

8. $U = \frac{1}{2}Li^2$; $P = \frac{dU}{dt}$

9. $e = -M\left(\frac{di}{dt}\right)$

10. $E = L \times \frac{di}{dt}$

11. $M = \frac{\mu_0 N_1 N_2 A_2}{l}$

12. $M = \frac{\mu_0 N_1 N_2 A_2}{l}$

13. $\tau = \frac{L}{R} = \frac{Li}{V}$

LEVEL - II (CW)

MAGNETIC FLUX & FARADY'S LAWS, INDUCED EMF,CURRENT AND CHARGE

1. **A circular coil of 'n' turns is kept in a uniform magnetic field such that the plane of the coil is perpendicular to the field. The magnetic flux associated with the coil is now ϕ. Now the coil is opened and made into another circular coil of twice the radius of the previous coil and kept in the same field such that the plane of the coil is perpendicular to the field. The magnetic flux associated with this coil now is**

   1) ϕ 2) 2ϕ 3) $\frac{\phi}{4}$ 4) $\frac{\phi}{2}$

2. **A physicist works in a laboratory where the magnetic field is 2 T. She wears a necklace** enclosing an area 100 cm^2 **of field and having a resistance of 0.1 Ω. Because of power failure, the field decays to 1 T in millisecond. The electric charge circulated in the necklace assuming that the magnetic field is perpendicular to area covered by the necklace is**

   1) 0.01 C 2) 0.001 C 3) 0.1 C 4) 1.0 C

3. A square loop PQRS of side 'a' and resistance 'r' is placed near an infinitely long wire carrying a constant current I. The sides PQ and RS are parallel to the wire. The wire and the loop are in the same plane. The loop is rotated by 180° about an axis parallel to the long wire and passing through the mid points of the side QR and PS. The total amount of charge which passes through any point of the loop during rotation is:

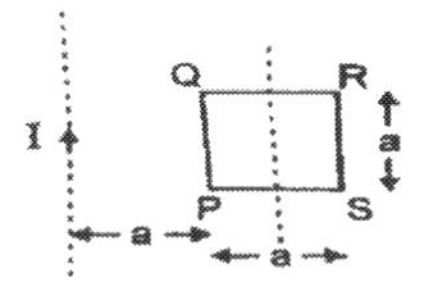

1) $\frac{\mu_0 Ia}{2\pi r}\ell n2$ 2) $\frac{\mu_0 Ia}{\pi r}\ell n2$ 3) $\frac{\mu_0 Ia^2}{2\pi r}$ 4) $\frac{\mu_0 i\, a}{2\pi r}$

MOTIONAL E.M.F

4. Two parallel rails of a railway track insulated from each other and with the ground are connected to a millivoltmeter. The distance between the rails is one metre. A train is traveling with a velocity of 72 kmph along the track. The reading of the millivoltmeter (in m V) is : (Vertical component of the earth's magnetic induction is 2×10^{-5}T)

1) 144 2) 0.72 3) 0.4 4) 0.2

5. A rod PQ is connected to the capacitor plates. The rod is placed in a magnetic field (B) directed downwards perpendicular to the plane of the paper. If the rod is pulled out of magnetic field with velocity $\vec{v}$ as shown in Figure.

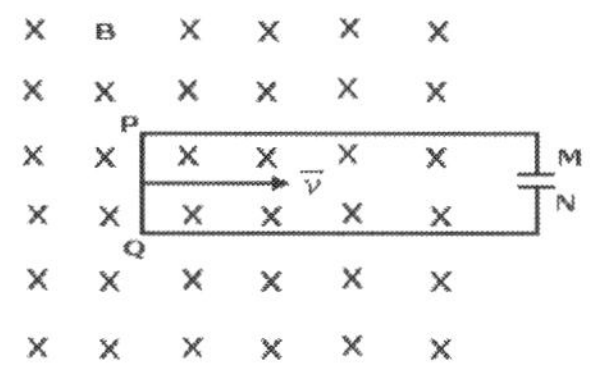

1) Plate M will be positively charged 2) Plate N will be positively charged
3) Both plates will be similarly charged 4) no charge will be collected on paltes.

6. A wire is sliding as shown in Figure. The angle between the acceleration and the velocity of the wire is

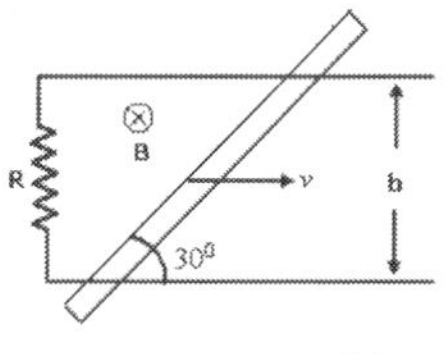

1) 30^0 2) 40^0 3) 120^0 4) 90^0

7. A conducting wire xy of length l and mass m is sliding without friction on vertical conduction rails ab and cd as shown in figure. A uniform magnetic filed B exists perpendicular to the plane of the rails, x moves with a constant velocity of

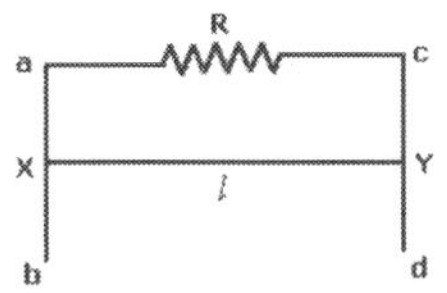

1) $\frac{mgR}{Bl}$ 2) $\frac{mgR}{Bl^2}$ 3) $\frac{mgR}{B^2l^2}$ 4) $\frac{mgR}{B^2l}$

8. A conducting rod AB of length $l = 1$ m is moving at a velocity v = 4 m/s making an angle 30° with its length. A uniform magnetic field B = 2T exists in a direction perpendicular to the plane of motion. Then :

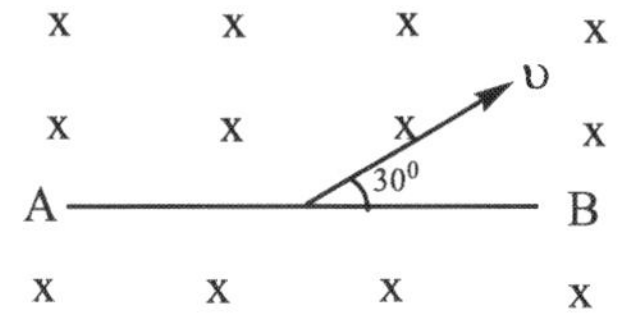

1) $V_A - V_B = 8V$ 2) $V_A - V_B = 4V$ 3) $V_B - V_A = 8V$ 4) $V_B - V_A = 4V$

9. A wire KMN moves along the bisector of the angle θ with a constant velocity v in a uniform magnetic field B perpendicular to the plane of the paper and directed inward. Which of the following is correct?

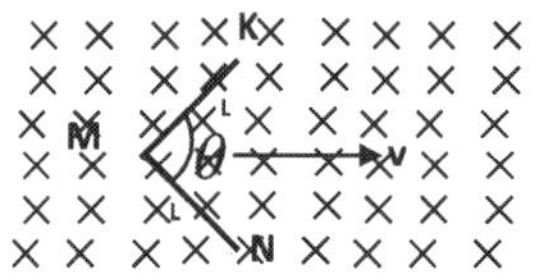

1) Effective length of the wire is $2L\sin\frac{\theta}{2}$

2) E.m.f induced between K and N is $2BLVSin\frac{\theta}{2}$

3) The shape of KMN is immaterial, only the end points K N are important.

4) All the above

10. A uniform magnetic field exists in region given by $\vec{B} = 3\hat{i} + 4\hat{j} + 5\hat{k}$. A rod of length 5 m is placed along y – axis is moved along x – axis with constant speed 1 m/sec. Then the magnitude of induced e.m.f in the rod is:

1) zero 2) 25 volt 3) 20 volt 4) 15 volt

11. A conducting rod PQ of length 1 m is moving with uniform velocity of 2 m/s in a uniform magnetic field of 2T directed into the plane of paper. A capacitor of capacity $c = 10\mu F$ is connected as shown. Then:

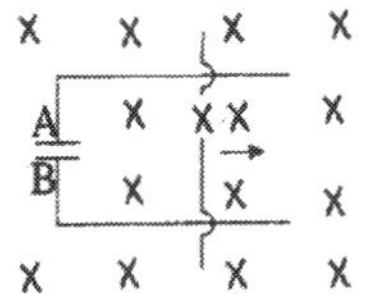

1) $q_A = +40\mu C,\ q_B = +40\mu C$ 2) $q_A = +40\mu C,\ q_B = -40\mu C$

3) $q_A = -40\mu C,\ q_B = +40\mu C$ 4) $q_A = q_B = 0$

12. A rectangular loop with a sliding connector of length $l = 1.0$ m is situated in a uniform magnetic field B = 2T perpendicular to the plane of loop. Resistance of connecter is as shown in figure. The external force required to keep the connector moving with a constant velocity $v = 2$ m/s is

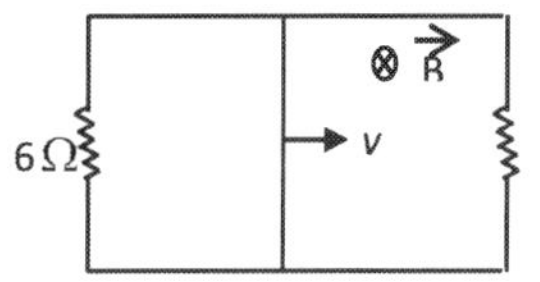

1) 6N 2) 4 N 3) 2 N 4) 1 N

13. **A straight rod of length l is rotating about axis passing through O is shown. A uniform magnetic field B exists parallel to the axis of rotation. E.m.f induced between P and Q is:**

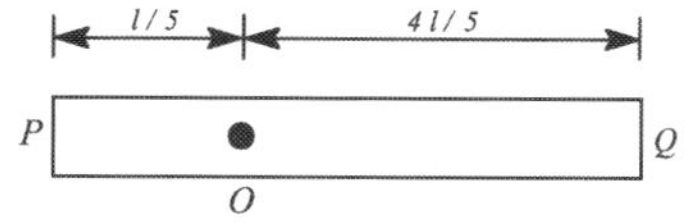

1) $\frac{8}{25}B\omega l^2$ 2) $\frac{3}{10}B\omega l^2$ 3) $\frac{7}{25}B\omega l^2$ 4) zero

14. **A rod of length 10 cm made up of conducting and non-conducting material (shaded part is non-conducting). The rod is rotated with cosntant angular velocity 10 rad / s about point O, in constant magnetic field of 2 T as shown in the figure. The induced emf between the point A and B of rod will be :**

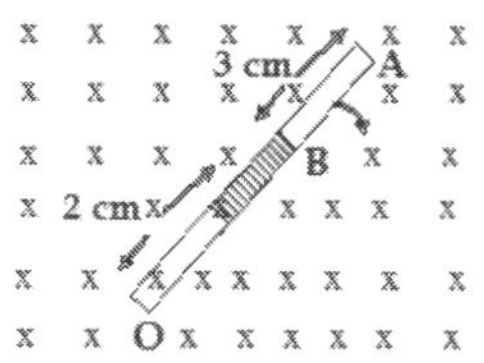

1) 0.029 V 2) 0.1 V 3) 0.051 V 4) 0.064 V

INDUCED ELECTRIC FIELD

15. **A time varying magnetic field is present in a cylindrical region of radius R as shown in the figure. A positive charge q is taken slowly from P to Q through POQ, the magnetic field** varies with time as $B = B_0 t$ **(where B_0 is a constant) and directed into the plane of the paper. If W is the workdone then W =**

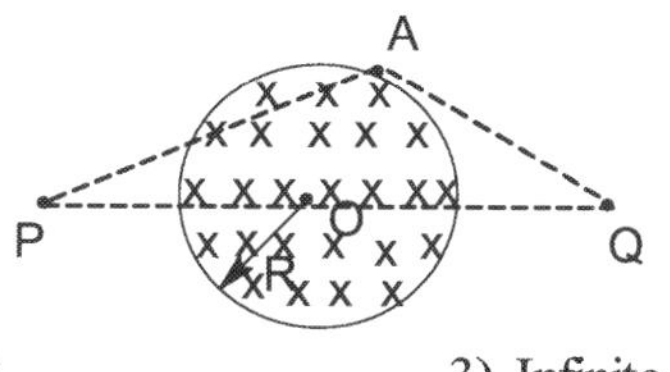

1) Zero 2) B_0 3) Infinite 4) $2B_0$

16. **A metallic square loop ABCD is moving in its own plane with velocity v in a uniform magnetic field perpendicular to its plane as shown in the figure. Electric field is induced:**

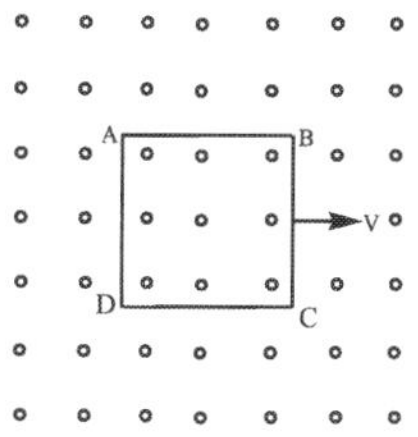

1) in AD, but not in BC
2) in BC, but not in AD
3) neither in AD nor in BC
4) in both AD and BC

17. **A triangular wire frame ABC (each side =2m) is placed in a region of time variant magnetic field having dB/dt = $\sqrt{3}$ T/s. The magnetic field is perpendicular to the plane of the triangle. The base of the triangle AB has a resistance 1 Ω while the other two sides have resistance 2 Ω each. The magnitude of potential difference between the points A and B will be**

1) 0.4 V 2) 0.6 V 3) 1.2 V 4) None

18. A uniform magnetic field of induction B is confined to a cylindrical region of radius R. The magnetic field is increasing at a constant rate of dB/dt (tesla/second). A charge q of mass m, placed at the point P on the periphery of the field experiences an acceleration:

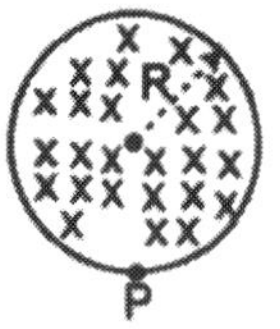

1) $\frac{1}{2}\frac{eR}{m}\frac{dB}{dt}$ toward left 2) $\frac{1}{2}\frac{eR}{m}\frac{dB}{dt}$ toward right 3) $\frac{eR}{m}\frac{dB}{dt}$ toward left 4) zero

SELF INDUCTION AND MUTUAL INDUCTION

19. The flux linked with a coil is 0.8 Wb when a 2 A current is flowing through it. If this current begins to increase at the rate of 400 A/s, the induced emf in the coil will be

1) 20 V 2) 40 V 3) 80 V 4) 160 V

20. A solenoid of self inductance 1.2 H is in series with a tangent galvanometer of reduction factor 0.9 A. They are connected to a battery and the tangent galvanometer shows a deflection of 53°. The energy stored in the magnetic field of the solenoid is (tan 53° = 4/3)

1) 0.864 J 2) 0.72 J 3) 0.173 J 4) 1.44 J

21. There are two batteries 'A' and 'B' having same emf. A has no internal resistance and B has some internal resistance. An inductance is connected first to 'A' and the energy in the uniform magnetic field setup inside is 'U'. It is now disconnected from 'A' and reconnected to 'B'. The energy stored in the uniform magnetic field will be

1) U 2) > U 3) < U 4) *zero*

22. An emf induced in a secondary coil is 10000 V when the current breaks in the primary. The mutual inductance is 5 H and the current reaches to zero in 10^{-4} s in primary. The maximum current in the primary before the break is

1) 0.2 A 2) 0.3 A 3) 0.4 A 4) 0.5 A

23. A mutual inductor consists of two coils X and Y as shwon in figure in which one quarter of the magnetic flux produced by X links with Y, giving a mutual inductance M. What will be the mutual inductance when Y is used as the primary?

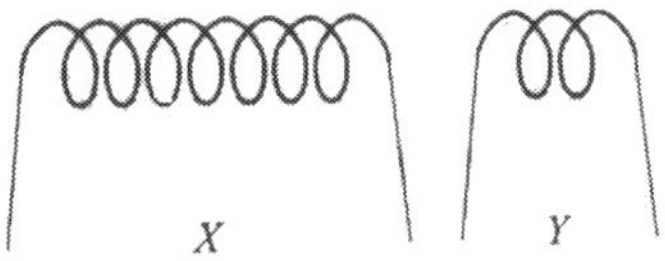

1) M/4 2) M/2 3) M 4) 2M

24. A long solenoid of length L, cross section A having N_1 turns has wound about its centre a small coil of N_2 turns as shown in figure. The mutual inductance of two circuit is

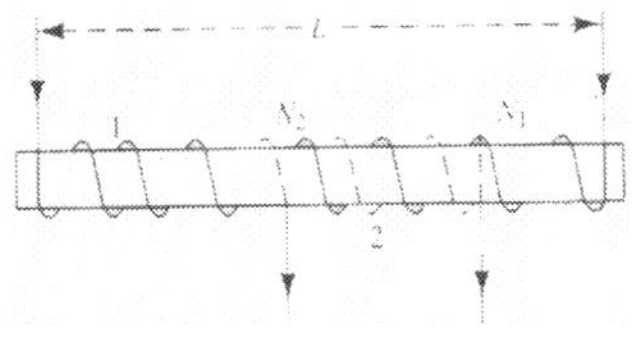

1) $\frac{\mu_0 A(N_1/N_2)}{L}$ 2) $\frac{\mu_0 A(N_1 N_2)}{L}$ 3) $\mu_0 A(N_1 N_2)L$ 4) $\frac{\mu_0 A(N_1^2 N_2)}{L}$

25. A small coil of radius r is placed at the centre of a large coil of radius R, where $R >> r$. The two coils are coplanar. The mutual inductance between the coils proportional to

1) r/R 2) r^2/R 3) r^2/R^2 4) r/R^2

26. The coefficient of mutual inductance of two circuits A and B is 3 mH and their respective resistances are 10 and 4Ω. How much current should change in 0.02 s in circuit A, so that the induced current in B should be 0.0060 A?

1) 0.24 A 2) 1.6 A 3) 0.18 A 4) 0.16 A

27. Two coaxial circular loops of radius 0.5 m and $5\times10^{-2}\,m$ are separated by a distance 0.5 m and carry currents $2A$ and $1A$ respectively. The force between the loops due to mutual induction is

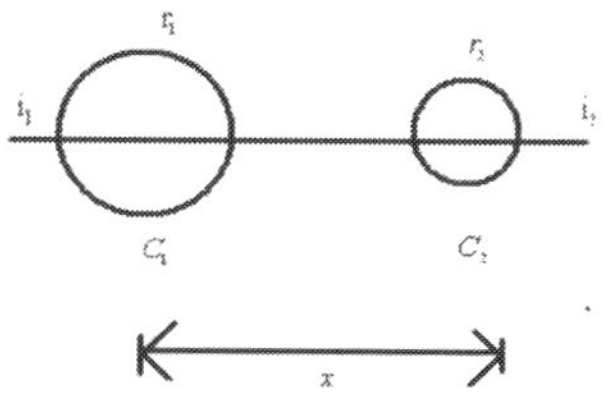

1) $2.09\times10^{-8}\ N$ 2) $1.06\times10^{-6}\ N$ 3) $4.18\times10^{-8}\ N$ 4) $8.3\times10^{-5}\ N$

28. The mutual inductance between the rectangular loop and the long straight wire as shown in figure is M.

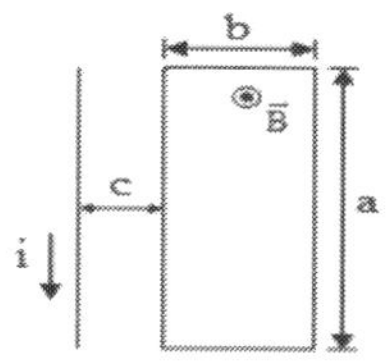

1) M = zero 2) $M = \frac{\mu_0 a}{2\pi} ln\left(1+\frac{c}{b}\right)$ 3) $M = \frac{\mu_0 a}{2\pi} ln\left(\frac{a+c}{b}\right)$ 4) $M = \frac{\mu_0 a}{2\pi} ln\left(1+\frac{b}{c}\right)$

29. An infinite long straight conducting cylindrical shell of radius a is surrounded by a thin coaxial infinite conducing cylindrical shell of radius b. Assuming current flows uniformly through the cylindrical shell and returns through the outer shell, the inductance per unit length for this arrangement is

1) $\frac{2\mu_0}{\pi}\ln\left(\frac{b}{a}\right)$ 2) $\frac{\mu_0}{\pi}\ln\left(\frac{b}{a}\right)$ 3) $\frac{\mu_0}{2\pi}\ln\left(\frac{b}{a}\right)$ 4) $\frac{\mu_0}{4\pi}\ln\left(\frac{b}{a}\right)$

D.C TRANSIENT CIRCUITS

30. The key 'K' is switched on at t = 0. Then the currents through battery at t = 0 and $t = \infty$ are

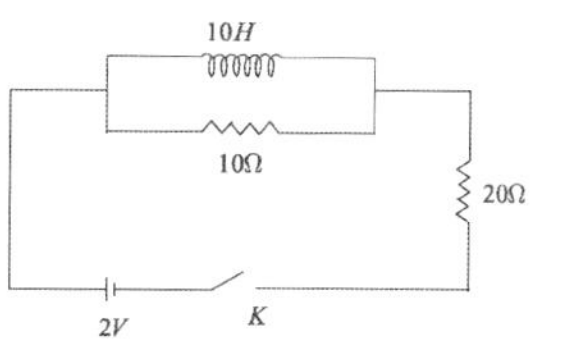

1) $\frac{1}{15}A, \frac{1}{10}A$ 2) $\frac{1}{10}A, \frac{1}{15}A$ 3) $\frac{2}{15}A, \frac{1}{10}A$ 4) $\frac{1}{15}A, \frac{2}{25}A$

31. **The current in an L - R circuit builds upto 3/4th of its steady state value in 4sec. Then the time constant of this circuit is**

1) $\frac{1}{\ln 2}$sec 2) $\frac{3}{\ln 2}$sec 3) $\frac{4}{\ln 2}$sec 4) $\frac{2}{\ln 2}$sec

32. **For the circuit shown, initially 'S' is closed for a long time so that steady state has been reached. Then at t = 0, 'S' is opened, due to which the current decays to zero. The heat generated in inductor is.**

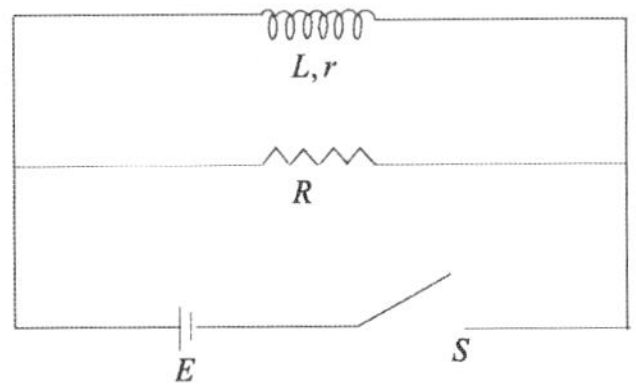

1) zero 2) $\frac{E^2L}{2(R+r)}$ 3) $\frac{E^2L}{2r(R+r)}$ 4) $\frac{E^2R}{2r(R+r)}$

33. **A coil of inductance 1 henry and resistance 10 Ω is connected to a resistance less battery of e.m.f 50 V at time t=0. Calculate the ratio of the rate at which magnetic energy is stored in the coil to the rate at which energy is supplied by the battery at t=0.1s**

1) e 2) $\frac{1}{e-1}$ 3) $1-\frac{1}{e}$ 4) $\frac{1}{e}$

34. A) 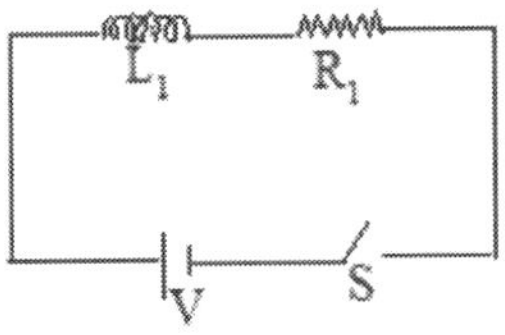

B) 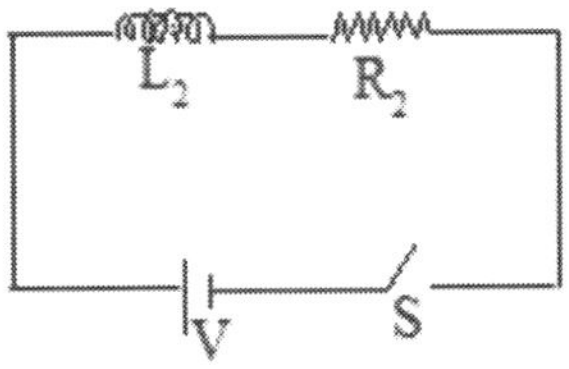

C) 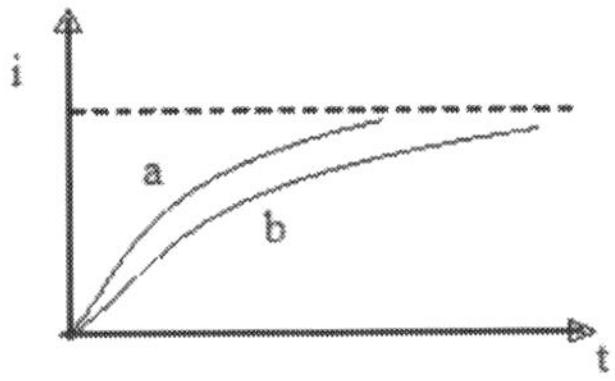

Current growth in two LR circuits (A) and (B) is as shown in figure (C). It follows that

1) $R_1 > R_2; L_1 < L_2$ 2) $R_1 > R_2; L_1 = L_2$ 3) $R_1 = R_2; L_1 < L_2$ 4) $R_1 < R_2; L_1 > L_2$

35. **In an LR circuit, current at t = 0 is 20 A. After 2 s it reduces to 18 A. The time constant of the circuit is :**

1) $\ln\left(\frac{10}{9}\right)$ 2) 2 3) $\frac{2}{ln\left(\frac{10}{9}\right)}$ 4) $2\ln\left(\frac{10}{9}\right)$

36. **A choke coil has an inductance of 4 H and a resistance of 2 Ω. It is connected to a battery of 12 V and negligible internal resistance. The time taken for the current to become 3.78 A is nearly**

1) 8 s 2) ½ s 3) 2 s 4) 4 s

37. **A coil of 40 H inductance is connected in series with a resistance of 8Ω and this combination is connected to the terminals of a 2V battery. The inductive time constant of the circuit is (in seconds)**

1) 40 2) 20 3) 5 4) 0.2

38. **In the circuit, the final current through 30Ω resistance when circuit is completed is**

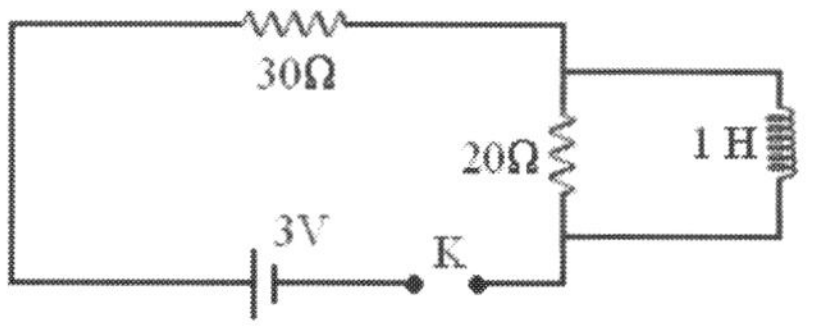

1) 3 A 2) 0.1 A 3) 5 A 4) 0.5 A

39. **In the circuit shown in figure, switch S is closed at time $t = 0$. The charge that passes through the battery in one time constant is**

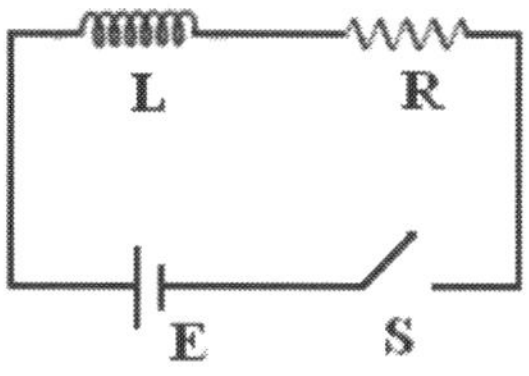

1. $\frac{eR^2E}{L}$ 2. $E\left(\frac{L}{R}\right)$ 3. $\frac{EL}{eR^2}$ 4. $\frac{eL}{ER}$

40. **A coil of wire having finite inductance and resistance has a conducting ring placed co-axially within it. The coil is connected a battery at time $t = 0$, so that a time dependent current $I_1(t)$ starts flowing through the coil. If $I_2(t)$ is the current induced in the ring and $B(t)$ is the magnetic field at the axis of the coil dut to $I_1(t)$ then as a function of time $(t > 0)$, the product $I_2(t)\ B(t)$**

1) increases with time 2) decreases with time

3) does not vary with time 4) passes through a maximum

KEY - LEVEL - II (CW)

01) 2	02) 3	03) 2	04) 3	05) 1	06) 3	07) 3	08) 2	9) 4	10) 2
11) 2	12) 2	13) 2	14) 3	15) 1	16) 4	17) 2	18) 1	19) 4	20) 1
21) 3	22) 1	23) 3	24) 2	25) 2	26) 4	27) 1	28) 4	29) 3	30) 1
31) 4	32) 3	33) 4	34) 3	35) 3	36) 3	37) 3	38) 2	39) 3	40) 4

HINTS - LEVEL - II (CW)

1. $\phi = BAn \quad n(2\pi r) = \text{constant}$

2. $q = \frac{N(B_1 - B_2)A}{R} = \frac{1(2-1)10^{-2}}{0.1} = 0.1C$

3. Conceptual

4. $e = B_v lu = 2\times10^{-5}\times1\times20 = 0.4mV$

5. Consider the force on an electron in PQ. This electron experiences a force towards Q. Free electron in PQ tend to move towards N. So M will be positively charged.

6. $\vec{F}_m \perp \vec{v}$

$\therefore \theta = 90^0 + 30^0 = 120^0$

7. $BIl = mg$ or $B\dfrac{Bvl}{R}l = mg$ or $v = \dfrac{mgR}{B^2l^2}$

8. $e = Blv\sin\theta$

9. $e = B\,l_{eff}\,v$

$l_{eff} = 2\left(l\sin\theta/2\right)$

10. $e = \vec{l}\cdot\left(\vec{v}\times\vec{B}\right)$

11. $e_{induced} = Blv$ and $q = Ce_{induced}$ Apply Right hand rule to know direction of induced current.

12. $F = \dfrac{B^2l^2v}{R_P}$ here $R_P = \dfrac{6\times3}{6+3} = 2\Omega$

13. $e = B\omega\int_{-l/5}^{4l/5} x\ dx = \dfrac{3B\omega l^2}{10}$

14. $e = B\omega\int_{7}^{10} x\ dx = 0.051V$

15. In taking charge from P to Q one has to perform work against the force experienced by charge due to the induced electric field. The induced electric field is perpendicular to line POQ and hence $W_1 = 0$

16. conceptual

17. $P.D$ in the loop $= \int \vec{E}\cdot d\vec{l} = \dfrac{d\phi}{dt} = A\dfrac{dB}{dt}$

$i = \dfrac{\Delta V}{R_T}$ $R_T = 5\ \Omega$

$V_{AB} = iR_{AB}$

18. $\int \vec{E}\cdot d\vec{l} = \dfrac{d\phi}{dt} = A\dfrac{dB}{dt}$

$\vec{a} = -\dfrac{q\vec{E}}{m}$ Direction of induced field can determined from Lenz's law

19. $\phi = Li \Rightarrow L = \frac{0.8}{2} = 0.4$

$e = L\frac{di}{dt} = 0.4 \times 400 = 160v$

20. $i = K\tan\theta = 0.9 \times \frac{4}{3} = 1.2A, U = \frac{1}{2}Li^2 = \frac{1}{2} \times 1.2 \times 1.2^2 = 0.864$

21. $U \propto i^2$

22. $e = L\frac{di}{dt} = 5\left(\frac{i_1 - 0}{10^{-4}}\right)$

23. The mutual inductance M remains the same whether X or Y is used as the primary.

24. $\phi_2 = N_2 B_1 A$ or $\phi_2 = N_2 \frac{\mu_0 N_1 I_1}{L} A$ or $\phi_2 = \frac{\mu_0 N_1 N_2 A}{L} I_1$

Comparing with $\phi_2 = MI_1$, we get $M_2 = \frac{\mu_0 N_1 N_2 A}{L}$

25. Magnetic field at the centre of a large coil $E \propto \frac{\mu_0 I}{2R}$

Magnetic flux linked with smaller coil $= \frac{\mu_0 I}{2R} \times \pi r^2$

$M = \frac{\phi}{I} = \frac{\mu_0 \pi r^2}{2R}$

$\therefore M \propto \frac{r^2}{R}$

26. Induced current in $B = 0.006A = 6 \times 10^{-3} A$

induced emf in $B = 6 \times 10^{-3} \times 4V = 24 \times 10^{-3} V$

Now, $M\frac{dI}{dt} = 24 \times 10^{-3}$

or $dI = \frac{24 \times 10^{-3} \times 0.02}{3 \times 10^{-3}} A = 0.16A$

27. $F = \frac{\mu_0}{4\pi}\frac{6M_1 M_2}{d^4}$

where $M_1 = n_1 i_1 A_1$, $M_2 = n_2 i_2 A_2$

28. $d\phi = B\,adx$

$d\phi = \frac{\mu_o i}{2\pi x} a\,dx$

$\phi = \frac{\mu_o ia}{2\pi}\ln\left(\frac{c+b}{c}\right) = \text{Mi}$

29. $\int_a^b \frac{\mu_0 i}{2\pi x} dx = \frac{\mu_0 i}{2\pi} \log\left(\frac{b}{a}\right)$

$$L = \frac{\mu_0}{2\pi} \log\left(\frac{b}{a}\right)$$

30. V = iR

31. $i = i_0\left(1 - e^{-\frac{Rt}{L}}\right),\ i = \frac{3i_0}{4}$ when t = 4 sec.

$$\frac{3i_0}{4} = i_0\left(1 - e^{-\frac{4R}{L}}\right),\ e^{-4/\tau} = \frac{1}{4}$$

$$e^{4/\tau} = 4$$

$$\frac{4}{2} = \ln 4, \tau = \frac{4}{2\ln 4} = \frac{2}{\ln 2}$$

32. Energy stored in indcutor $U = \frac{1}{2}Li_0^2, i_0 = \frac{E}{r}, U = \frac{1}{2}L\frac{E^2}{r^2}$

fraction of energy dissipated in 'r'

$$U = \frac{1}{2}\frac{LE^2}{r^2}\frac{r}{r+R}, U = \frac{LE^2}{2r(R+r)}$$

33. $\frac{\frac{dU}{dt}}{p} = \frac{1}{2}\frac{L(2i)\frac{di}{dt}}{Vi} = \frac{L}{V}i_0\frac{R}{L}e^{-\frac{R}{L}t} = \frac{R}{V}\left(\frac{V}{R}\right)e^{-1} = \frac{1}{e}$

34. $i_0 = \frac{V}{R_1} = \frac{V}{R_2} \Rightarrow R_1 = R_2$

$$\tau_1 < \tau_2 \Rightarrow \frac{L_1}{R_1} < \frac{L_2}{R_2} \Rightarrow L_1 < L_2$$

35. $I = I_0 e^{-t/\tau} \Rightarrow 18 = 20e^{-2/\tau} \Rightarrow \frac{10}{9} = e^{2/\tau}$

$$\Rightarrow \ln\frac{10}{9} = \frac{2}{\tau} \Rightarrow \tau = \frac{2}{\ln\left(\frac{10}{9}\right)}$$

36. $i = i_0\left[1 - e^{-\frac{R}{L}t}\right]$

37. $t = \frac{L}{R}$

38. The current is short cicuited through inductor. In stready state, current will pass through inductor instead of 20Ω. Then current through 30Ω is : $I = \frac{3}{30} = 0.1A$

39. The current at time t is given by $i = i_0\left(1 - e^{-t/\tau}\right)$

Here $i_0 = \frac{E}{R}$ and $\tau = \frac{L}{R}$

$$\therefore q = \int_0^{\tau} i dt = \int_0^{t} i_0\left(1 - e^{-t/\tau}\right)dt$$

$$= \frac{i_0\tau}{e} = \frac{\left(\frac{E}{R}\right)\left(\frac{L}{R}\right)}{e} = \frac{EL}{eR^2}$$

40. The equation of $I_1(t), I_2(t)$ and $B(t)$ will take the following forms

$I_1(t) = K_1\left(1 - e^{-k2t}\right) \rightarrow$ current growth in L-R circuit

$B(t) = K_3\left(1 - e^{-k2t}\right) \rightarrow B(t) \propto I_1(t)$

$I_2(t) = K_4 e^{-k2t}$

$$\left[I_2(t) = \frac{e_2}{R} and\ e_2 \propto = -M\frac{dI_1}{dt}\right]$$

Therefore, the product $I_2(t)B(t) = K_5 e^{-k2t}\left(1 - e^{-k2t}\right)$

The value of this product zero at $t = 0$ and $t = \infty$. Therefore, the product will pass through a maximum value.

LEVEL - II (HW)

MAGNETIC FLUX AND MOTIONAL EMF

1. A coil of 30 turns of wire each of 10 cm² area is placed with its plane perpendicular to a magnetic field of 0.1 T. When the coil is suddenly withdrawn from the field, a galvanometer connected in series with the coil indicated that a 10 μC charge passes around the circuit. The combined resistance of the coil and galvanometer is

   1) 3 Ω 2) 30 Ω 3) 300 Ω 4)3000 Ω

2. A square coil of side 0.5 m has movable sides. It is placed such that its plane is perpendicular to uniform magnetic field of induction 0.2 T. If all the sides are allowed to move with a speed of 0.1 m/s for 4 sec outwards, average induced emf is

   1) Zero 2) 0.01V 3) 0.028V 4) 0.072V

3. A uniform magnetic field exists in a regiion given by $\vec{B} = 3\hat{i} + 4\hat{j} + 5\hat{k}$. A rod of length 5 m along $y-axis$ moves with a constant speed of 1 m/s along $x-axis$. Then the induced emf in the rod will be

   1. 0 2. 25 V 3. 20 V 4. 15 V

4. Two identical conducting rings A and B of radius R are rolling over a horizontal conducting plane with same speed v but in opposite direction. A constant magnetic field B is present pointing into the plane of paper. Then the potential difference between the highest points of the two rings is

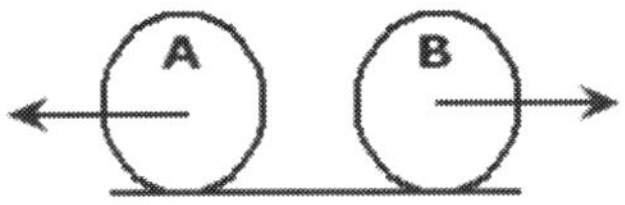

1. 0 2. $2BvR$ 3. $4BvR$ 4. none of these

5. A flexible wire loop in the shape of a circle has a radius that grows linearly with time. There is a magnetic field perpendicular to the plane of the loop that has a magnitude inversely proportional to the distance from the centre of the loop, $B(r) \propto \frac{1}{r}$ How does the emf E vary with time?

1. $E \propto t^2$ 2. $E \propto t$ 3. $E \propto \sqrt{t}$ 4. E is constant

SELF INDUCTION & MUTUAL INDUCTION

6. The e.m.f. induced in a secondary coil is 20000 V when the current breaks in the primary coil. The mutual inductance is 5H and the current reaches to zero in 10^{-4} sec in the primary. The maximum current in the primary before it breaks is

1) 0.1A 2) 0.4A 3) 0.6A 4) 0.8A

7. A small square loop of wire of side 'l' is placed inside a large square loop of side 'L' (L>>l). If the loops are coplanar and their centres coincide, the mutual induction of the system is directly proportional to :

1) $\frac{L}{l}$ 2) $\frac{l}{L}$ 3) $\frac{L^2}{l}$ 4) $\frac{l^2}{L}$

8. A straight solenoid of length 1 m has 5000 turns in the primary and 200 turns in the secondary. If the area of cross section is $4cm^2$, the mutual inductance will be

1. 503 H 2. 503 mH 3. 503 μH 4. 5.03 H

L-R AND C-R CIRCUITS (D.C.)

9. A $8\mu F$ capacitor is charged by a 400V supply through 0.1 $M\Omega$ resistance. The time taken by the capactor to develop a potential difference of 300V is: (Given $\log_{10} 4 = 0.602$)

1) 2.2 sec 2) 1.1 sec 3) 0.55 sec 4) 0.48 secs

10. A resistance with an inductor of 8 H has the same time constant as it has with a condenser of capacity 2 μF. The value of the resistance expressed in ohms is

1) 500 2) 250 3) 4000 4) 2000

11. The cell in the circuit shown in figure is ideal. The coil has an inductance of 4 mH and a resistance of $2m\Omega$. The switch is closed at $t = 0$. The amount of energy stored in the inductor at $t = 2$ s is (take $e = 3$)

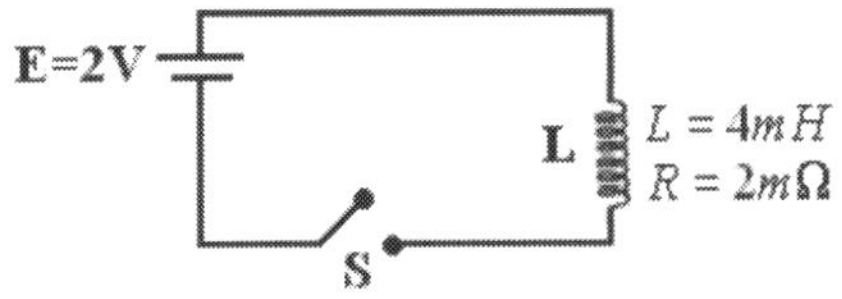

1. $\frac{4}{3}J$ 2. $\frac{8}{9}\times10^3 J$ 3. $\frac{8}{3}\times10^{-3} J$ 4. $2\times10^3 J$

KEY - LEVEL - II (HW)

1) 3 2) 2 3) 2 4) 3 5) 4 6) 2 7) 4 8) 3 9) 4 10) 2
11) 2

HINTS - LEVEL - II (HW)

1. $q = \frac{d\phi}{R} = \frac{NA(B_1 - B_2)}{R}$

$10\mu c = \frac{30\times10^{-3}(0.1-0)}{R}$

$R = 300\Omega$

2. $e = B\left(\frac{dA}{dt}\right)$

3. b

$e = (\vec{v}\times\vec{B}).\vec{l}$

$e = \left[\vec{i}\times(3\vec{i}+4\vec{j}+5\vec{k})\right].5\vec{j}$

$\Rightarrow e = -25\text{V}$

4. c

$V_2 - V_1 = \frac{1}{2}B\omega(2R)^2 = 2BR(\omega R) = 2BRv$

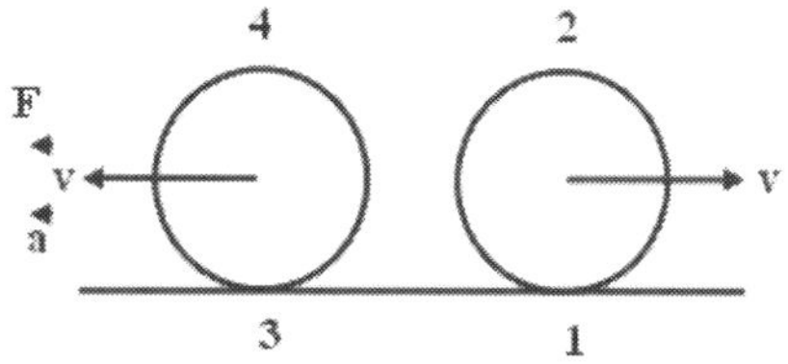

$V_3 - V_4 = 2BRv$ But $V_1 = V_3$ $\Rightarrow V_2 - V_4 = 4BRv$

5. d

Let radius of the loop is r at any time t and in further time dt, radius increases by dr.

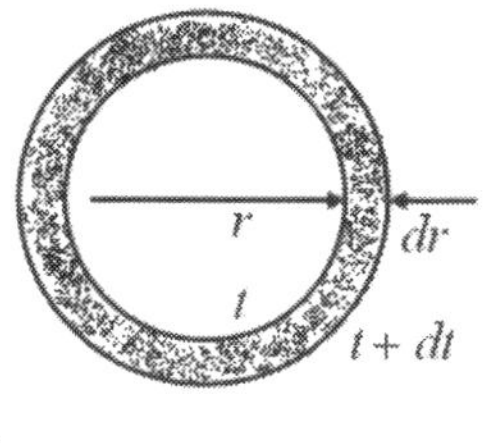

Then change in flux : $d\phi = (2\pi r dr)B$

$\Rightarrow e = \frac{d\phi}{dt} = 2\pi r\left(\frac{dr}{dt}\right)\frac{k}{r}$

$\Rightarrow e = 2\pi ck$ (constant) $\left[\because \frac{dr}{dt} = c, B = \frac{k}{r}\right]$

6. $e=-M.\frac{(i_2-i_1)}{time}$; $2\times10^4=\frac{-5(0-i_1)}{10^{-4}}$

7. Flux $= Mi = BA$

$Mi=\frac{4\mu o\times l^2 i}{\pi L\sqrt{2}}$

$M\propto\frac{l^2}{L}$

8. $M=\frac{\mu_0 N_1 N_2 A}{l}a$, Here l=1m

$=503\times10^{-6}$H

9. $\tau=\frac{L}{R}=RC$

$R=\sqrt{L/C}=\sqrt{8/2\times10^{-6}}$; $R=2\ K\Omega$

10. $q=q_0\left(1-e^{-t/T}\right)$

$V=\left(\frac{q}{C}\right)$

11. Use $i=\frac{E}{R}\left[1-e^{-t/\tau}\right]$ and $U=\frac{1}{2}Li^2$

LEVEL - III

MAGNETIC FLUX & FARADY'S LAWS, INDUCED EMF,CURRENT AND CHARGE

1. A magnetic field in a certain region is given by $B=\left(40\hat{i}-15\hat{k}\right)\times10^{-4}T$. The magnetic flux passes through a loop of area 5.0 cm^2 is placed flat on xy plane is

1) 750nWb 2) -750nWb 3) 360nWb 4) -360nWb

Motional E.M.F

2. A straight conducting rod of length 30cm and having a resistance of 0.2 ohm is allowed to slide over two parallel thick metallic rails with uniform velocity of 0.2 m/s as shown in the figure. The rails are situated in a horizontal plane if the horizontal component of earth's magnetic field is 0.3×10^{-4} T and a steady current of $3\mu A$ is induced through the rod. The angle of dip will be :

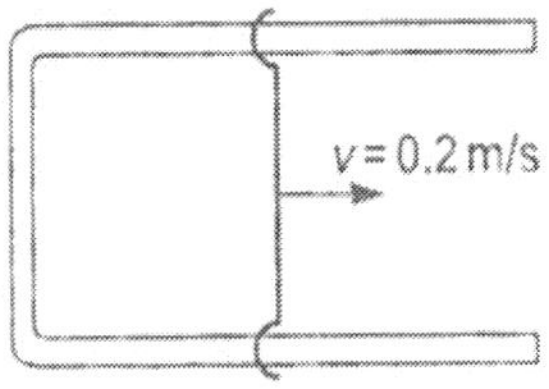

1) $\tan^{-1}\left(\frac{3}{4}\right)$ 2) $\tan^{-1}\left(\frac{1}{\sqrt{3}}\right)$ 3) $\tan^{-1}\left(\sqrt{3}\right)$ 4) $\tan^{-1}\left(\frac{1}{3}\right)$

3. **A 0.1 m long conductor carrying a current of 50 A is perpendicular to a magnetic field of 1.25 mT. The mechanical power to move the conductor with a speed of $1\,ms^{-1}$ is**

1) 0.25 mW 2) 6.25 mW 3) 0.625 mW 4) 1 W

4. **A circuit ABCD is held perpendicular to the uniform magnetic field of $B = 5\times10^{-2}\,T$ extending over the region PQRS and 3directed into the plane of the paper. The circuit is moving out of the field at a uniform speed of $0.2ms^{-1}$ for 1.5s. During this time, the current in the 5Ω resistor is**

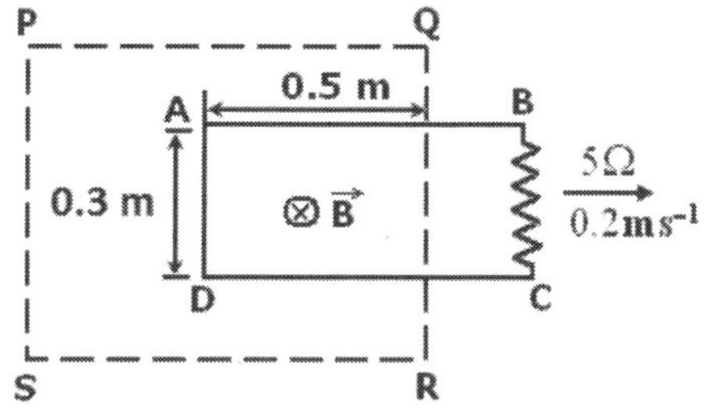

1) 0.6 mA from B to C 2) 0.9 mA from B to C
3) 0.9 mA from C to B 4) 0.6 mA from C to B

5. **Figure shows a copper rod moving with velocity v parallel to a long straight wire carrying current $=100A$. Calculate the induced emf in the rod, where $v = 5ms^{-1}$, $a = 1cm$, $b = 100cm$.**

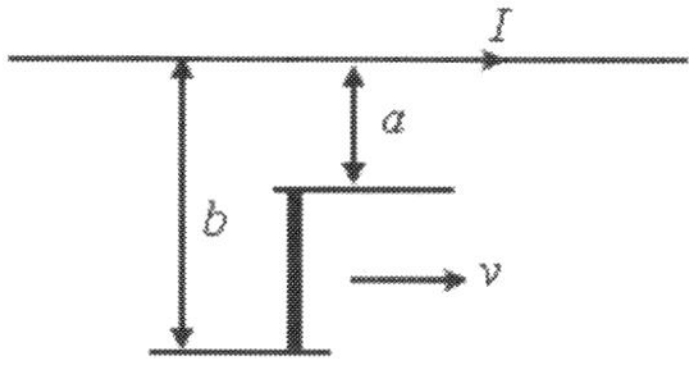

1) 0.23 mV 2) 0.46 mV 3) 0.16 mV 4) 0.32 mV

6. **A rectangular loop with a sliding conductor of length l is located in a uniform magnetic field perpendicular to the plane of the loop. The magnetic induction is B. The conductor has a resistance R_1 and R_2, respectively. Find the current through the conductor during its motion to the right with a constant velocity v.**

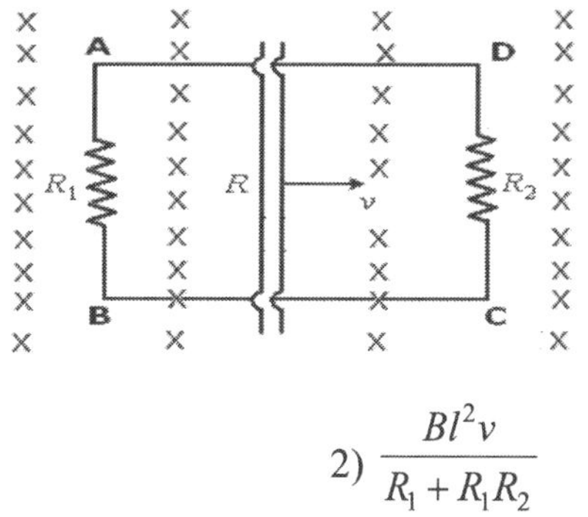

1) $\dfrac{Blv(R_1+R_2)}{R_1(R_1+R_2)}$ 2) $\dfrac{Bl^2v}{R_1+R_1R_2}$

3) $\dfrac{Blv(R_1+R_2)}{R_1R_2+R(R_1+R_2)}$ 4) $\dfrac{Bl^2v}{R_1R_2+R(R_1+R_2)}$

7. A conductor of length l and mass m can slide without any friction along the two vertical conductors connected at the top through a capacitor. A uniform magnetic field B is set up $\perp$ to the plane of paper. The acceleration of the conductor

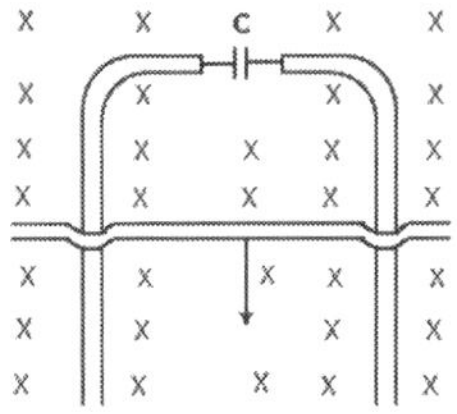

1) is constant 2) Increases 3) decreases 4) cannot say

8. A wheel has three spokes and is in a uniform magnetic field perpendicular to its plane, with the axis of rotation of the wheel parallel to the magnetic field. When the wheel rotates with a uniform angular velocity ω, the emf induced between the center and rim of the wheel is 'e'. If another wheel having same radius but with six spokes is kept in the same field and rotated with a uniform angular velocity '$\omega/2$', the emf induced between the center and the rim will be

1) e 2) e/2 3) 2e 4) e/4

9. A metal rod of length 1 m is rotated about one of its ends in a plane at right angles to a uniform magnetic field of induction $2.5\times10^{-3}Wbm^{-2}$. If it makes 1800 rpm, then the induced emf between its ends approximately is

1) 0.24 V 2) 0.12 V 3) 0.36 V 4) 0.48 V

10. A bicycle is resting on its stand in the east-west direction and rear wheel is rotating at 100 revolution per minute. Length of each spoke is 30 cm, and vertical component of earth's magnetism is 1.5×10^{-5} tesla. If the emf induced in the spokes is $3\pi\times10^{-6}V$, the angle of dip will be :

1) $\tan^{-1}\left(\frac{3}{4}\right)$ 2) $\tan^{-1}\left(\frac{1}{\sqrt{3}}\right)$ 3) $\tan^{-1}\left(\sqrt{3}\right)$ 4) $\tan^{-1}\left(\frac{1}{3}\right)$

11. Find the linear speed the bicycle required to power its head light by a generator, whose rubber shaft presses against the wheel of cycle of radius 0.33 m, turns at an angular speed of 44 times as great as the angular speed of the tire itself. The coil consists of 75 turns, has an area of 2.6 X 10^{-3} m^2, and rotates in a 0.10 T magnetic field. When the peak emf being generated is 6.0 V.

1) 2.5 m/s 2) 5 m/s 3) 2.3 m/s 4) 4.6 m/s

12. The coil of an ac generator rotates at a frequency of 60 Hz and develops an inducecd emf of 120 V (rms). The coil has an area of 3.0 x10^{-3} m^2 and consists of 500 turns. The magnitude of the magnetic field in which the coil rotates is

1) 0.30 T 2) 3.00 T 3) 0.03 T 4) 30 T

13. A conducting loop of radius R is present in a uniform magnetic field B perpendicular to the plane of ring. If radius R varies as a function of time t as $R = R_0 + t^2$. The emf induced in the loop is

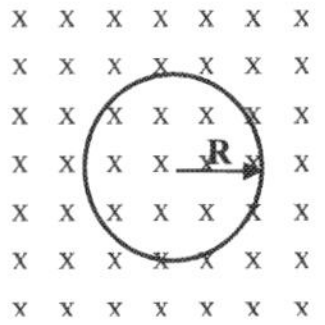

1) $2\pi Bt\left(R_0+t^2\right)$ Clockwise 2) $2\pi Bt\left(R_0+t^2\right)$ Anticlockwise

3) $4\pi Bt\left(R_0+t^2\right)$ Anticlockwise 4) $4\pi Bt\left(R_0+t^2\right)$ Clockwise

14. A conducting rod is free to slide along a pair of conducting rails, in a region where a uniform and constant (in time) magnetic field is directed into the plane of the paper, as the drawing illustrates. Initially the rod is at rest. There is no friction between the rails and the rod. What happens to the rod after the switch is closed? If any induced emf develops, be sure to account for its effect.

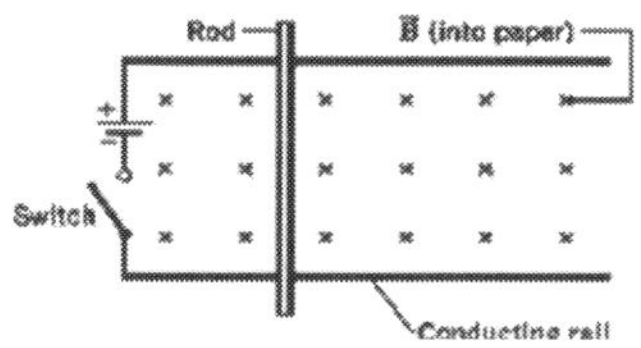

1) The rod accelerates to the right, its velocity increasing without limit.

2) The rod does not move.

3) The rod accelerates to the right for a while and then slows down and comes to a halt.

4) The rod accelerates to the right and eventually reaches a constant velocity at which it continues to move.

15. A rectangular loop has a sliding connector PQ of length l and resistance $R\Omega$ and it is moving with a speed v as shown. The set-up is placed in a uniform mangetic field going into the plane of the paper. The three currents I_1, I_2 and I are

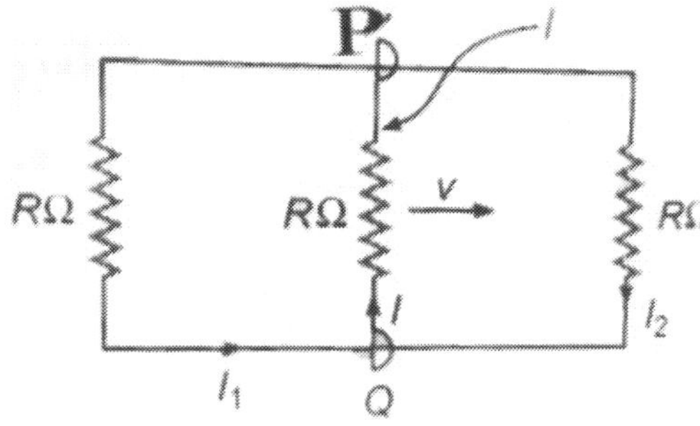

1) $I_1 = -I_2 = \frac{Blv}{R}, I = \frac{2Blv}{R}$

2) $I_1 = I_2 = \frac{Blv}{3R}, I = \frac{2Blv}{3R}$

3) $I_1 = I_2 = I = \frac{Blv}{R}$

4) $I_1 = I_2 = \frac{Blv}{6R}, I = \frac{Blv}{3R}$

INDUCED ELECTRIC FIELD

16. A specially uniform magnetic field of 0.080 T is directed into the plane of the page and perpendicular to it, as shown in the accompanying figure. A wire loop in the plane of the page has constant area 0.010 m^2. The magnitude of the magnetic field decreases at a constant rate of 3.0 X 10^{-4} T/s. The magnitude and direction of the induced emf is

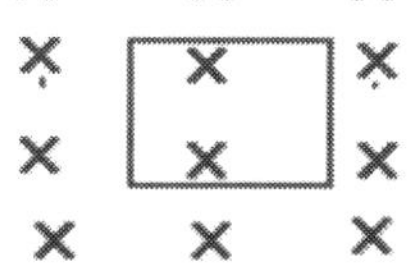

1) 3.0 X 10^{-6} V clockwise

2) 3.0 X 10^{-6} V anticlockwise

3) 2.4 X 10^{-5} V anticlockwise

4) 8.0 X 10^{-4} V clockwise

17. A magnetic field induction is changing in magnitude in a region at a constant rate dB/dt. A given mass m of copper drawn into a wire and formed into a loop is placed perpendicular to the field. If the values of specific resistance and density of copper are ρ and σ respectively, then the current in the loop is given by :

1) $\frac{4\pi m}{\rho\sigma}\frac{dB}{dt}$ 2) $\frac{m}{4\pi\rho\sigma}\frac{dB}{dt}$ 3) $\frac{m}{\rho\sigma}\frac{dB}{dt}$ 4) $\frac{2\pi m}{\rho\sigma}\frac{dB}{dt}$

18. Magnetic flux linked with a stationary loop of resistance R varies with respect to time during the time period T as follows: $\phi = at(T-t)$ the amount of heat generated in the loop during that time (inductance of the coil is negligible) is

1) $\frac{aT}{3R}$ 2) $\frac{a^2T^2}{3R}$ 3) $\frac{a^2T^2}{R}$ 4) $\frac{a^2T^3}{3R}$

19. A wire is bent to form the double loop shown in Figure. There is a uniform magnetic field directed into the plane of the loop. If the magnitude of this field is decreasing, the current will flow from

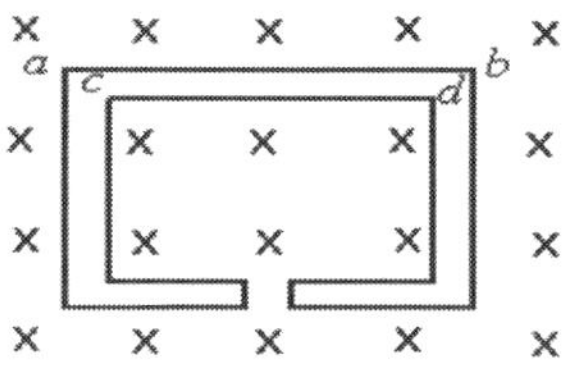

1) a to b and c to d 2) b to a and d to c 3) a to b and d to c 4) b to a and c to d

SELF INDUCTION AND MUTUAL INDUCTION

20. Two coils have self-inductance $L_1 = 4$ mH and $L_2 = 1$ mH respectively. The currents in the two coils are increased at the same rate. At a certain instant of time both coils are given the same power. If I_1 and I_2 are the currents in the two coils, at that instant of time respectively, then the value of (I_1/I_2) is:

1) 1/8 2) 1/4 3) 1/2 4) 1

21. The length of a wire required to manufacture a solenoid of length l and self-induction L is (cross-sectional area is negligible)

1. $\sqrt{\frac{2\pi Ll}{\mu_0}}$ 2. $\sqrt{\frac{\mu_0 Ll}{4\pi}}$ 3. $\sqrt{\frac{4\pi Ll}{\mu_0}}$ 4. $\sqrt{\frac{\mu_0 Ll}{2\pi}}$

22. The inductance L of a solenoid of length l, whose windings are made of material of density D and resistivity ρ, is (the winding resistance is R)

1) $\frac{\mu_0}{4\pi l}\frac{Rm}{\rho D}$ 2) $\frac{\mu_0}{4\pi r}\frac{lm}{\rho D}$ 3) $\frac{\mu_0}{4\pi l}\frac{R^2 m}{\rho D}$ 4) $\frac{\mu_0}{2\pi R}\frac{lm}{\rho D}$

23. Two concentric and coplanar coils have radii a and $b(>> a)$ as shown in figure. Resistance of the inner coil is R. Current in the outer coil is increased from 0 to i, then the total charge circulating the inner coil is

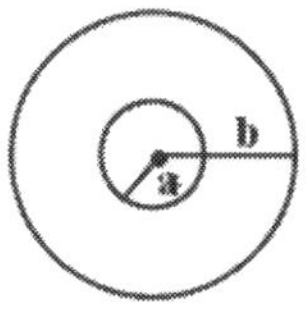

1) $\frac{\mu_0 i a^2\pi}{2Rb}$ 2) $\frac{\mu_0 iab}{2R}$ 3) $\frac{\mu_0 iab}{2a}\frac{\pi b^2}{R}$ 4) $\frac{\mu_0 ib}{2\pi R}$

24. The inner loop has an area $5\times10^{-4}m^2$ and a resistance of 2Ω. The larger circular loop is fixed and has a radius of 0.1m. Both the loops are concentric and coplanar. The smaller loop is rotated with an angular velocity ω rad s^{-1} about its diameter. The magnetic flux with the smaller loop is

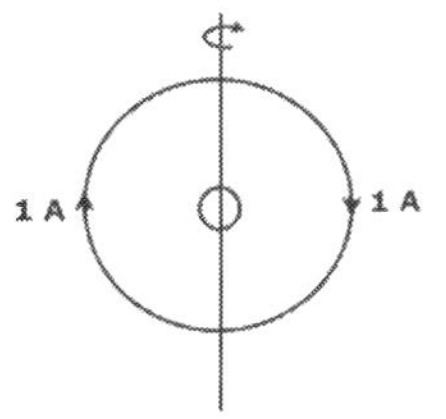

1) $2\pi\times10^{-6}$ weber 2) $\pi\times10^{-9}$ weber 3) $\pi\times10^{-9}\cos\omega t$ weber 4) zero

25. A circular wire loop of radius R is placed in the x-y plane centred at the origin O. A square loop of side $a\left(a<<R\right)$ having two turns is places with its centre at $z=\sqrt{3}R$ along the axis of the circular wire loop, as shown in figure. The plane of the squre loop makes an angle of 45^0 with respect to the z-axis. If the mutual inductance between the loops is given by $\frac{\mu_0 a^2}{2^{P/2}R}$, then the value of P is

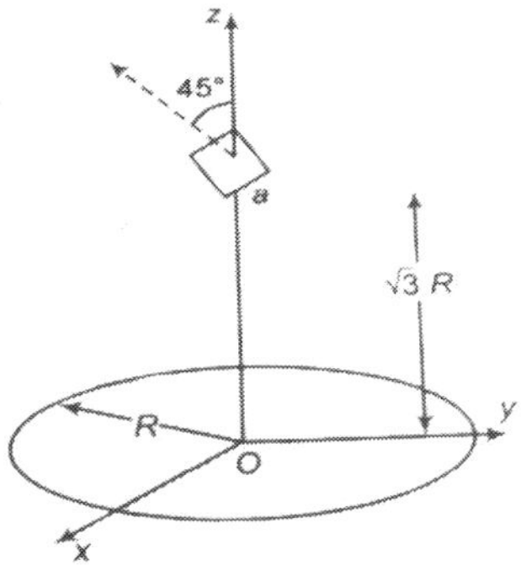

1) 3 2) 5 3) 7 4) 9

D.C Transient circuits

26. A coil of some internal resistance 'r' behaves like an inductance. When it is connected in series with a resistance R_1, the time constant is found to be τ_1. When it is connected in series with a resistance R_2, the time constant is found to be τ_2. The inductance of the coil is

1) $\frac{\tau_1\tau_2(R_1-R_2)}{(\tau_2-\tau_1)}$ 2) $\frac{(\tau_2-\tau_1)}{\tau_1\tau_2(R_1-R_2)}$ 3) $\frac{(R_1-R_2)}{(\tau_2-\tau_1)}$ 4) $\frac{(\tau_2-\tau_1)}{(R_1-R_2)}$

27. An inductance L and a resistance R are connected in series to a battery of voltage V and negligible internal resistance through a switch. The switch is closed at t = 0. The charge that passes through the battery in one time constant is [e is base of natural logarithms]

1) $\frac{eR^2V}{L}$ 2) $\frac{VL}{R}$ 3) $\frac{VL}{eR^2}$ 4) $\frac{eL}{VR}$

28. **A coil of inductance 300 mH and resistance 2Ω is connected to a source of voltage 2V. The current reaches half of its steady state value in (nearly)**

1) 0.05 s 2) 0.1 s 3) 0.15 s 4) 0.03 s

29. **In the circuit shown, X is joined to Y for a long time and then X is joined to Z. The total heat produced in R_2 is**

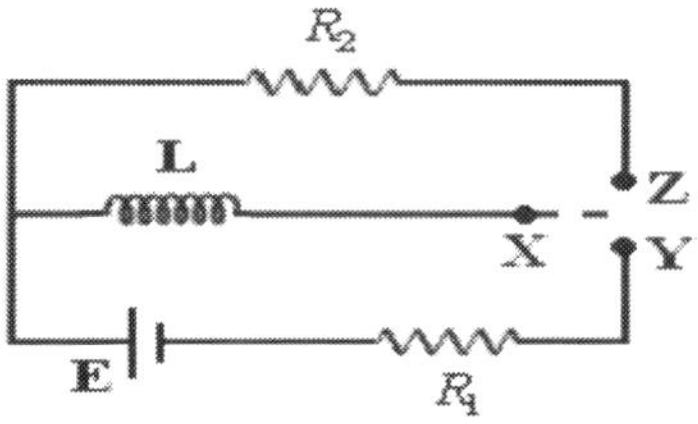

1. $\frac{LE^2}{2R_1^2}$ 2. $\frac{LE^2}{2R_2^2}$ 3. $\frac{LE^2}{2R_1R_2}$ 4. $\frac{LE^2R_2}{2R_1^3}$

30. **A condenser in series with a resistor is connected through a switch to a battery of negligible internal resistance and having an emf of 12 V. One second after closing the switch, the condenser is found to have a potential difference of 6 V. The time constant of the system is**

1) 2 s 2) $\frac{1}{\log_e 2}$ 3) $\log_e 2$ 4) $\log_e\left(\frac{1}{2}\right)$

31. **The time constant of the circuit shown is**

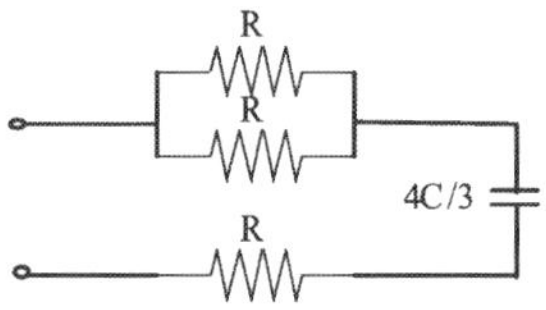

1) RC 2) 2RC 3) 3RC 4) 4RC

32. **In the circuit shown below. the key K is closed at $t=0$. The current through the battery is**

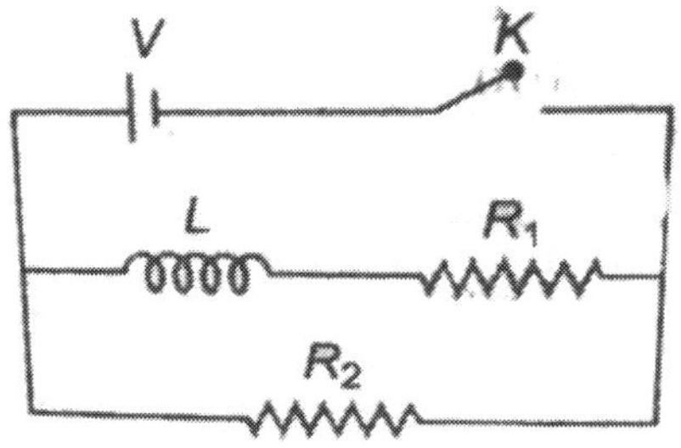

1) $\frac{VR_1R_2}{\sqrt{R_1^2+R_2^2}}$ at $t=0$ and $\frac{V}{R_2}$ at $t=\infty$

2) $\frac{V}{R_2}$ at $t=0$ and $\frac{V(R_1+R_2)}{R_1R_2}$ at $t=\infty$

3) $\frac{V}{R_2}$ at $t=0$ and $\frac{VR_1R_2}{\sqrt{R_1^2+R_2^2}}$ at $t=\infty$

4) $\frac{V(R_1+R_2)}{R_1R_2}$ at $t=0$ and $\frac{V}{R_2}$ at $t=\infty$

KEY - LEVEL - III

01) 2	02) 4	03) 2	04) 1	05) 2	06) 3	07) 1	08) 2	09) 1	10) 1
11) 3	12) 1	13) 3	14) 4	15) 2	16) 1	17) 2	18) 4	19) 3	20) 2
21) 3	22) 1	23) 1	24) 3	25) 3	26) 1	27) 3	28) 2	29) 1	30) 2
31) 2	32) 2								

HINTS - LEVEL - III

1. $\phi = \vec{B}.\vec{A}$

$= \left[\left(40\hat{i} - 15\hat{k}\right)10^{-4}\right] \cdot \left[5 \times 10^{-4}\hat{k}\right] = -750 \ nWb$

2. Only veritical lines of earth's magnetism will cut when the rod is allowed to slide over the rails. Let the vertical component of earth's magnetic field is B_v.

Induced

current $i = \dfrac{B_v l\upsilon - 0}{R} = \dfrac{B_v l\upsilon}{R}$

$B_V = 10^{-5}T$ Given $B_H = 0.3 \times 10^{-4}T$

The angle of dip θ is given by $\theta = \tan^{-1}\left(\dfrac{B_v}{B_H}\right)$

3. $P = Fv = BIlv = 1.25 \times 10^{-3} \times 50 \times 0.1 \times 1W$

$6.25 \times 10^{-3}\text{W} = 6.25\text{mW}$

An iterating alternating

$P = EI = (Blv)I$

4. $I = \dfrac{Blv}{R}$

$I = \dfrac{5 \times 10^{-2} \times 0.3 \times 0.2}{5}\text{A} = 0.6\text{mA}$

Area and flux are decreasing. So, current flows to increase the flux. Clearly, current should be clockwise. So, it flows from B to C through 5Ω.

5. Let there be an element dx of rod at a distance x from the wire.

emf developed in the element, $dE = Bdxv$

$\therefore dE = \left(\dfrac{\mu_0}{4\pi}\dfrac{2I}{x}\right)dxv$

$\therefore E = \dfrac{\mu_0 Iv}{2\pi}\int_a^b \dfrac{dx}{x} = \dfrac{\mu_0 Iv}{2\pi}\log_e \dfrac{b}{a}$

$\therefore E = \dfrac{4\pi \times 10^{-7} \times 100 \times 5}{2\pi}\log_e \dfrac{100}{1} = 4.6 \times 10^{-4}\text{V} = 0.46\text{mV}$

6. Induced emf $= Blv$. R is internal resistance of seat of emf, i.e., of rod

Total resistance of circuit $= R + \frac{R_1R_2}{R_1+R_2}$

$$\therefore I = \frac{Blv}{R + \frac{R_1R_2}{(R_1+R_2)}} = \frac{Blv(R_1+R_2)}{R_1R_2 + R(R_1+R_2)}$$

7. Let v is the velocity of conductor at any time, then induced emf: $e = Blv$ .... (i)

charge on capacitor : $q = Ce = CBlv$

Current in circuit : $I = \frac{dq}{dt} = CBl\frac{dv}{dt}$

for conductor $mg - IBl = \frac{mdv}{dt}$

$$\Rightarrow mg - CB^2l^2\frac{dv}{dt} = \frac{mdv}{dt} \Rightarrow \frac{dv}{dt} = \frac{mg}{m + CB^2l^2}$$

This is the acceleration of conductor which is constant.

8. $e = B\frac{dA}{dt} = \frac{1}{2}B\omega r^2$

$\therefore$ In every spokes same emf developed, and cycle spokes is equivalent to parallel combination of cells.

9. $\varepsilon = \frac{1}{2}B\omega l^2$

10. The bicycle is resting in east -west direction, therefore horizontal magneti field lines of earth's magnetisation will cut across the spokes of the wheel. Let this horizontal field of earth's magnetism is B_H. The induced emf between the axle and the rim of the wheel is $e = \frac{1}{2}B_H l^2\omega$

$\therefore B_H = 2\times10^{-5}T$

given $B_v = 1.5\times10^{-5}T$. Angle of dip $\theta = \tan^{-1}\left(\frac{B_v}{B_H}\right)$

11. $v = r\left(\frac{1}{44}\right)\omega_{coil}$ where $\omega_{coil} = \frac{e.m.f}{NAB}$

12. $B = \frac{\sqrt{2}V_{r.m.s}}{NA\omega}$

13. $e = -\frac{d\phi}{dt} = \pi B\frac{d}{dt}(R^2)$

14. conceptual

15. A moving conductor is equivalent to a battery of emf

$= vBl\,(motion\ emf)$

Equivalent circuit

$I = I_1 + I_2$

Applying Kirchhoff's law

$I_1R + IR - vBl = 0 - - - - - (i)$

$I_2R + IR - vBl = 0 - - - - - (ii)$

Adding Eqs. (i) and (ii)

$2IR + IR = 2vBl$

$I = \frac{2vBl}{3R}$

$B = \frac{\mu_0 i R^2}{(a^2 + R^2)^{3/2}}$

$\phi = BA\,Cos(45)$

$M = \frac{\phi}{i} = \frac{\mu_0 a^2}{2^{7/2} R^2}$

16. $e = A\frac{dB}{dt} = 10^{-2} \times 3 \times 10^{-4} = 3 \times 10^{-6}$

17. Let r is the radius of the loop. The magnetic flux, at a time t, through the loop is given by

$\phi = B(\pi r^2)$

$e = \frac{d\phi}{dt} = \pi r^2 (\frac{dB}{dt})$

(the numerical valu of induced emf)

Induced current in the loop is given by $i = \frac{e}{R} = \frac{\pi r^2}{R}\frac{dB}{dt}$ ....(i)

where R is the resistance of the wire let a is the corss-section area and l the length of the wire. Then

$R = \rho \frac{l}{\pi a^2}$ ....(ii)

where $l = 2\pi r$; $m = \pi a^2 l \sigma$

On putting the value of a^2 in eq. (ii), we get

$i = \frac{m}{4\pi\rho\sigma}\frac{dB}{dt}$

18. Given that $\phi = at(T - t)$

induced emf, $E = \frac{d\phi}{dt} = \frac{d}{dt}\left[at(T-t)\right]$

$= at(0-1) + a(T-t) \quad = a(T - 2t)$

So, indeced emf is also a function of time.

$\therefore$ Heat generated in time T is

$$H=\int_0^T \frac{E^2}{R}dt=\frac{a^2}{R}\int_0^T (T-2t)^2 dt=\frac{a^2T^3}{3R}$$

19. By Lenz's law, clockwise current is induced in both loops. Greater the area, large will be the induced emf Outer loop has greater area.

20. $i \propto \frac{1}{L};\ \frac{i_1}{i_2}=\frac{L_2}{L_1}=\frac{1}{4}$

21. $L=\frac{\mu_0 N^2 A}{l}$

If x is the length of the solenoid with r as radius, then $x=2\pi rN, A=\pi r^2$

$$\therefore L=\mu_0\left(\frac{x^2}{4\pi^2 r^2}\right)\frac{\pi r^2}{l}\quad \left[\therefore N=\frac{x}{2\pi r}\right]$$

$$\therefore x=\sqrt{\frac{4\pi Ll}{\mu_0}}$$

22. For a solenoid, $L=\mu_0 N^2\frac{A}{l}$. If x is the length of the wire and a is the area of cross-section, then

$R=\frac{\rho x}{a}$ and $m=axD$

$$Rm=\frac{\rho x}{a}axD, x=\sqrt{\frac{Rm}{\rho D}}$$

Also, $x=2\pi rN, N=\frac{x}{2\pi r}\ \left(\therefore L=\frac{\mu_0 N^2 A}{l}\right)$

$$\therefore L=\mu_0\left(\frac{x}{2\pi r}\right)^2\frac{\pi r^2}{l}=\frac{\mu_0}{4\pi l}\frac{Rm}{\rho D}$$

23. $\Delta q=\frac{\Delta\phi}{R}$

$$\phi_i=0;\ \phi_f=\left(\frac{\mu_0}{2}\frac{i}{b}\right)\left(\pi a^2\right)=\frac{\mu_0 i a^2\pi}{2b}$$

$$\therefore \Delta\phi=\phi_f-\phi_i=\frac{\mu_0 i a^2\pi}{2b}$$

So, $\Delta q=\frac{\mu_0 i a^2\pi}{2Rb}$

24. Magnetic field due to larger loop

$$= \frac{\mu_0 1}{2r} = \frac{4\pi \times 10^{-7} \times 1}{2 \times 0.1} \text{T} = 2\pi \times 10^{-6} \text{T}$$

Now, magnetic flux linked withthe smaller loop, ϕ

$$= NBA\cos\omega t = 1 \times 2\pi \times 10^{-6} \times 5 \times 10^{-4} \cos\omega t$$

25. $B = \dfrac{\mu_0 iR^2}{\left(a^2 + R^2\right)^{3/2}}$

$$\phi = B\,A Cos(45)$$

$$M = \frac{\phi}{i} = \frac{\mu_0 a^2}{2^{7/2} R^2}$$

26. $\tau_1 = \dfrac{L}{R_1 + r}; \tau_2 = \dfrac{L}{R_2 + r}$

27. $q = \int idt = \int i_0 \left(1 - e^{\frac{-R}{L}t}\right) dt$

28. $i = i_0 \left[1 - e^{-\frac{R}{L}t}\right]$

29. The current in L for steady state $= \dfrac{E}{R_1}$

$\therefore$ Energy stored in $L = \dfrac{1}{2} L I_0^2 = \dfrac{1}{2} L \dfrac{E^2}{R_1^2}$

30. $q = q_0 \left(1 - e^{-t/Rc}\right)$

t = 1 second

$$CV = CV_0\left(1 - e^{-t/Rc}\right); \frac{1}{2} = e^{-t/Rc}$$

$$\ln 2 = \frac{1}{Rc} \ln e$$

$$RC = \frac{1}{\ln 2}$$

31. $R^1 = \dfrac{R}{2} + R = \dfrac{3R}{2}$

$$i = RC = \frac{3R}{2} \times \frac{4C}{3} = 2RC$$

32. At $t=0$, inductor behaves like an infinite resistance

so at $t=0, i=\frac{V}{R_2}$

and at $t=\infty$, inductor behaves like a conducting wire

$$i=\frac{V}{R_{eq}}=\frac{V(R_1+R_2)}{R_1R_2}$$

LEVEL - IV

ASSERTION & REASON

Instructions for the Assertion & reason type questions:

A. Both Assertion and Reason are true and Reason is the correct explanation of Assertion

B. Both Assertion and Reason are true and Reason is not the correct explanation of Assertion

C. Assertion is true but reason is false

D. Assertion is false but reason is true.

1. **Assertion (A):** Whenever the magnetic flux linked with a closed coil changes there will be an induced emf as well as an induced current.

   Reason (R): According to Faraday, the induced emf is inversely proportional to the rate of change of magnetic flux linked with a coil.

   1) A   2) B   3) C   4) D

2. **Assertion (A):** Lenz's law is a violation of the law of conservation of energy.

   Reason (R): Induced emf always oppose the change in magnetic flux that has produced them.

   1) A   2) B   3) C   4) D

3. **Assertion(A):** An inductor in a D.C. circuit opposes both a steady current and a changing current

   Reason(R): Induced emf is generated only when the flux linked with the inductor remains unchanged.

   1) A   2) B   3) C   4) D

4. **Assertion(A):** The possibility of an electric bulb fusing is higher at the time of switching on and off.

   Reason(R): Inductive effects produce a large current at the time of switch-on and switch-off.

   1) A   2) B   3) C   4) D

5. **Assertion(A):** L/R and CR both have same dimensions.

   Reason(R): L/R and CR are time constants

   1) A   2) B   3) C   4) D

6. **Assertion(A):** When a charged condenser discharges through a resistor, the time taken for half the charge to be lost is always same, irrespective of the initial value of the charge.

   Reason(R): The rate of decay of charge in aCR circuit is a linear function of time.

   1) A   2) B   3) C   4) D

Match The Following

7. A square loop is placed near a long straight current carrying wire as shown, match the following table.

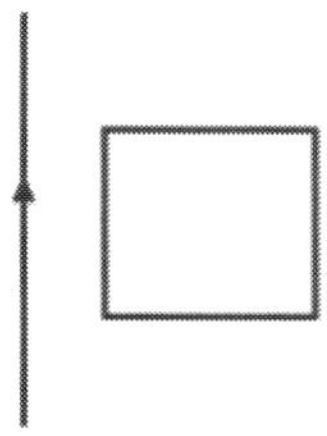

Table - 1	**Table - 2**
a) If current is increased	**p) Induced current in loop is clockwise**
b) If current is decreased	**q) Induced current in loop is anticlockwise**
c) If loop is moved away from the wire	**r) Wire will attract the loop**
d) If loop is moved to wards the wire	**s) Wire will repel the loop**

1) a-q,s; b-p,r; c-q,s; d-p,r
2) a-p,r; b-q,s; c-p,r; d-q,s
3) a-q,s; b-p,r; c-q,r; d-p,q
4) a-q,s; b-p,r; c-p,r; d-q,s

GRAPHS

8. An equilateral triangular loop ADC having some resistance is pulled with a constant velocity v out of a uniform magnetic field directed into the paper. At time $t = 0$, side DC of the loop is at edge of the magnetic field.

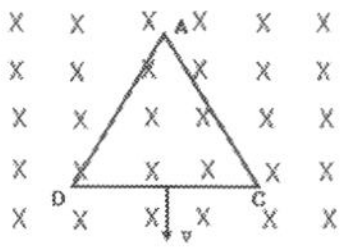

The induced current (i) versus time (t) graph will be as

1) 

2) 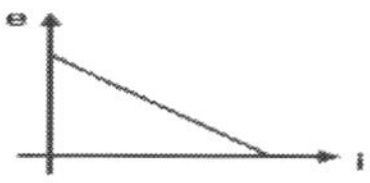

3) 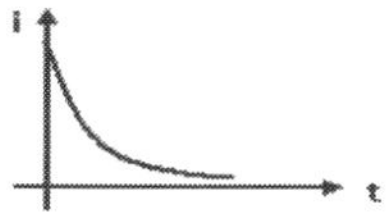

4) 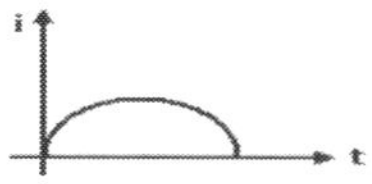

9. The current through the coil in figure(i) varies as shown in figure (ii). Which graph best shows the ammeter A reading as a function of time?

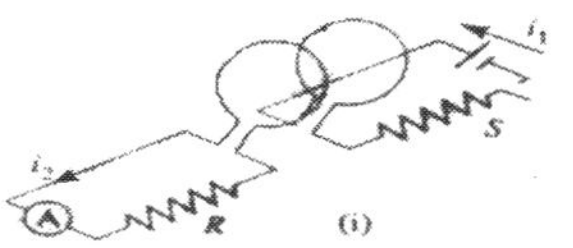

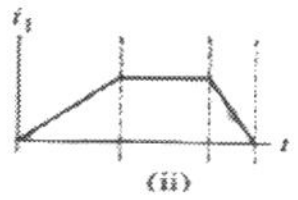

1) 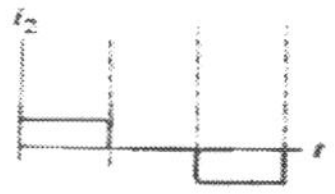

2) 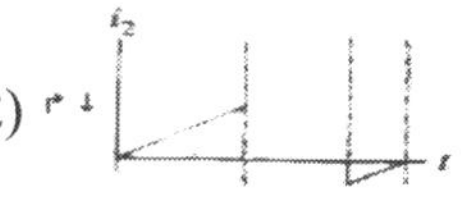

3) 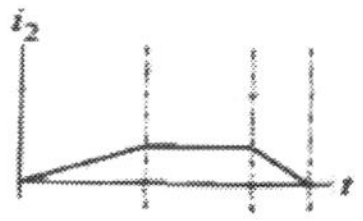

4) 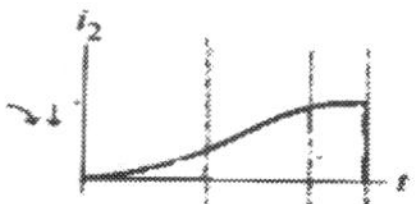

10. Switch S of the circuit shown in figure is closed at $t=0$

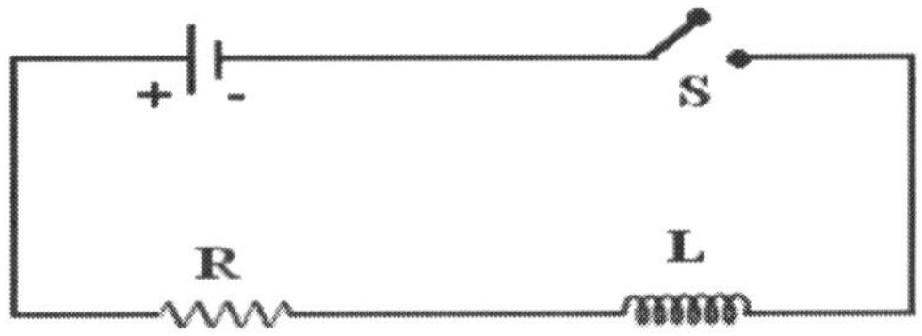

If e denotes the induced emf in L and i the current flowing through the circuit at time t, then which of the following graphs correctly represents the variation of e with i?

1) 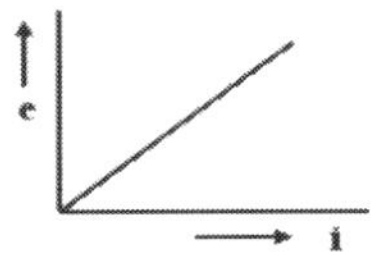

2) 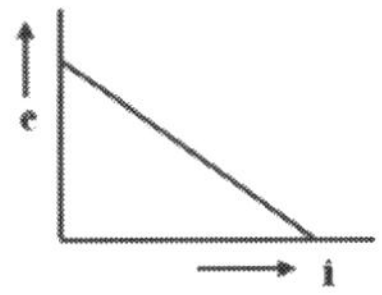

3) 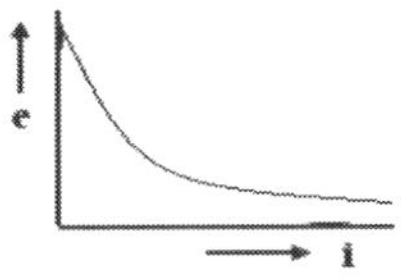

4) 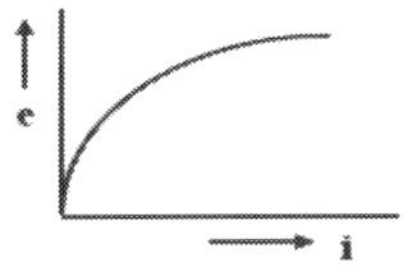

11. In an LR circuit connected to a battery, the rate at which energy is stored in the inductor is plotted against time during the growth of curretn in the circuit. Which of the following best represents the resulting curve?

1) 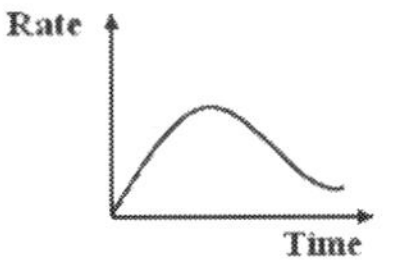

2) 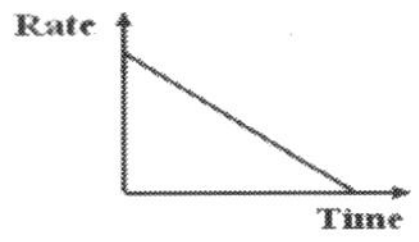

3) 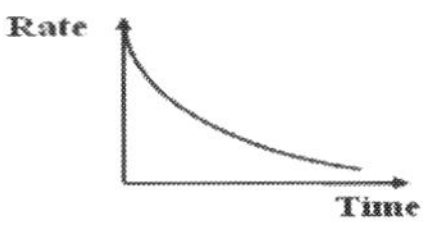

4) 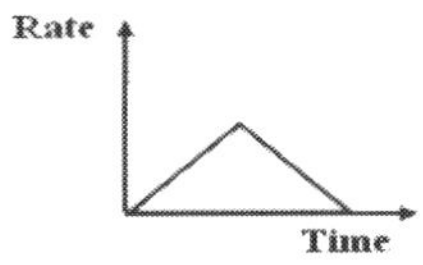

12. In the circuit shown, switch k_2 is open and switch k_1 is opened and switch k_2 is simultaneously closed. The variation of inductor current with time is

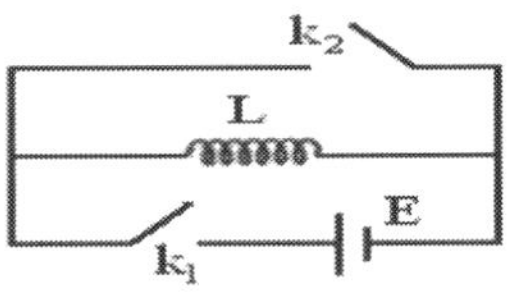

1) 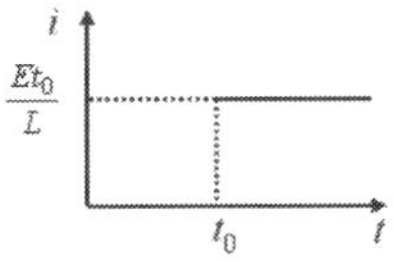

2) 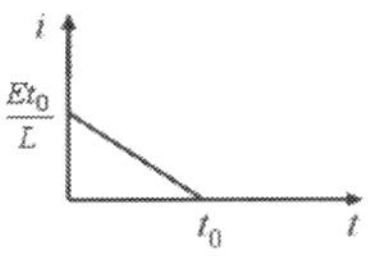

3) 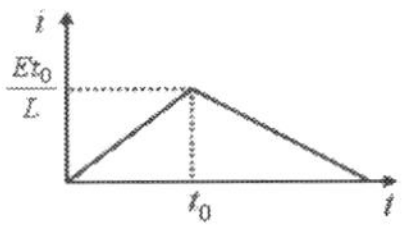

4) 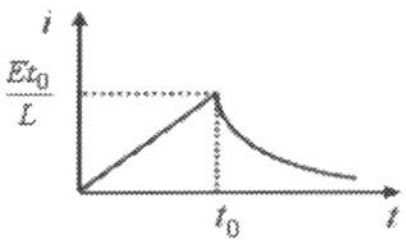

13. Switch S of the circuit shown in figure is closed at $t = 0$. If e denotes the induced emf in L and I is the current flowing through the circuit at time t, which of the following graphs is/are correct?

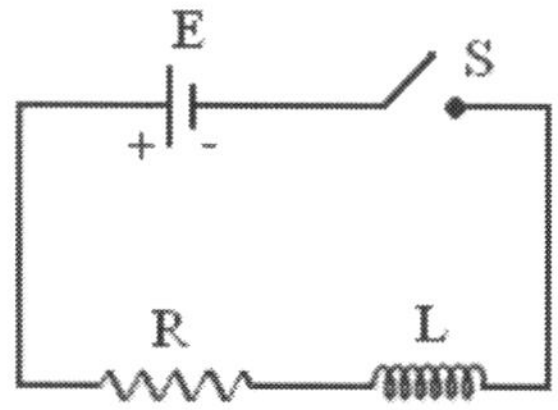

1) 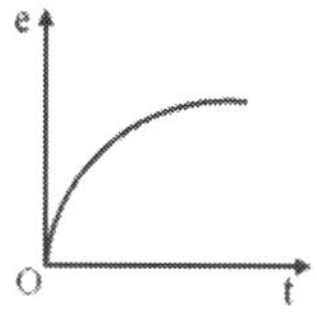

2) 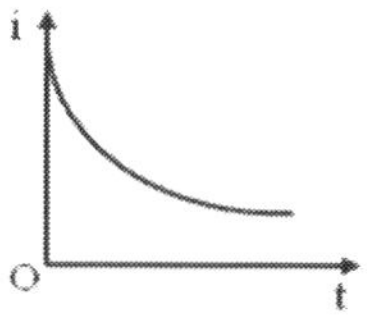

3) 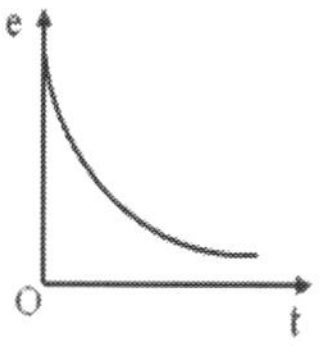

4) 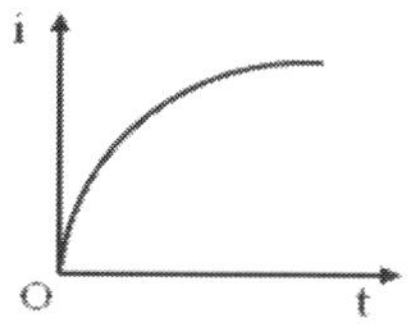

14. Two infinitely long conducting parallel rails are connected through a capacitor C as shown in figure. a conductor of length l is moved with constant speed v_0. Which of the following graph truly depicts the variation of current through the condctor with time?

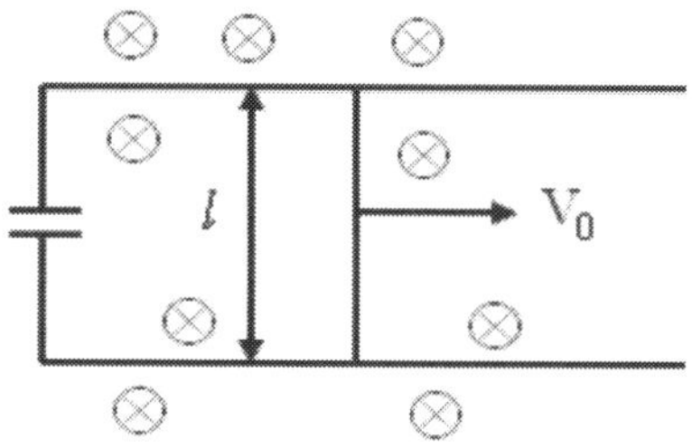

1) 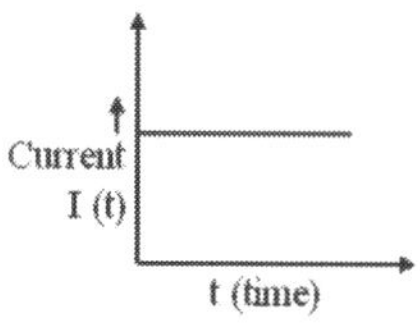

2) 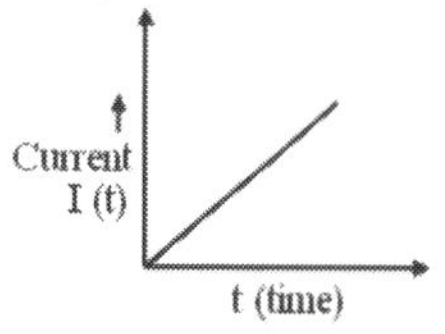

3) 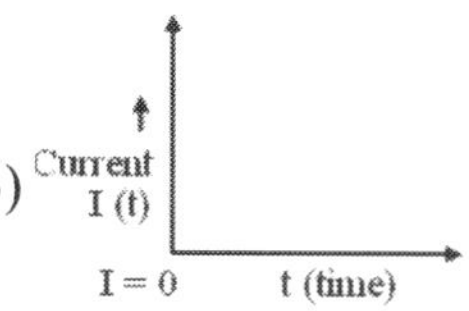

4) 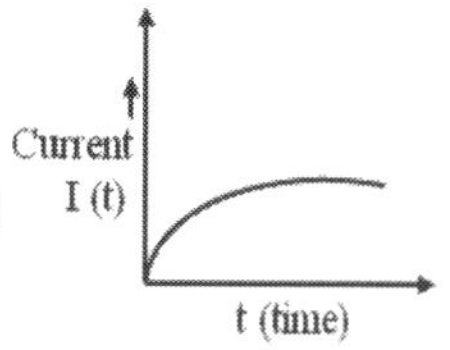

15. The figure shows an isosceles triangle wire frame with angle equal to $\frac{\pi}{2}$. The frame starts entering into the region of uniform magnetic field B with constant velocity v at $t = 0$. The longest side of the frame is perpendicular to the direction of velocity. If 'i' is the instantaneous

current through the frame then choose the alternative showing the correct variation of 'i' with time.

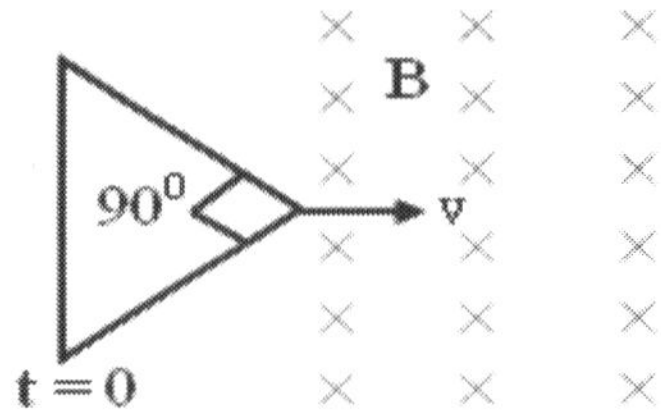

1) 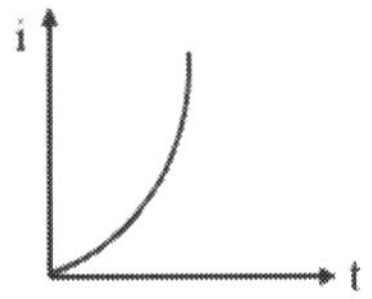

2) 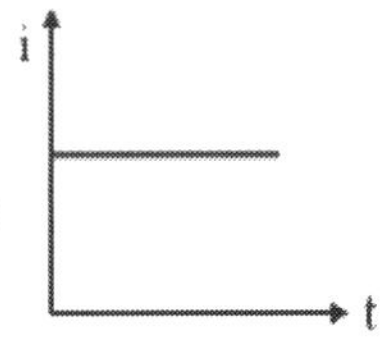

3) 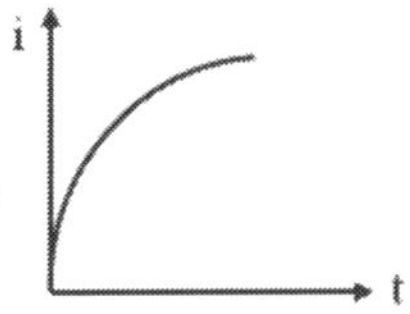

4) 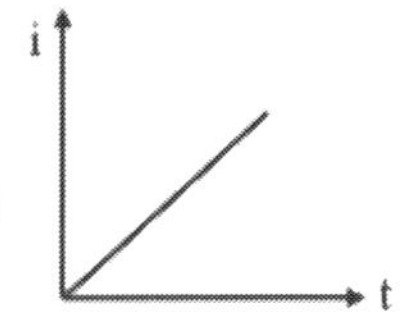

16. **A flexible conducting wire in the form of a circle is kept in a uniform magnetic field with its plane normal to the field. Radius of that circle changes with time as shown. Then which of the following graphs represents the variation of induced emf with time** $R = R_0, t < t_0; R = R_0 + t, t_0 < t < 2t_0$:

1) 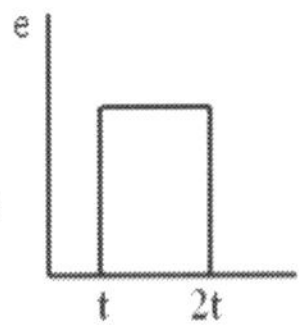

2) 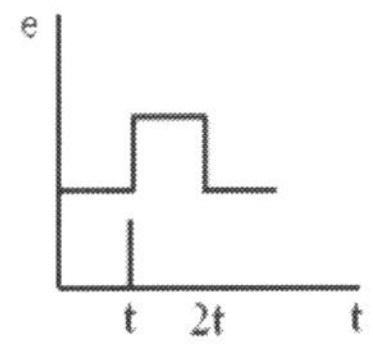

3) 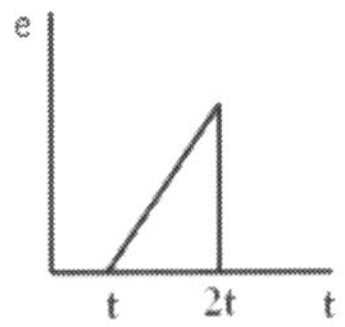

4) 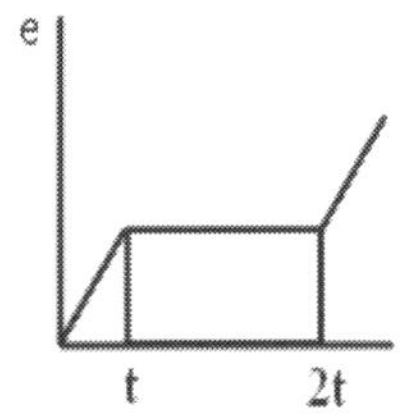

17. **A wire loop is placed in a region of time varying magnetic field which is oriented orthogonally to the plane of the loop as shown in figure. The graph shows the magnetic field variation as the function of time. Assume the positive emf is the one which drives a current in the clockwise direction and seen by the observer in the direction of B. Which of the following grapsh best represents the induced emf a function of time.**

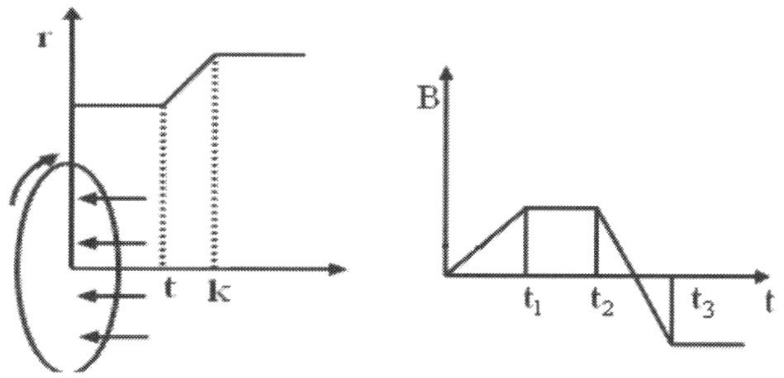

1) 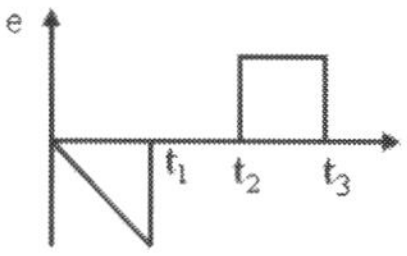

2) 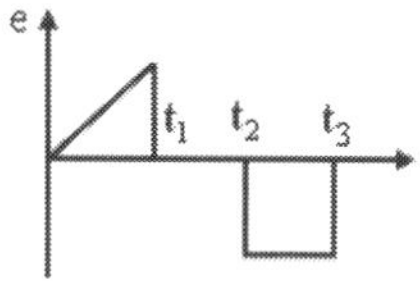

3) 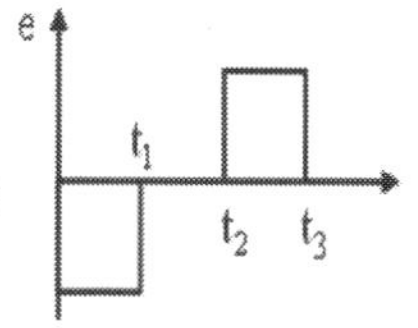

4) 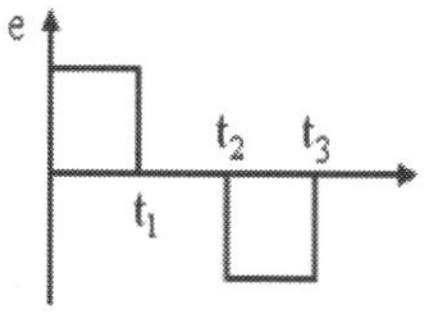

18. **The current is an induction coil varies with time t, according to the graph shown in figure. Which of the following graphs shows the induced emf (e) in the coil with time.**

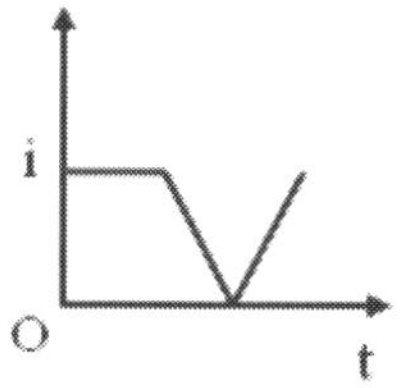

1) 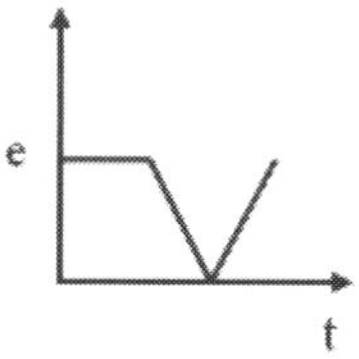

2) 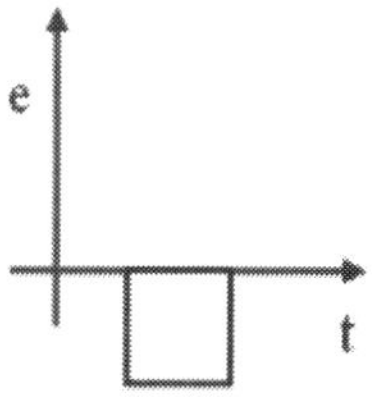

3) 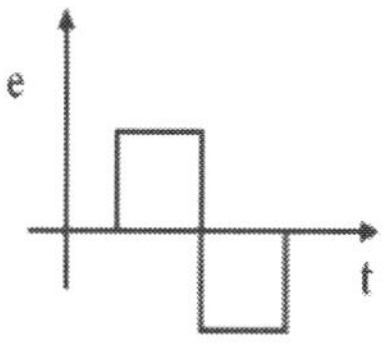

4) 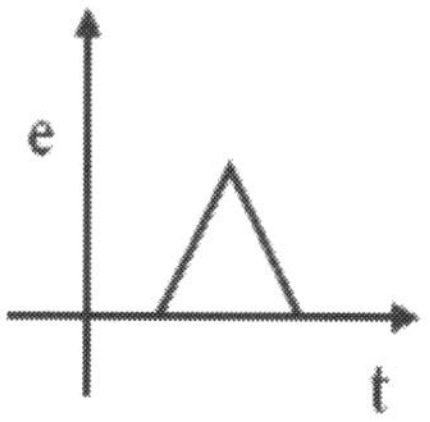

19. **A and B are two coils placed closely as shown. The current in coil A varies as shown in the graph.**

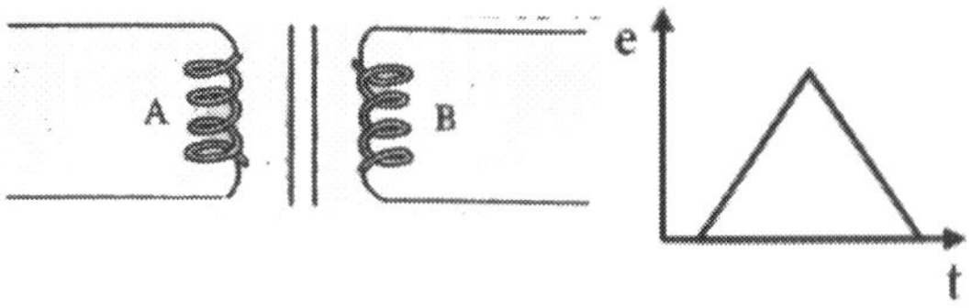

Then which of the following graphs gives the best representation of variation of induced emf in coil B?

1) 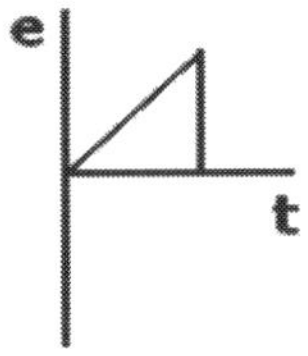

2) 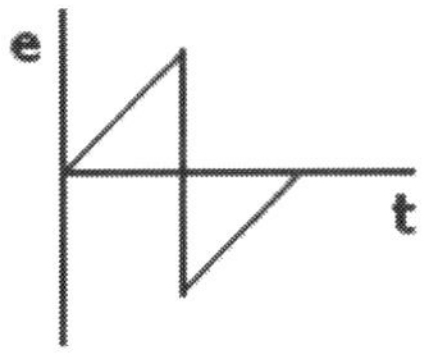

3) 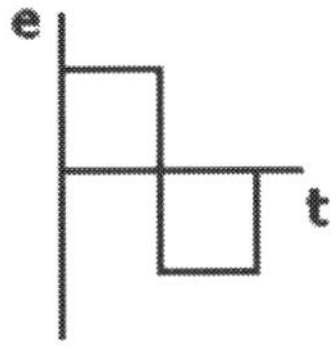

4) 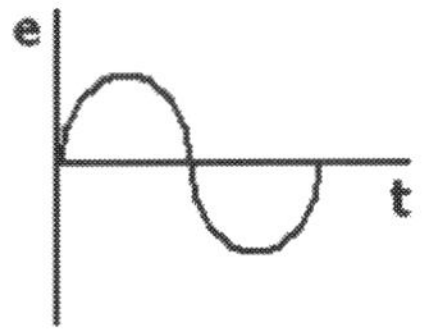

20. **A flexible wire bent in the form of a circle is place in a uniform mgnetic field perpendicularly to the plane of the coil. The radius of the coil changes as shown in Figure. The graph of magnitude of induced emf in the coil is represented by**

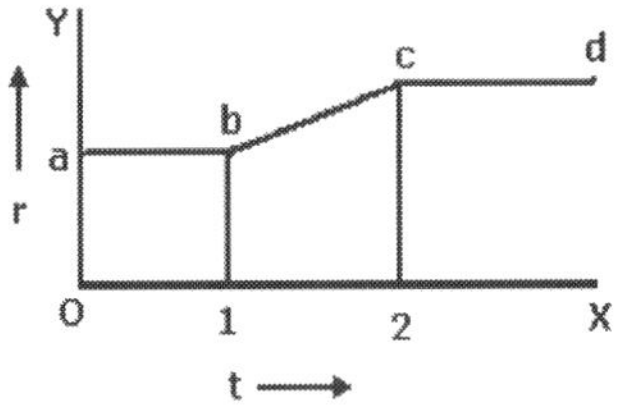

1. 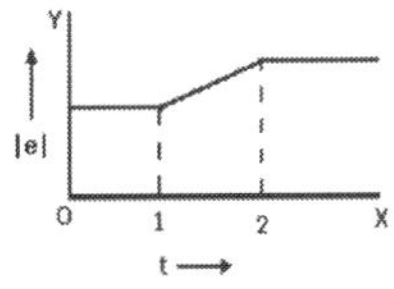

2. 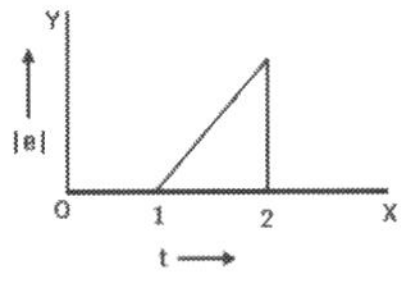

3. 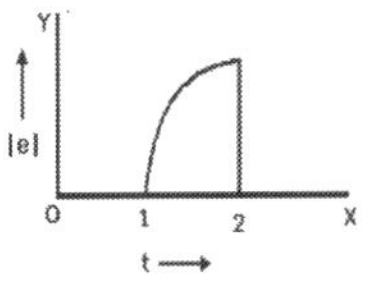

4. 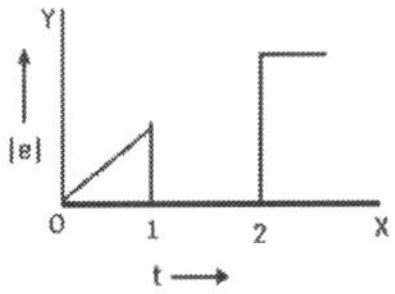

KEY - LEVEL - IV

01) 3	02) 4	03) 4	04) 1	05) 1	06) 3	07) 4	08) 2	09) 1	10) 3
11) 1	12) 1	13) 3,4	14) 3	15) 4	16) 3	17) 3	18) 3	19) 3	20) 2

Chapter - 3

ALTERNATING CURRENT

ALTERNATING CURRENT

SYNOPSIS

INTRODUCTION

ALTERNATING CURRENT (AC)

- Electric current, which keeps on changing in magnitude and direction periodically is defined as alternating current.
- It obeys Ohm's law and Joule's heating law.

  It is produced using the principle of electromagnetic induction.
- Graphical representations for alternating quantities can be represented in the form of the following graphs.

ALTERNATING VOLTAGE (A.V)

- The voltage, which changes in magnitude and direction with respect to time is defined as alternating voltage.
- The alternating voltage in general use is sinusoidal voltage. It is produced by rotating a coil in a uniform magnetic field with uniform angular velocity.

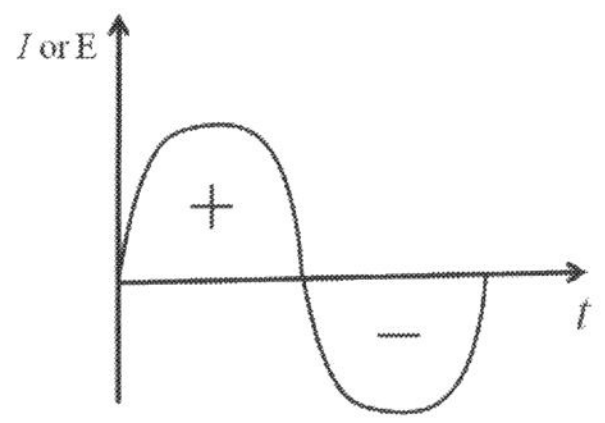

sinusoidal form of ac

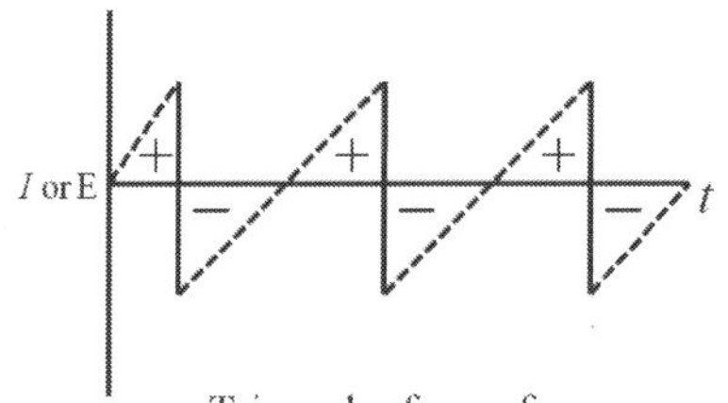

Triangular form of ac

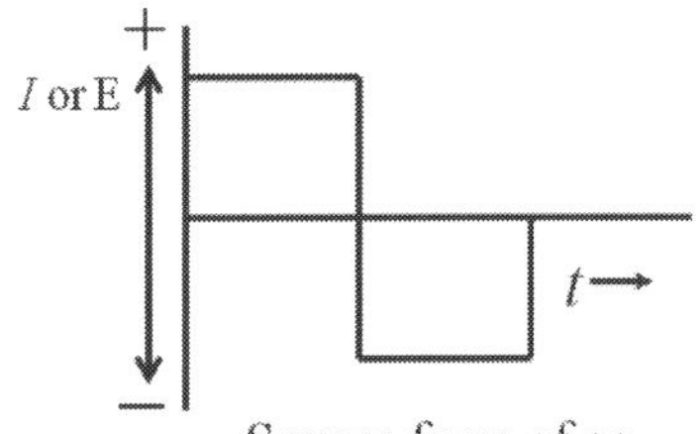

Square form of ac

ADVANTAGES OF ALTERNATING CURRENT OVER DIRECT CURRENT

- The cost of generation of **ac** is less than that of **dc**.
- **ac** can be conveniently converted into **dc** with the help of rectifiers.
- By supplying **ac** at high voltages, we can minimise transmission losses or line losses.
- **ac** is available in a wide range of voltages. These volatages can be easily stepped up or stepped down with the help of transformers.

DISADVANTAGES OF ALTERNATING CURRENT OVER DIRECT CURRENT:

- **ac** is more dangerous than **dc**.
- **ac** is transmitted more by the surface of the conductor. This is called skin effect . Due to this reason that several strands of thin insulated wire, instead of a single thick wire, need be used.
- For electrorefining,electro- typing electroplating,only dc can be used but not ac.

INSTANTANEOUS VALUE OF CURRENT OR VOLTAGE (I or E)

- The value of current or voltage in an **ac** circuit at any instant of time is called its instantaneous value.
- Instantaneous current,

  $I = I_0 \sin\omega t$ (or) $I = I_0 \sin(\omega t + \phi)$
- Instantaneous voltage,

  $E = E_0 \sin\omega t$ (or) $E = E_0 \sin(\omega t + \phi)$

  Where $(\omega t + \phi)$ is called phase

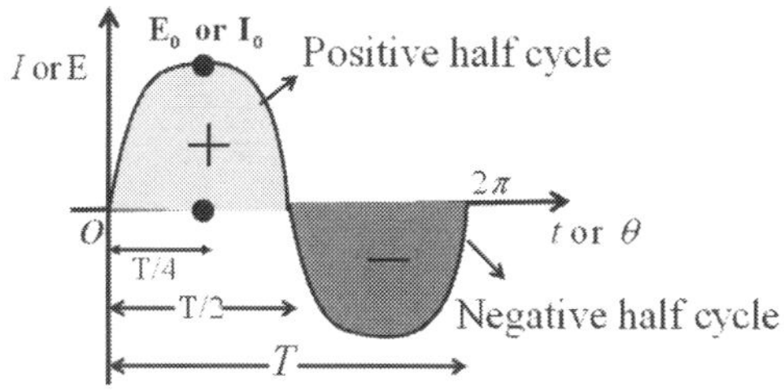

AMPLITUDE OF A.C. (PEAK VALUE) (I_0) or (I_m):

It is the maximum value of A.C. The value of A.C. becomes maximum twice in one cycle.

Note: Average value of a function $F(t)$ over a period of T is given by

$$<F(t)> = F_{avg} = \frac{\int_0^T F(t)dt}{\int_0^T dt} = \frac{1}{T}\int_0^T F(t)dt$$

Eg:- $<\sin^2\omega t> = \frac{1}{2}$: $<\cos^2\omega t> = \frac{1}{2}$

$<\sin 2\omega t> = 0 ; <\cos 2\omega t> = 0$

AVERAGE VALUE OF A.C. <I>

- The value of current at any instant 't' is given by $I = I_0 \sin\omega t$.
- The average value of a sinusoidal wave over one complete cycle is given by

$$I_{avg} = \frac{\int_0^T I.dt}{\int_0^T dt} = \frac{\int_0^T I_0 \sin\omega t.dt}{\int_0^T dt} = 0$$

For half cycle.

$$<I> = \frac{\int_0^{T/2} Idt}{\int_0^{T/2} dt} = \frac{\int_0^{T/2} I_0 \sin\omega t\, dt}{\int_0^{T/2} dt} = \frac{2I_0}{\pi} = 0.636 I_0$$

$I_{avg} = 63.7\%$ of I_0

- Similarly

$$E_{avg} = \frac{2E_o}{\pi} = 0.637E_0 = 63.7\%E_0$$

FREQUENCY OF A.C. (f)

- It is the number of cycles completed by A.C. in one second.

TIME PERIOD OF A.C. (T)

- It is the time taken by A.C. to complete one cycle.

f= 1/ T

MEAN SQUARE VALUE OF A.C. $<I^2>$

- $<I^2> = \frac{I_0^2}{2}$

R.M.S. VALUE (I_{rms}) or EFFECTIVE VALUE (I) or VIRTUAL VALUE OF A.C.:-

It is the square root of the average of squares of all the instantaneous values of current over one complete cycle. (or)

- It is equal to that direct current which produces same heating in a resistance as is produced by the A.C. in same resistance during same time.

$$I_{rms}^2 = \frac{\int_0^T I^2.dt}{\int_0^T dt} = \frac{\int_0^T I_0^2.\sin^2 \omega t.dt}{T}$$

$$= \frac{I_0^2}{T}\int_0^T \left[\frac{1-\cos 2\omega t}{2}\right]dt = \frac{I_0^2}{2T}\left[t - \frac{\sin 2\omega t}{2\omega}\right]_0^T = \frac{I_0^2}{2}$$

$$\therefore I_{rms} = \frac{I_0}{\sqrt{2}} = 0.707\ I_0$$

- It is equal to that direct current which produces same heating in a resistance as is produced by the A.C. in same resistance during same time.

MEAN SQUARE VALUE OF A.C. $<I^2>$

- $<I^2> = \frac{I_0^2}{2}$

FORM FACTOR

$$\text{Form factor} = \frac{\text{rms value}}{\text{average value over half cycle}}$$

$$Form\ factor = \frac{I_{rms}}{I_{avg}} = \frac{E_{rms}}{E_{avg}}$$

We know that $I_{rms} = \frac{I_0}{\sqrt{2}}$ and $I_{ave} = \frac{2I_0}{\pi}$

$\therefore$ Form factor $= \frac{I_0}{\sqrt{2}} \times \frac{\pi}{2I_0} = \frac{\pi}{2\sqrt{2}} = 1.11$

Note :

- **ac** ammeter and voltmeter read the r.m.s value i.e., effective value of alternating current and voltage respectively.
- **ac** can be measured by using hot wire ammeters or hot wire voltmeters because the heat generated is independent of the direction of current.
- **ac produces the same heating effects as that of dc of magnitude $i = i_{rms}$**
- **ac** is more dangerous than **dc** of same voltage.

  100V **ac means** $E_{rms} = 100V$, $E_0 = 100\sqrt{2}V$

  100V **dc is equivalent to** E_{rms}
- **ac** can be produced by the principle of electromagnetic induction.

WE.1 You have two copper cables of equal length for carrying current. One of them has a single wire of area of across section A, the other has ten wires each of cross section area $\frac{A}{10}$. Judge their suitability for transporting ac and dc.

Solution:

For transporting d.c.., both the wires are equally suitable, but for transporting a.c., we prefer wireof multiple strands.

ac is transmitted more by the surface of the conductor. This is called skin effect . Due to this reason that several strands of thin insulated wire, instead of a single thick wire, need be used.

W.E-2

If the voltage in an ac circuit is represented by the equation.

$V = 220\sqrt{2}\sin(314t - \phi)$

calculate (a) peak and rms value of the voltage, (b) average voltage, (c) frequency of ac.

solution.

(a) As in case of ac, $V = V_0 \sin(\omega t - \phi)$

The peak value $V_0 = 220\sqrt{2} = 311V$

and as in case of ac. $V_{rms} = \frac{V_0}{\sqrt{2}}; V_{rms} = 220V$

(b) In case of ac $V_{av} = \frac{2}{\pi}V_0 = \frac{2}{\pi} \times 311 = 198.17V$

(c) As $\omega = 2\pi f, 2\pi f = 314$

i.e, $f = \frac{314}{2 \times \pi} = 50Hz$

W.E-3

A current is made of two components a dc component i_1 = 3A and an ac component $i_2 = 4\sqrt{2}\sin\omega t$***. Find the reading of hot wire ammeter?***

solution.

$i = i_1 + i_2 = 3 + 4\sqrt{2}\sin\omega t$

$$i_{rms}^2 = \frac{\int_0^T i^2.dt}{\int_0^T dt} = \frac{\int_0^T \left(3+4\sqrt{2}\sin\omega t\right)^2 dt}{T}$$

$$i_{rms}^2 = \frac{1}{T}\int_0^T \left(9+24\sqrt{2}\sin\omega t + 32\sin^2\omega t\right)dt$$

$\therefore i_{rms} = 5A$

W.E-4

If a direct current of value **a** ampere is superimposed on an alternating current $I = b\sin\omega t$ flowing through a wire, what is the effective value of the resulting current in the circuit?

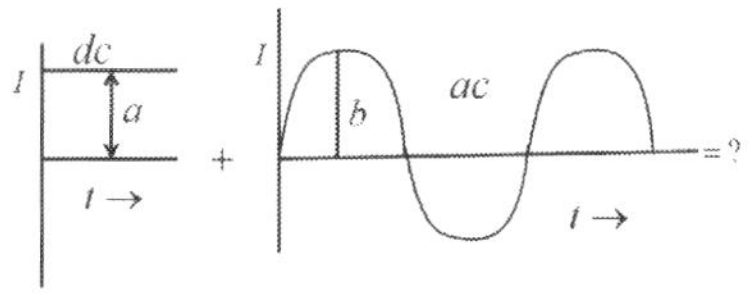

Solution.

As current at any instant in the circuit will be, $I = I_{dc} + I_{ac} = a + b\sin\omega t$

So,

$$I_{eff} = \left[\frac{\int_0^T I^2 dt}{\int_0^T dt}\right]^{\frac{1}{2}} = \left[\frac{1}{T}\int_0^T (a+b\sin\omega t)^2\, dt\right]^{\frac{1}{2}}$$

i.e,

$$I_{eff} = \left[\frac{1}{T}\int_0^T \left(a^2 + 2b\sin\omega t + b^2 + \sin^2\omega t\right)dt\right]^{\frac{1}{2}}$$

But as

$$\frac{1}{T}\int_0^T \sin\omega t\; dt = 0 \quad \text{and} \frac{1}{T}\int_0^T \sin^2\omega t\, dt = \frac{1}{2}$$

So, $I_{eff} = \left[a^2 + \frac{1}{2}b^2\right]^{1/2}$

- **RESISTANCE (R)**

  It is the opposition offered by a conductor to the flow of direct current.

- **IMPEDANCE (Z)**

  It is the opposition offered by a conductor to the flow of alternating current.

$$Z = \frac{|\text{alternating emf}|}{|\text{alternating current}|} = \frac{\text{peak value of alternating voltage}}{\text{peak value of AC}}$$

$$= \frac{\text{RMS value of alternating voltage}}{\text{RMS value of AC}}$$

- **Admittance(Y):**

  Reciprocal of impedance of a circuit is called admittance of the circuit.

  $\text{admittance (Y)} = \frac{1}{Z}$

  S.I. Unit: ohm^{-1} i.e. mho or siemen.

PHASE:-

- The physical quantity which represents both the instantaneous value and direction of A.C. at any instant is called its phase.
- It is dimensionless quantity and its unit is Radian

Phase Difference

- The diffrerence between the phases of current and voltage is called Phase difference.
- If alternating emf and current are $E = E_0 \sin(\omega t + \phi_1)$ and $i = i_0 \sin(\omega t + \phi_2)$

  then phase difference is $\phi = \phi_1 \sim \phi_2$

- **The quantity varies sinusoidally with time and can be represented as projection of a rotating vector, is called as phasor.**
- **A diagram, representing alternating emf and current (of same frequency) as rotating vectors (Phasors) with phase angle between them is called as phasor diagram.**

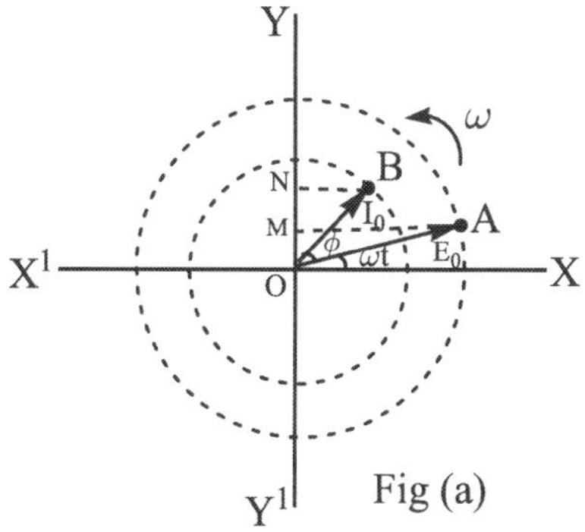

Fig (a)

- In figure (a), $\overline{OA}$ and $\overline{OB}$ represent two rotating vectors having magnitudes E_0 and I_0 in anti clock wise direction with same angular velocity 'ω'.
- OM and ON are the projections of $\overline{OA}$ and $\overline{OB}$ on Y-axis respectively.
- OM = E and ON = I, represent the instantaneous values of alternating emf and current.
- $\underline{|BOA} = \phi$ represents the phase angle by which current I_0 leads the alternating emf E_0.
- The phasor diagram, in a simple representation is

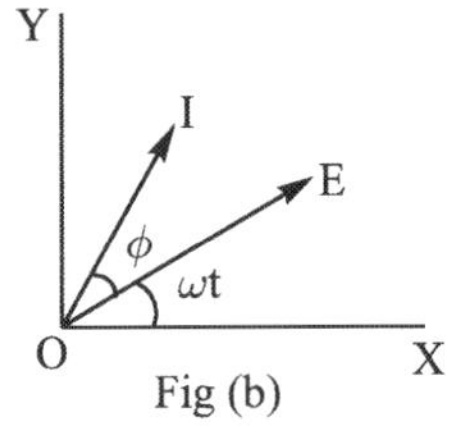

Fig (b)

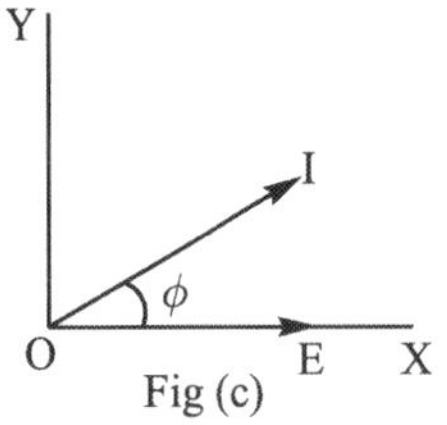

Fig (c)

WE - 5:

Use a phasor diagram to represent the sine waves in the following Figure.

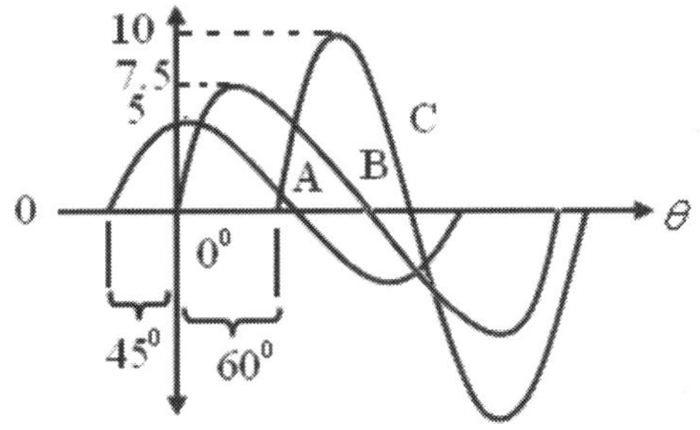

Sol:-The phasor diagram representing the sine waves is shown in figure.The length of each phasor represents the peak value of the sine wave.

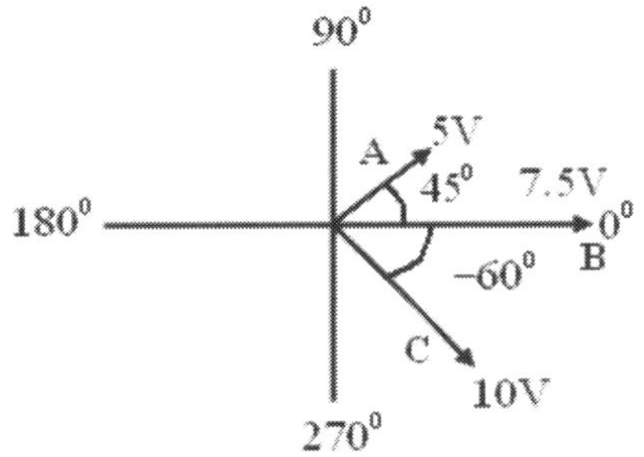

Note:-If e.m.f (or voltage) in A.C. is $E = E_0 \sin \omega t$ and the current $I = I_0 \sin(\omega t + \varphi)$ Where phase difference φ is Positive if current leads,Negative if current lags and zero if current is inphase with the emf (or voltage).

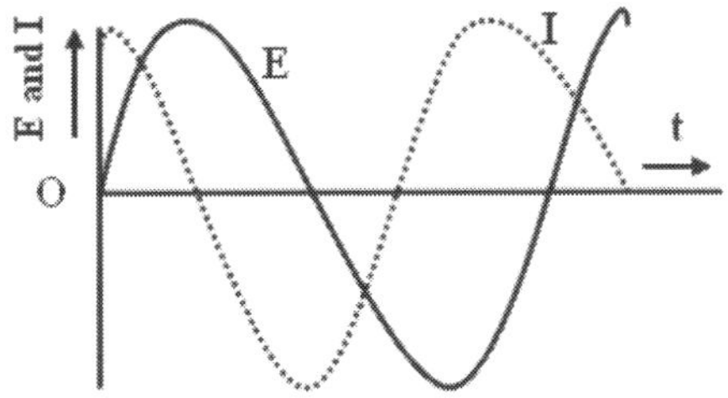

- instantaneous emf is $E = E_0 \sin \omega t$
- instantaneous current $I = I_0 \sin(\omega t + \phi)$

$$\text{where}\left[\phi = \frac{\pi}{2}\right]$$

Current leads emf by $\frac{\pi}{2}$

- 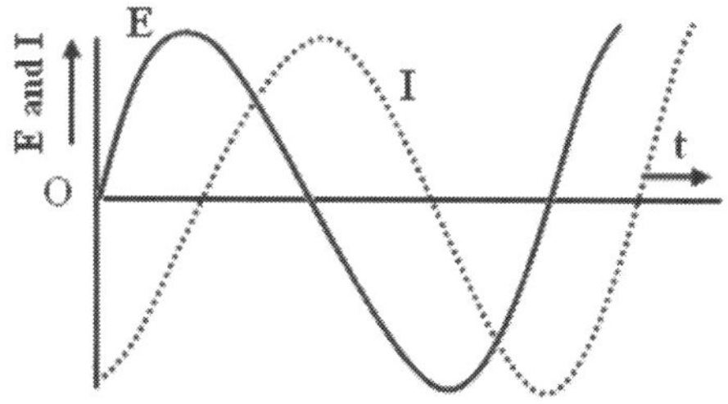

E=E_0 sin ωt

$I = I_0 \sin(\omega t - \phi)$ where $\phi = \frac{\pi}{2}$

- Current lags emf by $\frac{\pi}{2}$ **or** emf leads current by $\frac{\pi}{2}$.

A.C THROUGH A RESISTOR

- A pure resistor of resistance R is connected across an alternating source of emf

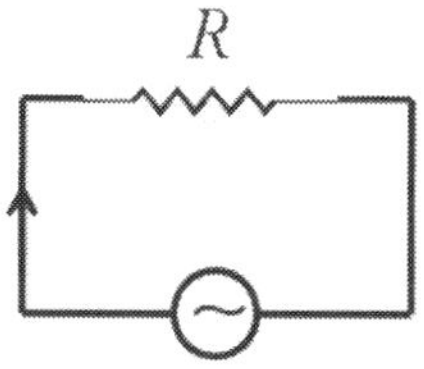

$E = E_0 \sin \omega t$

- The instantaneous value of alternating emf is $E = E_0 \sin \omega t$
- The instantaneous value of alternating current is

$I = \frac{E}{R} = \frac{E_0}{R}(\sin \omega t) = I_0 \sin \omega t$

- Peak value of current, $I_0 = \frac{E_0}{R}$
- **Phasor diagrams:-**

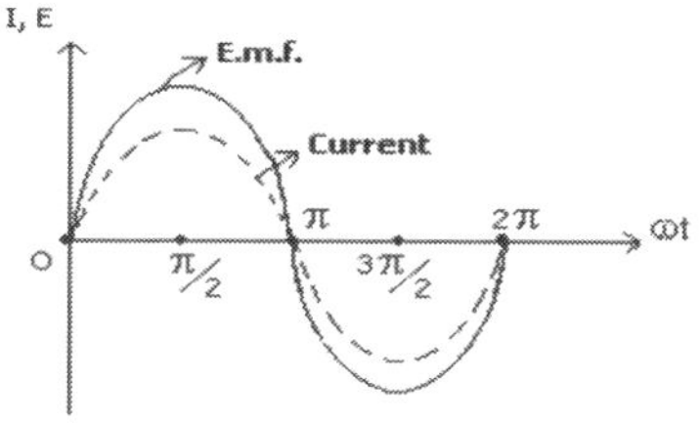

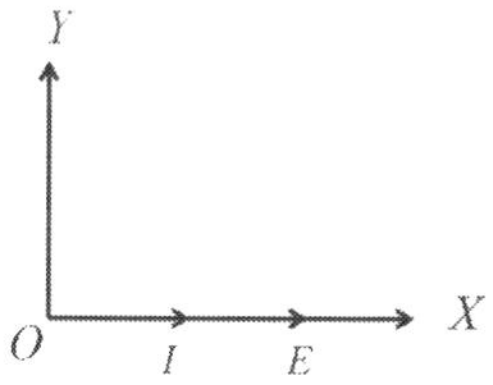

- emf and current will be in phase $(\Delta\phi = 0°)$
- emf and current have same frequency

- peak emf is more than peak current
- The value of impedance (Z) is equal to R and reactance (X) is zero
- Apart from instantaneous value, current in the circuit is independent of frequency and decreases with increase in R (similar to that in dc circuits).

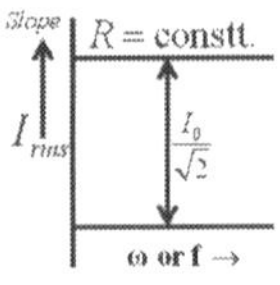

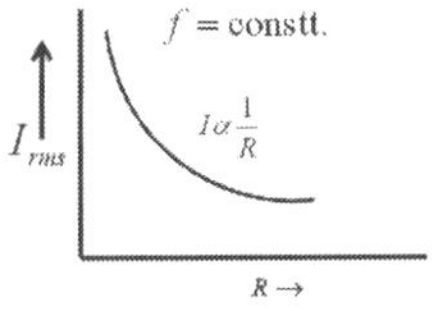

POWER

- power factor $\cos\phi = \cos 0° = 1$
- Instantaneous power $P_i = E_o I_o \sin^2 \omega t$
- Average power over time 'T' sec =

$$P_{avg} = E_{rms} I_{rms} \cos\phi = E_{rms} . I_{rms} = \frac{E_{rms}^2}{R}$$

- Apparent or virtual power : The product of apparent voltage and apparent current in an electric circuits is called apparent power. This is always positive.

$$P_{ap} = E_{rms} I_{rms} = \frac{E_0 I_0}{2}$$

A.C THROUGH AN INDUCTOR

- A pure inductor of inductance L is connected across an alternating source of emf E

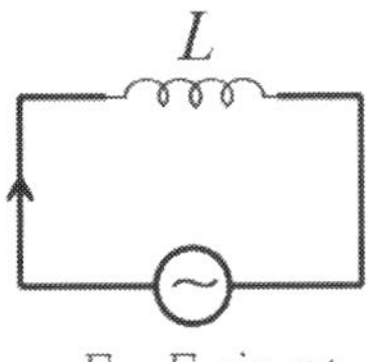

$E = E_0 \sin \omega t$

- The instantaneous value of alternating emf is $E = E_0 \sin \omega t$ ....(1)
- The induced emf across the inductor $= -L.\frac{dI}{dt}$ which opposes the growth of current in the circuit. As there is no potential drop across the circuit, so

$$E + \left(-L.\frac{dI}{dt}\right) = 0$$

or $L.\frac{dI}{dt} = E$

$$\frac{dI}{dt} = \frac{E_0}{L} \sin \omega t$$

On integrating $I = -\frac{E_0}{L\omega} \cos \omega t = I_0 \sin\left(\omega t - \frac{\pi}{2}\right)$ ....(2)

- The instantaneous value of alternating current is

$$\Rightarrow I = I_0 \sin\left(\omega t - \frac{\pi}{2}\right)$$

Where Peak value of current, $I_0 = \dfrac{E_0}{\omega L} = \dfrac{E_0}{X_L}$

From equation 1 & 2

- **Phase difference between alternating voltage and current is** $\dfrac{\pi}{2}$

- **The alternating current lags behind the emf** by a phase angle of $\dfrac{\pi}{2}$

- **Phasor diagram**

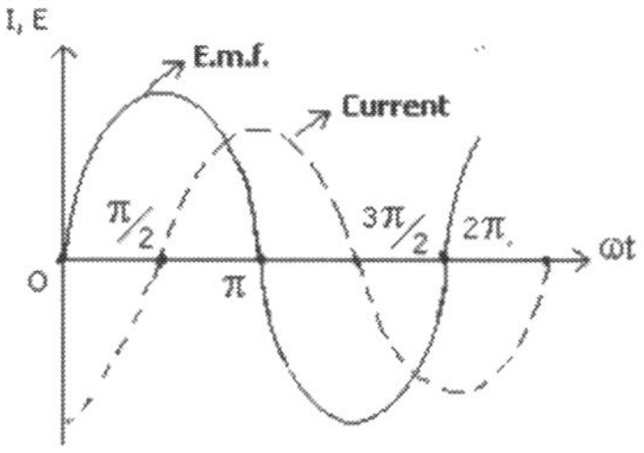

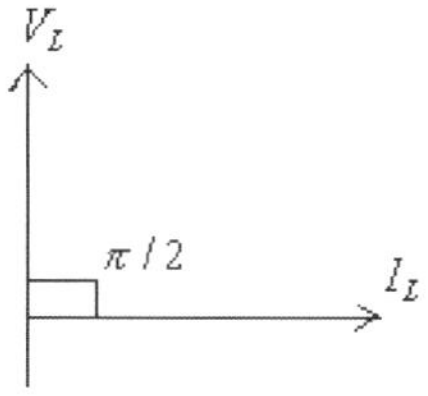

INDUCTIVE REACTANCE (X_L)

- The opposition offered by an inductor to the flow of **ac** is called an inductive reactance.
- The quantity ωL is analogous to resistance and is called reactance of Inductor represented by X_L .
- It allows D.C. but offers finite impedance to the flow of A.C.
- Its value depends on L and f.
- Inductance not only causes the current to lag behind emf but it also limits the magnitude of current in the circuit.
- $I_0 = \dfrac{E_0}{\omega L} \Rightarrow \omega L = \dfrac{E_0}{I_0} = X_L$,

$\therefore X_L = \omega L = 2\pi f L \Rightarrow X_L \alpha f$

$X_L - f$ curve $X_L - L$ curve

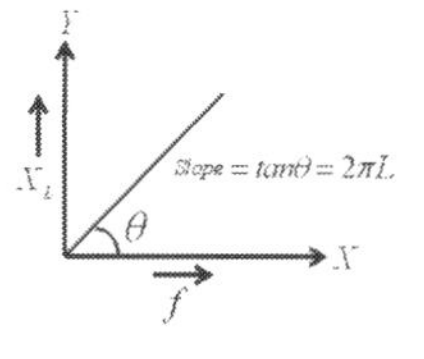

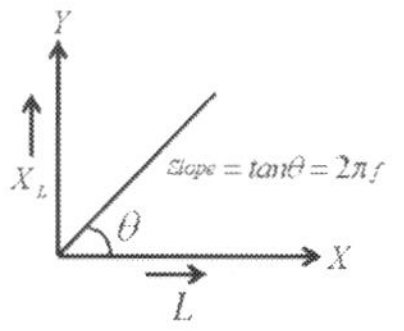

- For dc, $f = 0 \quad \therefore X_L = 0$

  For ac, high frequencies, $X_L = \infty$

  $\therefore$ dc can flow easily through inductor.

- Inductive reactance in terms of RMS value is $X_L = \omega L = \frac{E_{rms}}{I_{rms}}$

Power supplied to inductor

- The instantaneous power supplied to the inductor is $P_L = iv = i_0 \sin\left(\omega t - \frac{\pi}{2}\right) \times v_0 \sin(\omega t)$

$$= -i_0 v_0 \cos(\omega t)\sin(\omega t) \quad = -\frac{i_0 v_0}{2}\sin(2\omega t)$$

So, the average power over a complete cycle is

$$P_{avg} = E_{rms} \cdot I_{rms} \cos\phi = 0 \quad (\because \Delta\phi = 90°) \quad = -\frac{i_0 v_0}{2}\langle \sin(2\omega t)\rangle = 0$$

Since the average of $\sin(2\omega t)$ over a complete cycle is zero.

Thus, the ***average power supplied to an inductor over one complete cycle is zero.***

A.C THROUGH A CAPACITOR

- When an alternating emf is applied to a capacitor, then alternating current is constituted in the circuit. Due to this charge on the plates and electric field between the plates of capacitor vary sinusoidally with time.
- At any instant the potential difference between the plates of a capacitor is equal to applied emf at that time.

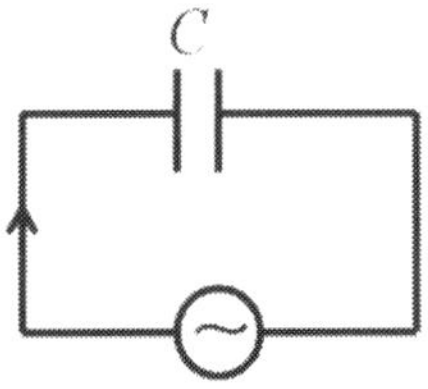

$E = E_0 \sin \omega t$

- A capacitor of capacity C is connected across an alternating source of emf
- The instantaneous value of alternating emf is $E = E_0 \sin \omega t$ ....(1)
- Let q be the charge on the capacitor at any instant.

  Accoding to kirchhoff's loop rule

$$E - \frac{q}{C} = 0 \Rightarrow q = CE_0 \sin \omega t$$

$$I = \frac{dq}{dt} = C\omega E_0 \cos\omega t = I_0 \sin\left(\omega t + \frac{\pi}{2}\right) \text{..............(2)}$$

- The instantaneous value of alternating current is

$$I = I_0 \sin\left(\omega t + \frac{\pi}{2}\right) \text{...............(2)}$$

where peak value of current, $I_0 = \dfrac{E_0}{\left(\dfrac{1}{\omega C}\right)}$

- From equation 1 & 2

current leads the emf by an angle $\frac{\pi}{2}$.

Phasor diagram

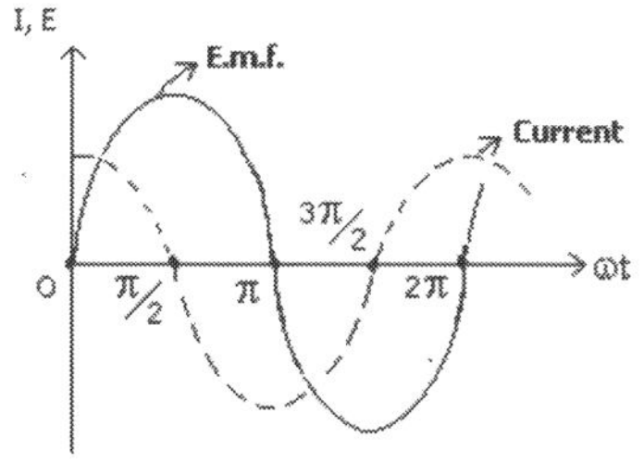

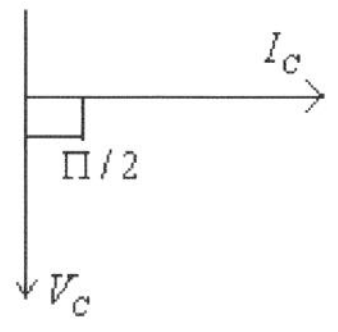

CAPACITIVE REACTANCE (X_C)

- The resistance offered by a capacitor to the flow of **ac** is called capacitive reactence.
- The quantity $\frac{1}{\omega C}$ is analogous to resistance and is called reactance of capacitor represented by X_C

$$I_0 = \frac{E_0}{\left(\frac{1}{\omega C}\right)} \Rightarrow X_C = \frac{1}{\omega C} = \frac{1}{2\pi fC} = \frac{E_0}{I_0} = \frac{E_{rms}}{I_{rms}}$$

- It is that part of impedance in which A.C. leads the A.V. by a phase angle of $\frac{\pi}{2}$.
- Its value is $X_c = \frac{1}{\omega C} = \frac{1}{2\pi fC}$.
- Its value depends on C and f.
- It bypasses A.C. but blocks D.C.
- It is produced due to pure capacitor or induced charge.
- $X_C - f$ curve $X_C - C$ curve

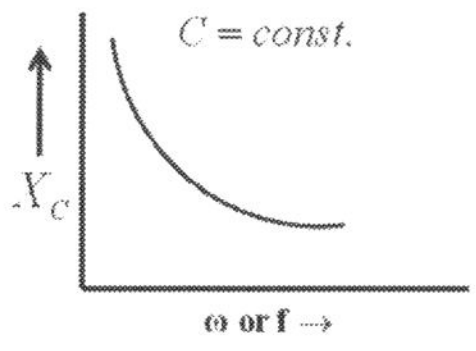

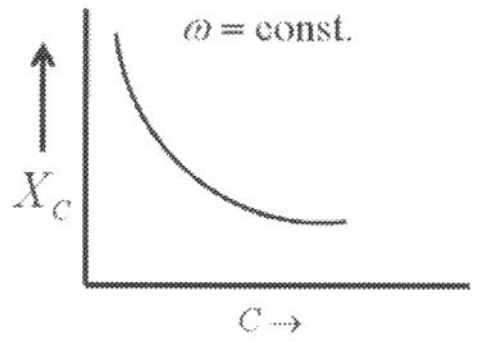

Note:

Resistance, Impedance, Reactance have the same units and Dimensional Formulae.

i.e. SI unit is ohm; Dimensional Formulae is $\left(ML^2T^{-3}A^{-2}\right)$

Power supplied to capacitor:

- The instantaneous power supplied to the capacitor is $P_c = iv = i_0\cos(\omega t)v_0\sin(\omega t)$

$= i_0v_0\cos(\omega t)\sin(\omega t) = \frac{i_0v_0}{2}\sin(2\omega t)$

So, the average power over a complete cycle is zero

since $<\sin(2\omega t)>= 0$ over a complete cycle.

$P_{avg} = V_{rms}.I_{rms}\cos\phi = V_{rms}.I_{rms}\cos 90° = 0$

$\therefore$ **no power is consumed in a purely capacitive circuit.**

WE: 6

An alternating voltage

$E = 200\sqrt{2}\sin(100t)$ **is connected to a** $1\mu F$ **capacitor through an ac ammeter. What will be the reading of the ammeter?**

Solution.

Comparing $E = 200\sqrt{2}\sin(100t)$ with

$E = E_0\sin\omega t$

$E_0 = 200\sqrt{2}V$ and $\omega = 100(rad/s)$

$$X_C = \frac{1}{\omega C} = \frac{1}{100\times10^{-6}} = 10^4\Omega$$

$$I_{rms} = \frac{E_{rms}}{Z} = \frac{E_0}{\sqrt{2}X_C} = \frac{200\sqrt{2}}{\sqrt{2}\times10^4} = 20\,mA$$

WE. 7

Find the maximum value of current when a coil of inductance 2H is connected to 150 V,50 cycles / sec supply.

Solution :

Here $L = 2H$, $E_{rms} = 150\text{ V}$, $\nu = 50\text{ Hz}$

$X_L = L\omega = L\times2\pi\nu = 2\times2\times3.14\times50 = 628\text{ ohm}$

RMS value of current through the inductor,

$$I_{rms} = \frac{E_{rms}}{X_L} = \frac{150}{628} = 0.24A$$

Maximum value (or peak value) of current is given by

$$I_{rms} = \frac{I_0}{\sqrt{2}}$$

or $I_0 = \sqrt{2}\,I_{rms} = 1.414\times0.24 = 0.339A$

WE:8

An inductor of 1 henry is connected across a 220 v, 50 Hz supply. The peak value of the current is approximately.

Solution

peak value of current

$$i_0 = \frac{E_0}{X_L} = \frac{\sqrt{2}E_{rms}}{\omega L} = \frac{\sqrt{2}E_{rms}}{2\pi fL}$$

$$= \frac{\sqrt{2}(220)}{2\pi \times 50 \times 1} = 0.99A$$

WE.:9

A capacitor of $2\mu F$ is connected in a radio circuit. The source frequency is 1000 Hz. If the current through the capacitor branch is 2 mA then the voltage across the capacitor is

Solution.

$$V_C = IX_C = I \times \frac{1}{\omega C} = \frac{I}{2\pi fC} = \frac{2\times10^{-3}}{2\pi\times10^3\times2\times10^{-6}} = 0.16V$$

A.C THROUGH LR SERIES CIRCUIT

- LR circuit consists of a resistor of resistance R and an inductor of inductance L in series with a source of alternating emf
- The instantaneous value of alternating emf is $E = E_0 \sin \omega t$

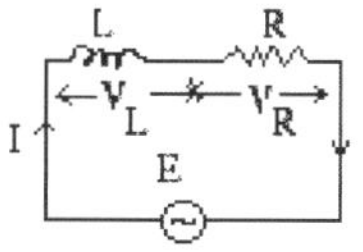

- The potential difference across the inductor is given by, $V_L = IX_L$ .....(1)
- The potential difference across the resistor,

$V_R = IR$ .....(2)

current I lags the Voltage V_L by an angle of $\frac{\pi}{2}$,

Therefore, the resultant of V_L and V_R is

$OC = \sqrt{OA^2 + OB^2}$ or $E = \sqrt{{V_R}^2 + {V_L}^2}$

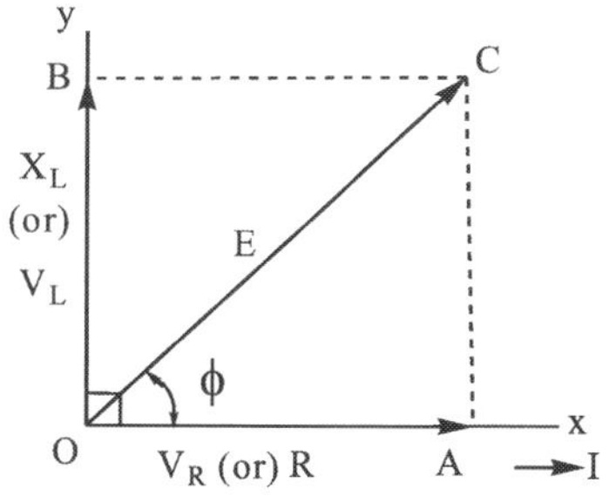

fig (b)

Using equations (1) and (2), we get

$$E = \sqrt{I^2R^2 + I^2X_L{}^2} = I\sqrt{R^2 + X_L{}^2}$$

where $X_L = \omega L$ is the inductive reactance.

$$\text{or } I = \frac{E}{\sqrt{R^2 + X_L{}^2}} \quad ...(3) \qquad I = \frac{E}{Z_{LR}}$$

$$Z_{LR} = \sqrt{R^2 + X_L^2} = \sqrt{R^2 + L^2\omega^2}$$

The effective opposition offered by LR circuit to **ac** is called the **impedance** of LR circuit.

Let ϕ be the angle made by the resultant of V_L and V_R with the X-axis, then from figure, we get

$$\tan\phi = \frac{AC}{OA} = \frac{OB}{OA} = \frac{V_L}{V_R} = \frac{IX_L}{IR}$$

$$\text{or } \tan\phi = \frac{X_L}{R} = \frac{\omega L}{R}$$

Note: In series LR circuit, emf leads the current or the current is said to lag behind the emf by an angle ϕ

$\therefore$ Current in L-R series circuit is given by

$$I = \frac{E}{Z_{LR}} = \frac{E_0}{Z_{LR}} \sin(\omega t - \phi)$$

(or) $I = I_0 \sin(\omega t - \phi)$

Note :

- $Z_{LR} = \sqrt{R^2 + L^2\omega^2} = \sqrt{R^2 + L^2 \times 4\pi^2 f^2}$.
  Thus Z_{LR} increases with the frequency of **ac**, so Z_{LR} is low for lower freqeuncy of **ac** and high for higher frequency of **ac**
- The phase angle between voltage and current increases with the increase in the frequency of **ac**

C-R SERIES CIRCUIT WITH ALTERNATING VOLTAGE

- Let an alternating source of emf $E = E_0 \sin\omega t$ is connected to a series combination of a pure capacitor of capacitance (C) and a resistor of resistance (R) as shown in figure (a)

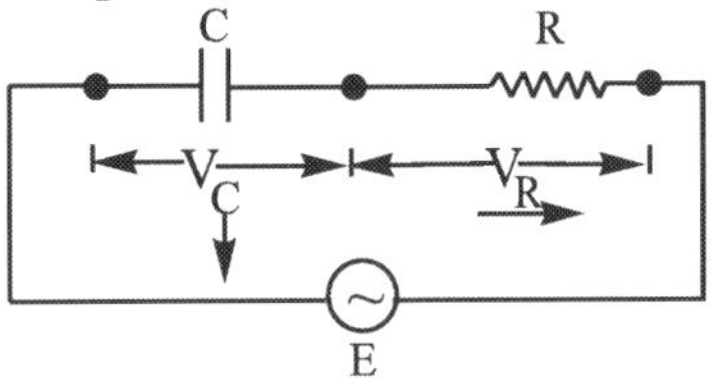

fig (a)

- Let I be the r.m.s value of current flowing through the circuit. The potential difference across the capacitor,
  $V_C = IX_C$ **.....(i)**
- The current leads emf by an angle $\frac{\pi}{2}$ when ac flows through capacitor.
- The potential difference across the resistor,
  $V_R = IR$ .....(ii)
- The emf and current are in phase when ac flows through resistor.

Phasor diagram.

- In figure V_C is represented by OB along negative Y - axis and the current I is represented along X - axis.
- V_R is represented by OA along X - axis.
- The resultant potential difference of V_C and V_R is represented by OC.

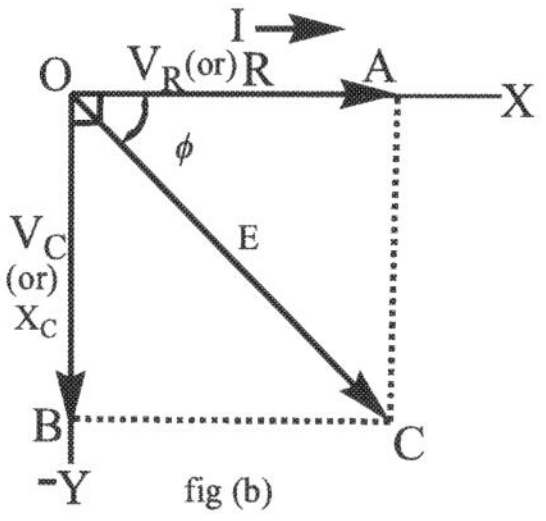

fig (b)

- Also, the emf and current are in phase when **ac** flows through the resistor. So, V_R is represented by OA along X-axis.
- Therefore, the resultant potential difference of V_C and V_R is represented by OC and is given by

$$OC = \sqrt{OA^2 + OB^2} \quad \text{or } E = \sqrt{V_R^{\ 2} + V_C^{\ 2}}$$

Using equations (i) and (ii), we get

$$E = \sqrt{I^2R^2 + I^2X_C^{\ 2}} = I\sqrt{R^2 + X_C^{\ 2}}$$

$$\text{or } I = \frac{E}{\sqrt{R^2 + X_C^{\ 2}}} = \frac{E}{Z_{CR}}$$

From the above equations of I and E, we have $Z_{CR} = \sqrt{R^2 + X_C^2} = \sqrt{R^2 + \left(\frac{1}{C\omega}\right)^2}$

Where Z_{CR} is the effective opposition offered by the CR circuit to **ac**, which is the **impedance** of CR circuit.

Let ϕ be the angle made by E with X-axis

$$\tan\phi = \frac{AC}{OA} = \frac{V_C}{V_R} = \frac{IX_C}{IR}$$

$$\text{or } \tan\phi = \frac{X_C}{R} = \frac{I}{C\omega R}$$

In series CR circuit, emf lags behind the current or in other words, the current is said to lead the emf by an angle ϕ given by the above equation.

$\therefore$ Current in C-R series circuit is given by

$$I = \frac{E}{Z_{CR}} = \frac{E_0}{Z_{CR}} sin(\omega t + \phi)$$

(or) $I = I_0 . sin(\omega t + \phi)$

Note:

- The resultant potential difference of V_C and V_R is represented by OC Impedance of CR circuit.

$$Z_{CR} = \sqrt{R^2 + X_C^{\ 2}} = \sqrt{R^2 + \frac{1}{C^2\omega^2}} = \sqrt{R^2 + \frac{1}{4\pi^2 f^2 C^2}}$$

Thus $Z_{CR} \propto \frac{1}{f}$

- The resultant potential difference of V_C and V_R is represented by OC For very high frequency (v) of **ac**. $Z \to R$ and for very low frequency of **ac**, $Z \to \infty$
- Phase angle between voltage and current is given by $\tan\phi = \frac{1}{C\omega R} = \frac{1}{2\pi v CR}$

As v increases, phase angle ϕ decreases.

L - C SERIES CIRCUIT WITH ALTERNATING VOLTAGE.

Let an alternating source of emf $E = E_0 \sin\omega t$ is connected to the series combination of a pure capacitor of capacitance (C) and an inductor of inductance (L) is shown in fig.

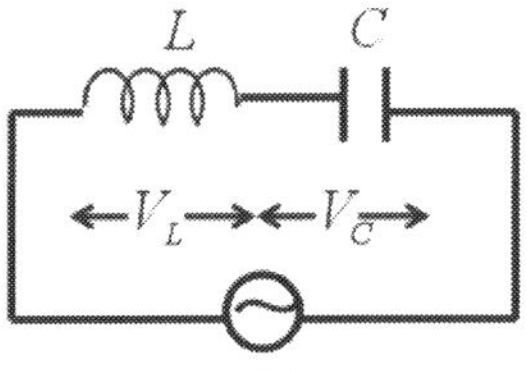

Let I be the rms value of current flowing the circut

The P.D across 'L' is $V_L = I.X_L$

The current I lags V_L by an angle $\pi/2$.

The P.D across capacitance is $V_C = I.X_C$.

The current I leads V_C by an angle $\pi/2$.

The voltage V_L and V_C are represented by OB and OC respecitvely.

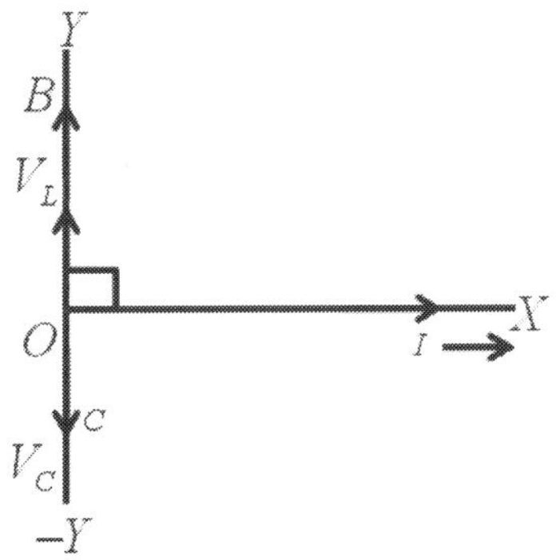

The resultant P.D of V_L and V_C is $V = V_L \sim V_C = I(X_L \sim X_C) = \left[\omega L \sim \frac{1}{\omega C}\right] = IZ_{LC}$

- From the above equations, **Impedance of L -C** circuit is $Z_{LC} = \left[(\omega L) \sim \frac{1}{\omega C}\right]$

- If $\omega L > \frac{1}{\omega C}$ i.e, $X_L > X_C$ then $V_L > V_C$ potential difference $V = V_L - V_C$.
- Now current lags behind voltage by $\pi/2$.
- If $\omega L < \frac{1}{\omega C}$ then $V_L < V_C$ resultant potential difference $(V) = V_C - V_L$

Now current leads emf by $\pi/2$.

If $\omega L = \frac{1}{\omega C}$ then $Z = \omega L - \frac{1}{\omega C} = 0$

Current $I = \frac{E}{Z} = \alpha$

In L - C, circuit, the phase difference between voltage and current is always $\pi/2$.

Power factor $\cos\phi = \cos\pi/2 = 0$.

So, power consumed in L - C circuit is $P = V_{rms} \times I_{rms} \times \cos\phi = 0$

$\therefore$ **In L - C circuit no power is consumed.**

Note:

In L - C, circuit, the impendence $Z = \left|\omega L - \frac{1}{\omega C}\right|$

Current $I = \frac{E}{Z}$.

So, the impedence and current varies with frequency.

At a particular angular frequency (ω_0) $\omega L = \frac{1}{\omega C}$

and current $I = \frac{E}{Z}$ becomes maximum (I_0) and resonance occurs.

At resonance $Z = 0$ and $I_0 = \frac{E_0}{Z} = \infty$.

Resonant angular frequency $\omega_0 = \frac{1}{\sqrt{LC}}$

Resonant frequecny $f_0 = \frac{1}{2\pi\sqrt{LC}}$.

WE:10

A 0.21 H inductor and a 12 ohm resistance are connected in series to a 220 V. 50 Hz ac source. Calculate the current in the circuit and the phase angle between the current and the source voltage .

Solution:

Here $X_L = \omega L = 2\pi fL = 2\pi \times 50 \times 0.21 = 21\pi\,\Omega$

$Z = \sqrt{R^2 + X_L^2} = \sqrt{12^2 + (21\pi)^2} = \sqrt{144 + 4348}$

$$Z = \sqrt{4492} \simeq 67.01\Omega$$

$$I = \frac{V}{Z} = \frac{220}{67.02} = 3.28A$$

$$\phi = \tan^{-1}\left(\frac{X_L}{R}\right) = \tan^{-1}\left(\frac{21\pi}{12}\right)$$

The current will lag the applied voltage by in $\tan^{-1}\left(\frac{21\pi}{12}\right)$ phase.

WE:11

When 100 volt dc is applied across a coil, a current of 1 amp flows through it; when 100 V ac of 50 Hz is applied to the same coil, only 0.5 amp flows. Calculate the resistance and inductance of the couil.

Solution:

In case of a coil, i.e, L - R circuit,

$$I = \frac{V}{Z} \text{ with } Z = \sqrt{R^2 + X_L^{\ 2}} = \sqrt{R^2 + (\omega L)^2}$$

So when dc is applied, $\omega = 0$, so $Z = R$

and hence $I = \frac{V}{R}, i.e, R = \frac{V}{I} = \frac{100}{1} = 100\Omega$

and when ac of 50 Hz is applied.

$$I = \frac{V}{Z}, \text{ i,e. } Z = \frac{V}{I} = \frac{100}{0.5} = 200\Omega$$

but $Z = \sqrt{R^2 + \omega^2 L^2}$, i.e, $\omega^2 L^2 = Z^2 - R^2$

i.e, $(2\pi fL)^2 = 200^2 - 100^2 = 3\times 10^4$

$$L = \frac{\sqrt{3}\times 10^2}{2\pi\times 50} = \frac{\sqrt{3}}{\pi} H = 0.55H$$

WE: 12

A $10\mu F$ capacitor is in series with a 50Ω resistance and the combination is connected to a 220V, 50 Hz line. Calculate (i) the capacitive reactance, (ii) the impedance of the circuit and (iii) the current in the circuit.

Solution :

Here, $C = 10\mu F = 10\times 10^{-6} = 10^{-5} F$

$R = 50$ ohm, $E_{rms} = 220V, \nu = 50Hz$,

(i) Capacitive reactance,

$$X_C = \frac{1}{\omega C} = \frac{1}{2\pi\nu C} = \frac{1}{2\times 3.14\times 50\times 10^{-5}} = 318.5\ \Omega$$

(ii) Impedance of CR circuit.

$$Z_{CR} = \sqrt{R^2 + X_C^{\ 2}} = \sqrt{(50)^2 + (318.5)^2} = 322.4\Omega$$

(iii) Current, $I_{rms} = \frac{E_{rms}}{Z_{CR}} = \frac{220}{322.4} = 0.68A$

WE :13

A coil has an inductance of 0.7 H and is joined in series with a resistance of 220 Ω. When an alternating e.m.f of 220 V at 50 cps is applied to it, then the wattless component of the current in the circuit is

Solution:

$$\tan\phi = \frac{X_L}{R} = \frac{\omega L}{R} = \frac{2\pi \times 50 \times 0.7}{220} = 1$$

$$\therefore \phi = 45°, Z = \sqrt{R^2 + X_L^2} = \sqrt{220^2 + 220^2}$$

$$= 220\sqrt{2}\Omega$$

Wattless component of currnet $= I_v \sin\phi$

$$= \frac{E_v}{Z}\sin 45° = \frac{220}{220\sqrt{2}} \times \frac{1}{\sqrt{2}} = 0.5A$$

A.C THROUGH LCR SERIES CIRCUIT

- A circuit containing pure inductor of inductance (L), pure capacitor of capacitance (C) and resistor of resistance (R), all joined in series, is shown in figure.
- Let E be the r.m.s value of the applied alternating emf to the LCR circuit.

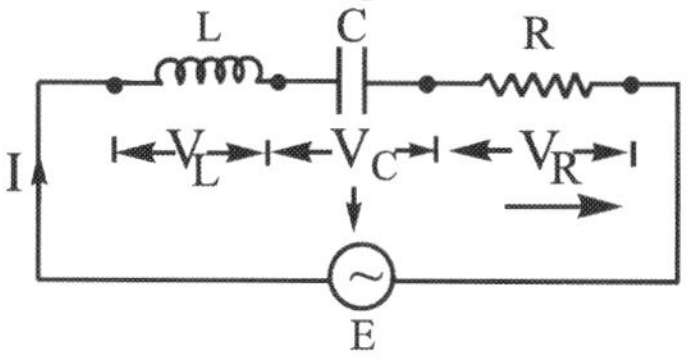

- The potential difference across L,

  $V_L = IX_L$ .....(i)
- The potential difference across C,

  $V_C = IX_C$ .....(ii)
- The potential difference across R,

  $V_R = IR$ .....(iii)

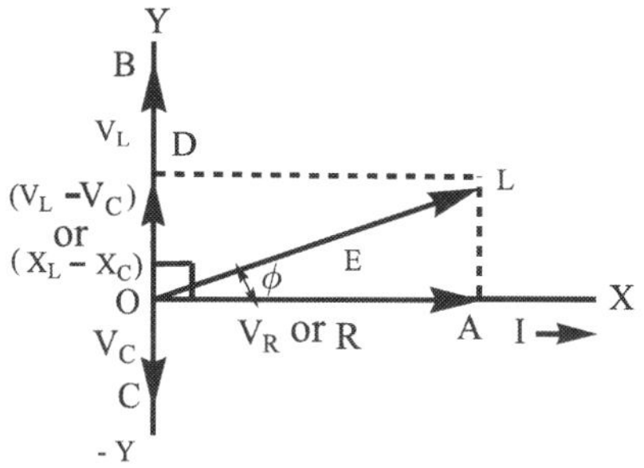

- Since V_L and V_C are in opposite phase, so their resultant $(V_L - V_C)$ is represented by OD (Here $V_L > V_C$)

  The resultant of V_R and $(V_L - V_C)$ is given by OL.

  The magnitude of OL is given by

$$OL = \sqrt{(OA)^2 + (OD)^2} = \sqrt{V_R^2 + (V_L - V_C)^2} = I\sqrt{R^2 + (X_L - X_C)^2}$$

$$Z = \frac{E}{I} = \sqrt{R^2 + (X_L - X_C)^2}$$

$\therefore$ Impedance (Z) of LCR circuit is given by

$$Z = \sqrt{R^2 + (X_L - X_C)^2}$$

$$\therefore I = \frac{E}{\sqrt{R^2 + (X_L - X_C)^2}} = \frac{E}{\sqrt{R^2 + \left(L\omega - \frac{1}{C\omega}\right)^2}}$$

- Let ϕ be the phase angle between E and I, then from above figure,

$$\tan\phi = \frac{V_L - V_C}{V_R} = \frac{IX_L - IX_C}{IR} = \frac{X_L - X_C}{R}$$

$$\tan\phi = \frac{\left(L\omega - \frac{1}{C\omega}\right)}{R}$$

$\therefore$ Current in L-C- R series circuit is given by

$$I = \frac{E}{Z} = \frac{E_0}{Z} sin(\omega t \pm \phi)$$

(or) $I = I_0 . sin(\omega t \pm \phi)$

- If X_L and X_C are equal then $Z = R$ i.e., expression for pure resistance circuit.
  If $X_L = 0$ then $Z = \sqrt{R^2 + X_C^2}$ i.e., expression for series RC circuit.
- Similarly if $X_C = 0$ then $Z = \sqrt{R^2 + X_L^2}$ i.e. expression for series RL circuit.

  Also, $\cos\phi = \frac{R}{Z}$

case (i) : If $X_L > X_C$ then ϕ is +ve. In this case the current lags behind the emf by a phase angle

$$\phi = Tan^{-1}\left(\frac{X_L - X_C}{R}\right)$$

case (ii) : If $X_L < X_C$ then ϕ is -ve. In this case the current leads the emf by a phase angle

$$\phi = Tan^{-1}\left(\frac{X_C - X_L}{R}\right)$$

case (iii) : If $X_L = X_C$ then ϕ is 0. In this case the current and emf are in phase.

- If $X_L > X_C$, then the circuit will be inductive
- If $X_L < X_C$, then the circuit will be capacitive
- If $X_L = X_C$, then the circuit will be purely resistive.
- The LCR circuit can be inductive or capacitive or purely resistive depending on the value of frequency of alternating source of emf.
- At some frequency of alternating source, $X_L > X_C$ and for some other frequency, $X_L < X_C$. There exists a particular value of frequency where $X_L = X_C$ (This situation is

explained under resonance of LCR series circuit)

Note:

Relation between applied pd & pd's across the components in L - C - R circuit

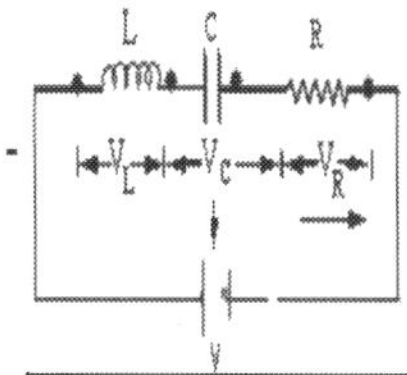

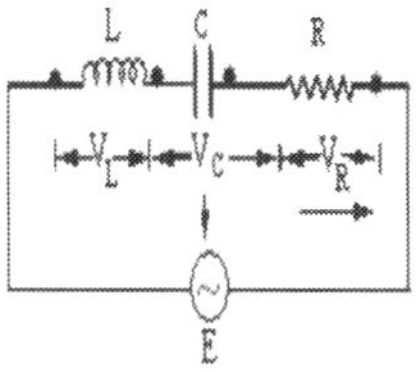

$\boxed{\mathbf{V = V_R + V_L + V_C}}$ $V = IZ$

(only before steady state) $= I\sqrt{R^2 + (X_L - X_C)^2} = \sqrt{(IR)^2 + (IX_L - IX_C)^2}$

$\mathbf{V^2 = V_R{}^2 + (V_L - V_C)^2}$ where $V_L = IX_L = I\omega L$ and $V_C = IX_C = \frac{I}{\omega C}$ and $V_R = IR$

Note:

Rules to be followed for various combinations of ac circuits

- Compute effective resistance of the circuit as R
- Calculate the net reactance of the circuit as $X = X_L - X_C$ where $X_L = \omega L$, $X_C = \frac{1}{\omega C}$.
- Resistance offered by all the circuited elements to the flow of **ac** is impedance (Z)

$$\therefore\ Z = \sqrt{R^2 + X^2} = \sqrt{R^2 + (X_L - X_C)^2}$$

- Calculate the peak value of current as $I_0 = \frac{E_0}{Z}$
- The phase difference between emf & current can be known by constructing an **ac** triangle as

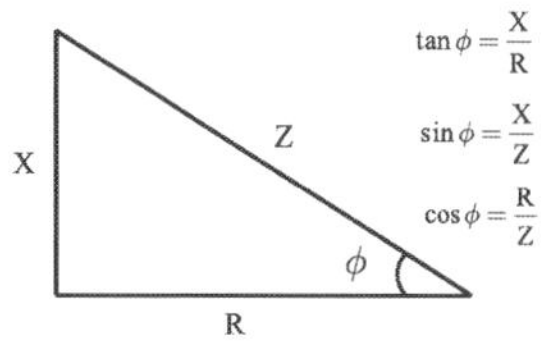

WE: 14

In a circuit L, C and R are connected in series with an alternating voltage source of frequency f. The current leads the voltage by $45°$. The value of C is :

Solution:

As current leads the voltage by $45°$,

$$\therefore\ \tan\theta = \frac{X_C - X_L}{R} = \tan 45° = 1$$

$$\therefore X_C - X_L = R \text{ or } X_C = X_L + R$$

$$or\ \frac{1}{\omega C} = \omega L + R \Rightarrow C = \frac{1}{\omega(\omega L + R)}$$

$$C = \frac{1}{2\pi f\,(2\pi fL + R)}$$

WE: 15

In a series LCR circuit, the voltage across the resistance, capacitance and inductance is 10V each. If the capacitance is short circuited the voltage across the inductance will be

Solution:

As $V_R = V_L = V_C$

$R = X_L = X_C$

$Z = R$

$V = IR = 10\,\text{volt}$

When capacitor is short circuited,

$Z = \sqrt{R^2 + X_L^2} = \sqrt{R^2 + R^2} = R\sqrt{2}$

New current, $I' = V/Z = \dfrac{V}{R\sqrt{2}} = \dfrac{10}{R\sqrt{2}}$

Potential drop across inductance

$= I'X_L = I'R = \dfrac{10 \times R}{R\sqrt{2}} = \dfrac{10}{\sqrt{2}}\,\text{volt}$

WE:16

An inductance of $\dfrac{200}{\pi}mH$ **. a capacitance of** $\dfrac{10^{-3}}{\pi}F$ **and a resistance of** 10Ω **are connected in series with an AC source of 220 V, 50 Hz. The phase angle of the circuit is**

Soution:

Here, $L = \dfrac{200}{\pi}mH = \dfrac{200 \times 10^{-3}}{\pi}H = \dfrac{0.2}{\pi}H$

$C = \dfrac{10^{-3}}{\pi}F, R = 10\Omega$

$E_v = 220V, n = 50Hz$

$X_L = \omega L = 2\pi nL = 2\pi \times 50 \times \dfrac{0.2}{\pi} = 20\,\Omega$

$X_C = \dfrac{1}{\omega C} = \dfrac{1}{2\pi nC} = \dfrac{\pi}{2\pi \times 50 \times 10^{-3}} = 10\Omega$

$\tan\phi = \dfrac{(X_L - X_C)}{R} = \dfrac{20-10}{10} = 1$

$\phi = \dfrac{\pi}{4}$

WE:17

In a series LCR circuit, $R = 200\Omega$ **and the voltage and the frequency of the main supply is 220 V and 50 Hz. respectively. On taking out the capacitance from the circuit. The current lags behind the voltage by** $30°$**. On taking out the inductor from the circuit,**

the current leads the voltage by $30°$. The power dissipated in the LCR circuit is

Solution:

Here, $R = 200\Omega, E_v = 220V$

In L -R circuit, $\tan 30° = \frac{X_L}{R}$

In C - R circuit, $\tan 30° = \frac{X_C}{R}$

$\therefore \frac{X_L}{R} = \frac{X_C}{R}$ or $X_L = X_C$

In L - C - R circuit, if θ is the phase difference between voltage and current, then

$$\tan\theta = \frac{X_L - X_C}{R} = \frac{0}{200} = 0 \Rightarrow \theta = 0°$$

i.e., current and voltage are in the same phase.

$\therefore$ **Average power** $= E_v \times I_v \cos\theta = \frac{E_v^2}{R}(\because \theta = 0) = \frac{(220)^2}{200} = 242W$

WE:18

An LCR circuit has L = 10 mH. R = 3 ohm and $C = 1\mu F$ connected in series to a source of 15 cos ωt volt. What is average power dissipated per cycle at a frequecny that is 10% lower than the resonant frequency?

Solution:

Here, $L = 10^{-2}H, R = 3\Omega, C = 10^{-6}F$

Resonant frequency,

$$\omega_0 = \frac{1}{\sqrt{LC}} = \frac{1}{\sqrt{10^{-2} \times 10^{-6}}} = 10^4 \text{ rad/s}$$

Actual frequency, $\omega = (90\%)\omega_0 = 9 \times 10^3$ rad/s

$$X_L = \omega L = 9 \times 10^3 \times 10^{-2} = 90\Omega$$

$$X_C = \frac{1}{\omega C} = \frac{1}{9 \times 10^3 \times 10^{-6}} = \frac{1000}{9}\Omega$$

$$Z = \sqrt{R^2 + (X_C - X_L)^2} = \sqrt{3^2 + \left(\frac{100}{9} - 90\right)^2} = 21.3\Omega$$

Power dissipated / cycle $= E_v I_v \cos\phi = E_0\left(\frac{E_v}{Z}\right)\frac{R}{Z} = \left(\frac{E_v}{Z}\right)^2 \times R$

$$= \left(\frac{15}{\sqrt{2} \times 21.3}\right)^3 \times 3 = 0.744W$$

WE:19

A 750 Hertz - 20 volt source is connected to a resistance of 100 ohm, an inductance of 0.1803 henry and a capacitance of $10\mu F$, all in series. What is the time in which the resistance(Thermal capacity $= 2\text{joule/°C}$) will get heated by $10°C$?

Soution:

Here, $v = 750Hz,\ E_v = 20V,\ R = 100\Omega$

$L = 0.1803\,H, C = 10\mu F = 10^{-5}F, t = ?$

$\Delta\theta = 10°C$, **thermal capacity** $= 2\text{J/°C}$

$X_L = \omega L = 2\pi vL = 2\times 3.14\times 750\times 0.1803$

$$X_C = \frac{1}{\omega C} = \frac{1}{2\pi\times 750\times 10^{-5}} = 21.2\Omega$$

$$Z = \sqrt{R^2 + (X_L - X_C)^2} = \sqrt{100^2 + (850 - 21.2)^2} = 835\Omega$$

Power dissipated $= E_v I_v \cos\phi = E_v\left(\frac{E_v}{Z}\right)\left(\frac{R}{Z}\right) = \frac{20^2\times 100}{(835)^2} = 0.0574\text{W}$

Heat produced in resistance $= 2\times 10 = 20J$

If t is the required time, then

$$P\times t = 20 \Rightarrow t = \frac{20}{P} = \frac{20}{0.0574} = 348\text{s}$$

RESONANT FREQUENCY:

At resonance, $X_L = X_C$

$$\Rightarrow \omega_0 L = \frac{1}{\omega_0 C} \quad \Rightarrow \omega_0 = \frac{1}{\sqrt{LC}}\frac{rad}{\sec}$$

$$\Rightarrow f_0 = \frac{1}{2\pi\sqrt{LC}} Hz(\text{or cps})$$

- Resonant frequency doesn't depend upon the resistance of the circuit.
- At resonance frequency, the value of I_0 is maximum which is given by $I_{0(\max)} = \frac{E_0}{R}$

Note :

At resonance power factor is maximum.

It is given by $\cos\phi = \frac{Z}{R} = \frac{R}{R} = 1$.

The variation of I_0 with ω

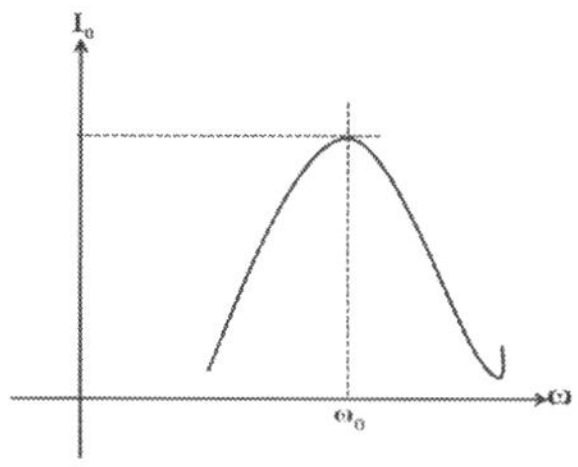

- The variation of ϕ and Z with angular frequency.

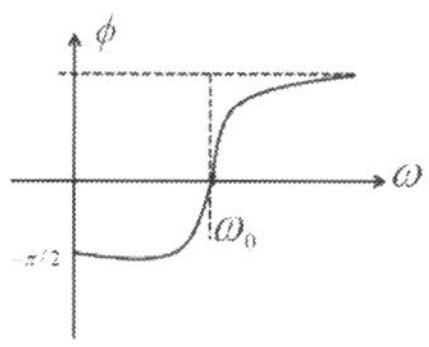

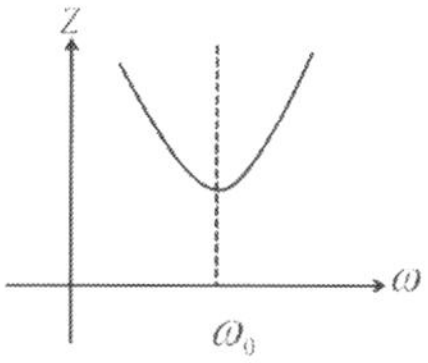

- The power factor is unity at resonance.
  Maximum power is dissipated during.
  resonance.

- 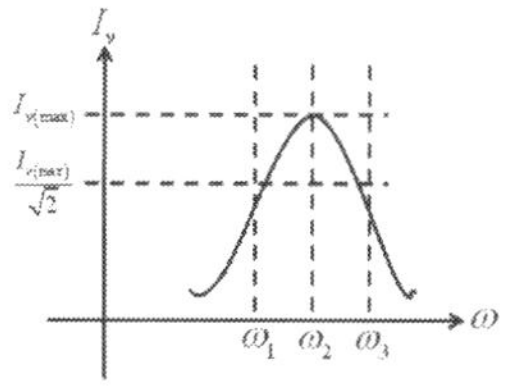

  These two frequencies

  $$\omega_1\omega_2 = \frac{1}{LC}$$

  band width $\omega_3 - \omega_1 = \frac{R}{L}$

  Half power frequencies and band width.

- The frequencies at which the power in the circuit is half of the maximum power (The power at resonance) are called half power frequencies.

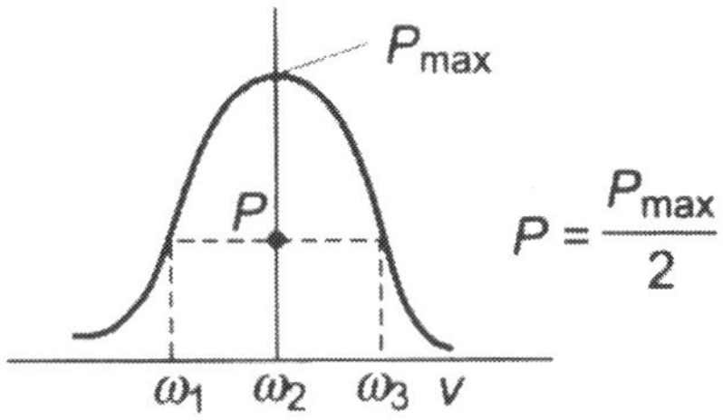

- The current in the circuit at half power frequecies (HPF) is $1\sqrt{2}$ or 0.707 or 70.7% of maximum current (current at resonance).
- There are two half power frequencies

$\omega_1 \rightarrow$ called lower half power frequency. At this frequency the circuit is capacitive.

$\omega_3 \rightarrow$ called upper half power frequency. It is greater than ω_2. At this frequency the circuit is inducitve.

- **Band width** $(\Delta\omega)$: The difference of half power frequencies ω_1 and ω_2 is called band width $(\Delta\omega)$ and $\Delta\omega = \omega_2 - \omega_1$.
- For series resonant circuit it can be proved $(\Delta\omega = R/L)$

QUALITY FACTOR (Q - FACTOR) OF SERIES RESONANT CIRCUIT.

- The characteristic of a series resonant circuit is determined by the quality factor $(Q-\text{factor})$ of the circuit.
- It defines sharpness of $i-v$ curve at resonance when $Q-$ factor is large, the sharpness of resonance curve is more and vice - versa.

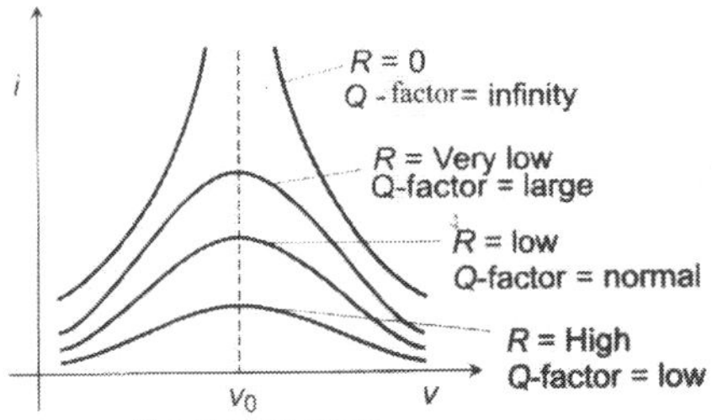

Resonance curve

- $Q-$ factor also defined as follows

$$Q-\text{factor} = 2\pi \times \frac{\text{Maximum energy stored}}{\text{energy dissipation}}$$

$$= \frac{2\pi}{T} \times \frac{\text{Maximum energy stored}}{\text{Mean power dissipated}} = \frac{\text{Resonant frequency}}{\text{Band width}} = \frac{\omega_0}{\Delta\omega}$$

$$Q-\text{factor} = \frac{V_L}{V_R} \text{ or } \frac{V_C}{V_R} = \frac{\omega_0 L}{R} \text{ or } \frac{1}{\omega_0 CR}$$

$$\Rightarrow Q-\text{factor} = \frac{1}{R}\sqrt{\frac{L}{C}}$$

Wattless Current:

In an ac circuit, $R = 0 \Rightarrow \cos\phi = 0$ so $P_{av} = 0$, i.e.., **in resistanceless circuit the power consumed is zero, Such a circuit is called the wattless circuit and the current flowing is called the wattless current. Or**

The component of current which does not contribute to the average power dissipation is called wattless current.

wattless current $= I_{rms} \sin\phi$

CHOKE COIL:

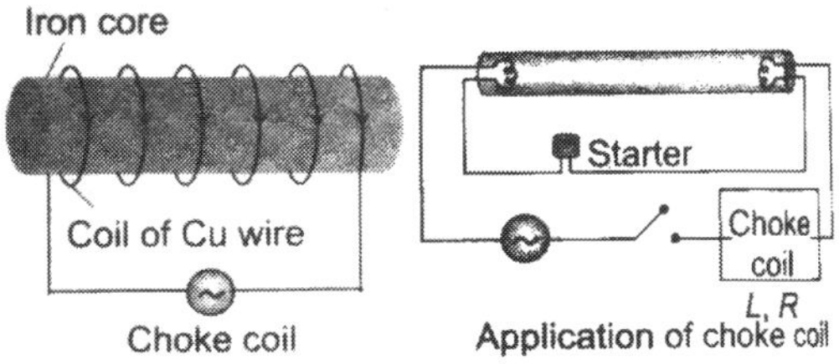

- Choke coil (or ballast) is a device having high inductance and negligible resistance.
- It is used to control current in ac circuits and is used in fluorescent tubes.
- The power loss in a circuit containing choke coil is least.
- In a dc circuit current is reduced by means of a rheostat.This resutls in a loss of electrical energy I^2R per sec.
- The current in an ac circuit can however be reduced by means of a device which involves very small amount of loss of energy.
- It consists of a copper coil wound over a soft iron laminated core. This coil is put in series with the circuit in which current is to be reduced.
- Soft iron is used to improve inductance (L) of the circuit.
- The inductive reactance or effective opposition of the choke coil is given by $X_L = \omega L = 2\pi v L$
- For an ideal choke coil $r = 0$, no electric cnergy is wasted, i.e., average power P = 0.
- In actual practice choke coil is equivalent to a $R-L$ circuit.
- Choke coil for different frequencies are made by using different substances in their core.
- For low frequency L should be large thus iron core choke coil is used. For high frequency ac circuit, L Should be small, so air cored choke coil is used.
- The choke coil can be used only in ac circuits not in dc circuits, because for dc frequency $v = 0$. Hence $X_L = 2\pi v L = 0$.
- Choke coil is based on the principle of wattless current.
- The current in the circuit $I = \frac{E}{Z}$ with

$$Z = \sqrt{(R+r)^2 + (\omega L)^2}.$$

- The power loss in the choke

$$p_{av} = V_{rms} I_{rms} \cos\phi \to 0$$

as $\cos\phi = \frac{r}{Z} = \frac{r}{\sqrt{r^2 + \omega^2 L^2}} = \frac{r}{\omega L} \to 0$

WE:20

An ideal choke coil takes a current of 8 ampere when connected to an AC supply of 100 volt and 50 Hz. A pure resistor under the same conditions takes a current of 10 ampere. If the two are connected to an AC supply of 150 volts and 40 Hz. then the current in a series combination of the above resistor and inductor is

Solution:

For pure inductor,

$$X_L = \frac{E_0}{I_v} = \frac{100}{8} = \frac{25}{2}\Omega$$

$$\omega L = \frac{25}{2}; L = \frac{25}{2\omega} = \frac{25}{2\times 2\pi \times 50} = \frac{1}{8\pi}H$$

$$R = \frac{V}{I} = \frac{100}{10} = 10\Omega$$

For the combination, the supply is 150 v, 40 Hz

$$\therefore X_L = \omega L = 2\pi \times 40 \times \frac{1}{8\pi} = 10\Omega$$

$$Z = \sqrt{X_L^2 + R^2} = \sqrt{10^2 + 10^2} = 10\sqrt{2}\text{ohm}$$

$$I_v = \frac{E_v}{Z} = \frac{150}{10\sqrt{2}}A = \frac{15}{\sqrt{2}}A$$

WE:21

An electric bulb has a rated power of 50 W at 100 V. If it is used on an AC source of 200 V, 50 Hz, a choke has to be used in series with it. This choke should have an inductance of

Soution:

Here, $P = 50W, V = 100\text{volt}$

$$I = \frac{P}{V} = \frac{50}{100} = 0.5A,\ R = \frac{V}{I} = \frac{100}{0.5} = 200\Omega$$

Let L be the inductance of the choke coil

$$\therefore I_v = \frac{E_v}{Z} = \text{or } Z = \frac{E_v}{I_v} = \frac{200}{0.5} = 400\,\Omega$$

Now $X_L = \sqrt{Z^2 - R^2} = \sqrt{400^2 - 200^2}$

$$\omega L = 100 \times 2\sqrt{3}$$

$$L = \frac{200\sqrt{3}}{\omega} = \frac{200\sqrt{3}}{2\pi v} = \frac{200\sqrt{3}}{100\pi} = \frac{2\times 1.732}{3.14} = 1.1H$$

L C Oscillations:-

- When a fully charged capacitor is connected across an inductor, the L-C oscillations are set up in the circuit. During the oscillations at any time 't'.

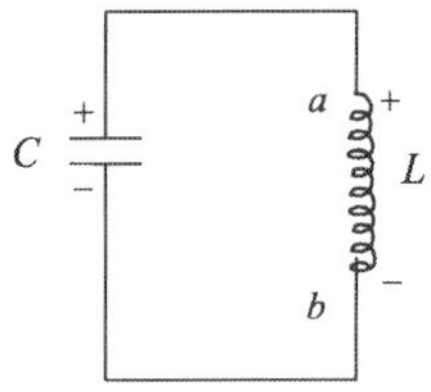

- The charge on the capacitor is given by $q = q_0 \cos \omega t$
- The current in the circuit is given by $I = -q_0\omega \sin \omega t$
- The rate of change of current in the circuit is given by $\frac{dI}{dt} = -q_0\omega^2 \cos \omega t$
- From Kirchhoff's Law $\frac{q}{C} - L\frac{di}{dt} = 0$

but $I = \frac{-dq}{dt}$

$$\frac{d^2q}{dt^2} + \frac{1}{LC}q = 0$$

This is similar to $\frac{d^2x}{dt^2} + \omega^2 x = 0$

$\therefore$ Charge oscillates with natural frequency $\omega_0 = \frac{1}{\sqrt{LC}}$

- If the oscillations are undamped the energy of the system oscillates between the capacitor and inductor. i.e. the maximum electrical energy in the capacitor $\left((U_E) = \frac{1}{2}\frac{q_m^2}{C}\right)$ equals the maximum magnetic energy $\left(U_m = \frac{1}{2}Li_m^2\right)$.
- But in reality, these oscillations are damped oscillations due to finite resistance of circuit i.e. every inductor has some resistance and energy radiate away from system in the form of electromagnetic energy.

TRANSFORMER

- A transformer works on the principle of mutual induction.
- It is a static device that is used to increase or decrease the voltage in an AC circuit.
- On a laminated iron core two insulated copper coils called primary and secondary are wound.
- Primary is connected to an alternating source of emf, By mutual induction, an emf is induced in the secondary.

VOLTAGE RATIO:

- If V_1 and V_2 are the primary and secondary voltages in a transformer, N_1 and N_2 are the number of turns in the primary and secondary coils of the transformer, then $\frac{V_1}{V_2} = \frac{N_1}{N_2}$.
- In a transformer the *voltage per turn* is the same in primary and secondary coils.
- The ratio N_2/N_1 is called transformation ratio.
- The voltage ratio is the same as the ratio of the number of turns on the two coils.

CURRENT RATIO:

- If the primary and secondary currents are I_1 and I_2 respectively, then for ideal transformer $\frac{V_2}{V_1} = \frac{I_1}{I_2} = \frac{N_2}{N_1}$.

- In an ideal transformer the *ampere turns* are the same in primary and secondary coils.
- If $N_s > N_P$ voltage is stepped up, then the transformer is called step - up transformer.
- If $N_s < N_P$ voltage is stepped down, then the transformer is called step - down transformer.
- In step - up transformer, $V_S > V_P$ and $I_S < I_P$
- In step - down transformer, $V_S < V_P$ and $I_S > I_P$
- Frequency of input a.c is equal to frequency of output a.c
- Transformation of voltage, is not possible with d.c

Efficiency of transformer (η)

Effeiciency is defined as the ratio of output power and input power.

$$\text{Efficiency} = \frac{\text{output power}}{\text{input power}}$$

$$\text{i.e., } \eta\% = \frac{P_{out}}{P_{in}} \times 100 = \frac{V_s i_s}{V_P i_P} \times 100$$

- For an ideal transformer $P_{out} = P_{in}$ so $\eta = 100\%$ (But efficiency of practical transformer lies between 70% - 90 %)

For practical transformer $P_{in} = P_{out} + P_{losses}$

$$\text{So } \eta = \frac{P_{out}}{Pin} \times 100 = \frac{P_{out}}{(P_{out} + P_L)} \times 100 = \frac{(P_{in} - P_L)}{P_{in}} \times 100$$

- In an ideal transformer the input power is equal to the output power. $V_1 I_1 = V_2 I_2$

The efficiency of an ideal transformer is 100%.

LOSSES IN A TRANSFORMER:

- The losses in a transformer are divided in to two types. They are *copper losses* and *iron losses*.
- The loss of energy that occurs in the copper coils of the transformer (i.e. primary and secondary coils) is called 'copper losses'. These are nothing but joule heating losses where electrical energy is converted in to heat energy.

The loss of energy that occurs in the iron core of the transformer (i.e. hysteresis loss and eddy current loss) is called 'iron losses'.

MINIMIZING THE LOSSES IN A TRANSFORMER:

- The core of a transformer is laminated and each lamination is coated with a paint of insulation to reduce the 'eddy current' losses.
- By choosing a material with narrow 'hysteresis loop' for the core, the hysteresis losses are minimized.

Uses of transformer:

- A transformer is used in almost all ac operations, e.g
- In voltage regulators for TV, refrigerator, computer, air conditioner etc.
- In the induction furnaces.
- Step down transformer is used for welding purposes.

- In the transmission of ac over long distnace.
- Step down and step up transformers are used in electical power distribution.
- Audio Frequency transformers are used in radiography, television, radio, telephone etc.
- Radio frequency transformers are used in radio communication.

WE: 22

A transformer having efficiency 90% is working on 100 V and at 2.0 kW power. If the current in the secondary coil is 5A, calculate (i) the current in the primary coil and (ii) voltage across the secondary coil.

Sol. Here $\eta = 90\% = \frac{9}{10}, I_s = 5A$, $E_p = 100V$,

(i) $E_p I_p = 2kW = 2000W$

$$I_p = \frac{2000}{E_p} \text{ or } I_p = \frac{2000}{100} = 20A$$

(ii) $$\eta = \frac{\text{Output power}}{\text{Input power}} = \frac{E_s I_s}{E_p I_p} \text{ or } E_s I_s = \eta \times E_p I_p$$

$$= \frac{9}{10} \times 2000 = 1800W$$

$$\therefore E_s = \frac{1800}{I_s} = \frac{1800}{5} = 360 \text{ volt}$$

WE:23

A step up transformer operates on a 230 V line and a load current of 2 ampere. The ratio of the primary and secondary windings is 1 : 25. What is the current in the primary ?

Sol. Using the relation

$$\frac{N_P}{N_S} = \frac{I_S}{I_P}; I_P = \frac{N_S I_S}{N_P}$$

Here $N_p / N_s = 1/25$ (or) $N_s / N_p = 25/1 = 25$ and $I_S = 2A$

Current in primary,

$I_P = 25 \times 2 = 50A$

SKIN EFFECT:

- A direct current flows uniformly throughout the cross section of the conductor.
- An alternaitng current, on the other hand, flows mainly along the surface of the conductor. This effect is known as skin effect.
- When alternating current flows through a conductor, the flux changes in the inner part of the conductor are higher.
- Therefore, the inductance of the inner part is higher than that of the outer part. Higher the frequency of alternating current, more is the skin effect.
- The depth upto which ac current flows through a wire is called skin depth(δ).

$$R = \frac{V_R^2}{P_R} \Rightarrow R \alpha V_R^2$$

$(V_R = \text{rated voltage, } P_R = \text{rated power})$

CONCEPTUAL QUESTIONS

INSTANTANEOUS, PEAK,R.M.S & AVERAGE VALUES OF A.C AND A.V

1. **In an ac circuit the current**
   1) is in phase with the voltage 2) leads the voltage
   3) lags the voltage 4) any of the above depending on the circumstances
2. **The average e.m.f during the positive half cycle of an a.c. supply of peak value E_0 is**
   1) E_0/π 2) $E_0/\sqrt{2}$ 3) $E_0/2\pi$ 4) $2E_0/\pi$
3. **Alternating current is transmitted to distant places at**
   1) high voltage and low current 2) high voltage and high current
   3) low voltage and low current 4) low voltage and high current
4. **In case of a.c circuit, Ohm's law holds good for**
   a) Peak values of voltage and current
   b) Effective values of voltage and current
   c) Instantaneous values of voltage and current
   1) only a is true 2) only a and b are true 3) only c is true 4) a, b and c are true
5. **In case of AC circuits the relation V = i Z, where Z is impedance, can directly applied to**
   1) peak values of voltage and current only 2) rms values of voltage and current only
   3) instantaneous values of voltage and current only 4) both 1 and 2 are true
6. **Alternating current can not be measured by direct current meters, because**
   1) alternating current can not pass through an ammeter
   2) the average value of current for complete cycle is zero
   3) some amount of alternating current is destroyed in the ammeter
   4) peak value of current is zero
7. **The r.m.s. value of potential due to superposition of given two alternating potentials $E_1 = E_0 \sin \omega t$ and $E_2 = E_0 \cos \omega t$ will be**
   1) E_0 2) $2E_0$ 3) $E_0\sqrt{2}$ 4) Zero
8. **If the instantaneous values of current is $I = 2\cos(\omega t + \theta)$ A in a circuit, the r.m.s. value of current in ampere will be**
   1) 2 2) $\sqrt{2}$ 3) $2\sqrt{2}$ 4) zero
9. **If a capacitor is connected to two different A.C. generators, then the value of capacitive reactance is**
   1) directly proportional to frequency 2) inversely proportional to frequency
   3) independent of frequency 4) inversely proportional to the square of frequency
10. **In general in an alternating current circuit**
    1) the average value of current is zero 2) the average value of square of the current is zero
    3) average power dissipation is zero 4) the phase difference between voltage and current is zero

A.C ACROSS R-L,R-C,L-C & L-C-R SERIES CIRCUIT

11. **The magnitude of induced e.m.f in an LR circuit at break of circuit as compared to its value at make of circuit will be**
    1) less 2) more
    3) some times less and some times more 4) nothing can be said

12. **The emf and current in a circuit are such that $E = E_0 \sin \omega t$ and $I = I_0 \sin(\omega t - \theta)$. This AC circuit contains.**

1) R and L 2) R and C 3) only R 4) only C

13. **The correct graph between the resistance of a conductor with frequency is**

1) 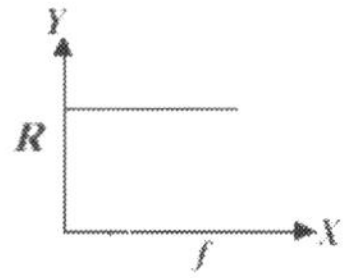

2) 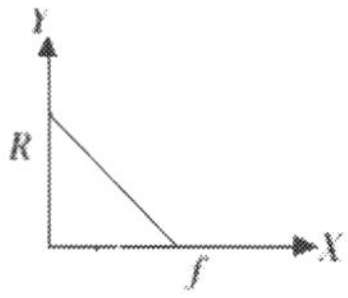

3) 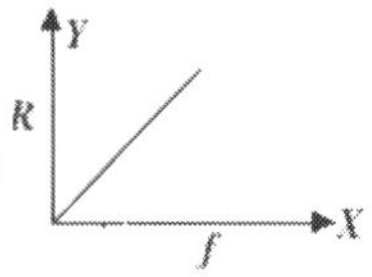

4) 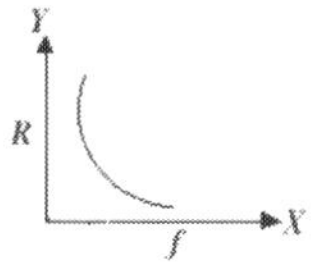

14. **Same current is flowing in two alternating circuits. The first circuit contains only inductance and the other contains only a capacitor. If the frequency of the e.m.f. is increased, the current will**

1) increase in first circuit and decrease in the other
2) increase in both circuits
3) decrease in both circuits
4) decrease in first circuit and increase in the other

15. **When an a.c source is connected across a resistor**

1) The current leads the voltage in phase
2) The current lags behind the voltage in phase
3) The current and voltage are in same phase
4) The current and voltage are out of phase

16. **The phase angle between current and voltage in a purely inductive circuit is**

1) zero 2) π 3) $\pi/4$ 4) $\pi/2$

17. **Ratio of impedence to capacitive reactance has**

1) no units 2) ohm 3) ampere 4) tesla

18. **An inductor coil having some resistance is connected to an AC source. Which of the following have zero average value over a cycle**

1) induced emf in the inductor only
2) current only
3) both 1 and 2
4) neither 1 nor 2

19. **The current does not rise immediately in a circuit containing inductance**

1) because of induced emf
2) because of high voltage drop
3) both 1 and 2
4) because of joule heating

20. **In an AC circuit containing only capacitance the current**

1) leads the voltage by 180º
2) lags the voltage by 90º
3) leads the voltage by 90º
4) remains in phase with the voltage

21. **A bulb is connected first with dc and then ac of same voltage. Then it will shine brightly with**

1) AC 2) DC 3) Equally with both 4) Brightness will be in ratio 1/14

22. **A capacitor of capacity C is connected in A.C. circuit. If the applied emf is $V = V_0 \sin \omega t$, then the current is**

1) $I = \frac{V_0}{L\omega} \sin \omega t$
2) $I = \frac{V_0}{\omega C} \sin\left(\omega t + \frac{\pi}{2}\right)$
3) $I = V_0 C\omega \sin \omega t$
4) $I = V_0 C\omega \sin\left(\omega t + \frac{\pi}{2}\right)$

23. **At low frequency a condenser offers**
1) high impedance
2) low impedance
3) zero impedance
4) impedance of condenser is independent of frequency

24. **Statement (A) : The reactance offered by an inductance in A.C. circuit decreases with increase of AC frequency.**
Statement (B) : The reactance offered by a capacitor in AC circuit increases with increase of AC frequency.
1) A is true but B is false
2) Both A and B are true
3) A is false but B is true
4) Both A and B are false

25. **Statement (A) : With increase in frequency of AC supply inductive reactance increases.**
Statement (B) : With increase in frequency of AC supply capacitive reactance increase
1) A is true but B is false
2) Both A and B are true
3) A is false but B is true
4) Both A and B are false

26. **In an A.C circuit having resistance and capacitance**
1) emf leads the current
2) current lags behind the emf
3) both the current and emf are in phase
4) current leads the emf.

27. **Select the correct options among the following: In an R-C circuit**
a) instantaneous A.C is given by $I = I_0 \sin(wt + \phi)$
b) the alternating current in the circuit leads the emf by a phase angle ϕ.
c) Its impedance is $\sqrt{R^2 + (\omega c)^2}$
d) Its capacitive reactance is ω c
1) a, b are ture
2) b, c, d are true
3) c, d are true
4) a, c are true

28. **If the frequency of alternating e.m.f. is f in L-C-R circuit, then the value of impedance Z will change with log (frequency) as**
1) increases
2) increases and then becomes equal to resistance, then it will start decreasing
3) decreases and when it becomes minimum equal to the resistance then it will start increasing
4) go on decreasing

29. **An inductance and resistance are connected in series with an A.C circuit. In this circuit**
1) the current and P.d across the resistance lead P.d across the inductance by $\frac{\pi}{2}$
2) the current and P.d across the resistance lags behind the P.d across the inductance by angle $\frac{\pi}{2}$
3) The current across resistance leads and the P.d across resistance lags behind the P.d across the inductance by $\frac{\pi}{2}$
4) the current across resistance lags behind and the P.d across the resistance leads the P.d across the inductance by $\frac{\pi}{2}$

30. **An LCR circuit is connected to a source of alternating current. At resonance, the applied voltage and the current flowing through the circuit will have a phase difference of**
1) $\pi/4$
2) zero
3) π
4) $\pi/2$

31. **The incorrect statement for L-R-C series circuit is**

1) The potential difference across the resistance and the appleid e.m.f. are always in same phase

2) The phase difference across inductive coil is 90^0

3) The phase difference between the potential difference across capacitor and potential difference across inductance is 90^0

4) The phase difference between potential difference across capacitor and potential difference across resistance is 90^0

32. **In series L - C - R resonant circuit, to increase the resonant frequency**

1) L will have to be increased
2) C will have to be increased
3) LC will have to be decreased
4) LC will have to be increased

33. **If in a series L - C - R ac circuit, the voltages across R, L, C are V_1, V_2, V_3 respectively. Then the voltage of applied AC source is always equal to**

1) $V_1+V_2+V_3$
2) $\sqrt{V_1^2+(V_2+V_3)^2}$
3) $V_1-V_2-V_3$
4) $\sqrt{V_1^2+(V_2-V_3)^2}$

34. **In non-resonant circuit, the nature of circuit for frequencies greater than the resonant frequency is**

1) resistive
2) capacitive
3) inductive
4) both 1 and 2

35. **The phase difference between voltage and current in an LCR series circuit is**

1) zero always
2) $\pi/4$ always
3) π
4) between 0 and $\pi/2$

36. **In an LCR a.c circuit at resonance, the current**

1) Is always in phase with the voltage
2) Always leads the voltage
3) Always lags behind the voltage
4) May lead or lag behind the voltage

37. **An inductance L and capacitance C and resistance R are connected in series across an AC source of angular frequency ω. If $\omega^2 > \frac{1}{LC}$ then**

1) emf leads the current
2) both the emf and the current are in phase
3) current leads the emf
4) emf lags behind the current

38. **Consider the following two statements A and B and identify the correct answer.**

A) At resonance of *L - C - R* series circuit, the reactance of circuit is minimum.

B) The reactance of a capacitor in an A.C circuit is similar to the resistance of a capacitor in a D.C. circuit

1) A is true but B is false
2) Both A and B are true
3) A is false but B is true
4) Both A and B are false

39. **Choose the wrong statement of the following.**

1) The peak voltage across the inductor can be less than the peak voltage of the source in an LCR circuit

2) In a circuit containing a capacitor and an ac source the current is zero at the instant source voltage is maximum

3) When an AC source is connected to a capacitor,then the rms current in the circuit gets increased if a dielectric slab is inserted into the capacitor.

4) In a pure inductive circuit emf will be in phase with the current.

40. **The essential difference between a d.c. dynamo and an a.c. dynamo is that**

1) a.c. has an electromagnet but d.c. has a permanent magnet
2) a.c. will generate a higher voltage
3) a.c.has slip rings but the d.c. has a commutator
4) a.c. dynamo has a coil wound on soft iron, but the d.c. dynamo has a coil wound on copper

41. The unit of impedence is

1) ohm 2) mho 3) ampere 4) volt

42. The power factor of a.c. circuit having L and R connected in series to an a.c. source of angular frequency ω is given by

1) $\frac{\sqrt{R^2+\omega^2L^2}}{R}$ 2) $\frac{R}{\sqrt{R^2+\omega^2L^2}}$ 3) $\frac{\omega L}{R}$ 4) $\frac{R}{\omega L}$

43. The capacitor offers zero resistance to

1) D.C. only 2) A.C. & D.C. 3) A.C. only 4) neither A.C. nor D.C.

44. Power factor is defined as

1) apparent power/true power
2) true power/apparent power
3) true power (apparent power)^2
4) true power x apparent power

TRANSFORMER

45. The core of a transformer is laminated so that

1) energy loss due to eddy currents may be reduced
2) rusting of the core may be prevented
3) change in flux may be increased
4) ratio of voltage in the primary to that in the secondary may be increased

46. A step up transformer is used to

1) increase the current and increase the voltage
2) decrease the current and increase the voltage
3) increase the current and decrease the voltage
4) decrease the current and decrease the voltage

47. A transformer changes the voltage

1) without changing the current and frequency
2) without changing the current but changes the frequency
3) without changing the frequency but changes the current
4) without changing the frequency as well as the current

48. A step up transformer is connected on the primary side to a rechargable battery which can deliver a large current. If a bulb is connected in the secondary, then

1) the bulb will glow very bright
2) the bulb will get fused
3) the bulb will glow, but with less brightness
4) the bulb will not glow

49. The ratio of primary voltage to secondary voltage in a transformer is 'n'. The ratio of the primary current to secondary current in the transformer is

1) n 2) $1/n$ 3) n^2 4) $1/n^2$

50. In a step down transformer, the number of turns in the primary is always

1) greater than the number of turns in the secondary
2) less than the number of turns in the secondary
3) equal to the number of turns in the secondary
4) either greater than or less than the number of turns in the secondary

51. The transformer ratio of a step up transformer is

1) greater than one
2) less than one
3) less than one and some times greater than one
4) greater than one and some times less than one

52. A stepup transformer develops 400V in secondary coil for an input of 200V A.C. Then the type of transformer is

1) Steped down 2) Steped up 3) Same 4) Same but with reversed direction

53. **Assertion(A) : If changing current is flowing through a machine with iron parts, results in loss of energy.**

Reason(R): Changing magnetic flux through an area of the iron parts causes eddy currents.

1)Both A and R are individually true and R is the correct explanation of A

2)Both A and R are individually true but R is not the correct explanation of A

3)A is true but R is false   4)Both A and R are false

54. **Transformers are used in**

1) d.c circuits only   2) a.c. circuits only

3) Both a.c and d.c circuits   4) Integrated circuits.

55. **The magnitude of the e.m.f. across the secondary of a transformer does not depend on**

1) The number of the turns in the primary   2) The number of the turns in the secondary

3) The magnitude of the e.m.f applied across the primary

4) The resistance of the primary and the secondary

56. **For an ideal transformer ratio of output to the input power is always**

1) greater than one   2) equal to one   3) less than one   4) zero

57. **Consider the following two statements A and B and identify the correct answer.**

A) In a transformer a large alternating current at low voltage can be transformed into a small alternating current at high voltage

B) Energy in current carrying coil is stored in the form of magnetic field.

1) A is true but B is false   2) Both A and B are true

3) A is false but B is true   4) Both A and B are false

58. **Statement (A) : Flux leakage in a transformer can be minimized by winding the primary and secondary coils one over the other.**

Statement (B) : Core of the transformer is made of soft iron

1) A is true but B is false   2) Both A and B are true

3) A is false but B is true   4) Both A and B are false

59. **Statement (A) : In high current low voltage windings of a transformer thick wire is used to minimize energy loss due to heat produced**

Statement (B) : The core of any transformer is laminated so as to reduce the energy loss due to eddy currents

1) A is true but B is false   2) Both A and B are true

3) A is false but B is true   4) Both A and B are false

60. **Statement (A) : Step up transformer converts low voltage, high current to high voltage, low current**

Statement (B) : Transformer works on both ac and dc

1) A is true but B is false   2) Both A and B are true

3) A is false but B is true   4) Both A and B are false

61. **To reduce the iron losses in a transformer, the core must be made of a material having**

1) low permeability and high resistivity   2) high permeability and high resistivity

3) low permeability and low resistivity   4) high permeability and low resistivity

62. **Maximum efficiency of a transformer depends on**

1) the working conditions of technicians.   2) weather copper loss =1/2 x iron loss

3) weather copper loss = iron loss   4) weather copper loss =2 x iron loss

63. **For a LCR series circuit with an A.C. source of angular frequency ω**

1) circuit will be capacitive if $\omega > \frac{1}{\sqrt{LC}}$
2) circuit will be inductive if $\omega = \frac{1}{\sqrt{LC}}$
3) power factor of circuit will be unity if capacitive reactance equals inductive reactance
4) current will be leading voltage if $\omega > \frac{1}{\sqrt{LC}}$

64. **The value of current in two series L C R circuits at resonance is same when connected across a sinusoidal voltage source. Then**

1) both circuits must be having same value of capacitance and inductance
2) in both circuits ratio of L and C will be same
3) for both the circuits X_L / X_C must be same at that frequency
4) both circuits must have same impedance at all frequencies

65. **When an AC source of emf $e = E_0 \sin(100t)$ is connected across a circuit, the phase difference betwen the emf e and the current i in the circuit is observed to be $\frac{\pi}{4}$ ahead,If the circuit consists possibly of $R-C$ or $R-L$ or $L-C$ in series, find the relationship between the two elements:**

1) $R = 1k\Omega,\ C = 10\mu F$
2) $R = 1k\Omega,\ C = 1\mu F$
3) $R = 1k\Omega,\ L = 10H$
4) $R = 1k\Omega,\ L = 1H$

66. **An AC voltage source of variable angular frequency ω and fixed amplitude V_0 is connected in series with a capacitance C and an electric bulb of resistance R (inductance zero). When ω is increased**

1) the bulb glows dimmer
2) the bulb glows brighter
3) total impendance of the circuit is unchanged
4) total impendance of the circuit increases

Assertion & Reason

1) Both Assertion and Reason are true and Reason is the correct explanation of Assertion.
2) Both Assertion and Reason are true but Reason is not the correct explanation of Assertion.
3) Assertion is true but Reason is false
4) Assertion is false but Reason is true

67. **Assertion (A): The average value of $<\sin^2 \omega t>$ is zero.**
Reason (R): The average value of function $F(t)$ over a period T is $<F(t)> = \frac{1}{T}\int_0^T F(t)dt$

68. **Assertion (A): If current varies sinusoidally the average power consumed in a cycle is zero.**
Reason (R): If current varies sinusoidally the average power consumed is zero

69. **Assertion (A) : The power consumed in an electric circuit is never negative**
Reason (R) : The average power consumed in an electric circuit is $P = \frac{V^2}{R} = I^2R$

70. **Assertion (A): The inductive reactance limits the current in a purely inductive circuit in the same way as the resistance circuit.**
Reason (R): The inductive reactance is directly proportional to the inductance and to the frequency of the varying current.

71. **Assertion (A) : An ac emf which oscillates symmetrically about zero, the current it sustains also oscillates symmetrically about zero.**

Reason (R): In any circuit element, current is always in the phase with voltage

72. **Assertion (A): A lamp is connected in series with a capacitor and ac source connected across their terminals consequently current flow in the circuit and the lamp will shine.**

Reaosn(R): capacitor block dc current and allow ac current

73. **Assertion (A): An electric lamp is connected in series with a long solenoid of copper with air core and then connected to AC source. If an iron rod is inserted in solenoid the lamp will become dim.**

Reason (R): If iron rod is inserted in solenoid, the induction of solenoid increases.

74. **An inductor, capacitor and resistance connected in series. The combination is connecte across AC source.**

Assertion (A): Peak current through each remains same

Reason (R) : Average power delivered by source is equal to average power consumed by resistance.

75. **Assertion (A): when frequency is greater than resonance frequency in a series LCR circuit, it will be an inductive circuit.**

Reason (R): Resultant voltage will lead the current

76. **Assertion (A): Maximum power is dessipated in a circuit (through R) in resonance**

Reason (R) : At resonance in a series LCR circuit, the voltage across indcutor and capacitor are out of phase.

KEY CONCEPTUAL

1) 4	2) 4	3) 1	4) 2	5) 4	6) 2	7) 1	8) 2	9) 2	10) 1
11) 2	12) 1	13) 1	14) 4	15) 3	16) 4	17) 1	18) 3	19) 3	20) 3
21) 3	22) 4	23) 1	24) 4	25) 1	26) 4	27) 1	28) 3	29) 2	30) 2
31) 3	32) 3	33) 4	34) 3	35) 4	36) 1	37) 1	38) 1	39) 4	40) 3
41) 1	42) 2	43) 4	44) 2	45) 1	46) 2	47) 3	48) 4	49) 2	50) 1
51) 1	52) 2	53) 1	54) 2	55) 4	56) 2	57) 2	58) 4	59) 2	60) 1
61) 2	62) 3	63) 3	64) 3	65) 1	66) 2	67) 4	68) 4	69) 1	70) 2
71) 4	72) 1	73) 1	74) 2	75) 1	76) 1				

HINTS

67. $< \sin^2 \omega t >= 1/2$

68. $< P >= \frac{1}{2} i_m^2 R$

69. I is scalar in Joules heating effect is independent an direction of current.

70. $I = \frac{V}{X_L} \& i = \frac{V}{R} \quad X_L = \omega L = 2\pi v L$

71. In inductor current lags the voltage by $\frac{\pi}{2}$

In capcitor current leads the voltage by $\frac{\pi}{2}$

72. $X_C = \dfrac{1}{2\pi fC}$

for dc $f = 0$ then $X_L = \infty$

for ac $f \neq 0$ then $X_C = finite$

73. $L \propto \mu_r$

more voltage is present across inductor

so less voltage across bulb

74. In series current is same, inductor and capacitor does not consume power

75. At resonance $X_L = X_C$ and frequency $f_0 = \dfrac{1}{2\pi}\sqrt{\dfrac{1}{LC}}$ If $f > f_0$ then $X_L > X_C$, so it will be an inductive circuit. AC current must lag AC voltage.

76. At resonance $P = I_{max}^2 R$ and V_L and V_C are out of phase. I_{max} is due to $Z_{min} = R$ which is due to out of phase of V_L and V_C.

LEVEL - I (CW)

INSTANTANEOUS, PEAK,R.M.S & AVERAGE VALUES OF A.C AND A.V

1. The r.m.s. value of an a.c. of 50 Hz is 10 A. The time taken by the alternating current in reaching from zero to maximum value and the peak value of current will be

1) 2×10^{-2} sec and 14.14 A
2) 1×10^{-2} sec and 7.07 A
3) 5×10^{-3} sec and 7.07 A
4) 5×10^{-3} sec and 14.14 A

2. An inductor has a resistance R and inductance L. It is connected to an A.C. source of e.m.f E_V and angular frequency ω, then the current I_v in the circuit is

1) $\dfrac{E_V}{\omega L}$
2) $\dfrac{E_V}{R}$
3) $\dfrac{E_V}{\sqrt{R^2+\omega^2L^2}}$
4) $\sqrt{\left(\dfrac{E_V}{R}\right)^2+\left(\dfrac{E_V}{\omega L}\right)^2}$

3. The peak voltage of 220 Volt AC mains (in Volt) is

1) 155.6
2) 220.0
3) 311
4) 440.0

4. The peak value of A.C. is $2\sqrt{2}$A. It's apparent value will be

1) 1A
2) 2A
3) 4A
4) zero

5. Alternating current in circuit is given by $I = I_0 \sin 2\pi nt$. Then the time taken by the current to rise from zero to r.m.s. value is equal to

1) 1/2n
2) 1/n
3) 1/4n
4) 1/8n

6. Using an A.C. voltmeter the potential difference in the electrical line in a house is read to be 234 volt. If the line frequency is known to be 50 cycles/second, the equation for the line voltage is

1) $V = 165\ \sin(100\pi\ t)$
2) $V = 331\ \sin(100\pi\ t)$
3) $V = 220\ \sin(100\pi\ t)$
4) $V = 440\ \sin(100\pi\ t)$

7. A mixer of 100Ω resistance is connected to an A.C. source of 200V and 50 cycles/sec. The value of average potential difference across the mixer will be

1) 308V
2) 264V
3) 220V
4) zero

A.C ACROSS PURE RESISTOR, INDUCTOR & CAPACITOR

8. **The equation of an alternating voltage is E=220 $\sin(\omega t + \pi/6)$ and the equation of the current in the circuit is I=10 $\sin(\omega t - \pi/6)$. Then the impedance of the circuit is**

1) 10 ohm   2) 22 ohm   3) 11 ohm   4) 17 ohm

9. **A steady P.D. of 10V produces heat at a rate 'x' in resistor. The peak value of A.C. voltage which will produce heat at rate of x/2 in same resistor is**

1) 5 V   2) $5\sqrt{2}$ V   3) 10 V   4) $10\sqrt{2}$ V

10. **An alternating voltage of $E = 200\sqrt{2}\sin(100t)V$ is connected to a condenser of 1 μF through an A.C. ammeter. The reading of the ammeter will be**

1) 10 mA   2) 40 mA   3) 80 mA   4) 20 mA

11. **The inductance of a coil is 0.70 henry. An A.C. source of 120 volt is connected in parallel with it. If the frequency of A.C. is 60Hz, then the current which is flowing in inductance will be**

1) 4.55 A   2) 0.355 A   3) 0.455 A   4) 3.55 A

TRANSFORMER

12. **A transformer steps up an A.C. voltage from 230 V to 2300 V. If the number of turns in the secondary coil is 1000, the number of turns in the primary coil will be**

1) 100   2) 10,000   3) 500   4) 1000

13. **The transformer ratio of a transformer is 5. If the primary voltage of the transformer is 400 V, 50 Hz, the secondary voltage will be**

1) 2000 V, 250 Hz   2) 80 V, 50 Hz   3) 80 V, 10 Hz   4) 2000 V, 50 Hz

14. **A step-up transformer works on 220V and gives 2 A to an external resistor. The turn ratio between the primary and secondary coils is 2:25. Assuming 100% efficiency, find the secondary voltage, primary current and power delivered respectively**

1) 2750 V, 25 A, 5500 W   2) 2750 V, 20 A, 5000 W

3) 2570 V, 25 A, 550 W   4) 2750 V, 20 A, 55 W

A.C ACROSS L-R, L-C & L-C-R SERIES CIRCUITS

15. **A coil of self - inductance $\left(\frac{1}{\pi}\right)$ H is connected in series with a 300 Ω resistance. A voltage of 200V at frequency 200Hz is applied to this combination. The phase difference between the voltage and the current will be**

1) $\tan^{-1}\left(\frac{4}{3}\right)$ 2) $\tan^{-1}\left(\frac{3}{4}\right)$ 3) $\tan^{-1}\left(\frac{1}{4}\right)$ 4) $\tan^{-1}\left(\frac{5}{4}\right)$

16. **A condenser of $10\mu F$ and an inductor of 1H are connected in series with an A.C. source of frequency 50Hz. The impedance of the combination will be (take $\pi^2 = 10$)**

1) zero   2) Infinity   3) 44.7 Ω 4) 5.67 Ω

17. **A 100 km telegraph wire has capacity of 0.02 $\mu F/km$, if it carries an alternating current of frequency 5 kHZ. The value of an inductance required to be connected in series so that the impedence is minimum.**

1) 50.7mH   2) 5.07mH   3) 0.507mH   4) 507mH

18. In an LCR series circuit the rms voltages across R, L and C are found to be 10 V, 10 V and 20 V respectively. The rms voltage across the entire combination is

1) 30 V 2) 1V 3) $20V$ 4) $10\sqrt{2}\,V$

19. In the circuit shown, a 30V d.c. source gives a current 2.0 A as recorded in the ammeter A and 30V a.c. source of frequency 100Hz gives a current 1.2A. The inductive reactance is

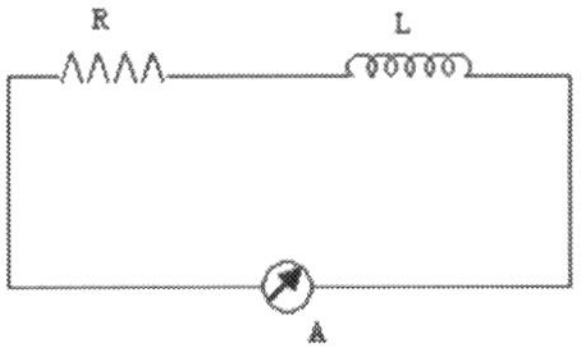

1) 10 ohm 2) 20 ohm 3) $5\sqrt{34}$ ohm 4) 40 ohm

20. A choke coil has negligible resistance. The alternating potential drop across it is 220 volt and the current is 5mA. The power consumed is

1) $220\times\frac{5}{1000}$ W 2) $\frac{220}{5}$ W 3) zero 4) 2.20 x 5W

21. In an A.C. circuit, the instantaneous values of e.m.f. and current are $E = 200 \sin 314t$ volt and $I = \sin(314t + \pi/3)$ ampere then the average power consumed in watts is

1) 200 2) 100 3) 0 4) 50

22. In a black box of unkown elements (L, C or R or any other combination) an AC voltage $E = E_0 \sin(\omega t + \phi)$ is applied and current in the circuit was found to be $i = i_0 \sin(\omega t + \phi + \pi/4)$. Then the unknown elements in the box may be

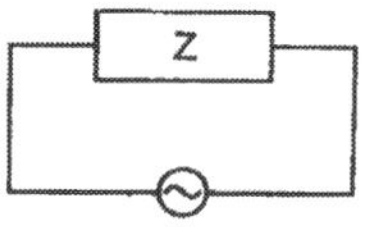

1) only capacitor 2) both inductor and resistor
3) either capacitor, resistor and inductor or only capacitor and resistor
4) only resistor

23. The voltage time (V - t) graph for triangular wave having peak value V_0 is as shown in figure.

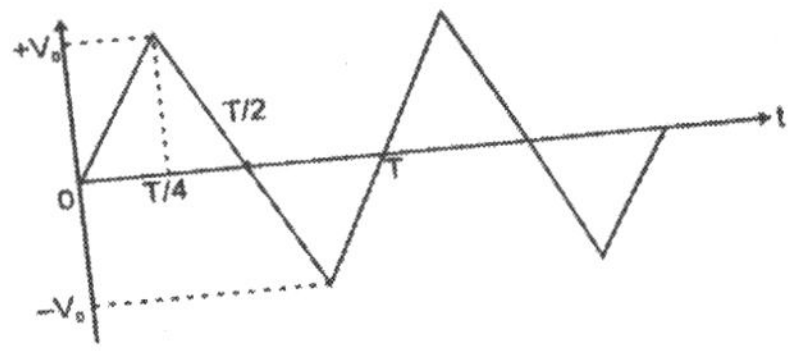

The rms value of V in time interval from $t = 0$ to $\frac{T}{4}$ is

1) $\frac{V_0}{\sqrt{3}}$ 2) $\frac{V_0}{2}$ 3) $\frac{V_0}{\sqrt{2}}$ 4) $2V_o$

KEY LEVEL - I (CW)

1) 4	2) 3	3) 3	4) 2	5) 4	6) 2	7) 4	8) 2	9) 3	10) 4
11) 3	12) 1	13) 4	14) 1	15) 1	16) 1	17) 3	18) 4	19) 2	20) 3
21) 4	22) 3	23) 1							

HINTS LEVEL - I (CW)

1. $i_0 = \sqrt{2} i_{rms} \quad , T = \frac{1}{f}, t = \frac{T}{4}$

2. $i = \frac{E_0}{\sqrt{R^2 + X_L^2}}, \; X_L = L\omega$

3. $V_0 = \sqrt{2}.V_{r.m.s.} = \sqrt{2} \times 200 = 311\, volt$

4. $I_{rms} = \frac{I_0}{\sqrt{2}}$

5. $t = \frac{T}{4} = \frac{1}{4f}$

6. $E = E_0 \sin \omega \text{ t}$; voltage read is r.m.s. value $E_0 = \sqrt{2} \times 234V = 331$ volt and
$\omega \text{ t} = 2\pi\, n\, t = 2\pi \times 50 \times t = 100\,\pi\, t$ Thus, the eqn of line voltage is given by $V = 331\ \sin(100\pi\ t)$

7. For one complete rotation, average voltage is zero

8. $Z = \frac{E_0}{I_0}$

9. $\frac{v^2}{R} = x, \frac{v_1^2}{R} = \frac{x}{2} \Rightarrow v_1 = \frac{v}{\sqrt{2}}$
$\therefore$ in the second case $V_{rms} = V_1$
$\therefore$ $V_0 = \sqrt{2} V_1$

10. $I_{rms} = \frac{E_{rms}}{X_C} = \frac{E_0 \omega C}{\sqrt{2}}$

11. $X_L = 2\pi fl = 6.28 \times 60 \times 0.70 = 263.76\Omega$
$I = \frac{V}{X_L} = \frac{120}{263.76} = 0.455A$

12. $\frac{n_s}{n_p} = \frac{V_s}{V_p}$

13. Frequency remains same. $\frac{V_S}{V_P} = 5$

14. $\frac{E_s}{E_P} = \frac{N_s}{N_P} = \frac{i_P}{i_s}, \quad P = E_s i_s$

15. $\tan\theta = \frac{2\pi fL}{R}$, $f = \frac{1}{2\pi\sqrt{LC}}$

16. $Z = \left(2\pi fL - \frac{1}{2\pi fC}\right)$

17. $\omega = \frac{1}{\sqrt{LC}} \Rightarrow L = \frac{1}{\omega^2 C} = \frac{1}{(2\pi n)^2 C}$

18. $V = \sqrt{V_R^2 + (V_L - V_C)^2}$

19. When d.c. source, $R = \frac{V}{I} = \frac{30}{2} = 15\Omega$

When a.c. source, $Z = \frac{30}{1.2} = 25\Omega$

$X_L = \sqrt{(25)^2 - (15)^2} = \sqrt{625 - 225} = 20\Omega$

20. Average power is zero

21. $P_{avg} = I_{rms} E_{rms} \cos\phi = \frac{1}{\sqrt{2}} \times \frac{200}{\sqrt{2}} \cos 60°$

50W

22. Here current leads the voltage. So, there is reactance which is capacitive

$\Rightarrow X = X_C - X_L$ *or* $X = X_C$ alone besides R

23. Ans : (a)

$$V = \frac{V_0 t}{T/4} = \frac{4V_0 t}{T}$$

$$V_{rms} = \sqrt{\langle V^2 \rangle} = \frac{4V_0}{T} \left\{ \frac{\int_0^{T/4} t^2 dt}{\int_0^{T/4} dt} \right\} = \frac{V_0}{\sqrt{3}}$$

LEVEL - I (HW)

INSTANTANEOUS, PEAK, R.M.S & AVERAGE VALUES OF A.C AND A.V

1. For a given AC source the average emf during the positive half cycle

1) depends on E_0 2) depends on shape of wave

3) both 1 and 2 4) depends only on peak value of E_0

2. An alternating emf given by $V = V_0 \, Sin \, \omega t$ has peak value 10 volt and frequency 50 Hz. The instantaneous emf at $t = \frac{1}{600} s$ is

1) 10 V 2) $5\sqrt{3}V$ 3) 5 V 4) 1V

3. The equation of A.C. of frequency 75Hz, if it's RMS value is 20A is

1) $I = 20 Sin(150\pi t)$ 2) $I = 20\sqrt{2} Sin(150\pi t)$ 3) $I = \frac{20}{\sqrt{2}} Sin(150\pi t)$ 4) $I = 20\sqrt{2} Sin(75\pi t)$

4. The voltage of an A.C. source varies with time according to the equation $V = 50 \sin 100\pi t \cos 100\pi t$, where 't' is in sec and 'V' is in volt. Then

1) The peak voltage of the source is 100 V 2) The peak voltage of the source is $100/\sqrt{2}V$

3) The peak voltage of the source is 25 V 4) The frequency of the source is 50 Hz

5. The form factor for a sinusoidal A.C. is

1) $2\sqrt{2} : \pi$ 2) $\pi : 2\sqrt{2}$ 3) $\sqrt{2} : 1$ 4) $1 : \sqrt{2}$

6. At resonance the peak value of current in L-C-R series circuit is

1) E_0/R 2) $\frac{E_0}{\sqrt{R^2 + \left(\omega L - \frac{1}{\omega C}\right)^2}}$

3) $\frac{E_0}{\sqrt{2}\sqrt{R^2 + \left(\omega^2 L - \frac{1}{\omega^2 C^2}\right)^2}}$ 4) $\frac{E_0}{\sqrt{2}R}$

7. In an AC circuit, the rms value of the current, I_{rms} is related to the peak current I_0 as

1) $I_{rms} = \frac{1}{\pi} I_0$ 2) $I_{rms} = \frac{1}{\sqrt{2}} I_0$ 3) $I_{rms} = \sqrt{2} I_0$ 4) $I_{rms} = \pi I_0$

8. A voltmeter connected in an A.C circuit reads 220V. It represents,

1) peak voltage 2) RMS voltage

3) Average voltage 4) Mean square voltage

9. If the instantaneous current in a circuit is given by $I = 2\cos(\omega t + \phi)$ A, the rms value of the current is

1) $2A$ 2) $\sqrt{2}A$ 3) $2\sqrt{2}A$ 4) zero

10. The time taken by an AC of 50 Hz in reaching from zero to its maximum value will be

1) 0.5 s 2) 0.005 s 3) 0.05 s 4) 5s

11. A generator produces a voltage that is given by V=240 sin 120t V, where t is in second. The frequency and r.m.s. voltage are respectively

1) 60Hz and 240V 2) 19Hz and 120V 3) 19Hz and 170V 4) 754Hz and 170V

A.C ACROSS PURE RESISTOR, INDUCTOR & CAPACITOR

12. A 220 V, 50 Hz AC supply is connected across a resistor of 50 $k\Omega$. The current at time t second, assuming that it is zero at $t = 0$, is

1) $4.4\sin(314t)mA$ 2) $6.2\sin(314t)mA$ 3) $4.4\sin(157t)mA$ 4) $6.2\sin(157t)mA$

13. A resistance of 20Ω is connected to a source of alternating current rated 110 V, 50 Hz. Then the time taken by the current to change from its maximum value to the r.m.s. value is

1) 2.5×10^{-3} sec 2) 2.5×10^{-2} sec 3) 5×10^{-3} sec 4) 25×10^{-3} sec

14. A condenser of capacity 1pF is connected to an A.C source of 220V and 50Hz frequency. The current flowing in the circuit will be

1) 6.9×10^{-8}A 2) 6.9A 3) 6.9×10^{-6}A 4) zero

15. In a circuit, the frequency is $f = \frac{1000}{2\pi}$ Hz and the inductance is 2 henry, then the reactance will be

1) 200Ω 2) $200\mu\Omega$ 3) 2000Ω 4) $2000\mu\Omega$

TRANSFORMER

16. The transformer ratio of a transformer is 10:1. The current in the primary circuit if the secondary current required is 100 A assuming the transformer be ideal, is

1) 500 A 2) 200 A 3) 1000 A 4) 2000 A

17 The transformer ratio of a transformer is 10:1. If the primary voltage is 440 V, secondary emf is

1) 44 V 2) 440V 3) 4400 V 4) 44000 V

A.C ACROSS L-R, L-C & L-C-R SERIES CIRCUITS

18. The frequency at which the inductive reactance of 2H inductance will be equal to the capacitive reactance of $2\mu F$ capacitance (nearly)

1) 80Hz 2) 40 Hz 3) 60Hz 4) 20Hz

19. In a series LCR circuit $R = 10\Omega$ and the impedance $Z = 20\Omega$. Then the phase difference between the current and the voltage is

1) 60° 2) 30° 3) 45° 4) 90°

20. In an L-C-R series circuit, $R = \sqrt{5}\Omega$, $X_L = 9\Omega$, $X_C = 7\Omega$. If applied voltage in the circuit is 50V then impedance of the circuit in ohm will be

1) 2 2) 3 3) $2\sqrt{5}$ 4) $3\sqrt{5}$

21. In an AC circuit the potential differences across an inductance and resistance joined in series are respectively 16 V and 20 V. The total potential difference across the circuit is

1) 20 V 2) 25.6 V 3) 31.9 V 4) 53.5 V

22. Current in an ac circuit is given by $i = 3\sin\omega t + 4\cos\omega t$ then

1) rms value of current is 5 A

2) mean value of this current in one half period will be $6/\pi$

3) if voltage applied is $V = V_m \sin\omega t$ then the circuit must be containing resistance and capacitance

4) if voltage applied is $V = V_m \sin\omega t$, the circuit may contain resistance and inductance

23. A fully charged capacitor C with initial charge q_0 is connected to a coil of self inductance L at $t = 0$. The time at which the energy is stored equally between the electric and the magnetic fields is

1) $\frac{\pi}{4}\sqrt{LC}$ 2) $2\pi\sqrt{LC}$ 3) $\sqrt{LC}$ 4) $\pi\sqrt{LC}$

LEVEL - I (HW) KEY

1) 3	2) 3	3) 2	4) 3	5) 2	6) 1	7) 2	8) 2	9) 2	10) 2
11) 3	12) 2	13) 1	14) 1	15) 3	16) 3	17) 3	18) 1	19) 1	20) 2
21) 2	22) 3	23) 1							

LEVEL - I (HW) HINTS

1. $E_{av} = \frac{2E_0}{T}\int_0^{T/2} \sin(\omega t)\,dt = \frac{2E_0}{\pi}$

2. $V = 10\sin(100\pi t)$; $t = \frac{1}{600}s$

3. $i = i_0 \sin\omega t = \sqrt{2}\, i_{rms}\, sin(2\pi ft)$

4. $V_0 = \sqrt{2}.V_{r.m.s.}$

5. Form factor $= \frac{\text{rms value}}{\text{avg value over half a cycle}}$

6. Conceptual

7. Conceptual

8. Conceptual

9. $I_{rms} = \frac{I_0}{\sqrt{2}}$

10. $t = \frac{T}{4} = \frac{1}{4f}$

11. $V = V_m \sin\omega t$ compare to given equation, we get

$V_m = 240$ and $\omega = 120$

$f = \frac{\omega}{2\pi} = \frac{120}{6.28} = 19H_2$ and voltage $= \frac{V_m}{\sqrt{2}} = \frac{240}{\sqrt{2}} = 170V$

12. $i = i_0 \sin \omega t$

$\omega = 2\pi f;\ i_0 = \frac{E_0}{R} = \frac{\sqrt{2} \times E_{rms}}{R}$

13. $E = E_0 \cos \omega t$, $i = i_0 \cos(2\pi ft)$

but $i = \frac{i_0}{\sqrt{2}}$

14. $i_{rms} = \frac{E_{rms}}{X_c}$

15. $X_L = \omega L = 2\pi fl = 2\pi \times \frac{1000}{2\pi} \times 2 = 2000\Omega$

16. $\frac{N_S}{N_p} = \frac{I_P}{I_S}$

17. $\frac{N_S}{N_p} = \frac{V_s}{V_P}$

18. $f = \frac{1}{2\pi\sqrt{LC}}$

19. $\cos\phi = \frac{R}{Z}$

20. Impedance, $Z = R + X_C + X_L = \left(\sqrt{5}i - 7j + 9j\right) = \sqrt{5}i + 2j$

$|Z| = \sqrt{5+4} = 3\Omega$

21. $V_{rms} = \sqrt{16^2 + 20^2} = \sqrt{656} \approx 25.6V$

22.. Ans : (c) $i = 5\left(\frac{3}{5} sin\,\omega t + \frac{4}{5} cos\,\omega t\right) = 5\,sin(\omega t + \delta)$

rms value is $\frac{5}{\sqrt{2}}$

Mean value can not be decided.

Here current leads voltage so, it is RC circuit

23. As initially charge is maximum

$q = q_0 \cos \omega t$

$\Rightarrow i = \frac{dq}{dt} = -\omega q_0 \sin \omega t$

Given $\frac{1}{2}Li^2 = \frac{q^2}{2C}$

$$\Rightarrow \frac{1}{2}L\left(\omega q_0 \sin\omega t\right)^2 = \frac{\left(q_0\cos\omega t\right)^2}{2C}$$

But, $\omega = \frac{1}{\sqrt{LC}} \Rightarrow \tan\omega t = 1$

$$\omega t = \frac{\pi}{4} \Rightarrow t = \frac{\pi}{4\omega} = \frac{\pi}{4}\sqrt{LC}$$

LEVEL - II (CW)

INSTANTANEOUS, PEAK,R.M.S & AVERAGE VALUES OF A.C AND A.V

1. **The average current of a sinusoidally varrying alternating current of peak value 5A with initial phase zero, between the instants t = T/8 to t = T/4 is (Where 'T' is time period)**

1) $\frac{10}{\pi}\sqrt{2}A$ 2) $\frac{5}{\pi}\sqrt{2}A$ 3) $\frac{20\sqrt{2}}{\pi}A$ 4) $\frac{10}{\pi}A$

A.C ACROSS L-R, L-C & L-C-R SERIES CIRCUITS

2. **A 100Ω resistance is connected in series with a 4H inductor. The voltage across the resistor is $V_R = 2\sin(1000t)V$. The voltage across the inductor is**

1) $80\sin\left(1000t+\frac{\pi}{2}\right)$ 2) $40\sin\left(1000t+\frac{\pi}{2}\right)$ 3) $80\sin\left(1000t-\frac{\pi}{2}\right)$ 4) $40\sin\left(1000t-\frac{\pi}{2}\right)$

3. **The reading of voltmeter and ammeter in the following figure will respectively be**

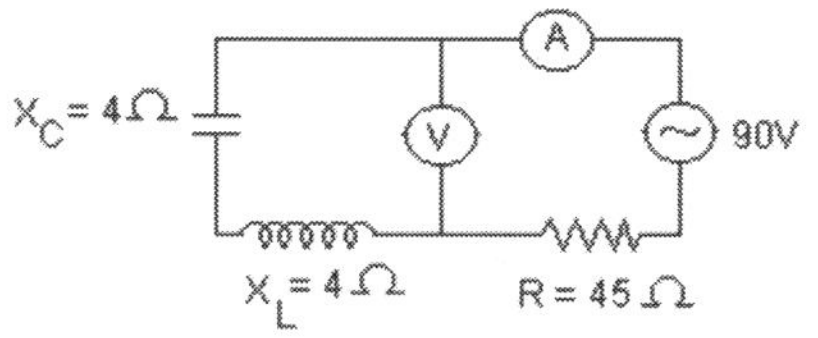

1) 0 and 2A 2) 2A and 0V 3) 2V and 2A 4) 0V and 0A

4. **In the following circuit, the values of current flowing in the circuit at f = 0 and f = ∞ will respectively be**

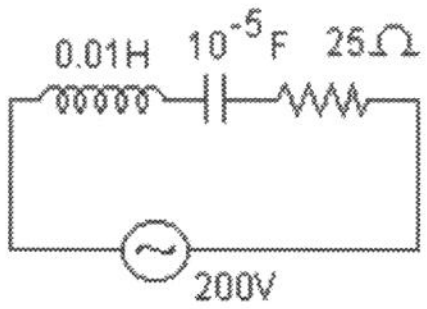

1) 8A and 0A 2) 0A and 0A 3) 8A and 8A 4) 0A and 8A

5. **In the series L-C-R circuit figure the voltmeter and ammeter readings are**

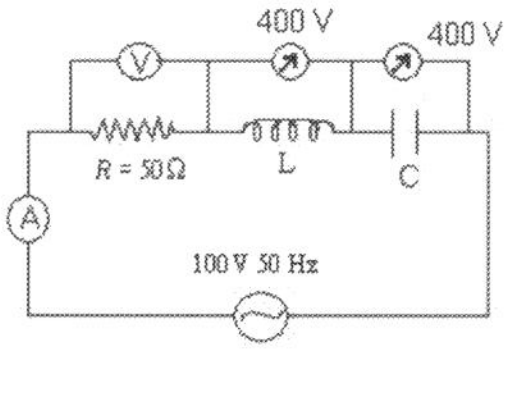

1) V = 100 volt, I = 2 A 2) V = 100 volt, I = 5 A
3) V = 1000 volt, I = 2 A 4) V = 300 volt, I = 1 A

6. The potential difference between the ends of a resistance R is V_R, between the ends of capacitor is $V_C = 2V_R$ and between the ends of inductance is $V_L = 3V_R$. Then the alternating potential of the source in terms of V_R will be

1) $\sqrt{2}V_R$ 2) V_R 3) $\frac{V_R}{\sqrt{2}}$ 4) $5V_R$

7. A 220V, 50Hz a.c. generator is connected to an inductor and a 50Ω resistance in series. The current in the circuit is 1.0A. The P.D. across inductor is

1) 102.2V 2) 186.4V 3) 213.6V 4) 302V

8. The figure shows variation of R, X_L and X_C with frequenc f in a series L, C, R circiut. Then for what frequency point, the circiut is inductive

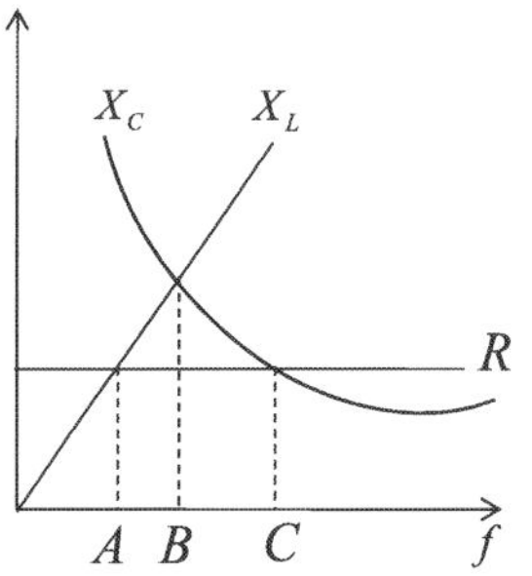

1) A 2) B 3) C 4) All points

9. A constant voltage at different frequencies is applied across a capacitance C as shown in the figure. Which of the following graphs correctly depicts the variation of current with frequency

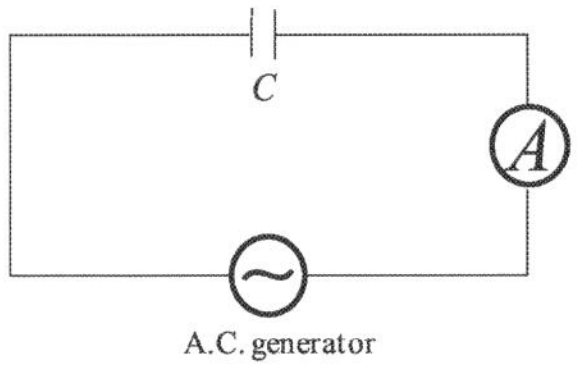

1) 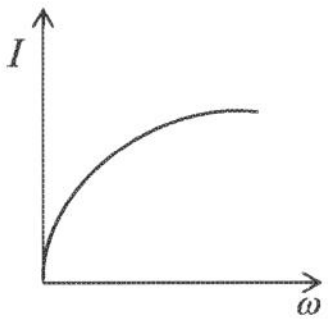

2) 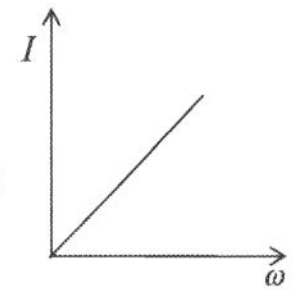

3) 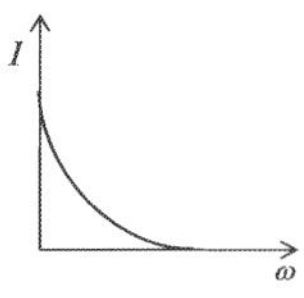

4) 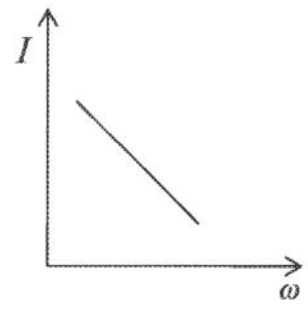

10. In a series $L-C-R$ circuit $R = 200\Omega$ and the voltage and the frequency of the main supply is 220 V and 50Hz respectively. On taking out the capacitance from the circuit the current lags behind the voltage by 30^0. On taking out the inductor from the circuit the current leads the voltage by 30^0. The power dissipated in the $L-C-R$ circuit is

1) 305 W 2) 210 W 3) zero 4) 242 W

11. In a series resonant LCR circuit, the voltage across R is 100V and $R = 1k\Omega$ with $C = 2\mu F$. The resonant frequency ω is 200 rad/s. At resonance the voltage across L is

1) $2.5\times10^{-2}V$ 2) 40 V 3) 250 V 4) $4\times10^{-3}V$

KEY LEVEL - II (CW)

1) 1 2) 1 3) 1 4) 2 5) 1 6) 1 7) 3 8) 3 9) 2 10) 1
11) 2

LEVEL - II (CW) HINTS

1. $<i>=\frac{\int_{T/8}^{T/4} idt}{\int_{T/8}^{T/4} dt}$

2. $i=\frac{(V_0)_R}{R}, V_L=(V_0)_L\sin\left(\omega t+\frac{\pi}{2}\right) and (V_0)_L=X_L i$

3. $I_{rms}=\frac{E_{rms}}{R}=2A$; $V_{rms}=I_{rms}(X_L-X_C)=0$

$\therefore$ circuit is at resonance

4. $I=\frac{E}{Z}=\frac{E}{\sqrt{R^2+\left[2\pi f L-\frac{1}{2\pi f C}\right]^2}}$

5. $I_{r.m.s.}=\frac{V_{r.m.s.}}{Z}=\frac{V_{r.m.s.}}{R}=\frac{100}{50}=2A$

$V=\sqrt{V_R^2+(V_L-V_C)^2}$

6. $\overline{V}_S=\overline{V_B}+\overline{V_C}+\overline{V_L}=V_R\hat{i}-2V_R\hat{j}+3V_R\hat{j}$

$=V_R\hat{i}+V_R\hat{j}$, $|\overline{V}|=\sqrt{2}V_R$

7. $I=\frac{E}{Z}$, $\therefore$ $I=\frac{220}{Z}$, $Z=220\Omega$

$Z^2=R^2+X_L^2$ $\therefore$ $X_L=\sqrt{Z^2-R^2}$

$L=\frac{1}{\omega}\sqrt{Z^2-R^2}$ $\therefore L=\frac{1}{2\pi f}\sqrt{Z^2-R^2}=0.68H$

$\therefore V_L=\omega LI=2\pi\times0.5\times0.68\times1=213.6V$

8. At A : $X_C>X_L$
At B : $X_C=X_L$
At C : $X_C<X_L$

9. For capacitive circuits $X_C=\frac{1}{\omega C}$

$\therefore i=\frac{V}{X_C}V\omega C\Rightarrow i\propto\omega$

10. The given circuit is under resonance as $X_L=X_C$

Hence, power dissipated in the circuit is $P = \frac{V^2}{R} = 242W$

11. At resonance, $\omega L = \frac{1}{\omega C}$

current flowing through the circuit $I = \frac{V_R}{R} = \frac{100}{1000} = 0.1A$

So, voltage across L is given by

$V_L = I X_L = I\omega L$ but $\omega L = \frac{1}{\omega C}$

$V_L = \frac{1}{\omega C} = \frac{0.1}{200 \times 2 \times 10^{-6}} = 250V$

LEVEL - II (HW)

INSTANTANEOUS, PEAK,R.M.S & AVERAGE VALUES OF A.C AND A.V

1. An alternating current 'i' is given by $i = i_0 \sin 2\pi(t/T + 1/4)$. Then the average current in the first one quarter time period is

1) $\frac{2i_0}{\pi}$ 2) $\frac{I_0}{\pi}$ 3) $\frac{I_0}{2\pi}$ 4) $\frac{3I_0}{\pi}$

A.C ACROSS L-R, L-C & L-C-R SERIES CIRCUITS

2. In an LR circuit, R = 10Ω and L = 2H. If an alternating voltage of 120V and 60Hz is connected in this circuit, then the value of current flowing in it will be _______ A (nearly)

1) 0.32 2) 0.16 3) 0.48 4) 0.8

3. The equation of an alternating current is $I = 50\sqrt{2}\sin 400\pi t$ A, then the frequency and the root mean square value of the current are respectively.

1) 200Hz, 50 A 2) 400Hz, $50\sqrt{2}$A 3) 200Hz, $50\sqrt{2}$A 4) 500Hz, 200A

4. A circuit operating at $\frac{360}{2\pi}Hz$ contains a $1\mu F$ capacitor and a 20Ω resistor. The inductor must be added in series to make the phase angle for the circuit zero is

1) 7.7 H 2) 10 H 3) 3.5 H 4) 15 H

5. A resistor R and capacitor C are connected in series across an AC source of rms voltage 5 V. If the rms voltage across C is 3 V then that across R is

1) 1V 2) 2 V 3) 3 V 4) 4 V

6. An *LCR* series circuit containing a resistance of 120Ω has angular resonance frequency 4×10^5 $rad\ S^{-1}$. At resonance the voltage across resistance and inductance are 60V and 40V respectively. Then the values of L and C are respectively.

1) $0.2mH, 1/32\mu F$ 2) $0.4mH, 1/16\mu F$ 3) $0.2mH, 1/16\mu F$ 4) $0.4mH, 1/32\mu F$

7. The natural frequency of an LC - circuit is 1,25,000 cycles per second. Then the capacitor C is replaced by another capacitor with a dielectric medium of dielectric contant k. In this case, the frequency decreases by 25 kHz. The value of k is

1) 3.0 2) 2.1 3) 1.56 4) 1.7

8. In the given figure, the instantaneous value of alternating e.m.f. is $e = 14.14 \sin \omega t$. The reading of voltmeter in volt will be

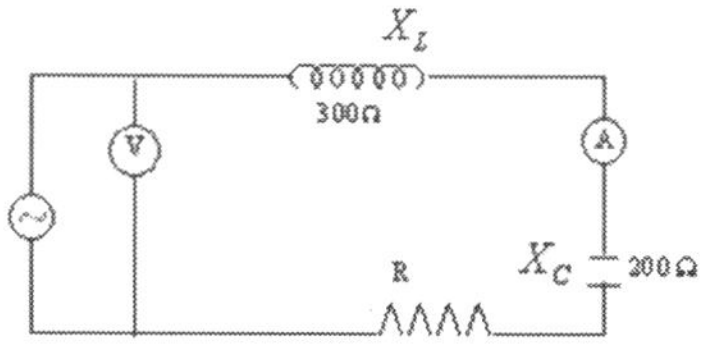

1) 141.4 2) 10 3) 200.0 4) 70.7

9. A coil of inductance 0.1H is connected to 50V, 100Hz generator and current is found to be 0.5A. The potential difference across resistance of the coil is

1) 15V 2) 20V 3) 25V 4) 39V

10. The voltage of A.C. source varies with time according equation. $V = 120 \sin 100\pi t \cos 100\pi t$. Then the frequency of source is

1) 50Hz 2) 100Hz 3) 150Hz 4) 200Hz

11. The current in a coil of self inductance 5 henry is increasing according to $i = 2\sin^2 t$. The amount of energy spent during the period when current changes from 0 to 2 amperes is

1) 10J 2) 5J 3) 100J 4) 2J

12. In an AC circuit the voltage applied is $E = E_0 \sin \omega t$. The resulting current in the circuit is $I = I_0 \sin\left(\omega t - \frac{\pi}{2}\right)$. The power consumption in the circuit is given by

1) $P = \frac{E_0 I_0}{\sqrt{2}}$ 2) P= zero 3) $P = \frac{E_0 I_0}{2}$ 4) $P = \sqrt{2} E_0 I_0$

TRANSFORMER

13. The efficiency of a transformer is 98%. The primary voltage and current are 200 V and 6 A. If the secondary voltage is 100 V, the secondary current is

1) 11.76 A 2) 12.25 A 3) 3.06 A 4) 2.94 A

KEY LEVEL - II (HW)

1) 1 2) 2 3) 1 4) 1 5) 4 6) 1 7) 3 8) 2 9) 4 10) 2
11) 4 12) 4 13) 3

LEVEL - II (HW) HINTS

1. $$\langle i \rangle = \frac{\int_0^{T/4} i\,dt}{\int_0^{T/4} dt}$$

2. $$I = \frac{E}{Z} = \frac{E}{\sqrt{R^2 + 4\pi^2 f^2 L^2}}$$

3. $I = 50\sqrt{2} \sin 400\pi t$; $I = I_0 \sin \omega t$

Comparing two equations, we get

$\omega = 2\pi f = 400\pi$; f= 200Hz and $I_0 = 50\sqrt{2}$

$I_{rms} = \frac{I_0}{\sqrt{2}} = 50A$

4. $\omega L = \frac{1}{\omega C}$

5. $E = \sqrt{E_R^2 + E_C^2}$

6. At resonance $i = \frac{V}{R} = \frac{V_L}{X_L}$; $X_L = \omega L,\ \omega = \frac{1}{\sqrt{LC}}$

7. $n = \frac{1}{2\pi\sqrt{LC}}; \frac{n_1}{n_2} = \sqrt{\frac{C_2}{C_1}} = \sqrt{\frac{Kc}{c}}; \frac{n_1}{n_2} = \sqrt{K}$

8. Reading of voltmeter in rms value $E = \frac{14.14}{\sqrt{2}} = 10V$

9. $I = \frac{E}{Z}$; $0.5 = \frac{50}{Z}$ $Z = 100\Omega$

$Z^2 = R^2 + \omega^2 L^2$, then R = 78Ω

Now $V_R = \sqrt{V_{LR}^2 - V_L^2} = 39V$

$\left[V_R^2 + V_L^2 = V_{LR}^2\right]$

10. V = 120sin $100\pi t \cos 100\pi t$

= 60 x 2 sin$100\pi t . \cos 100\pi t$ = 60sin $200\pi t$ and $\therefore \omega = 200\pi$

$\therefore n = 100 cps$

11. Energy E = 1/2LI^2

When I = change of current from minimum to maximum. = 2–0 = 2J.

12. For given circuit current is lagging the voltage by $\pi/2$ so circuit is purely inductive and there is no power consumption in the circuit. The work done by battery is stored as magnetic energy in the inductor.

13. $\eta = \frac{V_s I_s}{V_p I_p} \Rightarrow \frac{98}{100} = \frac{100 \times i_s}{200 \times 6}$

LEVEL - III

1. The emf of an A.C. source is given by $E = 8\sin\omega t + 6\cos\omega t$ volt. The rms value of emf is

1) $5\sqrt{2}V$ 2) $7\sqrt{2}V$ 3) 10 V 4) $10\sqrt{2}V$

2. The dielectric strength of air is $3\times10^6 V/m$. A parallel plate air capacitor has area $20cm^2$ and plate separation 1mm. Then maximum r.m.s. voltage of an A.C. source which can be safely connected to this capacitor.

1) 212.2 V 2) 21.22 V 3) 2.122 V 4) 2122 V

3. **A pure resistive circuit element 'x' when connected to an A.C. supply of peak voltage 100 V gives a peak current of 4 A which is in phase with the voltage. A second circuit element 'y' when connected to the same AC supply also gives the same value of peak current but the current lags behind by 90^0. If the series combination of 'x' and 'y' is connected to the same supply. R.M.S. value of current is**

1) $\frac{5}{\sqrt{2}}A$ 2) 2A 3) 1/2 A 4) $\frac{\sqrt{2}}{5}A$

4. **An ideal inductor takes a current of 10 A when connected to a 125 V, 50 Hz AC supply. A pure resistor across the same source takes 12.5 A. if the two are connected in series across a $100\sqrt{2}\,V$, 40 Hz supply, the current through the circuit will be**

1) 10 A 2) 12.5 A 3) 20 A 4) 25 A

5. **A circuit containing resistance R_1, Inductance L_1 and capacitance C_1 connected in series resonates at the same frequency 'n' as a second combination of R_2, L_2 and C_2. If the two are connected in series. Then the circuit will resonates at**

1) n 2) 2n 3) $\sqrt{\frac{L_2C_2}{L_1C_1}}$ 4) $\sqrt{\frac{L_1C_1}{L_2C_2}}$

6. **An AC source of variable frequency is applied across a series L-C-R circuit. At a frequency double the resonance frequency. The impedance is $\sqrt{10}$ times the minimum impedance. The inductive reactance is**

1) R 2) 2R 3) 3R 4) 4R

7. **A 20V, 750 HZ source is connected to a series combination of R = 100Ω, C = $10\,\mu$F and L = 0.1803 H. Calculate the time in which resistance will get heated by 10^oC. (If thermal capacity of the material = 2 J / ^oC)**

1) 328 sec 2) 348 sec 3) 3.48 sec 4) 4.32 sec

8. **An AC source of angular frequency ω is fed across a resistor R and a capacitor C in series. The current registered is I. If now the frequency of source is changed to $\omega/3$ (but maintaining the same voltage), the current in the circuit is found to be halved. The ratio of reactance to resistance at the original frequency ω is**

1) $\sqrt{\frac{3}{5}}$ 2) $\sqrt{\frac{5}{3}}$ 3) $\frac{3}{5}$ 4) $\frac{5}{3}$

9. **An LCR circuit has L = 10 mH, R = 3Ω, and C = 1 μF connected in series to a source of $15\cos\omega t$ volt. The current amplitude at a frequency that is 10% lower than the resonant frequency is**

1) 0.5 A 2) 0.7 A 3) 0.9 A 4) 1.1 A

10. **In the given circuit, *R* is a pure resistor, *L* is a pure inductor, S is a 100V, 50 Hz AC source, and A is an AC ammeter. With either K_1 or K_2 alone closed, the ammeter reading is *I*. If the source is changed to 100 V, 100 Hz, the ammeter reading with K_1 alone closed and with K_2 alone closed will be respectively.**

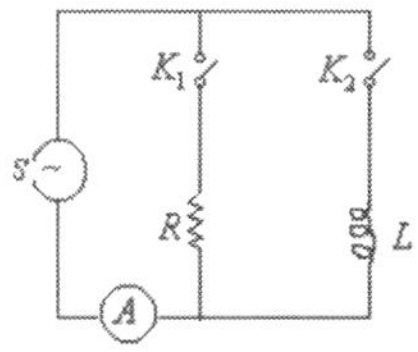

1) $I, I/2$ 2) $I, 2I$ 3) $2I, I$ 4) $2I, I/2$

11. A capacitor has a resistance of $1200\ M\Omega$ and capacitance of $22\ \mu F$. When connected to an a.c. supply of frequency 80 hertz, then the alternating voltage supply required to drive a current of 10 virtual ampere is

1) $904\sqrt{2}V$ 2) 904V 3) $904/\sqrt{2}V$ 4) 452V

12. A 120V, 60Hz a.c. power is connected 800Ω non-inductive resistance and unknown capcitance in series. The voltage drop across the resistance is found to be 102V, then voltage drop across capacitor is

1) 8V 2) 102V 3) 63V 4) 55V

13. An A.C. is given by equation $I = I_1 \cos \omega t + I_2 \sin \omega t$. The r.m.s. value of current is given by

1) $\frac{I_1 + I_2}{2}$ 2) $\frac{(I_1 + I_2)^2}{\sqrt{2}}$ 3) $\frac{1}{\sqrt{2}}\sqrt{I_1^2 + I_2^2}$ 4) $\frac{I_1^2 + I_2^2}{2}$

14. A coil has an inductance of 0.7H and is joined in series with a resistance of $220\,\Omega$. When an alternating e.m.f. of 220V at 50 c.p.s. is applied to it, then the wattless component of the current in the circuit is

1) 5 ampere 2) 0.5 ampere 3) 0.7 ampere 4) 7 ampere

15. Two alternating voltage generators produce emfs of the same amplitude E_0 but with a phase difference of $\frac{\pi}{3}$. The resultant e.m.f.is

1) $E_0 \sin\left(\omega t + \frac{\pi}{3}\right)$ 2) $E_0 \sin\left(\omega t + \frac{\pi}{6}\right)$ 3) $\sqrt{3}E_0 \sin\left(\omega t + \frac{\pi}{6}\right)$ 4) $\sqrt{3}E_0 \sin\left(\omega t + \frac{\pi}{2}\right)$

16. The potential difference across a 2H inductor as a function of time is shown in figure. At time t = 0, current is zero. Current t = 2 second is

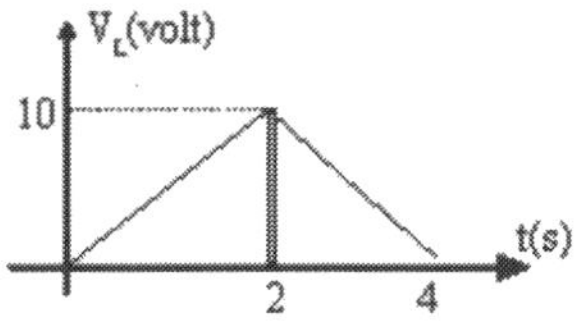

1) 1A 2) 3A 3) 4A 4) 5A

17. For the circuit shown in the figure the rms value of voltages across R and coil are E_1 and E_2, respectively.

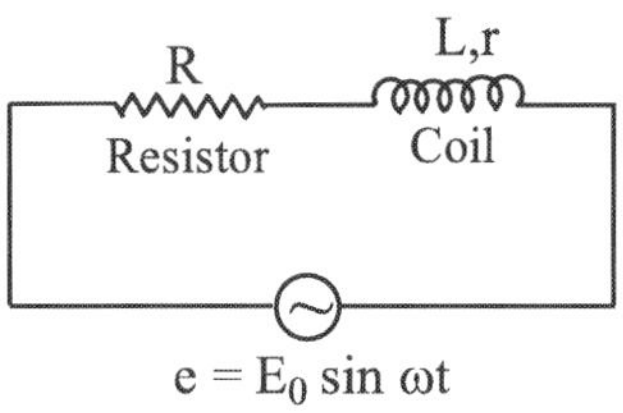

The power (thermal) developed across the coil is

1) $\frac{E - E_1^2}{2R}$ 3) $\frac{E - E_1^2 - E_2^2}{2R}$ 3) $\frac{E^2}{2R}$ 4) $\frac{(E - E_1)^2}{2R}$

18. **A bulb is rated at 100 V, 100 W, it can be treated as a resistor. Find out the inductance of an inductor (called choke coil) that should be connected in series with the bulb to operate the bulb at its rated power with the help of an ac source of 200 V and 50 Hz**

1) $\frac{\pi}{\sqrt{3}}H$ 2) 100 H 3) $\frac{\sqrt{2}}{\pi}H$ 4) $\frac{\sqrt{3}}{\pi}H$

19. **An ac source of angular frequency ω is fed across a resistor R and a capacitor C in series. The current registered is I. If now the frequency of source is changed to $\omega/3$ (but maintaining the same voltage), the current in the circcuit is found to be halved. Then the ratio of reactance to resistance at the original frequency ω is**

1) $\sqrt{3/5}$ 2) $\sqrt{5/3}$ 3) $\sqrt{2/3}$ 4) $\sqrt{3/2}$

20. **In the circuit diagram shown, $X_C = 100\Omega, X_L = 200\Omega \& R = 100\Omega$. The effective current through the source is**

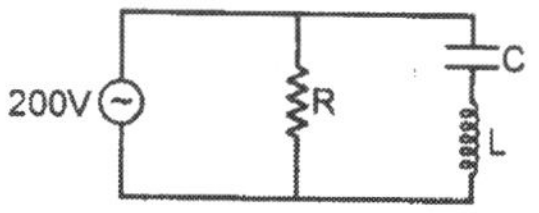

1) 2 A 2) $2\sqrt{2}A$ 3) 0.5 A 4) $\sqrt{0.4}A$

KEY LEVEL - III

1) 1	2) 4	3) 2	4) 1	5) 1	6) 4	7) 2	8) 1	9) 2	10) 1
11) 2	12) 3	13) 3	14) 2	15) 3	16) 4	17) 2	18) 4	19) 1	20) 2

HINTS LEVEL-III

1. $E_0 = \sqrt{8^2 + 6^2} = 10$

$E_{rms} = \frac{E_0}{\sqrt{2}}$

2. maximum voltage $V_o = E \times d$

maximum r.m.s. voltage $= V_o / \sqrt{2}$

3. $X_L = \frac{\varepsilon_o}{I_o} = 25\Omega$; $R = \frac{\varepsilon_o}{I_o} = 25\Omega$

$Z = \sqrt{R^2 + X_C^2}$;

$I_0^1 = \varepsilon_o / Z = 4/\sqrt{2}A$

$I_{r.m.s.} = I_0^1/\sqrt{2} = \frac{4/\sqrt{2}}{\sqrt{2}} = 2A$

4. For 50 Hz and 125 V supply

$X_L = \omega L = \frac{V}{i_L} \Rightarrow L = \frac{1}{8\pi}$, $R = \frac{V}{i_R} = 10\Omega$

For 40 Hz, $100\sqrt{2}\,V$ supply

$$i = \frac{V}{\sqrt{R^2 + X_L^2}} = \frac{V}{\sqrt{R^2 + 4\pi^2 f^2 L^2}}$$

5. $$n = \frac{1}{2\pi\sqrt{L_1C_1}} = \frac{1}{2\pi\sqrt{L_2C_2}}$$

$$L_1C_1 = L_2C_2$$

$$L_{net} = L_1 + L_2; C_{net} = \frac{C_1C_2}{C_1 + C_2}$$

$$L_{net}C_{net} = (L_1 + L_2)\left(\frac{C_1C_2}{C_1 + C_2}\right)$$

$$L_{net}C_{net} = L_2C_2$$

6. $$Z^2 = R^2 + (\omega L - 1/\omega C)^2$$

$$10R^2 = R^2 + (2\omega_o L - 1/2\omega_o C)^2$$

minimum impedance Zmin = R

$$\omega_o^2 LC = 1 \quad \text{-------} \quad (1)$$

$$2\omega_o L - \frac{1}{2\omega_o C} = 3R \quad \text{-------} \quad (2)$$

from(1) $\frac{1}{2\omega_o C} = R \quad \therefore\ X_C = R$

from(2) $X_C = 2\omega_o L = 3R + R = 4R$

7. $$X_C = \frac{1}{2\pi nc} = 21.2\Omega$$

$$Z = \sqrt{R^2 + (X_L - X_C)^2} = 835\Omega$$

$$I_V = Ev/z = 0.0239A$$

$$I_V^2 Rt = (ms)\Delta\theta \Rightarrow t = \frac{(ms)\Delta\theta}{I_V^2 R}$$

8. at frequency ω, $X_C = 1/\omega C$

at frequency $\omega/3$, $X_C' = \frac{3}{\omega C} = 3X_C$

$$I = \frac{V}{\sqrt{R^2 + X_C^2}}; \frac{I}{2} = \frac{V}{\sqrt{R^2 + 9X_C^2}}; \frac{X_C}{R} = \sqrt{\frac{3}{5}}$$

9. $c_v = \frac{90}{100} c_{v_0} = \frac{90}{100} x \frac{1}{\sqrt{LC}}$ =9000 rad/s

$$i_0 = \frac{E_0}{\sqrt{R^2 + \left(\omega L - \frac{1}{\omega C}\right)^2}}$$

10. In the second case induction reactance becomes 2 times thus current through L when K_2 is closed becomes $\frac{i}{2}$. But current through R when K_1 is closed does not change

11. f = 80Hz, I_V = 10A

Current through R, $I_R = \frac{E_V}{R} = \frac{E_V}{12 \times 10^8}$

Current through C $I_C = \frac{E_V}{X_C} = 2\pi fC \times E_V$

$= 2\pi \times 80 \times 22 \times 10^{-6} \times E_V$

$= 352\pi \times 10^{-5} \times E_V$ $\qquad I_V^2 = I_R^2 + I_C^2$

$(10^2) \frac{E_V^2}{(12 \times 10^8)^2} + (352 \times 10^{-5} \times E_V)^2$

$= E_V^2 \left(\frac{1}{144 \times 10^{16}} + 1.2 \times 10^{-4}\right)$

$E_V^2 = \frac{100 \times 10^4}{1.2}$ $\quad E_V \approx 904$ volt

12. $V^2 = V_R^2 + V_C^2$

$V_C^2 = V^2 - V_R^2$

$V_C^2 = (120)^2 - (102)^2$

$V_C = 63V$

13. $I = I_1 \cos \omega t + I_2 \sin \omega t$

$+2I_1 I_2 \cos \omega t . \sin \omega t$

$I_1^2 . \frac{1}{2} + I_2^2 . \frac{1}{2} + 2I_1 I_2 \times 0$

$I_{rms} = \frac{(I_1^2 + I_2^2)^{1/2}}{\sqrt{2}}$

14. Watt less component of

A.C. $= I_V \sin\theta = \frac{E_V}{Z} \sin\theta$

$= \frac{220}{\sqrt{R^2 + L^2\omega^2}} \times \frac{L\omega}{\sqrt{R^2 L^2 \omega^2}}$

$\therefore L0.7\times 2\pi \times 50$

$= \dfrac{220\times L\omega}{(R^2+L^2\omega^2)} = 0.7\times 2\times \dfrac{22}{7}\times 50$

$= \dfrac{220\times (0.7\times 2\pi \times 50)}{(220^2+220^2)} = 220\Omega$

$= \dfrac{220\times 220}{220^2(2)} = \dfrac{1}{2} = 0.5$

15. $E_1 = E_0 \sin\omega t;\; E_2 = E_0 \sin(\omega t + \pi/3)$

$E = E_2 + E_1$

$= E_0 \sin(\omega t + \pi/3) + E_0 \sin \omega t$

$= 2E_0 \sin(\omega t + \pi/6)\cos(\pi/6)$

$= \sqrt{3}E_0 \sin(\omega t + \pi/6)$

16. $|e| = L\dfrac{di}{dt} \Rightarrow |e|dt = L(i_2 - i_1)$

$|e|dt =$ area of Δ le for $t = 0$ to 2 sec.

17. Draw the phasor diagram.

$$E^2 = E_1^2 + E_2^2 + 2E_1E_2\cos\theta.$$

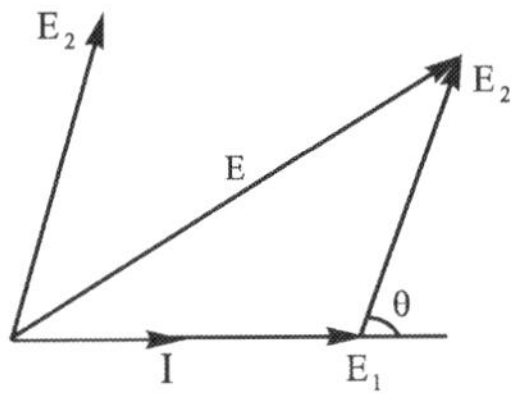

Thermal power developed in coil is

$$P = E_2 \cos\theta \times I \text{ and } I = \frac{E_1}{R} \Rightarrow P = \frac{E_1E_2}{R}\cos\theta = \frac{E^2 - E_1^2 - E_2^2}{2R}$$

18. Resistance of bulb is $R = \dfrac{100\times 100}{100} = 100\Omega$

Rated current is $\dfrac{100}{100} = 1A$

In ac, $I_{rms} = \dfrac{V_{rms}}{Z}$

$Z = 200\Omega$

$\sqrt{100^2 + (\omega L)^2} = 200 \Rightarrow \omega^2 L^2 = 30000$ and $L = \sqrt{\dfrac{30000}{(100\pi)^2}} = \dfrac{\sqrt{3}}{\pi}$ henry.

19. Ans : (a)

$$I = \frac{V}{Z_1} = \frac{V}{\sqrt{R^2 + X_c^2}}$$

$$\frac{I}{2} = \frac{V}{Z_2} = \frac{V}{\sqrt{R^2 + (3X_c)^2}}$$

$$\Rightarrow R^2 + (3X_c)^2 = 4\left(R^2 + X_c^2\right)$$

$$3R^2 = 5X_c^2 \Rightarrow \frac{Xc}{R} = \sqrt{\frac{3}{5}}$$

20. $I_R = \frac{V}{R} = \frac{200}{100} = 2A$

$$I_{LC} = \frac{200}{X_L \times X_C} = 2A$$

$$I = \sqrt{2^2 + 2^2} = 2\sqrt{2}\ A$$

as I_R inphase with V

I_{LC} lages behind V by $\pi/2$

LEVEL-IV

1. **At resonance, V_L and V_C are both very much greater than the applied potential, V itself. The quality factor for an LCR circuit in resonance is given by $Q=\frac{X_L}{R}$. In practice, $Q = 200$ has been achieved.**

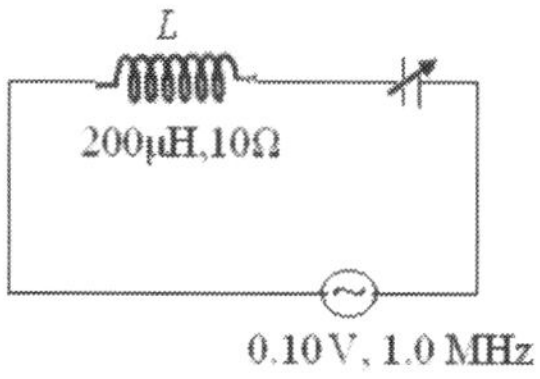

(a) **At resonance, the capacitor has been adjusted for**

1 $200\times10^{-6}\mu F$ 2. $0.00013\mu F$ 3. $0.0013\mu F$ 4. 0.0013F

(b) **At resonance, the potential difference across the inductance is**

1. 1.3 V 2. 13 V 3. 0.3 V 4. none of these

(c) **The potential across the capacitance at resonance is**

1. 13 V 2. > 13 V 3. < 13 V 4. none of these

(d) **The Q factor is**

1. $\frac{V_L}{V_C}$ 2. $\frac{V_C}{V_L}$ 3. $\frac{V_C}{V}$ 4. $\frac{V_L}{V}$

(e) **choose the right statement.**

1. V_L+V_C can be greater than $V_{applied}$ 2. $V_L+V_C=V_{applied}$

3. $V_L+V_C<V_{applied}$ 4. none of these

2. **A physics lab is designed to study the transfer of electrical energy from one circuit to another by means of a magnetic field using simple transformers. Each transformer has two coils of wire electrically insulated from each other but wound around a common core of ferromagnetic material. The two wires are close together but do not touch each other.**

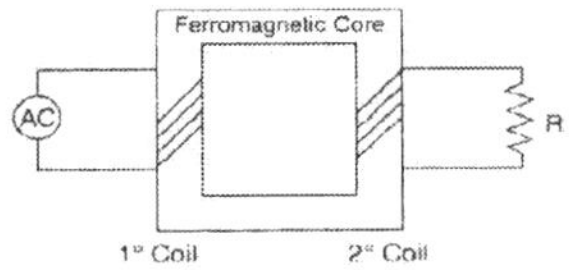

The primary coil is connected to a source of alternating (AC) current. The secondary coil is connected to a resistor such as a light bulb. The AC source produces an oscillating voltage and current in the primary coil that produces an oscillating magnetic field in the core material. This in turn induces an oscillating voltage and AC current in the secondary coil.

Student collected the following data comparing the number of turns per coil (N), the voltage (V) and the current (I) in the coils of three transformers

	Primary Coil			Secondary coil		
	N_1	V_1	I_1	N_2	V_2	I_2
Transformer 1	100	10V	10 A	20	20 V	5 A
Transformer 2	100	10V	10 A	50	5 V	20 A
Transformer 3	200	10V	10 A	100	5 V	20 A

a) **The primary coil of a transformer has 100 turns and is connected to a 120V AC source. How many turns are in the secondary coil if there is a 2400 V across it**

1) 5 2) 50 3) 200 4) 2000

b) **A transformer with 40 turns in its primary coil is connected to a 120 V AC source. If 20 W of power is supplied to the primary coil, the power developed in the secondary coil is**

1) 10 W 2) 20 W 3) 80 W 4) 160 W

c) **One of the following is a correct expression for R, the resistance of the load connected to the secondary coil (pick the correct one)**

1) $\left(\frac{V_{1^0}}{I_{1^0}}\right)\left(\frac{N_{2^0}}{N_{1^0}}\right)$ 2) $\left(\frac{V_{1^0}}{I_{1^0}}\right)\left(\frac{N_{1^0}}{N_{2^0}}\right)^2$ 3) $\left(\frac{V_{1^0}}{I_{1^0}}\right)\left(\frac{N_{1^0}}{N_{2^0}}\right)$ 4) $\left(\frac{V_{1^0}}{I_{1^0}}\right)\left(\frac{N_{1^0}}{N_{2^0}}\right)^2$

d) **A 12 V battery is used to supply 2.0 mA of current to the 300 turns in the primary coil of a given transformer. What is the current in the secondary coil if $N_2 = 150$ turns**

1) zero 2) 1.0 mA 3) 2.0 mA 4) 4.0 mA

<u>**KEY LEVEL-IV**</u>

1) a) 2 b) 2 c) 1 d) 3,4 e) 4

2) a) 4 b)2 c)1 d) 4

Chapter - 4

ELECTRO MAGNETIC WAVES

ELECTRO MAGNETIC WAVES

SYNOPSIS

➢ According to Maxwell, an accelerated charge produces a sinusoidal time-varying magnetic field, which in turn produces a sinusoidal time varying electric field. The two fields so produced are mutually perpendicular. They constitute electro magnetic waves which can propagate through empty space.

➢ **Displacement current:-** According to Ampere's circuital Law, the magnetic field B is related to steady current I as $\oint \vec{B}.\vec{dl} = \mu_0 I$ ....(i) where I is the current travelling through the surface bounded by closed loop.

In 1864, Maxwell showed that relation (i) is logically inconsistent. He accounted for this inconsistency as follows: Consider a parallel plate capacitor having plates P and Q being charged with battery B.

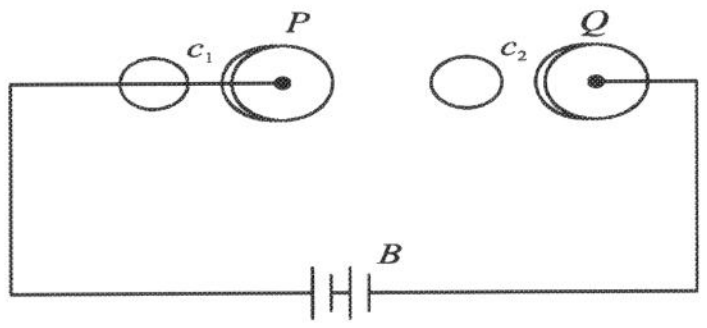

➢ During charging, a current I flows through the connecting wires which changes with time. This current will produce magnetic field around the wires which can be detected using a magnetic compass needle. Consider two loops c_1 and c_2 parallel to the plates P and Q of the capacitor. c_1 is enclosing only the connecting wire attached to the plate P of the capacitor and c_2 lies in the region between the two plates of capacitor. For the loop c_1, a current I is flowing through it, hence Ampere's circuital law for loop c_1 gives $\oint_{C_1} \vec{B}.\vec{dl} = \mu_0 I \quad ....(ii)$

➢ Since the loop c_2 lies in the region between the plates of the capacitor, no current flows in this region. Hence Ampere's circuital law for loop c_2 gives $\oint_{C_2} \vec{B}.\vec{dl} = 0 \quad ....(iii)$

➢ The relations (ii) and (iii) continue to be true even if two loops c_1 and c_2 are infinitesimally close to the plate P of the capacitor. In the other hand, as the loops c_1 & c_2 are infinitesimally close, it is expected that $\oint_{C_1} \vec{B}.\vec{dl} = \oint_{C_2} \vec{B}.\vec{dl} \quad ...(iv)$

Thus, relation (iv) is in contradiction with relations (ii) and (iii). This led Maxwell to point out that Ampere's circuital law as given by (i) is logically inconsistent.

➢ **Idea of Displacement Current :** Maxwell predicted that not only a current flowing in a conductor produces magnetic field but also a time-varying electric field (i.e., changing electric field) in a vacuum/ free space (or in a dielectric) produces a magnetic field. It means a changing electric field gives rise to a current which flows through a region so long as the electric field is changing there. Maxwell also predicted that this current produces the same magnetic field as a conduction current can produce. This current is known as 'displacement current'.

➢ Thus, displacement current is that current which comes into play in the region in which the electric field and hence the electric flux is changing with time.

Maxwell defined this displacement current in space where electric field is changing with time as

$$I_D = \varepsilon_0 \frac{d\phi_E}{dt} \quad ...(v)$$

where ϕ_E is the electric flux.

- Maxwell also found that conduction current (I) and displacement current (I_D) together have the property of continuity, although, individually, they may not be continuous.
- This idea led Maxwell to modify Ampere's circuital law in order to make the same logically consistent.

He states Ampere circuital law in the form, $\oint_C \vec{B}.\vec{dl} = \mu_0 (I + I_D)$

$$= \mu_0 \left(I + \varepsilon_0 \frac{d\phi_E}{dt} \right)$$

It is now called as Ampere-Maxwell's law.

- This means that out side the capacitor plates, we have only conduction current $i_c = i$ and no displacement current $i_d = 0$. On the other hand, inside the capacitor, there is no conduction current $i_c = 0$ and there is only displacement current $i_d = i$

MAXWELL'S EQUATIONS

- Maxwell, in 1862, gave the basic Laws of electricity and magnetism in the form of four fundamental equations which are known as Maxwell's equations. In the absence of any dielectric and magnetic material may be stated in the integral form as below.

1. **Gauss's Law for electrostatics :-**

This Law gives the total electric flux in terms of charge enclosed by the closed surface.

In the usual notations $\oint \vec{E}.\vec{dS} = \frac{q_{in}}{\varepsilon_0}$

This Law states that electric lines of force start from positive charge and end at negative charge i.e., electric lines force do not form closed paths.

2. **Gauss's Law for magnetism :-**

Mathematically $\oint \vec{B}.\vec{dS} = 0$

- This Law shows that the no. of magnetic lines of force entering a closed surface is equal to no. of magnetic lines of force leaving that closed surface.
- This law tells that the magnetic lines of force form a continuous closed path.
- This Law also predicts that the isolated magnetic monopoles does not exist.

3. **Faraday's Law of electro magnetic induction :-**

Mathematic cally $\oint \vec{E}.\vec{dl} = \frac{-d\phi_B}{dt}$ = induced emf.

- This law gives a relation between electric field and changing magnetic flux.
- This law tells that changing magnetic field is a source of electric field.

4. **Amper's-Maxwell's Law :-**

Mathematically $\oint \vec{B}.\vec{dl} = \mu_0 (i_c + i_d) = \mu_0 \left(i_c + \varepsilon_0 \frac{d\phi_E}{dt} \right)$

- This law states that magnetic field can be produced by a conduction current as well as by displacement current.
- At any instant in a circuit, conduction current is equal to displacement current.

5. **Lorentz Force :-** Force acting on a charge 'q' moving in a region where electric and magnetic fields similar to EM waves are existing simultaneously is $\vec{F} = q\left(\vec{E} + \vec{v} \times \vec{B}\right)$

W.E.-1: ***What is the instantanuous displacement current in space between plates of parallel plate capacitor of capacitor*** $1\mu F$ ***which is charging at rate of*** $10^6 V / S$

Sol. As $I_d = \varepsilon_0 \frac{d\phi_E}{dt} = \varepsilon_0 A \frac{d}{dt}(E) = \varepsilon_0 A \frac{d}{dt}\left(\frac{V}{d}\right)$

$$= \frac{\varepsilon_0 A}{d}\frac{dV}{dt} = C\frac{dV}{dt};\ I_d = 10^{-6} \times 10^6 = 1A$$

Source of EM waves :-

- Accelerated charges radiate energy in the form of EM waves. So it is source of EM waves
- An oscillating charge produces an oscillating electric field inturn which produces an oscillating magnetic field.
- The oscillating electric and magnetic fields regenerate each other and propagate through space as waves called EM waves.
- An electric charge oscillating harmonically with frequency 'υ' produces EM waves of same frequency.

Characteristics of EM waves :

1) EM waves are transverse in nature whose speed is same as that of speed of light
2) The two fields $\vec{E}$ and $\vec{B}$ have same frequency of oscillation and they are in phase with each other.
3) Keeping these features in mind, we can assume that if EM wave is travelling along positive direction along x-axis, the electric field is oscillating parallel to the y-axis and that magnetic field is parallel to z-axis, then we can write the electric and magnetic fields as sinusoidal functions of position 'x' and time 't'

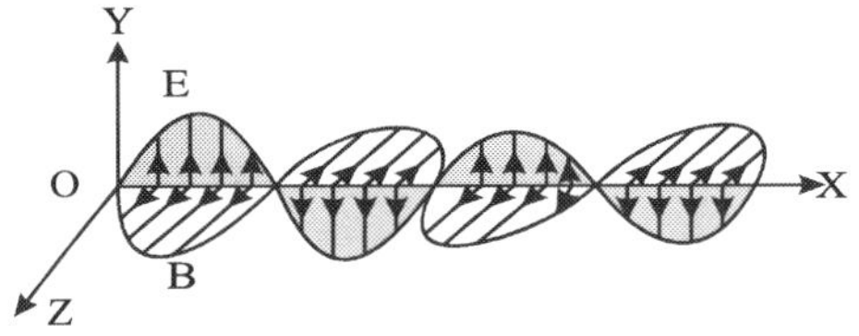

$E = E_0 Sin(kx - wt)$; $B = B_0 \sin(kx - wt)$

In this, E_0 & B_0 are the amplitudes of the fields

4) EM waves can be polarised.
5) EM waves are self-sustaining oscillations of electric and magnetic fields in free space or vaccum. EM waves travel through vaccum with speed of light 'C' where. $C = \frac{1}{\sqrt{\varepsilon_0 \mu_0}} = 3 \times 10^8 m / Sec$
6) The speed of EM waves in any other medium of permittivity ε and permeability μ is

$$C_{med} = \frac{1}{\sqrt{\mu\varepsilon}} = \frac{1}{\sqrt{\mu_r \mu_0 \varepsilon_r \varepsilon_0}} = \frac{C_0}{\sqrt{\mu_r \varepsilon_r}}$$

$\sqrt{\mu_r \varepsilon_r} = \frac{C_0}{C_{med}} = n$ R.I.of medium

7) In vaccum, EM waves are of different wavelengths, but velocity is same.

W.E.-2 : ***Electro magnetic waves travel in a medium with speed of*** $2\times10^8 m/\sec$. ***The relative permeability of the medium is 1 find relative permittivity.***

Sol. Give $C = 2\times10^8 m/\sec, \mu_r = 1$

Speed of EM waves in medium

$$C_{med} = \frac{1}{\sqrt{\mu_r \mu_0 \varepsilon_r \varepsilon_0}} = \frac{C_0}{\sqrt{\mu_r \varepsilon_r}} \;;\; \varepsilon_r = \frac{c_0^2}{c^2 \mu_r} = \frac{(3\times10^8)^2}{(2\times10^8)^2\times1} = 2.25$$

W.E-3: ***Suppose that the electric field amplitude of an EM wave is*** $E_0 = 120N/C$ ***and that its frequency*** $v = 50\mu HZ$. ***Determine (a)*** B_0, ω, λ ***and*** K ***(b) Find expressions for E and B***

Sol. a) i) Using $C = \frac{E_0}{B_0}$ we get

$$B_0 = \frac{E_0}{C} = \frac{120}{3\times10^8} = 4\times10^{-7}T = 400nT$$

ii) $\omega = 2\pi v = 2\times\pi\times50\times10^6 = 3.14\times10^8 rad/Sec$ iii) $c = \upsilon\lambda \Rightarrow \lambda = \frac{C}{\upsilon} = \frac{3\times10^8}{50\times10^6} = 6m$

iv) $K = \frac{2\pi}{\lambda} = \frac{2\pi}{6} = \frac{2\times3.14}{6} = 1.05m^{-1}$

b) $\vec{E} = E_0 \sin(kx - \omega t)$

$= 120\sin(1.05x - 3.14\times10^8 t)$

$B = B_0 \sin(kx - \omega t)$

$= 400\times10^{-9}\sin(1.05x - 3.14\times10^8 t)$

Energy density of EM waves:

➢ Consider a plane electro magnetic wave propagating along x-axis. The electric and magnetic fields in a plane EM wave can be given by $E = E_0 \sin(kx - \omega t)$ and $B = B_0 \sin(kx - \omega t)$

➢ In any small volume 'dV', the energy of electric field is $U_E = \frac{1}{2}\varepsilon_0 E^2 dV$ and energy of the magnetic field in volume 'dV' is $U_B = \frac{B^2}{2\mu_0}dV$

➢ Thus total energy of EM wave is $u = \frac{1}{2}\varepsilon_0 E^2 dV + \frac{B^2}{2\mu_0}dV$

➤ Energy density of EM wave is $U = \frac{1}{2}\varepsilon_0 E^2 + \frac{B^2}{2\mu_0}$

$$= \frac{1}{2}\varepsilon_0 E_0^2 \sin^2(kx - wt) + \frac{B_0^2}{2\mu_0}\sin^2(kx - \omega t)$$

If we take average over a long time, the $\sin^2$ terms have an average value of $\frac{1}{2}$

Thus $u_{aV} = \frac{1}{4}\varepsilon_0 E_0^2 + \frac{B_0^2}{4\mu_0}$

Now $E_0 = CB_0$ and $\mu_0\varepsilon_0 = \frac{1}{C^2}$

$$\therefore u_E = \frac{1}{4}\varepsilon_0 E_0^2 = \frac{1}{4}\varepsilon_0(C^2 B_0^2) = \frac{1}{4}\varepsilon_0 \frac{1}{\mu_0\varepsilon_0} B_0^2 = \frac{B_0^2}{4\mu_0} = u_B$$

Hence is an EM wave, average energy density of electric field is equal to average energy density of magnetic field.

The units of u_E & u_B are Jm^{-3}

∴ average energy density of EM wave

$$u = u_E + u_B = 2u_E = 2u_B = \frac{1}{2}\varepsilon_0 E_0^2 = \frac{B_0^2}{2\mu_0}$$

Intensity of electro magnetic wave:

Intensity of EM wave is defined as the energy crossing per second per unit area of a source perpendicular to the direction of propagation of the wave. It is denoted by I.

i.e. Intensity $I = \frac{total\ EM\ wave\ energy}{Surface\ area \times time}$

$$\frac{u_{aV} \times \Delta V}{A\Delta t} = \frac{u_{aV} \times Ac\Delta t}{A\Delta t}$$

Interms of electric fields $I = \frac{1}{2}\varepsilon_0 E_0^2 C$ ---------(1)

Interms of magnetic field $I = \frac{B_0^2}{2\mu_0}C$ ---------(2)

Either eq (1) or (2) may be used to find intensity of EM waves

ii) The intensity of EM radiation from an isotropic point source at a distance r is $I = \frac{P}{4\pi r^2}$ where P is power of source

Note: The rate of flow of energy crossing a unit area in an EM wave is described by the vector 'S' called poynting vector which is described by the expression.

$$\vec{S}=\frac{1}{\mu_0}(\vec{E}\times\vec{B})$$

Since $\vec{E}$ and $\vec{B}$ are mutually perpendicular

$$|\vec{E}\times\vec{B}|=EB$$

Thus magnitude of Poynting vector

$$S=\frac{EB}{\mu_0}=\frac{E^2}{\mu_0 C}$$

SI unit of S is $J\sec^{-1}m^{-2}\ (or)\ Wm^{-2}$

This relation shows that the value of electric vector at any instant in the EM wave is about 377 times the value of magnetic vector. It is because of this reason, optical properties of light is due to electric field. Average of Poynting vector is given by

$$I=S_{aV}=\frac{E_0B_0}{2\mu_0}=\frac{1}{2}\varepsilon E_0^2 C=\frac{CB_0^2}{2\mu_0}$$

W.E.-4 : ***The electric field of an electro magnetic wave is given by*** $E=50\sin\omega\left(t-\frac{x}{c}\right)N/C$. ***Find energy contained in a cylinder of cross-section*** $10cm^2$ ***and length*** $50cm$ ***along x-axis***

Sol: Average volume of energy density $u_{aV}=\frac{1}{2}\varepsilon_0E_0^2$

Total volume of cylinder $V=Al$

Total energy of contained in cylinder

$$U=(U_{aV})V=\left(\frac{1}{2}\varepsilon_0E_0^2\right)(Al)=5.5\times10^{-12}J$$

Momentum and Radiation Pressure

i) Electro magnetic waves have linear momentum as well as energy. When EM waves strike a surface, pressure is exerted on it, called radiation pressure.

ii) When EM waves are incident on a surface and the total energy transferred to the surface in a time t is U then magnitude of momentum transferred to surface is $p=\frac{U}{C}$ (total absorption)

$$\left[E=mC^2=(mC)C\Rightarrow mC=\frac{E}{C}\Rightarrow P=\frac{E}{C}\right]$$

iii) When radiation incident on a surface is entirely reflected back along its original path, magnitude of momentum delivered to the surface is $p=\frac{2U}{C}$ where 'C' is velocity of light.

iv) When the radiation incident on a surface (Perfect absorber) radiaton pressure

$$P_r=\frac{F}{A}=\frac{1}{A}\frac{dp}{dt}=\frac{1}{A}\frac{d}{dt}\left(\frac{U}{C}\right)=\frac{1}{AC}\frac{dU}{dt}=\frac{S}{C}$$

$\frac{dU/dt}{A}$ is called average value of Poynting vector.

If the surface is perfect reflector radiation pressure $P_r = \frac{2S}{C}$

W.E-5: ***Light with an energy flux of*** $18W/cm^2$ ***falls on a non reflecting surface at normal incidence. If the surface has an area of*** $20cm^2$ ***then find average force exerted on the surface during a 30 minute time span.***

Sol. Total energy falling on the surface is

$U = (18\times10^4)(20\times10^{-4})(30\times60) = 6.48\times10^5 J$

Total momentum delivered (complete absorption)

$p = \frac{U}{C} = \frac{6.48\times10^5}{3\times10^8} = 2.16\times10^{-3} kg\,m/\sec$ average force exerted

$F = \frac{p}{t} = \frac{2.16\times10^{-3}}{30\times60} = 1.2\times10^{-6} N$

W.E.-6: ***The rms value of electric field of light coming from sun is*** $720N/C$. ***Find average energy density of em wave.***

Sol: Total average energy density $= \frac{1}{2}\varepsilon_0 E_0^2 = \varepsilon_0 E_{rms}^2$

$\left(\therefore E_{rms} = \frac{E_0}{\sqrt{2}}\right) = 8.85\times10^{-12}\times(720)^2 = 4.58\times10^{-6} Jm^{-3}$

Electro magnetic spectrum :

➢ The array obtained on arranging all the electromagnetic waves in an order on the basis of their wavelength is called the electromagnetic spectrum

In the order of increasing wavelength, these waves are (i) Gamma rays, (ii) X-rays, (iii) Ultraviolet rays, (iv) Visible light, (v) infrared waves, (vi) Microwaves and (vii) Radio waves.

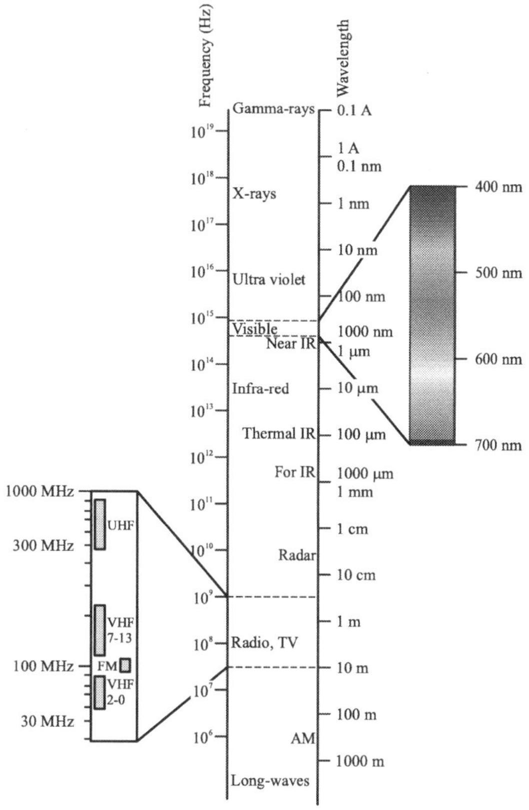

The figure illustrates the general spectrum of the electromagnetic radiations, in which the wavelength is expressed in metre.

1) **Gamma rays :** They were discovered by Becquerel and Curie in 1896. Their wavelength is of the order of 10^{-14} to $10^{-10}\,m$. The main sources are the natural and artificial radioactive substances. These rays affect the photographic plate. These rays are mainly used in the treatment of cancer disease.

2) **X-rays :** They were discovered by Roentgen in 1895. Their wavelength is of the order of $10^{-12}\,m$ to $10^{-18}\,m$. X-rays are produced when highly energetic cathode rays are stopped by a metal target of high melting point. They affect the photographic plate and can penetrate through the transparent materials. They are mainly used in detecting the fracture of bones, hidden bullet, needle, costly material, etc., inside the body and also used in the study of crystal structure.

3) **Ultraviolet rays :** They were discovered by Ritter in 1801. Their wavelength is of the order $10^{-9}\,m$ to $4\times10^{-7}\,m$. In the radiations received from sun, major part is that of the ultraviolet radiation. Its other sources are the electric discharge tube, carbon arc etc. These radiations are mainly used in excitation of photoelectric effect and to kill the bacteria of many diseases.

4) **Visible light :** This was first studied in 1666 by Newton. The radiations in the range of wavelength from $4\times10^{-7}\,m$ to $7\times10^{-7}\,m$ fall in the visible region. The wavelength of the light of violet colour is the shortest and that of red colour is the longest. Visible light is obtained from the glowing bodies, while they are white hot. The light obtained from the electric bulbs, sodium lamp, fluorescent tube is the visible light.

5) **Thermal or infrared waves:** They were discovered by Herchell in 1800. Their wavelength is of the order of $7\times10^{-7}\,m$ to $10^{-3}\,m$. A body on being heated, emits out the infrared waves. These radiations have the maximum heating effect. The glass absorbs these radiations, therefore for the study of these radiations rock salt prism is used instead of a glass prism. These waves are mainly used for therapeutic purpose by the doctors because of their heating effect.

6) Microwaves: They were discovered by Hertz in 1888. Their wavelength is in the range of nearly $10^{-4}m$ to 1m. These waves are produced by the spark discharge or magnetron valve. They are detected by the crystal or semiconductor detector. These waves are used mainly in radar and long distance communication.

7) Radiowaves: They were first discovered in 1895 by Marconi. Their wavelength is in the range of 0.1m to $10^5 m$. They can be obtained by the flow of high frequency alternating current in an electric conductor. These waves are detected by the tank circuit in a radio receiver or transmitter.

Application of EM waves

1) Radio and microwave radiations are used in radio and TV communication system. Microwave radiations are mainly used in radar and TV communication.

2) **Infrared radiations are used**

i) in green houses to keep the plants warm

ii) in revealing the secret writings on the ancient walls

iii) for looking through haze, fog and mist during war time, as these radiations can pass through them.

3) **Ultraviolet radiations are used**

i) in preserving the food stuffs.

ii) in the detection of invisible writing, forged documents, finger prints in forensic laboratory.

iii) Ultraviolet radiations are also used for knowing the structure of the molecules and arrangement of electrons in the external shells.

4) **X-rays many applications** these rays provide us valuable information

i) about the structure of atomic nuclei

ii) in the study of crystal structure

iii) in the fracture of bones etc.

5) $\gamma-$ **rays were used**

i) in treatment of cancer and tumours

ii) to produce nuclear reactions.

CONCEPTUAL QUESTIONS

1. If $\vec{E}$ and $\vec{B}$ are the electric and magnetic field vectors of electromagnetic waves then the direction of propagation of electromagnetic waves is along the direction of -

1. $\vec{E}$ 2. $\vec{B}$ 3. $\vec{E}\times\vec{B}$ 4. $\vec{B}\times\vec{E}$

2. The electromagnetic waves do not transport-

1. energy 2. charge 3. momentum 4. information

3. A capacitor is connected in an electric circuit. When key is pressed, the current in the circuit is-

1. zero 2. maximum

3. any transient value 4. depends on capacitor used

4. Displacement current is continuous-

1. when electric field is changing in the circuit 2. when magnetic field is changing in the circuit

3. in both types of fields. 4. through wire and resistance only

5. The conduction current is the same as displacement current when the source is

1. A.C. only 2. D.C. only

3. both A.C and D.C. 4. neither for A.C. nor for D.C.

Electro Magnetic Waves

6. **The Maxwells four equations are written as**

(i) $\oint \vec{E}.\overrightarrow{dS} = \frac{q_0}{\varepsilon_0}$ (ii) $\oint \vec{B}.\overrightarrow{dS} = 0$

(iii) $\oint \vec{E}.\overrightarrow{dl} = \frac{d}{dt}\oint \vec{B}.\overrightarrow{dS}$ (iv) $\oint \vec{B}.\overrightarrow{dl} = \mu_0\varepsilon_0 \frac{d}{dt}\oint \vec{E}.\overrightarrow{dS}$

The equations which have sources of $\vec{E}$ and $\vec{B}$ are

1. (i), (ii), (iii) 2. (i), (ii) 3. (i) and (iii) only 4. (i) and (iv) only

7. **Out of the above four equations, the equations which do not contain source field are -**

1. (i) and (ii) 2. (ii) only 3. all of four 4. (iii) only

8. **Out of the four Maxwell's equations above, which one shows non-existence of monopoles?**

1. (i) and (iv) 2. (ii) only 3. (iii) only 4. none of these

9. **Which of the above Maxwell's equations shows that elelctric field lines do not form closed loops?**

1. (i) only 2. (ii) only 3. (iii) only 4. (iv) only

10. **In an electromagnetic wave the average energy density is associated with -**

1. electric field only 2. magnetic field only
3. equally with electric and magnetic fields. 4. average energy density is zero.

11. **The displacement current flows in the dielectric of a capacitor when the potential difference across its plates-**

1. becomes zero 2. has assumed a constant value
3. is increasing with time 4. decreasing with time

12. **Select wrong statement from the following-Electromagnetic waves**

1. are transverse 2. travel with same speed in all media
3. travel with the speed of light 4. are produced by accelerating charge.

13. **The waves related to telecommuni-cation are-**

1. infra red 2. visible light 3. microwaves 4. ultraviolet rays

14. **The magnitudes of Poynting vector and electric field vector are respectively S and E, then**

1. $S = E^2\sqrt{\frac{\varepsilon_0}{\mu_0}}$ 2. $S = E^2\sqrt{\varepsilon_0\mu_0}$

3. $S = E^2\sqrt{\frac{\mu_0}{\varepsilon_0}}$ 4. $S^2 = \frac{E^2}{\mu_0}$

15. **The nature of electromagnetic wave is-**

1. longitudinal 2.longitudinal stationary
3. transverse 4. transverse stationary

16. **The frequencies of** x-rays , γ-rays **and ultraviolet rays are respectively a,b and c. Then:**

1. $a < b, b < c$ 2. $a > b, b > c$ 3. $a > b, b < c$ 4. $a < b, b > c$

17. **Electromagnetic radiation of frequency** 3×10^5MHz **can be produced by which of the following wave**

1. Radiowave 2. X-rays 3. Ultraviolet 4. Microwave

18. **Maxwell's equations describe the fundamental laws of**

1. electricity only 2. magnetism only 3. mechanics only 4. both 1 and 2

19. **Which of the following statements is not correct?**
1. photographic plates are sensitive to infrared rays
2. photographic plates are sensitive to ultraviolet rays
3. Infra-red rays are invisible but can cast shadows like visible light
4. infrared photons have more energy than photons of visible light

20. **Radio waves and visible light in vacuum have**
1. same velocity but different wavelength 2. continuous emission spectrum
3. band absorption spectrum 4. line emission spectrum

21. **Energy stored in electromagnetic oscillations is in the form of**
1. electrical energy 2. magnetic energy 3. both 1 and 2 4. neither of the above

22. **Whcih wave is not electromagnetic in nature?**
1. micro 2. radio 3. X-ray 4. audio

23. **Total energy of EM waves in free space is given by**
1. $\frac{E^2}{2\varepsilon_0}+\frac{B^2}{2\mu_0}$ 2. $\frac{\varepsilon_0 E^2}{2}+\frac{\mu_0 B^2}{2}$ 3. $\frac{E^2+B^2}{C}$ 4. $\frac{\varepsilon_0 E^2}{2}+\frac{B^2}{2\mu_0}$

24. **Which of the following waves have the maximum wavelength?**
1. Ultraviolet rays 2. I.R. rays 3. UV rays 4. radio waves

25. **Electromagnetic waves are transverse in nature is evident by**
1. polarization 2. interference 3. reflection 4. diffraction

26. **Which of the following are not electromagnetic waves?**
1. cosmic rays 2. gamma rays 3. $\beta - rays$ 4. X-rays

27. **Let $\vec{E},\vec{B}$ and $\vec{C}$ represent the electric field, magnetic field and velocity of an electromagnetic wave respectively. Their directions, at any instant point along the unit vectors given below in order. Which of the following cannot be true?**
1) $\hat{k},\hat{i},\hat{j}$ 2) $\hat{k},\hat{j},-\hat{i}$ 3) $\hat{i},\hat{j},\hat{k}$ 4) $-\hat{j},-\hat{k},-\hat{i}$

28. **A lamp radiates power P_0 uniformly in all directions, the amplitude of electric field strength E_0 at a distance r from it is**
1) $E_0=\frac{P_0}{2\pi\varepsilon_0 cr^2}$ 2) $E_0=\sqrt{\frac{P_0}{2\pi\varepsilon_0 cr^2}}$ 3) $E_0=\sqrt{\frac{P_0}{4\pi\varepsilon_0 cr^2}}$ 4) $E_0=\sqrt{\frac{P_0}{8\pi\varepsilon_0 cr}}$

29. **A radiation of energy E falls normally on a perfectly reflecting surface. The momentum transferred to the surface**
1) E/C 2) 2E/C 3) EC 4) E/C^2

30. **An electromagnetic wave passing through vacuum is described by the equation;**
$E=E_0\sin(kx-\omega t)$ and $B=B_0\sin(kx-\omega t)$; **then**
1) $E_0=B_0$ 2) $E_0\omega=B_0k$ 3) $E_0B_0=\omega k$ 4) $E_0k=B_0\omega$

31. **The frequency of visible light is of the order of**
1. 10^{15}Hz 2. 10^{10}Hz 3. 10^{6}Hz 4. 10^{4}Hz

32. Which of the following wavelength falls in X-ray region?

1. $1 A^0$ 2. $10 A^0$ 3. $10^{-2} A^0$ 4. $10^{-3} A^0$

33. An electro magnetic wave is vaccume has the electric and magnetic field $\vec{E}$ and $\vec{B}$ which are always perpendicular to each other. The direction of polarisation is given by $\vec{X}$ and that of wave propagation by $\vec{K}$ then

1) $\vec{X} \| \vec{B}$ and $\vec{K} \| \vec{B} \times \vec{E}$ 2) $\vec{X} \| \vec{E}$ and $\vec{K} \| \vec{E} \times \vec{B}$

3) $\vec{X} \| \vec{E}$ and $\vec{K} \| \vec{E} \times \vec{B}$ 4) $\vec{X} \| \vec{E}$ and $\vec{K} \| \vec{B} \times \vec{E}$

Note : Directions q.no. 34 to 46

1. Both Assertion and reason are true and the reason is correct explanation of the Assertion.
2. Both Assertion and reason are true, but reason is not correct explanation of Assertion.
3. Assertion is true, reason is false 4. Assertion is false, reason is true

34. Assertion: Displacement current arises on account of change in electric flux.

Reason: $I_d = \epsilon_0 \dfrac{d\phi_E}{dt}$

35. Assertion (A): In an e.m. wave, magnitude of magnetic field vector $\vec{B}$ is much smaller than the magnitude of vector $\vec{E}$

Reason(R): This is because in an e.m. wave E/B $= c = 3\times10^8\ m/s$.

36. Assertion(A): Microwaves have more energy than the radio waves

Reason(R): $E = h\upsilon = \dfrac{hc}{\lambda}$

37. Assertion: Displacement current decreases with the increase in frequency of a.c. supplied to a capacitor

Reason: Reactance due to capacitance is directly proportional to the frequency of a.c.

38. Assertion: The electrostatic field lines cannot form a closed path.

Reason: $\oint \vec{E}.\vec{dl} = 0$

39. Assertion: The magnetic flux through a closed surface is zero

Reason: Gauss's law applies in the case of electric flux only

40. Assertion: A changing electric-field produces a magnetic field.

Reason:A changing magnetic field produces an electric field.

41. Statement I: Sound waves are not electromagnetic waves.

Statemenet II: Sound waves require a material medium for propagation.

42. Statement I: Electromagnetic waves are transverse in nature.

Statement II: The electric and magnetic fields of an e.m. wave are perpendicular to each other and also perpendicular to the direction of wave propagation.

43. Statement I: Electromagnetic waves exert pressure, called radiation pressure.

Statement II: This is because they carry energy.

44. **Statment I: in an electric circuit a capacitor of reactance 100 Ω is connected across a 220 V source. The displacement current is 2.2 A.**
Statement II: The data is insufficient.

45. **Statement I: An e.m. radiation of energy 14.4 keV belongs to X-ray region.**
Statement II: $E = hv = hc/\lambda$

46. **Statement I : The velcoity of all electromagnetic waves in vacuum is different.**
Statement II: The different electromagnetic waves are of different frequency

47. **Column I**
(A) Aveage energy density of electric field in electromagnetic wave
(B) Average energy density of magnetic field in electromagnetic wave
(C) Total average energy density of electromagnetic wave
(D) Intensity of electromagnetic wave
Column II
(P) $\frac{1}{2}\epsilon_0 E_0^2 c$ (Q) $\frac{1}{2}\epsilon_0 E_0^2$ (R) $\frac{1}{4}\epsilon_0 E_0^2$ (S) $\frac{1}{4}\epsilon_0 \frac{B_0^2}{\mu_0}$
1. A-P, B-Q, C-R, D-S 2. A-Q, B-R, C-S, D-P
3. A-R, B-S, C-Q, D-P 4. A-Q, B-R, C-S, D-P

48. **Column I** / **Column II**
(A) Radiowaves (P) Vibrations of atoms and molecules
(B) Infrared radiations (Q) Arc lamp
(C)Ultraviolet rays (R) Nuclear origin
(D)Gamma rays (S) Oscillating circuit
1. A-P, B-Q, C-R, D-S 2. A-Q, B-R, C-S, D-P
3. A-R, B-S, C-P, D-Q 4. A-S, B-P, C-Q, D-R

CONCEPTUAL KEY

1) 3	2) 2	3) 2	4) 1	5) 2	6) 4	7) 2	8) 2	9) 1	10) 3
11) 3	12) 2	13) 3	14) 1	15) 3	16) 4	17) 4	18) 4	19) 4	20) 1
21) 3	22) 4	23) 4	24) 4	25) 1	26) 3	27) 4	28) 2	29) 2	30) 4
31) 1	32) 1	33) 2	34) 1	35) 1	36) 2	37) 3	38) 1	39) 3	40) 2
41) 1	42) 1	43) 2	44) 3 45) 2		46) 4	47) 3	48) 4		

LEVEL - I (CW)

DISPLACEMENT CURRENT

1. **A Parallel plate condenser of capacity 100 pF is connected to 230 V of AC supply of 300rad/sec frequency. The rms value of displacement current.**
1) $6.9\mu A$ 2) $2.3\mu A$ 3) $9.2\mu A$ 4) $4.6\mu A$

2. **A parallel plate capacitor of plate separation 2mm is connected in an electric circuit having source voltage 400V. If the plate area is $60\,cm^2$, then the value of displacement current for 10^{-6} sec. will be**
1) $1.062\,amp$ 2) 1.062×10^{-2} amp 3) $1.062\times10^{-3}\,amp$ 4) $1.062\times10^{-4}\,amp$

MAGNETIC FIELD PRODUCED BETWEEN PLATES OF PARALLEL PLATE CAPACITOR

3. **The magnetic field between the plates of a capacitor when $r > R$ is given by -**

1. $\frac{\mu_0 I_D r}{2\pi R^2}$ 2. $\frac{\mu_0 I_D}{2\pi R}$ 3. $\frac{\mu_0 I_D}{2\pi r}$ 4. zero

4. **A condenser is charged using a constant current. The ratio of the magnetic fields at a distance of $\frac{R}{2}$ and R from the axis is (R is the radius of plate)**

1) 1:1 2) 2:1 3) 1:2 4) 1:4

5. **The magnetic field between the plates of radius 12 cm separated by distance of 4mm of a parallel plate capacitor of capacitance 100 pF. along the axis of plates having conduction current of 0.15 A is**

1) zero 2) 1.5 T 3) 15 T 4) 0.15 T

WAVE EQUATION

6. **The wave function (in S.I unit) for an electromagnetic wave is given as - $\psi(x,t) = 10^3 \sin \pi \left(3\times10^6 x - 9\times10^{14} t\right)$**

The speed of the wave is

1. 9×10^{14} m/s 2. 3×10^8 m/s 3. 3×10^6 m/s 4. 3×10^7 m/s

7. **The velocity of all radiowaves in free space is 3×10^8 m/s, the frequency of a wave of wavelength 150 m is**

1. 45 MHz 2. 2 MHz 3. 2 KHz 4. 20 KHz

8. **The relative permeability of glass is 3/8 and the dielectric constant of glass is 8. The refractive index of glass is**

1) 1.5 2) 1.1414 3) 1.732 4) 1.6

9. **An electromagnetic wave of frequency 3 MHz passes from Vacuum into a dielectric medium with permittiviy $\in = 4.0$ Then**

1) Wave length doubled and frequency remains unchanged
2) wave length is doubled and frequency becomes half
3) wave length is halved and frequency remains unchanged
4) wave length and frequency both remain unchanged

RELATION BETWEEN B & E

10. **In a plane electromagnetic wave, the electric field oscillates sinusoidally at a frequency of 2×10^{10} Hz and amplitude 48V/m. The amplitude of oscillating magnetic field will be**

1. $\frac{1}{16}\times10^{-8}$ Wb/m^2 2. 16×10^{-8} Wb/m^2 3. 12×10^{-7} Wb/m^2 4. $\frac{1}{12}\times10^{-7}$ Wb/m^2

11. **In a plane electromagnetic wave, the electric field oscillates sinnusoidally at a frequency of 2×10^{10} Hz and amplitude 48V/m., the wavelength of the wave will be -**

1. 1.5 m 2. 66.6 m 3. 1.5 cm 4. 66.6 cm

12. In an apparatus the electric field was found to oscillate with an amplitude of $18Vm^{-1}$. The rms of the oscillating magnetic field is

1) $6\times10^{-8}T$ 2) $4.23\times10^{-8}T$ 3) $9\times10^{-8}T$ 4) $7.0\times10^{-8}T$

13. The amplitude of the sinusoidally oscillating electric field of a plane wave is 60v/m. Then the amplitude of the magnetic field is

1) $12\times10^{7}T$ 2) $6\times10^{7}T$ 3) $6\times10^{-7}T$ 4) $2\times10^{-7}T$

MOMENTUM AND FORCE

14. Light with energy flux of $18w/cm^2$ falls on a non reflecting surface of area $20cm^2$ at normal incidence the momentum delivered in 30 minutes is

1) $1.2\times10^{-6}kgms^{-1}$ 2) $2.16\times10^{-3}kgms^{-1}$ 3) $1.18\times10^{-3}kgms^{-1}$ 4) $3.2\times10^{-3}kgms^{-1}$

15. Light with energy flux of $24wm^{-2}$ is incident on a well polished disc of radius 3.5cm for one hour. The momentum transferred to the disc is

1) $1.1\mu\ kg\ ms^{-1}$ 2) $2.2\mu\ kg\ ms^{-1}$ 3) $3.3\mu\ kg\ ms^{-1}$ 4) $4.4\mu\ kg\ ms^{-1}$

ENERGY DENSITY

16. The maximum electric field of a plane electromagnetic wave is 88 V/m. The average energy density is

1) $3.4\times10^{-8}Jm^{-3}$ 2) $13.7\times10^{-8}Jm^{-3}$ 3) $6.8\times10^{-8}Jm^{-3}$ 4) $1.7\times10^{-8}Jm^{-3}$

17. The rms value of the electric field of a plane electromagnetic wave is 314V/m. The average energy density of electric field and the average energy density are

1) $4.3\times10^{-7}Jm^{-3};2.15\times10^{-7}Jm^{-3}$ 2) $4.3\times10^{-7}Jm^{-3};8.6\times10^{-7}Jm^{-3}$

3) $2.15\times10^{-7}Jm^{-3};4.3\times10^{-7}Jm^{-3}$ 4) $8.6\times10^{-7}Jm^{-3};4.3\times10^{-7}Jm^{-3}$

INTENSITY

18. If a source of power 4 kW produces 10^{20} photons/second, the radiation belong to a part of the spectrum called

1. X - rays 2. Ultraviolet rays 3. Microwaves 4. γ rays

19. The intensity of electromagnetic wave at a distance of 1 Km from a source of power 12.56 kw. is

1) $10^{-3}Wm^{-2}$ 2) $4\times10^{-3}Wm^{-2}$ 3) $12.56\times10^{-3}Wm^{-2}$ 4) $1.256\times10^{-3}Wm^{-2}$

20. The sun delivers $10^3W/m^2$ of electromagnetic flux incident on a roof of dimensions $8m\times20m$, will be

1) 6.4×10^3W 2) 3.4×10^4W 3) 1.6×10^5W 4) 3.2×10^5W

LEVEL-I (C.W) - KEY

1) 1	2) 2	3) 3	4) 3	5) 1	6) 2	7) 2	8) 3	9) 3	10) 2
11) 3	12) 2	13) 4	14) 2	15) 2	16) 3	17) 2	18) 1	19) 1	20) 3

LEVEL-I (H.W) - HINTS

1. $I_d = \frac{V}{Z} = \frac{V}{X_C} = VC\omega$

2. $I_d = \varepsilon_0 \frac{VA}{td}$

3. $r > R \quad B = \frac{\mu_0 I_d}{2\pi r}$

4. $r < R \, B = \frac{\mu_0 I_0 r}{2\pi R^2}$; $r = R \quad B = \frac{\mu_0 I_d}{2\pi R}$

5. $B = \frac{\mu_0 I_d r}{2\pi R^2}$; $r = 0$

6. $C = \frac{\omega}{K}$

7. $C = n\lambda$

8. RI of glass $= \frac{C_0}{C} = \sqrt{\frac{\mu\varepsilon}{\mu_0\varepsilon_0}} = \sqrt{\mu_r\varepsilon_r}$

9. $C = n\lambda$

10. $B = \frac{E}{C}$

11. $\lambda = \frac{C}{\nu}$

12. $C = \frac{E}{B}$

13. $C = \frac{E_0}{B_0}$

14. $p = \frac{U}{C} = \frac{IAt}{C}$

15. $P = \frac{2U}{C} = \frac{2IAt}{C}$

16. $U_{av} = \varepsilon_0 E_{Max}^2$

17. $(U_{av})_E = \frac{1}{2}\varepsilon_0 E_{Max}^2$; $U_{av} = \varepsilon_0 E_{\max}^2$

18. $P = \frac{nE}{t}$; $E = \frac{hc}{\lambda}$

19. $I = \frac{P}{4\pi r^2}$

20. $\phi = PA$

LEVEL - I

DISPLACEMENT CURRENT

1. The voltage between the plates of a parallel plate condenser of capacity $2.0\ \mu F$ is charging at a rate of $10Vs^{-1}$. The displacement current

1) $2mA$ 2) $2\mu A$ 3) $20\mu A$ 4) $2A$

2. A parallel plate condenser of capacity $10\mu F$ is charged with a constant charging current of 0.16A. The displacement current is

1) $0.16\mu A$ 2) $0.16A$ 3) $0.96A$ 4) $9.6A$

MAGNETIC FIELD PRODUCED BETWEEN PLATES OF PARALLEL PLATE CAPACITOR

3. The graph representing the variation of induced magnetic field in the gap of the condenser plates during its charging with the distance from the axis of the gap is

1)
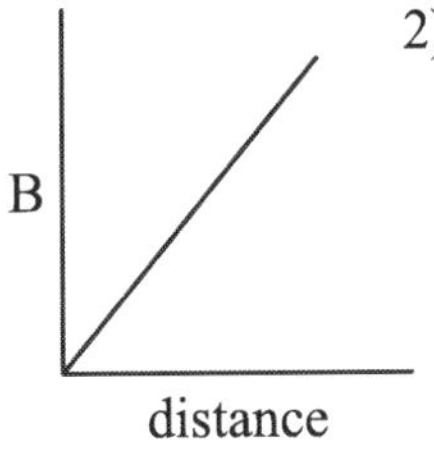

2)
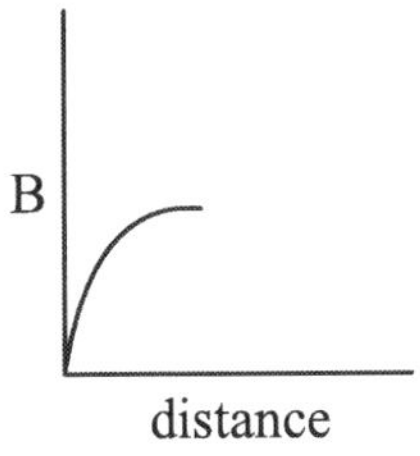

3)
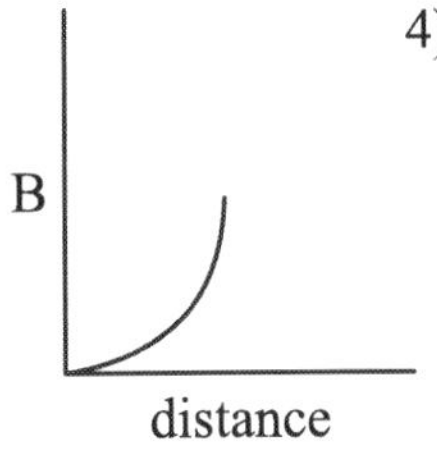

4)
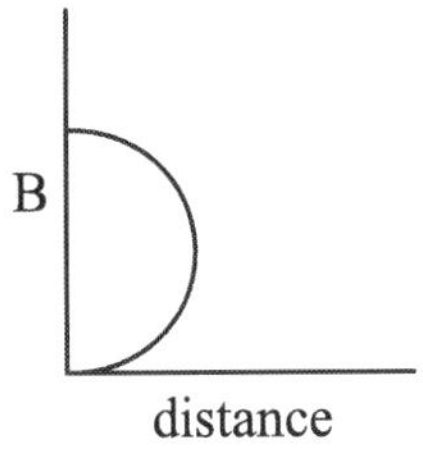

4. The electrical field in the gap of a condenser charges as $10^{12}Vm^{-1}s^{-1}$. If the radius of each plate of the condenser is 3cm, the magnetic field at the edge of plate in the gap is

1) $1.67\,mT$ 2) $0.167\,\mu T$ 3) $0.5\,\mu T$ 4) $5\,\mu T$

WAVE EQUATION

5. An LC circuit contains inductance $L = 1\mu H$ and capacitance $C = 0.01\mu H$. The wavelength of electromagnetic wave generated is nearly

1. 0.5 m 2. 5 m 3. 188 m 4. 30 m

6. The wave length of the Green light of mercury is 550nm. If the refractive index of the glass is 1.5, the time period of the electrical vector in glass nearly $\left(C_0 = 3\times10^8 mS^{-1}\right)$

1) $1.8\times10^{-9}S$ 2) $3.6\times10^{-15}S$ 3) $9\times10^{-15}S$ 4) $2.75\times10^{-15}S$

7. The all India Radio, station at Vijayawada transmits its signals at 840 K C/s. The length of the radio wave is

1) 35.7m 2) 357m 3) 35.7 km 4) 3.57 m

RELATION BETWEEN B & E

8. **A point source of electromagnetic radiation has an average power output of 800W, The maximum value of electric field at a distance 3.5 m from the source will be 62.6 V/m, the maximum value of magnetic field will be -**

1. 2.09×10^{-5}T 2. 2.09×10^{-6}T 3. 2.09×10^{-7}T 4. 2.09×10^{-8}T

9. **A plane E.M. wave of frequency 40 MHz travels along X-axis. At same point at same instant, the electric field E has maximum value of 750 N/C in Y-direction. The magnitude and direction of magnetic field is**

1) $2.5\mu T$ along X-axis 2) $2.5\mu T$ along Y-axis

3) $2.5\mu T$ aloing Z-axis 4) $5\mu T$ along Z-axis

10. **A plane electromagnetic wave of frequency 25 MHz travels in free space along the x-direction. At a particular point in space and time $\overline{E} = 6.3\hat{j}$. The magnetic field $\overline{B}$ at this point is**

1) $4.2\times10^{-8}\hat{k}T$ 2) $2.1\times10^{-8}\hat{k}T$ 3) $18.9\times10^{8}\hat{k}T$ 4) $2.1\times10^{8}\hat{k}T$

MOMENTUM AND FORCE

11. **Light with energy flux $36w/cm^2$ is incident on a well polished metal square plate of side 2cm. The force experienced by it is**

1) $0.96\mu N$ 2) $0.24\mu N$ 3) $0.12\mu N$ 4) $0.36\mu N$

ENERGY DENSITY

12. **The rms value of the electric field of the light coming from the sun is 720 N/C. The average total energy density of the electromagnetic wave is**

1) $3.3\times10^{-3} J/m^3$ 2) $4.58\times10^{-6} J/m^3$

3) $6.37\times10^{-9} J/m^3$ 4) $81.35\times10^{-12} J/m^3$

13. **In an electromagnetic wave, the amplitude of electric field is 1V/m. The frequency of wave is $5\times10^{14} Hz$. The wave is propagating along z-axis. The average energy density of electric field, in $Joule/m^3$, will be**

1) 1.1×10^{-11} 2) 2.2×10^{-12} 3) 3.3×10^{-13} 4) 4.4×10^{-14}

INTENSITY

14. **About 5% of the power of a 100 W light bulb is converted to visible radiation. The average intensity of visible raditaion at a distance of 1 m from the bulb:**

1. $0.4W/m^2$ 2. $0.5W/m^2$ 3. $0.6W/m^2$ 4. $0.8W/m^2$

15. **The sun radiates electromagnetic energy at the rate of $3.9\times10^{26} W$. Its radius is $6.96\times10^{8} m$. The intensity of sun light at the solar surface will be (in W/m^2)**

1) 1.4×10^4 2) 2.8×10^5 3) 64×10^6 4) 5.6×10^7

16. **The intensity of TV broad cast station of $E = 800\,Sin\left(10^9 t - kx\right)$ V/M is...... and the wave length in meter is ......**

1) $850\,wm^{-2};0.6\pi$ 2) $425\,wm^{-2};0.6\pi$ 3) $850\,wm^{-2};0.3\pi$ 4) $425\,wm^{-2};0.3\pi$

LEVEL-I - KEY

1) 3	2) 2	3) 1	4) 2	5) 3	6) 4	7) 2	8) 3	9) 3	10) 2
11) 1	12) 2	13) 2	14) 1	15) 3	16) 1				

LEVEL-I - HINTS

1. $I_d = C\dfrac{dv}{dt}$

2. $I_C = I_d$

3. $B = \dfrac{\mu_0 i}{2\pi r}$ and $i = \dfrac{r^2 i_0}{R^2}$

4. $\int \vec{B}.\vec{dl} = \mu_0 \varepsilon_0 \dfrac{d\phi_E}{dt}$; $B(2\pi r) = \dfrac{1}{C^2}\dfrac{dE}{dt}\pi r^2$

5. $n = \dfrac{1}{2\pi\sqrt{LC}}, \lambda = \dfrac{C}{n}$

6. $\mu = \dfrac{C_0}{C}, T = \dfrac{\lambda}{C}$

7. $\lambda = \dfrac{C}{n}$

8. $B_m = \dfrac{E_m}{C}$

9. $B = \dfrac{E}{C}$

10. $B = \dfrac{E}{C}$

11. $F = \dfrac{2U}{Ct} = \dfrac{2IAt}{Ct}$

$= \dfrac{2IA}{C}$

12. $U_{av} = \dfrac{1}{2}\varepsilon_0 E_{\max}^2$

13. $(V_{av})_E = \dfrac{1}{4}\varepsilon_0 E_{\max}^2$

14. $I = \dfrac{P}{4\pi r^2}$

15. $I = \dfrac{P}{4\pi r^2}$

16. $\lambda = \dfrac{c}{n} = c\left(\dfrac{2\pi}{\omega}\right)$

$I = \dfrac{E_0 B_0}{2\mu_0} = \dfrac{E_0^2}{2\mu_0 C}$

LEVEL - II

DISPLACEMENT CURRENT

1. **A parallel plate condenser consists of two circular plates each of radius 2cm separated by a distance of 0.1mm. A time varying potential difference of $5\times10^{13} v/s$ is applied across the plates of the condenser. The displacement current is**

   1) $5.50A$ 2) $5.56\times10^{2} A$ 3) $5.56\times10^{3} A$ 4) $2.28\times10^{4} A$

2. **A parallel plate condenser has conducting plates of radius 12cm separated by a distance of 5mm. It is charged with a constant charging current of 0.16 A, the rate at which the potential difference between the plate change is**

   1) $1\times10^{9} Vs^{-1}$ 2) $2\times10^{10} Vs^{-1}$ 3) $3\times10^{12} Vs^{-1}$ 4) $2\times10^{9} Vs^{-1}$

MAGNETIC FIELD PRODUCED BETWEEN PLATES OF PARALLEL PLATE CAPACITOR

3. **A condenser has two conducting plates of radius 10cm separated by a distance of 5mm. It is charged with a constant current of 0.15A. The magnetic field at a point 2cm from the axis in the gap is**

   1) $1.5\times10^{-6}T$ 2) $3\times10^{-8}T$ 3) $6\times10^{-8}T$ 4) $3\times10^{-6}T$

4. **An AC rms voltage of 2V having a frequency of 50 KHz is applied to a condenser of capacity of $10\mu F$. The maximum value of the magnetic field between the plates of the condenser if the radius of plate is 10cm is**

   1) $0.4 p\mu$ 2) $4\pi \mu T$ 3) $2\mu T$ 4) $40\pi \mu T$

WAVE EQUATION

5. **The wave emitted by any atom or molecule must have some finitetotal length which is known as the coherence length. For sodium light, this length is 2.4cm. The number of oscillations in this length will be Given $\lambda = 5900A^{0}$**

   1) $4.068\times10^{5} Hz$ 2) $4.068\times10^{4} Hz$ 3) $4.068\times10^{6} Hz$ 4) $4.068\times10^{8} Hz$

6. **A wave is propagating in a medium of dielectric constant 2 and relative permeability 50. The wave impedance is**

   1) 5Ω 2) 376.6Ω 3) 3776Ω 4) 1883Ω

RELATION BETWEEN B & E

7. **The magnetic field in travelling EM wave has a peak value of 20nT. The peak value of electric field strength is (AIE : 2013)**

   1) 6 V/m 2) 9 V/m 3) 12 V/m 4) 3 V/m

MOMENTUM AND FORCE

8. **A plane electromagnetic wave of wave intensity $6W/m^2$ stikes a small mirror of area $40\ cm^2$, held prependicular to the approaching wave. The momentum transfered by the wave to the mirror each second will be**

   1. 6.4×10^{-7}kg-m/s 2. 4.8×10^{-8}kg-m/s 3. 3.2×10^{-9}kg-m/s 4. 1.6×10^{-10}kg-m/s

9. **In the above question the radiation force on the mirror will be**

   1. 6.4×10^{-7}N 2. 4.8×10^{-8}N 3. 3.2×10^{-9}N 4. 1.6×10^{-10}N

ENERGY DENSITY

10. **A point source of electromagnetic radiation has an average power output of 800W. The maximum value of electric field at a distance 3.5 m from the source will be 62.6 V/m, the energy density at a distance 3.5 m from the source will be - (in joule/m³)**

    1. 1.73×10^{-5} 2. 1.73×10^{-6} 3. 1.73×10^{-7} 4. 1.73×10^{-8}

11. **An electromagnetic radiation has an energy 14.4 KeV. To which region of electromagnetic spectrum does it belong?**
1. Infra red region 2. Visible region 3. X-rays region 4. γ- ray region

INTENSITY

12. **A laser beam can be focussed on an area equal to the square of its wavelength. A He-Ne laser radiates energy at the rate of 1mW and its wavelength is 600 nm. The intensity of focussed beam will be**
1) $3.2\times10^{9}W/m^{2}$ 2) $2.8\times10^{13}W/m^{2}$ 3) $2.7\times10^{9}W/m^{2}$ 4) $3.2\times10^{13}W/m^{2}$

13. **The intensity of solar radiation at the earths surface is $1KW\,m^{-2}$. The power entering the pupil of an eye of diameter 0.5 cm is**
1. 39.2 mw 2) 19.6 mw 3) 9.8 mw 4) 4.9 mw

LEVEL-II (C.W.) - KEY

1) 3	2) 4	3) 3	4) 3	5) 2	6) 4	7) 1	8) 4	9) 4	10) 4
11) 4	12) 3	13) 2							

LEVEL-II (C.W.) - HINTS

1. $I_d = \frac{\varepsilon_0 A}{d}\times\frac{dv}{dt}$

2. $\frac{dv}{dt} = \frac{dI_d}{\varepsilon_0 A}$

3. $B = \frac{\mu_0 ir}{2\pi R^2}$

4. $B = \frac{\mu_0 C}{2\pi R}\left(\frac{dv}{dt}\right)$

5. No.of oscillations in this length $= \frac{l}{\lambda}$

6. Wave impedence $Z = \frac{E}{H} = \frac{CB}{H} = \sqrt{\frac{\mu}{\varepsilon}}$

7. $B_m = \frac{E_m}{C}$

8. $P = \frac{2U}{C} = \frac{2\delta_{av}A}{C}$

9. Momentum per sec is force

10. Energy density $U = \frac{1}{2}\varepsilon_0 E_m^2$

11. $\lambda = \frac{12,400}{14.4\times10^3} = \frac{hc}{E}$

12. $I = \frac{P}{A} = \frac{P}{\lambda^2}$

13. $P = IA = I\pi R^2$

LEVEL - II

DISPLACEMENT CURRENT

1. **The area of each plate of a parallel plated condenser is $144\,cm^2$. The electrical field in the gap between the plates changes at the rate of $10^{12}\,V\,m^{-1}\,s^{-1}$. The displacement current is**

1) $\frac{4}{\pi}A$ 2) $\frac{0.4}{\pi}A$ 3) $\frac{40}{\pi}A$ 4) $\frac{1}{10\pi}A$

2. **A condenser having circular plates having radius 2cm and separated by a distance of 3mm. It is charged with a current of 0.1 A. The rate at which the potential difference between the plates change is**

1) $9\times10^{10}V/S$ 2) $1.8\times10^{10}V/S$ 3) $2.7\times10^{6}V/S$ 4) $2.7\times10^{10}V/S$

3. **An AC source having a frequency of 50 Hz and voltage supply of 300v is applied directly to the condenser of capacity $100\mu F$. The peak and rms values of displacement current are**

1) $9.42\,A;\frac{9.42}{\sqrt{2}}A$ 2) $\frac{9.42}{\sqrt{2}}A;9.42\sqrt{2}A$ 3) $9.42\sqrt{2}A;9.42A$ 4) $9.42A;9.42A$

MAGNETIC FIELD PRODUCED BETWEEN PLATES OF PARALLEL PLATE CAPACITOR

4. **The capacity of a parallel plate condenser is 50 pF. A magnetic field of $4\times10^{-7}T$ is produced at a distance of 10cm from the axis of the gap. The charging current is**

1) 0.1A 2) 0.2 A 3) 0.3 A 4) 0.15 A

5. **The diameter of the condenser plate is 4cm. It is charged by an external current of 0.2A. The maximum magnetic field induced in the gap**

1) $2\mu T$ 2) $4\mu T$ 3) $6\mu T$ 4) $8\mu T$

6. **A condenser of capacity 50 p F is connected to an AC supply of 220 V 50 Hz. The rms value of magnetic field at a distance of 5cm from the axis is**

1) $22\pi\,x10^{-14}T$ 2) $22\pi\,x10^{-12}T$ 3) $44\pi\,x10^{-13}T$ 4) $\frac{11}{5}\pi\,x10^{-12}T$

WAVE EQUATION

7. **The velocity of an electromagnetic wave in a medium is $2\times10^{8}\,mS^{-1}$. If the relative permeability is 1 the relative permittivity of the medium is $\left(C_0 = 3\times10^{8}\,mS^{-1}\right)$**

1) 2.25 2) 1.5 3) 4/9 4) 2/3

RELATION BETWEEN B & E

8. **In a plane electromagnetic wave, the electric field oscillates sinusoidally at a frequency of $2\times10^{10}\,Hz$ and amplitude 48 V/m. The amplitude of oscillating magnetic field will be**

1) $\frac{1}{16}\times10^{-8}Wb/m^2$ 2) $16\times10^{-8}Wb/m^2$

3) $12\times10^{-7}Wb/m^2$ 4) $\frac{1}{12}\times10^{-7}Wb/m^2$

9. **In an apparatus, the electric field was found to oscillate with amplitude of 18 V/m. The amplitude of the oscillating magnetic field will be**

1) $4\times10^{-6}T$ 2) $6\times10^{-8}T$ 3) $9\times10^{-9}T$ 4) $11\times10^{-11}T$

MOMENTUM AND FORCE

10. **Light with energy flux $36Wm^{-2}$ is incident on a circular part of radius 1.4 m of a perfectly black body. The force experienced by the body and the momentum delivered in 10 minutes are**

1) $2.2\mu N; 7.2\mu\, kgms^{-1}$ 2) $3.5\mu N; 7.4\mu\, kgms^{-1}$

3) $0.74\mu N; 444\mu\, kgms^{-1}$ 4) $7.4\mu N; 2.2\mu\, kgms^{-1}$

11. **Light with energy flux $18\,wcm^{-2}$ is incident on a mirror of size $2cm\times2cm$ normally. The force experienced by it and momentumdelivered in one minute are**

1) $0.48\,\mu N; 28.8\,\mu\, kgms^{-1}$ 2) $48\,\mu N; 2.88\,\mu\, kgms^{-1}$

3) $28.8\,\mu N; 4.8\,\mu\, kgms^{-1}$ 4) $0.24\,\mu N; 28.8\,\mu\, kgms^{-1}$

12. **Electromagnetic radiation with energy flux $50\,W\,cm^{-2}$ is incident on a totally absorbing surface normally for 1 hour. If the surface has an area of $0.05\,m^2$, then the average force due to the radiation pressure, on it is;**

1) $8.3\times10^{-7}N$ 2) $8.3\times10^{-5}N$ 3) $1.2\times10^{-7}N$ 4) $1.2\times10^{-5}N$

ENERGY DENSITY

13. **In an electromagnetic wave in vacuum. The electrical and magnetic fields are $40\pi\, V/m$ and $0.4\times10^{-7}T$. The Poynting vector**

1) $4.4Wm^{-1}$ 2) $0.44Wm^{-1}$ 3) $5.65Wm^{-1}$ 4) $4.0Wm^{-1}$

INTENSITY

14. **The amplitude of magnetic field at a region carried by an electromagnetic wave is $0.1\mu T$. The intensity of wave is**

1) $4\mu Wm^{-2}$ 2) $1.2Wm^{-2}$ 3) $4Wm^{-2}$ 4) $1.2\mu Wm^{-2}$

LEVEL-II - KEY

1) 2 2) 4 3) 3 4) 2 5) 1 6) 2 7) 1 8) 2 9) 2 10) 3
11) 1 12) 2 13) 4 14) 2

LEVEL-II - HINTS

1. $I_d = \varepsilon_0 A \dfrac{dE}{dt}$ 2. $\dfrac{dv}{dt} = \dfrac{dI_d}{\varepsilon_0 A}$

3. $I_{rms} = \dfrac{E_{rms}}{X_C}$; $I_{rms} = \dfrac{I_0}{\sqrt{2}}$ 4&5. $B = \dfrac{\mu_0 i}{2\pi r}$

6. $i_{rms} = \dfrac{E_{rms}}{X_C}, B_{rms} = \dfrac{\mu_0 i_{rms}}{2\pi r}$ 7. $RI = \dfrac{1}{\sqrt{\mu_r \varepsilon_r}} = \dfrac{C_0}{C}$

8&9. $B_0 = \frac{E_0}{C}$

10,11. Momentum $P = \frac{U}{C} = \frac{IAt}{C}$; Force $F = \frac{IA}{C}$

12. Force $F = \frac{IA}{C}$

13. Poynting vector $= \frac{B_0 E_0}{\mu_0}$ 14. $I = \frac{B_0^2 C}{2\mu_0}$

LEVEL - III

1. **A parallel plate capacitor of plate seperation 2 mm is connected in an electric circuit having source voltage 400V. If the plate area is $60\,cm^2$, then the value of displacement current for 10^{-6} sec. will be-**

1. 1.062 amp. 2. 1.062×10^{-2}amp 3. 1.062×10^{-3}amp 4. 1.062×10^{-4}amp

2. **A long straight wire of resistance R, radius 'a' and length 'l' carries a constant current 'I'. The poynting vector for the wire will be-**

1. $\frac{IR}{2\pi al}$ 2. $\frac{IR^2}{al}$ 3. $\frac{I^2R}{al}$ 4. $\frac{I^2R}{2\pi al}$

3. **To establish an instantaneous displacement current of 2A in the space between two parallel plates of $1\,\mu F$ capacitor, the potential difference across the capacitor plates will have to be changed at the rate of**

1) $4\times10^4 V/s$ 2) $4\times10^6 V/s$ 3) $2\times10^4 V/s$ 4) $2\times10^6 V/s$

4. **The sun delivers $10^3 W/m^2$ of electromagnetic flux to the earth's surface. The total power that is incident on a roof of dimensions 8m × 20m, will be -**

1. 6.4×10^3 W 2. 3.4×10^4 W 3. 1.6×10^5 W 4. none of these

5. **The sun delivers $10^3 W/m^2$ of electromagnetic flux to the earth's surface. The total power that is incident on a roof of dimensions 8m × 20m is 1.6×10^5 W, the radiation force on the roof will be -**

1. 3.33×10^{-5}N 2. 5.33×10^{-4}N 3. 7.33×10^{-3}N 4. 9.33×10^{-2}N

6. **An electric field of 300V/m is confined to a circular area 10 cm in diameter. If the field is increasing at the rate of 20V/m-s, the magnitude of magnetic field at a point 15cm from the centre of the circle will be-**

1. 1.85×10^{-15}T 2. 1.85×10^{-16}T 3. 1.85×10^{-17}T 4. 1.85×10^{-18}T

7. **A lamp emits monochromatic green light uniformly in all directions. The lamp is 3% efficient in converting electrical power to electromagnetic waves and consumes 100W of power. The amplitude of the electric field associated with the electromagnetic radiation at a distance of 10m from tha lamp will be -**

1. 1.34 V/m 2. 2.68 V/m 3. 5.36V/m 4.9.37 V/m

8. **A flood light is covered with a filter that transmits red light. The electric field of the emerging beam is represented by a sinusoidal plane wave**

$E_x = 36\sin\left(1.20\times10^7 z - 3.6\times10^{15} t\right) V/m$

The average intensity of the sun will be-

1. $0.86 W/m^2$ 2. $1.72 W/m^2$ 3. $3.44 W/m^2$ 4. $6.88 W/m^2$

9. **A plane electromagnetic wave of frequency 40 MHz travels in free space in the X-direction. At some point and at some instant, the electric field $\vec{E}$ has its maximum value of 750 N/C in Y-direction. The wavelength of the wave is-**

1. 3.5 m 2. 5.5 m 3. 7.5 m 4. 9.5 m

10. **A plane electromagnetic wave propagating in the x-direction has a wavelength of 60 mm. The electric field is in the y-direction and its maximum magnitude is $33 Vm^{-1}$. The equation for the electric field as function of x and t is**

1. $11\sin\pi(t - x/c)$ 2. $33\sin\pi\times10^{10}(t - x/c)$

3. $33\sin\pi(t - x/c)$ 4. $11\sin\pi\times10^{10}(t - x/c)$

LEVEL-III - KEY

1) 2 2) 4 3) 4 4) 3 5) 2 6) 4 7) 1 8) 2 9) 3 10) 2

LEVEL-III - HINTS

1. $I_D = \varepsilon_0 \frac{\partial \phi_E}{\partial t} = \varepsilon_0 \frac{EA}{t} = \frac{\varepsilon_0 A}{t}\frac{v}{d}$ $I_D = \frac{8.85\times10^{-12}\times400\times60\times10^{-4}}{2\times10^{-3}\times10^{-6}} = 1.062\times10^{-2} amp$

Hence the corrent answer will be (4).

2. If V is the potential difference across the wire, then $E = \frac{V}{\ell} = \frac{IR}{\ell}$ and magnetic field at the surface of wire $B = \frac{\mu_0 I}{2\pi a}$. Hence poynting vector, directed radially inwards, is gives by $S = \frac{EB}{\mu_0} = \frac{IR}{\mu_0 \ell}\frac{\mu_0 I}{2\pi a} = \frac{I^2 R}{2\pi a l}$

Hence the correct answer will be (4).

3. $I_D = \varepsilon_0 \frac{d\phi_E}{dt} = \varepsilon_0 \frac{d}{dt}(EA) = \varepsilon_0 A \frac{d}{dt}\left(\frac{V}{d}\right)$

or $I_D = \frac{\varepsilon_0 A}{d}\frac{dV}{dt} = C\frac{dV}{dt}$

$\therefore \frac{dV}{dt} = \frac{I_D}{C} = \frac{2}{10^{-6}} = 2\times10^6 V/s$

Hence the correct answer will be (4).

4. $P = 10^3 \times 160 = 1.6\times10^5 W$

5. $F = PA = \frac{SA}{c} = \frac{1.6\times10^5}{3\times10^8} = 5.33\times10^{-4} N$

(or) F = total power / velocity of light

6. $B = \frac{\mu_0 \varepsilon_0}{2\pi R}\left(\frac{\pi d^2}{4}\right)\frac{dE}{dt}$

$= \frac{2\times10^{-7}\times8.85\times10^{-12}\times3.14\times0.01\times20}{4\times0.15}$

$=1.85\times10^{-18}T$

7. $S_{av} = \frac{P}{4\pi R^2} = \frac{1}{2}\varepsilon_0 c E_0^2 \qquad \therefore E_0 = \sqrt{\frac{P}{2\pi R^2 \varepsilon_0 c}}$

$= \sqrt{\frac{3}{2\times3.14\times100\times8.85\times10^{-12}\times3\times10^8}} = 1.34V/m$

8. $I_{av} = \frac{c\varepsilon_0 E_0^2}{2} = \frac{3\times10^8\times8.85\times10^{-12}\times36^2}{2} = 1.72W/m^2$

9. $\lambda = \frac{C}{f} = \frac{3\times10^8}{4\times10^7} = 7.5m$

10. $\omega = 2\pi v = \frac{2\pi c}{\lambda} = \frac{2\pi\times3\times10^8}{60\times10^{-3}} = 10^{10}\pi rad/s$

$E_y = E_0 \sin\omega\left(t-\frac{x}{c}\right) = 33\sin\pi\times10^{10}(t-x/c)$

Chapter - 5

DUAL NATURE OF MATTER

DUAL NATURE OF MATTER AND RADIATION

SYNOPSIS

Electron Emission :-

- The metals have free electrons and these normally cannot escape out of the metal surface.
- The free electron is held inside the metal surface by the attractive forces of the ions.
- A certain minimum amount of energy is requried to be given to an electron to pull it out from the surface of the metal and this energy is known as "Work Function".
- Work function (ϕ) = 5.65 eV, highest (for platinum) ϕ = 1.88 eV, lowest (for ceasium)

Metal	Work function ω_0 (eV)	Metal	Work function ω_0 (eV)
Cs	2.14	Al	4.28
K	2.30	Hg	4.49
Na	2.75	Cu	4.65
Ca	3.20	Ag	4.70
MO	4.17	Ni	5.15
Pb	4.25	Pt	5.56

- This minimum energy required for the electron emission can be supplied by any one of the following processes.

  a) Thermionic emission : "Sufficient thermal energy can be imported to free electrons" by suitably heating

  b) Field emission: "By applying a very strong electric field ($\approx 10^8 V/m$)".

  c) Photo electric emission: "By irradiating the metal surface with suitable E.M radiaton".

Photo electric effect :

- Photo electric effect was discovered by Hertz in 1887. In his experiments, Hertz observed that high voltage spark passes across the metal electrodes more easily when cathode is illuminated with ultra violet rays from an arc lamp.
- In 1888 Hallwachs under took the study further. He connected zinc plate to an electroscope. He found that when zinc plate is illuminated with ultra violet light it became positively charged. A positively charged zinc plate became more positively charged when it is further illuminated with ultra violet light .
- From these observations he concluded that negatively charged particles were emitted by the zinc plate under the action of ultra violet light. After the discovery of electron these particles were called as photo electrons.
- The emission of electrons from a metal plate when illuminated by electromagnetic radiation of suitable wavelength is called Photoelectric effect.

Lenard's Experimental Study of Photoelectric effect :

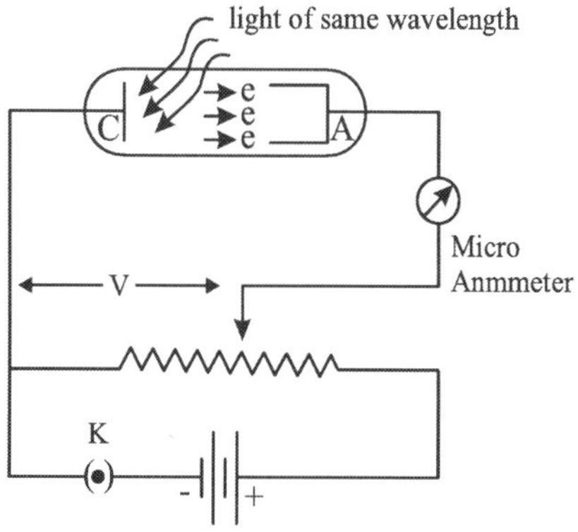

- The apparatus used for experimental study of photoelectric effect. A metal plate C called cathode (emitter) and a metal cup A called anode (collector) are sealed in a vacuum chamber.
- A beam of monochromatic light enters the window of a vacuum chamber and falls on cathode C. The photoelectrons emitted are collected by the anode A.
- When key K is open and monochromatic light is made incident on the cathode, then current is measured by the ammeter. i.e., even though applied voltage is zero current flows in the circuit.

  These photoelectrons emitted from the cathode C moves towards anode A. But less energetic electrons comes to rest before reaching the anode.
- When anode is given positive potential w.r.t the cathode, electrons in the space charge are attracted towards the anode so photocurrent increases. If potential of the anode is increased gradually the effect of space charge becomes negligible at some potential and then every electron that is emitted from the cathode will be able to reach the anode. The current then becomes constant even though voltage is increased and this current is called saturation photocurrent.
- When anode is given negative potential w.r.t the cathode, the photo electrons will be repelled by the anode and some electrons will go back to cathode so current decreases. At some negative potential anode current becomes zero.This potential is called stopping potential.
- The minimum negative potential(V_0) given to the collector with respect to the emitter for which 'photocurrent' becomes zero is called 'stopping potential'.
- Stopping potential is related to maximum kinetic energy of photoelectrons, because at this potential even the most energetic electron just fails to reach the anode.

  So work done by the stopping potential is equal to the maximum kinetic energy of the electrons.

  $$(-e)(-V_0)=\frac{1}{2}mv_{max}^2-0;\ \therefore\ eV_0=\frac{1}{2}mv_{max}^2$$
- A graph is plotted with current on y-axis and applied voltage on x-axis. It is as shown in below graph

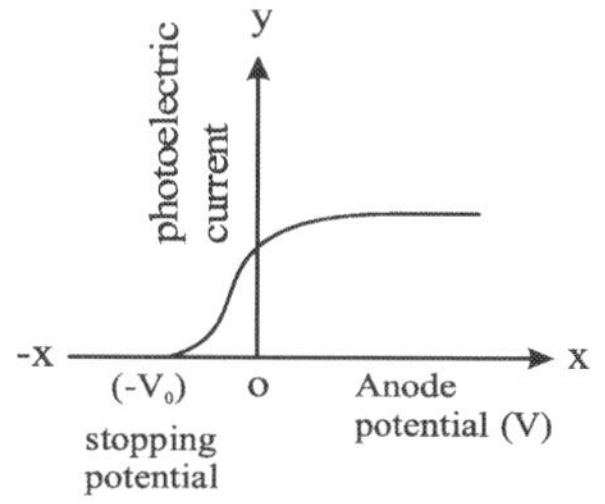

➢ **Variation of Photo current with intensity of incident light :** Keeping the frequency of incident light and nature of the cathode constant, for different intensities of incident light saturation photo current is measured.

When a graph is plotted with saturation photocurrent on y-axis and intensity of incident light on x-axis, it is as shown in figure.

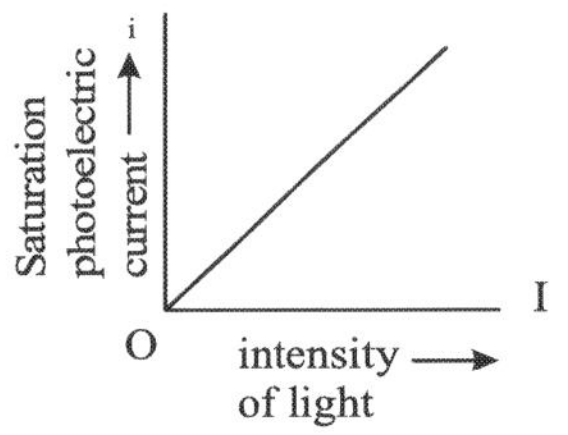

Observations :

(i) It is observed that saturation photocurrent (i) is proportional to the intensity (I) of incident light at a given frequency

➢ **Variation of saturation photo current with stopping potential at constant intensity :** Keeping the frequency of incident light and nature of the cathode constant, for different intensities of incident light photo current is measured.

When a graph is plotted with photocurrent on y-axis and applied voltage on x-axis. It is as shown in figure.

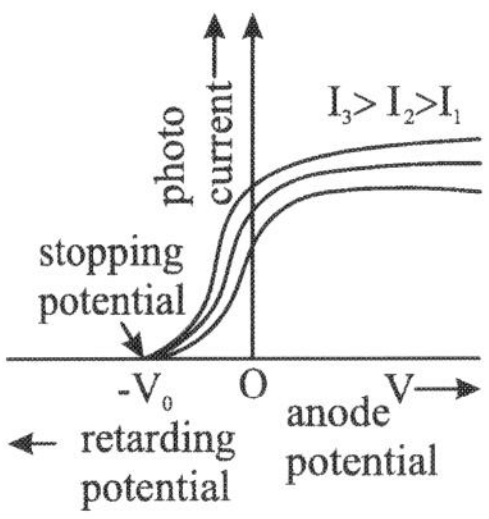

Observations :

(i) The value of stopping potential is independent of the intensity of incident light, if frequency is constant.

(ii) The magnitude of saturation current depends on the intensity of light. Higher the intensity, larger the saturation current.

➢ **Variation of frequency of incident light on stopping potential :**

Keeping the intensity of incident light and nature of the cathode constant, for different frequencies of incident light, photo current is measured.

When a graph is plotted with photocurrent on y-axis and applied voltage on x-axis. It is as shown in figure.

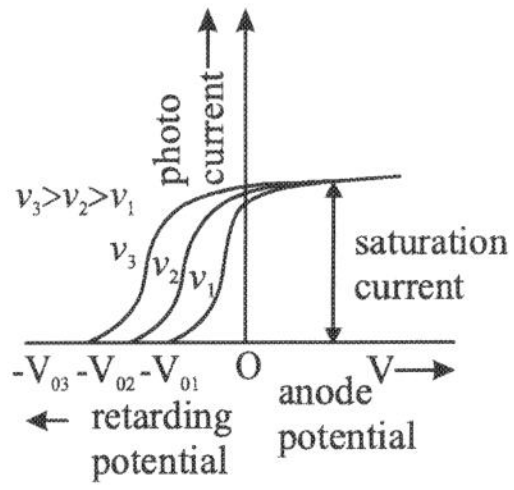

Observations :

(i) Larger the frequency of incident radiation, larger is the stopping potential.

So The maximum kinetic energy of the emitted electrons depends on the frequency of incident light and nature of the metal plate. Maximum kinetic energy of photo electrons is independent of the intensity of incident light.

(ii) The saturation photo current is independent of the frequency of incident radiation.

➢ **Variation of Stopping potential with frequency of incident light :**

When a graph is plotted with stopping potential on y-axis and frequency of incident radiation on x-axis, keeping the metal constant, then it is as shown in figure.

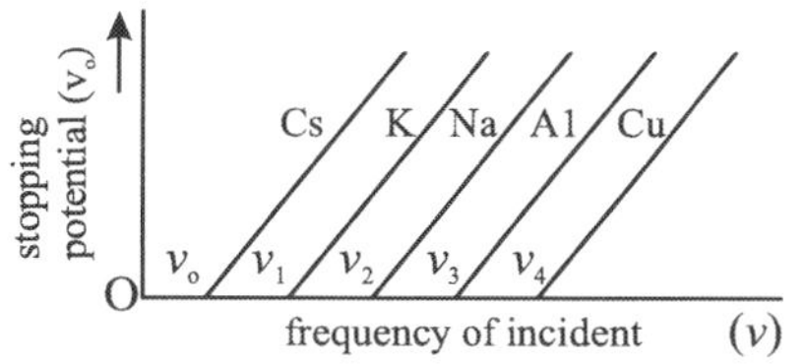

Observations :

(i) Threshold frequency (ν_0) is a characteristic of the metal plate and at this frequency, kinetic energy of the photo electrons is zero.

(ii) Above threshold frequency, kinetic energy of photo electrons range from zero to a maximum value.

(iii) Maximum kinetic energy and Stopping potential increases linearly with increasing frequency as shown in the above figure.

LAWS OF PHOTOELECTRIC EFFECT:

➢ If the frequency of incident radiation is less than a certain value called threshold frequency, electrons are not emitted from a given metal surface, whatever be the intensity of the incident radiation.

➢ The maximum kinetic energy of photoelectrons depends on the frequency of the incident radiation, but it is independent of the intensity of the radiation. The maximum kinetic energy of photoelectrons is a linear function of the frequency of the incident radiation.

➢ The photocurrent increases with intensity of incident radiation, but it is independent of the frequency of incident radiation.

➢ There is no time lag between the incidence of the incident radiation and the emission of photo electrons.

QUANTUM THEORY OF LIGHT:

➢ Since wave theory of light can not explain observed facts about photo electric effect, Einstein thought that light has got particle-like behaviour during its interaction with matter. According to this theory, light consists of particles associated with definite amount of energy and momentum. These particles were latter called as photons.

A photon is a packet of energy, given by $E = h\nu$

where h = 6.62×10^{-34} Js

(where Planck's constant, h = 6.62×10^{-34} Js

= 4.14×10^{-15} eVs)

ν = frequency of the wave associated with photon

then $C = \nu\lambda$

C = 3×10^8 ms^{-1} = velocity of light

λ = wavelength of the wave associated with photon $\therefore E = h\nu = \dfrac{hC}{\lambda}$

Conclusions :

(i) Energy of a photon is normally expressed in electron volt (eV)

(ii) Electron volt is the change in the energy of an electron when it passes through a potential difference of one volt. 1eV = $1.6\text{x}10^{-19}$ J

(iii) Charge of a photon is zero, so it can not be deflected by electric and magnetic fields.

(iv) Rest mass of a photon is zero. It is not a material particle, it is a bundle of energy.

(v) Effective mass of a photon is $m = \frac{E}{C^2} = \frac{h\nu}{C^2} = \frac{h}{\nu\lambda^2}$

(vi) A photon possesses momentum given by $p = \frac{h}{\lambda} = \frac{h\nu}{C} = \frac{E}{C}$

INTENSITY

➢ Intensity (I) of radiation at a given point is the energy transmitted through unit area perpendicular to that area in unit time. Intensity(I) of radiation at a distance 'r' from a monochromatic source of power 'P' is, $I = \frac{P}{4\pi r^2}$. If 'N' photons are emitted in time 't' by a monochromatic source of power 'P' then $P = \frac{Nh\nu}{t} = \frac{NhC}{\lambda t}$. So increase in intensity of monochromatic radiation means increase in the number of photons incident on unit area in unit time.

W.E-1: ***While working with light and X-rays, there is a useful relation between the energy of a photon in electron volts (eV) and the wavelength of the photon in angstom (A^0). Suppose the wavelength of aphoton is λA^0. Then energy of the photon is***

Sol: $E = hv = \frac{hc}{\lambda}$

Here wavelength =

$\lambda \times 10^{-10} m; h = 6.62 \times 10^{-34} Js, c = 3 \times 10^8 ms^{-1}$

$$\therefore E = \frac{(6.62 \times 10^{-34}) \times (3 \times 10^8)}{\lambda \times 10^{-10}}$$

$$= \frac{(6.62 \times 10^{-34}) \times (3 \times 10^8)}{(\lambda \times 10^{-10}) \times (1.6 \times 10^{-19})} eV = \frac{12400}{\lambda} eV$$

$$\therefore E = \frac{12400}{\lambda} eV$$

Note : (λ is taken in A^0 and 12400 in A^0 eV)

W.E-2: ***If wavelength of radiation is $4000 A^0 = 400 nm$ then the energy of the photon is***

Sol: $E = \frac{hC}{\lambda} = \frac{12400 eVA^0}{4000 A^0} = \frac{1240 eVnm}{400 nm} = 3.1 eV$

W.E-3: ***A monochromatic source of light operating at 200 W emits 4×10^{20} photons per second. Find the wagelength of the light.***

Sol: Power = $P = \frac{N}{t} h\upsilon$

Energy of photon = $E = \frac{P}{\left(\frac{N}{t}\right)} = \frac{200}{4\times10^{20}} = 5\times10^{-19}$

EINSTEIN'S PHOTO ELECTRIC EQUATION:

- For explaining photoelectric effect, Einstein postulated that light consists of particles called photons. Energy of a photon of frequency ν is $h\nu$.
- According to this theory the emission of a photoelectron was the result of the interaction of a single photon with an electron, in which the photon is completely absorbed by the electron.
- The minimum amount of energy required to eject an electron from a metal surface is called work function (W) of that metal. It is also called threshold energy.
- The minimum frequency of radiation required to eject an electron from a metal surface is called threshold frequency(ν_0) for that metal. $\therefore \quad W = h\nu_0$
- Work function of a metal depends on nature of the metal, it will not depend on frequency and intensity of the radiation.
- When a photon of energy $h\nu$ is absorbed by an electron, an amount of energy atleast equal to work function W (provided $h\nu$ >W) is used up in liberating the electron from the surface and the difference ($h\nu$ - W) is equal to the maximum kinetic energy.of that electron.

$$\therefore \quad \frac{1}{2}mV_{max}^2 = h\nu - W \qquad \rightarrow(1)$$

$$\therefore \quad h\nu = W + \frac{1}{2}mV_{max}^2 \qquad \rightarrow(2)$$

$$\therefore \quad h\nu = h\nu_0 + \frac{1}{2}mV_{max}^2 \qquad \rightarrow(3)$$

The above relation is called the Einstein's Photoelectric equation. Here 'm' is the mass of the electron and V_{max} is the maximum velocity of the photoelectrons. Infact, most of the electrons possess kinetic energy less than the maximum value, as they lose a part of their kinetic energy due to collisions before escaping from the metal.

Thus from the above discussion the laws of photoelectric effect from Einstein's Photoelectric equations are deduced.

i) From equation (1) maximum kinetic energy of photoelectrons is $KE_{max} = h\nu - h\nu_0$.

For photoelectric emission to take place kinetic energy of electrons must be positive. It follows that $h\nu > h\nu_0 \quad \Rightarrow \quad \nu > \nu_0$.

It proves that for photoelectric emission to take place, from a given metal the *frequency of the incident radiation must be greater than threshold frequency for that metal.*

If frequency of the incident radiation is less than threshold frequency then no photoelectric emission will take place, whatever be the intensity of the incident radiation, or how long it falls on the metal surface.

ii) From equation (1) it follows that maximum kinetic energy of photoelectrons depends linearly on the frequency. It proves that *the maximum kinetic energy of photoelectrons increases as frequency of incident radiation increases.*

Since Einstein's equation does not involve a factor representing intensity, it proves that *the maximum kinetic energy of emitted electrons is independent of the intensity of incident radiation.*

iii) According to Einstein, the photoelectric effect arises , when *a single photon is absorbed by a single electron.* So number of photoelectrons ejected will be large if intense radiation is incident. This is because intensity of radiation is proportional to number of photons per unit area per unit time. Hence if intensity of incident radiation is larger, then number photons incident is larger and number of electrons ejected is larger.

It proves *that number of photoelectrons ejected from a metal surface depends on intensity of incident radiation.* Further, there is no effect of frequency of incident radiation on number of photoelectrons emitted. It is because one photon is capable of ejecting only one electron, provided, $\nu > \nu_0$

iv) According to Einstein, the basic process in photoelectric emission is absorption of a photon of light by an electron. So *as the photon is absorbed, emission of electron takes place instantaneously* irrespective of intensity.

Conclusions :

(a) Alkali metals can cause photoelectric effect with visible light.

(b) Work function of Alkali metals is around 2eV.

(Cs=2.14eV, K= 2.3eV, Na=2.75eV, Al=4.28eV, Cu=4.65eV, Ag=4.7eV, Ni=5.15eV, Pt=5.65eV)

Among all metals work function is least for Cesium(2.14eV)

(c) Work function $W = h\nu_0 = \frac{hC}{\lambda_0}$

where ν_0=threshold frequency, λ_0=threshold wavelength

(d) Einstein's equation can be written as follows:

$KE_{max} = E - W$ (or) $KE_{max} = h\nu - h\nu_0$

(or) $KE_{max} = \frac{hC}{\lambda} - \frac{hC}{\lambda_0}$

(e) $\frac{1}{2}mV_{max}^2 = E - W$ (or) $\frac{1}{2}mV_{max}^2 = h\nu - h\nu_0$

(or) $\frac{1}{2}mV_{max}^2 = \frac{hC}{\lambda} - \frac{hC}{\lambda_0}$

(f) $eV_0 = E - W$ (or) $eV_0 = h\nu - h\nu_0$

(or) $eV_0 = \frac{hC}{\lambda} - \frac{hC}{\lambda_0}$

W.E-4: ***The work function of a metal is 3.0eV. It is illuminated by a light of wave length 3 x 10^{7}m. Calculate i) threshold frequency, ii) the maximum energy of photoelectrons, iii) the stopping potential. (h =6.63 x 10^{-34} Js and c = 3 x 10^{8}ms^{-1}).***

Sol. i) $W = h\nu_0 = 3.0eV = 3 \times 1.6 \times 10^{-19}J$

Threshold frequency

$$\nu_0 = \frac{W}{h} = \frac{3 \times 1.6 \times 10^{-19}}{6.63 \times 10^{-34}} = 0.72 \times 10^{15} Hz.$$

ii) Maximum kinetic energy $(E_{max}) = h(\nu - \nu_0)$

$$\lambda = 3 \times 10^{-7} m, \ \nu = \frac{c}{\lambda} = \frac{3 \times 10^8}{3 \times 10^{-7}} = 1 \times 10^{15} Hz$$

$K_{max} = h(\nu - \nu_0) = 6.63 \times 10^{-34} (1 - 0.72) \times 10^{15} J = 1.86 \times 10^{-19} J.$

iii) $K_{max} = eV_0$ where V_0 is stopping potential in volt and e is the charge of electron

$V_0 = \frac{K_{max}}{e}$. Here $K_{max} = 1.86 \times 10^{-19}$ J and

$e = 1.6 \times 10^{-19}$ C; $V_0 = \frac{1.86 \times 10^{-19} J}{1.6 \times 10^{-19} C} = 1.16V$

W.E-5: ***The work function of a photosensitive element is 2eV. Calculate the velocity of a photoelectron when the element is exposed to a light of wavelength*** $4\times10^3\overset{0}{A}$.

Sol. Einstein's photoelectric equation is

$$\frac{1}{2}mv^2=\frac{hc}{\lambda}-W_0$$

$$\frac{1}{2}mv^2=\frac{6.62\times3}{4\times10^3\times10^{-10}}\times10^{-26}-2\times1.6\times10^{-19}$$

$$v^2=\frac{1.765\times2}{9.1}\times10^{12}$$

$$v=\sqrt{\frac{1.765\times2}{9.1}}\times10^6=6.228\text{ x }10^5\text{ ms}^{-1}$$

W.E-6: ***A metal of work function 4eV is exposed to a radiation of wavelength 140 x 10^{-9}m. Find the stopping potential developed by it.***

Sol. $E=\frac{hc}{\lambda}$ $E=\frac{6.62\times10^{-34}\times3\times10^8}{140\times10^{-9}\times1.6\times10^{-19}}eV$ $=8.86\text{eV}$

work function $W_0=4\text{eV}$

$eV_0=E-W_0=8.86-4=4.86\text{eV}$

$\therefore$ Stopping potential $V_0=4.86V$

W.E-7: ***Radiations of wavelength 200nm propagating in the form of a parallel beam, fall normally on a plane metallic surface. The intensity of the beam is 5mW and its cross sectional area 1.0mm^2. Find the pressure exerted by the radiation on the metallic surface, if the radiation is completely reflected.[Roorkee 2001]***

Sol. $\therefore E=\frac{12400}{\lambda}=\frac{12400}{2000}=6.2eV\approx10^{-18}J$

Number of photons passing a point per second is $n=\frac{P}{E}=\frac{5\times10^{-3}}{10^{-18}}=5\times10^{15}$. momentum of each photon

$p=\frac{E}{C}=3.3\times10^{-27}J/s$. Change in momentum after each strike $=2p=6.6\text{ x }10^{-27}$ J/s

Total momentum change per second is $F=\frac{dp}{dt}=\frac{n\times2p}{t}=5\times10^{15}\times6.6\times10^{-27}=33\times10^{-12}N$

$\therefore$ pressure $\frac{F}{A}=33\times10^{-6}N/m^2$

W.E-8: ***In a photocell bi chromatic light of wave length 2480 A^0 and 6000A^0 are incident on a cathode whose workfunction is 4.8eV. If a uniform magnetic field of 3x10^{-5} T exists parallel to the plate, find the radius of the circular path described by the photoelectron. (mass of electron is*** $9x10^{-31}kg$***)***

Sol. $E_1=\frac{12400}{\lambda_1}=\frac{12400}{2480}=5eV$; $E_2=\frac{12400}{\lambda_2}=\frac{12400}{6000}=2.06eV$

As $E_2<W_0$ and $E_1>W_0$, photo electric emission is possible only with λ_1.

Maximum K.E of emitted photo electrons

$K=E_1-W_0=0.2$ eV.

Photo electrons experience magnetic force and move along a circular path of radius

$$r = \frac{mv}{Bq} = \frac{\sqrt{2mK}}{Bq} \Rightarrow r = 5cm.$$

W.E-9: ***A monochromatic light of wavelength λ is incident on an isolated metalic sphere of radius a. The threshold wavelength is λ_0 which is larger than λ. Find the number of photoelectrons emitted before the emission of photo electrons stops.***

Sol. As the metallic sphere is isolated, it becomes positively charged when electrons are ejected from it. There is an extra attractive force on the photoelectrons. If the potential of the sphere is raised to V, the electron should have a minimum energy W + eV to be able to come out. Thus, emission of photoelectrons will stop when

$$\frac{hc}{\lambda} = W + eV = \frac{hc}{\lambda_0} + eV \quad \text{or,} \quad V = \frac{hc}{e}\left(\frac{1}{\lambda} - \frac{1}{\lambda_0}\right).$$

The charge on the sphere needed to take its potential to V is $Q = (4\pi\varepsilon_0 a)V$

The number of electrons emitted is, therefore,

$$n = \frac{Q}{e} = \frac{4\pi\varepsilon_0 aV}{e} = \frac{4\pi\varepsilon_0 ahc}{e^2}\left(\frac{1}{\lambda} - \frac{1}{\lambda_0}\right)$$

W.E-10: ***A small metal plate (work function W) is kept at a distance d from a singly ionized, fixed ion. A monochromatic light beam is incident on the metal plate and photoelectrons are emitted. Find the maximum wavelength of the light beam so that some of the photoelectrons may go round the ion along a circle.***

Sol. Electron is moving around the ion in a Circle of radius 'd'. $\frac{1}{4\pi\varepsilon_0}\frac{e^2}{d^2} = \frac{mV^2}{d}$, $\therefore mV^2 = \frac{1}{4\pi\varepsilon_0}\frac{e^2}{d}$

$$\therefore K.E = \frac{1}{8\pi\varepsilon_0}\frac{e^2}{d} \text{ ----------(1)}$$

But $K.E_{max} = \frac{hc}{\lambda} - W$ ----------(2)

$$\therefore \lambda = \frac{hc\, 8\pi\varepsilon_0 d}{e^2 + 8\pi\varepsilon_0 dW}$$

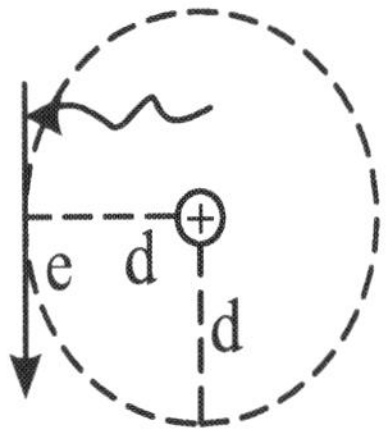

Photo electric cells :-

- A device which converts light energy into electrical energy is known as a photo electric cell or photo cell.
- It is a technological application of the photo electric effect.
- It is a device whose electrical properties are affected by light:
- It consists of a semi-cylindrical photosensitive metal plate (emitter) and a wire loop (collector).
- when light of suitable wavelength falls on the emitter ,photoelectrons are emitted and enters into the circuit giving a few amperes of current.

- It converts a change in intensity of illumination into a change in photocurrent.
- These are the devices used to convert light energy into electrical energy.
- In the outer photoelectric effect photo electrons liberated are emitted out of the metal surface.
- In the inner photoelectric effect photoelectrons liberated are in the metal surface.
- There are three types of photo cells.

  1) Photo emissive cells 2) Photo voltaic cells

  3) Photo conductive cells.
- Photo emissive cells depend upon outer photoelectric effect whereas photo voltaic and photo conductive cells depend upon inner photoelectric effect.
- Photo emissive cells are of two types.

  1. Vacuum type and 2. Gas filled type
- In the Vacuum type of photo emissive cell, an evacuated glass tube has its inner surface coated with an alkali metal.

  To study stellar spectra potassium hydride photosensitive surface is used.

  In the vacuum type, current is directly proportional to the intensity of incident radiation.
- In a gas filled emissive cell, an inert gas such as He, Ar or Ne at a low pressure, some tenths of mm of Hg, is filled. Gas filled cells produce much more intense photoelectric current due to ionization by collision in the gas. But there is no proportionality between the current and intensity of incident radiation.
- Photovoltaic cell is a true cell as it generates e.m.f. without the application of any external potential.

  When light is incident on a semi conductor coated on a metallic plate covered by a semi transparent film electrons are emitted and travel in the direction opposite to the light rays.

  The semi conductors used are cuprous oxide, selenium.

  The metal plates used are copper plate, iron plate.

  The semi transparent films used are silver, gold, platinum.

  For small load resistances the current is nearly proportional to the intensity of incident radiation.
- In a photo conductive cell, conductivity of a semi conductor is increased when light falls on it. But the response is slow. The current is not proportional to the intensity of light. Photo electric cells are used
- in exposure meters
- to compare intensities of illuminations of two light sources.
- in recording and reproduction of sound in films
- in video cameras
- to study stellar spectra
- in electronic relay circuits such as Burglar's alarm, counting devices, switching on and off street lights, etc.

DUAL NATURE OF MATTER - (de-BROGLIE HYPOTHESIS) :

- Photoelectric effect and Compton effect proves that radiation behaves like particles (photons), where as Interference and Diffraction proves that radiation behaves like waves.

  So 'radiation has dual nature' i.e., radiation behaves like particles when interacting with matter and radiation behaves like waves when propagating in a medium.

de Broglie Hypothesis

1) The universe consists of matter and radiation only.
2) Nature loves symmetry
3) If radiation has dual nature then matter also should have dual nature.

- According to de Broglie particles like electron, proton and neutron, also have both wave and particle properties. The waves associated with moving particle are called matter waves and the wavelength is called the de Broglie wavelength of a particle.

  For a photon Energy, $E = \frac{hC}{\lambda} = mC^2$

  where m = effective mass then wavelength $\lambda = \frac{h}{mC} = \frac{h}{p}$

  where p = momentum of the photon

  de Broglie extended the same for particles also.

  So if a particle of mass 'm' is moving with velocity 'v' then its momentum p = mv, hence de Broglie wave length of the matter wave associated with is given by $\lambda = \frac{h}{p} = \frac{h}{mv}$

  Davisson and Germer studied the scattering of electrons by a nikel target. The wavelength λ of diffracted electrons was determined by Davisson and Germer. The experimental values of wavelength λ were found to agree with the theoretical value $\lambda = \frac{h}{mv}$

  Hence it is concluded that electrons behaves like waves and undergo diffraction.

- For definite sized objects like a car the corresponding wavelength is very small to detect the wave properties. But the de-Broglie wavelength of the electron is large enough to be observed. Because of their small mass, electrons have a small momentum and hence large wavelength $\lambda = h/p$.

Conclusions :

(i) deBroglie wavelength $\lambda = \frac{h}{p} = \frac{h}{mv}$

Where momentum p = mv ; m=mass, v= velocity

(ii) deBroglie wavelength $\lambda = \frac{h}{\sqrt{2mK}}$

where kinetic energy, $K = \frac{p^2}{2m} \Rightarrow p = \sqrt{2mK}$

(iii) If a particle having charge q starting from rest is accelerated through a potential difference V then gain in kinetic energy, K=qV

so, deBroglie wavelength $\lambda = \frac{h}{\sqrt{2mqV}}$

(iv) For electron $\lambda = \frac{12.27}{\sqrt{V}} \overset{o}{A} = \sqrt{\frac{150}{V}} \overset{o}{A}$

(v) For proton $\lambda = \frac{0.286}{\sqrt{V}} \overset{o}{A} = \sqrt{\frac{0.082}{V}} \overset{o}{A}$

(vi) For dueteron $\lambda = \frac{0.202}{\sqrt{E}} \mathring{A}$

(vii) For α particle $\lambda = \frac{0.101}{\sqrt{V}} \mathring{A}$

(viii) for neutron $\lambda = \frac{0.286}{\sqrt{E}} \mathring{A}$

where E = kinetic energy in electron volts

(ix) The de-Broglie wavelength of a particle is independent of nature of the particle and these waves are not electromagnetic. Diffraction effects have been obtained with streams of electrons, protons, neutrons and alpha particles.

(x) de-Broglie explains Bohr's criterion to select the allowed orbits in which angular momentum of the electron is an integral multiple of $\frac{h}{2\pi}$. According to his hypothesis, an electron revolving round the nucleus is associated with certain wavelength 'λ' which depends on its momentum mv. It is given by

$$\lambda = \frac{h}{mv} = \frac{h}{p}$$

In an allowed orbit, an electron can have an integral multiple of this wavelength.

That is the nth orbit consists of n complete de-Broglie wavelengths $2\pi r_n = n\lambda_n$, where n is the principle quantum number.

where r_n is the radius of nth orbit and λ_n is the wavelength of electron in n^{th} orbit

$$\lambda_n = \frac{2\pi r_n}{n} \quad \text{(or)} \quad \lambda_n = \frac{2\pi}{n}(0.53 \times n^2)\,\mathring{A}.$$

$$\lambda_n = 2\pi \times 0.53 nA^\circ$$

(a) 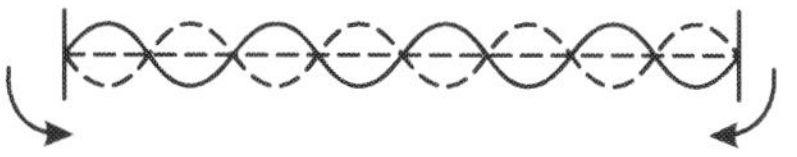

(b) 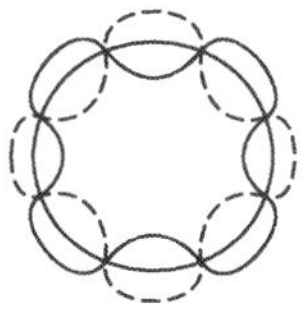

Figure (a) shows the waves on a string have a wavelength related to the length of the string allowing them to interfere constructively as shown If we imagine the string bent into a closed circle we get an idea of how electrons in circular orbits can interfere constructively as shown in figure(b). If the wavelength does not fit in to the circumference the electron interferes destructively, electron can not exist in such an orbit.

W.E-11: ***A particle of mass 'm' projected horizontally with velocity u. If it makes an angle θ, with the horizontal after some time, then at that instant, its de Broglie wavelength is***

Sol: For a projectile horizontal component of velocity is constant.

$\therefore v_x = u_x ; V\cos\theta = u$

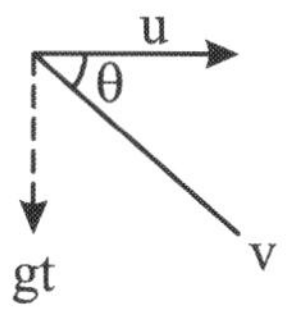

$\therefore$ de Broglie wavelength, $\lambda = \frac{h}{mv} = \frac{h\cos\theta}{mu}$

W.E-12: ***Electrons are accelerated through a potential difference of 150V. Calculate the de Broglie wavelength.***

Sol. V = 150V; h=6.62 x 10^{-34}Js, m=9.1 x 10^{-31}kg, e = 1.6 x 10^{-19}C

$$\therefore \lambda = \frac{h}{\sqrt{2Vem}} = \frac{6.62 \times 10^{-34}}{\sqrt{2 \times 9.1 \times 10^{-31} \times 1.6 \times 10^{-19} \times 150}} = 1\overset{0}{A}$$

W.E-13: ***Find the ratio of de Broglie wavelength of molecules of hydrogen and helium which are at temperatures 27°C and 127°C respectively***

Sol. Since, $\lambda = \frac{h}{m\upsilon} = \frac{h}{\sqrt{3mkT}}$; $\frac{\lambda_H}{\lambda_{He}} = \sqrt{\frac{m_{He}T_{He}}{m_H T_H}} = \sqrt{\frac{8}{3}}$

W.E-14: ***With what velocity must an electron travel so that its momentum is equal to that of a photon with a wavelength of*** $5000\overset{0}{A}$

$(h = 6.6x10^{-34} Js,\ m_e = 9.1x10^{-31} Kg)$

Sol. $mv = \frac{h}{\lambda} \Rightarrow v = \frac{6.6\times10^{-34}}{9.1\times10^{-31}\times5000\times10^{-10}} = 1450 m/s$

W.E-15: ***If 10,000V applied across an X-ray tube, what will be the ratio of deBroglie wavelength of the incident electrons to the shortest wavelength of X-ray produced (e/m of electron is*** $1.7x10^{11} C/Kg$***)***

Sol. Debroglie wave length of incident electron is $\lambda_1 = \frac{h}{\sqrt{2meV}}$ 1

Shortest wavelength of x ray photon is $\lambda_2 = \frac{hc}{Ve}$ 2

$\Rightarrow \frac{\lambda_1}{\lambda_2} = \frac{1}{c}\sqrt{\left(\frac{V}{2}\right)\left(\frac{e}{m}\right)} = 0.1$

W.E-16: ***Photons of energies 4.25eV and 4.7eV are incident on two metal surfaces A and B respectively. The maximum KE of emitted electrons are respectively*** T_A ***eV and*** $T_B = (T_A - 1.5)eV$***. The ratio de-Broglie wavelengths of photo electrons from them is*** $\lambda_A : \lambda_B = 1:2$***, then find the work function of Aand B***

Sol. Debroglie wavelength

$\lambda = \frac{h}{\sqrt{2km}} \Rightarrow \lambda \propto \frac{1}{\sqrt{k}}$ (k = k.E = T); $\frac{\lambda_B}{\lambda_A} = \sqrt{\frac{T_A}{T_B}}$

$2 = \sqrt{\frac{T_A}{T_A - 1.5}} \Rightarrow T_A = 2eV$

$\Rightarrow W_A = 4.25 - T_A = 2.25 eV$

$\Rightarrow T_B = T_A - 1.5 = 2 - 1.5 = 0.5eV$

$\Rightarrow W_B = 4.7 - T_B = 4.7 - 0.5 = 4.2 eV$

HEISEN-BERG UNCERTAINITY PRINCIPLE

- The matter-wave picture elegantly incorporated the Heisenberg's uncertainty principle. According to the principle, it is not possible to measure both the position and momentum of an electron (or any other particle) at the same time exactly. There is always some uncertainity (Δx) in the specification of position and some uncertainity (Δp) in the specification of momentum. The product of Δx and Δp is of the order of $\hbar$ (with $\hbar = \frac{h}{2\pi}$) i.e., $\Delta x \Delta p = \hbar$.

- Equation allows the possibility that Δx is zero, but then Δp must be infinite in order that the product is nonzero. Similarly, if Δp is zero, Δx must be infinite. Ordinarily, both Δx and Δp are nonzero such that their product is of the order of h.
- Now, if an electron has a definite momentum p, (i.e., Δp=0), by the de Broglie relation, it has a definite wavelength λ. A wave of definite (single) wavelength extends all over space. By Born's probability interpretation this mens that the electron is not localizedin any finite region of space. That is, its position uncertainity is infinite $(\Delta x \to \infty)$, which is consistent with the uncertainity principle.
- In general, the matter wave associated with the electron is not extended all over space. It is a wve packet extending over some finite region of space. In that case Δx is not infinite but has some finite value depending on the extension of the wave packet. Also, you must appreciate that a wave packet of finite extension does not have a single wavelength. It is built up wavelengths spread around some central wavelength.
- By de Broglie's relation, then, the momentum of the electron will also have a spread - an uncertainity Δp. This is as expected from the uncertainity principle. It can be shown that the wave packet description together with de Broglie relation and Born's probability interpretation reproduce the Heisenberg's uncertainity princile exactly.
- The de Broglie relation will be seen to justify bohr's postulate on quantisation of angular momentum of electron in an atom.

Figure shows a schematic diagram of (a) a localised wave packet, and (b) an extended wave with fixed wavelength.

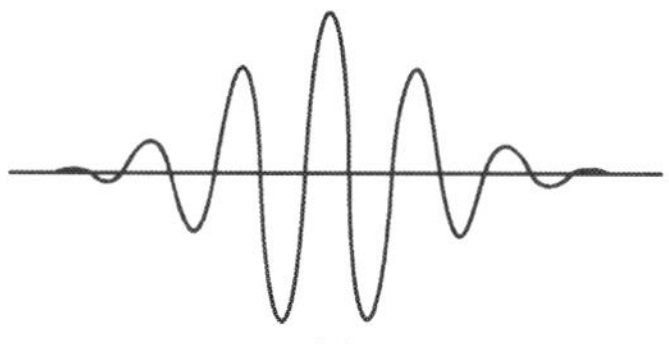
(a)

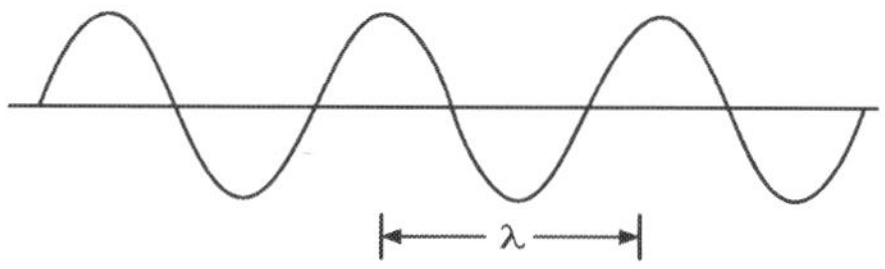

(b)

Figure (a) the wave packet description of an electron. The wave packet corresponds to a spread of wavelength around some central wavelength (and hence by de Broglie relation, a spread in momentum). Consequently, it is associated with an uncertainity in position (Δx) and an uncertainity in momentum (Δp). (b) the matter wave corresponiding to a definite momentum of an electron extends all over space. In this case, $\Delta p = 0$ and $\Delta x \to \infty$.

W.E-17: ***If the uncertainity in the position of proton is $6\times10^{8}\,m$, then the minimum uncertainity in its speed is***

Sol: $\Delta p = m\Delta v = \dfrac{h}{\Delta x}$

or $\Delta v = \dfrac{h}{m\Delta x} = \dfrac{1.034\times10^{-34}}{1.67\times10^{-27}\times6\times10^{-8}} = 1\ ms^{-1}$

W.E-18: ***The correctness of velocity of an electron movign with velocity 50 ms^{-1} is 0.005%. The accuracy with which its position can be measured will be***

Sol: Here, $\Delta v = \frac{0.005 \times 50}{100} = 0.0025 ms^{-1}$

$$\Delta x = \frac{h}{m\Delta v} = \frac{1.034 \times 10^{-34}}{9.1 \times 10^{-31} \times 0.0025}$$
$$= 4634 \times 10^{-5} m$$

DAVISSON AND GERMER'S ELECTRON DIFFRACTION EXPERIMENT

i) The first experimental evidence of matter wave was given by two American physicists, Davisson and Germer in 1927. They also succeeded in measuring the de - Broglie wave length associated with slow electrons.

ii) A beam of electron emitted by electron gun is made to fall on nikel crystal cut along cubical axis at a particular angle.

iii) Ni crystal behaves like a three dimensional diffraction grating and it diffracts the electron beam obtained from electron gun.

iv) The diffracted beam of electrons received by the detector which can be positioned at any angle by rotating about the point of incidence.

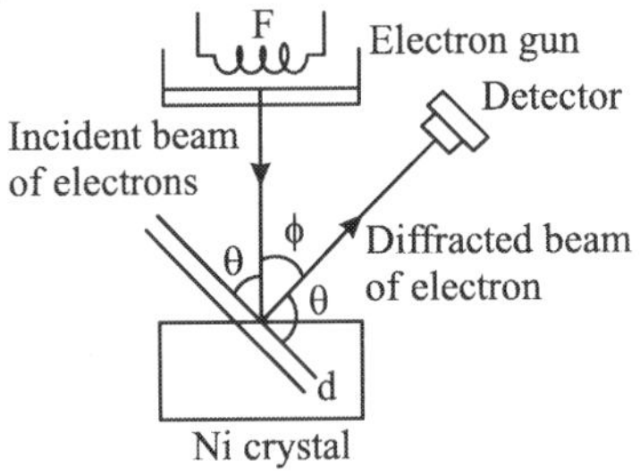

v) The energy of the incident beam of electron can also be varied by changing the applied voltage to the electron gun.

vi) According to classical physics, the intensity of scattered beam of electrons was not the same but different at different angles of scattering. It is maximum for diffracting angle 50^0 at 54 volt P.D.

vii) It is seen that a bump begins to appear in the curve for 44 volt electrons. With increasing potential in the bump moves upwards and becomes most prominent in the curve for 54 volt electrons at $\phi = 50^0$. At higher potential the bump gradually disappears.

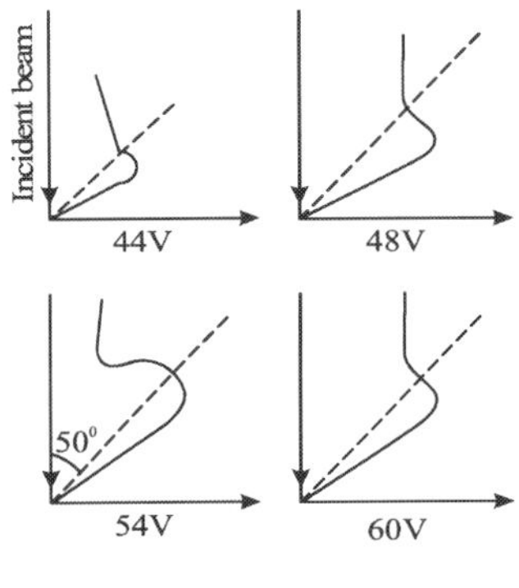

viii) If the de Broglie waves are associated with electron, then these should be diffracted like x - rays. using the Bragg's formula $2d\sin\theta = n\lambda$, we can determine the wavelength of these waves.

Where 'd' is the distance between the diffracting planes. $\theta = \left[\frac{180-\phi}{2}\right]$ = glancing angle for incident beam = Bragg's angle.

ix) The distance between diffracting planes in Ni - crystal for this experiment is d = 0.91A^0 and for n = 1; $\lambda = 2\times 0.91\times 10^{-10}$ sin 65 = 1.65A^0

Now debroglie wave length can also be determined using the formula ;

$$\lambda = \frac{12.27}{\sqrt{V}} = \frac{12.27}{\sqrt{54}} = 1.67A^0$$

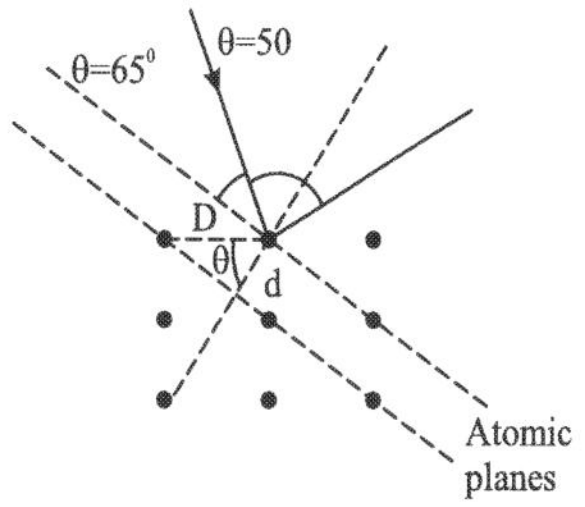

Thus the deBroglie hypothesis is verified.

x) The Bragg's formula can be rewritten in the form containing inter atomic distance D and scattering angle 'ϕ'.

$$\therefore \theta = 90 - \frac{\phi}{2} \text{ and } d = D\cos\theta = D\sin\frac{\phi}{2}$$

using $\sin\theta = \cos\frac{\phi}{2}$

$$\lambda = 2d\sin\theta = 2d\left(\sin\frac{\phi}{2}\right)\cos\frac{\phi}{2} = d\sin\phi$$

$$\boxed{\lambda = d\sin\phi}$$

ADDITIONAL

X-RAYS :

Roentgen discovered the X-rays.

i) Most commonly x-rays are produced by the deceleration of high energy electrons bombarding a hard metal target.

ii) The target should have

a) high atomic weight

b) high melting point

c) high thermal conductivity

iii) They are electromagnetic waves of very short wavelength. i.e., order of wavelength 0.1A° to 100A°, order of frequency 10^{16}Hz to 10^{19} Hz, order of energy 124eV to 124keV

iv) Most of the kinetic energy of electrons is converted into heat and only a fraction is used in producing x-rays (less than 1% x - rays and more than 99% heat).

v) Intensity of x-rays depends on the number of electrons striking the target which inturn depends on filament current.

vi) Quality of x - rays (hard /soft) depends on P.D applied to x - rays tube.

vii) high frequency x-rays are called hard x-rays

viii) low frequency x-rays are called soft x-rays

ix) Penetrating power of x-rays is a function of potential difference between cathode and target.

x) Interatomic distance in crystals is of the order of the wavelength of x-rays hence crystals diffract x-rays.

xi) Production of x-rays is converse of photoelectric effect.

X-RAY SPECTRUM

i) Continuous X-ray spectrum:

a) It is produced when high speed electrons are suddenly stopped by a metal target.

b) It contains all wave lengths above a minimum wavelength λ_m. ($\therefore$ continuous spectrum) For a given accelerating potential, λ_m is called cut off wave length.

c) Properties of continuous x - rays spectra are independent of nature of target metal and they depend only on accelerating potential.

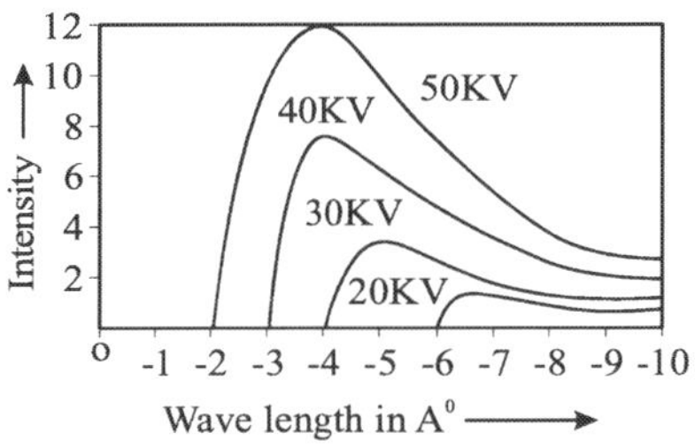

d) $$\boxed{\lambda_{min} = \frac{hc}{eV} = \frac{12400}{V} A^0}$$

$\therefore \lambda_{min} \alpha \frac{1}{V}$ it is Duane and Hunt's law

e) Maximum frequency of emitted x - ray photon is $\boxed{\nu_{max} = \frac{ev}{h}}$

f) In this spectrum intensity first increases, reaches a maximum value I_{max} and then decreases.

g) Every spetrum starts with certian minimum wave length called limiting wave length or cut off wave length λ_{min}.

h) With the increase in target potential, λ_{min} and wavelength corresponding to maximum intensity λ_0 shifts towards minimum wavelength side.

i) At a given potential the range of wave length of continous x - rays produced is λ_{min} to ∞.

j) Efficiency of x - ray tube

$$\eta = \frac{out\ put\ power}{input\ power} x100$$

input power P = VI. Where V is P.D applied to x - ray tube I = anode current

ii) **Characterstic X-ray spectrum:**

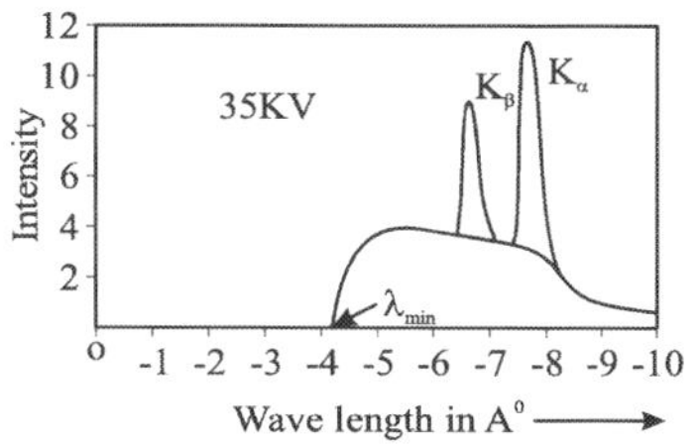

a) Produced due to transition of electrons from higher energy level to lower energy level in target atoms

b) Wavelengths of these x-rays depend only on atomic number of the target element and independent of target potential.

c) Characteristic x-rays of an element consists of K, L,M and N series.

d) K-series of lines are obtained when transition takes place from higher levels to k shell

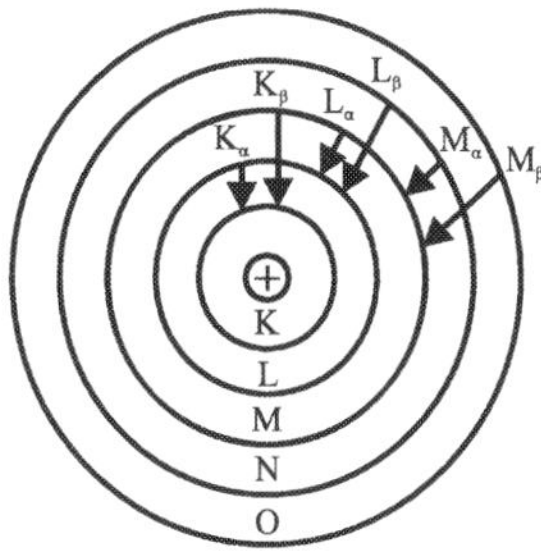

e) This spectrum is useful in identifying the elements by which they are produced.

f) Relation among the energies $E_{k\alpha} < E_{k\beta} < E_{k\gamma}$, $E_{k\alpha} > E_{L\alpha}$

g) Intensity of x - rays $I_{k\alpha} > I_{k\beta} > I_{k\gamma}$

h) Relation among frequences $\nu_{k\alpha}, \nu_{k\beta}$ and $\nu_{L\alpha}$ is $\nu_{k\beta} = \nu_{k\alpha} + \nu_{L\alpha} \Rightarrow \frac{1}{\lambda_{K\beta}} = \frac{1}{\lambda_{K\alpha}} + \frac{1}{\lambda_{L\alpha}}$

h) $E_K - E_L = h\nu_{K\alpha} = \frac{hc}{\lambda_{K\alpha}}$

$$E_K - E_M = h\nu_{K\beta} = \frac{he}{\lambda_{K\beta}}$$

$$E_L - E_M = h\nu_{L\alpha} = \frac{he}{\lambda_{K\alpha}}$$

iii) Intensity and wavelength $(I - \lambda)$ graph

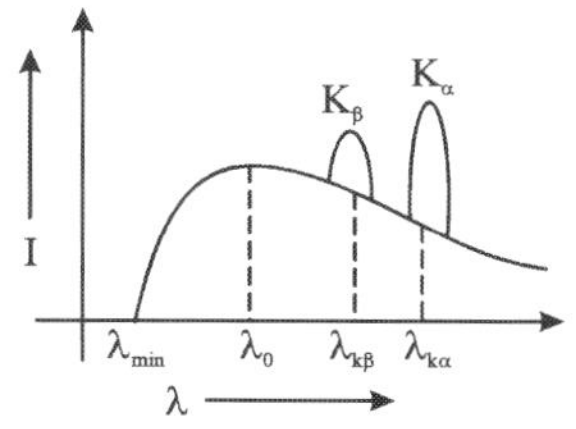

As target potential V is increased

a) $(\lambda_0 - \lambda_{min})$ decreases

b) Wavelength of k_α remains constant.

c) diffrence between λ_{k_α} and λ_{min} increases

d) diffrence between λ_{k_α} line and λ_{k_β} line remains constant.

e) Difference between $\lambda_{k\alpha} - \lambda_0$ increases.

MOSELEY'S LAW

i) "The square root of frequency (ν) of the spectral line of the characteristic x-rays spectrum is directly proportional to the atomic number(z) of the target element.

$\sqrt{\nu} \propto Z$ or $\sqrt{\nu}$ =a(Z-b)

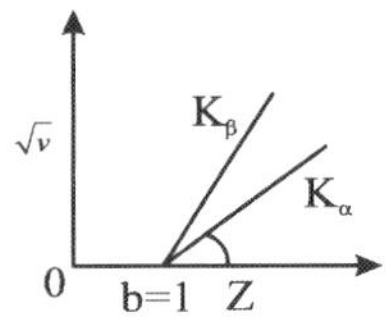

ii) The slope(a) of $\sqrt{\nu}$ -Z curve varies from series to series and also from line to line of a given series.

For K series $\sqrt{\frac{\nu_1}{\nu_2}} = \left(\frac{Z_1-1}{Z_2-1}\right)$

$\Rightarrow \sqrt{\frac{\lambda_2}{\lambda_1}} = \left(\frac{Z_1-1}{Z_2-1}\right)$

iii) $a_{k\gamma} > a_{k\beta} > a_{k\alpha}$

iv) The intercept on 'Z' axis gives the screening constant 'b' and it is constant for all spectral lines in given series but varies with the series.

b = 1 for k series ($k_\alpha, k_\beta, k_\gamma$)

b = 7.4 for L series

v) The wavelength of characteristic X-rays is given by $\frac{1}{\lambda}$=R(Z-b)^2 $\left[\frac{1}{n_1^2} - \frac{1}{n_2^2}\right]$

vi) Ratio of k_α and k_β lines from a given target is $\frac{\lambda_{k\alpha}}{\lambda_{k\beta}} = \frac{32}{27}$

vii) **Significance :**

a) The elements must be arranged in the periodic table as per their atomic numbers but not on their atomic weights.

b) Helped to discover new elements like masurium (43) and illinium (61) etc.

c) Decided the positions and atomic numbers of rare earth metals.

CONCEPTUAL QUESTIONS

PHOTO ELECTRIC EFFECT

1. **The rest mass of a photon is**

1) zero   2) $1.6\times10^{-19}\,kg$ 3) $3.1\times10^{-30}\,kg$ 4) $9.1\times10^{-31}\,kg$

2. **The mass of a photon in motion is (given its frequency = x)**

1) $\frac{hx}{c^2}$ 2) hx^3 3) $\frac{hx^3}{c^2}$ 4) zero

3. **Photoelectric effect supports the quantum nature of light because**

1) There is minimum frequency of light above which no photo electrons are emitted

2) The maximum kinetic energy of photo electrons depends on both frequency and intensity of light

3) Even when a metal surface is faintly illuminated, the photoelectrons do not leave the surface immediately

4) The maximum K.E. of photo electrons depends only on the frequency of light and not on intensity

4. **Which of the following statement is wrong?**

1) Einstein explained photo electric effect with the help of quantum theory

2) Millikan determined the value of planck's constant depending upon the property of photo electric effect

3) The maximum K.E. of the photo electrons depends upon the intensity of incident radiation

4) As the frequency of incident photon increases the corresponding stopping potential also increases

5. **In photoelectric emission, the energy of the emitted electron is**

1) larger than that of the incident photons   2) smaller than that of the incident photons

3) same at that of the incident photons   4) proportional to the intensity of the incident light

6. **A laser beam of output power 'P' consists only of wavelength λ. If Planck's constant is h and the speed of light is c, then the number of photons emitted per second is**

1) $P\lambda/hc$ 2) $P\lambda/h$ 3) $hc/P\lambda$ 4) hc/P

7. **In photoelectric effect, which of the following property of incident light will not affect the stopping potential**

1) Frequency   2) Wavelength   3) Energy   4) Intensity

8. **The best suitable metal for photo electric effect is**

1) Iron   2) Steel   3) Aluminium   4) Cesium

9. **Photo electric effect can be explained only by assuming that light**

1) is a form of transverse waves   2) is a form of longitudinal waves

3) can be polarized   4) consists of quanta

10. **When green light is incident on a metal, photo electrons are emitted by it but no photo electrons are obtained by yellow light. If red light is incident on that metal then**

1) No electron will be emitted   2) Less electrons will be emitted

3) More electrons will be emitted   4) we can not predict

11. **If the energy and momentum of a photon are E and P respectively, then the velocity of photon will be**

1) E/P   2) $(E/P)^2$ 3) EP   4) 3×10^7 m/s

12. **The photo electric effect proves that light consists of**

1) Photons   2) Electrons

3) Electromagnetic waves   4) Mechanical waves

13. **Intensity of light incident on a photo sensitive surface is doubled. Then**
1) the number of emitted electrons is tripuled
2) the number of emitted electrons is doubled
3) the K.E of emitted electrons is doubled
4) the momentum of emitted electrons is doubled

14. **A point source of light is used in a photoelectric effect. If the source is moved farther from the emitting metal, the stopping potential**
1) will increase
2) will decrease
3) will remain constant
4) will either increase or decrease

15. **If the frequency of light in a photoelectric experiment is doubled, the stopping potenital will**
1) be doubled
2) be halved
3) become more than double
4) become less than double

16. **With the decrease in the wave length of the incident radiation the velocity of the photoelectrons emitted from a given metal**
1) remains same
2) increases
3) decreases
4) increases first and then decreases

17. **Sodium surface is illuminated with ultraviolet light and visible radiation successively and the stopping potentials are determined. Then the potential**
1) is equal in both the cases
2) greater for ultraviolet light
3) more for visible light
4) varies randomly

18. **In photo electric effect, the slope of the straight line graph between stopping potential and frequency of the incident light gives the ratio of Planck's constant to**
1) charge of electron
2) work function
3) photo electric current
4) K.E. of electron

19. **From the graph shown, the value of Work function if the stopping potential (V), and frequency of the incident light, ν, are on y and x- axes respectively is**

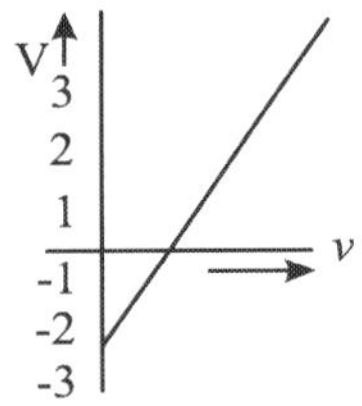

1) 1eV
2) 2eV
3) 4eV
4) 3eV

20. **A graph is drawn between frequency of the incident radiation (on X- axis) and stopping potential (on Y-axis). Then the slope of the straight line indicates**
1) h.e
2) h/e
3) e/h
4) (e-h)

21. **In an experiment of photo electric emission for incident light of 4000 A^0, the stopping potential is 2V. If the wavelength of incident light is made 3000 A^0, then the stopping potential will be**
1) Less than 2 volt
2) More than 2 volt
3) 2 volt
4) Zero

22. **Light of wavelength λ falls on a metal having work function hc / λ_o Photoelectric effect will take place only if**
1) $\lambda \geq \lambda_0$
2) $\lambda \geq 2\lambda_0$
3) $\lambda \leq \lambda_0$
4) $\lambda < \lambda_0/2$

23. **Emission of electrons in photoelectric effect is possible, if**
1) metal surface is highly polished
2) the incident light is of sufficiently high intensity
3) the light is incident at right angles to the surface
4) the incident light is of sufficiently low wavelength

24. **The work function of a metal**
1) is different for different metals
2) is the same for all the metals
3) depends on the frequency of the light
4) depends on the intensity of the incident light

25. **The process of photo electric emission depends on**
1) Temperature of incident light
2) Nature of surface
3) Speed of emitted photo electrons
4) Speed of the incident light

26. **The threshold wavelength of lithium is 8000 A^0. When light of wavelength 9000 A^0 is made to be incident on it, then the photo electrons**
1) Will not be emitted
2) Will be emitted
3) Will sometimes be emitted and sometimes not
4) Data insufficient

27. **The correct curve between the stopping potential (V_0) and intensity of incident light (I) is**

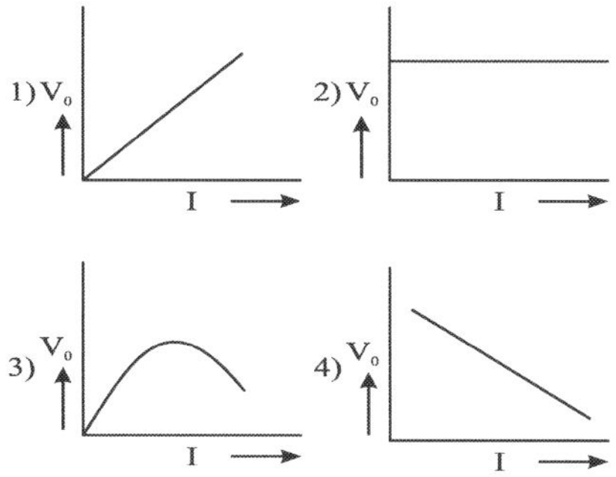

28. **The photo electrons emitted from the surface of sodium metal are**
1) Of speeds from 0 to a certain maximum
2) Of same de Broglie wavelength
3) Of same kinetic energy
4) Of same frequency

29. **The necessary condition for photo electric emission is**
1) $h\upsilon \leq h\upsilon_0$
2) $h\upsilon \geq h\upsilon_0$
3) $E_k > h\upsilon_0$
4) $E_k < h\upsilon_0$

30. **At stopping potential, the photo electric current becomes**
1) Minimum
2)Maximum
3) Zero
4) Infinity

31. **Stopping potential depends on**
1) Frequency of incident light
2) Intensity of incident light
3) Number of emitted electrons
4) Number of incident photons

32. **Work function is the energy required**
1) to excite an atom
2) to produce X-rays
3) to eject an electron just out of the surface
4) to explode the atom

33. **Threshold wavelength depends on**
1) frequency of incident radiation
2) work function of the substance
3) velocity of electrons
4) energy of electrons

34. **If the work function of a metal is ϕ_0, then its threshold wavelength will be**
1) hc ϕ_0
2) $\dfrac{c\phi_0}{h}$
3) $\dfrac{h\phi_0}{c}$
4) $\dfrac{hc}{\phi_0}$

35. **The work function of a metal is X eV When light of energy 2X eV is made to be incident on it then the maximum kinetic energy of emitted photo electron will be**
1) 2 eV
2) 2X eV
3) X eV
4) 3X eV

36. **If the distance of 100W lamp is increased from a photocell, the saturation current i in the photo cell varies with distance d as**

1) $i \propto d^2$ 2) $i \propto d$ 3) $i \propto \frac{1}{d}$ 4) $i \propto \frac{1}{d^2}$

37. **A source of light is placed at a distance 4m from a photocell and the stopping potential is then 7.7 volt. If the distance is halved, the stopping potential now will be**

1) 7.7 volt 2) 15.4 volt 3) 3.85 volt 4) 1.925 volt

38. **A milliammeter in the circuit of a photocell measures**

1) number of electrons released per second 2) energy of photon
3) velocity of photoelectrons 4) momentum of the photo electrons

39. **The Einstein's photoelectric equation is based upon the conservation of**

1) Mass 2)momentum 3) angular momentum 4) energy

40. **The stopping potential of the photocell is independent of**

1) wavelength of incident light 2) nature of the metal of photo cathode
3) time for which light is incident 4) frequency of incident light

41. **The maximum energy of emitted photo electrons is measured by**

1) the current they produce 2) the potential difference they produce
3) the largest potential difference they can transverse 4) the speed with which they emerge

42. **Three metals have work functions in the ratio 2:3:4. Graphs are drawn for all between the stopping potential and the incident frequency. The graphs have slopes in the ratio**

1) 2: 3: 4 2) 4: 3: 2 3)6: 4: 3 4)1: 1: 1

43. **The curve between current (I) and potential difference (v) for a photo cell will be**

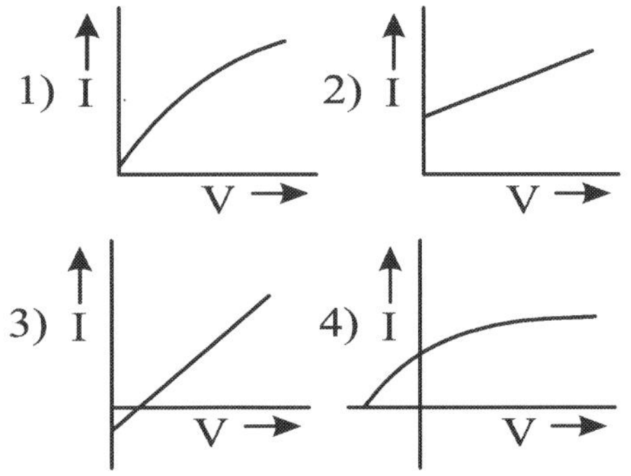

44. **Which conservation law is obeyed in Einstein's photo electric equation?**

1) Charge 2) Energy 3) Momentum 4) Mass

45. **In photo electric effect, the photo electric current**

1) increases when the frequency of incident photon increases
2) decreases when the frequency of incident photon decreases
3) does not depend upon the photon frequency but depends on the intensity of incident beam
4) depends both on the intensity and frequency of the incident beam.

46. **The photoelectric current can be increased by**

1) increasing frequency 2) increasing intensity
3) decreasing intensity 4)decreasing wavelength

47. The threshold wavelength for sodium is 5×10^{-7} m. Photoemission occurs for light of

1) Wavelength of 6×10^{-7} m and above
2) Wavelength of 5×10^{-7} m and below
3) Any wavelength
4) All frequencies below 5×10^{14} Hz

48. If Planck's constant is denoted by h and electronic charge by e, then photoelectric effect allows determination of:

1) Only h
2) Only e
3) Both h and e
4) Only h/e

49. The electron behaves as waves because they can

1) be diffracted by a crystal
2) ionise a gas
3) be deflected by magnetic fields
4) be deflected by electric fields

50. A nonmonochromic light is used in an experiment on photoelectric effect. The stopping potential

1) is related to the mean wavelength
2) is related to the longest wavelength
3) is related to the shortest wavelength
4) is not related to the wavelength

51. The incident photon involved in the photoelectriceffect experiment

1) completely disappears
2) comes out with increased frequency
3) comes out with a decreased frequency
4) comes out with out change in frequency

52. In a photoelectric experiment, the maximum velocity of photoelectrons emitted

1) depends on intensity of incident radiation
2) does not depend on cathode material
3) depends on frequency of incident radiation
4) does not depend on wavelength of incident radiation

53. The number of electrons emitted by a surface exposed to light is directly proportional to

1) Frequency of light
2) Work function
3) Thereshold wavelength
4) Intensity of light

54. Emission of electrons in photo electric effect is possible, if

1) metal surface is highly polished
2) the incident light is of sufficiently high intensity
3) the light is incident at right angles to the surface
4) the incident light is of sufficiently low wavelength

55. When orange light falls on a photo sensitive surface the photocurrent begins to flow. The velocity of emitted electrons will be more when surface is hit by

1) red light
2) violet light
3) thermal radiations
4) radio waves

56. When the amplitude of the light wave incident on a photometal sheet is increased then

1) the photoelectric current increases
2) the photoelectric current remains unchanged
3) the stopping potential increases
4) the stopping potential decreases

57. Which of the following is dependent on the intensity of incident radiation in a photoelectric experiment

1) work function of the surface
2) amount of photoelectric current
3) stopping potential
4) maximum kinetic energy

58. **Maximum kinetic energy (E_k) of a photoelectron varies with the frequency (ν) of the incident radiation as**

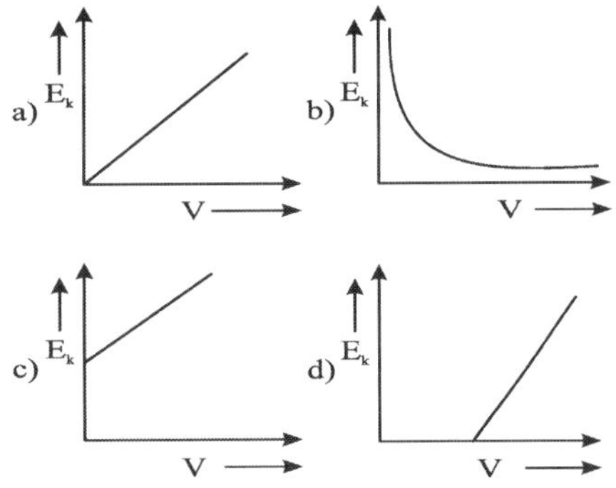

1) a　　2) b　　3) c　　4) d

59. **Which one of the following is true in photoelectric emission**

1) photoelectric current is directly proportional to the amplitude of light of given frequency

2) photoelectric current is directly proportional to the intensity of light of given frequency at moderate intensities

3) above the threshold frequency the maximum kinetic energy of photoelectrons is inversely proportional to the frequency of incident light

4) the threshold frequency depends on the intensity of incident light

60. **If the work function of the metal is W and the frequency of the incident light is υ, then there is no emission of photoelectrons if**

1) $\nu < W/h$　　2) $\nu > W/h$　　3) $\nu \geq W/h$　　4) $\nu \leq W/h$

61. **Kinetic energy with which the electrons are emitted from a metal surface due to photoelectric effect is**

1) Dependent of the intensity of illumination　　2) Dependent on the frequency of light

3) Inversely proportional to the intensity of illumination

4) Directly proportional to the intensity of illumination

62. **When ultraviolet radiation is incident on a surface, no photoelectrons are emitted. If a second beam causes emission of photoelectrons, it may consist of :**

1) radio waves　　2) infrared rays　　3) visible light rays　　4) X-rays

63. **The maximum kinetic energy (E_k) of emitted photoelectrons against frequency ν of incident radiation is plotted as shown in fig. The slope of the graph is equal to**

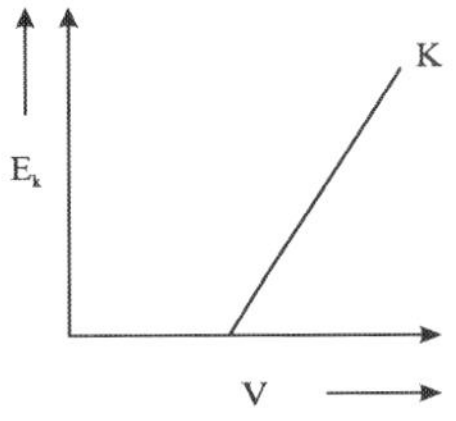

1) charge on electron　　2) work function of emitter

3) Planck's constant　　4) ratio of Planck's constant and chargeon electron

64. **Einstein's photoelectric equation states that $E_k = h\nu - W$, In this equation E_k refers to :**

1) kinetic energy of all ejected electrons　　2) mean kinetic energy of emitted electrons

3) minimum kinetic energy of emitted electrons　　4) maximum kinetic energy of emitted electrons

65. **The function of photoelecrtic cell is**
1) to convert electrical energy into light energy. 2) to convert light energy into electrical energy
3) to convert mechanical energy into electrical energy 4) to convert DC into AC.

66. **Photoelectric effect can be explained only by assuming that light:**
1) is a form of transverse waves 2) is a form of longitudinal waves
3) can be polarised 4) consists of quanta

67. **When light falls on a photosensitive surface, electrons are emitted from the surface .The kineticenergy of these electrons does not depend on the:**
1) Wave length of light 2) thickness of the surface layer
3) type of material used for the layer 4) intensity of light.

68. **Photoelectric effect is described as the ejection of electrons form the surface of a metal when:**
1) it is heated to a high temparature 2) light of a suitable wave lenght is incident on it
3) electrons of a suitable velocity impinge on it 4) it is placed in a strong electric field

69. **Though quantum theory of light can explain a number of phenomena observed with light, it is necessary to retain the wave-nature of light to explain the phenomena of:**
1) photoelectric effect 2) diffraction 3) compton effect 4) black body radiation

70. **In the following diagram if $V_2 > V_1$ then**

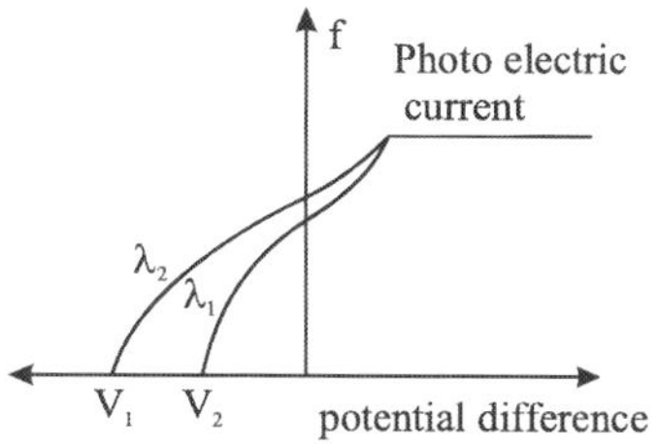

1) $\lambda_1 = \sqrt{\lambda_2}$ 2) $\lambda_1 < \lambda_2$ 3) $\lambda_1 = \lambda_2$ 4) $\lambda_1 > \lambda_2$

71. **When an X-ray photon collides with an electron and bounces off, its new frequency**
1) is lower than its original frequency 2) is same as its original frequency
3) is higher than its original frequency 4) depends upon the electron's frequency

72. **A point source of light is used in a photoelectric effect. If the source is removed farther from the emitting metal, the stopping potential**
1) will increase 2) will decrease 3) will remain constant
4) will either increase or decrease

73. **De-Broglie wavelength depends on**
1) mass of the particle 2) size of the particle
3) material of the particle 4) shape of the particle

74. **The deBroglie wavelength associated with a particle of mass m, moving with a velocity v and energy E is given by**
1) h/mv^2 2) mv/h^2 3) $h/\sqrt{2mE}$ 4) $\sqrt{2mE}/h$

75. **Choose the correct statement**
1) Any charged particle in rest is accompanied by matter waves
2) Any uncharged particle in rest is accompanied by matter waves
3) The matter waves are waves of zero amplitude
4) The matter waves are waves of probability amplitude

76. **An electron of charge e and mass m is accelerated from rest by a potential difference V. The de Broglie wavelength is**

1) Directly proportional to the square root of potential difference.

2) Inversely proportional to the square root of potential difference

3) Directly proportional to the square root of electron mass

4) Inversely proportional to the cube root of electron mass

77. **Which of the following particles - neutron, proton, electron and deuteron has the lowest energy if all have the same de Broglie wavelength**

1) neutron 2) proton 3)electron 4) deuteron

78. **The momentum of a proton is p. The corresponding wavelength is**

1) h/p 2) h p 3) p/h 4) $\sqrt{hp}$

79. **A wave is associated with matter when it is**

1) stationary 2) in motion with a velocity

3) in motion with speed of light

4) in motion with speed greater than that of light

80. **An electron of mass 9.1 x 10^{-31}kg and charge 1.6 x 10^{-19} C is accelerated through a potential difference of V volt. The de Broglie wavelength (λ) associated with the electron is**

1) $\frac{12.27}{\sqrt{V}}$ A^0 2) $\frac{12.27}{V}$ A^0

3) $12.27\sqrt{V}$ A^0 4) $\frac{1}{12.27\sqrt{V}}$ A^0

81. **The de Broglie wavelength of a molecule of thermal energy KT (K is Boltzmann constant and T is absolute temperature) is given by**

1) $\frac{h}{\sqrt{2mKT}}$ 2) $\frac{h}{2mKT}$ 3) $h\sqrt{2mKT}$ 4) $\frac{1}{h\sqrt{2mKT}}$

82. **The wavelengths of a proton and a photon are same. Then**

1) Their velocities are same 2) Their momenta are equal

3) Their energies are same 4) Their speeds are same

83. **If the value of Planck's constant is more than its present value, then the de Broglie wavelength associated with a material particle will be**

1) More 2) Less 3) Same

4) More for lighter particles and less for heavy particles

84. **The wavelength of matter waves does not depend on**

1) Momentum 2) Velocity 3) Mass 4) Charge

85. **The wave nature of matter is not observed in daily life because their wave length is**

1) Less 2) More 3) In infrared region 4) In ultraviolet region

86. **The ratio of the wavelengths of a photon and that of an electron of same energy E will be [m is mass of electron]**

1) $\sqrt{\frac{2m}{E}}$ 2) $\sqrt{\frac{E}{2m}}$ 3) $C\sqrt{\frac{2m}{E}}$ 4) $\sqrt{\frac{EC}{2m}}$

87. **One of the following figures respesents the variation of particle momentum with associated de Broglie wavelength**

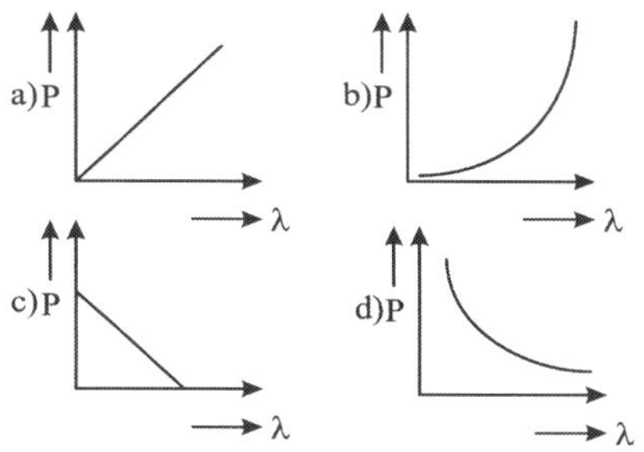

1) a 2) b 3) c 4) d

88. **A point source causes photoelectric effect from a small metal plate. Which of the following curves may represent the saturation photocurrent as a function of the distance between the source and the metal?**

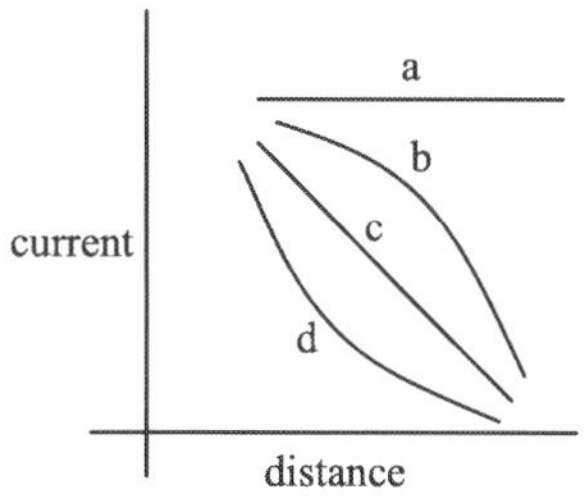

1) a 2) b 3) c 4) d

89. **Matter waves are:**

1) electromagnetic waves
2) mechanical waves
3) either mechanical or electromagnetic waves
4) neither mechanical nor electromagnetic waves

90. **The incorrect statement is**

1) Material wave (de-Broglie wave) can travel in vacuum
2) Electromagnetic wave can travel through vacuum
3) The velocity of photon is the same as light passes through any medium
4) Wavelength of de-Broglie wave depends upon velocity

91. **The magnitude of the de-Broglie wavelength (λ) of an electron (e),proton(p),neutron (n) and α - particle (α) all having the same energy of MeV, in the increasing order will follow the sequence:**

1) $\lambda_e, \lambda_p, \lambda_n, \lambda_\alpha$ 2) $\lambda_\alpha, \lambda_n, \lambda_p, \lambda_e$ 3) $\lambda_e, \lambda_n, \lambda_p, \lambda_\alpha$ 4) $\lambda_p, \lambda_e, \lambda_\alpha, \lambda_n$

92. **Moving with the same velocity ,one of the following has the longest deBroglie wavelength**

1) β -particle 2) α -particle 3) proton 4) neutron

93. **Debroglie wavelength of a particle at rest position is**

1) zero 2) finite 3) infinity 4) cannot be calculated

94. **Debroglie wavelength of protons accelerated by an electric field at a potential difference v is**

1) $\frac{0.108}{\sqrt{V}}$ 2) $\frac{0.202}{\sqrt{V}}$ 3) $\frac{0.286}{\sqrt{V}}$ 4) $\frac{0.101}{\sqrt{V}}$

95. Debroglie wavelength of uncharged particles depends on

1) mass of particle
2) kinetic energy of particle
3) nature of particle
4) All above

96. Debroglie wavelength of a moving gas molecule is

1) proportional to temperature
2) inversely proportional to temperature
3) independent of temperature
4) inversely proportional to square root of temperature

97. The particles that can be accelerated by an electric field is

1) proton
2) electron
3) alpha particle
4) all above

98. If a proton and an electron are confined to the same region, then uncertainity in momentum

1) for proton is more, as compared to the electron
2) for electron is more, as compared to the proton
3) same for both the particles
4) directly proportional to their masses

99. Which phenomenon best supports the theory that matter has a wave nature ?

1) electron momentum
2) electron diffraction
3) photon momentum
4) photon diffraction

100. The wavelength of de-Broglie wave associated with a thermal neutron of mass m at absolute temperature T is given by (Here, k is the Boltzmann constant)

1) $\frac{h}{\sqrt{2mkT}}$
2) $\frac{h}{\sqrt{mkT}}$
3) $\frac{h}{\sqrt{3mkT}}$
4) $\frac{h}{2\sqrt{mkT}}$

ASSERTION & RESONING

In each of the following questions, a statement is given and a corresponding statement or reason is given just below it. In the statements, mark the correct answer as

1) If both Assertion and Reason are true and Reason is correct explanation of Assertion.

2) If both Assertion and Reason are true but Reason is not the correct explanation of Assertion.

3) If Assertion is true but Reason is false.

4) If both Assertion and Reason are false.

101. Assertion (A) : For a fixed incident photon energy, photoelectrons have a wide range of energies ranging from zero to the maximum value K_{max}

Reason (R) : Initially, the electrons in the metal are at different energy level.

102. Consider the following statements A and B, identify the correct choice in the given answers.

A) Tightly bound electrons of target material scattered X-ray photon,resulting in the Compton effect.

B) Photoelectric effect takes place with free electrons.

103. The frequency and intensity of a light source are both doubled. Consider the following statements.

(A) The saturation photocurrent remains almost the same.

(B) The maximum kinetic energy of the photoelectrons is doubled.

1) Both A and B are true
2) A is true but B is false
3) A is false but B is true
4) Both A and B are false

STATEMENT TYPE

In each of the following questions, a statement is given and a corresponding statement or reason is given just below it. In the statements, mark the correct answer as

1) Statement I is true, Statement II is true; statement II is a correct explanation of statement I.

2) Statement I is true, Statement II is true, Statement II is NOT a correct explanation for statement I.

3) Statement I is true, Statement II is false

4) Statement I is false, Statemenet II is true.

104. A proton and an electron both have energy 50 eV.

Statement-I: Both have different wavelengths

Statement-II: Wavelength depends on energy and not on mass.

105. Statement I: Though light of a single frequency (monochromic light) is incident ona metal, the energies of emitted photoelectrons are different.

Statement II: The energy of electrons just after they absorb photons incident on the metal surface may be lost in collision with other atoms in the metal before the electron is ejected out of the metal.

106. Statement I: The de Broglie wavelength of a molecule (in a sample of ideal gas) varies inversely as the square root of absolute temperature.

Statement II: The deBroglie wavelength of a molecule (in sample of ideal gas) depends on temperature.

107. Statement-I: A metallic surface is irradiated by a monochromatic light of frquency $v > v_0$ (the threshold frequency). The maximum kinetic energy and the stopping potential are K_{max} and v_0 are also doubled.

Statement-II: The maximum kinetic energy and he stopping potential of photoelectrons emitted from a surface are linearly dependent on the frquency of incident light. **[AIEEE-2011]**

108. Statement I: Davisson-Germer experiment established the wave nature of electrons

Statement II: If electrons have wave nature, they can interface and show diffraction. **[AIEEE-2012]**

CONCEPTUAL KEY

1) 1	2) 1	3) 4	4) 3	5) 2	6) 1	7) 4	8) 4	9) 4	10) 1
11) 1	12) 1	13) 2	14) 3	15) 3	16) 2	17) 2	18) 1	19) 4	20) 2
21) 2	22) 3	23) 4	24) 1	25) 2	26) 1	27) 2	28) 1	29) 2	30) 3
31) 1	32) 3	33) 2	34) 4	35) 3	36) 4	37) 1	38) 1	39) 4	40) 3
41) 3	42) 4	43) 4	44) 2	45) 3	46) 2	47) 2	48) 4	49) 1	50) 3
51) 1	52) 3	53) 4	54) 4	55) 2	56) 1	57) 2	58) 4	59) 2	60) 1
61) 2	62) 4	63) 3	64) 4	65) 2	66) 4	67) 4	68) 2	69) 2	70) 4
71) 1	72) 3	73) 1	74) 3	75) 4	76) 2	77) 4	78) 1	79) 2	80) 1
81) 1	82) 2	83) 1	84) 4	85) 1	86) 3	87) 4	88) 4	89) 4	90) 3
91) 2	92) 1	93) 3	94) 3	95) 4	96) 4	97) 4	98) 3	99) 2	100) 3
101) 1	102) 4	103) 2	104) 3	105)1	106) 2	107) 3	108) 1		

LEVEL - I

PHOTO ELECTRIC EFFECT

1. The frequency of a photon associated with an energy of 3.31 eV is (given h = 6.62 x 10^{-34} Js)

1) 0.8 x 10^{15} Hz 2) 1.6 x 10^{15} Hz 3) 3.2 x 10^{15} Hz 4) 8.0 x 10^{15} Hz

2. A radiation of wave length 2500 A^0 is incident on a metal plate whose work function is 3.5 eV. Then the potential required to stop the fastest photo electrons emitted by the surface is (h = 6.63×10^{-34}Js & c= 3×10^8 m/s)

1) 1.86V 2) 3.00 V 3) 1.46V 4) 2.15 V

3. The work function of a metal is 2.5 eV. The maximum kinetic energy of the photoelectrons emitted if a radiation of wavelength 3000 A^0 falls on it is
($h = 6.63\times10^{-34}$Js and $c = 3\times10^8$ m/s)
1) 1.12×10^{-19}J 2) 4.8×10^{-19}J 3) 3.2×10^{-19}J 4) 2.61×10^{-19}J

4. The work function of a substance is 4.0 eV. The longest wavelength of light that can cause photoelectric emission from this substance is approximately
1) 220 nm 2) 310 nm 3) 540 nm 4) 400 nm

5. A laser used to weld detached retains emits light with a wavelength 652 nm in pulses that are of 20ms duration. The average power during each pulse is 0.6W. The energy in each pulse and in a single photon are
1) $7.5\times10^{15}eV, 2.7eV$ 2) $6.5\times10^{16}eV, 2.9eV$
3) $6.5\times10^{16}eV, 2.7eV$ 4) $7.5\times10^{16}eV, 1.9eV$

6. Electrons ejected from the surface of a metal, when light of certain frequency is incident on it, are stopped fully by a retarding potential of 3 volts. Photo electric effect in this metallic surface begins at a frequency $6\times10^{14}s^{-1}$. The frequency of the incident light in s^{-1} is [$h=6\times10^{-34}$J-sec;charge on the electron=1.6×10^{-19}C]
1) 7.5×10^{13} 2) 13.5×10^{13} 3) 14×10^{14} 4) 7.5×10^{15}

7. The threshold wavelength for emission of photoelectrons from a metal surface is 6×10^{-7}m. The work function of the material of the metal surface is .
1) 3.3×10^{-19}J 2) 6.67×10^{-19}J 3) 1.23×10^{-19}J 4) 2.37×10^{-19}J

8. The maximum velocity of an electron emitted by light of wavelength λ incident on the surface of a metal of workfunction ϕ is where h = Planck's constant, m = mass of electron and c=speed of light
1) $\left[\frac{2(hc+\lambda\phi)}{m\lambda}\right]^{1/2}$ 2) $\frac{2(hc-\lambda\phi)}{m}$ 3) $\left[\frac{2(hc-\lambda\phi)}{m\lambda}\right]^{1/2}$ 4) $\left[\frac{2(hc\lambda-\phi)}{m}\right]^{1/2}$

9. The work function of nickle is 5eV. When light of wavelength 2000A^0 falls on it, emits photoelectrons in the circuit. Then the potential difference necessary to stop the fastest electrons emitted is (given $h=6.67\times10^{-34}$Js)
1) 1.0V 2) 1.75V 3) 1.2V 4) 0.75V

MATTER WAVES

10. If an electron and a proton have the same KE, the ratio of the de Broglie wavelengths of proton and electron would approximately be
1) 1 : 1837 2) 43 : 1 3) 1837 : 1 4) 1 : 43

11. If electron is having a wavelength of 100 A^o, then momentum is (gm cm s^{-1}) units
1) 6.6×10^{-32} 2) 6.6×10^{-29} 3) 6.6×10^{-25} 4) 6.6×10^{-21}

12. The de-broglie wavelength of an electron and the wavelength of a photon are same. The ratio between the energy of the photon and the momentum of the electron is [M 2006]
1) h 2) c 3) 1/h 4) 1/c

13. A proton and an alpha particle are accelerated through the same potential difference. The ratio of wavelengths associated with proton and alpha particle respectively is
1) $1:2\sqrt{2}$ 2) 2:1 3) $2\sqrt{2}:1$ 4) 4:1

14. Ratio of debroglie wavelengths of uncharged particle of mass m at 27^0C to 127^0C is nearly
1) 1.16 2) 0.16 3) 1.33 4) 0.8

15. A particle is projeted horizontally with a velocity 10m/s. What will be the ratio of de-Broglie wvelengths of the particle, when the velocity vector makes an angle 30^0 and 60^0 with the horizontal

1) $\sqrt{3}:1$ 2) $1:\sqrt{3}$ 2) $2:\sqrt{3}$ 4) $\sqrt{3}:2$

16. A positron and a proton are accelerated by the same accelerating potential. Then the ratio of the associated wavelengths of the positron and the proton will be [M = mass of proton, m = mass of positron]

1) $\frac{M}{m}$ 2) $\sqrt{\frac{M}{m}}$ 3) $\frac{m}{M}$ 4) $\sqrt{\frac{m}{M}}$

LEVEL-I - KEY

1) 1	2) 3	3) 4	4) 2	5) 4	6) 3	7) 1	8) 3	9) 3	10) 4
11) 4	12) 2	13) 3	14) 1	15) 1	16) 2				

LEVEL-I - HINTS

1. E = hv

2. $V_0 e = \left[\frac{12400}{\lambda\, in\, A^0} - \omega_0\right] eV$

3. $K.E. = \left[\frac{12400}{\lambda\, in\, A^0} - \omega_0\right] eV$

4. λ in $A^0 = \frac{12400}{E}$ in eV

5. $E = pt; E(\text{in each photon}) = \frac{hc}{\lambda}$

6. K.E. = $V_0 e$ and $V_0 e = h[\upsilon - \upsilon_0]$

where V_0 is the stopping potential and υ_0 is the Threshold frequency

7. $\omega_0 = \frac{hc}{\lambda_0}$ 8. $E = \phi + \frac{1}{2}mv^2$ 9. $V_0 = \frac{E - \omega_0}{e}$

10. $\lambda = \frac{h}{\sqrt{2mE}}$ 11. $P = \frac{h}{\lambda}$

12. $\lambda_e = \lambda_{ph}; \frac{E_{ph}}{m_e V} = \frac{E_{ph}}{\left(\frac{h}{\lambda_e}\right)} = \frac{\frac{hc}{\lambda_{ph}}}{\frac{h}{\lambda_e}} = C$

13. $\frac{\lambda_p}{\lambda_\alpha} = \sqrt{\frac{q_\alpha}{q_p} \times \frac{m_\alpha}{m_p}}$ 14. $\lambda \alpha \frac{1}{\sqrt{T}}$

15. $\lambda = \frac{h}{P}, P = m \times v$ Take v component

16. $\lambda = \frac{h}{p}, \lambda = \frac{h}{\sqrt{2Vmq}}$

LEVEL - I -2

PHOTO ELECTRIC EFFECT

1. **The threshold wavelength for a surface having a threshold frequency of 0.6×10^{15} Hz is (given $c = 3 \times 10^8$ m/s)**
   1) 4000 A° 2) 6000 A° 3) 5000A° 4) 3500A°

2. **Two photons of energies twice and thrice the work function of a metal are incident on the metal surface. Then the ratio of maximum velocities of the photoelectrons emitted in the two cases respectively, is**
   1) $\sqrt{2}:1$ 2) $\sqrt{3}:1$ 3) $\sqrt{3}:\sqrt{2}$ 4) $1:\sqrt{2}$

3. **The photo electric work function for a metal surface is 4.125 eV. The cut-off wavelength for this surface**
   1) 4125 A° 2) 2062.5 A° 3) 3006.06 A° 4) 6000A°

4. **The energy of emitted photoelectrons from a metal is 0.9 eV. The work function of the metal is 2.2 eV. Then the energy of the incident photon is**
   1) 0.9 eV 2) 2. 2 eV 3) 4. 4 eV 4) 3.1 eV

5. **A photoelectron is moving with a maximum velocity of 10^6 m/s. Given $e=1.6\times10^{-19}$ c, and $m = 9.1\times 10^{-31}$ kg, the stopping potential is**
   1) 2.5 V 2) 2.8 V 3) 2.0 V 4) 1.4 V

6. **A metal of work function 4eV is exposed to a radiation of wavelength 140×10^{-9}m.Then the stopping potential developed by it ($h = 6.63\times10^{-34}$Js and $c= 3\times10^8$ m/s)**
   1) 6.42 V 2) 2.94 V 3) 4.86V 4) 3.2 V

7. **Threshold wavelength for a metal having work function w_o is λ. Then the threshold wavelength for the metal having work function 2 w_o is**
   1) 4λ 2) 2λ 3) $\lambda/2$ 4) $\lambda/4$

8. **The work function of metals A and B are in the ratio 1:2. If light of frequencies f and 2f are incident on metal surfaces A and B respectively, the ratio of the maximum kinetic energies of the photo electrons emitted is (2000 M)**
   1) 1:1 2) 1:2 3) 1:3 4) 1:4

9. **The threshold wave length for photo electric emission from a material is 5,200A^0, photo electrons will be emitted when this material is illuminated with mnochromaic radiation from a**
   1) 50 watt infrared lamp 2) 1 watt infrared lamp
   3) 1 watt ultraviolet lamp 4) 50 watt sodium vapour lamp

MATTER WAVES

10. **A particle having a de Broglie wavelength of 1.0 A° is associated with a momentum of (given $h = 6.6 \times 10^{-34}$ Js)**
    1) 6.6×10^{-26} kg m/s 2) 6.6×10^{-25} kg m/s
    3) 6.6×10^{-24} kg m/s 4) 6.6×10^{-22} kg m/s

11. **The de Broglie wavelength of an electron having 80 eV of energy is nearly** ($1\ eV = 1.6x10^{-19} J$, **Mass of electron =** 9×10^{-31}kg, **Planck's constant** $= 6.6\times10^{-34}$ Js) **(nearly) (2001 E)**
    1) 140 A° 2) 0.14 A° 3) 14 A° 4) 1.4 A°

12. **Electrons are accelerated through a p.d. of 150V. Given m = 9.1x10^{-31}kg,e =1.6x10^{-19} c, h = 6.62x10^{-34} Js, the de Broglie wavelength associated with it is**

1) 1.5 A° 2) 1.0 A° 3) 3.0 A° 4) 0.5 A°

13. **If accelerating potential of an alpha particle is doubled than its new debrolgie wavelength becomes**

1) $\frac{1}{\sqrt{2}}$ times of initial 2) $\sqrt{2}$ times of initial

3) $\frac{1}{2}$ times of initial 4) 2 times of initial

14. **The ratio of the deBroglie wavelenths of proton, deuteron and alpha particle accelerated through the same potential difference 100V is**

1) $2:2:1$ 2) $1:2:2\sqrt{2}$ 3) $1:2:2\sqrt{2}$ 4) $2\sqrt{2}:2:1$

LEVEL - I -2 KEY

1) 3 2) 4 3) 3 4) 4 5) 2 6) 3 7) 3 8) 2 9) 3 10) 3
11) 4 12) 2 13) 1 14) 4

LEVEL-I -2 - HINTS

1. $\lambda = \frac{c}{v}$

2. $\frac{V_1}{V_2} = \sqrt{\frac{E_1 - \omega_0}{E_2 - \omega_0}}$

3. λ in $A^0 = \frac{12400}{E}$ in eV

4. $E = K.E + \omega_0$

5. $\frac{1}{2}mv^2 = eV_0$

6. $V_0 = \frac{\left[\frac{12400}{\lambda\, in\, A^0} - \omega_0\right]}{e}$

7. $w_0 = \frac{hc}{\lambda_0}$

8. $\frac{KE_1}{KE_2} = \frac{E_1 - W_1}{E_2 - W_2} = \frac{hv_1 - w_1}{hv_2 - w_2} = \frac{hf - w}{2hf - 2w} = \frac{1}{2}$

9. $P = \frac{NE}{t}, E = \frac{hc}{\lambda}$

10. $P = \frac{h}{\lambda}$

11. $\lambda = \frac{h}{\sqrt{2mE}}$

12. $\lambda = \sqrt{\frac{150}{V}}$

13. $\lambda \alpha \frac{1}{\sqrt{V}}$

14. $\lambda = \frac{h}{P}, \frac{P^2}{2m} = V \times q$

LEVEL - II

PHOTO ELECTRIC EFFECT

1. **A photometal is illuminated by lights of wavelengths λ_1 and λ_2 respectively. The maximum kinetic enegies of electrons emitted in the two cases are E_1 and E_2 respectively. The work function of metal is.**

1) $\frac{E_2\lambda_1 - E_1\lambda_2}{\lambda_1}$ 2) $\frac{E_1\lambda_1 - E_2\lambda_2}{\lambda_1 + \lambda_2}$ 3) $\frac{E_1\lambda_1 + E_2\lambda_2}{\lambda_1 - \lambda_2}$ 4) $\frac{E_2\lambda_2 - E_1\lambda_1}{\lambda_1 - \lambda_2}$

2. **Light of wavelength λ strikes a photo sensitive surface and electrons are ejected with kinetic energy E. If the kinetic energy is to be increased to 2E, the wavelength must be changed to λ' where**

1) $\lambda' = \frac{\lambda}{2}$ 2) $\lambda' = 2\lambda$ 3) $\frac{\lambda}{2} < \lambda' < \lambda$ 4) $\lambda' > \lambda$

3. **Ultraviolet light of wavelength 300 nm and intensity 1.0 W/m² falls on the surface of a photoelectric material. If one percent of the incident photons produce photoelectrons, then the number of photoelectrons emitted from an area of 1.0 cm² of the surface is nearly (in per second)**

1) 9.61×10^{14} 2) 4.12×10^{13} 3) 1.51×10^{12} 4) 2.13×10^{11}

4. **Light rays of wavelengths 6000 A° and of photon intensity 39.6 watts/m² is incident on a metal surface. If only one percent of photons incident on the surface emit photo electrons, then the number of electrons emitted per second per unit area from the surface will be**

[Planck constant = 6.64×10^{-34} J - S; Velocity of light = 3×10^8 ms^{-1}]

1) 12×10^{18} 2) 10×10^{18} 3) 12×10^{17} 4) 12×10^{15}

5. **Light of wavelength 4000 A° is incident on a metal surface of work function 2.5 eV. Given h=6.62×10^{-34} Js, c = 3×10^8 m/s, the maximum KE of photoelectrons emitted and the corresponding stopping potential are respectively**

1) 0. 6 eV, 0.6 V 2) 2.5 eV, 2.5 V 3) 3.1 eV, 3.1 V 4) 0.6 eV, 0.3 V

6. **The K.E of the electron is E when the incident wavelength is λ. To increase the K.E of the electron to 2E, the incident wavelength must be**

1) 2λ 2) $\frac{\lambda}{2}$ 3) $\frac{hc\lambda}{E\lambda + hc}$ 4) $\frac{2hc\lambda}{E\lambda + hc}$

7. **A photon of energy 15 eV collides with H-atom. Due to this collision, H-atom gets ionized .The maximum kinetic energy of emitted electron is :**

1) 1.4 eV 2) 5 eV 3) 15eV 4) 13.6eV

8. **The anode voltage of a photocell is kept fixed. The wavelength λ of the light falling on the cathode in gradually changed. The plate current I of the photocell varies as follows :**

1)
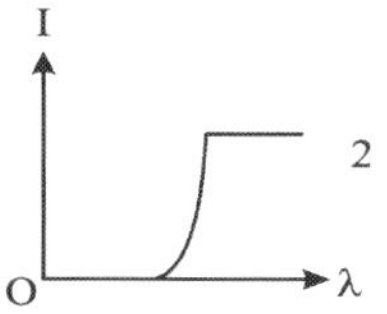

2)
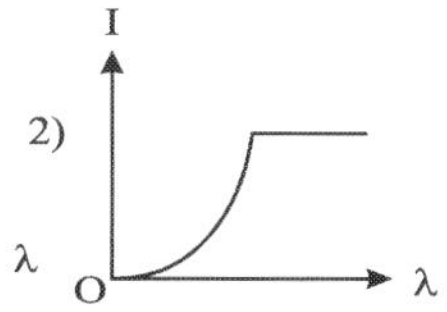

3)
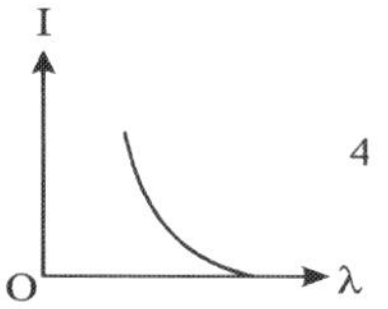

4)
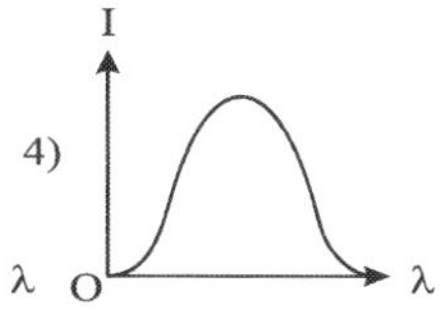

9. **According to Einstein's photoelectric equation, the graph between the kinetic energy of photoelectrons ejected and the frequency of incident radiation is :**

1)
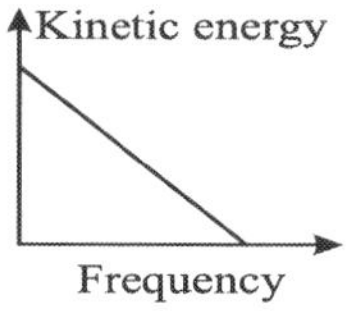

2)
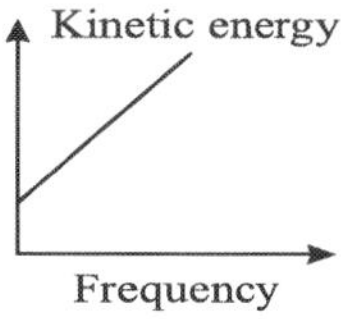

3)
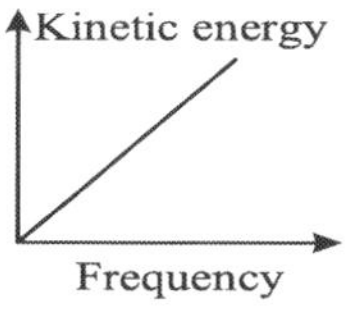

4)
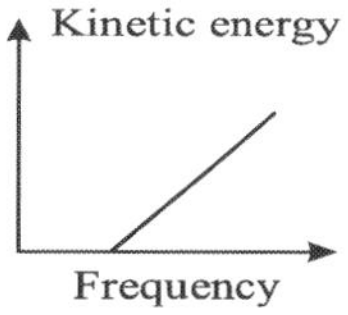

10. **The graph shown in figure show the variation of photoelectric current (i) and the applied voltage (V) for two different materials and for two different intensities of the incident radiation.**

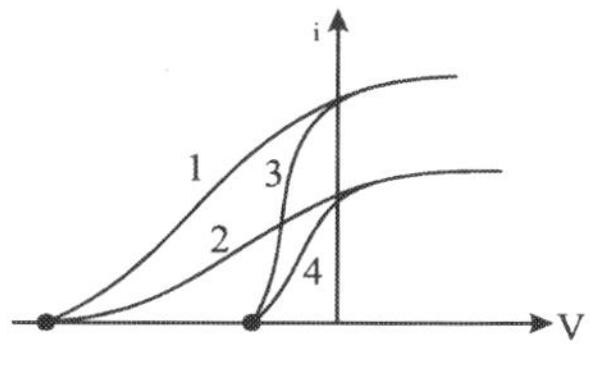

Identify the pairs of curves that correspond to (a) different material (b) same intensity of incident radiations.

1) Curve 1 and 3, Curve 2 and 4
2) Curve 1 and 2, Curve 3 and 4
3) Curve 1 and 4, Curve 2 and 3
4) Curve 1 only, Curve 2 and 4

MATTER WAVES

11. **A proton when accelerated through a p.d. of V volt has a wavelength λ associated with it. An α - particle in order to have the same wavelength λ must be accelerated through a p.d. of**

1) V/8 volt 2) V/4 volt 3) V volt 4) 2V volt

12. **An electron of mass m and charge e initially at rest gets accelerated by a constant electric field E. The rate of change of de-Broglie wavelength of this electron at time t ignoring relativistic effects is**

1) $\frac{-h}{eEt^2}$ 2) $\frac{-eEt}{E}$ 3) $\frac{-mh}{eEt^2}$ 4) $\frac{-h}{e.E}$

13. **If the velocity of a particle is increased three times, then the percentage decrease in its de Broglie wavelength will be**

1) 33.3% 2) 66.6% 3) 99.9% 4) 22.2%

14. **If the momentum of an electron is changed by p_m, then the de Broglie wavelength associated with it changes by 0.5%. The initial momentum of electron will be**

1) $p_m/200$ 2) $p_m/100$ 3) $200p_m$ 4) $100p_m$

15. **When the mass of an electron becomes equal to thrice its rest mass, its speed is**

1) $\frac{2\sqrt{2}}{3}c$ 2) $\frac{2}{3}c$ 3) $\frac{1}{3}c$ 4) $\frac{1}{4}c$

16. **Which of the following figures represents the variation of particle momentum with the associated de Broglie wave-length ?**

1) 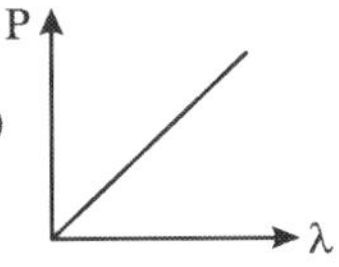

2) 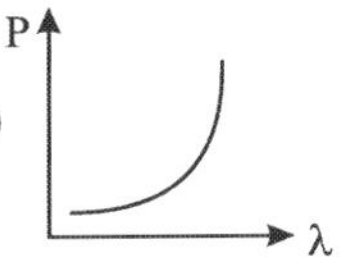

3) 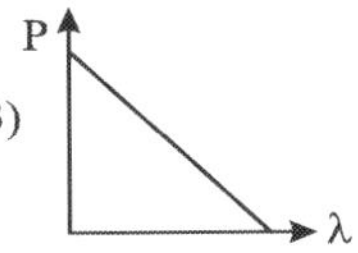

4) 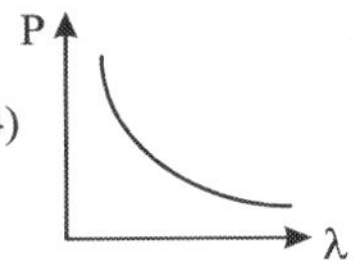

17. **The de Broglie wave present in fifth Bohr orbit is :**

1) 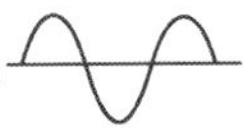

2) 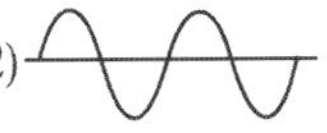

3) 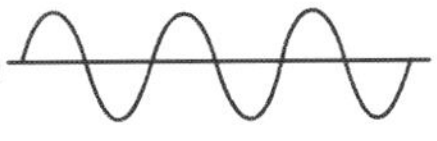

4) 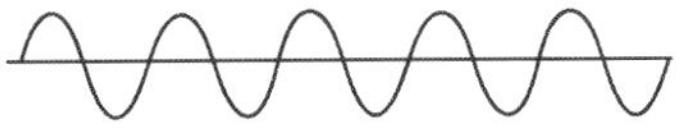

HEISEN-BERG UNCERTAINITY PRINCIPLE AND DAVISSON-GERMER EXPERIMENT

18. **The correctness of velocity of an electron moving with velocity 50 ms^{-1} is 0.005%. The accuracy with which its position can be measured will be**

1) $4634\times10^{-3}m$ 2) $4634\times10^{-5}m$ 3) $4634\times10^{-6}m$ 4) $4634\times10^{-8}m$

19. **If the uncertainity in the position of an electron is 10^{-10} m, then the value of uncertainity in its momentum (in kg-ms^{-1}) will be**

1) 3.33×10^{-24} 2) 1.03×10^{-24} 3) 6.6×10^{-24} 4) 6.6×10^{-20}

20. a) Name the experiment for which the adjacent graph, showing the variation of intensity of scattered electrons with the angle of scattering (θ) was obtained.

b) Also name the important hypothesis that was confirmed by this experiment.

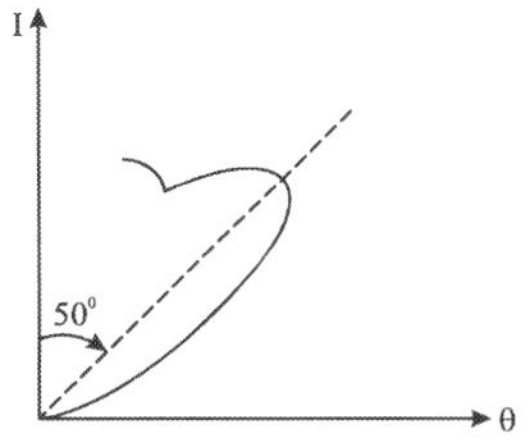

1) A) Davission and Germer experiment B) de Broglie hypothesis
2) A) Photo electric effect B) de Broglie hypothesis
3) A) Thermionic emission B) de Broglie hypothesis
4) A) Photocell B) de Broglie hypothesis

LEVEL-II - KEY

1) 4	2) 3	3) 3	4) 3	5) 1	6) 3	7) 1	8) 3	9) 4	10) 1
11) 1	12) 1	13) 2	14) 3	15) 1	16) 4	17) 4	18) 2	19) 2	20) 2

LEVEL-II - HINTS

1. $\frac{hc}{\lambda_1}=W+E_1;\frac{hc}{\lambda_2}=W+E_2$

2. use $\frac{hc}{\lambda_1}>\frac{hc}{\lambda}$ and $\frac{2hc}{\lambda}>\frac{hc}{\lambda_1}$

3&4. $P=\left(\frac{n}{t}\right)\left(\frac{hc}{\lambda}\right)$

5. $K.E=\left[\frac{12400}{\lambda\, in\, A^0}-\omega_0\right]$

6. $\frac{\lambda^1}{\lambda}=\frac{\omega+E}{\omega+2E}$

7. Energy of photon ionization energy $+K.E_{max}$ =15ev - 13.6ev=1.4ev

10. (a) Curves 1 and 3 (different materials due to different stopping potentials)
(b) Curves 2 and 4 (same intensity due to same current)

11. $\lambda=\frac{h}{p}=\frac{h}{\sqrt{2mvq}}$ (λ = constant);

$v_1q_1m_1=v_2q_2m_2$

12. Here $u=0, a=\frac{eE}{m}$'$v=?.t=t$

$v = u + at = 0, \frac{eE}{m} t$

de-Broglie wavelength,

$\lambda = \frac{h}{mv} = \frac{h}{m(eEt/m)} = \frac{h}{eEt}$

Rate of change of de-Broglie wavelength

$\frac{d\lambda}{dt} = \frac{h}{eE}\left(-\frac{1}{t^2}\right) = \frac{-h}{eEt^2}$

13. $\lambda \alpha \frac{1}{V}$

14. $\frac{\Delta\lambda}{\lambda} = \frac{-\Delta P}{P}$

15. $m = \frac{m_0}{\sqrt{1 - \frac{v^2}{c^2}}}$

18. Here, $\Delta v = \frac{0.005 \times 50}{100} = 0.0025 ms^{-1}$

$\Delta x = \frac{h}{m\Delta v} = \frac{1.034 \times 10^{-34}}{9.1 \times 10^{-31} \times 0.0025} = 4634 \times 10^{-5} m$

19. $\Delta p = \frac{h}{\Delta x} = \frac{1.034 \times 10^{-34}}{10^{-10}} = 1.034 \times 10^{-24} kg - ms^{-1}$

20. A) Davission and Germer experiment
B) de Broglie hypothesis

LEVEL - II - 1
PHOTO ELECTRIC EFFECT

1. When a metal surface is illuminated by a monochromatic light of wave - length λ, then the potential difference required to stop the ejection of electrons is 3V. When the same surface is illuminated by the light of wavelength 2 λ, then the potential difference required to stop the ejection of electrons is V. Then for photoelectric effect, the threshold wavelength for the metal surface will be

1) 6 λ 2) $4\lambda/3$ 3) 4 λ 4) 8 λ

2. If U.V. Light of wavelengths 800 A° and 700 A° can liberate electrons with kinetic energies of 1.8eV and 4 eV respectively from hydrogen atom in ground state, then the value of planck's constant is

1) 6.57 x 10^{-34} Js 2) 6.63 x 10^{-34} Js 3) 6.66 x 10^{-34} Js 4) 6.77 x10^{-34} Js

3. In a photoelectric effect experiment, photons of energy 5 eV are incident on a metal surface. They liberate photoelectrons which are just stopped by an electrode at a potential of -3.5 V with respect to the metal. The work function of the metal is

1) 1.5 eV 2) 3.5 eV 3) 5. 0 eV 4) 8.5 eV

4. **The number of photons emitted per second by a 62W source of monochromatic light of wavelength 4800 A° is**

1) 1.5×10^{19} 2) 1.5×10^{20} 3) 2.5×10^{20} 4) 4×10^{20}

5. **Photons of frequencies 2.2×10^{15} Hz and 4.6×10^{15} Hz are incident on a metal surface. The corresponding stopping potentials were found to be 6.6 V and 16.5 V respectively. Given $e = 1.6 \times 10^{-19}$ c, the value of universal planck's constant is**

1) 6.6×10^{-34} Js 2) 6.7×10^{-34} Js 3) 6.5×10^{-34} Js 4) 6.8×10^{-34} Js

6. **If stopping potentials corresponding to wavelengths 4000A^0 and 4500A^0 are 1.3V and 0.9V respectively, then the work function of the metal is**

1) 0.3eV 2) 1.3eV 3) 1.8eV 4) 5eV

7. **In a photoelectric experiment anode potential is plotted against plate current**

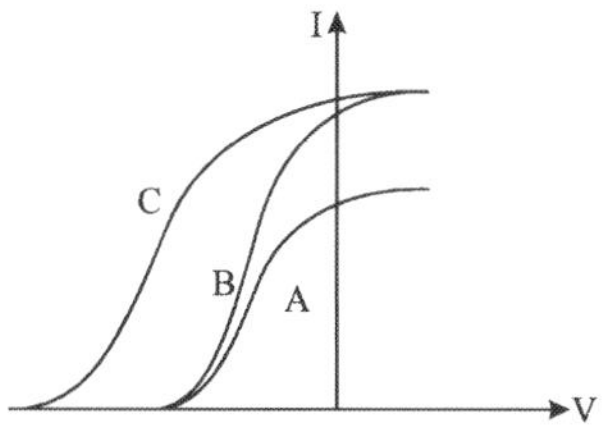

1) A and B will have same intensities while B and C will have different frequencies
2) B and C will have different intensities while A and B will have different frequencies.
3) A and B will have different intensities while B and C will have equal frequencies.
4) B and C will have equal intensities while A and B will have same frequencies.

MATTER WAVES

8. **An electron moves with a speed of $\frac{\sqrt{3}}{2}$ c. Then its mass becomes....times its rest mass.**

1) 2 2) 3 3) 3/2 4) 4

9. **Photons of energy 2.0 eV fall on a metal plate and release photoelectrons with a maximum velocity V. By decreasing λ by 25% the maximum velocity of photoelectrons is doubled. The work function of the metal of the material plate in eV is nearly**

1) 2.22 2) 1.985 3) 2.35 4) 1.80

10. **A proton when accelerated through a p.d of V volt has wavelength λ associated with it .An electron to have the same λ must be accelerated through a p.d of**

1) $\frac{V}{8}$ volt 2) 4V volt 3) 2V volt 4) 1838V volt

11. **The momentum aphoton of electromagnetic radiation is 3.3×10^{-29} kgms^{-1}. The frequency of these waves is:**

1) 3.0×10^3 Hz 2) 6.0×10^3 Hz 3) 7.5×10^{12} Hz 4) 1.5×10^{13} Hz

12. **If the energy of a particle is reduced to one fourth, then the percentage increase in its de Broglie wavelength will be**

1) 41% 2) 141% 3) 100% 4) 71%

13. **The de Broglie wavelength associated with an electron of velocity 0.3 c and rest mass 9.1×10^{-31}kg is**

1) 7.68×10^{-10} m 2) 7.68 x 10-12 m 3) 5.7×10^{-12} m 4) 9.1×10^{-12} m

14. The two lines A and B shown in figure are the graphs of the de Broglie wavelength λ as a function of $\frac{1}{\sqrt{V}}$ (V is the accelerating potential) for two particles having the same charge.

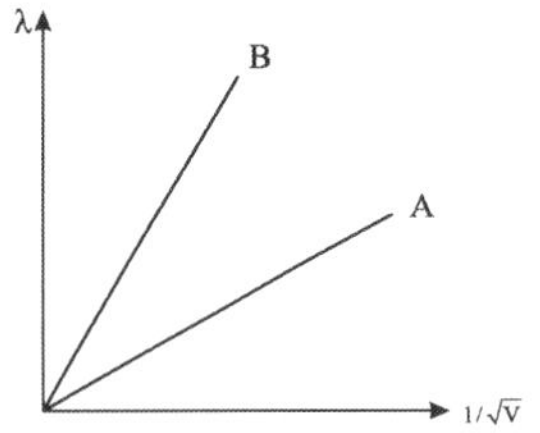

Which of the two represents the particle of heavier mass ?

1) A 2) B 3) Both A and B 4) Data insufficient

HEISEN-BERG UNCERTAINITY PRINCIPLE AND DAVISSON-GERMER EXPERIMENT

15. The uncertainity in the position of a particle is equal to the de-Broglie wavelength. The uncertainity in its momentum will be

1) $\frac{h}{\lambda}$ 2) $\frac{2h}{3\lambda}$ 3) $\frac{\lambda}{h}$ 4) $\frac{3\lambda}{2h}$

16. If the uncertainity in the position of proton is $6\times10^{8}m$, then the minimum uncertainity in its speed will be

1) $1\ cms^{-1}$ 2) $1\ ms^{-1}$ 3) $1\ mms^{-1}$ 4) $100\ ms^{-1}$

17. From Davisson-Germer experiment an α particle and a proton are accelerated through the same pd V. Find the ratio of the de Broglie wavelengths associated with them

1) $1:2\sqrt{2}$ 2) $2\sqrt{2}:1$ 3) $1:\sqrt{2}$ 4) $\sqrt{2}:1$

LEVEL-II -2 - KEY

1) 3	2) 1	3) 1	4) 2	5) 1	6) 3	7) 4	8) 1	9) 4	10) 4
11) 4	12) 3	13) 2	14) 1	15) 1	16) 2	17) 1			

LEVEL-II -2 - HINTS

1. $\frac{hc}{\lambda}=\frac{hc}{\lambda_0}+eV_0$ or $hc\left(\frac{1}{\lambda}-\frac{1}{\lambda_0}\right)=eV_0$

2. $\frac{hc}{\lambda}=w+eV_0$

3. $E=w+eV_0$

4. $P=\left(\frac{n}{t}\right)\left(\frac{hc}{\lambda}\right)$

5. $hv=w+eV_0$

6. $h\gamma=$ w+K.E $\Rightarrow h\gamma=$ w+V_0e

8. $m=\frac{m_0}{\sqrt{1-\frac{v^2}{c^2}}}$

9. $\frac{1}{2}mv_{max}^{2} = h\nu - w$ and $h\nu = \frac{hc}{\lambda}$

10. $V_1 \times q_1 \times m_1 = V_2 \times q_2 \times m_2$

11. $\upsilon = \frac{pc}{h}$

12. $\lambda \, \alpha \frac{1}{\sqrt{E}}$

13. $\lambda = \frac{h\sqrt{1-\frac{v^2}{c^2}}}{m_0 V}$

14. Slope $\alpha \frac{1}{\sqrt{m}}$

15. $\delta p = \frac{h}{\Delta x} = \frac{h}{\lambda}$

16. $\Delta p = m\Delta v = \frac{h}{\Delta x}$

or $\Delta v = \frac{h}{m\Delta x} = \frac{1.034 \times 10^{-34}}{1.67 \times 10^{-27} \times 6 \times 10^{-8}} = 1 \ ms^{-1}$

17. $\lambda = \frac{h}{p} = \frac{h}{\sqrt{2mqV}}$

LEVEL - III

1. **When a surface 1 cm thick is illuminated with light of wave lenght λ the stopping potential is V_0, but when the same surface is illuminated by light of wavelength 3λ, the stopping potential is $\frac{V_0}{6}$. The threshold wavelength for metallic surface is:**

   1) 4λ 2) 5λ 3) 3λ 4) 2λ

2. **A photon of energy 2.5 eV and wavelength λ falls on a metal surface and the ejected electrons have velocity 'v'. If the λ of the incident light is decreased by 20%, the maximum velocity of the emitted electrons is doubled. The work function of the metal is**

   1) 2.6 eV 2) 2.23 eV 3) 2.5 eV 4) 2.29 eV

3. **When a metal surface is illuminated by light of wavelengths 400 nm and 250 nm, the maximum velocities of the photoelectrons ejected are V and 2V respectively. The work function of the metal is**

   1) 2hc x 10^6 J 2) 1.5hc x 10^6 J 3) hc x 10^6 J 4) 0.5hc x 10^6 J

4. **A source of light is placed above a sphere of radius 10cm. How many photoelectrons must be emitted by the sphere before emission of photoelectrons stops? The energy of incident photon is 4.2 eV and the work function of the metal is 1.5 eV.**

   1) 2.08 x 10^{18} 2) 1.875 x 10^8 3) 2.88 x 10^{18} 4) 4 x 10^{19}

5. **Figure shows the variation of the stopping potential (V_0) with the frequency (v) of the incident radiations for two different photosensitive material M_1 and M_2. What are the values of work functions for M_1 and M_2 respectively**

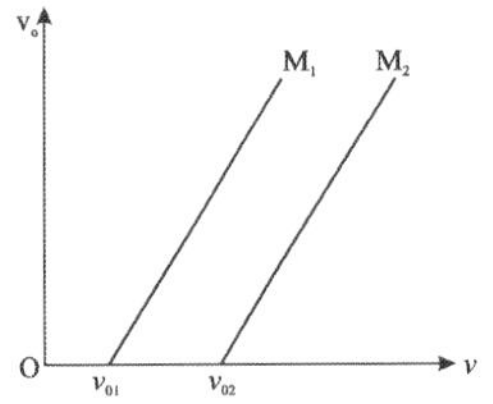

1) hv_{01}, hv_{02} 2) hv_{02}, hv_{01} 3) hv_{01}, hv_{01} 4) hv_{02}, hv_{02}

6. **From the above figure the values of stopping potentials for M_1 and M_2 for a frequency $v_3 (> v_{02})$ of the incident radiatioins are V_1 and V_2 respectively. Then the slope of the line is equal to**

1) $\dfrac{V_2 - V_1}{v_{02} - v_{01}}$ 2) $\dfrac{V_1 - V_2}{v_{02} - v_{01}}$

3) $\dfrac{V_2}{v_{02} - v_{01}}$ 4) $\dfrac{V_1}{v_{02} - v_{01}}$

7. **Photoelectric effect experiments are performed using three different metal plates p, q and r having work functions $\phi_p = 2.0eV$, $\phi_q = 2.5eV$ and $\phi_r = 3.0eV$ respectively. A light beam containing wavelengths of 550 nm, 450 nm and 350 nm with equal intensities illuminates each of the plates. The correct I-V graph for the experiment is : [Take $hc = 1240$ eV nm]**

1)
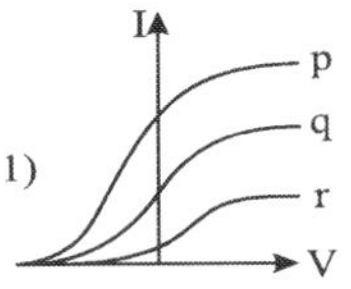

2)
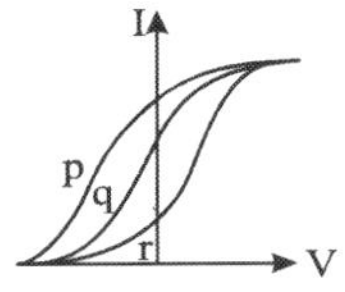

3)
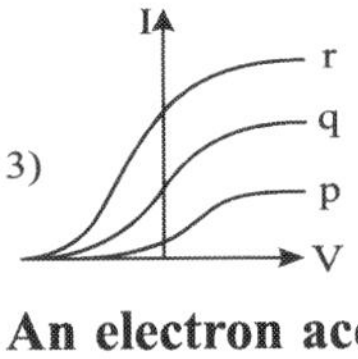

4)
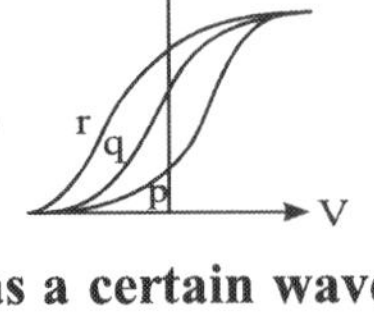

8. **An electron accelerated under a p.d. of V volt has a certain wavelength λ. Mass of the proton is 2000 times the mass of an electron. If the proton has to have the same wavelength λ, then it will have to be accelerated under p.d. of (volts)**

1) 100 2) 2000 3) V/2000 4) $\sqrt{2000}$

9. **The graph between the stopiing potential (V_0) and $(1/\lambda)$ is shown in figure, ϕ_1, ϕ_2 and ϕ_3 are work functions. Which of the following is correct**

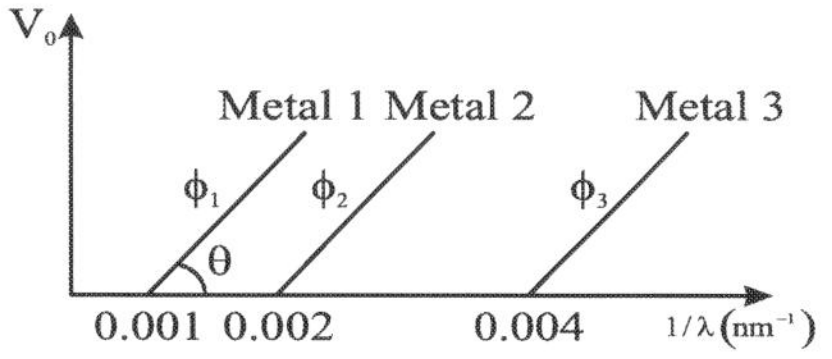

1) $\phi_1 : \phi_2 : \phi_3 = 1:2:3$
2) $\phi_1 : \phi_2 : \phi_3 = 4:2:1$
3) $\tan\theta$ is directly proportional to hc/e, where h is Planck's constant and c is the speed of light
4) ultraviolet light can be used to emit photoelectrons from metal 2 and metal 3 only.

10. **For certain photosensitive material, a stopping potential of 3.0 V is required for light of wavelength 300 nm, 2.0 V for 400 nm and 1.0V for 600nm. The work function of the material is (nearly)**

1) 2.5 ev
2) 1.5 ev
3) 2.0 ev
4) 1.0 ev

11. **An electron (mass m) with an initial velocity $v = v_0\hat{i}$ $(v_0 > 0)$ is in an electric field $E = -E_0\hat{i}$ (E_0 = constant > 0). Its de Broglie wavelength at time t is given by**

1) $\dfrac{\lambda_0}{\left(1+\dfrac{eE_0t}{mv_0}\right)}$
2) $\lambda_0\left(1+\dfrac{eE_0t}{mv_0}\right)$
3) λ_0
4) $\lambda_0 t$

12. **An electron (mass m) with an initial velocity $v = v_0\hat{i}$ is in an electric field $E = E_0\hat{j}$.**

If $\lambda_0 = \dfrac{h}{mv_0}$, its de Broglie wavelength at time t is given by

1) λ_0
2) $\lambda_0\sqrt{1+\dfrac{e^2E_0^2t^2}{m^2v_0^2}}$
3) $\dfrac{\lambda_0}{\sqrt{1+\dfrac{e^2E_0^2t^2}{m^2v_0^2}}}$
4) $\dfrac{\lambda_0}{\left(1+\dfrac{e^2E_0^2t^2}{m^2v_0^2}\right)}$

13. **The kinetic energy of α-particles at a distance 5×10^{-14} m from the uranium nucleurs will be (in joules). Which is moving in a field of 1mega volt potential difference**

1) 6.4×10^{-13}
2) 4.3×10^{-13}
3) 2.1×10^{-13}
4) 3.4×10^{-14}

14. **The stopping potential for the photoelectric emitted from a metal surface of work function 1.7 eV is 10.4 V. Find the wavelength of the radiation used. Also identify the energy levels in hydrogen atom, which will emit this wavelength**

1) 1024 $A^0, n=3$ to $n=1$
2) 1024 $A^0, n=2$ to $n=1$
3) 2044 $A^0, n=2$ to $n=1$
4) 2044 $A^0, n=3$ to $n=1$

15. A graph regarding photoeletric effect is shown between the maximum kinetic energy of electrons and the frequency of the incident light. On the basis of data as shown in the graph, calculate the work fucntion

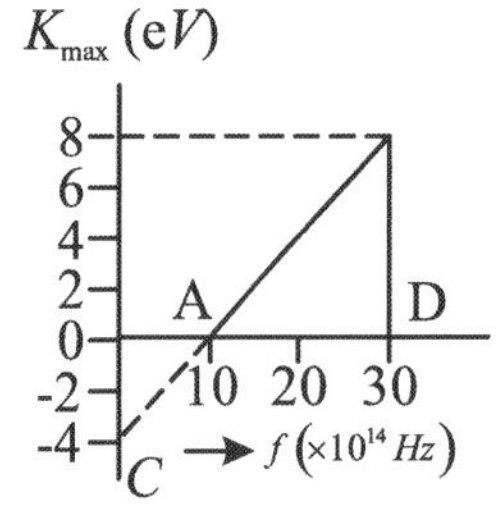

1) 2 eV　　2) 4 eV　　3) 4.2 eV　　4) 2.5 eV

16. Light of wavelength 180 nm ejects photoelectrons from a plate of metal whose work fucntion is 2 eV. If a unifrom magnetic of 5×10^{-5} T be applied parallel to the plate, what would be the radius of the path followed by electrons ejected normally from the plate with maximum energy

1) 0.148 m　　2) 0.2 m　　3) 0.25 m　　4) 0.3 m

17. Light described at a place by the equation

$E=(100\,V/M)\times\left[\sin\left(5\times10^{15}s^{-1}\right)t+\sin\left(8\times10^{15}s^{-1}\right)t\right]$

falls on a metal surface having work fucntion 2.0 eV. Calculate the maximum kinetic energy of the photoelectrons

1) 3.27 eV　　2) 5 eV　　3) 1.27 eV　　4) 2.5 eV

18. The electric field associated with a light wave is given by

$E=E_0\times\sin\left[\left(1.57\times10^{7}m^{-1}\right)(x-ct)\right]$. **Find the stopping potential when this light is used in an experiment on photoelectric effect with the similar having work function 1.9 eV**

1) 1.2 V　　2) 1.1 V　　3) 2 V　　4) 2.1 V

19. Electrons with de-Broglie wavelength λ fall on the target in an X-ray tube. The cut-off wavelength λ_0 of the emitted X-rays is

1) $\lambda_0=\frac{2mc\lambda^2}{h}$　　2) $\lambda_0=\frac{2h}{mc}$　　3) $\lambda_0=\frac{2m^2c^2\lambda^2}{h^2}$　　4) $\lambda_0=\lambda$

20. A photocell is illumuniated by small bright source placed 1m away. When the same source of light is placed 1/2m away, the number of electrons emitted by photocathode would

1) increase by a factor of 2　　2) decrease by a factor of 2

3) increase by a factor of 4　　4) decrease by a factor of 4

LEVEL-III - KEY

1) 2	2) 4	3) 1	4) 2	5) 1	6) 2	7) 1	8) 3	9) 3	10) 4
11) 1	12) 3	13) 3	14) 1	15) 2	16) 1	17) 1	18) 1	19) 1	20) 3

LEVEL-III - HINTS

1. $eV_0 = hc\left(\frac{1}{\lambda}-\frac{1}{\lambda_0}\right)$ $\frac{eV_0}{6} = hc\left(\frac{1}{3\lambda}-\frac{1}{\lambda_0}\right)$

2. $\frac{v_1^2}{v_2^2} = \frac{\frac{hc}{\lambda_1}-\omega}{\frac{hc}{\lambda_2}-\omega}$ use $E = \frac{12400}{\lambda\left(A^0\right)}$

3. $\frac{hc}{\lambda} = w + \frac{1}{2}mv^2$

4. Stopping potential energy $= eV_0 = E - \omega$

   $V_0 = \frac{E-\omega}{e} = \frac{9\times10^9 ne}{r}$; n = no of electrons

5. $W = hv$

6. $hv_{01} + eV_1 = hv_{02} + eV_2$ $e(V_1 - V_2) = h(V_{02} - V_{01})$; $\frac{h}{e} = \frac{(V_1 - V_2)}{(V_{02} - V_{01})}$

7. Explain based on graph between V & I for different metals and light of different wave lengths.

8. $V_1 \times q_1 \times m_1 = V_2 \times q_2 \times m_2$

9. $W = hv$

10. $V_s e = hv - w = \frac{12400}{\lambda \ in \ A^0} - w$; w = 1ev

11. $V = V_0 + \frac{eE_0 t}{m}$; $\lambda = \frac{h}{mv}$

12. $V_x = V_0$; $V_y = V_0 + at = V_0 + \frac{eE_0}{m}t$

    $\Rightarrow V = \sqrt{V_x^2 + V_y^2}$

    Resultant velocity is $V = \sqrt{V_0^2 + \frac{e^2E_0^2t^2}{m^2v_0^2}}$ $\lambda = \frac{h}{mv}$

13. K.E = qv- $\frac{2Ze^2}{4\Pi\varepsilon_0 r}$

14. E = $K.E_{max} + w$; $\lambda = \frac{hc}{E}$

15. From the graph

    threshold frquency $(f_0) = 10\times10^{14}\,Hz$

$$h=\frac{8\times1.6\times10^{-19}}{20\times10^{14}}=6.4\times10^{-34}J$$

work function $= hf_0 = 4eV$

16. $E=\frac{hc}{\lambda}=6.9eV$; $K.E_{\max}=4.9eV$

$$Bqv=\frac{mv^2}{r} \Rightarrow Bq=\frac{P}{r} \Rightarrow Bq=\frac{\sqrt{2KEm}}{r}$$

$$\therefore K.E_{\max}=\frac{B^2q^2r^2}{2m} \quad ; \quad r=0.148m$$

17. $E=100\sin 5\times10^{15}t+100\sin 8\times10^{15}t$

$$v_{\max}=\frac{8\times10^{15}}{2\pi}; K.E_{\max}=hv_{\max}-w; K.E_{\max}=3.27eV$$

18. $v=\frac{1.57\times10^{7}\times3\times10^{8}}{2\pi}=0.75\times10^{15}Hz$

$$E=\frac{6.62\times10^{-34}\times0.75\times10^{15}}{1.6\times10^{-19}}\ eV=3.1eV$$

$eV_0=E-w$; $V_0=1.2V$

19. $\lambda=\frac{h}{p}=\frac{h}{\sqrt{2mE_k}}=E_k=\frac{h^2}{2m\lambda^2}$

also $\frac{hc}{\lambda_0}=E_k=\frac{h^2}{2m\lambda^2}$ (or) $\lambda_0=\frac{2mc\lambda^2}{h}$

20. $I\alpha\frac{1}{r^2}$

LEVEL - IV

MULTIPLE ANSWER TYPE

1. **Photoelectric effect supports quantum nture of light because :**

   1) there is a minimum frequency of light below which no photoelectrons are emitted

   2) the maximum kinetic energy of photoelectrons depends only on the frequency of light and not on its intensity.

   3) even when the metal surface is faintly illuminated, the photoelectrons leave the surface immediately

   4) electric charge of the photoelectrons is quantized.

2. **If the wavelength of light in an experiment on photoelectric effect is doubled :**

   1) the photoelectric emission will not take place

   2) the photoelectric emission may or may not take place

   3) the stopping potential will increase 4) the stopping potential will decrease

3. **The frequency and intensity of light source are both doubled. Which of the following statement (statements) is (are) true ?**

1) The saturation photocurrent gets doubled.

2) the saturation photocurrent remains almost the same

3) the maximum KE of the photoelectrons is more than doubled.

4) the maximum KE of the photoelectrons get doubled.

4. **In which of the following situations, the heavier of the two particles has smaller de Broglie wavelength ? The two particles :**

1) move with same speed
2) move with the same linear momentum
3) move with the same kinetic energy
4) have fallen through the same height

5. **When a monochromatic point source of light is at a distance of 0.2m from a photoelectric cell, the cut-off voltage and the saturation current are respectively 0.6V and 18.0 mA. If the same source is placed 0.6m away from the photoelectric cell, then :**

1) the stopping potential will be 0.2 V
2) the stopping potential will be 0.6 V
3) the saturation current will be 6.0 mA
4) the saturation current will be 2.0 mA

6. In a photoelectric experiment the wavelength of the incident light is decreased from 6000A^0 **to 4000A^0 while the intensity of radiation remains the same. Choose the correct statement(s)**

1) the cut-off potential will increase
2) the cut-off potential will decrease
3) the photoelectric current will increase
4) the kinetic energy of the emitted photoelectrons will increase

COMPREHENSION TYPE

Passage I:

Photoelectric threhold of silver is $\lambda = 3800\,\overset{o}{A}$. ultraviolet light of $\lambda = 2600\,\overset{o}{A}$ is incident on silver surface. (Mass of the electron $9.11 \times 10^{-31} kg$)

7. **Calculate the value of work function in eV.**

1) 1.77 2) 3.27 3) 5.69 4) 2.32

8. **Calculate the maximum kinetic energy (in eV) of the emitted photoelectrons.**

1) 1.51 2) 2.36 3) 3.85 4) 4.27

9. **Calculate the maximum velocity of the photoelectrons.**

1) 72.89×10^8 2) 57.89×10^8 3) 42.93×10^8 4) 68.26×10^8

Passage II:

A 100 W point source emits monochromatic light of wavelength $6000\,\overset{0}{A}$.

10. **Calculate the total number of photons emitted by the source per second.**

1) 5×10^{20} 2) 8×10^{20} 3) 6×10^{21} 4) 3×10^{20}

11. **Calculate the photon flux (in SI unit) at a distance of 5 m from the source. Given $h = 6.6 \times 10^{-34}$ Js and $c = 3 \times 10^8\, ms^{-1}$.**

1) 10^{15} 2) 10^{18} 3) 10^{20} 4) 10^{22}

12. **1.5 mW of $4000\overset{0}{A}$ light is directed at a photoelectric cell. if 0.10 per cent of the incident photons produce photoelectrons, find current in the cell. [Given $h = 6.6\times10^{-34}\ ms^{-1}$ and $e = 1.6\times10^{-19}\ C$]**

1) $0.59\ \mu A$ 2) $1.16\ \mu A$ 3) $0.48\ \mu A$ 4) $0.79\ \mu A$

Passage III:

When a particle is restricted to move along x-axis between x=0 and x=a, where a is of nanometer dimension, its energy can take only certain specific values. The allowed energies of the particle moving in such a restricted region, corrospond to the formation of standing waves with nodes at its ends x = 0 abd x = a. The wavelength of this standing wave is related to the linear momentum p of the particle according to the de Broglie relation. The energy of the particle of mass m is related to its linear momentum as $E = p^2/2m$. Thus, the energy of the particle can be denoted by a quantum number 'n' taking values 1,2,3,..........(n=1, called the ground state) corrosponding to the number of loops in the standing wave. Use the model described above to answer the following three questions for a particle moving along the line from x=0 to x=a.

13. **The allowed energy for the particle for a particular value of n is proportional to**

1) a^{-2} 2) $a^{-3/2}$ 3) a^{-1} 4) a^2

14. **If the mass of the particle is $m = 1.0\times10^{-30}$ kg and a=6.6 nm, the energy of the particle in its ground state is closest to**

1) 0.8 meV 2) 8 meV 3) 80 meV 4) 800 meV

15. **The speed of the particle that can take discrete values is proportional to**

1) $n^{-3/2}$ 2) n^{-1} 3) $n^{1/2}$ 4) n

16. **Statement-I: When ultraviolet light is incident on a photo cell, its stopping potential is v_0 and the maximum kinetic energy of the photoelectrons is K_{max} increase.**

Statement-II: Photoelectrons are emitted with speeds ranging from zero to a maximum value because of the range of frequencies present in the incident light.

1) Statement I is true, Statement II is true; statement II is a correct explanation of statement I.
2) Statement I is true, Statement II is true, Statement II is NOT a correct explanation for statement I.
3) Statement I is false, Statement II is true
4) Statement I is true, Statemenet II is false

17. **Wavelengths associated with different particles are given in Column - I. Match these wavelengths with their values given in Column-II.**

Column-I

A) Wavelength associated with an electron accelerated through a pd of 1V
B) Wavelength associated with an α -particle accelerated through a pd of 1V
C) Wavelength associated with a proton accelerated through a pd of 1V
D) Wavelength associated with a photon of energy 124.2 eV

Column - II

p) 10nm q) $0.10\ A^0$
r) $0.286\ A^0$ s) $12.27\ A^0$

18. In a permoting photoelectric experiment to study photoelectric effect, intensity of radiation (I), frequency of radiation (v), work function (ϕ_0) of the photosensitive emitter, distance (d) between emitter and collector are changed or kept constant. Match the changes given in Column - I to their effect given in Column - II.

Column - I

A) ϕ_0 is decreased, keeping v and I constant

B) d is increased, keeping I, v, ϕ_0 constant

C) v is increased, keeping I, ϕ_0, d constant

D) I is increased, keeping v, ϕ_0, d constant

Column - II

p) Saturation photoelectric current increases

q) stopping potential (V_0) increases

r) Maximum KE (K_{max}) of photoelectrons increases

s) Stopping potential remains the same

LEVEL-IV - KEY

1) 1,2,3	**2) 2,4**	**3) 1,3**	**4) 1,3,4**	
5) 2, 4	6) 1, 4	7) 2	8) 1	9) 1
10) 4	11) 2	12) 3	13) 1	14) 2
15) 4	16) 4			

17) A-s, B-q, C-r, D-p 18) A-q,r, B-s, C-q,r, D-p,s

LEVEL-IV HINTS

5. (b) Stopping potential remains the same as it depends on the frequency of incident radiation.

(D) Saturation current α intensity of incident radiation $\alpha \frac{1}{r^2}$. Since r becomes three times $\left(\frac{0.6m}{0.2m}\right)$,

saturation current becomes $\frac{18.0mA}{(3)^2} = 2.0mA$.

6. As $K_{max} = eV_0 = \frac{hc}{\lambda} - \phi_0$, when λ decreases, V_0 and K_{max} increase.

7. $E = hv$

$\frac{hc}{\lambda}$

8. $E = W.E. + K.E.$

9. $K = \frac{1}{2}mv^2$

Chapter - 6

ATOMS

ATOMS

SYNOPSIS

- At ordinary pressures and ordinary voltages gases are bad conductors of electricity.
- The p.d. required to start a spark through a gas is known as sparking potential. It is much greater than the p.d. required to maintain the spark once started.
- At ordinary pressures for moderately great spark lengths, the sparking potential is nearly proportional to the spark length, being about 30,000 V per cm in air between spherical electrodes of 1 cm diameter and less in case of pointed electrodes.
- According to Paschen's law, the sparking potential is directly proportional to the pressure (p) of the gas between the electrodes and the distance (d) between them.
- At ordinary pressures, the current is due to the presence of positive and negative ions as in the case of electrolytic conductivity of a solution.
- At low pressures, discharge (the passage of current through gases) occurs at much lower voltages than at high pressures.
- At lower pressures negative ion throws off its attendant atom or molecule and the resultant negatively charged particle, the electron travels free and faster than the positive ions.
- At low pressures in the discharge tube, as the pressure is reduced different phenomena such as sparking, positive column, glow of cathode, Faraday's dark space, Crooke's dark space and the striations of positive column are gradually observed.
- Finally at a pressure of about 0.01mm of Hg pressure, Crooke's dark space completely fills the discharge tube with the walls of the tube glowing with light i.e., producing fluorescence on the walls of the tube. This is due to some radiation emitted from the cathode surface to which the name cathode rays has been given.
- Cathode rays were first observed by Plucker. J J Thomson after studying their properties called them "Streams of negative corpuscles", while Johnson stoney who having found from the electrolysis that electricity was atomic in nature, suggested the name "electron".

J.J.Thomson's experimental determination of specific charge of an electron:

- Cathode rays can move undeflected through mutually perpendicular electric and magnetic fields acting in same region.

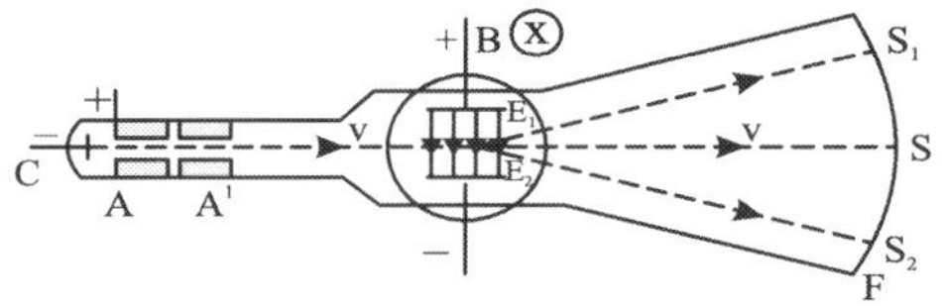

- If plate E_1 is connected to the positive terminal and E_2 to the negative terminal of a battery, vertically downward electric field (E) is produced perpendicular to the direction of motion of electrons. Then electrons will be deflected upwards striking the screen at S_1.
- If the magnetic field (B) alone is applied by using clockwise current carrying circular coil the direction of magnetic field is directed into the plane of the paper. Then by Fleming's left hand rule electrons will be deflected downwards striking the screen at S_2.
- Then both electric and magnetic fields are applied simultaneously and their values are adjusted so that the electron beam can move undeflected and strikes the screen at S.
- When the electron beam enters the region of crossed electric and magnetic fields with velocity v horizontally,

- If only electric field is switched on,
  Electric force on the electron which is vertically upwards is, $F_e = eE \rightarrow (1)$

  Where Electric field $E = \frac{V^1}{d}$

  V^1 = potential difference between the plates E_1 and E_2
  d = distance between the plates E_1 and E_2.
- If only magnetic field is switched on,

  Magnetic force acting on the electron, which is vertically downwards is, $F_m = Bev \rightarrow (2)$
  Where B = magnetic induction field
  e = charge of electron, v = velocity of electron
- If both electric and magnetic fields are applied simultaneously and the electrons moves undeflected, then magnetic and electric forces acting on the electrons are equal and opposite.
  $\therefore F_m = F_e$ Bev = eE
  $\therefore$ Velocity of electron $v = \frac{E}{B} \rightarrow (3)$
  As the electron is accelerated from cathode to anode, the workdone on the electron by the accelerating voltage is equal to the increase in kinetic energy of the electron, K.E of electron at anode = work done on the electron $\frac{1}{2}mv^2 = eV_0 \rightarrow (4)$; $\therefore \frac{e}{m} = \frac{v^2}{2V_0} \rightarrow (5)$

  Where V_0 = potential difference between cathode and Anode.
  m = mass of electron, e = charge of electron
  v = velocity of electron.
  substituting equation (3) in (5) we get

  $$\frac{e}{m} = \frac{E^2}{2V_0B^2} \rightarrow (6)$$

  Since we know E, B and V_0, specific charge of electron can be calculated.
- The value of specific charge of electron as measured by J.J.Thomson is
  1.7588×10^{11} C kg^{-1}

  Note :

(a) Specific charge of electron is found to be independent of nature of the gas.
(b) Specific charge of electron is found to be independent of nature of the electrodes.
(c) Specific charge of electron decreases with increase in velocity of electron, because charge is independent of velocity and mass increases with increase in velocity, since $m = \frac{m_0}{\sqrt{1 - \frac{v^2}{c^2}}}$

Where m_0 = rest mass of electron,
m = Relativistic mass of electron and
v = Velocity of electron; c = Velocity of light

Note : If cathode rays are moving along +ve x-axis and electric field is applied along +ve y-axis then magnetic field should be along +ve z-axis so that cathode rays moves undeflected. This is because net force on the electron

$$\vec{F} = e(\vec{E} + \vec{v} \times \vec{B}) = \vec{0} \Rightarrow \vec{E} = -(\vec{v} \times \vec{B})$$

W.E-1: ***An electron beam moving with a speed of $2.5 \times 10^{7}\ ms^{-1}$ enters into the magnetic field $4 \times 10^{-3}\ Wb/m^2$ directed perpendicular to its direction of motion. Find the intensity of the electric field required so that the electron moves undeflected.***

Sol. When electron goes undeflected, force due to electric field is equal and opposite to force due to magnetic field i.e., $Ee = Bev$

$B = 4 \times 10^{-3}\ Wb/m^2$ and $v = 2.5 \times 10^{7}\ ms^{-1}$

$E = Bv = 4 \times 10^{-3} \times 2.5 \times 10^{7} = 10^{5}\ V/m$

W.E-2: ***A mono energetic electron beam with a speed of $5.2 \times 10^{6}\ ms^{-1}$ enters into a magnetic field of induction 3×10^{-4} T, directed normal to the beam. Find the radius of the circle traced by the beam (Take $e/m = 1.76 \times 10^{11}\ Ckg^{-1}$)***

Sol. From the equation, $Bev = \frac{mv^2}{r}$

Radius of circle $r = \frac{mv}{Bq} = \frac{v}{B(e/m)}$

$$\therefore r = \frac{5.2 \times 10^{6}}{3 \times 10^{-4}\left(1.76 \times 10^{11}\right)} = 0.1m$$

W.E-3 : ***The deflecting plates in a Thomson's setup are x meters long. Intensity of electric field applied between the plates is E. The plates are maintained at a p.d of V volts. Electrons accelerated through a p.d of V volts enter from one edge of the plates midway in a direction parallel to the plates. Find the deflection at the other edge of the plates.***

Sol. Motion of a charged particle projected perpendicular to a uniform electric field is a parabola. Horizontal distance travelled by the electron in time t is $x = vt \Rightarrow t = \frac{x}{v}$

Deflection of the electron in y - direction is $y = \frac{1}{2}at^2$, $F = ma = Ee$; $a = \frac{Ee}{m}$; $y = \frac{1}{2}\left(\frac{Ee}{m}\right)\frac{x^2}{v^2}$

(since $u_y = 0$) $\Rightarrow y = \frac{Ee}{4 \times \left(\frac{1}{2}mv^2\right)}x^2$ ...... (1)

We know that, $\frac{1}{2}mv^2 = eV$ ...... (2)

From (1) and (2) $\Rightarrow y = \frac{E}{4V}x^2$

W.E-4: ***The specific charge of a proton is 9.6×10^{7} C/Kg. The specific charge of an alpha particle is***

Sol: Specific change $= \frac{q}{m}$; $\frac{S_\alpha}{S_p} = \frac{q_\alpha}{q_p}\frac{m_p}{m_\alpha}$

$$\frac{S_\alpha}{9.6 \times 10^{7}} = \frac{2}{1}.\frac{1}{4} = \frac{1}{2};\ S_\alpha = 4.8 \times 10^{7}\ C/kg$$

W.E-5: ***A beam of electrons enter at right angles to an uniform electric field with a velocity 3×10^{7} m/s, E= 1800 v/m while travelling through a distance of 10cm, the beam deflected by 2mm then specific charge of electron is***

Sol: $t = \frac{x}{u} = \frac{0.1}{3 \times 10^{7}} = \frac{1}{3} \times 10^{-8}s$; $y = \frac{1}{2}at^2$ $a = E\frac{q}{m}$

$$2\times10^{-3}=\frac{1}{2}\left(1800\,\frac{q}{m}\right).\left(\frac{1}{3}\times10^{-8}\right)^2;\frac{q}{m}=2\times10^{11}\,C/kg$$

W.E-6: ***In Thomson's experiment for determining e/m the potential difference between the cathode and the anode (in accelerating column) is same as that between the dflecting plates (in the region of crossed fields). If the p.d is doubled, by what factor should the magnetic field be increased to ensure that the electron beam remains undeflected***

Sol: $\frac{1}{2}m\upsilon^2=V.q$; $\frac{1}{2}m\left(\frac{E}{B}\right)^2=V.q$; $\frac{1}{2}m\frac{V^2}{d^2.B^2}=V.q$ $\quad\frac{q}{m}=\frac{V}{2d^2B^2}$; $B^2\alpha\ V\rightarrow B\alpha\ \sqrt{V}$;

$B^1\alpha\ \sqrt{2V}$

$B^1=\sqrt{2}B$

W.E-7: ***In a cathode ray tube, a p.d of 3000V is maintained between the deflector plates whose seperation is 2cm. A magnetic field of 2.5 x 10^{-3} T at right angles to the electric field gives no deflection of electron beam, which recieved an initial acceleration by a p.d of 10,000V. The e/m of electron is***

Sol: V_1 is p.d between the deflector plates and V_2 be the potential diference through which electrons are accelerated

$$\frac{1}{2}m\upsilon^2=V_2q\qquad\frac{1}{2}m.\frac{V_1^2}{d^2B^2}=V_2q;\frac{q}{m}=\frac{V_1^2}{2d^2\,B^2V_2}$$

$$\frac{q}{m}=\frac{9\times10^6}{2\times4\times10^{-4}\times6.25\times10^{-6}\times10^4}=1.8\times10^{11}c/kg$$

Determination of charge of an electron : Millikan's oil drop method.[Additional]

- Millikan's method to find charge of an oil drop is based on the measurement of terminal velocity of the charged oil drop under the action of gravity and under the combined action of gravity and electric field.

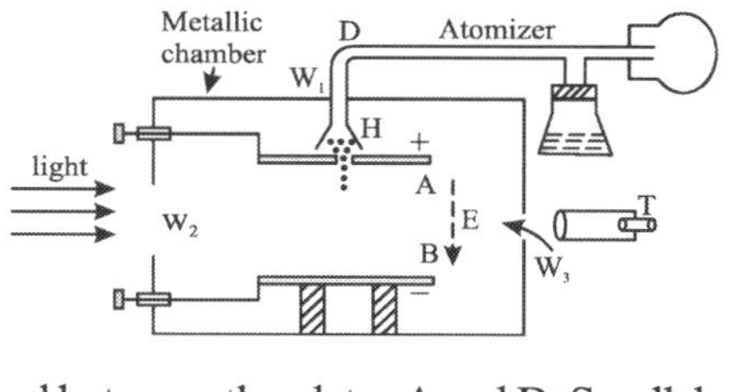

- Electric field can be produced between the plates A and B. Small droplets of non volatile oil (clock oil or apiezon oil) are sprayed using an atomizier above the hole in plate A. Some of the droplets acquire charge due to friction. The chamber is illuminated by sending light horizontally through it. The drops can be seen by using a telescope placed perpendicular to the light beam. A drop looks like a bright star moving upwards or downwards.
- An electric field is produced between the plates A and B by connecting upper plate to the positive terminal and lower plate to the negative terminal of a battery. The field is adjusted so that an oil drop slowly moves upwards. The drop experiences electric force and buoyancy in the upward direction where as weight and viscous force in the downward direction.
- The uniform velocity attained by the drop V_e is measured by measuring the distance travelled and time taken for it. Now electric field is switched off and the drop is allowed to fall freely. The oil drop now experiences viscous force and buoyancy directed upwards and weight downwards. The drop soon moves with terminal velocity V_g downwards. V_g can be measured by measuring distance travelled by the drop and time taken for it.

- Charge on the drop can be calculated using

  the formula $q = \frac{6\pi\eta(V_g + V_e)}{E}\left(\frac{9\eta V_g}{2(d - d^1)g}\right)^{1/2}$

  where η = Coefficient of viscosity of air

  d = density of oil; d^1 = density of air

  E = Intensity of electric field between the plates

  g = acceleration due to gravity

  V_e = terminal velocity of the oil drop in electric field

  V_g = terminal velocity of the oil drop in gravitational field

- After large number of observations Millikan found that the value of 'q' is an integral multiple of a common value 'e' such that q = ne. Where n is an integer and e is the charge of an electron. It is found that charge of electron, $e = 1.6 \times 10^{-19}$ C.

- When the electric field is applied between the plates A and B, suppose a negatively charged drop is moving upwards with constant velocity V_e.

  upward forces are electric force (F_e) and Buoyancy (F_b) while downward forces are viscous force (F_v) and weight (F_g).

  Net force on the drop is zero.

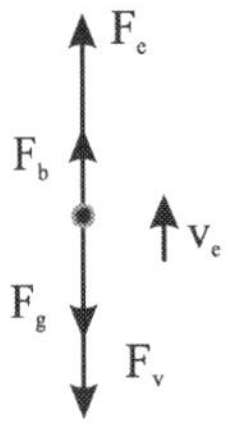

$\therefore F_e + F_b = F_g + F_v$

$$qE + \frac{4\pi}{3}R^3 d^1 g = \frac{4\pi}{3}R^3 dg + 6\pi\eta R V_e \rightarrow \quad (1)$$

R - Radius of the drop

If electric field is switched off, suppose the drop falls downwards with terminal velocity V_g.

upward forces are Buoyancy and viscous force while downward force is weight.

Net force on the drop is zero.

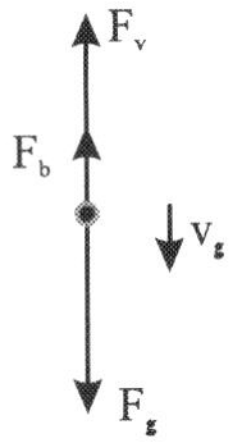

$\therefore F_b + F_v = F_g$

$$\frac{4\pi}{3}R^3 d^1 g + 6\pi\eta R V_g = \frac{4\pi}{3}R^3 dg \rightarrow \quad (2)$$

subtracting (2) from (1)

$$qE - 6\pi\eta R V_g = 6\pi\eta R V_e$$

$$q = \frac{6\pi\eta R(V_e + V_g)}{E} \rightarrow \quad (3)$$

from equation (2) $\frac{4\pi}{3}R^3(d - d^1)g = 6\pi\eta R V_g$

$$\therefore R^2 = \frac{9\eta V_g}{2(d-d^1)g} ; \ R = \sqrt{\frac{9\eta V_g}{2(d-d^1)g}} \rightarrow (4)$$

$$\therefore \ q = \frac{6\pi\eta(V_g + V_e)}{E}\left(\frac{9\eta V_g}{2(d-d^1)g}\right)^{1/2} \rightarrow (5)$$

Case-1 *:* In Millikan's oil drop method, if the drop is falling under gravity in the absence of electric field then, upward forces are Buoyancy (F_b) and viscous force (F_v) while downward force is weight (F_g). Net force on the drop is zero.

$$\therefore F_b + F_v = F_g$$

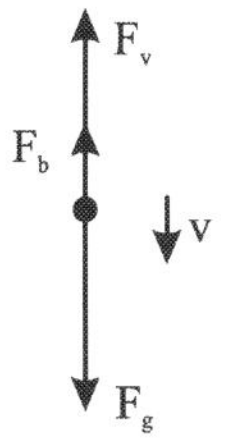

$$\frac{4\pi}{3}R^3 d^1 g + 6\pi\eta R V = \frac{4\pi}{3}R^3 dg$$

Case-2*:* In Millikan's oil drop method if a positively charged drop is at rest in the presence of electric field directed upwards then, upward forces are Buoyancy (F_b) and electric force (F_e) while downward force is weight (F_g).
Net force on the drop is zero.

$$\therefore F_e + F_b = F_g$$

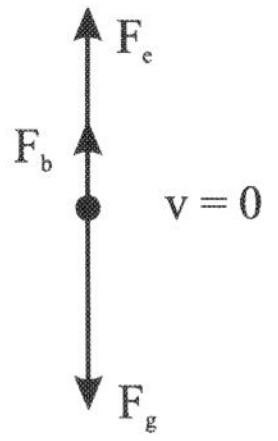

$$qE + \frac{4\pi}{3}R^3 d^1 g = \frac{4\pi}{3}R^3 dg$$

Case-3*:* In Millikan's oil drop method, if a positively charged drop is moving upwards with constant velocity V in the presence of electric field directed upwards then, upward forces are buoyancy (F_b) and electric force (F_e) while downward forces are weight (F_g) and viscous force (F_v). Net force on the drop is zero.

$$\therefore F_e + F_b = F_g + F_v$$

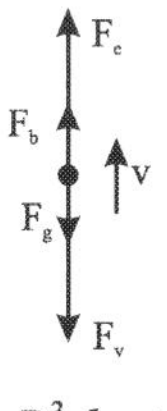

$$qE + \frac{4\pi}{3}R^3 d^1 g = \frac{4\pi}{3}R^3 dg + 6\pi\eta RV$$

Case-4*:* In Millikan's oil drop method if a positively charged drop is moving downwards with constant velocity V in the presence of electric field directed upwards then, upward forces are Buoyancy (F_b), electric force (F_e) and viscous force (F_v) while downward force is weight (F_g) .
Net force on the drop is zero.

$$\therefore F_e + F_b + F_v = F_g$$

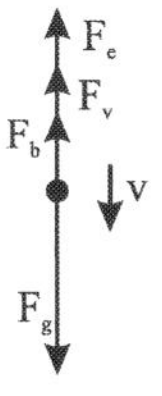

$$qE + \frac{4\pi}{3}R^3 d^1 g + 6\pi\eta RV = \frac{4\pi}{3}R^3 dg$$

Case-5 **: A charged oil drop having charge q_1 is falling freely under gravity in the absence of electric field with a velocity 'V_1'. It is held stationary in an electric field. As it acquires a charge it moves up with a velocity V_2. Then**

i) The new charge on the drop is $q_2 = (k+1)q_1$

ii) Additional charge aquired is $\Delta q = kq_1$

Where $k = \frac{V_2}{V_1}$

In the absence of electric field, drop is falling downwords with velocity V_1.

$$F_g = F_b + F_v$$
$$F_g - F_b = F_v$$
$$F_g - F_b = 6\pi\eta RV_1 \quad -(1)$$

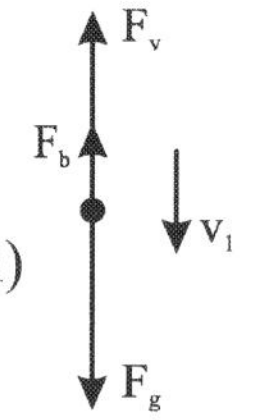

In the presence of electric field when the drop is at rest.

$$F_g = F_b + F_e$$
$$F_g - F_b = q_1 E \quad -(2)$$

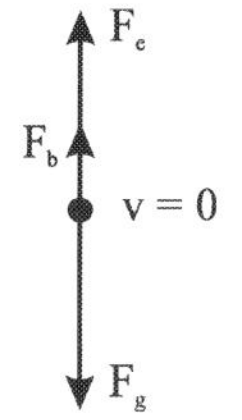

In the electric field, after acquiring final charge q_2 drop is moving upwards with velocity V_2. Then

$$F_g + F_v = F_b + F_e$$
$$F_g - F_b = q_2 E - 6\pi\eta RV_2 \quad but, V_2 = kV_1$$
$$F_g - F_b = q_2 E - k6\pi\eta RV_1 \quad -(3)$$

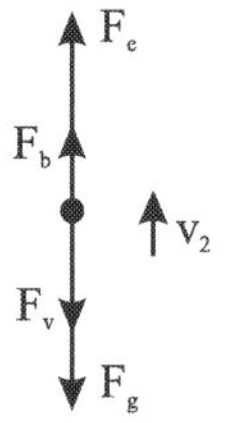

$$q_1 E = q_2 E - kq_1 E$$

So, the final charge on the drop is $q_2 = (k+1)q_1$

Additional charge aquired is $\Delta q = kq_1$

Case-6 **: In the Millikan's oil drop experiment the oil drop is subjected to a horizontal electric field E and the drop moves with a constant velocity making an angle θ to the horizontal. If the weight of the drop is W, neglecting buoyancy, the charge on the drop is $\frac{W\cot\theta}{E}$**

$$F_V \cos\theta = qE$$
$$F_V \sin\theta = F_g = W$$
$$\therefore q = \frac{W\cot\theta}{E}$$

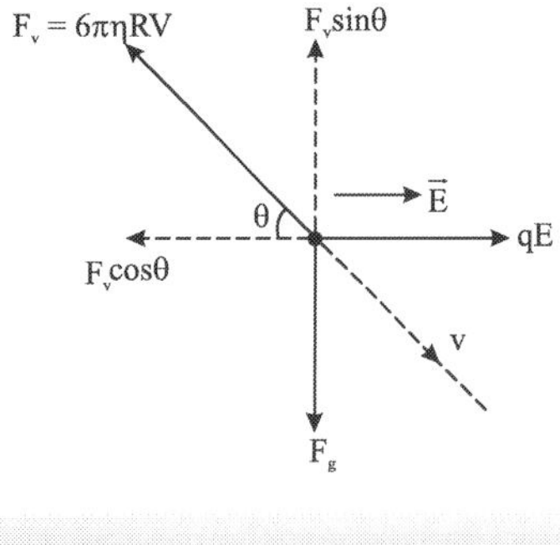

W.E-8: ***In Millikan's oil drop experiment an oil drop of radius r and charge Q is held in equilibrium between the plates of a charged parallel plate capacitor when the potential difference is V. To keep another drop of same oil whose radius is 2r and carrying charge 2Q in equilibrium between the plates, find the new potential difference required.***

Sol. Since drop is at rest QE = mg

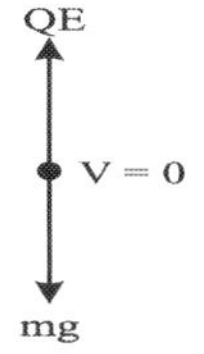

$$Q\frac{V}{d}=\frac{4\pi}{3}r^3\rho g \ ; \ V\propto\frac{r^3}{Q} ; \ \frac{V_2}{V_1}=\left(\frac{r_2}{r_1}\right)^3\frac{Q_1}{Q_2}$$

$$\frac{V_2}{V}=8\times\frac{1}{2} \quad ; \quad \therefore V_2=4V$$

W.E-9: ***A charged oil drop of charge q is falling under gravity with terminal velocity v in the absence of electric field. A electric field can keep the oil drop stationary. If the drop acquires an additional charge, it moves up with velocity 3v in that field. Find the new charge on the drop.***

Sol. In the absence of electric field

$mg=6\pi\eta RV\rightarrow$ (1)

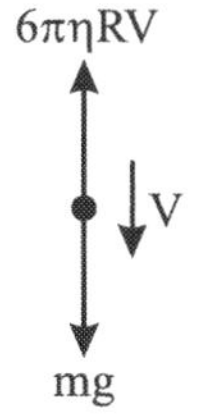

when electric field is applied

$mg=qE\rightarrow$ (2)

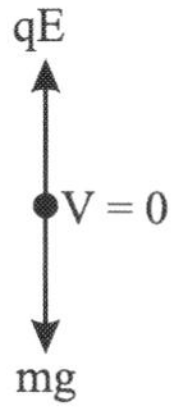

If q^1 = new charge and drop is moving up then

$q^1E=mg+3\times6\pi\eta RV\rightarrow$ (3)

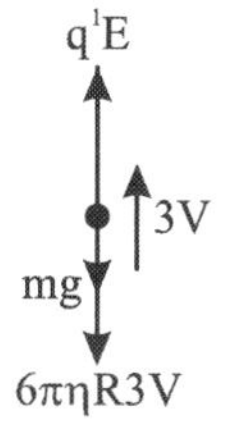

$q^1E=4mg=4qE \ \therefore q^1=4q$

W.E-10: ***In Millikan's method of determining the charge of an electron, the terminal velocities of oil drop in the presence and in the absence of an electric field are xcm/s upwards and y cm/s downwards respectively. Find the ratio of electric force to gravitational force on the oil drop. (Neglect Buoyancy)***

Sol. In Gravitational field, weight = viscous force

$W = 6\pi\eta ry$........(1)

In electric field,

Electric force = weight + viscous force

$Eq = W + 6\pi\eta rx$ ....(2)

Substitute (1) in (2) then $Eq = 6\pi\eta r(y+x)$ ....(3)

$$\frac{Electric\ force}{Gravitational\ force} = \frac{x+y}{y}$$

W.E-11: ***In a Millikan's experiment an oil drop of radius 1.5 x 10^{-6} m and density 890 kg/m^3 is held stationary between two condenser plates 1.2 cm apart and kept at a p.d of 2.3 kV. If upthrust due to air is ignored, then the number of excess electrons carries by the drop will be***

Sol: Eq = mg; $\frac{V}{d}.ne = \frac{4}{3}\pi r^3.\rho g$; $n = \frac{4}{3}\pi \times r^3 \rho g \frac{d}{Ve}$

$$n = \frac{4}{3} \times \frac{\pi \times 3.375 \times 10^{-18} \times 890 \times 9.8 \times 1.2 \times 10^{-2}}{2.3 \times 10^3 \times 1.6 \times 10^{-19}}$$

$n \simeq 4$

W.E-12: ***A charged oil drop is of charge q is falling freely under gravity in the absence of electric field with a velocity 'v'. It is held stationary in an electric field, as it acquires a charge it moves up with a velocity '3v. Now the charge on the drop is***

Sol: $mg = 6\pi\eta r\upsilon$, Eq = mg

$Eq^1 = mg + 6\pi\eta r(3\upsilon)$, $Eq^1 = Eq + 3Eq$

$Eq^1 = 4Eq \Rightarrow q^1 = 4q$

W.E-13: ***A charged oil drop falls with a terminal velocity V in the absence of electric field. An electric field E keeps the oil drop stationary in it. When the drop acquires a charge 'q' it moves up with same velocity. Find the initial charge on the drop.***

Sol: $mg = 6\pi\eta r\upsilon$,Eq = mg

$E(Q+q) = mg + 6\pi\eta r\upsilon$,$E(Q+q) = 2EQ$

$Q + q = 2Q \Rightarrow q = Q$

W.E-14: ***Two oil drops in Millikan's experiment are falling with terminal velocities in the ratio 1:4. The ratio of their de-Broglie wave length is***

Sol: $\upsilon_T \propto r^2$ $\qquad r \propto \sqrt{V_T}$

$$\frac{r_1}{r_2} = \sqrt{\frac{V_1}{V_2}} = \frac{1}{2} \qquad \frac{m_1}{m_2} = \frac{1}{8}$$

$$\therefore \frac{\lambda_1}{\lambda_2} = \frac{m_2}{m_1}.\frac{\upsilon_2}{\upsilon_1} = \frac{8}{1}.\frac{4}{1} = 32 : 1$$

Rutherford's α -particle Scattering Experiment:

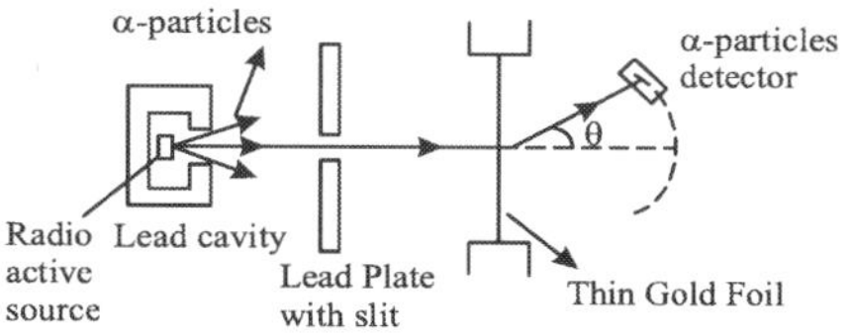

Experimental Observations:

a) Most of the α-particles were found to pass through the gold- foil without being deviated from their paths.

b) Some α-particles were found to be deflected through small angles $\theta < 90°$.

c) Few α-particles were found to be scattered at fairly large angles from their initial path $\theta > 90°$

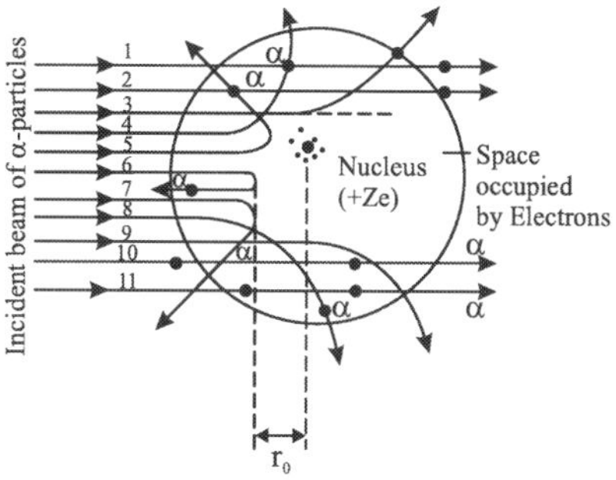

d) A very small number of α-particles about 1 in 8000 practically retracted their paths or suffered deflections of nearly $180°$.

➢ The observation (a) indicates that most of the portion of the atom is hollow inside.

➢ Because α-particle is positively charged, from the observations (b), (c) and (d) atom also have positive charge and the whole positive charge of the atom must be concentrated in small space which is at the centre of the atom is called nucleus. The remaining part of the atom and electrons are revolving around the nucleus in circular objects of all possible radii. The positive charge present in the nuclei of different metals is different . Higher the positive charge in the nucleus, larger will be the angle of scattering of α-particle.

Distance of Closest Approach :

➢ An α-particle which moves straight towards the nucleus in head on direction reaches the nucleus i.e, it moves close to a distance r_0 as shown the figure.

➢ As the α-particle approaches the nucleus, the electrostatic repulsive force due to the nucleus increases and kinetic energy of the alpha particle goes on converting into the electrostatic potential energy. When whole of the kinetic energy is converted into electrostatic potential energy, the α-particle cannot further move towards the nucleus but returns back on its initial path i.e α-particle is scattered through an angle of $180°$. The distance of α-particle from the nucleus in this stage is called as the distance of closest approach and is represented by $r_{0.}$

➢ Let m_α and v_α be the mass and velocity of the α-particle directed towards the centre of the nucleus. Then kinetic energy of the α-particle $K = \left(\frac{1}{2}\right) m_\alpha v^2{}_\alpha$

Because the positive charge on the nucleus is Ze and that on the α -particle 2e, hence the electrostatic potential energy of the α-particle, when at a distance r_0 from the centre of the nucleus, is given by

$$U = \frac{1}{4\pi \in_0} \cdot \frac{(2e)(Ze)}{r_0}$$

Because at $r = r_0$ kinetic energy of the α-particle appears as its potential energy, hence, K=U

$$\frac{1}{4\pi \in_0} \cdot \frac{(2e)(Ze)}{r_0} = \frac{1}{2} m_\alpha v_\alpha^2$$

$$r_0 = \frac{1}{4\pi \in_0} \frac{4Ze^2}{m_\alpha {v_\alpha}^2}$$

Note:

- When a mono energitic beam of α particles is projected towards a thin metal foil, some of the particles are found to deviate from their original path. This phenomenon is called α ray scattering
- It is caused by coulomb repulsive force between α particles and positive charges in atom.
- The number of α -particles scattered at an angle θ is given by $N = \dfrac{Q n t z^2 e^4}{(8\pi\varepsilon_0)^2 r^2 E^2 \sin^4\left(\frac{\theta}{2}\right)}$

where

$Q \rightarrow$ Total number of α particles striking the foil

$n \rightarrow$ number of atoms per unit volume of the foil

$r \rightarrow$ distance of screen from the foil

$t \rightarrow$ thickness of the foil

$z \rightarrow$ Atomic number of the foil atoms

$\theta \rightarrow$ angle of scettering

$E \rightarrow$ kinetic energy of α particles

$N \alpha t$; $N \alpha z^2$; $N \alpha \dfrac{1}{\sin^4 \frac{\theta}{2}}$

$N \alpha \dfrac{1}{E^2}$ or $N \alpha \dfrac{1}{\upsilon^4}$

where ϑ is the velocity of α particles falling on the foil.

- **Impact Parameter(b):** The perpendicular distance of the initial velocity vector of the α-particle from centre of the nucleus is called "impact parameter".

$$b = \frac{Ze^2 \cot\left(\frac{\theta}{2}\right)}{4\pi \in_0 \times \frac{1}{2} mv^2}$$

Bohr's model of Hydrogen like atoms :

- Electron can revolve round the nucleus only in certain allowed orbits called stationary orbits and the Coulomb's force of attraction between electron and the positively charged nucleus provides necessary centripetal force.

$$\frac{k(Ze)e}{r^2} = \frac{mV^2}{r}$$

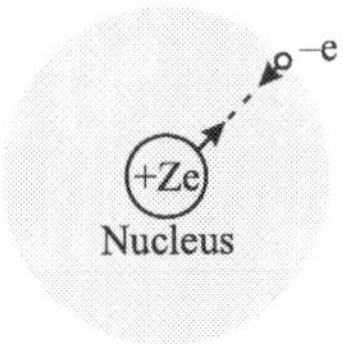

- Suppose m is the mass of electron, V is the velocity and 'r' is the radius of the orbit, then in stationary orbits the angular momentum of the electron is an integral multiple of $\frac{h}{2\pi}$, where h is the Planck's constant. The angular momentum $L = I\omega = mVr = n\frac{h}{2\pi}$ where n is called principal quantum number.
- An electron in a stationary orbit has a definite amount of energy. It posses kinetic energy because of its motion and potential energy on account of the attraction of the nucleus. Each allowed orbit is therefore associated with a certain quantity of energy called the energy of the orbit, which equals the total energy of the electron in it. In these allowed orbits electrons revolve without radiating energy.
- Energy is radiated or absorbed when an electron jumps from one stationary orbit to another stationary orbit. This energy is equal to the energy difference between these two orbits and emitted or absorbed as one quantum of radiation of frequency ν given by Planck's equation $E_2 - E_1 = h\nu = \frac{hc}{\lambda}$. This is called Bohr's frequency condition.
- **Radius of Bohr's orbit :** When mass of the nucleus is large compared to revolving electron, then electron revolves around the nucleus in circular orbit. According to first postulate

$$\frac{k(Ze)e}{r^2} = \frac{mV^2}{r} \left(\text{where } k = \frac{1}{4\pi \in_0}\right) \text{......(1)}$$

According to second postulate

$$mVr = n\frac{h}{2\pi} \text{ where } n = 1,2,3,4.............$$

$$\text{(or) } V = \frac{nh}{2\pi mr} \text{ .... (2)}$$

After solving the equations, radius of the orbit $r = \frac{n^2h^2}{4\pi^2kZme^2}$

For n^{th} orbit $r_n = \frac{h^2}{4\pi^2ke^2}\cdot\left(\frac{n^2}{mZ}\right)$ ....(3)

For hydrogen atom Z = 1, radius of the first orbit (n = 1) is given by $r_1 = 0.529\times10^{-10}\text{m} \simeq 0.53\text{Å}$ This value is called as Bohr's radius and the orbit is called Bohr's orbit. In general, the radius of the nth orbit of a hydrogen like atom is given by $r_n = 0.53\left(\frac{n^2}{Z}\right)\text{Å}$ where n = 1,2,3,....... (4)

Velocity of the Electron in the orbit :

The velocity of an electron in n^{th} orbit $V_n = \frac{nh}{2\pi mr_n}$ hence

$$V_n = \frac{2\pi ke^2}{h}\cdot\left(\frac{Z}{n}\right)\left(\because r_n = \frac{n^2h^2}{4\pi^2kZme^2}\right) \text{..(5)}$$

i.e the velocity of electron in any orbit is independent of the mass of electron. The above equation can also be written as

$$\therefore V_n = \left(\frac{c}{137}\right)\cdot\frac{Z}{n} \text{ m/s .... (6)}$$

Where 'c' is the speed of light in vacuum.

Time period of electron in the orbit :

Angular velocity of electron in n^{th} orbit

$\omega_n = \frac{V_n}{r_n} = \frac{\omega_0 Z^2}{n^3}$ where $\omega_0 = \frac{8\pi^3k^2e^4m}{h^3}$ ......(7) is the angular velocity of electron in first Bohr's orbit. The time period of rotation of electron in n^{th} orbit $T = \frac{2\pi}{\omega_n} = \frac{n^3}{2\pi\omega_0 Z^2}$ .... (8) i.e $T \propto \frac{n^3}{Z^2}$.

The time period of rotation increases as n increases and is independent on the mass of the electron.

Energy of the electron in the orbit :

- The kinetic energy of the electron revolving round the nucleus in n^{th} orbit is given by

$$K_n = \frac{1}{2}mV^2 = \frac{1}{2}m\left[\frac{2\pi k e^2}{h}.\frac{Z}{n}\right]^2$$

$$K_n = \frac{2\pi^2 k^2 e^4}{h^2}.\left(\frac{mZ^2}{n^2}\right) ....(9);\ K_n\ \alpha\ \frac{mZ^2}{n^2}$$

If the reference level (zero potential energy level) is at infinity then the electrostatic potential energy is given by

$$U_n = -\frac{k(Ze)e}{r_n} = -kZe^2\left[\frac{4\pi^2 kmZe^2}{n^2h^2}\right]$$

$$U_n = -\frac{4\pi^2k^2e^4}{h^2}\left(\frac{mZ^2}{n^2}\right) ....\ (10)$$

Total energy of the electron in n^{th} orbit

$$E_n = K_n + U_n = -\frac{2\pi^2k^2mZ^2e^4}{n^2h^2}$$

$$E_n = -\frac{2\pi^2k^2e^4}{h^2}\left(\frac{mZ^2}{n^2}\right) ....\ (11)$$

The expression of total energy for hydrogen like atom may be simplified as

$$E_n = -13.6\frac{Z^2}{n^2}\ \ eV,\ n = 1,2,3.....\ (12)$$

Where -13.6 eV is the total energy of the electron in the ground state of an hydrogen atom.

From the equations (9),(10)&(11) it is clear that

PE : K.E : T.E = -2 : 1 : -1

i.e $U = -2K$ and $E = -K$ (or) $E = -K = U/2$

- The state n = 1 is called ground state and n > 1 states are called excited states. When electron go from lower orbit to higher orbit speed and hence kinetic energy decrease, but both potential energy and total energy increases. $E\alpha\frac{1}{n^2}$ tells us that the energy gap between the two successive levels decreases as the value of n increases. At infinity level the total energy of the atom becomes zero.

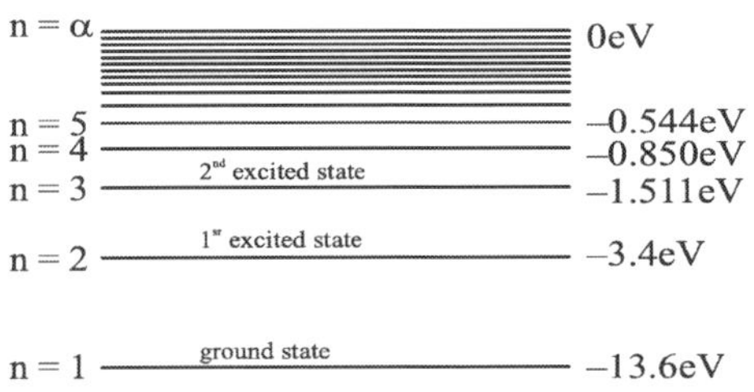

Energy level diagram of hydrogen atom (Z = 1) for normal and excited states as shown the figure.

The energy level diagram of hydrogen like atom with atomic number Z for normal and excited states as shown in Figure.

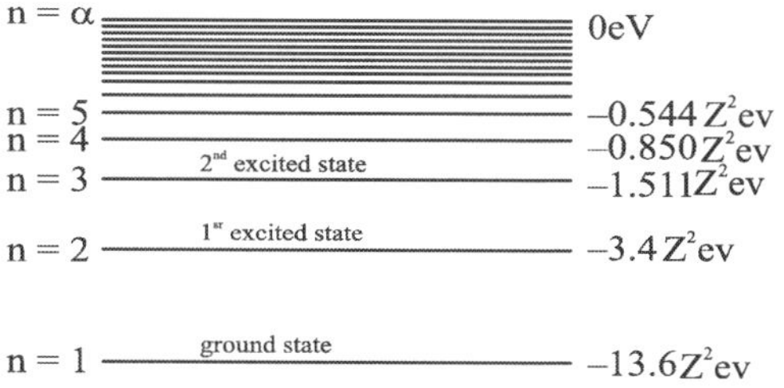

The total energy of the electron is negative implies the atomic electron is bound to the nucleus. To remove the electron from its orbit beyond the attraction of the nucleus, energy must be required.

The minimum energy required to remove an electron from the ground state of an atom is called its ionization energy and it is 13.6 Z^2 eV.

➢ In hydrogen atom the ground state energy of electron is –13.6 eV, so 13.6 eV is the ionization energy of the Hydrogen atom.

Emission of radiation:

➢ When an electron jumps from higher energy level n_2 to a lower energy level n_1 in stationary atom, the difference in energy is radiated as a photon whose frequency ν is given by Planck's formula.

$$E_{n_2} - E_{n_1} = h\nu$$

(or) $hv = E_2 - E_1 = 13.6Z^2\left[\frac{1}{n_1^2} - \frac{1}{n_2^2}\right]$ e.V

$\left(\therefore E_n = -\frac{13.6Z^2}{n^2}\ \text{e.V}\right)$ since $1\text{eV} = 1.6\times10^{-19}$J hence $h\frac{c}{\lambda} = (12.8\times10^{-18})Z^2\left[\frac{1}{n_1^2} - \frac{1}{n_2^2}\right]$ J

(or) wave number $\bar{v} = \frac{1}{\lambda} = RZ^2.\left[\frac{1}{n_1^2} - \frac{1}{n_2^2}\right]$ m^{-1}

where R is called for "Rydberg constant", when the nucleus is infinitely massive as compared to the revolving electron. In other words the nucleus is considered to be stationary. The numerical value of R is 1.097×10^7 m^{-1}.

Emission Spectrum of Hydrogen atom :

Electron in hydrogen atom, can be in excited state for very small time of the order of 10^{-8} second. This is because in the presence of conservative force system particles always try to occupy stable equilibrium position and hence minimum potential energy, which is least in ground state. Because of instability, when an electron in excited state makes a transition to lower energy state, a photon is emitted. Collection of such emitted photon frequencies is called an emission spectrum. This is as showing in figure.

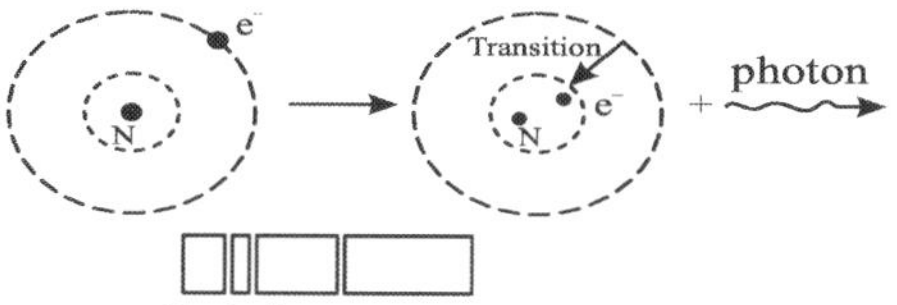

Emission spectrum

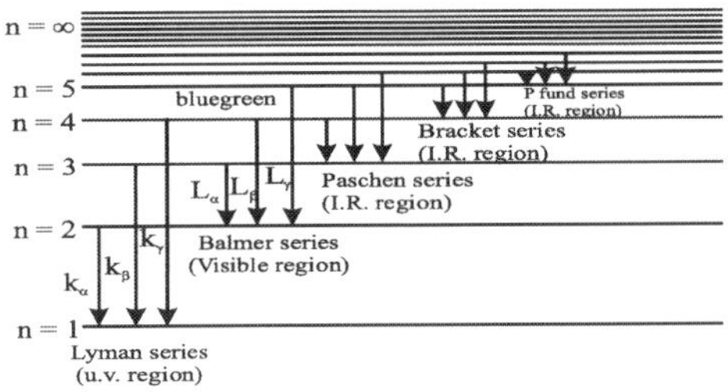

The Spectral Series of Hydrogen Atom as shown in figure, are explained below.

(a) Lyman Series : Lines corresponding to transition from outer energy levels $n_2 = 2,3,4,.........\infty$ to first orbit ($n_1 = 1$) constitute Lyman series. The wave numbers of different lines are given by,

$$\bar{v} = \frac{1}{\lambda} = R\left[\frac{1}{1^2} - \frac{1}{n_2^2}\right]$$

➢ Line corresponding to transition from $n_2 = 2$ to $n_1 = 1$ is first line; its wavelength is maximum.

$\frac{1}{\lambda_{max}} = R\left[\frac{1}{1^2} - \frac{1}{2^2}\right] = 1.1 \times 10^7\left[\frac{1}{1} - \frac{1}{4}\right] \therefore \lambda_{max} = 1212Å$ Similarly transition from $n_2 = \infty$ to $n_1 = 1$ gives line of minimum wavelength. $\frac{1}{\lambda_{min}} = R\left[\frac{1}{1^2} - \frac{1}{\infty^2}\right] = 1.1 \times 10^7 \quad \therefore \lambda_{min} = 912Å$

➢ Lyman series lies in ultraviolet region of electro magnetic spectrum.

➢ Lyman series is obtained in emission as well as in absorption spectrum.

(b) Balmer Series: Lines corresponding to $n_2 = 3,4,5,........\infty$ to $n_1 = 2$ constitute Balmer series. The wave numbers of different lines are given by, $\bar{v} = \frac{1}{\lambda} = R\left[\frac{1}{2^2} - \frac{1}{n_2^2}\right]$

➢ Line corresponding to transition n_2=3 to n_1=2 is first line, wavelength corresponding to this transition is maximum. Line corresponding to transition $n_2 = \infty$ to $n_1 = 2$ is last line; wavelength of last line is minimum.

$$\frac{1}{\lambda_{max}} = R\left[\frac{1}{2^2} - \frac{1}{3^2}\right] \quad \therefore \lambda_{max} = 6568Å \qquad \frac{1}{\lambda_{min}} = R\left[\frac{1}{2^2} - \frac{1}{\infty^2}\right] \quad \therefore \lambda_{min} = 3636Å$$

➢ Balmer series lies in the visible region of electromagnetic spectrum. The wavelength of L_α line is 656.8 nm (red). The wavelength of L_β line is 486 nm (blue green). The wavelength of L_γ line is 434 nm (violet). The remaining lines of Balmer series closest to violet light wavelength. The speciality of these lines is that in going from one end to other, the brightness and the separation between them decreases regularly.

➢ This series is obtained only in emission spectrum. Absorption lines corresponding to Balmer series do not exist, except extremely weakly, because very few electrons are normally in the state n = 2 and only a very few atoms are capable of having an electron knocked from the state n = 2 to higher states. Hence photons that correspond to these energies will not be strongly absorbed. In highly excited hydrogen gas there is possibility for detecting absorption at Balmer-line wavelengths.

(c) Paschen Series: Lines corresponding to $n_2 = 4,5,6,.......\infty$ to $n_1 = 3$ constitute Paschen series. The wave number of different lines are given by $\bar{v} = R\left[\frac{1}{3^2} - \frac{1}{n_2^2}\right]$

➢ Line corresponding to transition $n_2 = 4$ to $n_1 = 3$ is first line, having maximum wavelength. Line corresponding to transition $n_2 = \infty$ to $n_1 = 3$ is last line, having minimum wavelength

$$\frac{1}{\lambda_{max}} = R\left[\frac{1}{3^2} - \frac{1}{4^2}\right] \therefore \lambda_{max} = 18747Å$$

$$\frac{1}{\lambda_{min}} = R\left[\frac{1}{3^2} - \frac{1}{\infty}\right] = 1.1\times10^7 \times\left[\frac{1}{9} - 0\right]$$

$\therefore \lambda_{min} = 8202Å$

- Paschen series lies in the infrared region of electromagnetic spectrum.
- This series is obtained only in the emission spectrum.

(d) Bracket Series: The series corresponds to transitions from $n_2 = 5, 6, 7 ...., \infty$ to $n_1 = 4$. The wave number are given by, $\bar{v} = \frac{1}{\lambda} = R\left[\frac{1}{4^2} - \frac{1}{n_2^2}\right]$

- Line corresponding to transition from n_2=5 to n_1=4 has maximum wavelength and $n_2 = \infty$ to $n_1 = 4$ has minimum wavelength.

$$\frac{1}{\lambda_{max}} = R\left[\frac{1}{4^2} - \frac{1}{5^2}\right] \therefore \lambda_{max} = 40477Å$$

$$\frac{1}{\lambda_{min}} = R\left[\frac{1}{4^2} - \frac{1}{\infty^2}\right] \therefore \lambda_{min} = 14572Å$$

- This series lies in the infrared region of electromagnetic spectrum.

(e) Pfund Series: This series corresponds to transitions from $n_2 = 6, 7, 8, ..., \infty$ to $n_1 = 5$. The wave numbers are given by $\bar{v} = \frac{1}{\lambda} = R\left[\frac{1}{5^2} - \frac{1}{n_2^2}\right]$

- Line corresponding to transition from $n_2 = 6$ to $n_1 = 5$ has maximum wavelength and $n_2 = \infty$ to $n_1 = 4$ has minimum wavelength. $\frac{1}{\lambda_{max}} = R\left[\frac{1}{5^2} - \frac{1}{6^2}\right]$

$\therefore \lambda_{max} = 74563Å$ $\frac{1}{\lambda_{min}} = R\left[\frac{1}{5^2} - \frac{1}{\infty^2}\right]$

$\therefore \lambda_{min} = 22768Å$

- This series lies in infrared region of electromagnetic spectrum.

Note **:** In an atom emission transition may start from any higher energy level and end at any energy level below of it. Hence in emission spectrum the total possible number of emission lines from some excited state n_2 to another energy state $n_1 (< n_2)$ is $\frac{(n_2 - n_1)(n_2 - n_1 + 1)}{2}$

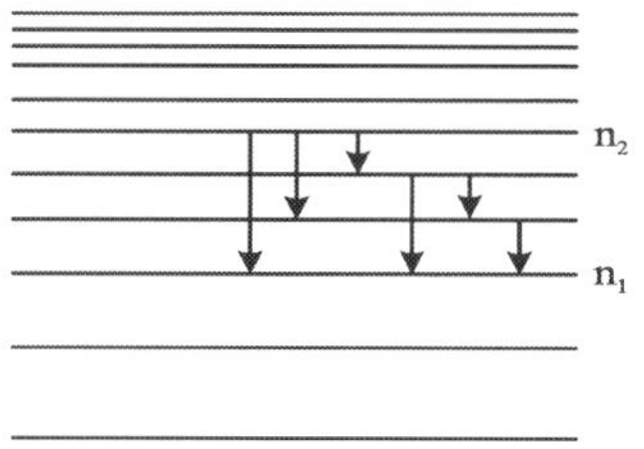

Note 1: for $n_2 = 4$, and $n_1 = 1$, the number of possible lines are 6.

Note 2 : If ΔE is the energy difference between two given energy states, then due to transition between these two states wavelength of emitted photon is $\lambda(Å) = \frac{12400}{\Delta E(eV)}$

Limitation of Bohr's model :

Despite its considerable achievements, the Bohr's model has certain short coming.

- It could not interpret the details of optical spectra of atoms containing more than one electron.
- It involves the concept of orbit which could not be checked experimentally
- It could be successfully applied only to single-electron atoms (e.g., H, He^+, Li^{2+}, etc.)
- Bohr's model could not explain the binding of atoms into molecules.
- No justification was given for the "principle of quantization of angular momentum".
- Bohr's model could not explain the reason why atoms should combine to form chemical bonds and why do the molecules become more stable on such combinations.
- Bohr had assumed that an electron in the atom is located at definite distance from the nucleus and is revolving with a definite velocity around it. This is against the Heisenberg uncertainty principle. With the advancements in quantum mechanics, it be came clear that there are no well defined orbits; rather there are clouds of negative charge.

S.No	Name of the series	Final State (n_1)	Initial State (n_2)	Formula	Series limit	Maximum wavelength	Region
1.	Lyman	$n_1 = 1$	$2,3,4,....\infty$	$\frac{1}{\lambda}=R\left(\frac{1}{1^2}-\frac{1}{n_2^2}\right)$	$\lambda=\frac{1}{R}=911A^\circ$	$\lambda=\frac{4}{3R}$	UV
2.	Balmer	$n_1 = 2$	$3,4,5....\infty$	$\frac{1}{\lambda}=R\left(\frac{1}{2^2}-\frac{1}{n_2^2}\right)$	$\lambda=\frac{4}{R}$	$\lambda=\frac{36}{5R}$	Visible
3.	Paschen	$n_1 = 3$	$4,5,6....\infty$	$\frac{1}{\lambda}=R\left(\frac{1}{3^2}-\frac{1}{n_2^2}\right)$	$\lambda=\frac{9}{R}$	$\lambda=\frac{144}{7R}$	Near IR
4.	Brackett	$n_1 = 4$	$5,6,7....\infty$	$\frac{1}{\lambda}=R\left(\frac{1}{4^2}-\frac{1}{n_2^2}\right)$	$\lambda=\frac{16}{R}$	$\lambda=\frac{400}{9R}$	Middle IR
5.	Pfund	$n_1 = 5$	$6,7,8....\infty$	$\frac{1}{\lambda}=R\left(\frac{1}{5^2}-\frac{1}{n_2^2}\right)$	$\lambda=\frac{25}{R}$	$\lambda=\frac{9000}{11R}$	Far IR

W.E-15: ***The electron in a hydrogen atom makes a transition $n_1 \to n_2$ where n_1 and n_2 are the principal quantum numbers of the two states. Assume the Bohr model to be valid. The time period of the electron in the initial state is eight times that in the final state. What are the possible values of n_1 and n_2?***

Sol. Since, $T \propto n^3$; $\frac{T_1}{T_2}=\frac{n_1^3}{n_2^3}$, As $T_1 = 8T_2$, $\left(\frac{n_1}{n_2}\right)^3 = 8$ (or) $n_1 = 2n_2$. Thus the possible values of n_1 and n_2 are $n_1 = 2, n_2 = 1, n_1 = 4, n_2 = 2, n_1 = 6, n_2 = 3$; and so on.

W.E-16: ***Find the kinetic energy, potential energy and total energy in first and second orbit of hydrogen atom if potential energy in first orbit is taken to be zero.***

Sol. E_1 = -13.60eV; $K_1 = -E_1$ = 13.60eV

$U_1 = 2E_1$ = -27.20eV $E_2 = -3.40eV$ K_2 = 3.40eV and U_2 = -6.80eV

Now, $U_1 = 0$, i.e., potential energy has been increased by 27.20eV. So, we will increase U and E in all energy states by 27.20eV while kinetic energy will remain unchanged.

Hence K(eV), U(eV), E(eV)

First orbit are	13.6,	0,	13.6
in Second orbit	3.40,	20.40,	23.80

W.E-17: ***A small particle of mass m moves in such a way that the potential energy $U = ar^2$ where a is a constant and r is the distance of the particle from the origin. Assuming Bohr's model of quantization of angular momentum and circular orbits, find the radius of nth allowed orbit.***

Sol. The force at a distance r is, $F = -\frac{dU}{dr} = -2ar$ Suppose r be the radius of nth orbit. Then the necessary centripetal force is provided by the above force. Thus, $\frac{mv^2}{r} = 2ar$ -------- (i)

Further, the quantization of angular momentum gives, $mvr = \frac{nh}{2\pi}$ -------- (ii)

Solving Eqs. (i) and (ii) for r, we get $r = \left(\frac{n^2h^2}{8am\pi^2}\right)^{1/4}$

W.E-18: ***Consider a hydrogen-like atom whose energy in n^{th} excited state is given by $E_n = -\frac{13.6Z^2}{n^2}$ when this excited atom makes transition from excited state to ground state most energetic photons have energy $E_{max} = 52.224eV$ and least energetic photons have energy $E_{min} = 1.224eV$. Find the atomic number of atom and the state of excitation.***

Sol. Maximum energy is liberated for transition $E_n \to 1$ and minimum energy for $E_n \to E_{n-1}$

Hence, $\frac{E_1}{n^2} - E_1 = 52.224eV$ ... (1)

and $\frac{E_1}{n^2} - \frac{E_1}{(n-1)^2} = 1.224\,eV$ ...... (2)

Solving above equations simultaneously, we get

$E_1 = -54.4eV$ and $n = 5$ Now $E_1 = -\frac{13.6Z^2}{1^2} = -54.4eV$. Hence, Z = 2 i.e, gas is helium originally excited to n = 5 energy state.

W.E-19: ***A hydrogen-like atom (atomic number Z) is in a higher excited state of quantum number n. This excited atom can make a transition to the first excited state by successively emitting two photons of energies 10.20 eV and 17.00 eV respectively. Alternatively the atom from the same excited state can make a transition to the second excited state by successively emitting two photons of energies 4.25 eV and 5.95 eV respectively. Determine the values of n and Z (ionization energy of hydrogen atom = 13.6 eV)***

Sol. The electronic transitions in a hydrogen-like atom from a state n_2 to a lower state n_1 are given by $\Delta E = 13.6Z^2\left(\frac{1}{n_1^2} - \frac{1}{n_2^2}\right)$. For the transition from a higher state n to the first excited state n_1 = 2, the total energy released is (10.2 + 17.0) eV or 27.2eV. Thus $\Delta E = 27.2$ eV, $n_1 = 2$ and $n_2 = n$.

We have $27.2 = 13.6Z^2\left[\frac{1}{4} - \frac{1}{n^2}\right]$ ...... (1)

For the eventual transition to the second excited state $n_1 = 3$, the total energy released is (4.25 + 5.95) eV or 10.2eV.

Thus $10.2 = 13.6Z^2\left[\frac{1}{9} - \frac{1}{n^2}\right]$ .....(2)

Dividing the Eq. (1) by Eq. (2) we get $\frac{27.2}{10.2} = \frac{9n^2 - 36}{4n^2 - 36}$.

Solving we get $n^2 = 36$ or $n = 6$

Substituting n = 6 in any one of the above equations, we obtain $Z^2 = 9$ (or) Z = 3, Thus n=6 and Z=3.

W.E-20: ***A doubly ionized lithium atom is hydrogen like with atomic number Z = 3. Find the wavelength of the radiation required to excite the electron in Li^{2+} from the first to the third Bohr orbit. Given the ionization energy of hydrogen atom as 13.6 eV.***

Sol. The energy of nth orbit of a hydrogen-like atom is given as $E_n = -\frac{13.6Z^2}{n^2}$ Thus for Li^{2+} atom, as Z = 3, the electron energies of the first and third Bohr orbits are For n = 1, $E_1 = -122.4$eV, for n = 3, $E_3 = -13.6$eV. Thus the energy required to transfer an electron from E_1 level to E_3 level is, $E = E_3 - E_1 = -13.6 - (-122.4) = 108.8\text{eV}$. Therefore, the radiation needed to cause this transition should have photons of this energy. hv = 108.8 eV. The wavelength of this radiation is or $\lambda = \frac{hc}{108.8\text{eV}} =$ 114.25 Å

W.E-21: ***A hydrogen atom in a state of binding energy 0.85 eV makes a transition to a state of excitation energy of 10.2 eV.***

(i) What is the initial state of hydrogen atom?

(ii) What is the final state of hydrogen atom ?

(iii) What is the wavelength of the photon emitted ?

Sol. (i) Let n_1 be initial state of electron. Then $E_1 = -\frac{13.6}{n_1^2}$eV Here E_1 = - 0.85 eV, therefore $-0.85 = -\frac{13.6}{n_1^2}$ or $n_1 = 4$

(ii) Let n_2 be the final excitation state of the electron. Since excitation energy is always measured with respect to the ground state, therefore

$\Delta E = 13.6\left[1 - \frac{1}{n_2^2}\right]$ here $\Delta E = 10.2\text{eV}$, therefore,

$10.2 = 13.6\left[1 - \frac{1}{n_2^2}\right]$ or $n_2 = 2$ Thus, the electron jumps from $n_1 = 4$ to $n_2 = 2$.

(iii) The wavelength of the photon emitted for a transition between $n_1 = 4$ to $n_2 = 2$, is given by

$$\frac{1}{\lambda} = R_\infty\left[\frac{1}{n_2^2} - \frac{1}{n_1^2}\right] \text{(or)} \frac{1}{\lambda} = 1.09 \times 10^7\left[\frac{1}{2^2} - \frac{1}{4^2}\right] = 4860\text{Å}.$$

W.E-22: ***A hydrogen atom initially in the ground level absorbs a photon, which excites it to the n = 4 level. Determine the wavelength and frequency of photon. To find the wavelength and frequency of photon use the relation of energy of electron in hydrogen atom is $E_n = -\frac{13.6}{n^2}$eV.***

Sol. For ground state $n_1 = 1$ to $n_2 = 4$.

Energy absorbed by photon, $E = E_2 - E_1$

$$= +13.6\left(\frac{1}{n_1^2} - \frac{1}{n_2^2}\right) \times 1.6 \times 10^{-19} \text{ J}$$

$$= 13.6\left(\frac{1}{1} - \frac{1}{4^2}\right) \times 1.6 \times 10^{-19}$$

$= 13.6\times1.6\times10^{-19}\left(\frac{15}{16}\right) = 20.4 \times 10^{-19}$.

or $E = h\nu = 20.4 \times 10^{-19}$

Frequency $\nu = \frac{20.4\times10^{-19}}{h} = \frac{20.4\times10^{-19}}{6.63\times10^{-34}}$

$= 3.076 \times 10^{15} = 3.1 \times 10^{15}$ Hz.

Wavelength of photon $\lambda = \frac{c}{\nu} = \frac{3\times10^{8}}{3.076\times10^{15}} = 9.74\times10^{-8}$ m. Thus, the wavelength is 9.7×10^{-8} m and frequency is 3.1×10^{15} Hz.

W.E-23: ***(a) Using the Bohr's model calculate the speed of the electron in a hydrogen atom in the n = 1, 2 and 3 levels.***

(b) Calculate the orbital period in each of these levels.

Sol. **(a)** Speed of the electron in Bohr's nth orbit

$\nu = \frac{c}{v}\alpha$ where, $\alpha = \frac{2\pi Ke^2}{ch}$

$\alpha = 0.0073 \qquad \therefore \quad v = \frac{c}{n}\times0.0073$

For $n = 1$, $\quad v_1 = \frac{c}{1}\times0.0073$

$= 3 \times 10^{8} \times 0.0073 = 2.19 \times 10^{6}$ m/s

For $n = 2$ $\quad v_2 = \frac{c}{2}\times0.0073$

$= \frac{3\times10^{8}\times0.0073}{2} = 1.095 \times 10^{6}$ m/s

For $n = 3$ $\quad v_3 = \frac{c}{3}\times0.0073$

$= \frac{3\times10^{8}\times0.0073}{3} = 7.3 \times 10^{5}$ m/s.

(b) Orbital period of electron is given by $T = \frac{2\pi r}{v}$

Radius of nth orbit $r_n = \left(\frac{n^2}{m}\right)\left(\frac{h}{2\pi}\right)^2\frac{4\pi\varepsilon_0}{e^2}$

$\therefore\ r_1 = \frac{(1)^2\times(6.63\times10^{-34})^2}{4\times9.87\times(9\times10)^9\times9\times10^{-31}\times6.6\times10^{-19})} = 0.53 \times 10^{-10}$ m.

For n = 1 $\quad T_1 = \dfrac{2\pi r_1}{v_1}$

$= \dfrac{2\times3.14\times0.53\times10^{-10}}{2.19\times10^{6}} = 1.52 \times 10^{-16}$ s

For n = 2, radius $r_n = n^2 r_1$

$r_2 = 2^2 . r_1 = 4\times0.53\times10^{-10}$ m and

velocity $v_n = \dfrac{v_1}{n}$

$\therefore\ v_2 = \dfrac{v_1}{2} = \dfrac{2.19\times10^{6}}{2}$

Time period $T_2 = \dfrac{2\times3.14\times4\times0.53\times10^{-10}\times2}{2.19\times10^{6}}$

$= 1.216 \times 10^{-15}$ s.

For n = 3, radius $r_3 = 3^2$

$r_1 = 9\ r_1 = 9 \times 0.53 \times 10^{-10}$ m

and velocity $v_3 = \dfrac{v_1}{3} = \dfrac{2.19\times10^{6}}{3}$ m/s

Time period $T_3 = \dfrac{2\pi r_3}{v_3}$

$= \dfrac{2\times3.14\times9\times0.53\times10^{-10}\times3}{2.19\times10^{6}} = 4.1 \times 10^{-15}$ s.

W.E-24: ***The radius of the innermost electron orbit of a hydrogen atom is 5.3 × 10^{-11} m. What are the radii of the n = 2 and n = 3 orbits?***

Sol. Given, the radius of the innermost electron orbit of a hydrogen $r_1 = 5.3 \times 10^{-11}$ m.

As we know that $\quad r_n = n^2 r_1$

For n = 2, radius $r_2 = 2^2$

$r_1 = 4 \times 5.3 \times 10^{-11} = 2.12 \times 10^{-10}$ m.

For n = 3, radius $r_3 = 3^2$

$r_1 = 9 \times 5.3 \times 10^{-11} = 4.77 \times 10^{-10}$ m.

W.E-25: ***A 12.5 eV electron beam is used to bombard gaseous hydrogen at room temperature. What series of wavelength will be emitted?***

Sol. Energy of electron beam E=12.5 eV = $12.5 \times 1.6 \times 10^{-10}$ J

Planck's constant h=6.63×10^{-34} J-s

Velocity of light c=3×10^{8} m/s

Using the relation $E = \dfrac{hc}{\lambda} = \dfrac{6.62\times10^{-34}\times3\times10^{8}}{12.5\times1.6\times10^{-19}}$

$= 0.993 \times 10^{-7}$ m $= 993 \times 10^{-10}$ m $= 993$ A^0

This wavelength falls in the range of Lyman series (912 A^0 to 1216A^0)

thus, we conclude that Lyman series of wavelength 993 A^0 is emitted.

W.E-26:*In accordance with the Bohr's model, find the quantum number that characterises the earth's revolution around the sun in an orbit of radius 1.5×10^{11} m with orbital speed 3×10^4 m/s. (Mass of earth = 6.0×10^{24} kg).*

A. Given, radius of orbit $r = 1.5 \times 10^{11}$ m

Orbital speed $v=3\times10^4$m/s;

Mass of earth $M=6\times10^{24}$kg

Angulalr momentum, $mvr = \frac{nh}{2\pi}$ or $n = \frac{2\pi vrm}{h}$

[where, n is the quantum number of the orbit]

$$= \frac{2\times3.14\times3\times10^4\times1.5\times10^{11}\times6\times10^{24}}{6.63\times10^{-34}}$$

$= 2.57 \times 10^{74}$ or $n = 2.6 \times 10^{74}$. Thus, the quantum number is 2.6×10^{74} which is too large. The electron would jump from n=1 to n=3

$$E_3 = \frac{-13.6}{3^2} = -1.5 \text{ eV.}$$

So, they belong to Lyman series.

CONCEPTUAL

THOMSON'S METHOD

1. **If 'C' denotes the velocity of light, velocity of cathode rays is**

1) equal to C 2) greater than C 3) less than C
4) either greater or less than C

2. **An electron has no initial velocity in a direction different from that of an electric field. The path of the electron is**

1) a straight line 2) a circle 3) a parabola 4) an ellipse

3. **Cathode rays enter an electric field normal to the field. Then their path in the electric field is**

1) a parabola 2) a circle 3) a straight line 4) an ellipse

4. **When the electron in the discharge tube is accelerated to high speeds**

1) The charge on the electron will decrease 2) The specific charge will decrease
3) The charge of the electron will increase 4) The value of e/m will increase

5. **The discovery of electron was a consequence of study of the**

1) discharge of electricity through atmosphere. 2) discharge of electricity through rarefied gases.
3) photoelectric effect. 4) nuclear fission.

6. **If a proton and an electron are accelerated through the same potential difference**

1) both the proton and electron have same K.E
2) both the proton and electron have same momentum
3) both the proton and electron have same velocity
4) both the proton and electron have same temperature

7. **A cathode ray tube has a potential difference of V between the cathode and anode. The speed of the cathode rays is given by**

1) $v \propto V$ 2) $v \propto V^{-1}$ 3) $v \propto \sqrt{V}$ 4) $v \propto V^2$

8. **In a region of space, cathode rays move along positive z-axis and a uniform magnetic field is applied along x-axis. If cathode rays pass undeviated, the direction of electric field will be along**

1) Negative x-axis 2) Positive y-axis 3) Negative y-axis 4) Positive z-axis

9. **A cathode ray particle is accelerated from rest through a potential difference of V volt. The speed of the particle is**

1) $\sqrt{\frac{2eV}{m}}$ 2) $\frac{m}{eV}$ 3) $\sqrt{meV}$ 4) $\sqrt{\frac{4eV}{m}}$

10. **In J.J.Thomson's method, electric field of intensity E, magnetic field of induction B and velocity V of the electrons were in mutually perpendicular directions. The condition for velocity is**

1) $V = E/B$ 2) $V = B/E$ 3) $V = BE$ 4) $V = \sqrt{B/E}$

11. **A positive ion of mass M kg and charge e coulomb travels from rest through a p.d. of V volt. The final K.E. is**

1) e V joule 2) MeV joule 3) Me/V joule 4) eV/M joule

12. **The increasing order of specific charge for electron(e), proton(p), neutron(n) and alpha particle(α) is**

1) e, p, n, α 2) n, α, p, e 3) n, p, e, α 4) n, p, α, e

13. **A uniform electric field and a uniform magnetic field are produced pointing in the same direction. An electron is projected with its velocity pointed in the same direction.**

1) the electron will turn to its right 2) the electron will turn to its left
3) the electron velocity will increase in magnitude 4) the electron velocity will decrease in magnitude

14. **When a positively charged particle enters a uniform magnetic field with uniform velocity its trajectory can be**

a) straight line b) a circle c) a helix
1) a only 2) a or b 3) a or c 4) any one of a,b and c

15. **The ratio e/m of electron is independent of**

a) Nature of cathode, anode **b) Nature of gas in discharge tube**
c) Applied voltage **d) Size of discharge tube**
1) a ,b,c are only correct 2) a,b,d are only correct
3) b,c,d are only correct 4) a,b,c,d are correct

16. **An electron is not deflected on passing through a certain region, because**

1) there is magnetic field in that region and the electron enters into it in any direction.
2) there may be magnetic field but the velocity of electron may be parallel to the direction of magnetic field
3) electron is a chargeless particle
4) there is electric field and the electron enters into it in any direction.

17. **In Thomson's experiment, when the electron strikes the undeflected spot, then it moves with**

1) constant acceleration 2)non uniform velocity
3) constant velocity 4)constant retardation

18. **An electron enters a magnetic field with a speed of 10^8 cm/s. The particle experiences a force due to the magnetic field and the speed of the electron**

1) will decrease 2) will increase

3) will remain constant 4) may increase or decrease

19. **A neutron, a proton, an electron and an alpha particle enter a region of constant magnetic field with equal velocities. The magnetic field is along the inward normal to the plane of the paper. The tracks of the particles are labled in the figure. The electron follows ____ track and alpha particle follows track ______**

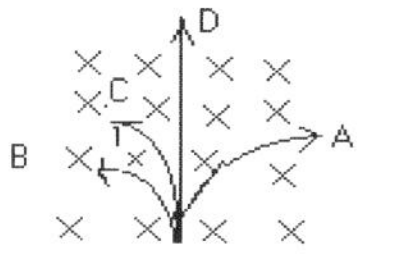

1) C,D 2) B, A 3) A,C 4) A,D

20. **When a charged particle moves through a magnetic field, the quantity which is not affected in the magnetic field is**

1) particle velocity 2)particle acceleration

3) linear momentum of the particle 4) Kinetic energy of the particle

21. **A negatively charged electroscope with zinc disc discharges when irradiated by an ultraviolet lamp. What caused this?**

1) α - particles from the source combine with electrons of the disc

2) electrons escape from the disc when ultraviolet radiation falls on it

3) ultraviolet rays ionize the air surrounding the electroscope

4) the disc becomes hot and thermionic emission takes place

22. **The force felt by an electron on entering into a magnetic field is independent of its**

1) Charge 2) Strength of the field

3) Mass 4) Direction of its velocity

23. **When an electron moves through a magnetic field, its speed will**

1) decrease 2) increase

3) remain the same 4) increase first and then decrease

24. **The direction of deflection of a cathode ray particle passing through a magnetic field can be found by**

1) Fleming's left hand rule 2) Laplace's law

3) Maxwell's cork screw rule 4) Ampere's rule

25. **Which of the following particles cannot be deflected by magnetic field**

1) electrons 2) neutrons 3) α - particles 4) protons

26. **A β - particle enters a magnetic field making an angle of 45^0 with the field lines. The path of the particle is**

1) circular 2) elliptical 3) spiral 4) a straight line

27. **An electron and a proton are injected into a uniform magnetic field at right angle to its direction with the same momentum. Then**

1) electron's path is less curved than proton's path

2) proton's path will be less curved than electron's path

3) the paths of both will be equally curved 4) both the trajectories will be straight

28. **A proton and an electron simultaneously enter into a region in which a uniform magnetic field acts normal to the motion of both the particles. The frequency of revolution of**

1) the proton is greater than that of the electron
2) the electron is greater than that of the proton
3) the proton is equal to that of the electron
4) both are having same frequency, but revolve in opposite direction

29. **You are sitting in a room in which uniform magnetic field is present in vertically downward direction. When an electron is projected in horizontal direction, it will be moving in circular path with constant speed**

1) clockwise in vertical plane
2) clockwise in horizontal plane
3) anticlockwise in vertical plane
4) anticlockwise in horizontal plane

30. **A charged particle of charge Q and mass m moves with velocity v in a circular path due to transverse magnetic field, B, then its frequency is**

1) $\frac{QB}{2\pi m}$
2) $\frac{QvB}{2\pi m}$
3) $\frac{QmvB}{2\pi}$
4) $\frac{vB}{2\pi Qm}$

31. **A proton and an electron simultaneously enter a region in which a uniform magnetic field acts normal to the motion of both the particles. The frequency of revolution of**

1) proton is greater than that of electron
2) electron is greater than that of proton
3) proton is equal to that of electron
4) proton depends on its velocity

32. **Imagine that you are sitting in a room with your back to one wall and that an electron beam traveling horizontally from the back wall towards the front one is deflected towards the right. What is the direction of the magnetic induction field that exists in the room?**

1) vertically upwards
2) vertically downwards
3) horizontal and perpendicular to the direction of motion of the electron beam
4) horizontal and parallel to the direction of motion of the electron beam

33. **Cathode rays are made to pass between the poles of a magnet as shown in figure. The effect of magnetic field is**

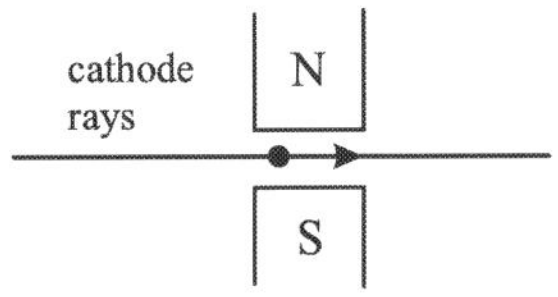

1) to deflect them towards the south pole
2) to deflect them perpendicular to the plane of the paper and towards the observer
3) to deflect them towards the north pole
4) to increase the velocity of the rays

34. **In which of the following fields cathode rays show minimum deflection**

1) Electric field
2) Magnetic field
3) Plasma field
4) Gravitational field

MILLIKAN'S OIL DROP EXPERIMENT

35. **An oil drop of mass m and charge +q is balanced in vacuum by a uniform electric field of intensity E. The direction of this field should be**

1) vertically up
2) vertically down
3) horizontal
4) inclined at 45^0 to the horizontal

36. **An oil drop of mass m falls through a medium that offers a viscous drag force F. If the velocity of the drop is constant it means that**

1) $F > mg$
2) $F < mg$
3) $F > mg$
4) $F = mg$

37. **An oil drop of mass m falls through a viscous medium. The viscous drag force (F) is proportional to the velocity of the drop. At the instant it begins to fall the net force that acts on the oil drop is (neglect buoyancy)**

1) mg 2) mg-F 3) F-mg 4) F

38. **An oil drop of mass m falls through air with a terminal velocity in the presence of upward electric field of intensity E. The drop carries a charge +q. R is the viscous drag force and F is the buoyancy force. Then for the motion of the drop.**

1) there is a net force directed upwards 2) there is a net force directed downwards.
3) mg = Eq + F + R 4) mg + Eq = F + R

39. **If g_E and g_M are the accelerations due to gravity on the surfaces of the earth and the moon respectively and if Millikan's oil drop experiment could be performed on the two surfaces, one will find the ratio $\frac{\text{electronic charge on the moon}}{\text{electronic charge on the earth}}$ to be**

1) g_E/g_M 2) 1 3) 0 4) g_M/g_E

40. **In Millikan's oil drop method, electron charge is accurately measured by conducting an experiment in which**

1) Apparent weight of drop is balanced by electric force on it
2) Apparent weight of drop is balanced by magnetic force
3) Electric force of drop is equal to magnetic force
4) Apparent weight of drop is balanced by force due to surface tension

41. **In Millikan's experiment, the oil drops acquire charge by**

1) induction 2) Friction 3) electric field 4) magnetic field

42. **According to Stoke's law, the viscous drag force on an oil drop is proportional to**

1) $\sqrt{V}$ 2) V^2 3) V^{-1} 4) V

43. **An oil drop of radius r carrying a charge q remains stationary in the presence of electric field of intensity E. If the density of oil is ρ, then**

1) $E=\frac{4}{3}\pi r^3\rho gq$ 2) $E=\frac{4}{3}\pi r^3\rho g$ 3) $E=\frac{4}{3}\pi r^3\rho g/q$ 4) $E=\frac{4}{3}\pi r^3\rho/g^3$

44. **The important conclusion given by Millikan's experiment about the charge is**

1) Charge is never quantised 2) Charge has no definite value
3) Charge is quantised 4) Charge on an oil drop always increases

ALPHA RAY SCATTERING

45. **'Coulomb's law correctly describes the electric force is that (pick the wrong statement)**

1) binds the electrons and neutrons in the nucleus of an atom.
2) binds electrons to nucleus
3)binds atoms together to form molecules 4)binds atoms and molecules to from solids

46. **Is the probability of backward scattering (i.e.scattering of α - particals at angles greater than 90^0) predicted by Thomson's model much less, about the same, or much greater than that predicted by Ruthorford's model?**

1) much greater 2) much less 3) same 4) slightly greater

47. An electron with KE 6 eV is incident on a hydrogen atom in its ground state.The collision.

1) must be elastic
2) may be partially elastic
3) must be completly inelastic
4) may be partially inelastic.

48. The angular momentum of the α-particles which are scattered through large angles by the heavier nuclei,is conserved because of the

1) nature of repulsive forces
2) conservation of kinetic energy
3) conservation of potential energy
4) there is no external torque

49. The Incorrect statement regarding Rutherford's atomic model is

1) Atom contains nucleus
2) Size of nucleus is very small in comparison to that of atom
3) Nucleus contains about 90 % mass of the atom
4) Electrons revolve round the nucleus with uniform speed

50. α-particles are

1) helium nucleii
2) sodium nucleii
3) ionised nucleii
4) hydrogen nucleii

51. In scattering experiment,the force that scatters particles is

1) nuclear force
2) coulomb force
3) Both(1)and(2)
4) gravitational force

BOHR'S THEORY

52. Bohr's atom model assumes

1) the nucleus is of infinite mass and is at rest
2)electron in a quantized orbit will not radiate energy.
3) mass of the electron remains constant.
4) 1,2,3 are correct

53. The ratio of magnetic dipole moment to angular momentum in a hydrogen like atom is

1) $\frac{e}{m}$
2) $\frac{e}{2m}$
3) $\frac{e}{3m}$
4) $\frac{2e}{m}$

54. The radius of hydrogen atom,when it is in its second excited state, becomes:

1) half
2) double
3) four times
4) nine times

55. For electron moving in n^{th} orbit of the atom, the angular velocity is proportional to:

1) n
2) $\frac{1}{n}$
3) n^3
4) $\frac{1}{n^3}$

56. The order of size of nucleus and Bohr's radius of an atom respectively are:

1) $10^{-14}m, 10^{-10}m$
2) $10^{-10}m, 10^{-8}m$
3) $10^{-20}m, 10^{-16}m$
4) $10^{-8}m, 10^{-6}m$

57. In the lowest energy level of hydrogen atom,electron has the angular momentum:

1) $\frac{\pi}{h}$
2) $\frac{h}{\pi}$
3) $\frac{h}{2\pi}$
4) $\frac{2\pi}{h}$

58. Atomic hydrogen is excited to the nth energy level.The maximum number of spectral lines which it can emit while returning to the ground state ,is:

1) $\frac{1}{2}n(n-1)$
2) $\frac{1}{2}n(n+1)$
3) $n(n+1)$
4) $n(n+1)$

59. As the orbit number increases ,the distance between two consecutive orbits in an atom or ion having single electron:

1) increases
2)decreases
3) remains the same
4) first increases and then becomes constant

60. Ionization energy of a hydrogen -like ion A is greater than that of another hydrogen like ion B. Let r, u, E and L represents the radius of the orbit, speed of the electron, energy of the atom and orbital angular momentum of the electron respectively in ground state, then

1) $r_A > r_B$ 2) $u_A > u_B$ 3) $E_A > E_B$ 4) $L_A > L_B$

61. The classification of discrete energy levels in atoms was first given experimentally by

1) Thomson's experiment 2) Millikan's oil drop experiment.
3) Frank - Hertz experiment 4) Leonard experiment

62. An atomic nucleus contains

1) only electrons 2) only protons 3) only neutrons
4) both protons and neutrons

63. On decreasing principal quantum number n, the values of r and v will

1)decrease 2) increase
3) r will increase but v will decrease. 4) r will decrease but v will increase

64. The possible values of principal quantum number can be

1) 1, 2, 3...8. 2) 0, 1, 2...8. 3) Only zero 4) only odd numbers

65. According to Sommerfeld, an electron revolves round a nucleus in

1) Circular orbits. 2) Elliptical orbits. 3) Hyperbolic orbits 4) parabolic

66. The main defect of Bohr atom model is

1) mixing of classical and quantum theories 2) exclusion of nuclear motion
3) failed to explain the fine structure of spectral lines 4) failed to explain other atoms

67. According to Bohr hypothesis, discrete quantity is

1) Momentum 2) Angular velocity 3) Potential energy 4) Angular momentum

68. If E_n and are the magnitude of total energy and angular momentum of electron in the nth Bohr orbit respectively, then

1) $E_n \propto J_n^2$ 2) $E_n \propto \frac{1}{J_n^2}$ 3) $E_n \propto J_n$ 4) $E_n \propto \frac{1}{J_n}$

69. If the radius of first Bohr orbit is r, then radius of second orbit will be

1) 2r 2) $\frac{r}{2}$ 3) 4r 4) $\sqrt{2r}$

ATOMIC SPECTRA

70. According to quantum mechanics, one of the following is wrong about spin of electron

1) it is related to intrinsic angular momentum 2) spin is rotation of electron about its own axis
3) value of spin quantum number must not be 1
4) +1/2 value of spin quantum number represents up spin

71. According to classical theory, the path of electron in Rutherford atom nuclei is

1) straight line 2) spiral 3) circular 4) parabolic

72. One of the following radiations are not emitted by electron transition in atoms, choose the option

1) ultra violet 2) infrared radiations
3) visible rays 4) α -rays

73. **The energy emitted by a source is in the form of**
1)Photons 2)electrons 3)protons 4)neutrons
74. **The unit of Planck's constant is equivalent to that of**
1) energy 2) angular momentum 3) velocity 4) force
75. **Hydrogen atom will be in its ground state,if its electron is in**
1) any energy level 2) the lowest energy state
3) the highest energy state 4) the intermediate state
76. **The wavelengths involved in the spectrum of deuterium ($_1D^2$) are slightly different from that of hydrogen spectrum, because**
1) the size of the nuclei are different 2) the nucelar forces are different in two cases
3) the masses of the two nuclei are different
4) the attraction between electron and the nucelus is different in the two cases
77. **When an electron jumps from n_1 th orbit to n_2 th orbit then the formula for energy radiated out is**
1) $E_1 - E_2 = h\nu$ 2) $E_2 - E_1 = h\nu$ 3) $E_2 + E_1 = h\nu$ 4) $E_2 - 2E_1 = h\nu$
78. **Which of the following parameters are the same for all hydrogen -like atoms and ions in their ground states?**
1) radius of the orbit 2) speed of the electron
3) energy of the atom 4) orbital angular momentum of the electron
79. **The fine structure of hydrogen spectrum can be explained by**
1) the presence of neutrons in the nucleus 2) the finite size of nucleus.
3) the orbital angular momentum of electrons 4) the spin angular momentum of electrons
80. **The visible region of hydrogen spectrum was first studied by**
1) Lyman 2) Balmer 3) Pfund 4) Brackett
81. **With the increase in quantum number the energy difference between consecutive energy levels**
1) remains constant 2) decreases. 3) increases
4)sometimes increases sometimes decreases
82. **Hydrogen atom does not emit X-rays because:**
1) its energy levels are too close to each other 2) its energy levels are too far apart
3) it has a very small mass 4) it has a single electron
83. **An electron makes transition from n= 3 , n=1 state in a hydrogen atom. The maximum possible number of photons emitted will be**
1) 1 2) 2 3) 3 4) 6

ASSERTION & REASONING

Choose the correct answer :

1) A and R are true and R is the correct explanation of A
2) A and R are true and R is not the correct explanation of A
3) A is true and R is false. **4) A is false but R is true**

84. **Assertion (A) :** Rydberg's constant varies with mass number of a given element.
Reason (R): The reduced mass of the electron is dependent on the mass of the nucleus only

85. **Assertion (A) :** Study of discharge of electricity through gases at low pressures resulted in the discovery of cathode rays.

Reason (R): Cathode rays are deflected by both magnetic and electric fields and the direction of deflection shows that they are negatively charged.

86. **Assertion (A) :** If electrons travel undeflected by the electric field E and magnetic field B then the velocity of electrons is given by $V = \frac{E}{B}$.

Reason (R) : When both electric and magnetic fields are applied simultaneously on electron beam. If force due to electric field is equal and opposite to force due to magnetic field then they travel undeflected.

87. **Assertion (A) :** In the Thomsons e/m experiment of electrons, the specific charge of electrons is independent of nature of the gas in the discharge tube.

Reason (R): Charge of a body is quantized.

88. **Assertion (A):** Between any two given energy levels, the number of absorption transition is always less than number of emission transition.

Reason (R): Absorption transitions start from the lowest energy level only and may end at any higher energy level. But emission transitions may start from any higher energy level and end at any energy level below it.

MATCH THE FOLLOWING

89. **In each situation of Column I, a physical quantity related to orbiting electron in hydrogen-like atom is given. The terms 'Z' and 'n' given in Column II have usual meaning in Bohr's theory. Match the quantities in Column I with the terms they depend on it Column II**

Column-I	**Column-II**
a. Frequency of orbiting electron	p. is directly proportional to n
b. Angular momentum of electron	q. is inversely proportional to n
c. Magnetic moment of electron	r. is inversely proportional to n^3
d. The average current due to orbiting of electron	s. is independent of Z

1) a -p,r ; b -q,s; c -q,s; d -q,r 2) a -q,s ; b -p,r; c-q,s; d -q,r
3) a-q,r ; b-q,s; c-q,s; a-p,r 4) a-p,r; b-q,r; c-q,s; d-q,s

90. **Column-I**

a. Radius of orbit is related with atomic number(Z)
b. Current associated due to orbital motion of electron with atomic number(Z)
c. Magnetic field at the center due to orbital motion of electron related with Z
d. Velocity of an electron related with atomic number(Z)

Column-II

p. is proportional to Z q. is inversely proportion to Z
r. is proportional to Z^2 s. is proportional to Z^3

1) a-q b-r c-r d-p 2) a-q b-s c-r d-p 3) a-p b-r c-s d-q 4) a-p b-s c-s d-q

91. **The spectral lines of hydrogen-like atom fall within the wavelength range from $950A^0$ to $1350A^0$ Then, match the following.**

Column-I

a. If it is atomic hydrogen atom and energy $E - = 0.85\ eV$
b. If it is atomic hydrogen atom and energy atom and energy $E - = 3.4\ eV$
c. If it is double ionized lithium atom, then
d. If it is singly ionized helium atom, then

Column-II

p. λ =1212A^0 and it corresponds to transition from 2 to 1

q. λ =134A^0 and it corresponds to transition from 2 to 1

r. λ =303A^0 and it corresponds to transition from 2 to 1

s. λ =970A^0 and it corresponds to transition from 4 to 1

1) a-s b-p c-q d-r 2) a-s b-q c-p d-r 3)a-p b-q c-r d-s 4) a-q b-r c-s d-p

92. Excitation energy of hydrogen atom is 13.6 eV. Match the following.

Column-I	**Column-II**
a. Energy of second excited state of hydrogen	p. – 3.4 eV
b. Energy of fourth state	q – 13.6eV of He^+
c. Energy of first excited	r. – 1.5 eV state of Li^{+2}
d. Energy of third excited state of Be^{+3}	s.None

1) a-r b-p c-s d-q 2)a-r b-s c-p d-q 3)a-s b-p c-q d-r 4)a-q b-p c-s d-r

93. Match the column–I with column–II

Column – I	**Column – II**
(a) Radius of orbit depends on principal quantum number as	(p) Increase
(b) Due to orbital motion of electrons, magnetic field arises at the centre of nucleus is proportional to principal quantum number as	(q) Decrease
(c) If electron is going from lower energy level to higher energy level than velocity of electron will	(r) Is proportional to $\frac{1}{n^2}$
(d) If electron is going from lower energy level to higher energy level than total energy of electron will	(s) Is proportional to n^2
	(t) Is proportional to $\frac{1}{n^5}$

1) a-s b-t c-q d-p 2) a-t b-s c-p d-q 3)a-p b-q c-t d-s 4) a-p b-q c-s d-t

94. Match the column–I with column–II

A certain amount of He+ ions are subjected to excitation and subsequently the emission spectrum for the same is observed. Column – I shows four different series to which the spectral lines belong to and column – II shows five different energies of the photons emitted during the process

Column – I	**Column – II**
(a) Lyman series	(p) 1.33 eV
(b) Balmer series	(q) 12.09 eV
(c) Paschen series	(r) 51 eV
(d) pfund sereis	(s) 5.5 eV
	(t) 3.87 eV

1) a-r b-q c-s,t d-p 2) a-q b-r c-p,t d-s 3)a-r b-p c-q d-t,s 4) a-s b-q c-p d-t

CONCEPTUAL - KEY

1) 3	2) 1	3) 1	4) 2	5) 2	6) 1	7) 3	8) 3	9) 1	10) 1
11) 1	12) 2	13) 4	14) 4	15) 4	16) 2	17) 3	18) 3	19) 3	20) 4
21) 2	22) 3	23) 3	24) 1	25) 2	26) 3	27) 3	28) 2	29) 2	30) 1
31) 2	32) 2	33) 2	34) 4	35) 1	36) 2	37) 1	38) 3	39) 2	40) 1
41) 2	42) 4	43) 3	44) 3	45) 1	46) 2	47) 1	48) 4	49) 3	50) 1
51) 2	52) 4	53) 2	54) 4	55) 4	56) 1	57) 3	58) 1	59) 1	60) 2
61) 3	62) 4	63) 4	64) 1	65) 2	66) 1	67) 4	68) 2	69) 3	70) 2
71) 2	72) 4	73) 1	74) 2	75) 2	76) 3	77) 1	78) 4	79) 4	80) 2
81) 2	82) 1	83) 3	84) 4	85) 2	86) 1	87) 2	88) 1	89) 1	90) 1
91) 1	92) 1	93) 1	94) 1						

LEVEL - I -1

J.J.THOMSONS METHOD

1. **An electron passes undeflected through perpendicular electric and magnetic fields of intensity 3.4×10^3 V/m and 2×10^{-3} Wb/m^2 respectively. Then its velocity in m/s is**

   1) 1.7×10^6 2) 6.8×10^6 3) 6.8 4) 1.7×10^8 m/s

2. **The ratio of specific charges of an electron to that of a hydrogen ion is**

   1) 2:1 2) 1:1 3) 1: 1840 4) 1840:1

3. **An α - particle and a proton are subjected to the same electric field, then the ratio of the forces acting on them is**

   1) 2: 3 2) 1: 2 3) 3 :2 4) 2 :1

4. **An electron is accelerated in an electric field of 40 V cm^{-1}. If e/m of electron is 1.76×10^{11} Ckg^{-1}, the acceleration is**

   1) 14.0×10^{14}ms^{-2} 2) 14.0×10^{10}ms^{-2} 3) 7.0×10^{10}ms^{-2} 4) 7.04×10^{14}ms^{-2}

5. **An electron beam moving with a speed of 2.5×10^7 ms^{-1} enters into the magnetic field directed perpendicular to its direction of motion. The magnetic induction of the field is 4×10^{-3} wb/m^2. The intensity of the electric field applied so that the electron is undeflected due to the magnetic field is.**

   1) 10^4N/C 2) 10^5N/C 3) 10^7N/C 4) 10^3N/C

6. **A particle carrying a charge moves perpendicular to a uniform magnetic field of induction B with a momentum p, then the radius of the circular path is**

   1) Be/p 2) pe/B 3) p/Be 4) Bep

MILLIKAN OIL DROP EXPERIMENT

7. **A water drop of mass 3.2×10^{-18} kg and carrying a charge of $1.6\times10^{-19} C$ is suspended stationary between two plates of an electric field. Given g=10 m/s^2, the intensity of the electric field required is**

   1) 2 V/m 2) 200 V/m 3) 20 V/m 4) 2000 V/m

8. **In a Millikan's oil drop experiment, an oil drop of mass 0.64×10^{-14} kg, carrying a charge 1.6×10^{-19} C remains stationary between two plates seperated by a distance of 5 mm. Given g=9.8 m/s^2; the voltage that must be applied between the plates being**

   1) 980 V 2) 1960 V 3) 3920V 4) 2880V

ALPHA RAY SCATTERING

9. $\alpha-$particles are projected towards the nuclei of the different metals, with the same kinetic energy. The distance of closest approach is minimum for

1) Cu(Z=29) 2) Ag(Z=47) 3) Au(Z=79) 4) Pd(Z=46)

10. In Rutherford experments on $\alpha-$ray scattering the number of particles scattered at 90^0 be 28 per minute. Then the number of particles scattered per minute by the same foil but at 60^0 are

1) 56 2) 112 3) 60 4) 120

11. For a given impact parameter (b), if the energy increase then the scattering angle (θ) will

1) Decrease 2) increase 3) become zero 4) become

BOHR'S MODEL OF ATOM

12. Find the frequency of revolution of the electron in the first stationary orbit of H-atom

1) $6\times10^{14}Hz$ 2) $6.6\times10^{10}Hz$ 3) $6.6\times10^{-10}Hz$ 4) $6.6\times10^{15}Hz$

13. Let the potential energy of a hydrogen atom in the ground state be zero. Then its energy in the first excited state will be

1) 10.2eV 2) 13.6eV 3) 23.8eV 4) 27.2eV

14. According to bohr model, the diameter of first orbit of hydrogen atom will be

1) $1.A^0$ 2) $0.529A^0$ 3) $2.25A^0$ 4) $0.725A^0$

15. The angular momentum of electron is J.its magnetic moment will be

1) $\frac{mJ}{2e}$ 2) $\frac{eJ}{2m}$ 3) $\frac{2m}{eJ}$ 4) $\frac{emJ}{2}$

16. The radius of shortest orbit in one electron system is 18 pm.It may be.

1) ${}_1^1H$ 2) ${}_1^2H$ 3) He^+ 4) Li^+

17. In the Bohr model of a hydrogen atom, the centripetal force is furnished by the coulomb attraction between the proton and the electron. If a_0 is the radius of the ground state orbit, m is the mass, e is the charge on the elctron and ε_0 is the vacuum permittivity, the speed of the electron is: [CBSE 1998]

1) Zero 2) $\frac{e}{\sqrt{\varepsilon_0 a_0 m}}$ 3) $\frac{e}{\sqrt{4\pi\varepsilon_0 a_0 m}}$ 4) $\frac{\sqrt{4\pi\varepsilon_0 a_0 m}}{e}$

18. The energy necessary to remove the electron from n=10 state in hydrogen atom will be

1)1.36 eV 2)0.0136 eV 3)13.6 eV 4) 0.136 eV

19. The ratio of energies of first two excited states hydrogen atom is

1) 3/1 2) 1/4 3) 4/9 4) 9/4

ATOMIC SPECTRA

20. The number of different wavelengths may be observed in the spectrum from a hydrogen sample if the atoms are excited to third excited state is

1) 3 2)4 3)5 4)6

21. The ratio of the frequencies of the long wavelength limits of the Balmer and Lyman series of hydrogen is

1)27:5 2)5:27 3)4:1 4)1:4

22. When an electron jumps from higher orbit to the second orbit in hydrogen ,the radiation emitted out will be in $(R=1.09\times10^{7}m^{-1})$:

1)ultraviolet 2) visible region 3)infrared region 4)X-ray region

23. The energy required to seperate a hydrogen atom into a proton and an electron is 13.6eV. Then the velocity of electron in a hydrogen atom is

1) $2.2\times10^{4}m/s$ 2) $2.2\times10^{2}m/s$ 3) $2.2\times10^{6}m/s$ 4) $2.2\times10^{10}m/s$

LEVEL-I -1 - KEY

1) 1	2) 4	3) 4	4) 4	5) 2	6) 3	7) 2	8) 2	9) 1	10) 2
11) 1	12) 4	13) 3	14) 1	15) 2	16) 4	17) 3	18) 4	19) 4	20) 4
21) 1	22) 2	23) 3							

LEVEL-I -1 - HINTS

1. $V=\frac{E}{B}$

2. $\frac{specific\ charge\ of\ electron}{specific\ charge\ of\ hydrogen\ ion}=\frac{q_1}{q_2}\times\frac{m_2}{m_1}$

3. F = Ee

4. $a=\frac{Ee}{m}$

5. $V=\frac{E}{B}$

6. $P=mv, r=\frac{mv}{Be}$

7. mg = qE

8. $Eq=mg\Rightarrow\left(\frac{V}{d}\right)q=mg$

9. $r=\frac{1}{4\pi\varepsilon_0}\frac{q_1q_2}{(KE)}\Rightarrow r\alpha q_2$

10. No. of particles scattered at angle θ is $N\alpha\frac{1}{\sin^4\left(\frac{\theta}{2}\right)}$

12. $T=\frac{2\pi r}{V}$ where $V=\frac{3\times10^8}{137}$; $r=0.53\times10^{-10}m$

13. $P.E_2=-\frac{27.2}{4}+27.2=20.4$ eV

$K.E_1=13.6eV,\ K.E_2=3.4eV$

$\therefore T.e_2=20.4+3.4\quad=23.8eV$

14. Diameter $= 2r_0 = 2 \times 0.529 = 1.058\ \overset{0}{A}$

15. $\dfrac{\text{magnetic moment}}{\text{angular moment}} = \dfrac{e}{2m}$

16. $r_n = \dfrac{n^2 r_B}{Z} \Rightarrow \dfrac{0.53A^\circ}{Z} = 0.18A^\circ \Rightarrow Z = 3$

17. $\dfrac{1}{4\pi\varepsilon_0}\dfrac{e^2}{a_0^2} = \dfrac{mv^2}{a_0}$

18. Energy requried $= E_\alpha - E_{10} = \dfrac{13.6}{10^2} = 0.136eV$

19. $E_n = -13.6Z^2 / n^2$

20. N$=\dfrac{n(n-1)}{2}$: where n=4

21. $\lambda_B = \dfrac{36}{5R} : \lambda_2 = \dfrac{4}{3R}$, $\quad \dfrac{\lambda_B}{\lambda_c} = \dfrac{27}{5}$:

22. $For\ \lambda_{\max}$:
$n_1 = 2\ \& n_2 = 3;$

23. Velocity of revolving electron is given as

$\dfrac{mv^2}{r} = \dfrac{1}{4\pi\varepsilon_0}\dfrac{e^2}{r^2}$, $\quad \therefore v = \dfrac{e}{\sqrt{4\pi\varepsilon_0 mr}}$

LEVEL - I -2

J.J.THOMSONS METHOD

1. **A cathode emits 1.8 x 10^{17} electrons/s and all the electrons reach the anode when it is given a positive potential of 400 V. Given e=1.6 x 10^{-19}C, the maximum anode current is**
1) 2.88 mA 2) 28.8 mA 3) 7.2 mA 4) 6.4 mA

2. **An electron of mass 9 x 10^{-31} kg moves with a speed of 10^7 m/s. It acquires a K.E. of**
1) 562.50 eV 2) 1125 eV 3) 1250 eV 4) 281.25 eV

3. **Two electron beams having velocities in the ratio 1:2 are subjected to the same transverse magnetic field. The ratio of the deflections is**
1) 1:2 2) 2:1 3) 4:1 4) 1:4

4. **The velocity of electrons accelerated by potential difference of 1×10^4 V (The charge of the elctron is 1.6×10^{-19} C and mass is 9.11×10^{-31}kg) is**
1) 5.93×10^7 ms^{-1} 2) 2.94×10^7 ms^{-1} 3) 6.87×10^7 ms^{-1} 4) 3.98×10^7 ms^{-1}

5. **Cathode ray tube is operating at 5 KV. Then the K.E. acquired by the electrons is**
1) 5 eV 2) 5 MeV 3) 5 KeV 4) 5 V

6. **A steam of similar negatively charged particals enters an electrical field normal to the electric lines of force with a velocity of $3\times10^7 m/s$. The electric intensity is 1800 V/m. While through a distance of 100 m, the electron beam is deflected by 2mm. Then the specific charge value of in c/kg is**

1) 2×10^{10} 2) 2×10^{7} 3) 2×10^{11} 4) 2×10^{4}

MILLIKAN OIL DROP EXPERIMENT

7. **An oil drop having a mass 4.8×10^{-10}g. and charge 2.4×10^{-18}C stands still between two charged horizantal plates seperated by a distance of 1cm. If now the polarity of the plates is changed the instantaneous acceleration of the drop is (in ms^{-2}) ($g=10ms^{-2}$)**

1) 5 2)10 3) 20 4) 40

8. **In millkan's oil drop experiment a charged drop of mass 1.8 X 10^{-14}kg is stationary between the plates. The distance between the plates is 0.9cm and potential difference is 2000 V. The number of electrons in the drop is ($g = 10ms^{-2}$)**

1) 2 2) 4 3) 5 4) 1

ALPHA RAY SCATTERING

9. **An α – particle accelerated through V volt is fired towards a nucleus.Its distance of closest approach is r.If a proton is accelerated through the same potential and fired towards the same nucleus,the distance of closest approach of proton will be:**

1) r 2) 2r 3) r/2 4) r/4

10. **In α – ray scattering, the scattering angle (θ) for impact parameter (b) to become zero is**

1) 0° 2) 90° 3) 180° 4) 45°

11. **The impact parameter at which the scattering angle is 90°, $z = 79$ and initial energy 10 MeV is**

1) $1.137\times10^{-14}m$ 2) $1.137\times10^{-16}m$ 3) $2.24\times10^{-17}m$ 4) $2.24\times10^{-18}m$

BOHR'S MODEL OF ATOM

12. **When a hydrogen atom emits a photon of energy 12.1 eV,its orbital angular momentum changes by**

1) $1.05\times10^{-34}J-s$ 2) $2.11\times10^{-34}J-s$ 3) $3.16\times10^{-34}J-s$ 4) $4.22\times10^{-34}J-s$

13. **An electron is in an excited state in a hydrogen like atom. It has a total energy -3.4eV. The kinetic energy of the electron is E and its de broglie wavelength is λ.**

1) $E=6.8eV.\lambda=6.6\times10^{-10}m$ 2) $E=3.4eV,\lambda=6.6\times10^{-10}m$

3) $E=3.4eV,\lambda=6.6\times10^{-11}m$ 4) $E=6.8eV,\lambda=6.6\times10^{-11}m$

14. **The ionisation potential of hydrogen atom is**

1) $12.97V$ 2) $10.2V$ 3) $13.6V$ 4) $27.2V$

15. **The energy of electron in an excited hydrogen atom is -3.4 eV. Its angular momentum according to bohr's theory will be**

1) $\frac{h}{\pi}$ 2) $\frac{h}{2\pi}$ 3) $\frac{3h}{2\pi}$ 4) $\frac{3}{2\pi h}$

16. **The velocity of an electron in its fifth orbit,if the velocity of an electron in the second orbit of sodium atom (atomicnumber=11)is v ,will be:**

1) v 2) $\frac{22v}{5}$ 3) $\frac{5}{2}v$ 4) $\frac{2}{5}v$

17. **The ratio of the kinetic energy and the potential energy of electron in the hydrogen atom will be**

1)1:2 2) -1:2 3) 2:1 4) -2:1

18. **If the potential energy of a H-atom in the ground state be zero then its potential energy in the first excited state will be**

1) 10.2 ev 2)20.4 eV 3) 23.8 eV 4)27.2 eV

ATOMIC SPECTRA

19. **The value of wavelength radiation emitted due to transition of electrons from $n = 4$ to $n = 2$ state in hydrogen atom will be**

1) $\frac{5R}{36}$ 2) $\frac{16}{3R}$ 3) $\frac{36}{5R}$ 4) $\frac{3R}{16}$

20. **The maximum number of photons emitted by an H-atom,if atom is excited to states with principal quantum number four is:**

1)4 2) 6 3) 2 4) 1

21. **For certain atom,there are energy levels A,B,C corresponding to energy values $E_A < E_B < E_c$.**

Choose the correct option if $\lambda_1, \lambda_2, \lambda_3$ are the wavelength of radiations corresponding to the transition from C to B,B to A and C to A respectively:

1) $\lambda_3 = \lambda_1 + \lambda_2$ 2) $\lambda_3 = \frac{\lambda_1\lambda_2}{\lambda_1 + \lambda_2}$ 3) $\lambda_1 + \lambda_2 + \lambda_3 = 0$ 4) $3\lambda_2 = \lambda_3 + 2\lambda_2$

22. **If 13.6 eV is the energy required to saperate a hydrogen atom into a proton and an electron then its orbital radius is**

1) 5.3×10^{-11}m 2) 5.3×10^{-12}m 3) 7.6×10^{-13}m 4) 7.6×10^{-14}m

LEVEL - I -2- KEY

1) 2	2) 4	3) 2	4) 1	5) 3	6) 1	7) 3	8) 3	9) 1	10) 3
11) 1	12) 2	13) 2	14) 3	15) 1	16) 4	17) 2	18) 2	19) 2	20) 2
21) 2	22) 1								

LEVEL - I -2- HINTS

1. $i = \frac{ne}{t}$

2. K.E. $= \left(\frac{\frac{1}{2}mv^2}{e}\right)$ in ev where e is charge of electron

3. $\frac{y_1}{y_2} = \frac{v_2}{v_1}$

4. $v=\sqrt{\frac{2V_0 e}{m}}$ Where V_0 is the P.D

5. K.E. = Ve Where V is P.D

6. $y=\frac{1}{2}\frac{F}{m}t^2=\frac{1}{2}\left(\frac{Eq}{m}\right)\left(\frac{x}{v}\right)^2$

7. mg = Eq and F = (mg+Eq) $\Rightarrow$ ma = 2mg $\Rightarrow$ a=2g

8. $Eq = mg \Rightarrow \left(\frac{V}{d}\right)ne = mg$

9. KE =PE $\Rightarrow q_1 V = \frac{1}{4\pi\varepsilon_0}\frac{q_1 q_2}{r}$

Independent of charge of the particle projected

11. $b=\frac{ze^2\cot(\theta/2)}{4\pi\varepsilon_0\times E}$

12. $\Delta E=-13.6eV\left(\frac{1}{n_1^2}-\frac{1}{n_2^2}\right)$; $\Delta L=\frac{h}{2\pi}(n_1-n_2)$

13. $|T.E|=K.E;\lambda=\frac{h}{\sqrt{2mK.E}}$

14. $E_{ionisation}=E_\alpha - E_1$

15. $.E_n=-\frac{13.6}{n^2}=-3.4\Rightarrow n=2;L=\frac{nh}{2\pi}=h/\pi$

16. $\upsilon_n \alpha \frac{1}{n}$:

17. P.e =-2 K.E :

18. $P.E.=0=-27.2+27.2=0$

$P.E.=-\frac{27.2}{4}+27.2=20.4$

19. $\frac{1}{\lambda}=RZ^2\left(\frac{1}{n_1^2}-\frac{1}{n_2^2}\right)$:

20. For maximum number. $N=\frac{(n-1)}{2}n$

21. $\frac{1}{\lambda_3}=\frac{1}{\lambda_1}+\frac{1}{\lambda_2}$ $\Rightarrow \lambda_3=?$

22. Total energy of electron in hydrogan atom,

$E=-13.6eV=-13.6\times1.6\times10^{-19}J$

$=\frac{-e^2}{8\pi\varepsilon_0 r}$ $\therefore r=\frac{e^2}{8\pi\varepsilon_0 E}$

LEVEL - II -1

J.J.THOMSONS METHOD

1. **Two ions having masses in the ratio 1:1 and charges 1:2 are projected into uniform magnetic field perpendicular to the field with speeds in the ratio 2:3 .The ratio of the radii of circular paths along which the two particles move is**

1) 4:3 2) 2:3 3) 3:1 4) 1:4

2. **In Thomson's experiment, a magnetic field of induction 10^{-2} wb/m^2 is used. For an undeflected beam of cathode rays, a p.d. of 500 V is applied between the plates which are 0.5 cm apart. Then the velocity of the cathode ray beam is ...... m/s.**

1) 4×10^7 2) 2×10^7 3) 2×10^8 4) 10^7

3. **A cathode ray beam is bent into an arc of a circle of radius 0.02 m by a field of magnetic induction 4.55 milli tesla. The velocity of electrons is (given e = 1.6×10^{-19} c and m = 9.1×10^{-31} kg)**

1) 2×10^7m/s 2) 3×10^7m/s 3) 1.6×10^7 m/s 4) 3.2×10^7m/s

4. **When two electrons enter into a magnetic field with different velocities, they deflect in different circular parts, in such a way that the radius of one path is double that of the other. 1×10^7 ms^{-1} is the velocity of electron in smaller circle of radius 2×10^{-3} m. The velocity of electron in the other circular path is :**

1) 4×10^7 ms^{-1} 2) 4×10^6 ms^{-1} 3) 2×10^7 ms^{-1} 4) 2×10^6 ms^{-1}

MILLIKAN OIL DROP EXPERIMENT

5. **A charged oil drop falls with terminal velocity V_0 in the absence of electric field. An electric field E keeps it stationary. The drop acquires additional charge q and starts moving upwards with velocity V_0. The initial charge on the drop was**

1) 4q 2) 2q 3) q 4) q/2

6. **A charged oil drop is moving with a velocity V_1.As it acquires an additional charge it moves up with the velocity V_2 in the same electric filed. It fall freely with a Velocity 'V' in the absence of electric field . The ratio of the charges before and after acquiring additional charge is**

1) $\frac{V_1 + V}{V_2 - v_1}$ 2) $\frac{V_1 + V_2}{2V}$ 3) $\frac{V + V_1}{V + V_2}$ 4) $\frac{V_2 - 2V_1}{2V_1 + V}$

ALPHA SCATTERING

7. **A proton strikes another proton at rest with speed V_o . Assume impact parameter to be zero. Their closest distance of approach is (mass of proton is m)**

1) $\frac{e^2}{4\pi\varepsilon m\upsilon_o^{\,2}}$ 2) $\frac{e^2}{\pi\varepsilon_0 m\upsilon_o^{\,2}}$ 3) $\frac{e^2}{m\upsilon_0^{\,2}}$ 4) zero

8. **A closest distance of approach of an α particle travelling with a velocity v towards Al_{13} nucleus is d. The closest distance of approach of an alpha particle travelling with velocity 4V towards Fe_{26} nucleus is**

1) d/2 2) d/4 3) d/8 4) d/16

BOHR'S THEORY

9. **The energy required to excite an electron from n=2 to n=3 energy state is 47.2 e V. The charge number of the nucleus, around which the electron revolving will be**

1) 5 2) 10 3)15 4) 20

10. **The debroglie wave length of an electron in the first Bohr orbit is**
1) Equal to the circumference of the first orbit
2) 1/2 th circumference of the first orbit
3) 1/4 th circumference of the first orbit
4) 3/4th circumference of the first orbit

11. **If the radius of first Bohr's orbit is x, then de-Broglie wavelength of electron in 3rd orbit is nearly:**
1) $2\pi x$ 2) $6\pi x$ 3) $9x$ 4) $x/3$

12. **The angular momentum of the electron in third orbit of hydrogen atom, if the angular momentum in the second orbit of hydrogen atom is L is**
1) L 2) 3L 3) (3/2) L 4) 2/3 L

13. **The de-Broglie wave length of the electron in the second Bohr orbit is (given r_0=0.53A^0)**
1) $3.33A^o$ 2) $6.66A^o$ 3) $9.99A^o$ 4) $1.06A^o$

ATOMIC SPECTRA

14. **The maximum wavelength of Brackett series of hydrogen atom will be ______ A^0**
1) 35,890 2) 14,440 3) 62,160 4) 40,477

15. **When a hydrogen atom emits a photon in going from n = 5 to n = 1, its recoil speed is almost**
1) $10^{-4} m/s$ 2) $2\times10^{-2} m/s$ 3) $4m/s$ 4) $8\times10^{-2} m/s$

16. **An orbital electron in the ground state of hydrogen has the magnetic moment μ_1. This orbital electron is excited to 3rd excited state by some energy transfer to the hydrogen atom. The new magnetic moment of the electron is μ_2, then**
1) $\mu_1 = 2\mu_2$ 2) $2\mu_1 = \mu_2$ 3) $16\mu_1 = \mu_2$ 4) $4\mu_1 = \mu_2$

LEVEL-II-1-KEY

1) 1 2) 4 3) 3 4) 3 5) 3 6) 3 7) 2 8) 3 9) 1 10) 1
11) 2 12) 3 13) 2 14) 4 15) 3 16) 4

LEVEL-II-1-HINTS

1. $\frac{r_1}{r_2} = \frac{m_1}{m_2} \times \frac{v_1}{v_2} \times \frac{q_2}{q_1}$

2. $V = \frac{E}{B}$

3. $v = \frac{rBe}{m}$

4. $r = \frac{mv}{Bq}$

5. $6\pi \eta r V_0 = mg; Eq = mg$
$E(q^1 + q) = 6\pi \eta r V_0 + mg = 2mg = 2Eq$

6. $Eq_1 - mg = 6\pi\eta a V_1 \rightarrow 1$ $\quad mg = 6\pi\eta a V$

$Eq_2 - mg = 6\pi\eta a V_2 \rightarrow 2$

7. Use:LCLM and conservation of energy.

ie: $m\upsilon_0 = 2m\upsilon - (1)$

$$\frac{m\upsilon_o^2}{2} = 2\times\frac{1}{2}m\left(\frac{\upsilon_o}{2}\right)^2 = \frac{e^2}{4\pi\varepsilon r}$$

8. $\frac{1}{2}mv^2 = \frac{1}{4\pi\varepsilon_0}\times\frac{q_1 q_2}{r}$; $r\alpha\frac{z^2}{v^2}$

9. $E = E_3 - E_2 = \frac{13.6Z^2}{4} - \frac{13.6Z^2}{9} = 47.2$

$\Rightarrow Z = ?$

10. $mvr = \frac{nh}{2\pi}$; $2\pi = \frac{h}{mv}$; $2\pi r = \lambda$

11. $mvr = \frac{nh}{2\pi}\left(r = \frac{r_0 n^2}{Z}\right) r_0 = x$

$2\pi r = n\lambda \Rightarrow \lambda = \frac{2\pi r}{n} = 6\pi x$

12. $L = \frac{nh}{2\pi}$

13. $2\pi r = n\lambda$;

$$\therefore \lambda = \frac{2\pi r}{n} = \frac{2\pi r_0 n^2}{n} = 2\pi r_0 n$$

14. $RZ^2\left(\frac{1}{n_1^2} - \frac{1}{n_2^2}\right) = \frac{1}{\lambda}$ *Where* $n_1 = 4; n_2 = 5$

15. From law of conservation of momentum

$\therefore m\upsilon = p = \frac{h}{\lambda} \Rightarrow \upsilon = \frac{h}{m\lambda}$

m $\rightarrow$ mass of proton :

$\lambda \rightarrow$ wave length of emitted photon

16. $\mu = niA = iA$, $= \frac{qv}{2\pi r}\times\pi r^2$

$M\alpha vr$ where $v\alpha\frac{z}{n}$ & $r\alpha\frac{n^2}{z}$

$\therefore \mu\alpha n$

LEVEL - II -2

J.J.THOMSONS METHOD

1. **A proton and an α-particle enter a magnetic field in a direction perpendicular to it. If the force acting on the proton is twice that acting on the α-particle, the ratio of their velocities is**
   1) 4 : 1 2) 1. : 4 3) 1 : 2 4) 2 : 1
2. **A proton, a deuteron and an α -particle are accelerated through the same p.d. of V volt. The velocities acquired by them are in the ratio**
   1) $1 : 1 : \sqrt{2}$ 2) $1 : \sqrt{2} : 1$ 3) 1 : 1 : 1 4) $\sqrt{2} : 1 : 1$
3. **An electron starts from rest and travels**
   0.9 m in an electric field of 200 V/m. After this, it enters a magnetic field at right angles to its direction of motion. If the radius of circular path of the electron is 9 cm, the magnetic field induction is (Given e=1.6 x 10^{-19}C, m=9x10^{-31}kg)
   1) 5 x 10^{-4} wb/m^2 2) 5 x 10^{-5} wb/m^2 3) 5 x 10^{-3} wb/m^2 4) 5 x 10^{-2} wb/m^2

MILLIKAN OIL DROP EXPERIMENT

4. **In the Millikan's experiment, the oil drop is subjected to a horizontal electric field of 2 N/C and the drop moves with a constant velocity making an angle of 45^0 with the horizontal. If the weight of the drop is W, then the electric charge, in coulomb, on the drop is**
   1) W 2) W/2 3) W/4 4) W/8

ALPHA SCATTERING

5. **An α -Particle of energy 5 Mev is scattered through 180^0 by a fixed uranium nuclens.The closest distance is in the order of:**
   1) $1A^o$ 2) $10^{-10}cm$ 3) $10^{-12}cm$ 4) $10^{-16}cm$
6. **An alpha nucleus of energy $\frac{1}{2}mv^2$ bombards a heavy nuclear target of charge Ze. Then the distance of closest approach for the alpha nucleus will be proportional to (AIEEE-2006)**
   1) $\frac{1}{v}$ 2) $\frac{1}{ze}$ 3) v^2 4) $\frac{1}{m}$

BOHR'S THEORY

7. **In the Bohr model of hydrogen atom, the ratio of the kinetic energy and total energy of electron in the nth quantum state will be**
   1) 1 2) -1 3) 2 4) -12
8. **The number of revolutions done by an electron 'e' in one second in the first orbit of hydrogen atom is**
   1) 6.57x10^{15} 2) 6.57 x 10^{13} 3) 1000 4) 6.57 x 10^{14}
9. **If $\left(\frac{0.51\times10^{-10}}{4}\right)$ m is the radius of smallest electron orbit in hydrogen like atom ,then this atom is :**
   1) hydrogen atom 2) He^+ 3) Li^{2+} 4) Be^{3+}
10. **In Bohr's orbit of hydrogen atom m kg is mass of an electron and e coulomb is the charge on it .The ratio (in SI units) of magnetic dipole moment to that of the angular momentum of electron is:**
    1) e/2m 2) e/m 3) 2e/m 4)2e/3m

11. In a sample of hydrogen like atoms all of which are in ground state, a photon beam contain photons of various energies is passed. In absorption spectrum, five dark line are observed. The number of bright lines in the emission spectrum will be(assume that all transitions take place)

1) 21 2) 10 3) 15 4) 12

ATOMIC SPECTRA

12. Let f_1 be the frequency of the series limit of the lyman series, f_2 be the frequency of the first line of the lyman series,and f_3 be the frequency of the series limit of the balmer series .

1) $f_1 - f_2 = f_3$ 2) $f_2 - f_1 = f_3$ 3) $f_3 = \frac{1}{2}(f_1 + f_2)$ 4) $f_1 + f_2 = f_3$

13. Ratio of difference of spacing between the energy levels with n=3 and n=4 and the spacing between the energy levels with n=8 and n=9 for a hydrogen like atom or ion is

1) 0.71 2) 0.41 3) 2.43 4) 14.82

14. A stationary hydrogen atom emits photon corresponding to the first line of Lyman series. If R is the Rydberg's constant and M is the mass of the atom, then the velocity acquired by the atom is (neglect energy absorbed by the photon)

1) $\frac{3Rh}{4M}$ 2) $\frac{4M}{3Rh}$ 3) $\frac{Rh}{4M}$ 4) $\frac{4M}{Rh}$

LEVEL-II -2- KEY

1) 1 2) 4 3) 1 4) 2 5) 3 6) 4 7) 2 8) 1 9) 4 10) 1
11) 3 12) 1 13) 2 14) 1

LEVEL-II -2- HINTS

1. $F = Bqv$

2. $v = \sqrt{\frac{2eV}{m}} \Rightarrow v \propto \sqrt{\frac{e}{m}}$

3. $V^2 - U^2 = 2as$; $V^2 = 2\frac{Eq}{m}S \Rightarrow V = \sqrt{2\frac{Eq}{m}S}$

$r = \frac{mV}{Bq} \Rightarrow B = \frac{mV}{rq}$

4. Tan $45^0 = \frac{Eq}{w}$ Eq = w

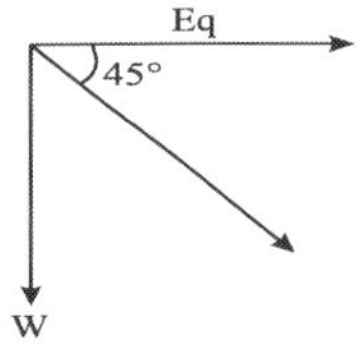

5. $\frac{1}{2}m\upsilon_o^2 = \frac{q_1 q_2}{4\pi\varepsilon_o r}$

6. $\frac{1}{2}mv^2 = \frac{1}{4\pi\varepsilon_0}\frac{(2e)(ze)}{r}$

7. $K.e = +\frac{13.6Z^2}{n^2}$ $\therefore K.e = |T.e|$ &

$T.e = -13.6z^2/n^2$ $T.e = -K.e$

8. $\upsilon_1(t) = n.2\pi r_1 \Rightarrow n = ?$

9. $r_n = \frac{n^2}{Z} r_o$:

10. Magnetic dipole moment $= \frac{er^2w}{2}$

Angular momentum $= mr^2w$

11. $\frac{n(n-1)}{2}$ Here $n = 6$

12.

f_1 relates $\Rightarrow n_1 = 1; n_2 = \alpha$

f_2 relates $\Rightarrow n_1 = 1; n_2 = 2$

f_3 relates $\Rightarrow n_1 = 2 : n_2 = \alpha$

13. $\frac{r_4 - r_3}{r_9 - r_8} = \frac{4^2 - 3^2}{9^2 - 8^2} = 0.41$

14. $p = -mv$; $\frac{h}{\lambda} = mv \Rightarrow v = \frac{h}{m}\left(\frac{1}{\lambda}\right)$ & $\frac{1}{\lambda} = \frac{3R}{4}$

LEVEL - III

1. Magnetic moment due to the motion of the electron in nth energy state of hydrogen atom is proportional to

1) n 2) n^0 3) n^5 4) n^3

2. The ratio between total accelertion of the electron in singly ionized helium atom and hydrogen atom (both in ground state) is

1) 1 2) 8 3) 4 4) 16

3. The shortest wavelength of the Brackett series of a hydrogen like atom (atomic number =Z) is the same as the shortest wavelength of the Balmar series of hydrogen atom. The value of Z is

1) 2 2) 3 3) 4 4) 6

4. According to Bohr's theory of hydrogen atoms, the product of the binding energy of the electron in the nth orbit and its radius in the n the orbit

1) is proportional to n^2 2) is inversely proportional to n^3

3) has a constant value of 10.2eV-A^0 4) has constant value 7.2 eV-A^0

5. If an electron drops from 4 th orbit to 2nd orbit in an H-atom,then

1) it gains 2.55 eV of potential energy 2) it gains 2.55 eV of total energy

3) it emits a 2.55 eV electron 4) it emits a 2.55 eV photon

6. The energy of an atom or ion in the ground state is -54.4 eV .it may be:

1) He^+ 2) Li^{2+} 3) hydrogen 4)deuterium

7. **An atom absorbs 2 eV energy and is excited to next energy state. The wavelength of light absorbed will be**

1) 2000 A° 2) 4000 A°. 3) 8000 A° 4) 6206 A°

8. **When an electron in the hydrogen atom in ground state absorbs a photon of energy 12.1 eV, its angular momenturm**

1) decreases by $2.11\times10^{-34} J-s$ 2) decreases by $1.055\times10^{-34} J-s$

3) increases by $2.11\times10^{-34} J-s$ 4) increases by $1.055\times10^{-34} J-s$

9. **Magnetic field at the centre (at nucleus) of the hydrogen like atoms (atomic number =Z) due to the motion of electron in n^{th} orbit is proportional to**

1) $\frac{n^3}{Z^5}$ 2) $\frac{n^4}{Z}$ 3) $\frac{Z^2}{n^3}$ 4) $\frac{Z^3}{n^5}$

10. **A neutron moving with a speed v makes a head on collision with a hydrogen atom in ground state kept at rest. The minimum kinetic energy of neutron for which inelastic collision will take place is (assume that mass of proton is nearly equal to the mass of neutron)**

1) 10.2 eV 2) 20.4eV 3) 12.1eV 4) 16.8ev

11. **A charged particle is moving in a uniform magnetic field in a circular path. The energy of the particle is doubled. If the initial radius of the circular path was R, the radius of the new circular path after the energy is doubled will be**

1) R/2 2) $\sqrt{2}R$ 3) 2R 4) R/$\sqrt{2}$

12. **An electron in hydrogen atom after absorbing an energy photon jumps from energy state n_1 to n_2. Then it returns to ground state after emitting six different wavelengths in emission spectrum. The energy of emitted photons is either equal to, less than or greater than the absorbed photons. Then n_1 and n_2 are**

1) $n_2=4, n_1=3$ 2) $n_2=5, n_2=3$ 3) $n_2=4, n_1=2$ 4) $n_2=4, n_1=1$

13. **The photon radiated from hydrogen corresponding to 2nd line of Lyman series is absorbed by a hydrogen like atom 'X' in 2nd excited stae. As a result then , the hydrogen-like atom 'X' makes a transition of nth orbit.**

1) X=He^+, n=4 2) X=Li^{++},n=6 3) X=He^+, n=6 4) X=Li^{++},n=9

14. **In a hypothetical system, a particle of mass m and charge -3q is moving around a very heavy particle charge q. Assume that Bohr's model is applicable to this system.then velocity of mass m in first orbit is**

1) $\frac{3q^2}{2\epsilon_0 h}$ 2) $\frac{3q^2}{4\epsilon_0 h}$ 3) $\frac{3q}{2\pi\epsilon_0 h}$ 4) $\frac{3q}{4\pi\varepsilon_0 h}$

15. **Consider a hydrogen-like atom whose energy in n^{th} excited state is given by $E_n = \frac{13.6Z^2}{n^2}$**

When this excited atom makes a transition from excited state to ground state, most energetic photons have energy E_{max} = 52.224 eV and least energetic photons have energy E_{min} = 1.224eV. Find the atomic number of atom and the state of excitation.

1) Z=2; n=5 2) Z=2, n=4 3) Z=3, n=6 4) Z=4, n =6

16. **29 electrons are removed from Zn-atom (Z=30) by certain means .The minimum energy needed to remove the 30th electron ,will be:**

1)12.24 keV 2)408 keV 3)0.45 keV 4)765keV

17. **The ionisation energy of Li^{2+} atom in ground state is :**

1) 13.6 x 9 eV 2) 13.6J 3) 13.6 erg 4)13.6x10^{-19} J

18. **A photon of energy 15 eV collides with H-atom. Due to this collision, H-atom gets ionized .The maximum kinetic energy of emitted electron is :**

1)1.4 eV 2) 5 eV 3)15eV 4) 13.6eV

19. **Monochromatic radiation of wavelength λ is incident on a hydrogen sample in ground state.Hydrogen atoms absorb a fraction of light and subsequently emit radiation of six different wavelengths. Find the value of λ.**

1) 80 nm 2) 97.5 nm 3) 105 nm 4) 60 nm

20. **The figure shows energy levels of a certain atom, when the system moves from level 2E to E, a photon of wavelength λ is emitted. The wavelength of photon produced during its transition from level 4/3 E to E level is:**

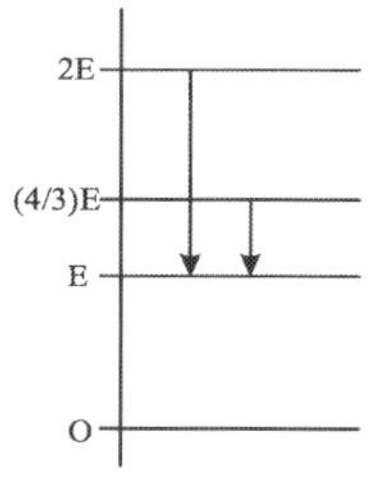

1) 3λ 2)3/4λ 3) λ/4 4)2λ

21. **When the electron in hydrogen atom jumps from the second orbit to the first orbit, the wavelength of the radiation emitted is λ. When the electron jumps from the third to the first orbit, the wavelength of the radiation emitted is:**

1) $\frac{9}{4}\lambda$ 2) $\frac{4}{9}\lambda$ 3) $\frac{27}{32}\lambda$ 4) $\frac{32}{27}\lambda$

22. **The ratio of the largest to shortest wavelengths in Balmer series of hydrogen spectra is:**

1) $\frac{25}{9}$ 2) $\frac{17}{6}$ 3) $\frac{9}{5}$ 4) $\frac{5}{4}$

23. **An electron in a hydrogen atom makes a transition $n_1 \rightarrow n_2$, where n_1 and n_2 are principal quantum numbers of the states. Assume the Bohr's model to be valid. The time period of the elctron in the initial states is eight times to that of final state. What is ratio of $\frac{n_1}{n_2}$**

1) 8:1 2) 4:1 3) 2:1 4) 1:2

24. **Any radiation in the ultra violet region of Hydrogen spectrum is able to eject photoelectrons from a metal. Then the maximum value of threshold frequency of the metal is, nearly**

1) 3.3x10^{15} HZ 2) 2.5x10^{15} HZ 3) 4.6 x 10^{14} HZ 4) 8.2x10^{14} HZ

25. **A hydrogen atom emits a photon corresponding to an electron transition from n=5 to n=1. The recoil speed of hydrogen atom in nearly equal to**

1) 10^{-4} m/s 2) 2x10^{-2} m/s 3) 4 m/ 4) 8x10^{-2} m/s

26. **The wave number of energy emitted when electron jumps from fourth orbit to second orbit in hydrogen in 20,497 cm^{-1}. The wave number of energy for the same transition in He^+ is**

1) 5,099 cm^{-1} 2) 20,497 cm^{-1} 3) 40,994 cm^{-1} 4) 81,988 cm^{-1}

27. **In a Bohr atom the electron is replaced by a particle of mass 150 times the mass of the electron and the same charge. If a_0 is the radius of the first Bohr orbit of the orbital atom, then that of the new atom will be**

1) 150 a_0 2) $\sqrt{150}\, a_0$ 3) $\frac{a_0}{\sqrt{150}}$ 4) $\frac{a_0}{150}$

28. **If the wavelength of first member of Balmer series of hydrogen spectrum is 6564 A^0, the wavelength of second member of Balmer series will be:**

1) 1215 A^0 2) 4848 A^0 3) 6050 A^0

4) data given is insufficient to calculate the value

29. **A hydrogen like atom (atomic number Z) is in a higher excited state of quantam number 'n'. This excited atom can make a transition to the first excited state by emitting a photon of energy 27.2eV. Alternatively the atom from the same excited state can make a transition to second excited state by emitting a photon of energy 10.20 eV. The value of n and z are given (Ionization energy of hydrogen atom is 13.6eV)**

1) n = 6 and z =3   2) n = 3 and z = 6   3) n = 8 and z = 4   4) n = 4 and z = 8

30. **Photons from n=2 to n=1 in Hydrogen atom is made to fall on a metal surface with work function 1.2ev. The maximum velocity of photo electrons emitted is nearly equal to**

1) $6x10^5$ m/s   2) $3x10^5$ m/s   3) $2x10^5$ m/s   4) $18x10^5$ m/s

31. **Let ϑ_1 be the frequency of the series limit of the Lyman series and ϑ_2 be the frequency of the first line of the Lyman series and ϑ_3 be the frequency of the series limit of Balmar series, then**

1) $\vartheta_1-\vartheta_2=\vartheta_3$ 2) $\vartheta_2-\vartheta_1=\vartheta_3$ 3) $2\vartheta_3=\vartheta_1+\vartheta_2$ 4) $\vartheta_1+\vartheta_2=\vartheta_3$

32. **a)Find the wavelength of the radiation required to excite the electron in Li from the first to the third Bohr orbit,**

b) How many spectral lines are observed in the emission spectrum of the above excited system?

1) 108.8 eV, 3   2) 13.6 eV, 4   3) 54.4 eV, 2   4) 10.2 eV, 3

33. **Find the wavelengths in a hydrogen spectrum between the range 500nm to 700nm.**

1) 540 nm   2) 580 nm   3) 654 nm   4) 696 nm

34. **The largest wavelength in the ultraviolet region of the hydrogen spectrum is 122 nm. The smallest wavelength in the infrared region of the hydrogen spectrum (to the nearest integer) is**

1) 802 nm   2) 823 nm   3) 1882 nm   4)1648 nm

35. **The electric potential between a proton and an electron is given by $V = V_0 ln\frac{r}{r_0}$, where r_0 is a constant. Assuming Bohr's model to be applicable, write variation of r_n with *n*, *n* being the principal quantum number**

1) $r_n \propto n$ 2) $r_n \propto 1/n$ 3) $r_n \propto n^2$ 4) $r_n \propto 1/n^2$

36. **If elements of quantum number greater than *n* were not allowed, the number of possible elements in nature would be ?**

1) $\frac{1}{2}n(n+1)$ 2) $\left\{\frac{n(n+1)}{2}\right\}^2$ 3) $\frac{1}{2}n(n+1)(2n+1)$ 4) $\frac{1}{3}n(n+1)(2n+1)$

37. **Magnetic field at the centre (at nucleus) of the hydrogen like atoms (atomic number $= z$) due to the motion of electron in n^{th} orbit is praportional to**

1) $\frac{n^3}{z^5}$ 2) $\frac{n^4}{z}$ 3) $\frac{z^2}{n^3}$ 4) $\frac{z^3}{n^5}$

38. **The recoil speed of a hydrogen atom after it emits a photon in going from $n = 5$, state to $n = 1$ state is (in ms^{-1})**

1) 4.718 2) 7.418 3) 4.178 4) 7.148

39. **The binding energy of an electron in the ground state of He-atom is $E_0 = 24.7eV$. The energy required to remove both the electrons from the atom is**

1) 24.6eV 2) 79.0eV 3) 54.4eV 4) none of these

40. **In hydrogen atom, the radius of nth Bohr orbit is V_n. The graph between $\log\left(\frac{r_n}{r_1}\right)$ and $\log n$ will be**

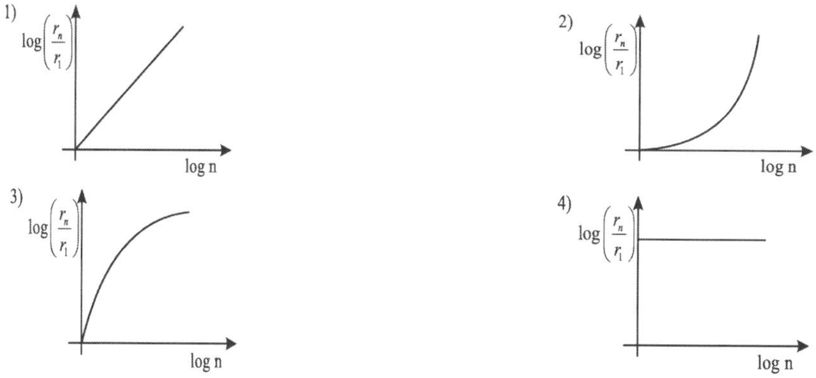

41. **In hydrogen atom, the area enclosed by n^{th} orbit is A_n. The graph between $\log\left(\frac{A_n}{A_1}\right)$ and $\log n$ will be**

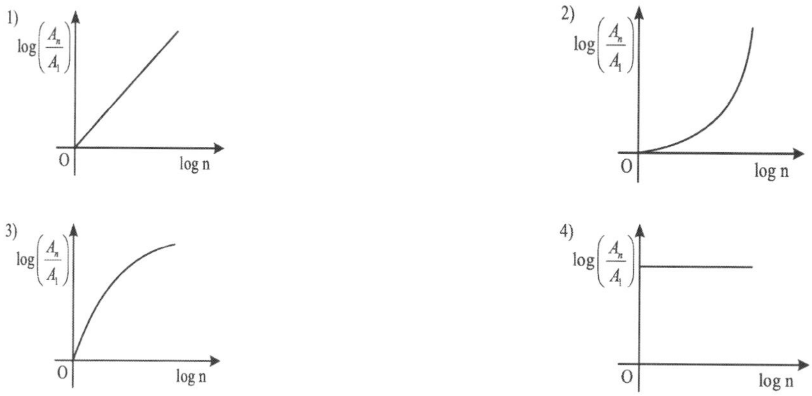

42. **An electron in the ground state of hydrogen atom is revolving in anticlockwise direction in the cicular orbit of radius R. The atom is placed in a uniform magnetic induction B such that the plane normal of electron orbit makes an angle 30° with B, as shown in figure. The torque experienced by electron willbe**

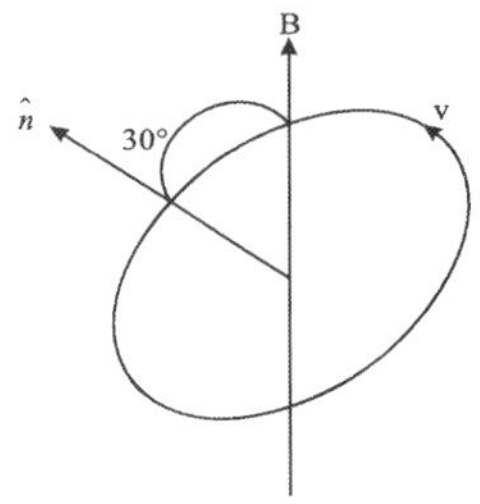

1) $\frac{ehB}{8\pi m}$ 2) $\frac{eh}{8\pi Bm}$ 3) $\frac{eB}{8\pi mh}$ 4) $\frac{hB}{8\pi em}$

43. **If we assume only gravitational attraction between proton and electron in hydrogen atom and the Bohr's quantization rule to be followed, then the expression for the ground state energy of the atom will be (the mass of proton is M and that of electron is m)**

1) $\frac{G^2M^2m^2}{h^2}$ 2) $\frac{2\pi^2G^2M^2m^3}{h^2}$ 3) $\frac{2\pi^2GM^2m^3}{h^2}$ 4) $\frac{h^2}{G^2M^2n^2}$

LEVEL - III - KEY

1) 1	2) 2	3) 1	4) 4	5) 4	6) 1	7) 4	8) 3	9) 4	10) 2
11) 2	12) 3	13) 4	14) 1	15) 1	16) 1	17) 1	18) 1	19) 2	20) 1
21) 3	22) 3	23) 3	24) 1	25) 3	26) 4	27) 4	28) 2	29) 1	30) 4
31) 1	32) 1	33) 3	34) 2	35) 1	36) 4	37) 4	38) 3	39) 2	40) 1
41) 1	42) 1	43) 2							

LEVEL - III - HINTS

1. $\frac{Magnetic\ Moment}{Angular\ Momentum} = \frac{e}{2m}$

$\therefore$ Magnetic Moment $\propto$ angular momentum

$M \propto n \left(L = \frac{nh}{2\pi} \right)$

2. $a = \frac{v^2}{r}$ $\therefore a \propto \frac{(z)^2}{(1/z)}$ or $\therefore a \propto z^3$ $\therefore \frac{a_1}{a_2} = \left(\frac{2}{1}\right)^3 = 8$

3. $(Z^2)\left(\frac{13.6}{16}\right) = \left(\frac{13.6}{4}\right)$; $Z^2 = 4$ or Z=2

4. $E_n \propto \frac{1}{n^2}$ and $r_n \propto n^2$; $\therefore E_n r_n$ is independent of n

Hence, $E_1 r_1 = (13.6eV)(0.53\overset{0}{A}) = 7.2eV - \overset{0}{A}$ =constant

5. $E_n = -\frac{13.6Z^2}{n^2} eV; \Delta E = E_4 - E_2$

6. $E = \frac{-13.6z^2}{n^2}$

7. $E = \frac{hc}{\lambda} \Rightarrow \lambda = ?$

8. After absorbing a photon of energy 12.1 eV electron jumps from ground state (n=1) to second excited state (n=3). Therefore change in angular momentum $\Delta L = L_3 - L_1$ $= 3\left(\frac{h}{2\pi}\right) - \frac{h}{2\pi} = \frac{h}{\pi}$

$= \frac{6.6 \times 10^{-34}}{3.14} J - s = 2.11 \times 10^{-34} J - s$

9. $B_n = \frac{\mu_0 I_n}{2r_n}$ (or) $B_n \propto \frac{I_n}{r_n} \propto \frac{(f_n)}{r_n}$

$B_n \propto \frac{(v_n / r_n)}{r_n} \propto \frac{v_n}{(r_n)^2}$; $\propto \frac{(Z/n)}{(n^2/Z)^2} \propto \frac{Z^3}{n^5}$

10. Let v=speed of neutron before collision
v_1= speed of neutron after collision
v_2= speed of proton or hydrogen atom after collision and ΔE=energy of exitation.
From conservation of linear momentum

$mv = mv_1 + mv_2$ ....(1)

From conservation of energy

$\frac{1}{2}mv^2 = \frac{1}{2}mv_1^2 + \frac{1}{2}mv_2^2 + \Delta E$ ....(2)

from 1 and 2 As $v_1 - v_2$ must be real

$v^2 - 4\frac{\Delta E}{m} \geq 0$

$\Rightarrow \frac{1}{2}mv^2 = 2\Delta E = 2 \times 10.2 = 20.4 eV$

11. $r \propto \sqrt{KE} \Rightarrow \frac{r_2}{r_1} = \sqrt{\frac{KE_2}{KE_1}}$

12. From $n_2 = 4$, six lines are obtained in emission spectrum. Now: $E_{4-2} = E_{absorbed}$

$E_{4\to3} < E_{absorbed}$ and $E_{4-1} E_{3-1}, E_{2-1} > E_{absorbed}$

Hence, $n_1 = 2$ and $n_2 = 4$

13. Energy of nth state in hydrogen is same as energy of 3n[th] state in Li^{++}.

$\therefore$ $3 \to 1$ transition in H would give same energy as the $9 \to 3$ transition in Li^{++}.

14. $\frac{mv^2}{r} = \frac{1}{4\pi \in_o}\frac{3q^2}{r^2}$; $mvr = \frac{nh}{2\pi}$

using above equations, putting n = 1; $v = \frac{3q^2}{2\in_o h}$

15. Maximum energy is liberated for transition $E_n \rightarrow 1$ and minimum energy for $E_n \rightarrow E_{n-1}$

Hence $\Delta E = (Rhc)\, Z^2\left[\frac{1}{n_1^2} - \frac{1}{n_2^2}\right]$

16. After removing 29 electrons, Zn atom will become Zn^{+29} $\therefore$ E= 13.6 Z^2 eV=12,240 eV =12.24KeV

17. (E)= Li^{+2} ; $E = Z^2(E)_H ; E = 9\times 13.6\,eV$

18. Energy of photon ionization energy + $K.E_{max}$ =15ev - 13.6ev=1.4ev

19. The atom is excited to n = 4 state

$$\frac{hc}{\lambda} = E_4 - E_1 = 12.75\,eV \Rightarrow \lambda = 97.5\,nm$$

20. $\frac{hc}{\lambda_1} = 2E - E = E \rightarrow 1$; $\frac{hc}{\lambda_2} = \frac{4}{3}E - E = \frac{E}{3} \rightarrow 2$

$$\frac{\lambda_2}{\lambda_1} = \frac{E}{\frac{E}{3}} = \frac{3}{1}$$

21. $\frac{1}{\lambda} = RZ^2\left(\frac{1}{n_1^2} - \frac{1}{n_2^2}\right)$

22. $\frac{1}{\lambda_{max}} = RZ^2\left(\frac{1}{n_1^2} - \frac{1}{n_2^2}\right)$ *Where* $n_1 = 2; n_2 = 3$

$\frac{1}{\lambda_{min}} = RZ^2\left(\frac{1}{n_1^2} - \frac{1}{n_2^2}\right)$ *Where* $n_1 = 2; n_2 = \alpha$

23. Time period of revolution of electron in the nth orbit is $T \,\alpha\, n^3$

24. $\therefore \upsilon = \frac{c}{\lambda}$

25. $v = \frac{h}{m\lambda}$; $\frac{1}{\lambda} = R\left[1 - \frac{1}{25}\right]$

26. $\bar{v} = RZ^2\left(\frac{1}{n_1^2} - \frac{1}{n_2^2}\right)$; $\bar{v}\alpha Z^2$

27. $r_n \alpha \frac{1}{m}$

28. $\frac{1}{\lambda} = RZ^2\left(\frac{1}{n_1^2} - \frac{1}{n_2^2}\right)$ for balmer series

n_1=2 and n_2= 3,4,5....

29. $\Delta E_1 = 27.2 = (13.6)\left[\frac{1}{4} - \frac{1}{n^2}\right]z^2$

$\Delta E_2 = 10.2 = (13.6)\left[\frac{1}{9} - \frac{1}{n^2}\right]z^2$

30. E= E_2-E_1; $E_n = -\frac{13.6}{n^2}; \frac{1}{2}mv^2 = E - W$

31.

$\frac{1}{\lambda} = R\left(\frac{1}{1^2} - \frac{1}{\alpha^2}\right)$ series limit for lyman series

$\frac{1}{\lambda} = R\left(\frac{1}{1^2} - \frac{1}{2^2}\right)$ series for lyman first line

$\frac{1}{\lambda} = R\left(\frac{1}{2^2} - \frac{1}{\alpha^2}\right)$ series limit for balmer series

32. (a) The energy in the first orbit=E_1=Z^2E_0 where E_0=-13.6eV is the energy of a hydrogen atom in ground state. Thus for Li^{++},

$E_1 = 9E_0 = 9\times(-13.6eV)$

The energy in the third orbit is

$E_3 = \frac{E_1}{n^2} = \frac{E_1}{9} = -13.6eV$

Thus, $E_3 - E_1 = 108.8\ eV$; $\frac{hc}{\lambda} = E_a - E_1$

33. The energy of a photon of wavelength 500 nm is

$\frac{hc}{\lambda} = \frac{1242eV - nm}{500nm} = 2.48eV$

The energy of a photon of wavelength 700nm is

$\frac{hc}{\lambda} = \frac{1242eV - nm}{700nm} = 1.77eV$

The energy difference between the states involved in the transition should, therefore, be between 1.77 eV and 2.44 eV. $n = 3 \rightarrow n = 2$ falls in this range $\lambda = \frac{hc}{\Delta E} = \frac{1242eV - nm}{1.9eV} = 654nm$

34. $\frac{1}{\lambda} = RZ^2\left(\frac{1}{n_1^2} - \frac{1}{n_2^2}\right)$

35. $U = eV = eV_0 \ln\left(\frac{r}{r_0}\right)$

$$F = \frac{-dU}{dr} \Rightarrow F = -\frac{eV_0 r_0}{r} \Rightarrow F = -\frac{K}{r}$$

$$\frac{mv^2}{r_n} = \frac{K}{r_n} \quad \Rightarrow mv = \sqrt{mK}\ ;\ mvr_n = \frac{nh}{2\pi}$$

$$\sqrt{mK}\, r_n = \frac{nh}{2\pi} \quad \Rightarrow r_n \propto n$$

36. For each principal quantum number n, number of electrons permitted equals the number of elements corresponding to the quantum number.

$$\begin{pmatrix} Total\ number \\ of\ electrons \end{pmatrix} = \Sigma 2n^2 = \frac{n(n+1)(2n+1)}{3}$$

37. $B_n = \frac{\mu_0 I_n}{2r_n}$ or $B_n \propto \frac{I_n}{r_n}; \propto \frac{(f_n)}{r_n}$

$$\therefore\ B_n \propto \frac{(v_n / r_n)}{r_n} \propto \frac{v_n}{(r_n)^2} \propto \frac{(z/n)}{(n^2/z)^2} \propto \frac{z^3}{n^5}$$

38. $E = R_\infty hc\left(1 - \frac{1}{25}\right)$ Momentum of photon emitted is $p = \frac{E}{c} = R_\infty h\left(\frac{24}{25}\right)$ Recoil momentum of H atom will also be p. $mv = p$

$$v = \frac{p}{m} = \frac{(1.097 \times 10^7)(6.626 \times 10^{-34})24}{(25)(1.67 \times 10^{-27})}$$

$\therefore v = 4.178\, ms^{-1}$

39. The total energy required to make the electron free from nucleus is the sum of the energy required to separate the electrons from the influence of each other and the energy required to separate the electrons from the influence of nucleus, i.e., Total required energy = BE of electron in He atom + ionization energy of He atom

$$= (24.6 + 13.6 \times 2^2)eV = (24.6 + 54.4)eV = 79eV$$

40. $r_n \propto n^2$

41. $A_n \propto n^4$

42. $\tau = M \times B$ and $M = iA$

Where $i = \frac{ev}{2\pi r}$ $\quad A = \pi r^2$

$$M = \frac{ev}{2\pi r} \times \pi r^2 = \frac{(mvr)e}{2m} = \frac{(nh)e}{2\pi(2m)}$$

$$\therefore M = \frac{eh}{4\pi m} \quad \tau = MB\sin 30^0$$

43. $mvr = \frac{nh}{2\pi}$; $\frac{mv^2}{r} = \frac{GMm}{r^2}$

Energy $E = \frac{1}{2}mv^2 - \frac{GMm}{r}$

LEVEL - IV

COMPREHENSION TYPE

Passage-I:

A particle of charge equal to that of an electron, -e, and mass 208 times the mass of electron (called μ -meson)moves in a circular orbit around a nucleus of charge+3e.(Take the mass of the nucleus to be infinite). Assuming that Bohr model of the atom is applicable to this system:

1. **Derive an expression for the radius of the nth Bohr orbit**

1) $\frac{\varepsilon_0 n^2 h^2}{208\pi m_e e^2}$ 2) $\frac{\varepsilon_0 n^2 h^2}{3\pi m_e e^2}$ 3) $\frac{\varepsilon_0 n^2 h^2}{624\pi m_e e^2}$ 4) $\frac{\varepsilon_0 n^2 h^2}{64\pi m_e e^2}$

2. **Find the value of n for which the radius of the orbit is approximately the same as that of the first Bohr orbit for the hydrogen atom.**

1) 10 2)15 3) 25 4) 30

3. **Find the wavelength of the radiation emitted when the μ – meson jumps from the third orbit to the first**

1) 0.4500 A^0 2) 0.5500 A^0 3) 0.6500 A^0 4) None of these

(Rydberg's constant=$1.097 \times 10^7 m^{-1}$)

Passage-II

Phtoelectrons are emitted when 4000 A^0 radiation is incident on a surface of work function 1.9eV. These photoelectrons pass through a region has a-particles to form He^+ ion, emitting a single photon in this process He^+ ions thus formed are in their fourth excited state.

4. **Energy of the fourth excited state is approx**

1) -4.2eV 2) -2.2eV 3) -3.2eV 4) -1.2eV

5. **Energy released during the combination of He^+ ions is.**

1) 5.38 eV 2) 3.38eV 3) 2.38eV 4) 1.38eV

6. **Energy of emitted photon in range of 3eV& 4eV after combination is**

1) 3.86eV 2) 3.24eV 3) 3.29eV 4) 5.24eV

Passage - III:

A sample of hydrogen gas in its ground state is irradiated with photons of 10.2eV energies. The radiation from the above sample is used to irradiate two other samples of excited ionized He^+ and excited ionized Li^{2+}, respectively. Both the ionized samples absorb the incident radiation.

7. **How many spectral lines are obtained in the spectra of Li^{2+}**

1) 10 2) 15 3) 20 4) 17

8. **Which is the smallest wavelength that will be observed in spectra of He^+ ion?**

1) 24.4nm 2) 28.8nm 3) 22.2nm 4) 30.6nm

9. **How many spectral lines are observed in spectra of He^+ ion?**

1) 2 2) 4 3) 6 4)8

10. **Which is the smallest wavelength observed in the spectra of Li^{2+}?**
1) 8.6nm 2) 10.4nm 3) 12.8nm 4)14.6nm

11. **Consider the spectral line resulting from the transition n = 2 to n = 1 in the atoms and the ions given below. The shortest wavelength is produced by**
1) hydrogen atom 2) deuterium atom
3) singly ionized helium 4) doubly ionized lithium

12. **A gas of monoatomic hydrogen is bombared with a stream of electron that have been accelerated from rest through a potential difference of 12.75 volt. In the emission spectrum, one cannot observe any line of**
1) Lyman series 2) Balmer series 3) Paschen series 4) Pfund series

13. **An electron jumps from the 4th orbit to 2nd orbit of hydrogen atom. Given the Rydberg's constant $r = 10^5\ cm^{-1}$, the frequency in hertz of the emitted radiation will be**
1) $\frac{3}{16}\times10^5$ 2) $\frac{3}{16}\times10^{15}$ 3) $\frac{9}{16}\times10^{15}$ 4) $\frac{3}{4}\times10^{15}$

14. **With increasing quantum number, the energy difference between adjacent energy levels in atoms**
1) decreases 2) increases 3) remains constant
4) decreases for low z and increases for high z

15. **Which of the following statements are true regarding Bohr's model of hydrogen atom?**
(I) orbiting speed of electron decreases as it shifts to discrete orbits away from the nucleus
(II) radii of allowed orbits of electron are proportional to the principal quantum number
(III) frequency with which electron orbits around the nucleus in discrete orbits is inversely proportional to the principal quantum number.
(IV) binding force with which the electrons are bound to the nucleus increases as it shifts to outer orbits.
Select the correct answer
1) I and III 2) II and IV 3) I, II and III 4) II, III and IV

16. A Hydrogen atom and a Li^{++} **ion are both in the second excited state. If l_H and l_{Li} are their respective electronic angular momenta, and E_H and E_{Li} their respective energies, then**
1) $l_H > l_{Li}$ and $|E_H|>|E_{Li}|$ 2) $l_H = l_{Li}$ and $|E_H|<|E_{Li}|$
3) $l_H = l_{Li}$ and $|E_H|>|E_{Li}|$ 4) $l_H < l_{Li}$ and $|E_H|<|E_{Li}|$

17. **Consider the spectral line resulting from the transition n = 2 → n = 1 in the atoms and ions given below, the shortest wavelength is produced by**
1) Hydrogen atom 2) deuterium atom
3) Singly ionized helium 4) doubly ionized lithium

18. **For hydrogen like system,**
(I) Ratio of magnetic moment to angular momentum is e/2m.
(II) energy of the electron is directly proportional to $1/n^2$.
(III) angular momentum varies inversely with n.
(IV) radius of an orbit is inversely related to n.
1) I, II 2) II, IV 3) I, II, IV 4) III, IV

19. **As per Bohr model,**

(I) minimum energy required to remove an electron from ground state of doubly ionized Li atom(z = 3) is 122.4 eV.

(II) energy of transition n = 3 to m = 2 is less than that of m = 2 to n = 1.

(III) minimum energy required to remove an electron from ground state of singly-ionised He atom (z = 2) is 27.2 eV.

(IV) A transition from state n = 3 to n = 2 in a hydrogen atom results in U-V radiations

1) II, III, IV 2) I, II 3) II, III 4) I, III, IV

20. **Assertion: A discharge tube appears dark, when evacuated to very low pressures.**
Reasoning: No colour is left at such low pressures.

1) If both assertion and reason are true and reason is the correct explanation of assertion.
2) If both assertion and reason are true but reason is not the correct explanation of assertion.
3) If assertion is true and reason is false. 4) If assertion is false but reason is true

21. **Assertion: An electron in hydrogen atom passes from n = 4 to n = 1 level. The maximum number of photons that can be emitted is 6.**
Reasoning: maximum number of photons omitted can only be 4.

1) If both assertion and reason are true and reason is the correct explanation of assertion.
2)If both assertion and reason are true but reason is not the correct explanation of assertion.
3)If assertion is true and reason is false. 4)If assertion is false but reason is true

LEVEL - IV - KEY

1) 3	2) 3	3) 2	4) 3	5) 2	6) 2	7) 2	8) 1	9)3	10) 2
11) 4	12) 4	13) 3	14) 1	15) 1	16) 2	17) 4	18) 1	19) 2	20) 3
21) 3									

LEVEL - IV - HINTS

1. $r_n = \dfrac{n^2 h^2 \in_o}{\pi m e^2 Z}$; Put $Z = 3$; $m = 208\, m_e$

2. $r_n = a_o \Rightarrow \dfrac{n^2 h^2 \in_o}{624\pi m_e e^2} = \dfrac{\in_o h^2}{\pi m_e e^2} \Rightarrow$ n= 25

3. $\dfrac{hc}{\lambda} = \Delta E$

4. $K_{\max} = \dfrac{hc}{\lambda} - \phi = 1.18 eV$; $E = -(13.6 eV)\dfrac{Z^2}{n^2}$

5. Difference of energy when photoelectron recombines with α particle =1.18-(-2.2)=3.38eV

6. For de-excitation from n = 5 to n = 3; E= 3.86 eV

7. Li^{+2} is excited from $n = 3$ to $n = 6$ So, $N = 15$

8. He^{+} is excited from n = 2 to n = 4

$\lambda_{\min} = \dfrac{hc}{\Delta E} = 24.4\, nm$

9. $N = \dfrac{4x(4-1)}{2} = 6$

10. $\lambda_{\min} = \dfrac{hc}{\Delta E} = 10.4\, nm$

Chapter - 7

NUCLEI

NUCLEI

SYNOPSIS

NUCLEUS:

- The nucleus of an atom is at the centre. Most of the mass of an atom is at the centre. The entire positive charge of an atom lies in the nucleus.
- All atomic nuclei are made up of elementary particles called protons and neutrons. Proton is the nucleus of the hydrogen atom. It has a positive charge of 1.6×10^{-19} C having a mass of 1.6726×10^{-27} kg. This is nearly equal to 1836 times the electron mass. Neutron is electrically neutral (i.e. neutron carries no charge). Mass of neutron is slightly greater than that of the proton (1.6750×10^{-27} kg). Both the proton and neutron together constitute the nucleus. They are called nucleons.
- Generally, atomic number is denoted by Z and mass number is denoted by A and (A-Z) gives number of neutrons (N) in the nucleus.

  $\therefore N = A - Z$; $A = Z + N$
- Nucleus is positively charged and its shape is considered as spherical.

TYPES OF NUCLEI:

- **ISOTOPES:** Atomic nuclei having same atomic number but different mass numbers are known as isotopes. They occupy same position in the periodic table and possess identical chemical properties. They have same proton number.

  Ex: 1) $_3Li^6, _3Li^7$ 2) $_1H^1, _1H^2, _1H^3$
- **ISOTONES :** Atomic nuclei having same number of neutrons are called isotones.

  Ex: 1) $_{17}Cl^{37}, _{19}K^{39}$, 2) $_7N^{17}, _8O^{18}, _9F^{19}$
- **ISOBARS:** Atomic nuclei having same mass number but different atomic numbers are called Isobars. They have same number of nucleons.

  Ex:–1) $_{18}Ar^{40}, _{20}Ca^{40}$, 2) $_{32}Ge^{76}, _{34}Se^{76}$
- **ISOMERS:** Atomic nuclei having same mass number and same atomic number but different nulear properties are called isomers.

  Ex:- $m_{35}Br^{80}$ metastable Bromine and $g_{35}Br^{80}$ ground state Bromine are two isomers with different half lives
- **ISODIAPHERS:** Nuclei having different Atomic number (Z) and mass number (A) but with same excess number of neutrons over protons (A-2Z) are called isodiaphers. Ex:- $_{11}Na^{23}$, $_{13}Al^{27}$

SIZE OF THE NUCLEUS:

- Nuclear sizes are very small and are measured in fermi (or) femtometer. 1 fermi=10^{-15}m
- Radius of the nucleus depends on number of nucleons. $R = R_0 A^{1/3}$

  above equation does not apply to heavy nucleides

  Value of $R_o = 1.4 \times 10^{-15}$m
- Radius of the nucleus is in the order of 10^{-15} m.
- Size of an atom is in the order of 10^{-10}m.
- If an α -particle with an initial kinetic energy E

approaches a target of atomic number Z, if the distance of closest approach is "d" then $\frac{1}{4\pi \epsilon_0}\frac{2Ze^2}{d} = E$

(Where 'e' is charge of an electron) If "v" represents the initial velocity of α particle, (m is mass of "α" particle) then $\frac{1}{4\pi \epsilon_0}\frac{2Ze^2}{d} = \frac{1}{2}mv^2$

Note : If a particle of charge q, mass m is projected towards a nucleus of charge Q with velocity v from infinity then the distance of closest approach d is give by $\frac{1}{4\pi \epsilon_0}\frac{qQ}{d} = \frac{1}{2}mv^2$

Note : If R, S and V be the Radius, surface area and volume of a nucleus with mass number A then

$$R \propto A^{\frac{1}{3}} \Rightarrow \frac{R_1}{R_2} = \left(\frac{A_1}{A_2}\right)^{\frac{1}{3}}; S \propto R^2 \propto A^{\frac{2}{3}} \Rightarrow \frac{S_1}{S_2} = \left(\frac{A_1}{A_2}\right)^{\frac{2}{3}}$$

$$V \propto R^3 \propto A \Rightarrow \frac{V_1}{V_2} = \frac{A_1}{A_2}$$

Note : If a stationary nucleus splits in to two lighter nuclei with mass numbers A_1 and A_2 then according to law of conservation of linear momentum, the two lighter nuclei move in opposite directions with equal momenta hence $\mathbf{m_1v_1 = m_2v_2}$

Ratio of velocities of the two nuclei

$$\frac{v_1}{v_2} = \frac{m_2}{m_1} = \frac{A_2}{A_1} = \left(\frac{R_2}{R_1}\right)^3 \quad (\because m \propto A \propto R^3)$$

Ratio of kinetic energy of the two nuclei

$$\frac{KE_1}{KE_2} = \frac{m_2}{m_1} = \frac{A_2}{A_1} = \left(\frac{R_2}{R_1}\right)^3$$

$$\left(\because KE = \frac{p^2}{2m} \ \& \ KE \propto \frac{1}{m} \text{ when p is constant}\right)$$

DENSITY OF THE NUCLEUS:

- Density of nucleus is independent of mass number of the atom.
- Density of the nucleus is 1.45×10^{17} Kgm^{-3}.
- The density is maximum at the centre and gradually falls to zero as we move radially out wards.
- Radius of the nucleus is taken as the distance between the centre and the point where the density falls to half of its value at the centre.
- Density of nucleus is of the order of $10^{14}\, gm/cc = 10^{17}\, kg/m^3$

W.E-1: ***Compare the radii of the nuclei of mass numbers 27 and 64.***

A. The ratio of the radii of the nuclei is

$$\frac{R_1}{R_2} = \left(\frac{A_1}{A_2}\right)^{\frac{1}{3}} = \left(\frac{27}{64}\right)^{\frac{1}{3}} (\because R = R_0A^{1/3}) = \frac{3}{4}$$

W. E-2 : ***The radius of the oxygen nucleus $^{16}_{8}O$ is 2.8×10^{-15}m. Find the radius of Lead nucleus***

A. $R_0 = 2.8 \times 10^{-15}$ m, $A_0 = 16$, $A_{Pb} = 205$ $R \propto A^{1/3}$

$$\frac{R_0}{R_{Pb}} = \left(\frac{A_0}{A_{Pb}}\right)^{1/3} = \frac{2.8\times10^{-15}}{R_{Pb}} = \left(\frac{16}{205}\right)^{1/3}$$

$R_{Pb} = 6.55 \times 10^{-15}$m.

ATOMIC MASS UNIT (A.M.U):

i) The masses of atoms, nuclei, sub atomic particles are very small. Hence, a small unit is used to express these masses. This unit is called as atomic mass unit (amu). ***1 amu is equal to one twelth part of the mass of carbon ($_6C^{12}$) isotope.***
Mass of $_6C^{12}$ is exactly 12 amu

ii) Now, the mass of 1 gm -mole of carbon is 12 gm and according to Avogadro's Hypothesis it has N (Avogadro's Number) atoms. Thus, the mass of one atom of carbon is (12/N) gm. According to the definition.

$$1amu = 1u = \frac{1}{12}\times(mass\ of\ one\ carbon\ atom)$$

$$= \frac{1}{12}\times\frac{12}{N} = \frac{1}{N}gm = \frac{1}{6.023\times10^{23}}gm$$

$$= 1.660565\times10^{-24}gm = 1.660565\times10^{-27}Kg$$

MASS - ENERGY EQUIVALENCE : According to Einstein's mass-energy equivalence principle, mass is another form of energy. Mass can be converted into energy & energy can be converted into mass according to the equation $E= mC^2$

Here m is the mass that disappears and E is the energy liberated. C is the velocity of light in vacuum.

When 1 amu of mass is converted in to energy

Energy liberated is given by

$E = (1.660565 \times 10^{-27}) \times 9 \times 10^{16}$ J= 931.5 MeV

hence 1 amu of mass is equivalent to 931.5 MeV of energy $\therefore$ $\boxed{1\ amu = 931.5\ MeV/C^2}$

The masses of electron, proton and neutron in terms of various units are :

Mass of the electron = $m_e = 9.1095 \times 10^{-31}$ kg

$= 0.000549$ u $= 0.511$ MeV/C^2

Mass of the proton = $m_p = 1.6726 \times 10^{-27}$ kg

$= 1.007276$ u $= 938.28$ MeV/C^2

Mass of the neutron = $m_n = 1.6750 \times 10^{-27}$ kg

$= 1.008665$ u $= 939.573$ MeV/C^2 .

NUCLEAR FORCES :

The attractive force which holds the nucleons together in the nucleus is called nuclear force.

Properties of nuclear forces :

1) ***Nuclear forces are strongest forces in nature.*** Nuclear forces are about 10^{38} times as strong as gravitational forces.The relative strengths of the gravitational, Coulomb's and nuclear forces are

$$F_g : F_e : F_n = 1 : 10^{36} : 10^{38}$$

2) ***Nuclear forces are short range forces .***

3) ***Nuclear forces are basically strong attractive forces, but contain a small component of repulsive forces.***

4) ***Nuclear forces are saturated forces.***
5) ***Nuclear forces are charge independent.***
6) ***Nuclear forces are spin-dependent.***
7) ***Nuclear forces are exchange forces.***
8) ***Nuclear forces are non-central forces.***

MASS DEFECT, BINDING ENEGRY, EINSTEIN'S MASS ENERGY RELATION

➢ When matter is completely annihilated, energy released is $E = mc^2$

➢ The energy equivalent to 1 amu is 931.5 M eV = $1.4925 \times 10^{-10} J$.

MASS DEFECT: Atomic mass is always less than the sum of the masses of constituent particles. The difference between the total mass of the nucleons and mass of the nucleus of an atom gives mass defect.

$$\Delta m = \left[\left[ZM_p + (A - Z)M_n\right] - M_{nucleus}\right]$$

Z = Atomic number; M_p = Mass of proton

M_n = Mass of neutron; A=Mass number

$M_{nucleus}$ = Mass of nucleus.

BINDING ENERGY: The energy required to bring the nucleons from infinity to form the nucleus is called binding energy or it is the energy required to split a nuclens into nucleons.

It is energy equivalent of mass defect BE= $[\Delta m]C^2$

NOTE: BE = mass defect x 931.5 MeV if mass is expressed in a.m.u.

B.E. per nucleon = Binding fraction

$$\frac{Binding\ Energy}{Mass\ Number} = \frac{\Delta m \times 931\ Mev}{A}$$

Averge Binding energy or Binding energy fraction: It is the Binding energy per nucleon (or) the average energy needed to separate a nuclei in to its individual nucleons.

➢ Binding energy is not a measure of stability of a nucleus.

PACKING FRACTION OF A NUCLEUS :

Packing fraction : It is defined as the mass defect per nucleon. Packing fraction $= \frac{\Delta m}{A} = \frac{M - A}{A}$

If the packing fraction is negative then the nucleus is more stable.

If the packing fraction is positive then the nucleus is unstable.

Packing fraction is zero for ${}_6C^{12}$

➢ Packing fraction measures the stability of a nucleus. Smaller the value of packing fraction, large is the stability of the nucleus.

VARIATION OF B.E. PER NUCLEON WITH MASS NUMBER

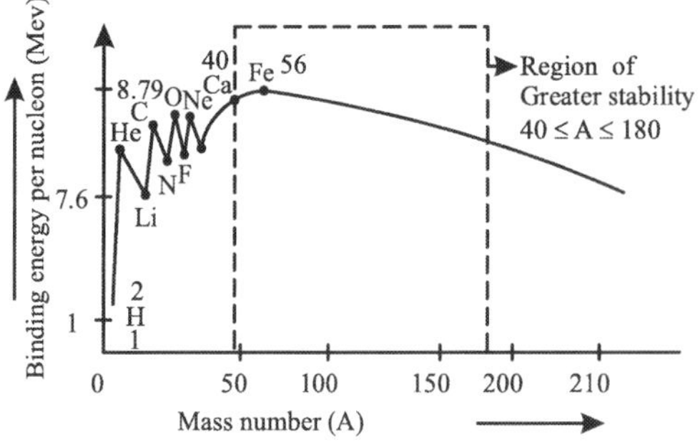

The main features of binding energy curve shown in figure are :

(1) The minimum value of binding energy per nucleon is in the case of deuteron (1.11MeV).

(2) The maximum value of $\frac{BE}{A}$ is 8.7MeV for the nucleus $_{28}Fe^{56}$ (iron) which is the most stable.

(3) Binding energy is high in the range 28<A<138. The binding energy of these nuclei is very close to 8.7 MeV.

(4) Further increase in the mass number, binding energy per nucleon decreases and consequently for the heavy nuclei like uranium it is 7.6 MeV.

(5) In the region of smaller mass numbers, the binding energy per nucleon curve shows the characteristic minima and maxima. Minima are associated with nuclei containing an odd number of protons and neutrons such as $_{3}^{6}Li, _{5}^{10}B, _{7}^{14}N$ and the maxima are associated with nuclei having an even number of protons and neutrons such as $_{2}^{4}He, _{6}^{12}C, _{8}^{16}O$.

(6) Nuclei with A> 220 are distinctly unstable. That means from A> 220 single heavy nucleus breaks into two nearly equal nuclei with mass number A< 150 and so which are most stable. This process takes at right of the BE curve as shown in figure . This process explains the nuclear fission.

(7) Light nuclei such as hydrogen combine to form heavy nucleus to form helium for greater stability. This process takes at left of the BE curve as shown in figure. This process explains the nuclear fusion.

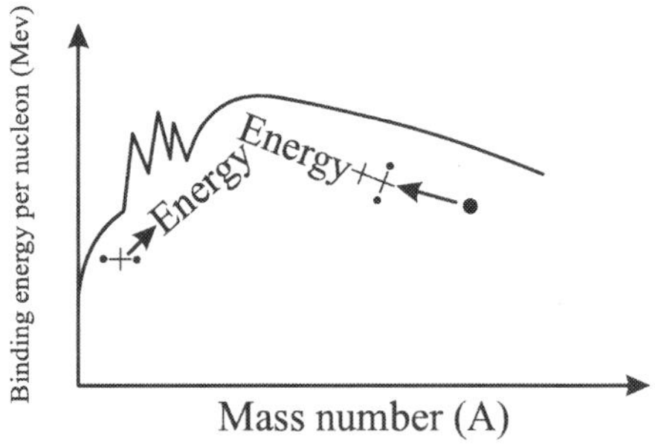

Note : Iron $\left(_{28}Fe^{56}\right)$ whose binding energy per nucleon stands maximum at 8.7 MeV is most stable and will undergo neither fission nor fusion.

Exo-ergic Reaction : ***The reaction in which energy will be released is called exo-ergic Reaction.*** $A + B \rightarrow C + D + Q$

Here A and B are called Reactants

C and D are called Products

Q is the amount of energy released

In an Exo - ergic Reaction

Mass of reactants > Mass of products

$$\Delta m = M_R - M_P = (M_A + M_B) - (M_C + M_D)$$

Energy Released $Q = \Delta m \times C^2$ joule [Δm is in kg] $= \Delta m \times 931.5$ MeV (Δ m is in amu)

If Binding energies are given then for Exo-ergic reactions.

(B.E) Products > (B.E) Reactants

Energy released $Q = (B.E)_P - (B.E)_R = [(B.E)_C + (B.E)_D] - [(B.E)_A + (B.E)_B]$

Endo-ergic Reaction : ***The reaction in which energy will be absorbed is called Endo-ergic Reaction.*** $A + B \rightarrow C + D - Q$

Here A and B are called Reactants

C and D are called Products

Q is the amount of energy absorbed

In an Endo - ergic Reaction

mass of reactants < Mass of products

$$\Delta m = M_P - M_R = (M_C + M_D) - (M_A + M_B)$$

Energy absorbed $Q = \Delta m \times C^2$ joule [Δm is in kg]

$\Rightarrow \Delta m \times 931.5$ MeV (Δ m is in amu)

If Binding energies are given then for

Endo-ergic reaction.

(B.E) Products < (B.E) Reactants

Energy absorbed $Q = (B.E)_R - (B.E)_P = [(B.E)_A + (B.E)_B] - [(B.E)_C + (B.E)_D]$

Note: A nuclear reaction can occur only if certain conservation laws are followed. These are :

1. Conservation of mass number A.
2. Conservation of charge.
3. Conservation of energy, linear momentum and angular momentum.

W.E-3 : ***Find the binding energy of $^{56}_{26}Fe$. Atomic mass of Fe is 55.9349u and that of Hydrogen is 1.00783u and mass of neutron is 1.00876u***

Sol. Mass of the hydrogen atom $m_H = 1.00783$u; Mass of neutron $m_n = 1.00867$ u; Atomic number of iron Z = 26; mass number of iron A = 56; Mass of iron atom $M_a = 55.9349$u

Mass defect $\Delta m = [Zm_H + (A-Z)m_n] - M_a$

= [26 x 1.00783+(56-20)1.00867]-55.93493

u = 0.5287 u.

$\therefore$ Binding energy = $(\Delta m)c^2$ = (0.52878)

c^2 = (0.52878)(931.5MeV) = 492.55 MeV

W.E-4 : ***Find the energy required to split $^{16}_{8}O$ nucleus into four α - particles. The mass of an α -particle is 4.002603u and that of oxygen is 15.994915u.***

Sol. Mass of a-particle = 4.002603 u

Mass of oxygen = 15.994915u

B.E= [Mass of 4 particles-Mass of oxygen] x 931.5MeV

B.E = [4 x 4.002603 - 15.994915] x 931.5 MeV

= (16.010412-15.994915) x 931.5 MeV = 0.015497 x 931.5 ; B.E = 14.43 MeV

W.E-5 : ***Calculate the binding energy per nucleon of $^{40}_{20}Ca$. Given that mass of $^{40}_{20}Ca$ nucleus = 39.962589 u, mass of proton = 1.007825 u. mass of Neutron = 1.008665 u and 1 u is equivalent to 931 MeV.***

Sol. A =40, Z = 20, A -Z = 20

$\Delta m = \{Zm_p + (A-Z)m_n\} - M_n$

= { (20 x 1.007825+(20x1.008665)} - 39.962589

= 40.329800 - 39.962589 ; $\Delta m = 0.367211$

Binding energy per nucleon =

$$\frac{\Delta m \times 931}{A} = \frac{0.367211 \times 931}{40} = 8.547 MeV.$$

W.E-6 : ***The binding energies per nucleon for deuterium and helium are 1.1 MeV and 7.0 MeV respectively. What energy in joules will be liberated when 2 deuterons take part in the reaction.***

Sol. $^2_1H + ^2_1H \rightarrow ^4_2He + Q$

Binding energy per nucleon of helium (^4_2He) =7 MeV

Binding energy = 4 x7 = 28 MeV

Binding energy per nucleon of deuterium (^2_1H)=1.1MeV

Binding energy = 2 x 1.1 = 2.2 MeV

Energy liberated (Q) = (28 -(2.2)2] = 23.6 Mev.

i.e. Q=23.6x10^6x1.6x10^{-19}; $Q = 37.76 \times 10^{-13} J$

W.E-7: ***The kinetic energy of α -particles emitted in the decay of $_{88}Ra^{226}$ into $_{86}Rn^{222}$ is measured to be 4.78 MeV. What is the total disintegration energy or the 'Q-value of this process'?***

Sol. The standard relation between the kinetic energy of the α -particle (KE_α) and the Q-value (or total disintegration energy) is

$$KE_a = \left(\frac{A-4}{A}\right).Q \qquad Q = \left(\frac{A}{A-4}\right).KE_a$$

$$= \left(\frac{226}{226-4}\right) \times 4.78\text{MeV} \quad = \frac{226}{222} \times 4.78\text{MeV}$$

Q = 4.865MeV $\approx$ 4.87MeV

W.E-8: ***A nucleus X-initially at rest, undergoes alpha-decay, according to the equation.***

$$^A_{92}X \rightarrow ^{228}_{z} Y + \alpha$$

The α -particle in the above process is found to move in a circular track of radius $1.1 \times 10^2 m$ ***in a uniform magnetic field of*** $3.0 \times 10^3 T$.

The energy (in MeV***) released during the process and binding energy of the parent nucleus X, respectively.***

Given: $m_y = 228.03 amu$; $m_\alpha = 4.003 amu$, $m(^1_0n) = 1.009 amu$; $m(^1_0H) = 1.008 amu$,

$1 amu = 1.66 \times 10^{-27}$; $kg \equiv 931.5 MeV/c^2$

Sol: The given equation is ${}_{92}^{A}X \rightarrow {}_{z}^{228}Y + {}_{2}^{4}He$

$A = 228 + 4 = 232$; $92 = z + 2 \qquad \therefore z = 90$

$$\frac{m_\alpha v_\alpha^2}{r} = qv_\alpha B \; ; \; v_\alpha = \sqrt{\frac{rqB}{m_\alpha}}$$

$$= \sqrt{\frac{1.1\times10^2 \times 2\times1.6\times10^{-19}\times3\times10^3}{4.003\times1.66\times10^{-27}}}$$

$$= 4.0\times10^6 m/s$$

From conservation of linear momentum, $m_\alpha v_\alpha = m_y v_y$

$$v_y = \frac{m_\alpha v_\alpha}{m_y} = \frac{(4.003)(4.0\times10^6)}{(228.03)} = 7.0\times10^4 m/s$$

There fore, energy released during the process $= \frac{1}{2}\left[m_\alpha v_\alpha^2 + m_y v_y^2\right] = \frac{(1.66\times10^{-27})}{(2\times1.6\times10^{-13})}$

$$\left[(4.003)(4.0\times10^6)^2 + (228.03)(7.0\times10^4)^2\right] MeV$$

$$= 0.34\, MeV = \frac{0.34}{931.5} amu = 0.000365 amu$$

Therefore, mass of ${}_{92}^{232}X = m_y + m_\alpha + 0.000365 = 232.033365\,u$

Mass defect $\quad \Delta m = 92(1.008) + (232 - 92)$

$(1.009) - 232.033365$.

$\therefore$ Binding energy $= 1.962635 \times 931.5\, MeV = 1828.2\, MeV$

NATURAL RADIO ACTIVITY :

Spontaneous decay of naturally occurring unstable nuclei by emission of certain sub particles (like α, β, and γ radiation) is called natural radio activity.

The emission of these rays takes place because of the instability of the nucleus. In the process of emitting these rays a nucleus tries to attain the stability.

In general natural radioactivity takes place in heavy nuclei beyond lead in the periodic table. There are also naturally radioactive light nuclei, such as potassium isotope ${}_{19}K^{40}$, the carbon isotope ${}_{6}C^{14}$ and the rubidium isotope ${}_{37}Rb^{87}$.

Regarding radioactivity.

(i) It is completely uneffected by the physical and chemical conditions to which the nucleus is subjected i.e we cannot change the radio activity by applying high temperature, high pressure and strong electric field etc.

(ii) The nucleus can disintegrate immediately (or) it may take infinite time.

(iii) The energy liberated during the radioactive decay comes from individual nuclei.

MODES OF DECAY :

The radioactive nucleus before decay is called a parent nucleus, the nucleus resulting from its decay by particles (Radiation) emission is called daughter nuclei.

This daughter nuclei may be stable (or) unstable.

$$\underset{\text{Parent}}{X} \longrightarrow \underset{\text{Daughter}}{Y} + \underset{\text{Radiation}}{R} + \underset{\text{Energy}}{Q}$$

Here R may be either α particle (or) β particle (or) γ radiation. Q is the energy of the emitted particles (or radiation).

α **-DECAY :** ***When a nucleus disintegrates by radiating α -rays, it is said to undergo α -decay.*** An α - particle is a helium nucleus. Thus a nucleus emitting an α particle losses two protons and two neutrons, as a result its atomic number Z decreases by 2, the mass number A decreases by 4 and the neutron number N decreases by 2.

$${}_{Z}X^{A} \rightarrow {}_{Z-2}Y^{A-4} + {}_{2}He^{4} + Q$$

Ex: ${}_{88}Ra^{226} \rightarrow {}_{86}Rn^{222} + {}_{2}He^{4} + 4.87MeV$

Both electric charge and nucleon number are conserved in the process of α decay.

Application : When a stationary Radio active nucleus x decays into another nucleus y by emitting an α -particle. $x \rightarrow y + \alpha$ particle $+Q$

Applying LCLM if α particle moves forward with a momentum 'P' then daughter nucleus y recoils with same momentum 'P' so that total momentum of the system is zero. Hence $\boxed{P_y = P_\alpha}$

The energy released 'Q' is in the form of K.E of daughter nucleus 'y' and 'α' particle.

$$\boxed{Q = KE_y + KE_\alpha}$$

Ratio of kinetic energies $\boxed{\dfrac{KE_y}{KE_\alpha} = \dfrac{M_\alpha}{M_y}}$

$\left(\because KE = \dfrac{P^2}{2m} \text{ and } KE \ \alpha \ \dfrac{1}{m} \text{ when 'P' is same}\right)$

$$1 + \frac{KE_y}{KE_\alpha} = 1 + \frac{M_\alpha}{M_y}; \quad \frac{KE_\alpha + KE_y}{KE_\alpha} = \frac{M_y + M_\alpha}{M_y}$$

$$\boxed{KE_\alpha = Q\left(\frac{M_y}{M_\alpha + M_y}\right)}; \quad \boxed{KE_y = Q\left(\frac{M_\alpha}{M_\alpha + M_y}\right)}$$

Notice that KE_α is very close to (but smallerthan) Q.

β **-DECAY : When a nucleus disintegrates by radiating $\beta -$ rays, it is said to undergo $\beta -$ decay.**

i) β particles are nothing but electrons. Hence when a nucleus emits a β particle, the atomic number (Z) increases by 1 unit, but the mass number does not change.

The general form of $\beta-$decay can be written as

$$_{Z}X^{A} \rightarrow _{Z+1}Y^{A} + _{-1}e^{0}.$$

Ex: $_{90}Th^{234} \rightarrow _{91}Pa^{234} + _{-1}e^{0}$

Both electric charge and nucleon number are conserved in β decay also.

γ -DECAY: When a nucleus disintegrates by radiating $\gamma-$rays, it is said to undergo $\gamma-$decay.

Gamma rays are nothing but electromagnetic radiations of short wavelengths (not exceeding 10^{-10}m.)

The emission of $\gamma-$rays from the nucleus does not alter either atomic number Z or mass number A. It just results in the change of the energy state of a nucleus.

When a parent nucleus emits an α or a β particle, the daughter nucleus may be formed in one of excited states. Such a nucleus will eventually comes to the ground state. In this process $\gamma-$radiation will be emitted.

$$_{Z}X^{A} \rightarrow _{Z}X^{A} + \gamma \; Photon(s)$$

Example: $_{38}{}^{*}Sr^{87} \rightarrow _{38}Sr^{87} + \gamma$.

$_{38}Sr$ is isomer of $_{38}$Sr.

Note : When a Radio active nucleus emits an α - particle followed by two β - particles, its isotope is formed.

$$_{Z}X^{A} \xrightarrow{\alpha} _{Z-2}Y^{A-4} \xrightarrow{2\beta^{-}} _{Z}X^{A-4}$$

Note : When a Radio active nucleus emits a β - particle its isobar is formed.

$$_{Z}X^{A} \xrightarrow{\beta^{-}} _{Z+1}Y^{A}$$

Note: When a Radio active nucleus emits a γ - particle its isomer is formed

$$_{Z}X^{A} \xrightarrow{\gamma} _{Z}X^{A}$$

Deflection of Radioactive radiations in electric and magnetic fields :

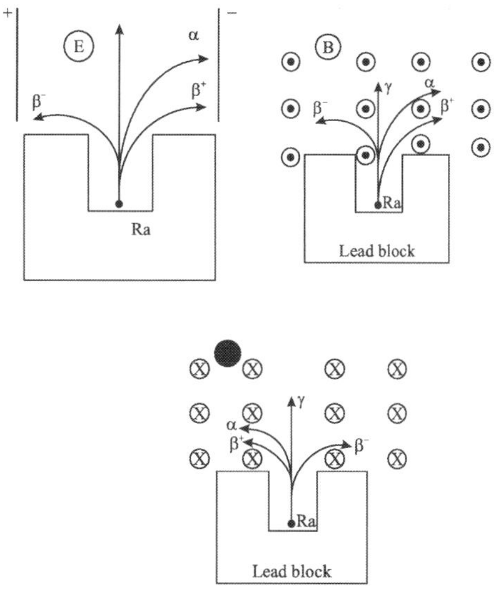

W.E-9: ***The nucleus $^{23}_{10}Ne$ decays by β^- emission. Write down the β-decay equation and determine the maximum kinetic energy of the electrons emitted. Given that :***
$m(^{23}_{10}Ne) = 22.994466u$; $m(^{23}_{11}Na) = 22.989770\ u$

Sol. $^{23}_{10}Ne \rightarrow ^{23}_{11}Na + \bar{e} + \bar{v} + Q$

For β^- - decay, Q = [M(x) - M(y)]C^2 = [22.994466-22.989770]931.5

= 0.004696 x 931.5 = 4.37 MeV

W.E-10: ***Calculate the binding energy of an α -particle. Given that mass of proton = 1.0073u, mass of neutron = 1.0087u. and mass of α -particle = 4.0015u.***

Sol. m_P = 1.0073u, m_N = 1.0087u, M = 4.0015u

N = A -Z =4 -2 =2 $\quad (\because {}_2He^4 = {}_ZX^A)$

B.E = Δm x 931.5 MeV $= \{[Zm_p + (A-Z)m_n] - M\} \times 931.5$

$[[(2\times1.0073)+(2\times1.0087)-4.0015]]\times931.5MeV$

= 0.0305 $\times$ 931.5 MeV ; B.E = 28.4 MeV

W.E-11: ***How many α and β - particles are emitted when uranium nucleus $({}_{92}U^{238})$ decay to ${}_{82}Pb^{214}$?***

Sol. Let n be the number of α - particles and m be the number of β - particles emitted.

${}_{92}U^{238} \rightarrow {}_{82}Pb^{214} + n\,{}_2He^4 + m\ {}_{-1}e^0$.

As mass is conserved, 238 = 214 + 4n + m (0)

= 214 + 4n ; 4n = 24; n = 6

As charge is conserved , 92 = 82 + 2n + m (-1)

10 = 2(6) - m $(\because n = 6)$; m = 2.

$\therefore 6\alpha$ - particles and 2 β - particles are emitted

RADIOACTIVE DECAY LAW:

Based on their experimental observations and analysis of certain radioactive materials Rutherford and Soddy formulated a theory of radioactive decay. According to them

After decay of a nucleus the new product (daughter) of nucleus has totally different physical as well as chemical properties.

The rate of radioactive decay (or) the number of nuclei decaying per unit time at any instant is directly proportional to the number of nuclei (N) present at that instant and is independent of the external physical conditions like temperature, pressure etc.

Let 'N' be the number of radioactive atoms present at a time 't' and N_0 is the initial number of radio active nuclei. Let dN atoms disintegrate in time 'dt'. According to the law of radioactive decay

$$\left(\frac{dN}{dt}\right) \propto N ; \left(\frac{dN}{dt}\right) = -\lambda N \ .... (1)$$

The proportionality constant λ is called decay constant (or) disintegration constant. The negative sign indicates that as time increases N decreases.

From eqn (1) $\frac{dN}{N} = -\lambda\, dt$ ....(2)

Integrating eq (2) on both sides $\int \frac{dN}{N} = -\lambda \int dt$

$\log_e N = -\lambda t + C$ .....(3)

Here C is the constant of integration

At t = O, N= N_0 Substituting in eqn (3),

we get, $\log_e N_0 = C$

$\therefore \log_e N = -\lambda t + \log_e N_0$;

$\therefore \log_e N - \log_e N_0 = -\lambda t$

$\therefore \log_e \left(\frac{N}{N_0}\right) = -\lambda t$; $\frac{N}{N_0} = e^{-\lambda t}$; $\boxed{N = N_0 e^{-\lambda t}}$ ... (4)

This shows that the number of radioactive nuclei decreases exponentially with time.

Above equation is known as the decay law (or) the law of radio-active decay. It is an exponential law.

taking logarithm on both sides for the above equation. $\log_e N = \log_e N_o - \lambda t$; $\lambda t = \log_e \frac{N_0}{N}$

$\therefore \boxed{t = \frac{1}{\lambda} \ln\left(\frac{N_0}{N}\right)}$

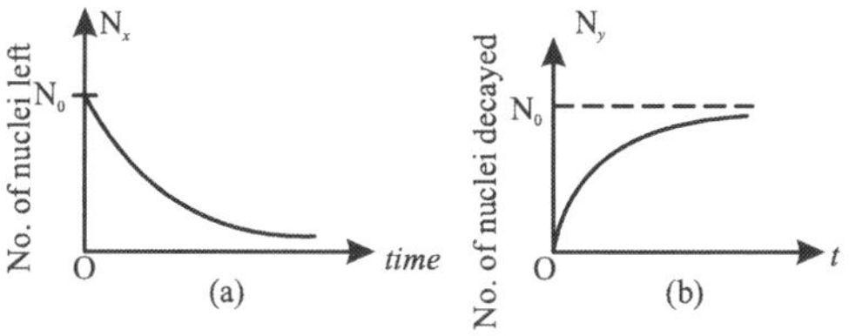

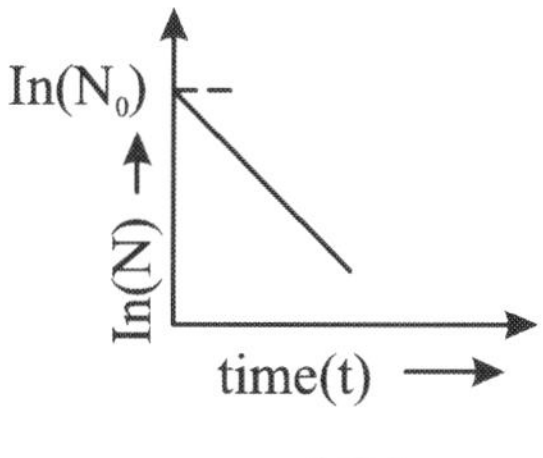

(C)

ACTIVITY (R) :

The number of decays per unit time (or) decay rate is called activity (R)

$|R| = \left|\frac{dN}{dt}\right| = \frac{d}{dt}\left(N_0 e^{-\lambda t}\right)$ (or) $R = \lambda N = \lambda N_0 e^{-\lambda t}$ (or)

$R = R_0 e^{-\lambda t}$, where $R_0 = \lambda N_0$ is the decay rate at t = 0, called initial activity.

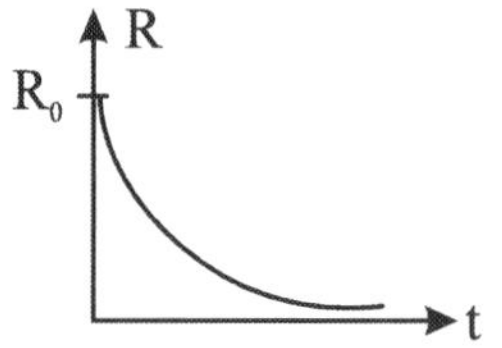

If a nucleus can decay simultaneously by n processes, which have activities R_1, R_2, ... and R_n. Then the resultant activity $R = R_1 + R_2 + ... + R_n$. **If nucleus decays simultaneously more than one process is called parallel decay.**

The S.I unit of activity is **Becquerel (Bq)** and other units are **curie (Ci) and Rutherford (Rd).**

1 Bq = 1 decay per second,

1 Rd = 10^6 decays per second.

1 Ci = 3.7×10^{10} decays per second.

Note : Curie is approximately equal to the activity of one gram of pure radium.

DECAY CONSTANT (λ) : It gives the ability of a nucleus to decay. ***The decay constant λ for a given radio active sample is defined as the reciprocal of the time during which the number of nuclei decreases to $\frac{1}{e}$ times their original value.***

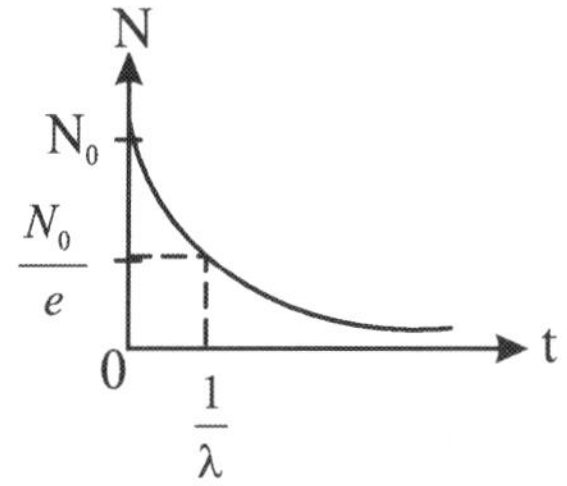

1) **Larger value of λ corresponding to decay in smaller time and vice versa.**
2) $\lambda = 0$ for stable nuclei.
3) Decay constant is the characteristic of the sample taken and does not vary with time.
4) If a nucleus can decay simultaneously by more than one process (say n), which have decay constants $\lambda_1, \lambda_2 .... and\ \lambda_n$, then the effective decay constant is $\lambda = \lambda_1 + \lambda_2 + ... + \lambda_n$. This is called parallel decay.

HALF LIFE (T) : As the name suggests, the half life of a radioactive sample is defined as ***"The time interval during which the activity of a radio active sample falls to half of its value, (or) The time interval during which the number of radio active nuclei of a sample disintegrate to half of its original number of nuclei"*** Half lives vary from isotope to isotope. While T may be as small as 10^{-16} s, its largest value may be as big as 10^9 years.

Eg: Half-life of uranium $\left({}^{238}_{92}U\right)$ is 4.47×10^9 years

half-life of krypton $\left({}^{89}_{36}Kr\right)$ is 3.16 minutes.

Relation between decay constant (λ) and half life period (T).

From Law of Radioactive decay $\frac{N}{N_0} = e^{-\lambda t}$

when $N = \frac{N_0}{2}$, $t=T$ $\therefore \frac{1}{2}=e^{-\lambda T}$ or $2 = e^{\lambda T}$

taking logirthms on both sides $\ln 2 = \lambda T$

(or) $\log_e 2=\lambda T$ $\therefore T = \frac{2.303\log_{10}2}{\lambda} = \frac{0.693}{\lambda}$

$$\therefore \boxed{T=\frac{\ln 2}{\lambda}=\frac{2.303\log 2}{\lambda}=\frac{0.693}{\lambda}}$$

The above relation establishes that the half - life (T) depends upon the decay constant λ of the radioactive substance. The value of λ is different for different radioactive substances.

Note :

i) Half life is the characteristic property of the sample and T cannot be changed by any known method.

ii) At any given instant whatever be the amount of the undecayed sample, it will be reduced to exactly half its value after a time equal to the half life of the sample.

iii) In parallel decay $\lambda = \lambda_1 + \lambda_2 + ...\lambda_n$ hence $\frac{1}{T}=\frac{1}{T_1}+\frac{1}{T_2}+........\frac{1}{T_n}$, where T is the equivalent half-life and $T_1, T_2 ....T_n$ are the half-lives in individual decay.

Application :

In a radioactive sample the number of nuclides undecayed after n-half lives (i.e., $t = nT$) is

$$t = nT = \frac{1}{\lambda}\ln\left(\frac{N_0}{N}\right) \text{ or } \frac{n(\ln 2)}{\lambda}=\frac{1}{\lambda}\ln\left(\frac{N_0}{N}\right)$$

$$\text{or } 2^n = \frac{N_0}{N}; \text{ or } \boxed{N = N_0\left(\frac{1}{2}\right)^n}$$

Note: The number of nuclei remain in the sample after **half of half life period (t=1/2T) is** given by

$$N = N_0\left(\frac{1}{2}\right)^n \text{ here } n = \frac{1}{2} \text{ then } N = N_0\left(\frac{1}{2}\right)^{\frac{1}{2}}$$

$$\therefore \boxed{N = \frac{N_0}{\sqrt{2}}} \text{taking } N_0 = 100, N = 50\sqrt{2} = 70.7$$

70.7% of nuclei remain and 29.3% of nuclei decayed.

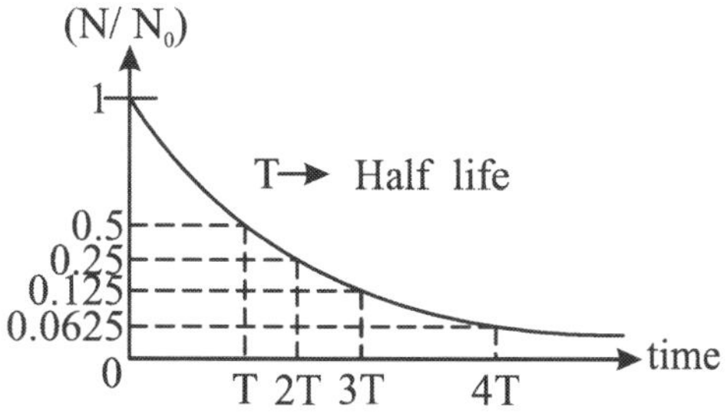

AVERAGE LIFE (OR) MEAN LIFE :

The phenomenon of radioactivity is random because we just can't predict which of the atoms in a given sample will decay first and when. Hence radioactivity process totally depends on chance. In decay process some of the atoms of the given sample may have very short life span, and others may not decay even after a very large span of time. So to determine the ability of the nucleus to decay it would be useful to calculate the average life. Hence average life is defined as ***the total life time of all the nuclei divided by the total number of original nuclei.***

i.e $\tau = \frac{\sum \text{life span of individual nucleus}}{\text{Total number of original nuclei}} = \frac{\sum t}{N_0}$

Let N_0 be the radio active nuclei that are present at $t = 0$ in the radioactive sample.

The number of nuclei which decay between t and (t + dt) is dN i.e the life time of these nuclei is 't'.

The total life time of these dN nuclei is (t dN)

$\therefore$ The total life time of all the nuclei present initially in the sample $= \int_{t=0}^{t=\infty} t\,dN$ $[\because N = 0 \text{ at infinity}]$

Average life time $\tau = \frac{\int t\,dN}{N_0}$ But $\frac{-dN}{dt} = \lambda N$

$dN = -\lambda N dt = -\lambda N_0 e^{-\lambda t} dt \quad (\because N = N_0 e^{-\lambda t})$

$\tau = \int_0^{\infty} t \frac{\lambda N_0 e^{-\lambda t}}{N_0} dt \;;\; \boxed{\tau = \frac{1}{\lambda}}$

The mean life (or) average life of a radio active sample is reciprocal to decay constant.

We know that $N = N_0 e^{-\lambda t}$; When $t = \tau$, $\quad N = N_0 e^{-\frac{1}{\tau} x \tau} = \frac{N_0}{e} = 0.37 N_0 = 37\%$ of N_0

Hence average life period of a radio active sample can also be defined as ***"The time interval during which 63% of sample decays or sample reduces to 37% of its original amount"***.

RELATION BETWEEN HALF LIFE PERIOD AND AVERAGE LIFE PERIOD

We know that $T = \frac{0.693}{\lambda}$ & $\tau = \frac{1}{\lambda}$

Hence $T = 0.693\tau$ (or) $\boxed{\tau = \frac{T}{0.693} = 1.443T}$

From the above equation it is clear that average life period is 44.3% greater than half life period.

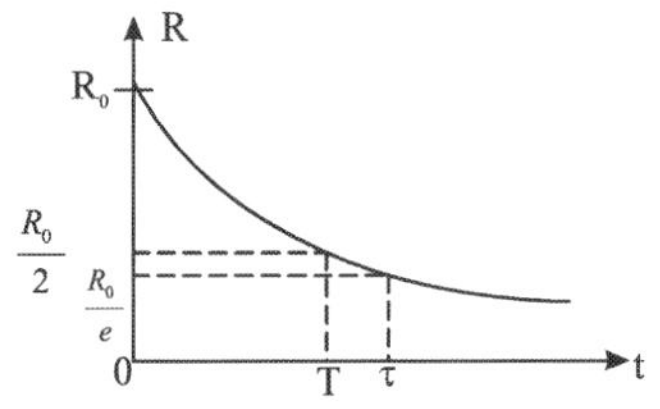

Determination of decay constant (λ) and half life period (T) of a radioactive sample graphically

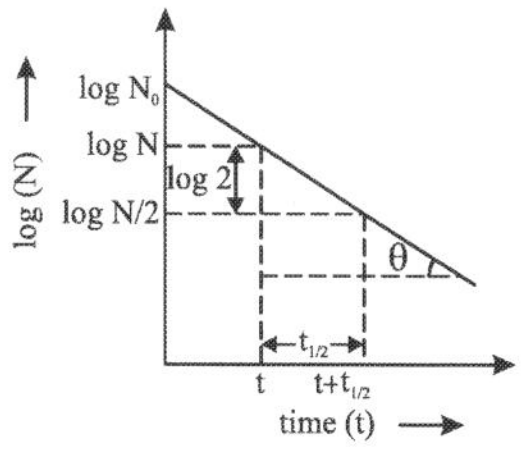

If N_0 and N be the number of atoms present undecayed initially and after a time t, then

We know that $N=N_0\, e^{-\lambda t}$ taking log on both sides

$$\log_e N = \log_e N_0 - \lambda t \Rightarrow \log N = \log N_0 - \frac{\lambda t}{2.303}$$

$$\log N = \left(\frac{-\lambda}{2.303}\right)t + \log N_0$$

Slope of the graph $m = -\tan\theta = \frac{-\lambda}{2.303}$

$\Rightarrow \boxed{\lambda = 2.303 \tan\theta}$

Half life period $T = \frac{2.303 \log 2}{\lambda}$

$T = \frac{2.303 \log 2}{2.303 \tan\theta}$ $\quad\therefore \boxed{T = (\log 2)\cot\theta}$

Note : In radioactive sample decay

1) The probability survival of nucleus after time $P_s = \frac{N}{N_0} = e^{-\lambda t}$.

2) The probability of nucleus to disintegrate in time t is $P_d = 1 - P_s = 1 - e^{-\lambda t}$.

W.E-12: ***A radioactive sample has an activity of 5.13 x10^7 Ci. Express its activity in 'becqueral' and 'rutherford'.***

Sol. Since 1 Ci = 3.7×10^{10} decays per second,

activity = 5.13×10^7 Ci

$= 5.13 \times 10^7 \times 3.7 \times 10^{10}$ Bq $= 1.9 \times 10^{18}$ Bq

Since , 1×10^6 decay per second = 1Rd

Activity $= 1.9 \times 10^{18}$ Bq $= \frac{1.9\times10^{18}}{1\times10^6} Rd = 1.9 \times 10^{12}$ Rd.

W.E-13: ***A radioactive substance has 6.0$\times$ 10^{18} active nuclei initially. What time is required for the active nuclei of the same substance to become 1.0$\times$10^{18} if its half-life is 40 s.***

Sol. The number of active nuclei at any instant of time t,

$$\frac{N_0}{N} = e^{\lambda t}; \quad \log_e\left(\frac{N_0}{N}\right) = \lambda t$$

$$\therefore t = \frac{\log_e\left(\frac{N_0}{N}\right)}{\lambda} = \frac{2.303 \log_{10}\left(\frac{N_0}{N}\right)}{\lambda}$$

In this problem, the initial number of active nuclei,

$N_0 = 6.0 \times 10^{18}$; $N = 1.0\times10^{18}, T = 40s,$

$$\lambda = \frac{0.693}{T} = \frac{0.693}{40} = 1.733\times10^{-2}\,s^{-1}.$$

$$t = \frac{2.303\log_{10}\left(\frac{6.0\times10^{18}}{1.0\times10^{18}}\right)}{1.733\times10^{-2}} = \frac{2.303\log_{10}(6)}{1.733\times10^{-2}} = \frac{2.303\times0.7782}{1.733\times10^{-2}} = 103.4s.$$

W.E-14: A radioactive sample can decay by two different processes. The half-life for the first process is T_1 and that for the second process is T_2. Find the effective half-life T of the radioactive sample.

Sol. Let N be the total number of atoms of the radioactive sample initially. Let $\frac{dN_1}{dt}$ and $\frac{dN_2}{dt}$ be the initial rates of disintegrations of the radioactive sample by the two processes respectively. Then

$$\frac{dN_1}{dt} = \lambda_1 N \text{ and } \frac{dN_2}{dt} = \lambda_2 N$$

Where λ_1 and λ_2 are the decay constants for the first and second processes respectively.

The initial rate of disintegrations of the radioactive sample by both the processes

$$= \frac{dN_1}{dt} + \frac{dN_2}{dt} = \lambda_1 N + \lambda_2 N = (\lambda_1 + \lambda_2)N.$$

If λ is the effective decay constant of the radioactive sample, its initial rate of disintegration.

$$\frac{dN}{dt} = \lambda N$$

But $\frac{dN}{dt} = \frac{dN_1}{dt} + \frac{dN_2}{dt}$

$$\lambda N = (\lambda_1 + \lambda_2)N$$

$$\lambda = \lambda_1 + \lambda_2$$

$$\frac{0.693}{T_1} + \frac{0.693}{T_2} = \frac{0.693}{T}$$

$$\frac{1}{T} = \frac{1}{T_1} + \frac{1}{T_2}; T = \frac{T_1T_2}{T_1 + T_2}.$$

W.E-15: ***Plutonium decays with a half life of 24,000 years. If plutonium is stored for 72,000 years, what fraction of it remains?***

A. $T_{1/2} = 24,000$ years

Duration of time (t) = 72,000 years

Number of half lifes (n) $= \frac{t}{T_{1/2}} = \frac{72000}{24000} = 3$

$$\therefore\ 1g \xrightarrow{1} \frac{1}{2}g \xrightarrow{2} \frac{1}{4}g \xrightarrow{3} \frac{1}{8}g$$

$\therefore$ Fraction of plutonium remains $= \frac{1}{8}g$

W.E-16: ***A certain substance decays to 1/32 of its initial activity in 25 days. Calculate its half-life.***

Sol. $1g \xrightarrow{1} \frac{1}{2}g \xrightarrow{2} \frac{1}{4}g \xrightarrow{3} \frac{1}{8}g \xrightarrow{4} \frac{1}{16}g \xrightarrow{5} \frac{1}{32}g \therefore n = 5$

$(n) = \frac{t}{t_{1/2}} \Rightarrow t_{1/2} = \frac{t}{n} = \frac{25}{5}$; $t_{1/2} = 5$ days

W.E-17: ***The half-life period of a radioactive substance is 20 days. What is the time taken for 7/8th of its original mass to disintegrate?***

Sol. Let the initial mass be one unit.

Mass reamaining $= 1 - \frac{7}{8} = \frac{1}{8}$

A mass of 1 unit becomes $\frac{1}{2}$ unit in 1 half life

$\frac{1}{2}$ unit becomes $\frac{1}{4}$ unit in 2nd half life

$\frac{1}{4}$ unit becomes $\frac{1}{8}$ unit in 3rd half life

$\therefore$ Time taken = 3 half lifes = 3 x 20 = 60 days

W.E-18: ***How many disintegrations per second will Occur in one gram of $^{238}_{92}U$, if its half-life against α-decay is 1.42 x 10^{17}s?***

Sol. Given Half-life period (T) $= \frac{0.693}{\lambda} =$

1.42×10^{17} s

$$\lambda = \frac{0.693}{1.42 \times 10^{17}} = 4.88 \times 10^{-18}$$

Avagadro number (N) = 6.023×10^{23} atoms

n = Number of atoms present in 1 g of $^{238}_{92}U = \frac{N}{A}$

$$= \frac{0.623 \times 10^{23}}{238} = 25.30 \times 10^{20}$$

Number of disintegrations $= \frac{dN}{dt} = \lambda n$

$= 4.88 \text{ x} 10^{-18} \text{ x } 25.30 \text{ x } 10^{20} = 1.2346 \text{ x } 10^{4}$ disintegrates/sec

W.E-19: ***One gram of radium is reduced by 2 milligram in 5 years by α-decay. Calculate the half-life of radium.***

Sol: Initial mass = 1 g, t = 5 years

$$\text{Reduced mass} = 2\text{mg} = 2 \times 10^{-3}\,\text{g} = \frac{2}{1000}\,\text{g}$$

$$\text{Remaining mass} = 1 - \frac{2}{1000} = \frac{998}{1000}$$

$$\frac{N}{N_0} = \frac{998}{1000} \left(\because \textit{Mass} \propto \textit{Number of atoms}\right)$$

$$\frac{N}{N_0} = e^{-\lambda t} \; ; \; \frac{998}{1000} = e^{-\lambda t}$$

$$\frac{1000}{998} = e^{\lambda t} = e^{5\lambda} \Rightarrow \log_e\left(\frac{1000}{998}\right) = 5\lambda$$

$$2.303\,(3.0000 - 2.9991) = 5\,\lambda$$

$$\lambda = \frac{2.303 \times 1 \times 0.0009}{5}$$

$$(T_{1/2}) = \frac{0.693}{\lambda} = \frac{0.693 \times 5}{2.303 \times 0.0009} = 1671.7 \text{ years}$$

W.E-20: ***The half-life of a radioactive substance is 5000 years. In how many years, its activity will decay to 0.2 times of its initial value? Given $\log_{10}5 = 0.6990$.***

Sol. T = 5000 years,

$$\frac{N}{N_0} = 0.2 = \frac{2}{10} = \frac{1}{5}$$

$$\lambda = \frac{0.693}{T} = \frac{0.693}{5000}$$

$$\frac{N}{N_0} = e^{-\lambda t}$$

$$\frac{1}{5} = \frac{1}{e^{\lambda t}} \Rightarrow 5 = e^{\lambda t}$$

$$\log_e^5 = \lambda t$$

$$2.303 \times 0.6990 = \lambda t$$

$$t = \frac{2.303 \times 0.6990 \times 5000}{0.693}$$

$$t = 11614.6\text{years} = 1.1615 \times 10^4 \text{ years}$$

W.E-21: ***Obtain the amount of $^{60}_{27}Co$ necessary to provide a radioactive source of 8.0 mCi strength. The half-life of $^{60}_{27}Co$ is 5.3 years.***

Sol. Half-life of $^{60}{}_{27}Co$ = 5.3 years

$= 5.3 \times 365 \times 24 \times 60 \times 60\ \text{S} = 5.3 \times 3.15 \times 10^7\ \text{s}$

Now `x` gm of ${}^{60}{}_{27}Co$ contains $\frac{10^{-3}x}{60}$ k - mole = $\frac{x\times10^{-3}}{60}\times6.025\times10^{26}$

N = 0.1004x x 10^{23} = 1.004 x x 10^{22} atoms

Required strength of ${}^{60}{}_{27}Co$ = 8 m Ci

= 8 x 10^{-3} x 3.7 x10^{10} dis/s

We know that, decay rate (R) = λN

$$N=\frac{R}{\lambda}=\frac{8\times3.7\times10^{7}}{0.693/T_{1/2}}$$

$$=\frac{29.6\times10^{7}}{0.693}\text{x}5.3\text{x}3.15\text{x}10^{7}=713.0909\text{ x }10^{14}$$

1.004x x 10^{22} = 713.0909 x 10^{14}

x = 710.2499 x 10^{-8} kg = 7.1 x 10^{-6} g

ARTIFICIAL TRANSMUTATION OF ELEMENTS

The conversion of one element into another by artificial means is called artificial transmutation of the element. Rutherford performed number of experiments in which the atoms of different stable elements, such as nitrogen, aluminium, phosphorus, etc, were bombarded by high speed α – particles from natural radioactive substances. Finally in 1919, he discovered the phenomenon of artificial transmutation.

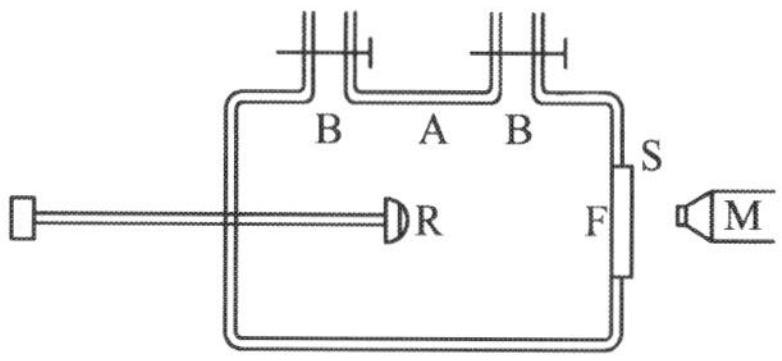

The apparatus used by Rutherford is as shown in Fig.

i. It consists of a chamber A provided with an adjustable rod, carrying a radio - active substance R (Radium C).
ii. The side of the glass tube facing 'R' is covered by metal plate with a central hole which is closed by a thin silver foil 'F'.
iii. A screen 'S', coated with a fluorescent material like zinc sulphide is arranged infront of the silver foil and the scintillations produced on it can be observed through the microscope 'M'.
iv. The side tubes B, B were used to fill various gases in the chamber.
v. The source of α - particles, Ra was placed on a small disc at R. Its distance from F was adjustable.
vi. The radio-active substance emits α -particles whose range in air was found to be about 7cm.
vii. When the glass tube is filled with nitrogen gas, scintillations are observed, even when 'R' is at a distance of 40cm from the foil.
viii. These particles producing scintillations can not be α -particles as they can not have such a long range.
ix. Rutherford concluded that nitrogen nucleus hit by an α (${}_2He^4$)-particle transmutes into oxygen nucleus along with a proton (${}_1H^1$).
x. The nuclear reaction causing artificial transmutation can be represented as ${}_7N^{14}+{}_2He^4\rightarrow{}_8O^{17}+{}_1H^1$
Thus an atom of nitrogen is transformed into an isotope of oxygen. This process is called transmutation of elements.

* High energy α - particles were used in the discovery of artificial transmutation and neutron because α - particles produce intense ionisation of the medium through which they pass and can be stopped after travelling a few mm in air.

Significance :

i) It leads to the discovery of proton and neutron.

ii) It helps to produce radio isotopes.

iii) It helps to produce transuranic elements.

RADIO ISOTOPES AND THEIR USES:

Radio isotopes have very short half lives and hence used for various purposes.

1) Medical applications :

a) $_{13}Al^{27} +_0 n^1 \rightarrow_{11} Na^{24*} +_2 He^4$

Radio – sodium is used to find out how a given medicine is circulated in the body. It is also used to find out circulatory disorders in blood vessels.

b) $_{15}P^{31} +_1 H^2 \rightarrow_{15} P^{32*} +_1 H^1$

Radio – phosphorus is used in the treatment of skin diseases. It is also used for the treatment of blood disorders.

c) $_{53}I^{127} +_0 n^1 \rightarrow_{53} I^{128*} + \gamma - \text{rays}$

Radio – iodine is used in the treatment of thyroid glands. Radio - iodine (I^{131}) is used for diagnosis and treatment of brain tumor and for the study of pumping condition of heart.

d) $_{27}Co^{59} +_0 n^1 \rightarrow_{27} Co^{60*} + \gamma - \text{rays}$ Radio – cobalt is used in the detection and treatment of cancer

e) Radio-iron is used to detect anemia and treat anemia.

2) In Geology:

a) Radio carbon(C^{14}) is used to determine the age of fossils by radio - carbon dating

b) Radio isotopes are used to determine the age of rocks by the ratio of U^{238} to Pb^{206}

3) In industry :

a) Radio – isotopes are used to find the wear and tear of machine parts

b) Radio isotopes are used to detect flaws in metal structures

c) Radio isotopes are used for treatment of alloys such as quenching , annealing and hardening.

d) Radio isotopes are used in the selection of appropriate lubricants.

4) In research: Radio - isotopes are used in the study of nuclear disintegrations of elements.

5) In food preservation: By exposing vegetables and other food stuffs to radiations from radio - active isotopes, their shelf life can be increased.

6) In agriculture:

a) Radio phosphorus (P^{32}) is used to study the uptake of phosphorus by plants using.

b) Radio sulphur (S^{34}) is used to study the transport of minerals in plants .

c) Radio zinc is used to develop new species of plants by causing genetic mutation.

d) Irradiation by γ - radiations of seeds to improve yields.

7) In Chemistry :

a) Radio oxygen (O^{18}) is used to study the mechanisms of photosynthesis and hydrolysis of ester

b) Radio isotopes are used in the chemical analysis of solubility of sparingly soluble salts such as $PbSO_4$ and AgCl and determination of trace amounts of elements in industrial raw materials and products.

NEUTRON

- It is electrically neutral and its mass is slightly greater than that of proton. It was discovered by chadwick
- Bothe – Becker equation:

$$_4Be^9 + _2He^4 \rightarrow \left[_6C^{13} \right] \rightarrow _6C^{12} + _0n^1$$

- Neutron is unstable outside the nucleus.

$_0n^1 \rightarrow _1H^1 + _{-1}e^0 + \bar{v}$ (anti neutrino)

- It has high penetrating power and low ionizing power.
- Slow moving neutrons are called thermal neutrons. Fast moving neutrons convert into thermal neutrons when they pass through a substance called moderator.
- Thermal neutrons have an average energy of nearly 0.025 eV. Fast moving neutrons have an average energy of 2 MeV.

NUCLEAR FISSION

- Nuclear Fission is a nuclear reaction in which a heavy atomic nucleus like U^{235} splits into two approximately equal parts, emitting neutrons and liberating large amount of energy.
- Bohr and Wheeler proposed liquid drop model to explain this fission process.
- Nucleus of U^{235} undergoes fission when it is struck by slow neutrons. This fission is not due to the impact of neutron.
- Energy of about 200 MeV is released during one fission reaction of $_{92}U^{235}$. The most probable nuclear fission reaction is

$$_{92}U^{235} + _0n^1 \rightarrow _{56}Ba^{141} + _{36}Kr^{92} + 3\,_0n^1 + \text{energy}$$

There is no guarantee that U^{235} always breaks into Barium and Krypton.

- On an average, in the fission of U^{235}, 2.5 neutrons are emitted per fission when fission occurs due to slow neutrons. U^{235} undergoes fission with fast neutrons also. But this probability is minimum.
- Fission fragments are unstable and emit neutrons some time after fission reaction which are called "delayed neutrons"
- 99% of neutrons emitted during fission process are prompt.
- Delayed neutrons play an important role in chain reaction

CHAIN REACTION:

- If the mass of fissionable material exceeds a critical value, chain reaction or self propagating fission reaction takes place.
- The rate of reaction increases in geometric progression during uncontrolled chain reaction.
- Chain-reaction : The process of continuation of nuclear fission which when once started continues spontaneously without the supply of additional neutrons from outside is defined as chain reaction.

Reproduction factor (K): "It is the ratio of number of neutrons in any particular generation to the number of neutrons in the preceeding generation.

Case(i): K<1 ; Chain reaction is not maintained.(sub-critical state)

Case (ii): K=1 : Chain reaction is maintained at steady rate. (critical state). In the state electricity is produced in the reactors at steady rate

Case (iii): K>1 : Chain reaction becomes self sustained and lead to atomic explosion(super critical state)

- Uncontrolled chain reaction takes place in atom bomb.

NUCLEAR REACTOR OR ATOMIC PILE:

- Nuclear reactor is a device in which nuclear fission is produced by controlled self sustaining chain reaction. And is used for the production of nuclear power (energy).
- The essential parts of a nuclear reactor are (i) the fuel, (ii) moderator, (iii) control rods, (iv) coolant, (v) radiation shields.
- **THE FUEL:** The common fuels used are uranium (U^{238}), enriched uranium (U^{235}) and plutonium (Pu^{236}) and Th^{232}.

MODERATOR:

- The function of a moderator is to slow down the fast moving neutrons to increase the rate of fission.
- The commonly used moderators in the order of efficiency are (i) Heavy water (ii) graphite, (iii) Berillium and Berillium Oxide
- Heavy water is a best moderator
- A good moderator should have
  (1) low atomic mass
  (2) poor absorption of neutrons
  (3) good scattering property.
  (4) The size of moderator atom should be nearly of same size as that of the size of a prompt neutron.

CONTROL RODS:

- The function of a control rod is to absorb (capture) the neutrons.
- Cadmium, Boron and steel rods are used as control rods in a nuclear reactor.
- Cadmium rods are best control rods
- They regulate the net rate of neutron production and hence they control the intensity of fission process.

COOLANT:

- The function of a coolant is to keep the reactor temperature at a low value so that there may not be any danger of heat damage to the reactor.
- Air and CO_2 are used as gaseous coolants. Water, Organic liquids, Helium, Liquid Sodium are used as liquid coolants. Liquid sodium is best coolant.
- **Protective shield:** The process of preventing radioactive effect around nuclear reactor is called Protective Shield.
- During the working of a nuclear reactor dangerous radiations such as high energy neutrons, gamma rays and thermal radiations are produced. To protect the persons working there, the reactor is thoroughly shielded with concrete wall of several feet thick and lined with metals like lead.

POWER OF A NUCLEAR REACTOR

In the nuclear reactor, large amount of heat will be generated in the core. These reactors have elaborate cooling systems that use water. This water absorbs the heat and produces steam. This steam in turn is used to run the steam turbines which ultimately generate electric power. Such reactors are called power reactors.

The power generated by a nuclear reactor is $\boxed{P = \frac{nE}{t}}$ here $\frac{n}{t}$ be the number of fissions per second and E be the energy released in each fission

$E = 200\ \text{MeV} = 200\times10^{6}\times1.6\times10^{-19}\ \text{J}$

$= 3.2 \times 10^{-11}\ \text{J}$

Note : Number of fissions per sec in a reactor of power 1 W is given by $\frac{n}{t}=\frac{P}{E}$

$$=\frac{1}{3.2\times10^{-11}}=3.125 \times 10^{10} \text{ fissions per sec}$$

Note : If only x% of energy released in fission is converted into electrical energy then out put power of reactor is $P=\frac{x}{100}\left(\frac{nE}{t}\right)$

Note: If 'x' gm of fuel with mass number 'A' completely undergo nuclear fission in time t sec in a reactor then its power is given by

Number of moles in x gm of fuel $=\frac{x}{A}$

Number of atoms (nuclei) present in x gm of fuel $n=\left(\frac{x}{A}\right)N_A$. Where N_A is Avogadro number

$\therefore$ power $P=\frac{nE}{t} \Rightarrow \boxed{P=\frac{xN_AE}{At}}$

Uses of Nuclear Reactors:

1) To generate electric power.

2) To produce nuclear fuel plutonium -239 and other radioactive materials which have a wide variety of applications in the fields of medicine, industry and research.

USES OF ATOMIC ENERGY :

1. **Generation of electric power** : The coolant in a nuclear reactor absorbs the heat generated as a result of the chain reaction and it releases the heat to the water which is converted into high pressure steam. This steam is used to drive turbine and operate the electric generator.
2. **Production of radio isotopes** : A small amount of the pure element is placed in an aluminium container and the container is placed in the reactor for a few days. The element absorbs neutrons and the element becomes radioactive isotope.

   $_{53}I^{127}+_0n^1\rightarrow_{53}I^{128*}+\gamma$.

   Radioiodine obtained in this way can be used to treat the thyroid gland. These radio isotopes have a number of applications in the field of medicine , agriculture, industry and basic research.
3. **Source of neutrons :** A large number of neutrons are produced in a reactor. They are used in research . The effect of neutrons on biological tissues is studied. A new branch of physics called Neutron Physics has come up.
4. Atomic energy is used to create artificial lakes, to divert the course of a river , to make tunnels for laying new railway tracks etc.
5. Atomic energy is used for driving automobiles, submarines and war - planes.
6. Atomic energy is used in war - fare for creating destructive atom bombs and hydrogen bombs.

NUCLEAR FUSION:

➢ The phenomenon in which two lighter nuclei combine to form a heavier nucleus of mass less than the total mass of the combining nuclei is called nuclear fusion. This mass defect appears as energy.

- At temperatures of about 10^7K, light nuclei combine to give heavier nuclei. Hence, fusion reactions are called thermo nuclear reactions.
- Nuclear fusion takes place in the sun and other stars.
- Energy produced in a single fission of $_{92}U^{235}$ is larger than that in a single fusion of Hydrogen into Helium.
- But fusion produces more energy than fission per nucleon.
- In fission, 0.09% of mass is converted into energy. In fusion 0.66% of mass is converted into energy.
- Hydrogen bomb is a fission – fusion bomb.

STELLAR AND SOLAR ENERGY:

Stellar and solar energy is due to fusion.

The cycles that occur are.

Proton - Proton Cycle & Carbon - Nitrogen Cycle

PROTON - PROTON CYCLE:

The Thermonuclear reactions involved are:

$$2(_1H^1)+2(_1H^1)\rightarrow 2(_1H^2)+2(_{+1}e^0)+Q_1$$

$$2(_1H^2)+2(_1H^1)\rightarrow 2(_2He^3)+Q_2$$

$$_2He^3+{}_2He^3\rightarrow(_2He^4)+2(_1H^2)+Q_3$$

On adding up these reactions, we obtain.

$$4(_1H^1)\rightarrow {}_2He^4+2\,{}_{+1}e^0+Q$$

Where $Q=Q_1+Q_2+Q_3$ is the total energy evolved in the fusion of 4 hydrogen nuclei (protons) to form Helium nucleus. The value of Q as calculated from mass defect comes out to be 26.7MeV

CARBON - NITROGEN CYCLE:

Proposed by bethe. It consists of following reactions.

$$_6C^{12}+{}_1H^1\rightarrow {}_7N^{13}+Q_1\;;\; {}_7N^{13}\rightarrow {}_6C^{13}+(_1e^0)+Q_2$$

$$_6C^{13}+{}_1H^1\rightarrow {}_7N^{14}+Q_3;\; {}_7N^{14}+{}_1H^1\rightarrow {}_8O^{15}+Q_4$$

$$_8O^{15}\rightarrow {}_7N^{15}+(_{+1}e^0)+Q_5;$$

$$_7N^{15}+{}_1H^1\rightarrow {}_6C^{12}+{}_2He^4+Q_6$$

On adding all these 6 equations, we get

$$4\,{}_1H^1\rightarrow {}_2He^4+2(_{+1}e^0)+Q$$

Where $Q=Q_1+Q_2+Q_3+Q_4+Q_5+Q_6$ The value of Q as calculated from mass defect is 26.7 Mev.

- In the sun both proton - proton cycle and carbon- nitrogen cycles occur with equal probabilities. In stars, whose interior temperatures are less than that of the sun, proton -proton cycle dominates the energy generation. Again in stars, whose interior temperatures are more than that of the sun, the energy generation is mainly due to carbon- nitrogen cycle.
- The core temperature of heavier stars may be larger than that of the sun and much larger nuclei may be formed.

NOTE: Due to enormous energy released in Sun and Stars the atmosphere of them will be in ionised state which is called Plasma (Which contains fast moving neutrons and electrons). Nuclear fusion can not be controlled.

Nuclear Fission	Nuclear Fusion
1) Neutrons are required for it	1) Protons are required for it
2) It is possible at normal pressure and temperature	2) It is possible at high pressure and temperature
3) Energy released per nucleon ≅0.9MeV	3) Energy released per nucleon ≅ 6MeV
4) % of mass getting converted into energy =0.1%	4) % of mass getting converted into energy = 0.7%
5) Fissionable materials are expensive.	5) Fusion materials are cheap
6) Harmful reactions are produced.	6) Harmful reactions are not produced.

W.E-22: ***An explosion of atomic bomb releases an energy of 7.6 x 10^{13} J. If 200 MeV energy is released on fission of one ^{235}U atom calculate (i) the number of uranium atoms undergoing fission. (ii) the mass of uranium used in the atom bomb***

Sol: E=7.6x10^3J;Energy released per fission = 200 MeV

$= 200 \times 10^6 \times 1.6 \times 10^{-19} = 3.2 \times 10^{-11}$ J

$$\text{Number of uranium atoms (n)} = \frac{\text{Total energy}}{\text{Energy per fission}}$$

$$n = \frac{7.6 \times 10^{13}}{3.2 \times 10^{-11}} = 2.375 \times 10^{24} \text{ atoms}$$

Avagadro number (N) = 6.023×10^{23}atoms

Mass of uranium =

$$\frac{n \times 235}{N} = \frac{2.375 \times 10^{24} \times 235}{6.023 \times 10^{23}} = 92.66\text{g}$$

W.E-23: ***Calculate the energy released by fission from 2 g of $^{235}_{92}U$ in kWh. Given that the energy released per fission is 200 MeV.***

Sol. Mass of uranium = 2g

Energy released per fission = 200 MeV

$= 200 \times 10^6 \times 1.6 \times 10^{-19} = 3.2 \times 10^{-11}$ J

Number of atoms in 2 gram of uranium is

$$n = \frac{2 \times 6.023 \times 10^{23}}{235} = 5.125 \times 10^{21} \text{ atoms}$$

Total energy released = No. of atoms x energ released per fission

$= 5.125 \times 10^{21} \times 3.2 \times 10^{-11} = 16.4 \times 10^{10}$J

$$\therefore \text{Energy in Kwh} = \frac{16.4 \times 10^{10}}{36 \times 10^5} \text{Kwh}$$

$= 0.455 \times 10^5$ Kwh $= 4.55 \times 10^4$ Kwh

W.E-24: ***200 Mev energy is released when one nucleus of ^{235}U undergoes fission. Find the number of fissions per second required for producing a power of 1 megawatt.***

Sol. Energy released = 200MeV

$= 200 \times 10^6 \times 1.6 \times 10^{-19} = 3.2 \times 10^{-11} J$

P = 1 mega watt = 10^6 watts.

$$\text{No.of fissions per second (n)} = \frac{\text{Total energy}}{\text{Energy per fission}}$$

$$n = \frac{10^6}{3.2 \times 10^{-11}} = 3.125 \times 10^{16} \text{ Fissions}$$

W.E-25: ***How much ^{235}U is consumed in a day in an atomic power house operating at 400 MW, provided the whole of mass ^{235}U is converted into energy?***

Sol. Power = 400 MW = 400 x 10^6 W;

time = 1 day = 86, 400 s.

Energy produced , E = power $\times$ time = 400 $\times$ 10^6 $\times$ 86,400 = 3.456 $\times$ 10^{13} J.

As the whole of mass is converted into energy, by Einstein's mass -energy relation.

$E = Mc^2$

$$\frac{E}{c^2} = \frac{3.456 \times 10^{13}}{(3 \times 10^8)^2} = 3.84 \times 10^{-4}\ kg = 0.384\ g.$$

W.E-26: ***How long can an electric lamp of 100W be kept glowing by fusion of 2.0 kg of deuterium ? Take the fusion reaction as $^2_1H + ^2_1H \to ^3_2He + n + 3.27$ MeV***

Sol. $_1H^2 +_1 H^2 \to^3_2 He + n + 3.27MeV$

No. of atoms in 2 kg of $_1H^2$ = 2/2 x 6.023 x 10^{26} = 6.023 x 10^{26} atoms

In the above reaction two deuterium nuclei are combined

Power (p) = w x rate of fusion.

$$= 3.27 \text{ MeV x } \frac{\textit{Number of atoms}}{\textit{Time expended}}$$

$$100 = 3.27 \times 10^6 \times 1.6 \times 10^{-19} \times \frac{6.023 \times 10^{26}}{2x}$$

$$\therefore x = \frac{3.27 \times 1.6 \times 6.023 \times 10^{+11}}{2} = 15.756 \times 10^{11} S$$

$$= \frac{15.756 \times 10^{11}}{365 \times 24 \times 60 \times 60} = \frac{15.756 \times 10^{11}}{3.15 \times 10^7} = 5 \times 10^4 \text{ years}$$

W.E-27: ***Suppose India had a target of producing by 2020 AD, 200,000 MW of electric power, ten percent of which was to be obtained from nuclear power plants. Suppose we are given that, on an avedrage, the efficiency of utilization (i.e conversion to electric energy) of thermal energy produced in a reactor was 25%. How much amount of fissionable uranium would our country need per year by 2020 ? Take the heat energy per fission of ^{235}U to be about 200 MeV.***

Sol. Required power from nuclear plants

= 10% of 2,00,000 MW = 2 x 10^{10}w

Required electric energy from nuclear plants in one year = 2 x 10^{10} x 365 x24 x 60 x 60

= 2 x 10^{10} x 3.15 x 10^{7} = 6.30 x10^{7} J

Available electric energy per fission = 25% of 200 MeV = 50 MeV = 8 x 10^{-12}J

Req. no. of fissions per year = $\frac{6.30\times10^{17}}{8\times10^{-12}}$ =0.7875x10^{29}

Req. no. of moles of U^{238} = $\frac{0.7875\times10^{29}}{6.023\times10^{23}}$ =0.1307 x10^{6}

Required mass of U^{238} = 0.1307 x 235 x10^{6} g

= 30.71 x10^{6} gm = 30.71 x10^{6} x 10^{-3} kg

= 0.03071 x 10^{6} kg = 3.071 x 10^{4} kg

W.E-28: Calculate the energy released by the fission 1 g of ^{235}U in joule, given that the energy released per fission is 200 MeV.
(Avogadro's number = 6.023 $\times$ 10^{23})

Sol. The number of atoms in 1 g of ^{235}U

$$=\frac{Avogadro's\ number}{Mass\ \ number}=\frac{6.023\times10^{23}}{235}=2.563\times10^{21}$$

Energy released per fission = 200 MeV

$$=200\times10^{6}\times1.6\times10^{-19}=3.2\times10^{-11}J.$$

Energy released by 1 g of ^{235}U

= Number of atoms $\times$ energy released per fission

$$=2.563\times10^{21}\times3.2\times10^{-11}J=8.202\times10^{10}J$$

W.E-29: ***In the process of nuclear fission of 1 gram uranium, the mass lost is 0.92 milligram. The efficiency of power house run by it is 10%. To obtain 400 megawatt power from the power house, how much uranium will be required per hour? (c = 3 x 10^{8} ms^{-1})***

Sol. Power to be obtained from power house = 400 mega watt\ Energy obtained per hour = 400 meagwatt x 1hour = (400 x 10^{6} watt) x 3600 second = 144 x 10^{10} joule

Here only 10% of input is utilised. In order to obtain 144 x 10^{10} joule of useful energy, the output energy from the power house $\frac{10E}{100}$ = 144 x 10^{10}J

E = 144 x 10^{11} joule

Let, this energy is obtained from a mass–loss of Δm kg.

Then $(\Delta m)c^2=144\times10^{11}$joule

$$\Delta m=\frac{144\times10^{11}}{\left(3\times10^{8}\right)^{2}}=16\times10^{-5}kg \quad =0.16\ \text{g}$$

Since 0.92 milli gram (= 0.92 x 10^{-3} g) mass is lost in 1 g uranium, hence for a mass loss of 0.16g, the uranium required is $=\frac{1\times0.16}{0.92\times10^{-3}}=174\ g$

Thus to run the power house, 174 gm uranium is required per hour.

➢ **PAIR AND PRODUCTION AND PAIR ANNIHILATION :** When an energetic γ - photon falls on a heavy nucleus, it is absorbed by the nucleus and a pair of electron and positon is produced. This phenomenon is called as *pair production* and can be represented by the following equation:

$$\underset{(\gamma-photon)}{hv} = \underset{(Positron)}{{}_{+1}\beta^0} + \underset{(electron)}{{}_{-1}\beta^0}$$

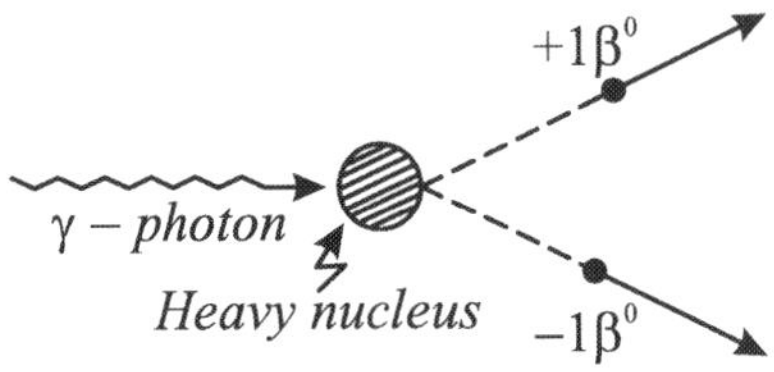

The rest mass energy of electron or positron is:

$$E_0 = m_0c^2 \quad = \left(9.1\times10^{-31}\right)\times\left(3\times10^8\right)^2$$

$$= 8.2\times10^{-14}J \;\simeq 0.51MeV.$$

Hence for pair production, the minimum energy of γ -photon must be $2\times0.51 = 1.02MeV$. If the energy of γ -photon is less than this, there may be Compton's effect. If energy of γ -photon is greater than E_0, then extra energy will become kinetic energy of the particles. If E is the energy of γ -photon, then kineric energy of each particle will be, $K_{electron} = K_{positron} = \dfrac{E-2E_0}{2}$

The inverse process of pair production is called *pair annihilation*. According to it when electron and a positron come close to each other, annihilate each other and produces minimum two γ -photons. Thus

$$\underset{(Positron)}{{}_{+1}\beta^0} + \underset{(electron)}{{}_{-1}\beta^0} = \underset{(\gamma-photon)}{2hf}$$

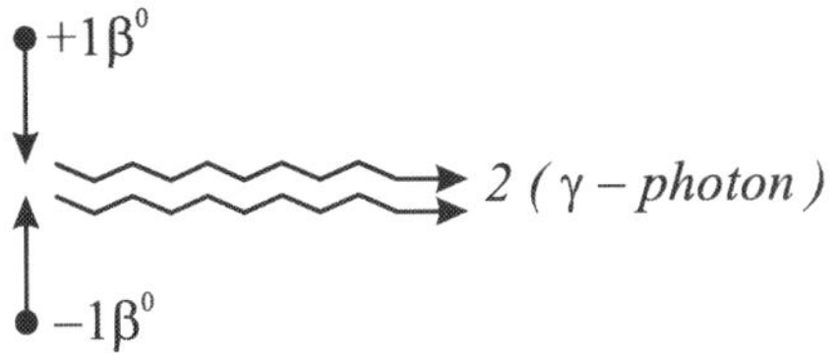

ADDITIONAL INFORMATION

ELEMENTARY PARTICLES :

We have realized, so far that there are only four fundamental constituents of matter. We can describe various physical processes involving atoms, molecules and nuclei in terms of electrons, protons, neutrons and photons. The first three are the building blocks of atoms and hence matter. The fourth one (i.e photon) is the quantized energy which is exchanged whenever electronic or nucleonic transition is involved.

Subsequently many more elementary particles and antiparticles have been discovered, using giant and modern accelerating machines.

The particles which are not constituted by any other particles are called Elementary particles. A brief discussion of important fundamental particles is as follows.

i) **Electron :** It was discovered in 1897 by Thomson. Its charge is –e and mass is 9.1 X 10^{-31} kg. Its symbol is e^- (or $_{-1}\beta^0$). It is a stable particle having spin $= \frac{1}{2}$

ii) **Proton :** It was discovered in 1919 by Rutherford in artificial nuclear disintegration. It has a positive charge +e and its mass is 1836 times (1.673 X 10^{-27}kg) the mass of electron. In free state, the proton is a stable particle. Its symbol is P^+. It is also written as $_1H^1$. It is a stable particle having spin = 1/2.

iii) **Neutron :** It was discovered in 1932 by Chadwick. Electrically it is a neutral particle. Its mass is 1839 times (1.675 X 10^{-27} kg) the mass of electron. In free state the neutron is unstable. Inside the nucleus the neutron is stable. Its symbol is n (or) $_0n^1$.

iv) **Positron :** It was discovered by Anderson in 1932. It is the antiparticle of electron, i.e., its charge is +e and its mass is equal to that of electron. Its symbol is e^+ (or $_{+1}\beta^0$)

v) **Antiproton :** It is the antiparticle of proton. It was discovered in 1955. Its charge is –e and its mass is equal to that of proton. Its symbol is P^-.

vi) **Antineutron :** It was discovered in 1956. It has no charge and its mass is equal to the mass of neutron. **The only difference between neutron and antineutron is that their magnetic momenta will be equal in magnitude and opposite in direction**. The symbol for antineutron is $\overline{n}$.

vii) **Neutrino and antineutrino :** The existence of these particles was predicted in 1930 by Pauli while explaining the emission of β - particles from radioactive nuclei, but these particles were actually observed experimentally in 1956. Their rest mass and charge are both zero but they have energy and momentum. These are mutually antiparticles of each other. They have the symbol ν and $\overline{\nu}$

viii) **Pi - Mesons :** The existence of pi - mesons was predicted by Yukawa in 1935, but they were actually discovered in 1947 in cosmic rays. Nuclear forces are explained by the exchange of pi-mesons between the nucleons. pi - mesons are of three types, positive π - mesons (π^+), negative pi-mesons (π^-) and neutral π - mesons (π^0). Charge on $\pi^\pm$ is $\pm$e. Whereas mass of $\pi^\pm$ is 274 times the mass of electron. π^0 has mass nearly 264 times the electronic mass. These are unstable having half life 10^{-8} sec and spin = 0

ix) **Mu-Mesons :** These were discovered in 1936 by Anderson and Neddermeyer. These are found in abundance in the cosmic rays at the ground level. There are two types of mu-mesons. Positive mu-meson (μ^+) and negative mu-meason (μ^-). There is no neutral mu-meson. Both the mu-mesons have the same rest mass 207 times the rest mass of the electron. These are unstable having half life 10^{-6} sec and spin = 1/2.

x) **Photon :** These are bundles of electromagnetic energy and travel with the speed of light. Energy and momentum of a photon of frequency ν are $h\nu$ and $\frac{h\nu}{c}$ respectively. They posses no charge. no mass and spin = 1 and are stable.

xi) **Gravitons :** Hypothetical particles that carry gravitational energy are called Gravitons. They possess no mass, no charge and spin = 2 as proposed by Dirac.

Antiparticles : An antiparticle is a form of matter that has the same mass as the particle but carries an opposite charge and / or a magnetic moment that is oriented in an opposite direction relative to the spin.

Name of the particle	Antiparticle
Electron (e^{-1})	Positron (e^{+1})
Proton (P^+)	Anti proton (P^-)
Neutron (n)	Anti Neutron $(\bar{n})$
Neutrino (ν)	Anti Neutrino $(\bar{\nu})$
Positive Pi-Meson (π^+)	Negative Pi-Meson(π^-)
Positive Mu-Meson (μ^+)	Negative Mu-Meson(μ^-)

Note: A few electrically neutral particles, like the photon and neutral π meson are their own antiparticles. A collision between a particle and an antiparticle results in annihilation of matter.

Classification of particles based on spin

1) **Bosons :** These particles have spin in the integral multiples of unity
2) **Fermions :** These particles have spin in the integral multiples of 1/2.

Classification of particles based on rest mass

1) **Photons :** Particles with zero rest mass
2) **Leptons :** Lighter particles
3) **Mesons :** Particles with intermediate mass
4) **Baryons :** Heavier particles

Classification of particles based on interaction

1) **Photons:** representing electromagnetic interactions.
2) **Leptons:** representing weak interactions.
3) **Hadrons:** representing strong interactions.
4) **Gravitons:** representing gravitational interactions.

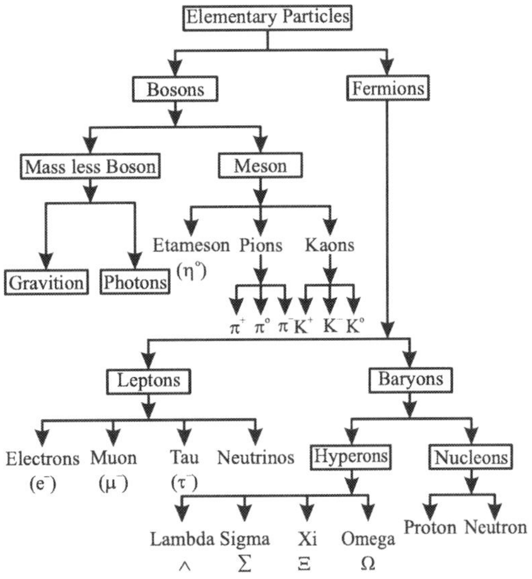

THE QUARK MODEL:

According to this model, each baryon and meson particle is made up of quarks. Quarks are supposed to be the ultimate building blocks of these particles and hence matter. They are point particles with no internal structure. Unlike any other charged particles, quarks are supposed to have fractional charges and spin 1/2.

Name	Quarks Symbol	Quarks Charge	Antiquarks Symbol	Antiquarks Charge
Up	u	$+\frac{2}{3}e$	$\bar{u}$	$-\frac{2}{3}e$
Down	d	$-\frac{1}{3}e$	$\bar{d}$	$+\frac{1}{3}e$
Strange	s	$-\frac{1}{3}e$	$\bar{s}$	$+\frac{1}{3}e$
Charmed	c	$+\frac{2}{3}e$	$\bar{c}$	$-\frac{2}{3}e$
Top	t	$+\frac{2}{3}e$	$\bar{t}$	$-\frac{2}{3}e$
Bottom	b	$-\frac{1}{3}e$	$\bar{b}$	$+\frac{1}{3}e$

These quarks combine to form some well known Mesons and Baryons. A meson consists of a quark and an antiquark whereas a baryon contains three quarks.

Particle	Quarks	charge	spin
proton	(u, u, d)	e	$\frac{1}{2}\uparrow$
neutron	(u d d)	0	$\frac{1}{2}\downarrow$
pi-meson π^+	$(u\bar{d})$	e	0

Photons and leptons are considered to be elementary particles and they are not composed of quarks.

Note : Isolated quark doesn't exist.

W.E-30: ***An electron-positron pair is produced when a γ-ray photon of energy 2.36MeV passes close to a heavy nucleus. Find the kinetic energy carried by each particle produced, as well as the total energy with each.***

Sol. The reaction is represented by

$$\gamma \rightarrow ({}_{-1}e^0) + ({}_{+1}e^0), \text{ so that}$$

$$E = m_0C^2 + K.E_{electron} + m_0C^2 + K.E_{positron} \quad 2.36\text{MeV} = 2m_0.C^2 + \text{K.E}_{(electron)} + \text{K.E}_{(positron)}$$

$$= 1.02 \text{ MeV} + K.E_{(e^-)} + K.E_{(e^+)}$$

$$\text{K.E. of } (e^-) = K.E_{(e^+)} = \frac{1}{2}(2.36 - 1.02)MeV,$$

(K.E. carried each) = 0.67 MeV (motional energy)

Total energy shared by each particle is obviously $m_0C^2 + K.E = 0.51MeV + 0.67MeV = 1.18$MeV.

W.E-31: ***A gamma ray photon of energy 1896 MeV annihilates to produce a proton-antiproton pair. If the rest mass of each of the particles involved be 1.007276 a.m.u approximately, find how much K.E these will carry?***

Sol. Working on the same lines as an electron-positron pair production, we notice that the reaction. $\gamma \rightarrow$ proton + antiproton, has the energy balance

$E = m_{0\,(proton)} C^2 + K.E_{(proton)} +$
$m_{0\,(antiproton)} C^2 + K.E_{(antiproton)}$

But m_0C^2 = energy equivalent of 1.007276 a.m.u $\approx$ 938 MeV. [$\because$ 1.007276 x 931 $\approx$ 938 MeV]
Thus K.E of each particle

$= \frac{1}{2}$ [1896 MeV - 2 $\times$ 938 MeV] = 10MeV.

W.E-32: Obtain the maximum kinetic energy of β-particles, and the radiation frequencies of γ decays in the decay scheme shown in Fig. 14.6. You are given that
$m(^{198}Au) = 197.968233$ u; $m(^{198}Hg) = 197.966760$ u

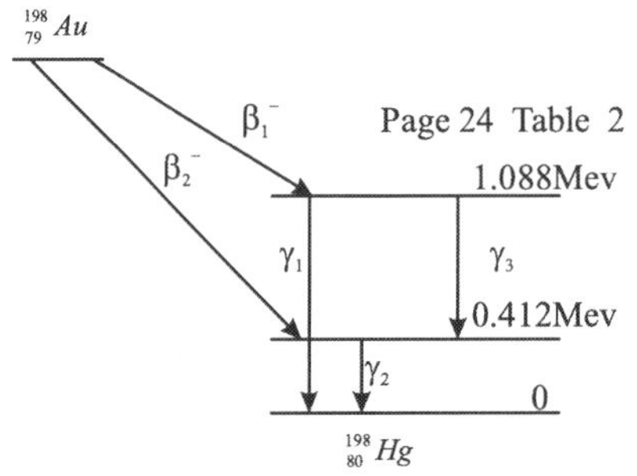

Sol. γ-rays are electro magnetic radiations having energy $E = h\upsilon \Rightarrow \upsilon = \frac{E}{h}$ where h = plank`s constant = 6.625 x 10^{-34} J.S

1) Frequencies of γ_1, γ_2 and γ_3 are calculated as follows

$$\gamma_1 = \frac{\Delta E}{h} = \frac{(1.088-0)MeV}{6.625\times10^{-34} J.s} = \frac{1.088\times10^6\times1.6\times10^{-19}}{6.625\times10^{-34}}$$

= 0.2627 x 10^{21} = 2.627 x 10^{20} Hz

$$\gamma_2 = \frac{\Delta E}{h} = \frac{(0.412-0)MeV}{6.625\times10^{-34} J.s} = \frac{0.412\times1.6\times10^{-19}\times10^6}{6.625\times10^{-34}}$$

= 0.0995 x10^{21} = 9.95 x 10^{19}Hz

$$\gamma_3 = \frac{\Delta E}{h} = \frac{(1.088-0.412)\times10^6\times1.6\times10^{-19}}{6.625\times10^{-34}}$$

= 0.1632 x 10^{21} = 1.632 x 10^{20} Hz

2) Now maximum K.E of β_1^- = [M($^{198}_{79}$Au)

M($^{198}_{80}$Hg)- $\frac{1.088}{931.5}$] c^2

($\because$ 1 amu = 931.5 MeV $\Rightarrow$ 1MeV = $\frac{1}{931.5}U$)

$= [197.968233 - 197.966760 - 0.001168]931.5$ MeV

$= 0.000305 \times 931.5 = 0.284$ MeV

Maximum K.E of $\beta_2^- = [M(^{198}_{79}Au) - M(^{198}_{80}Hg) - \frac{0.412}{931.5}]c^2$

$= [197.968233 - 197.966760 - 0.000442]931.5$

$= 0.001031 \times 931.5 = 0.9603$ MeV

CONCEPTUAL

COMPOSITION OF NUCLEUS, ISOTOPES, ISOBARS, ISOTONES SIZE OF THE NUCLEUS

1. **The particle A is converted to C via following reactions then**

   $A \rightarrow B + _2He^4$ $\quad$ $B \rightarrow C + 2\,_{-1}e^0$

   1) A and C are isobars $\quad$ 2) A and C are isotopes

   3) A and B are isobars $\quad$ 4) A and B are isotopes

2. **The particles which can be added to the nucleus of an atom without changing its chemical properties are**

   1) electrons $\quad$ 2) protons $\quad$ 3) neutrons $\quad$ 4) positron

3. **The radius of the nuclues is proportional to, (if A is the atomic mass number)**

   1) A $\quad$ 2) A^3 $\quad$ 3) $A^{1/3}$ $\quad$ 4) $A^{2/3}$

4. **The radius of a nucleus mainly depends on**

   1) Proton number $\quad$ 2) Electron Number $\quad$ 3) Mass number $\quad$ 4) Neutron number

5. **The nuclei $_6C^{13}$ and $_7N^{14}$ can be described as**

   1) isotones $\quad$ 2) isobars $\quad$ 3) isomers $\quad$ 4) isotopes

6. **The graph of ℓn (R/R_0) versus ℓn A (R =radius) of nucleus. A is atomic number is**

   1) Straight line $\quad$ 2) Parabola $\quad$ 3) Ellipse $\quad$ 4) Circle

7. **The nucleus of $_{56}Ba^{141}$ contains**

   1) 85 protons, 56 neutrons $\quad$ 2) 55 protons ,86 neutrons

   3) 56 protons,85 neutrons $\quad$ 4) 86 protons,55neutrons.

8. **The nuclear size is measured in units of**

   1) Angstrom $\quad$ 2) Fermi $\quad$ 3) Bar $\quad$ 4) Light-year

9. **Nuclides which have the same mass number are called**

   1) Isotopes $\quad$ 2) Isobars $\quad$ 3) Isotones $\quad$ 4) Isomars

10. **Observe the following statements regarding isotones**

    i) $^{39}K_{19}$ and $^{40}Ca_{20}$ are isotones

    ii) Nuclides having different atomic numbers (z) and mass number (A) but same number of neutrons (n) are called Isotones

    iii) $^{19}F_9$ and $^{23}Na_{11}$ are isotones

    The correct answer is

    1) i,ii and iii are correct $\quad$ 2) only i and ii are correct

    3) only i and iii are correct $\quad$ 4) only ii and iii are correct

11. A and B are isotopes. B and C are isobars. All three are radioactive. Which one of the following is true.

1) A, B and C must belong to the same element 2) A, B and C may belong to the same element

3) It is possible that A will change to B through a radioactive-decay process

4) It is possible that B will change to C through a radioactive-decay process

MASS DEFECT, BINDING ENEGRY, PACKING FRACTION AND MASS ENERGY RELATION

12. M,M_n & M_p denotes the masses of a nucleus of $_ZX^A$ a neutron, and a proton respectively. If the nucleus is separated in to its individual protons and neutrons then

1) $M=(A-Z)M_n+ZM_p$ 2) $M=ZM_n+(A-Z)M_p$

3) $M>(A-Z)M_n+ZM_p$ 4) $M<(A-Z)M_n+ZM_p$

13. The difference between the mass of a nucleus and the combined mass of its nucleons is

1) zero 2) positive 3) negative 4) zero, positive or negative

14. The mass number of a nucleus is

1) Always less than atomic number 2) Always more than atomic number

3) Equal to atomic number 4) Sometimes more or equal to atomic number.

15. If M is atomic weight , A is mass number then (M-A)/A represent

1) Mass defect 2) Packing fraction 3) Binding Energy 4) Chain Reaction

16. The difference between mass of the nucleus and total mass of its constituents is called

1) Packing fraction 2) Mass defect 3) Binding energy 4) Binding energy per neucleon

17. The parameter used to measure the stability of the nucleus is

1) Average binding energy 2) No of protons

3) No of neutrons 4) No of electrons

18. When the number of nucleons in a nuclues increases the binding energy per nucleon

1) Increase continuously with mass number 2) Decreases continuously with mass number

3) Remains constant with mass number

4) First increases and then decreases with increase in mass number

19. Maximum value of binding energy per nucleon is

1) 8MeV 2) 8.8MeV 3) 7.6MeV 4) 1.1MeV

20. The binding energy per nucleon is maximum at A=56 and its value is around_Mev/ Nucleon

1) 8.4 2) 8.7 3) 9 4) 7.8

21. Average binding energy per nucleon over a wide range is

1) 8MeV 2) 8.8MeV 3) 7.6MeV 4) 1.1MeV

22. The wrong statement about binding energy is

1) It is the sum of the rest mass energies of nucleons minus the rest mass energy of the nucleus.

2) It is the energy released when the nucleons combine to form a nucleus.

3) It is the energy required to break a given nucleus into its constituent nucleons.

4) It is the sum of the kinetic energies of all the nucleons in the nucleus.

23. The binding energies of a deutron and an α -particle are 1.125, 7.2 MeV/nucleon respectively. The more stable of the two, is

1) deutron 2) α -particle

3) both 4) sometimes deutron and sometimes α -particle

24. **Mass defect of an atom refers to**
1) inaccurate measurement of mass of neutrons
2) mass annihilated to produce energy to bind the nucleons
3) packing fraction
4) difference in the number of neutrons and protons in the nucleus

25. **The stability of a nucleus can be measured by**
1) Average binding energy
2) Packing fraction
3) Ratio of number of neutrons and protons
4) All the above.

26. **In a nuclear reaction some mass converts into energy. In this reaction total B.E of reactants when compared with that of products is**
1) always greater
2) always les
3) either greater or less
4) always equal

RADIOACTIVITY, α, β, γ RAYS, HALF LIFE, MEAN LIFE, DECAY CONSTANT

27. **The age of pottery is determined by archeologists using a radioisotope of**
1) carbon 2) cobalt 3) iodine 4) phosphorus

28. **During an artificial transmutation the nucleus emits**
1) β-particles
2) α-particles
3)always neutrons
4) may emit protons or neutrons

29. **When two deuterium nuclei fuse together to form a tritium nucleus, we get a**
1) neutron 2) deuteron 3) alpha particle 4) proton

30. **Identify the correct statement/statements**
a) Radiation causes genetic mutation
b) Restriction in blood circulation can be detected using radio-iodine
c) Hydrocarbon plastics are used as moderators in a nuclear reactor
d) The damage caused due to α-radiation is small due to its small penetrating power
1) a,b,c 2) a,c,d 3) b,c,d 4) a,b,d

31. **Identify the correct ascending order of α, β and γ with reference to their ionizing power**
I) α-ray **II) γ-ray** **III) β-ray**
1) II,III,I 2) I,III,II 3) II,I,III 4) I,II,III

32. **Two identical nuclei A and B of the same radioactive element undergo β^- decay. A emits a β^- particle and changes to A'. B emits a β^- particle and then a γ–photon immediately afterwards, and changes to B'.**
1) A' and B' have the same atomic number and mass number
2) A' and B' have the same atomic number and different mass numbers
3) A' and B' have different atomic numbers but the same mass number
4) A' and B' are isotopes

33. **Identify the correct ascending order of α, β and γ with reference to their penetrating power**
I) β-ray **II) α-ray** **III) γ-ray**
1) II,III,I 2) II,I,III 3) I,II,III 4) III,I,II

34. **If a beam consisting of<,< and < radiation is passed through an electric field perpendicular to the beam, the deflections suffered by the components, in decreasing order, are**

1) <,<ς   2) <ς ,<   3) <ς<,   4) <ς<ς

35. **Decrease in atomicnumber is observed during**

A) < -emission   B) < -emission   C)Positron emi ssion   D) electron capture

1) B is correct   2) A and B are correct
3) A,C and D are correct   4) Only C

36. **When $_{15}P^{30}$ decays to become $_{14}Si^{30}$,the particle released is**

1)electron   2) < -particle   3) neutron   4)positron

37. **During $<^{<}$ - decay, a neutron inside nucleus converts into proton, electron and x. Then the particle x is**

1) $<^{<}$ - meson   2)neutrons   3)anti-neutrino   4) $<^{<}$ -meson

38. **If a nucleus emits a gamma-ray, its atomic and mass number____ but there will be ___in the energy of the nucleus.Select suitable pair**

1)Remain same, increase   2)Remain same, decrease
3)Decrease, increase   4)increase, decrease

39. **In the followingnuclear reaction $_{13}Al^{27}+{}_{2}He^{4} <\ {}_{15}P^{30}$+X,X will be**

1)Proton   2) Electron   3)Neutron   4) < -particle

40. **In nuclear reaction $_{4}Be^{9}+{}_{2}He^{4}$-->$_{6}C^{12}$+X,X will be**

1)Proton   2)Neutron   3) < -particle   4) < -particle

41. **In nuclear reaction** $_{2}He^{4} <{}_{z}X^{A} <\ {}_{z<2}Y^{A<2}$ R **R denotes**

1)electron   2)positron   3)proton   4) neutron

42. **A positron is emitted by radioactive nucleus of proton no 90. The product nucleus will have proton number**

1)91   2)90   3)89   4)88

43. **$_{13}Al^{27}$ +< -particle < neutron +'X' then 'X' is**

1) $_{15}p^{31}$ 2) $_{14}Si^{30}$ 3) $_{15}p^{30}$ 4) $_{15}Si^{30}$

44. **The penentrating power of beta particle compared to alpha particle is**

1) Less   2) More   3) Equal   4) Can be more or less

45. **In a nuclear reactor , heavy water is used as a**

1)Controlling material   2) Moderator   3)Fuel   4)Heat exchanger

46. **The units of radioactivity is**

1)Fermi   2)Farad   3)Curie   4)Hertz

47. **The half-life of a radioactive isotope is 3 hours.The value of its disintegration constant is**

1)0.3 hour^{-1} 2)0.693 hour^{-1} 3)0.231 hour^{-1} 4)0.231 min^{-1}

48. $_{92}U^{238} <\ {}_{82}Pb^{206} < 8{}_{2}^{4}He$. **The number of < particles released in this reaction is**

1)6   2)3   3)1   4)10

49. **The radioactivity in anynucleus is measured in**

1)Curie   2)Rutherford   3)Both 1 &2   4)Newton

50. **The < -particles are**
1)high energy electrons
2)positively charged hydrogen ions
3)high energy< -radiation
4)doubly positively charged helium nuclei

51. **< -particles carries**
1)Ma ss 1 2)Ma ss 2 3)Ma ss 3 4)Ma ss 4

52. **At a specific instant, the emission of radio active compound is deflected in a magnetic field. The compound can emit**
i)electrons ii)Positrons iii) $He^{<2}$ iv) neutrons
1) i,ii,iii 2) i,ii,iii,iv 3)iv 4) ii,iii

53. **The atomicnumber (A) and mass number (M) of the nuclide formed where three alpha <<‹ and two ‹<‹ particles are emitted from $^{238}_{92}U$**
1) $A < 87, M < 233$ 2) $A < 86, M < 226$ 3)A =88, M = 227 4)A = 88, M = 226

54. **Element $_ZM^A$ emits one < (alpha) particle followed by two < (beta) particles. Among the following the daughter element is**
1) $_{Z<2}M^{A<4}$ 2) $_{Z<2}M^{A}$ 3) $_{Z}M^{A<4}$ 4) $_{Z<2}M^{A<4}$

55. **Particles which can be added to the nucleus of an atom without changing its chemical properties are called.**
1)Neutrons 2)Electrons 3)Protons 4) Alpha Particles

56. **An Electric field can deflect**
1) < - particles 2)X - rays 3)Neutrons 4) < - rays

57. **On the bombardment of Boron with neutron,- particle is emitted and product nucleus formed is ....**
1) $_6C^{12}$ 2) $_2Li^6$ 3) $_3Li^8$ 4) $_4Be^9$

58. **The one has maximum radioactive activity**
1)Uranium 2) Plutonium 3)Radium 4) Thorium

59. **Among the following one is not emitted by a radioactive substance**
1)Electrons 2) < - rays 3) Positron 4) Protons

60. $^{228}_{90}Th$ < $^{212}_{83}Bi$ ≪ <. **the number of < and < particals given out during the process are**
1) 4<,7< 2) 4<,1< 3) 4< 4) 7<

61. **The reciprocal of radioactive decay constant is called**
1)Half life period
2)Whole life period
3) Average life period
4) Avagadro number

62. **The missing particle in the reaction** ^1_1p ≪ < $^0_{<1}e$
1)deuteron 2)proton 3) neutron 4)< - particle

63. **In the Radioactive transformation**
R ≪≤ X ≪≤ Y ≪≤ Z; **the neuclii R and Z are**
1)Isotopes 2)Isobars 3)Isomers 4)Isotones

64. In the reaction $_{15}P^{30} < _{14}Si^{30}$, The change requires the emission of

1) < - particle 2)<-particle 3) neutron 4)positron

65. When a radioactive substance is subjected to a vacuum, the rate of disintegration per second

1)increases considerably 2)is not affected

3)increases only if the products are gases 4)suffers a slight decrease

66. A radioactive nudide can decay simultaneously by two different processes which have decay constants $<_1$ and $<_2$. The effective decay constant of the nuclide is $<$.

1) $< < <_1 < <_2$ 2) $< < \frac{1}{2} < <_1 < <_2 <$ 3) $\frac{1}{<} < \frac{1}{<_1} < \frac{1}{<_2}$ 4) $< < \sqrt{< <_1 <_2}$

67. A sample of radioactive material is used to provide desired doses of radiation for medical purposes. The total time for which the sample can be used will depend

1)only on the number of times radiation is drawn from it

2) only on the intensity of doses drawn from it

3)on both (a) and (b) 4)neither on (a) nor on (b)

68. A fraction f_1 of a radioactive sample decays in one mean life and a fraction f_2 decays in one half-life.

1) $f_1 < f_2$ 2) $f_1 < f_2$ 3) $f_1 < f_2$

4)May be (a), (b) or (c) depending on the values of the mean life and half-life.

NUCLEAR FORCES

69. The short range attractive nuclear forces that areresponsible for the binding of nucleons in a nucleus are supposed to be caused by the role played by the particles called

1)Positron 2)m-Meson 3)K-Meson 4) < - Meson

70. The strong interaction exists in

1)Gravitational foces 2)Electrostatic foce of attraction

3)Nuclear forces 4)Magnetic force on a moving charge

71. Nuclear forces are

1)Non-central forces 2) saturated 3)Spin dependent 4) All the above

72. Identify the correct statement/statements

a)At greater distances nuclear forces are negligible

b)Nuclear forces are non central forces

c)Nuclear forces are weakest in nature

d)Nuclear forces are charge dependent forces

1)a, b 2)b, c 3)c, d 4)a, d

73. Which of the following is not correct about nuclear forces?

1)They are short range attractive forces 2)They are independent of charge

3)They change to repulsion at very close distance 4) They obey inverse square law

74. Among the following, short range, charge independent and spin dependent forces are

1)Gravitational foces 2)Nuclear forces

3)Electromagnetic foces 4)Weak forces

75. **Let F_{pp},F_{pn} and F_{nn} denote the magnitudes of the nuclear force by a proton on a proton ,by a proton on a neutron and by a neutron on a neutron respectively when the separation is less than one fermi, then (2008 E)**

1) $F_{pp}>F_{pn}=F_{nn}$ 2) $F_{pp}=F_{pn}=F_{nn}$ 3) $F_{pp}>F_{pn}>F_{nn}$ 4) $F_{pp}<F_{pn}=F_{nn}$

76. **Two protons are kept at a seperation of 10nm.Let F_n be the nuclear force and F_e be electrostatic force between them. Then**

1)$F_e=F_n$ 2)$F_e>>F_n$ 3)$F_e<<F_n$ 4)$F_n= 3F_e$

77. **Two nucleons are at a separation of $1x10^{-15}$m. The net force between them is F_1 if both are neutrons, F_2 if both are protons and F_3 if one is a proton and other is a neutron. In such a case**

1)$F_2>F_1>F_3$ 2)$F_1=F_2>F_3$ 3)$F_1=F_2=F_3$ 4)$F_1=F_3>F_2$

78. **Two protons attract each other when**

1)the distance between them is 10^{-10} m 2)the distance between them is 10^{-1} m

3)the distance between them is 10^{-15} m 4)the distance between them is 10^{-6} m

79. **Among gravitational,electrostatic and nuclear forces,the two attractive forces between two neutrons are**

1)Electrostatic and nuclear 2)Electrostatic and gravitational

3)Gravitational and nuclear 4)Electrostatic

80. **Among the following interactions one is of least significant in nuclear physics is**

1)nuclear interaction 2)gravitational interaction

3)electrostatic interaction 4) electromagnetic interaction

81. **The origin of nuclear force between nucleons is due to the exchange of**

1) Mesons 2)Photons 3)Positrons 4)Electrons

DISCOVERY OF NEUTRON

82. **Which of the following particles is most unstable?**

1)Neutron 2)Proton 3) Electron 4) < particle

83. **A free neutron decays spontaneously into:**

1)a proton, an electron and an anti-neutrino 2)a proton, an electron and a neutrino

3)a proton and electron 4)a proton, an electron, a neutrino and an anti-neutrino

84. **Neutron was discovered by the experiment of**

1) Artificial transmutation of $_4Be^{<}$ by < - particles

2) Artificial transmutation of $_7N^{11}$ by < -particles

3)Rutherfored scattering of alpha particles by heavy nuclei 4)Bequerel with radio activity

85. **The average life of an isolated neutron is**

1) 1500 s 2) 1000 s 3) 1200 s 4)3 minutes

86. **Thermal neutrons energy is**

1) < 1 ev 2) > 1 ev 3) = 2 Mev 4)= 4 Mev

87. **A nucleus with an excess of neutrons may decay with the emission of**

1)a neutron 2)a proton 3)an electron 4)a positron

88. **The most penetrating atom smashing particle is**

1) neutron 2)proton 3)alpha particle 4)deuteron

89. Which of the fdlowigis formed bydecayofafree neutron?

1) A number of electrons | 2) Two Protons
3) A proton and an electron | 4) An < - particle

90. In neutron discovery experiment Beryllium target is bombarded by

1)Protons | 2) Alpha particles | 3)Neutrons | 4)Deutrons

91. Slow neutrons are sometimes refer to as thermal neutrons bcause

1)they are sort of heat radiations | 2)they are in thermal eqilibrium
3)they are capable of generatig heat
4)their energies are of same order as that of molecular energies at ambient temperatures.

92. Thermal neutrons are

1)Prompt neutrons | 2)Slow neutrons
3)Neutrons which are in the nucleus | 4)Neutrons from the sun

93. In neutron discovery exprriment, Be is bombarded with

1)Proton | 2)Deutrons | 3) < -particle | 4) < -particle

94. The process of producing a new stablenucleus from the other stablenucleus is called

1) Nuclear reaction | 2) Artificial transmutation
3)Nuclear fusion | 4)Nucl ear fission

NUCLEAR FISSION

95. At least how many thermal neutrons should be available to start a fission reaction

1)2 | 2)3 | 3)1 | 4)4

96. Which of the following changes in the artificial transmutation of elements?

1) number of neutrons | 2)number of electrons
3)atomic weight | 4)nucleus

97. During the fission process of Uranium,the amount of energy liberated per fission is nearly

1)100MeV | 2)200MeV | 3)150MeV | 4)300MeV

98. The number of neutrons that are released on an average during the fision of U^{235}nucleus is

1)3 | 2)1 | 3)2.5 | 4)5

99. Nuclear fisssion can be explained by

1)Optical model of the nucleus | 2)Shell model of nucleus
3)Collective model of the nucleus | 4)Liquid drop model of the nucleus

100. Percentage of mass lost during the fission of $_{92}U^{235}$ approximately is

1)0.01% | 2)0.1% | 2)0.7% | 4)0.9%

101. Most of energy released in the fission is carried by

1)neutrons | 2)fission fragments
3)neutrons and fragmentscarry equally | 4)positrons

102. Regarding Prompt neutrons

1)They are highly energetic | 2)They constitute 99%
3)Cannot initiate chain reaction | 4)1,2,3 are correct

103. Nuclear reactions obey the law of conservation of

1)Mass and energy | 2)Charge | 3)Momentum | 4) All theabove

104. The critical mass of a fissionable material is

1)0. 1kg equivalent
2) The minimum mass needed for chain reaction
3)The rest mass equivalent to 1020 joule
4)0.5kg

105. For fast chain reaction, the size of U^{235} block, as compared to its critical size, must be

1)greater 2)smaller 3)same 4)anything.

106. The critical mass of fissionable uranium-235 can be reduced by

1)adding impurities
2)heating material
3)surrounding it by a neutron-reflecting material
4) surrounding it by a neutron-absorbing material

107. Nuclear energy is released in fission since binding energy per nucleon is

1) smaller for fission fragments than for parent nucleus
2)the same for fission fragments and parent nucleus
3) larger for fission fragments than for parent nucleus
4)sometimes larger and sometimes smaller

108. In a critical chain reaction

1)energy is released at increasing rate
2)energy is released at steady rate
3)energy is released at decreasing rate
4)energy is not released

109. Among the following one is wrong

1)The energy of thermal neutrons is about 25 meV
2) In a nuclear reactor, when neutrons multiplication factor, K = 1 then the reaction is said to be critical
3) ${}_{92}U^{235}$ undergoes fission by bombardment of high energy neutron
4)On average 2.5 neutrons are emitted per fission of ${}_{92}U^{235}$

110.When 1gm of U^{235} is completely annihilated energy liberated is E_1 and when 1 gm of U^{235} completely undergoes fission the energy liberated is E_2 , then

1)$E_1 > E_2$ 2)$E_1 = E_2$ 3)$E_1 < E_2$ 4) $E_1 \leq E_2$

111. In the process of fission, the binding energy per nucleon

1)Increases 2)Decreases 3)Remains unchanged
4)Increases for mass number A < 56 nuclei but decreases for mass number A> 56

112. Assertion (A) : Fragments produced in the fission of U^{235} are radioactive.

Reason (R) : The fragments have abnormally high proton to neutron ratio

1)Both A and R are true and R is correct explanation of A
2)Both A and R are true and R is not correct explanation of A
3) A is true but R is false 4) A is false but R is true

113. The products of the fission of U^{235} by thermal neutron are

1)Ba^{141} and Kr^{92} and 3 neutron always
2)Xe^{140}, Sr^{94} and $2\,{}_0n^1$ always
3)can be different in each fission
4)should have same mass number

114. Consider the following statements A and B. Identify the correct in the given answer.

A)p-n; p-p; n-n forces between nucleons are not equal and charge dependent.

B) In nuclear reactor the fission reaction will be in accelerating state if the value of neutron reproduction factor k>1.

1)Both A and B are correct
2)Both A and B are wrong
3) A is wrong B is correct
4) A is correct B is wrong

115. The process of fission is responsible for the release of energy in

1)The hydrogen bomb 2) The atom bomb 3)The sun 4)The star

116.The working principle in atom bomb is

1)under-critical chain reaction
2)Critical chain reaction
3)super-critical chain reaction
4) All the above

117. Heavy water is

1)Water at 4^0C
2)Watercontaining various salts
3)Compound of heavy oxygen and hydrogen
4)Compound of oxygen and deuterium.

118. Nuclear reactor is surrounded by concrete walls to

1)Strengthen the construction
2)Control the chain reaction
3)form a protective shield
4)as moderator

119. The operation of a nuclear reactor is said to be critical, if the multiplication factor (K) has a value

1)1 2)1.5 3)2.1 4)2.5

120. Cadmium and Boron rods are used in a nuclear reactor to

1)Slow down the neutrons
2) Absorb excess number of thermal neutrons
3)speed up neutrons
4)absorb fast neutrons

121. The coolant in thenuclear reactor is

1)Liquid sodium 2)cadmium 3)Deuterium 4)Liquid hydrogen

122. Substance used to slow down the fast neutrons releasedduring nuclear fission is called.

1)Fuel 2)Moderator 3)Controlling rods 4)Reflecting rods

123. If the neutronreproduction factor K is

a)greater than 1 the fission reaction is accelerated
b)less than 1 the fission reaction retards
c)equal to 1 the fision reaction is at stady state

1)only a,b are true 2)only b,c are true 3)only a,c are true 4)only a,b,c true

124. Unstable fission fragments decay by emitting neutrons and electrons, neutrons so emitted are called

1)prompt nuetrons 2)delayed neutrons 3)stray neutrons 4)sustained neutrons

125.Chain reaction can be initiated by

1)prompt neutrons 2)delayed neutrons 3) slowed prompt neutrons 4) 2 or 3

126. The man-made element which was made in the nuclear reactor is

1)polonium 2) plutonium 3) thorium 4) uranium

127. In a fast breeder reactor, the main charm is that thenuclear ash is that it is

1)more fissile than parent fuel
2)not dangerous as a potential pollutant
3)easily disposed off
4)stable in terms of further decay

128. If "x" gm of anuclear fuel of mass number "A" undergoes fission inside a reactor then the number of fissions will be (N-Avagadro number)

1)NA/x 2)NAx 3)Nx/A 4) Ax/N

129. The reactor which produces power due to fission by fast neutron and at the same time regenerates more fissionable material than it consumes is

1)Thermal reactor 2)Breeder reactor 3)Both the above 4)Neither 1 & 2

130. The reactor in which thenumber of fissionable nuclides produced are more than the used is called

1)breeder reactor
2)Pressurised reactor
3)Heterogeneous reactor
4)Homogenerous reactor

131. A good moderator should

1)be a gas
2)have appetite for neutrons
3)be lighter in mass number
4)heavier in mass number

132. Who designed the atomic reactor?

1) Wilson 2) Fermi 3) Rutherford 4) Teller

133. From the following that are conserved in nuclear reactions are

1) mass number and energy
2) mass number and charge number
3) charge number and mass
4) mass number, charge number and energy

134. (A) Fission is a thermonuclear process **(B) Fusion is a thermonuclear process**
(C) Fusion is exothermic **(D) Fission is exothermic**
Choose the correct answer

1) A and B are correct
2) B and C are correct
3) A and C are correct
4) B,C and D are correct

135. Consider the following statements A and B. Identify the correct choice in the given answer.
(A) p-p,p-n,n-n forces between nucleons are not equal and charge dependent
(B) In nuclear reactor the fission reaction will be in accelerating state if the value of neutron reproduction factor k >1

1) Both A and B are correct
2) Both A and B are wrong
3) A is wrong and B is correct
4) A is correct and B is wrong.

136. A chain reaction in fission of Uranium is possible because

1) Large amount of energy is released
2) Two intermediate size nuclear fragments are formed
3) More than one neutron is given out in each fission
4) Fragments in fission are radioactive

137. A slow neutron can cause fission in

1) U^{238} 2) U^{235} 3) Pb^{206} 4) Sr^{90}

138. Heavy water is used as moderator in a nuclear reactor. The function of the moderator is

1) To slow down the neutrons to thermal energies
2) To control the energy released in the reactor
3) To cool the reactor faster
4) To absorb neutrons and stop chain reaction

139. To control fission process of the reactor , the following material is used_______

1) Graphite 2) Cadmium 3) Gold 4) Uranium

140. Nuclear fission is caused by

1) fast protons 2) fast neutrons 3) Slow protons 4) slow neutrons

141. The liquid drop model of nucleus was proposed by

1) Bohr, Wheeler 2) Fermi 3) Rutherford 4) Chadwick

142. In nuclear reactor, Cadmium rods are used as

1) Fuel 2) Moderator 3) Control Rods 4) None

143. The material used to slow neutrons in a reactor is called

1) Controlrod 2) Moderator 3) Fuel 4) Heat exchanger

NUCLEAR FUSION

144. Atomic mass of the most useful material for fusion reaction is

1) 1 2) 4 3) 235 4) 292.

145. Average *K.E* of thermal neutron is of the order of (in KeV)

1) 3.0 2) 0.03 3) 0.3 4) 0.003

146. Inside the sun

1) Four nuclei of hydrogen combine to form two nuclei of helium
2) Four nuclei of hydrogen combine to form four nuclei of helium
3) Four nuclei of hydrogen combine to form one nucleus of helium
4) Four nuclei of hydrogen is transformed into one nucleus of helium

147. As the age of star increases

1) Helium quantity increases
2) Helium quantity decreases
3) Helium quantity does not change
4) Helium, Hydrogen both quantities increases

148. In the carbon cycle of nuclear fusion carbon acts like a

1) Moderator 2) Activator 3) Catalyst 4) Controller

149. In a fusion process a proton and neutron combine to give a deuterium nucleus. If m_o and m_p be the mass of neutron and proton respectively the mass of deuterium nucleus is

1) equal to m_o+m_p 2) more than m_o+m_p 3) less than m_o+m_p
4) can be less than or more than (m_o+m_p)

150. The binding energies of the atoms of elements P and Q are E_P and E_Q respectively. Three atoms of elements Q fuse to form one atom of element P. In this process the energy released is e. The correct relation between E_P, E_Q and e will be

1) $E_Q=3E_P+e$ 2) $E_Q=3E_P-e$ 3) $E_P=3E_Q+e$ 4) $E_P=3E_Q-e$

151. The $\frac{B.E.}{A}$ for deutron and an α - particle are X_1 and X_2 respectively. The energy released in the fusion of deuterium into α -particle is

1) $4(X_2-X_1)$ 2) $2(X_2-X_1)$ 3) $4(X_2+X_1)$ 4) $\frac{X_2-X_1}{4}$

152. If Q_1 and Q_2 are the energies released in the fusion of hydrogen in Carbon - nitrogen cycle and proton - proton cycle respectively then

1) $Q_1 > Q_2$ 2) $Q_1 = Q_2$ 3) $Q_1 < Q_2$ 4) $Q_1 > Q_2$

153. Fusion reaction is initiated with the help of

1) low temperature 2) high temperature 3) neutrons 4) any particle

154. In an exo-ergic reaction the binding energies of reactants and products are E_1, E_2 respectively then

1) $E_1<E_2$ 2) $E_1=E_2$ 3) $E_1>E_2$ 4) $E_1 \geq E_2$

155. In an endo-ergic reaction the binding energies of reactants and products are E_1, E_2 respectively

1) $E_1<E_2$ 2) $E_1=E_2$ 3) $E_1>E_2$ 4) $E_1 \geq E_2$

156. Among the following reactions which is impossible

1) ${}_2He^4+{}_4Be^9={}_0n^1+{}_6C^{12}$
2) ${}_2He^4+{}_7N^{14}={}_1H^1+{}_8O^{17}$
3) $4({}_1H^1)={}_2He^4+2({}_{-1}e^0)$
4) ${}_3Li^7+{}_1H^1={}_4Be^8$

157. If the nuclei of masses X and Y are fused together to form a nucleus of mass m and some energy is released, then

1) X+Y=m 2) X+Y<m 3) X+Y> m 4) X-Y=m

158. Fusion reactions take place at about

1) 3×10^2 K 2) 3×10^3 K 3) 3×10^4 K 4) 3×10^6 K

159. The percentage of mass lost during nuclear fusion is

1)0.1% 2)0.4% 3)0.5% 4)0.65%

160. Nuclear fusion requires high temperature because

1) All nuclear reactions absorb heat

2)The particles can not come together unless they are moving rapidly

3)The binding energy must be supplied from an external source

4)The mass defect must be supplied

161. Among the following true option is

1)Energy released per nucleon is same in both fission and fusion reactions

2)Energy released per nucleon is more in fission than in fusion reaction

3)Energy released per nucleon is less in fission than in fusion reaction

4)No energy is released in fusion reaction

162. Fusion reaction take place at high temperature because

1)atoms are ionized at high temperature 2)molecules break up at high temperatures

3)nuclei break up at high temperature

4)kinetic energy is high enough to overcome repulsion between nuclei

163. In the carbon cycle from which stars hotter than the sun obtain their energy the $_6C^{12}$ isotope

1)splits up into three alpha particles 2) fuses with another $_6C^{12}$ nucleus to form $_{12}Mg^{24}$

3)is completely converted into energy 4)is regenerated at the end of the cycle

164. Source of solar energy can be said to be due to natural fusion in which hydrogen gets converted into helium with carbon serving as a natural catalyst. This carbon cycle was proposed by

1)Bethe 2) Yukawa 3)Fermi 4)Soddy

165. In Carbon-Nitrogen fusion cycle , protons are fused to form a helium nucleus, positrons and release some energy.The number of protons fused and the number of positrons released in this process respectively are

1)4,4 2)4,2 3)2,4 4)4,6

166. Nuclear fission and fusion can be explained on the basis of ____

1)Einstein theory of relativity 2)Einstein specific heat equation

3)Einstein mass-energy relation 4)Einstein photo electric equation

167. Energy in the sun is due to

1)Fossil fuels 2) Radioactivity 3)Fission 4)Fusion

168. The overall process of carbon nitrogen fusion cycle results in the fusion of 4 protons to yield helium nucleus and ---

1)positron 2)two electrons 3)two positrons 4) An electron.

169. The nucleus finally formed in fusion of protons in proton-proton cycle is that of

1)Heavy hydrogen 2)Carbon 3) Helium 4)Lithium

170. $4\,_1H^1 < \,_2He^4 < 2e^{<} < 26MeV$ **. The above reaction represents ......**

1)Fusion 2)Fission 3)b-decay 4)g-decay

171. The source of stellar energy is __process

1)Nuclear fission 2)Nuclear fusion

3)Nuclear fission&fusion 4)Nuclear decay

172. Fusion reaction takes place at veryhigh temperature because.

1)atoms are ionised at high temperatures. 2)molecules breakup at high temperature.

3) nuclei break up at high temperature.

4)kinetic energy is high enough to overcome repulsion between nuclei

173. In the carbon cycle, from which stars hotter than the sun obtain their energy the $_6C^{12}$ isotope

1)splits into three alpha particles

2)fuse with another $_6C^{12}$ nucleus to form $_{12}Mg^{24}$

3)is completely converted into energy. 4)is regenerated at the end of the cycle.

PAIR PRODUCTION & PAIRANNIHILATION

174. The phenomenon of pair production is

1)The production of an electron and a positron from < radiations

2)Ejection of an electron from a metal surface when exposed to ultraviolet light

3)Ejection of an electron from a nucleus 4)Ionization of a neutral atom

175. When the particle and its antiparticle unite, the result is

1)a heavier particle 2)two or more smaller particles

3)photons 4)partly matter and partly photons.

176. Particles and their antiparticles have

1)The same massesbut opposite spins

2) The same masses but opposite magnetic moment

3)The same masses and same magnetic moment 4)Opposite spins and same magnetic moment

177. To produce pair production, the minimum energy of <-ray should be (in MeV)

1)0.15 2)1 3)1.02 4)1.5

178. The rest mass energy of electron or positron is (in MeV)

1)0.51 2)1 3)1.02 4)1.5

179. A positron and an electron come close togetherto give a neutral one called

1) Electronium 2)Positronium 3) <-photon 4) <-particle

180. Positronium is converted into

1)2 Photons each of energy 0.51MeV 2)1 Photon of energy 1.02 MeV

3)2 Photons each of energy 1.02MeV 4)One Photon of energy 0.51MeV

181. In pair annihilation two <-ray photons are produced it is due to

1)Law of conservation of energy 2)Law of conservation of mass

3)Law of conservation ofmomentum 4)Law of conservation of angular momentum

182. In pair annihilation the least number of <- ray photons produced is

1)2 2)3 3)4 4)1

183. The number of protons, electrons and neutrons in the nucleus of $_{13}Al^{27}$ is

1)13, 13, 14 2)13, 0, 14 3)14,14, 13 4)14, 0, 13

184. $^{39}_{19}K$ and $^{40}_{20}Ca$ are

1)Isotopes 2)Isobars 3)Isotones 4)Isodiaphers

185. K^{40},Ar^{40},Ca^{40} are

1)Isobars 2)Isotopes 3)Isotones 4)Isogonals

186. Of the followingatoms $_6C^{14}, _7N^{13}, _{88}Ra^{236}, _7N^{14}, _8O^{16}$ and $_{86}Rn^{232}$ a pair of isobars is :

1) $_6C^{11}, _7N^{13}$ 2) $_7N^{13}, _7N^{14}$ 3) $_6C^{14}, _7N^{14}$ 4) $_6C^{14}, _8O^{16}$

187. Of the following a pair of isotones is

1) $_6C^{11}, _7N^{13}$ 2) $_7N^{13}, _7N^{14}$ 3) $_6C^{14}, _7N^{14}$ 4) $_6C^{14}, _8O^{16}$

188. Of the following a pair of isotopes is

1) $_6C^{11}, _7N^{13}$ 2) $_7N^{13}, _7N^{14}$ 3) $_6C^{14}, _7N^{14}$ 4) $_6C^{14}, _8O^{16}$

189. Of the following a pair of isodiaphers is

1) $_{88}Ra^{236}, _{86}Ra^{232}$ 2) $_7N^{13}, _7N^{14}$ 3) $_6C^{14}, _7N^{14}$ 4) $_6C^{14}, _8O^{16}$

ASSERTION & REASONS

These Questions consist of two statements each printed as A ssretion and Reason. While answering these questions you are required to choose any one of the following fourresponses

1) A and R are true and R is the correct explanation of A.

2) A and R are true and R is not the correct explanation of A.

3) A is true, R is false. 4) A is false, R is true.

190. (A): Free Neutron decays into proton, electron and antinuetrino

(R): Neutron is unstable outide thenucleus

191. (A): Nuclear forces arise from strong coulombic interactions between protons and neutrons

(R): Nulear forces are independent of charge of thenucleons

MATCHING TYPE

192. Match list I with list II

List I	List II
a)Red dwarfs cycle	e)Carbon nitrogen
b)stars having higher temperatures	f)Isotope of carbon
c)Blood circulation problems	g)proton - proton cycle
d)Radio carbon dating	h)radio sodium

1)a- e , b - g, c - h , d - f 2)a - g, b - e , c - f, d- h

3)a - g, b - e , c - h, d - f 4)a - f, b - e, c - g , d - h

193. Matching

List – I	List – II
A)K < 1	**e)Critical state**
B)K = 1	**f)S ub critical state**
C)K > 1	**g)Super critical**

1) A< f , B< e , C< g 2) A< e , B< f , C< g

3) A< g , B< e , C< f 4) A< e, B< g C < f

194. Matching.

List - I	List - II
A)Radio Iodine	e)Lukamia
B)Radio sodium	f)age of ancient objects
C)Radio phosphorous	g)Restriction in bloodcirculation
D)Radio carbon	h)F unctioning of thyroid gland

1) A< h , B< g , C< e , D< f 2) A< g , B< h , C< f , D< e

3) A< h , B < e , C< g , D< f 4) A< h , B< g , C< f , D< e

195. Match the following.

List – I	**List – II**
A)Moderator	e) Absorbs heat
B) Control rods	f)Prevent neutrons exposed outside
C)Radiation shielding	g) Absorb neutrons
D)Coolant	h)Slow down neutrons

1) A – h, B – g, C – f, D – e
2) A – g, B – h, C – f, D – e
3) A – h, B – g, C – e, D – f
4) A – h, B – f, C – e, D – g

196. Thisquestion contains Statement and Statement -2.Of the four choice given after the statements, choose the one that best decribes the two statements.

Statement - 1: Energy is released when heavy nuclei undergo fission or light nuclei undergo fusion. and

Statement - II : For heavy nuclei,binding energy per nucleon increases with increasing Z while for lightnuclei it decrease with increasing Z.

1)Statement - 1 is false,Statement -2 is true

2)Statement - 1is true, Statement -2 is true ; Statement -2 is correct explanation for Statement -1.

3)Statement - 1 is true, Statement -2 is true ; Statement -2 is *not* a correct explanation for statement -1

4)Statement -1 is true, Statement -2 is False.

CONCEPTUAL - KEY

1)2	2)3	3)3	4)3	5)1	6) 1	7)3	8)2	9)2	10)2
11)4	12) 4	13) 3	14) 4	15) 2	16) 2	17) 1	18) 4	19) 2	20) 2
21)1	22) 4	23) 2	24) 2	25) 4	26) 4	27) 1	28) 4	29) 4	30) 2
31)1	32) 1	33) 2	34) 3	35) 3	36) 4	37) 3	38) 2	39) 3	40) 2
41)4	42) 3	43) 3	44) 2	45) 2	46) 3	47) 3	48) 1	49) 3	50) 4
51)4	52) 1	53) 4	54) 3	55) 1	56) 1	57) 4	58) 3	59) 4	60) 2
61)3	62) 3	63) 1	64) 4	65) 2	66) 1	67) 4	68) 1	69) 4	70) 3
71)4	72) 1	73) 4	74) 2	75) 2	76) 3	77) 4	78) 3	79) 3	80) 2
81)1	82) 1	83) 1	84) 1	85) 2	86) 1	87) 3	88) 1	89) 3	90) 2
91)4	92) 2	93) 3	94) 2	95) 3	96) 4	97) 2	98) 3	99) 4	100) 2
101)2	102) 4	103) 4	104) 2	105)1	106) 3	107) 3	108) 2	109) 3	110) 1
111)1	112) 3	113) 3	114) 3	115) 2	116) 3	117) 4	118) 3	119) 1	120) 2
121)1	122) 2	123) 4	124) 2	125) 4	126)2	127) 1	128) 3	129) 2	130) 1
131)3	132) 2	133)4	134) 4	135) 3	136) 3	137) 2	138) 1	139) 2	140)4
141)1	142) 3	143) 2	144) 1	145) 2	146) 3	147)1	148) 3	149) 3	150) 3
151)1	152) 2	153) 2	154)1	155) 3	156) 3	157) 3	158) 4	159) 4	160) 2
161)3	162) 4	163) 4	164) 1	165) 2	166) 3	167) 4	168)3	169) 3	170) 1
171)2	172) 4	173) 4	174) 1	175)3	176) 2	177) 3	178) 1	179) 2	180) 1
181)3	182)1	183) 2	184) 3	185) 1	186) 3	187) 4	188) 2	189)1	190) 1
191)4	192) 3	193) 1	194) 1	195) 1	196)4				

LEVEL - I-1

SIZE OF THENUCLEUS

1. **The density of a nucleus in which mass of each nucleon is $1.67<10^{<27}$ kg and $R_0<1.4<10^{<15}m$ is**

   1) $1.453<10^{17}kg/m^3$
   2) $1.453<10^{16}kg/m^3$
   3) $1.453<10^{21}kg/m^3$
   4) $1.453<10^{10}kg/m^3$

2. **r_1 and r_2 are the radii of atomic nuclei of mass numbers 64 and 27 respectively. The ratio $<r_1/r_2<$ is**

   1) 64 / 27 2) 27 / 64 3) 4 / 3 4) 1

3. **The mass number of a nucleus is 216. The size of an atom without changing its chemical properties are called**

   1) $7.2<10^{<13}cm$ 2) $7.2<10^{<11}cm$ 3) $7.2<10^{<10}cm$ 4) $3.6<10^{<11}cm$

MASS DEFECT, BINDING ENERGY, PACKING FRACTION AND MASS ENERGY RELATION

4. **Energy released as mass of 2 amu is converted into energy is**

   1) 1.5×10^{-10} J 2) 3×10^{-10} J 3) 1863 J 4) 931.5 Mev

5. **A 1 MeV positron encounters a 1 MeV electron travelling in opposite direction. The total energy released is (in MeV)**

   1) 2 2) 3.02 3) 1.02 4) 2.04

6. **The binding energies of the nuclei A and B are E_a and E_b respectively. Three nuclei of the element B fuse to give one nucleus of element 'A' and an energy 'Q' is released. Then E_a, E_b, Q are related**

   1) $E_a<3E_b<Q$ 2) $3E_b<E_a<Q$ 3) $E_a<3E_b<Q$ 4) $E_b<3E_a<Q$

7. **The binding energies per nucleon for deuterium and helium are 1.1 MeV and 7.0 MeV respectively. The energy in joules will be liberated when 10^6 deuterons take part in the reaction**

   1) $18.88<10^{<3}J$ 2) $18.88<10^{<5}J$ 3) $18.88<10^{<7}J$ 4) $18.88<10^{<10}J$

8. **1 Kg of iron (specific heat 120 Cal $kg^{-1}C^{-1}$) is heated by 1000C. The increase in its mass is**

   1) Zero 2) 5.6×10^{-8} Kg 3) 5.6×10^{-16} Kg 4) 5.6×10^{-12} Kg

9. **In nuclear fission, 0.1% mass is converted into energy. The energy released in the fission of 1Kg mass is**

   1) 2.5×10^5 KWH 2) 2.5×10^7 KWH 3) 2.5×10^9 KWH 4) 2.5×10^{-7} KWH

<-DECAY, << DECAY, << DECAY

10. **After the emission of one <-particle followed by two <-particles from $^{238}_{92}U$, the number of neutrons in the newly formed nucleus is**

    1) 140 2) 142 3) 144 4) 146

11. **A radioactive nucleus undergoes a series of decays according to the sequence $A << A_1 << A_2 < A_3$. If the mass number and atomic number of A_3 are 172 and 69 respectively, then the mass number and atomic number of A is**

    1) 56, 23 2) 180, 72 3) 120, 52 4) 84, 38

12. The number of < and < particles emitted in the conversion of $_{90}Th^{232}$ to $_{82}Pb^{208}$ are

1)6, 4 2)4, 6 3)8, 6 4)6, 8

RADIOACTIVITY, HALF LIFE, MEAN LIFE, DECAY CONSTANT

13. The decay constant of a radio active element, which disintegrates to 10 gms from 20 gms in 10 mi nutes is

1)0.693 $min^{<1}$ 2)6.93 $min^{<1}$ 3)0.693 $sec^{<1}$ 4)0.0693 $min^{<1}$

14. Half life period of radium is 1600 years. 2 gm of radium undergoes decay and gets reduced to 0.125 gms in

1)3200 years 2)25600 years 3)8000 years 4)6400 years

15. After a certain lapse of time, the fraction of radioactive poloniumundecayed is found to be 12.5% of initialquantity. Theduration of this time lapse is (if the halflife of polonium is 138 days)

1)414days 2)407 days 3)421 days 4)410 days

16. Two radioactive substances X and Y initially contain an equal number of atoms. Their half-lives are 1 hour and 2 hours respectively. Then the ratio of their rates of disintegration after two hours is

1)1:1 2)2:1 3)1:2 4)2:3

17. 1 g of a radio active s ubstance disintegrates at the rate of 3.7 < 10^{10} disintegrations per second. The atomic mass of the substance is226. Calculate its mean life.

1) $1.2 <10^{5} s$ 2) $1.39<10^{11} s$ 3) $2.1 <10^{5} s$ 4)7.194 < 10^{10}S

NUCLEAR FISSION-NUCLEAR REACTOR

18. No. of uranium 235 nuclei required to undergo fission to give 9 x 10^3 joule of energy is

1)2.8125 x 10^{24} 2)28.125 x 10^{24} 3)281.25 x 10^{24} 4)28215 x 10^{24}

19. The energy supplied by a power plant is 40 billion kilowatt hour. It is supplied by annihilation of matter, the mass that is annihilated is.

1)1.6 gm 2)1.6 kg 3)1.6 mg 4)1.6 amu.

NUCLEAR FUSION

20. The amount of energy released in the fusion of two $_1H^2$ to form a $_2He^4$ nucleus will be

[Binding energy per nucleon of $_1H^2$ = 1.1 Mev Binding energy per nucleon of $_2He^4$ =7 Mev]

1)8.1 MeV 2)5.9 MeV 3)23.6 MeV 4)2 MeV

PAIR PRODUCTION & PAIRANNIHILATION

21. The minimum amount of energy released imahilation of electron-positron is

1)1.02MeV 2)0.58MeV 3)185MeV 4)200MeV

LEVEL - I-1 - KEY

1)1 2)3 3)1 4)2 5)2 6)1 7)3 8)4 9)2 10)2
11)2 12) 1 13) 4 14) 4 15) 1 16) 3 17) 4 18) 1 19) 1 20) 3
21)1

LEVEL - I-1 - HINTS

1. $d < \frac{Am}{\frac{4}{3}<R^3} < \frac{3Am}{4<\left(R_0 A^{\frac{1}{3}}\right)^3} < \frac{3Am}{4<R_0^3 A} < \frac{3m}{4<R_0^3}$

2. As $\frac{r_1}{r_2} < \left(\frac{A_1}{A_2}\right)^{1/3} < \left(\frac{64}{27}\right)^{1/3} < \frac{4}{3}$

3. $R < R_0 A^{1/3} < 1.2<10^{<13}<(216)^{1/3} < 7.2<10^{<13}$ cm

4. $E < MC^2$

5. Total energy=rest mass energy+KE

6. $3E_b < E_a < Q$

7. $E < 4\left(\frac{BE}{A}\right)$ of He $<\left(\frac{BE}{A}\right)$ of Deuterium

8. $<mc^2 < J<MS<t<$

9. $E < \frac{0.1}{100} mc^2$; inkwh $< \frac{E}{36<10^5}$

10. ${}_{92}U^{238} < {}_{92}X^{234} < {}_{2}He^{4} < 2{}_{<1}e^{0}$

no of neutrons=234-92=142

11. $A << A_1 << A_2 < A_3$

${}_{72}X^{180} << {}_{<1}e^{0}$ ${}_{73}X^{180}$ ${}_{2}He^{4}$ ${}_{71}X^{176}$

${}_{71}X^{176} << He^{4}$ ${}_{69}X^{172}$

12. ${}_{90}Th^{232} < {}_{82}Pb^{208} < 6{}_{2}He^{4} < 4{}_{<1}e^{0}$

13. $<< 0.693/T_{\frac{1}{2}}$

14. $N < N_0 e^{<<t}$

15. $\frac{dN}{dt} < <N$, here $N <$ avagadrono/238,

$<< 0.693/T_{\frac{1}{2}}$

16. $A < <N$

17. $\frac{N}{N_0} < \frac{1}{2^n}$

18. $200MeV <$ 1 fission

$9<10^{13} J <$?

Number of fissions $= \frac{9<10^{13}}{200<10^6<1.6<10^{<19}}$

19. $E < 40<10^6 < 36<10^5 J$; $E < mc^2$

$40<10^6 < 36<10^5 < m< 9< 10^{16}$

20. ${}_1H^2 <_1 H^2 < {}_2 He^4 < Q$

E = B.E. of products - B.E. of reactants

= 4(7) - 4(1.1) = 23.6 MeV

21. 0.51 MeV + 0.51 MeV = 1.02 MeV

LEVEL - I-2

SIZE OF THE NUCLEUS

1. Assume that the nuclear mass is of the order of 10^{-27}kg and the nuclear radius is of the order of 10^{-15}m. The nuclear density is of the order of

1)10^2Kg/m^3 2)10^{10}kg/m^3 3)10^{17}Kg/m^3 4)10^{31} kg/m^3

2. Highly energetic electrons are bombarded on a target of an element containing 30 neutrons. The ratio of radii of nucleus

1)25 2) 26 3) 56 4) 30

MASS DEFECT, BINDING ENEGRY, PACKING FRACTION AND MASS ENERGY RELATION

3. Sun radiates energy at the rate of 3.6x10^{26}J/ s. The rate of decrease in mass of sun is (Kgs^{-1}).

1)12 x 10^{10} 2)1.3 x 10^{20} 3)4 x 10^9 4)3.6 x 10^{36}

4. A slow neutron strikes a nucleus of ${}^{235}_{92}U$ splitting it into lighter nuclei of ${}^{141}_{56}Ba$ and ${}^{92}_{36}Kr$ along with three neutrons. The energy released in this reaction is (The masses of uranium, barium and krypton of this reaction are 235.043933, 140.917700 and 91.895400 u respectively. The mass of a neutron is 1.008665u)

1)740.69 MeV 2)156.9 MeV 3)186.9 MeV 4)209.8 MeV

5. The energy required to separate the typical middle mass nucleus ${}^{120}_{50}Sn$ into its constituent nucleons ($Mass of$ ${}^{120}_{50}Sn < 119.902199u$ **; mass of proton=1.007825 u and mass of neutron =** 1.008665 u)

1)951 MeV 2)805 MeV 3)1021MeV 4)1212 MeV

6. The mass defect in a nucleus is 3.5 amu. Then the binding energy of the nucleus is

1)32.58MeV 2)325.85 MeV 3)3260.25MeV 4)3.258 MeV

7. Consider the nuclear reaction:

$X^{200} < A^{110} < B^{90} <$ energy **The binding energy per nucleon for X^{200}, A^{110} and B^{90} is respectively 7.4MeV, 8.2MeV and 8.2Mev. The energy released is**

1)200MeV 2)160MeV 3) 110MeV 4)90MeV

< -DECAY, < < DECAY, < < DECAY

8. The isotope ${}^{238}_{92}U$ decays successively to form ${}^{234}_{90}Th$, ${}^{234}_{91}Pa$, ${}^{234}_{92}U$, ${}^{230}_{90}Th$ and ${}^{226}_{88}Ra$, The radiations emitted in these five steps are

1) <,<,<, <,< 2) <,<, <,<,< 3) <,<,<,<,<, 4) <,<,<,<,<

9. **The nuclide which disintegrates by emitting a $<$- particle to form $^{14}_{7}N$ contains**

1)8 neutrons 2)10 neutrons 3)7 neutrons 4)6 neutrons.

10. **A nucleus X initia lly at rest, undergoes alpha d ecay according to the equation $^{232}_{z}X < ^{A}_{90}Y<<$ What fraction of the total energy released in the decay will be the kinetic energy of the al pha particle**

1) $\frac{90}{92}$ 2) $\frac{228}{232}$ 3) $\sqrt{\frac{228}{232}}$ 4) $\frac{1}{2}$

RADIOACTIVITY, HALF LIFE, MEAN LIFE, DECAY CONSTANT

11. **A radio active sample contains 600 radio active atoms. Its half life period is 30 minutes. The no. of radio active atoms remaining, if the d ecay o ccurs for 90 mi nutes is**

1)300 2)200 3)400 4)75

12. **Radio active carbon - 14, in a wood sample decays w ith a half life of 5700 year s. T he fraction of the radio active carbon - 14, that remains after a dec ay period of 17,100 years is**

1)1/4 2)3/4 3)1/8 4)7/8

13. **The half-life of ^{238}U for $<<$ decay is $4.5<10^{9}$ years. The number of disintegrations per second occur in 1 g of ^{238}U is (Avogadro'snumber= $6.023<10^{23}mol^{<1}$)**

1) $1.532<10^{4}s^{<1}$ 2) $1.325<10^{4}s^{<1}$ 3) $1.412<10^{4}s^{<1}$ 4) $1.235<10^{4}s^{<1}$

14. **A certain substance decays to 1/32 of its initial activity in 25 days. The half-life is**

1)1 day 2)3 days 3)5 days 4)7 days

NUCLEAR FISSION-NUCLEAR REACTOR

15. **The energy released by the fission of 1 g of ^{235}U in joule, given that the energy released per fission is 200 MeV. (Avogadro'snumber $< 6.023<10^{23}$)**

1) $8.202<10^{12}$ 2) $8.202<10^{8}$ 3) $8.202<10^{10}$ 4) $8.202<10^{14}$

NUCLEAR FUSION

16. **The ratio of the amounts of energy released as a result of the fusion of 1 kg hydrogen $<E_1<$ and fission of 1 kg of $_{92}U^{236}$ $<E_2<$ will be**

1)1.28 2)3.28 3)5.28 4)7.28

LEVEL - I-2 - KEY

1)3 2)2 3)3 4)4 5)3 6)3 7)2 8)3 9)1 10)2
11)4 12) 3 13) 4 14) 3 15) 3 16) 4

LEVEL - I-2 - HINTS

1. $d < \frac{3M_n}{4<R^3}$ $R < radius$

2. As $R < R_0A^{1/3} < \frac{R_1}{R_2}< \left(\frac{<A_1}{<A_2}\right)^{1/3} < \frac{R}{R_{He}}< \left(\frac{<A}{<4}\right)^{1/3}$

$< \triangleleft 4^{1/3} < \left(\frac{<A<}{<4<}\right)^{1/3} < A < 56$ (or) $Z < A < N$

$Z < 56 < 30 < 26$

3. $\frac{dE}{dt} < \frac{dm}{dt}c^2$

4. $BE << mc^2 << m < 931.5MeV$

5. $<E << m\ 931.5MeV$

6. Binding energy $=<m< 931.5MeV$

7. Energy released=BE of products-BE of reactants

9. ${}_{7}N^{14} < {}_{6}X^{14} < {}_{<1}e^{0}$

10. $\frac{K.E<}{K.E} < \frac{A<4}{A}$ <Where A $< 232<$

11. $\frac{N}{N_0} < \left(\frac{<1}{<2}\right)^n, n < \frac{t}{T_{\frac{1}{2}}}$

12. $n < \frac{t}{T_{\frac{1}{2}}}; \frac{N}{N_0} < \left(\frac{<1}{<2}\right)^n$

13. $N < \frac{N_0}{2^n}$

14. $\frac{dN}{dt} < <N$ here $N <$ avagadrono/ 226, $< < 1/ <$

15. $E <$ (avagadrono./235)X-200x10^6x1.602x10^{19}J

16. Energy relesed for nucleon in fusion = 7.28 (energy released per nucleon in fission E_1) $< 7.28E_2$

LEVEL - II-1

SIZE OF THENUCLEUS

1. A nucleus x^{235} splits into twonuclei having the massnumbers in the ratio 2:1. The ratio of the radii of those twonuclei is

1)2:1 2)1:2 3)$2^{1/3}$:1 4)1:$2^{1/3}$

2. A match box of 5 cm x 5cm x 1 cm dimensions filled with nuclear matter. Its weight is in the order of

1)10g 2)10^{8} g 3)10^{12}g 4)10^{15}g

MASS DEFECT, BINDING ENEGRY, PACKING FRACTION AND MASS ENERGY RELATION

3. If the speed of light were 2/3 of its present value, the energy released in a given atomic explosion will be decreased by a fraction of

1)2/3 2)4/9 3)4/3 4)5/9

4. **The binding energy per nucleon of C^{12} is 7.68 MeV and that of C^{13} is 7.47 Mev. The energy required to remove one neutron from C^{13} is**

1)495 MeV 2)49.5 MeV 3)4.95 MeV 4)0.495 MeV

5. **The binding energy per each nucleon in the neighbourhood of medium nuclei is 8 MeV and the binding energy per each nucleon is about 7 MeV in the neighbourhood of Uranium. The energy released in the fission of U^{236} is**

1)212 eV 2)212 MeV 3) 2.12 MeV 4)0.9 MeV

< - DECAY,ς < DECAY,ς < DECAY

6. **^{22}Ne nucleus, after absorbing energy, decays into two < - particals and an unknown nucleus. The unknown nucleus is**

1)Carbon 2)Nitrogen 3)Boron 4)oxygen

RADIOACTIVITY, HALF LIFE, MEAN LIFE, DECAY CONSTANT

7. **The mass of one curie of U^{234} is**

1) $3.7 < 10^{10} g$ 2) $3.7 < 10^{<10} g$ 3) $6.25 < 10^{<34} g$ 4) $1.438 < 10^{<11} g$

8. **A radio active isotope having a half life of 3 days was received after 9 days. It was found that there was only 4 gms of the isotope in the container. The initial weight of the isotope when packed was**

1)8 g 2)64 g 3)48 g 4)32 g

9. **Half life of a radio active element is 5 min. 10 s ec. Time taken for 90% of it to disintegrate is nearly**

1)100 min 2) 1000sec 3) 10^4 sec 4)10^4 min

10. **The half life of $^{238}_{92}U$ undergoing << decay is $4.5 < 10^9$ years. Its activity of 1 g sample is**

1) $1.23 < 10^4$ Bq 2) $2.4 < 10^5$ Bq 3) $1.82 < 10^6$ Bq 4) $4.02 < 10^8$ Bq

NUCLEAR FISSION-NUCLEAR REACTOR

11. **In a thermo nuclear reaction $10^{<3} Kg$ of hydrogen is converted into $0.99 < 10^{<3} kg$ of helium. If the efficiency of the generator is 50%, the electrical energy generated in KWH is**

1) 10^5 2) $1.5 < 10^5$ 3) $1.25 < 10^5$ 4) $1.3 < 10^5$

12. **A nuclear reactor generated power at 50% efficiency by fission of $_{92}U^{235}$ into equal fragment of $_{46}Pd^{116}$ with the emission of two <-rays of 5.2 MeV each and three neutrons. The average B.E. per particle of U^{235} and Pd^{116} is 7.2 MeV and 8.2 MeV respectively. The amount of U^{235} consumed per hour to produce 1600Mw power is**

1)128 *gm* 2)1.4 *kg* 3)140.5 *gm* 4)281 *gm*

NUCLEAR FUSION

13. **In nuclear fusion,One gram hydrogen is converted into 0.993gm.If the efficiency of the generator be 5%,then the energy obtained in KWH is**

1) $8.75 < 10^3$ 2) $4.75 < 10^3$ 3) $5.75 < 10^3$ 4) $3.73 < 10^3$

PAIR PRODUCTION & PAIRANNIHILATION

14. A photon of energy 1.12 Mev splits into electron positron pair. The velocity of electron is (Neglect relativistic correction)

1)3 x 10^8 ms^{-1} 2)1.33 x 10^8 ms^{-1} 3)6 x $10^8 ms^{-1}$ 4)9 x $10^8 ms^{-1}$

LEVEL - II (C.W.) - KEY

1)3 2)4 3)2 4)3 5)2 6)1 7)4 8)4 9)2 10) 1
11)3 12) 1 13) 1 14) 2

LEVEL - II (C.W.) - HINTS

1. $R < A^{1/3}$

2.D= $\frac{M}{V}$

3. $E < mc^2$

4.Energy required =13 <7.47-12<7.68 = 4.95 MeV

5. $Q < BE_{product} < BE_{reactants}$

6. $^{22}_{10}Ne < 2\,{}_2He^4 < {}_6c^{12}$

7. $N < 234gm$

$3.7 < 10^{10} < xgm\,x < \frac{3.7 < 10^{10} < 234}{N}$

8. $n < \frac{t}{t_{\frac{1}{2}}} < \frac{9}{3} < 3; W < \frac{W_0}{2^n}$

$W_0 < 2^n, W < 2^3 < 4 << 32gm$

9. $t < \frac{t_{\frac{1}{2}}}{\log 2} \log \frac{< N_0 <}{< N <}$

10. Activity or rate of decay is $R << N$ Where $<< \frac{0.693}{T_{1/2}}$ and $N <$ No.of atom is 1 gm of

$^{238}_{92}U < \frac{1 < 6.025 < 10^{23}}{238} < 25.3 < 10^{20}$ atoms ; $R << N$

11. $\frac{E}{t} < \frac{mc^2}{t}$; effieiency $= \frac{\text{output}}{\text{input}}$

12. $\frac{50P}{100} < 1600MW; P < \frac{< BE_{product} < BE_{reactant} <}{t}$

13. Efficency $= \frac{output}{< mc^2}$

14. $E_< < \frac{<1}{<2} mv^2 \,\frac{<}{<} 2 < 2E_0$

LEVEL - II-2

SIZE OF THENUCLEUS

1. **A nucleus splits into two nuclear parts having radii in the ratio 1:2. Their velocities are in the ratio**

   1)8:1 2)6:1 3)4:1 4)2:1

MASS DEFECT, BINDING ENEGRY, PACKING FRACTION AND MASS ENERGY RELATION

2. **The atomic mass of $_7N^{15}$ is 15.000108 amu and that of $_8O^{16}$ is 15.994915 amu. The minimum energy required to remove the least tightly bound proton is (mass of proton is 1.007825 amu)**

   1)0.01 3018 amu 2)12.13 MeV 3)13.018 meV 4)12.13 eV

3. **Assume that a neutron breaks into a proton and electron. The energy released during this process is (Mass of neutron=$1.6725<10^{-27}$ Kg =mass of proton=$1.6725<10^{-27}$ Kg, mass of electron = $9<10^{-31}$ Kg.)**

   1)0.73 MeV 2)7.10 MeV 3) 6.30 MeV 4)5.4 MeV

< - DECAY,ς < DECAY,ς < DECAY

4. **A nucleus with mass number 220 initially at rest emits an << particle. If the Q value of the reaction is 5.5MeV then the kinetic energy of the << particle is**

   1)4.4 MeV 2)5.4 MeV 3)5.6 MeV 4)6.5 MeV

RADIOACTIVITY, HALF LIFE, MEAN LIFE, DECAY CONSTANT

5. **If the activity of ^{108}Ag is 3 micro curie, the number of atoms present in it are (<=0.005 sec^{-1})**

   1)2.2×10^7 2)2.2×10^6 3)2.2×10^5 4)2.2×10^4

6. **The half life period of Pb^{210} is 22 years. If 2g of Pb^{210} is taken, then after 11 years the amount of Pb^{210} will be present is**

   1)0.1414 g 2) 1.414 g 3) 2.828 g 4) 0.707 g

7. $^{221}_{87}Ra$ **undergoes radioactive decay with a half life of 4 days. The probability that a Ra nucleus will disintegrate in 8 days is**

   1)1 2)1/2 3)1/4 4) 3/4

NUCLEAR FISSION-NUCLEAR REACTOR

8. **When $_{92}U^{235}$ undergoes fission. About 0.1% of the original mass is conveted into energy..**

   Then the amount of $_{92}U^{235}$ should undergo fission per day in a nuclear reactor so that it provides energy of 200 mega watt electric power is

   1) $9.6<10^{<2}\,kg$ 2) $4.8<10^{<2}\,kg$ 3) $19.2<10^{<2}\,kg$ 4) $1.2<10^{<2}\,kg$

PAIR PRODUCTION & PAIR ANNIHILATION

9. **A <-ray photon creates an electron, positron pair. The rest mass of electron is 0.5 MeV. KE of the electron - positron pair system is 0.78 MeV. Then the energy of <-ray photon is (in MeV)**

   1)1.78 2)0.28 3)1.28 4)0.14

LEVEL-II -2- KEY

1)1 2)2 3)1 4)2 5)1 6)2 7)4 8)3 9)1

LEVEL-II-2 - HINTS

1. $m_1 v_1 < m_2 v_2 ; \frac{m_1}{m_2} < \frac{<R_1 <^3}{<R_2 <} < \frac{V_2}{V_1}$

2. $<m < m_p < m_{nitrogen} < m_{oxygen}$

   energy required $= <m < 931.5$ Mev

3. $<m < <m_p < m_e << m_n$;Energy released$= <$ m×c^2

4. $K.E_< < \frac{<A < 4 <}{< A <} Q < \frac{216}{220} < 5.5 < 5.4$MeV

5. $< < \frac{<\frac{dN}{dt}<}{N}$

6. $t < \frac{t_{\frac{1}{2}}}{\log 2} \log \frac{<N_0 <}{<N <}$

7. $n < \frac{t}{t_{1/2}} < \frac{8}{4} < 2 ;\ N_0 < 2^n N ;\ \frac{N_0}{N} < 2^2 < 4$

   Remaining atoms $= N < \frac{N_0}{4}$

   Dacayed atoms $< N_0 < N < N_0 < \frac{N_0}{4} < \frac{3N_0}{4}$

   $<$ Probability of decay $< \frac{N_0 < N}{N_0} < \frac{3}{4}$

8. E=MC^2

9. $E_< < 2 < E_K < E_0 <$

LEVEL - III

1. 50% of a radio active substance decays in 5 hours. The time required for the 87.5% decay is

1)1 0hours 2)1 5hours 3)12.5hours 4)17.5hours

2. 4 grams of radioactive substan ce A left 1/2 gm a fter some time. 1 gram of another radioactive substance B left 1/4 gm in the same p eriod. If half life of B is 2 hour s, the half life of A is

1)3/4 hours 2)4/3 hours 3)1/4 hours 4)1/2 hours

3. One mole of a < emitter of half life e qual to 2days was pla ced in a sealed t ube for 4 days at S.T.P. volume of helium collected is

1)22.4 lit 2)16.8 lit 3)11.2 lit 4)5.6 lit

4. **3 rutherfords of a radio active isotope of half- life equal to 3 days was received after 12 days. Initial isotope packed was**
1) 48 rutherfords 2)12 rutherfords 3) 25 rutherfords 4)36 rutherfords

5. **The half life of a radio active substance is 6 hours. The amount of the substance undergone disintegration when 36 gms of it undergoes decay for 18 hours is**
1) 31.5 gm 2)4.5 gm 3)18 gm 4)9 gm

6. **The radio active nuclides A and B have half lives t and 2t respectively. If we start an experiment with one mole of each of them, the mole ratio after time interval of 6t will be**
1)1 : 2 2)1 : 8 3)1 : 6 4)1 : 1

7. **20% of a radio active element disintegrates in 1hr. The percentage of the radio active element disintegrated in 2hrs will be**
1)36% 2) 64% 3) 60% 4)40%

8. **The C^{14} to C^{12} ratio in a certain piece of wood is 25% of that in atmosphere. The half life period of C^{14} is 5,580 years. The age of wood piece is**
1)5,580 years 2)2790 years 3)1395 years 4)11,160 years

9. **A radioactive sample can decay by two dfferentprocesses. The half-life for the first process is T_1 and that for the second process is T_2. The effective half-life T of the radioactive sample is**
1) $T < T_1 < T_2$ 2) $\frac{1}{T} < \frac{1}{T_1} < \frac{1}{T_2}$ 3) $T < \frac{T_1 < T_2}{T_1 T_2}$ 4) $T < \frac{T_1 < T_2}{T_1 T_2}$

10. **The age of the wood if only 1/16 part of original C^{14} is present in its piece is (T of C^{14} is 5,580 years)**
1)5580 years 2)11,160 years 3)22320 years 4)16740 years

11. **A piece of wood is found to have the $\frac{C^{14}}{C^{12}}$ ratio to be 0.5 times of that in a living plant The number of years back the plant died will be (T of C^{14} = 5,580 years)**
1)2,790 years 2)5,580 years 3)11,160 years 4)27,900 years

12. **A piece of wood collected from cro-Magnon caves gave 4 disintegrations / min. A freshly cut wood of the same weight gives 16 d.p.m. The cro-magnon man lived about (Half life of C^{14} is 5760 years. Assume the activity is due to C^{14} only)**
1)5700 years ago 2)2900 years ago 3)11520 years ago 4)1400 years ago

13. **The number of U^{238} nuclei in a rock sample qual to thenumber of Pb^{206} atoms. The halflife of U^{238} is $4.5<10^9$ years.The age of the rock is**
1) $4.5<10^9$ y 2) $9<10^9$ y 3) $13.5<10^9$ y 4) $18<10^9$ y

14. **Equal masses of two samples A and B of charcoal are burnt and the activity of resulting carbon-di-oxide from two samples is measured. The gas from sample A gives 10^4 counts per month and that from sample B gives $2.5<10^3$ counts per month. The age difference of two sample is (half life of C^{14} is 5730 years)**
1)5730y 2)11460y 3)17190y 4)22920y

15. **The half life of a radioactive substance is 20 minutes. The approximate time interval $<t_2 < t_1<$ between the time t_2, when $\frac{2}{3}$ of it has decayed and time t_1 and $\frac{1}{3}$ of it had decayed is:**

1) 14 minutes 2) 20 minutes 3) 28 minutes 4) 7 minutes

16. **A charged capacitor of capacitance C is discharged through a resistance R. A radio active sample decays with an average life t. Find R interms of C and t in order that the ratio of the electrostatic energy stored in the capacitor to the activity of the radio active sample remains constant with time**

1) $\frac{2t}{C}$ 2) $\frac{C}{2t}$ 3) 2t C 4) t C

17. **Uranium-238 decays to thorium-234 with half-life $5<10^9\ yr$. The resulting nucleus is in the excited state and hence further emits $<<$ rays to come to the ground state. It emits 20< rays per second. the emission rate will drop to 5<< rays per second in**

1) $1.25<10^9\ yr$ 2) $10^{10}\ yr$ 3) $10^{<8}\ yr$ 4) $1.25<10^{<9}\ s$

18. **A sample of radioactive material has mass m, decay constant $<$ and molecular weight M. Avagadro constant = N_A. The initial activity of the sample is**

1) $<m$ 2) $\frac{<m}{M}$ 3) $\frac{<mN_A}{M}$ 4) $mN_A e^{<}$

19. **A sample of radioactive material has mass m, decay constant $<$ and molecular weight M. Avagadro constant = N_A. The activity of the sample after time t will be**

1) $\left(\frac{<mN_A}{M}\right) e^{<<t}$ 2) $\left(\frac{<mN_A}{M}\right) e^{<<t}$ 3) $\left(\frac{<mN_A}{M}\right) e^{<<t}$ 4) $\frac{m}{<}\left(1<e^{<<t}\right)$

20. **In moon rock sample the ratio of the number of stable argon-40 atoms present to the number of radioactive potassium-40 atoms is 7:1. Assume that all the argon atoms were produced by the decay of potassium atoms, with a half-life of $2.5<10^9\ yr$. The age of the rock is**

1) $2.5<10^9\ yr$ 2) $5.0<10^9\ yr$ 3) $7.5<10^9\ yr$ 4) $10^{10}\ yr$

21. **The half-life of a radioactive sample is T. If the activities of the sample at time t_1 and t_2 ($t_1 < t_2$) are R_1 and R_2 respec tively, then the number of atoms disintegrated in time $t_2 < t_1$ is proportional to**

1) $(R_1 < R_2)T$ 2) $(R_1 < R_2)T$ 3) $\frac{R_1R_2}{R_1 < R_2}T$ 4) $\frac{R_1 < R_2}{T}$

22. **Considering a hypothetical annihilation of a stationary electron with a stationary positron, the wavelength of resulting radiation? (with usual notations)**

1) $\frac{h}{2m_0C}$ 2) $\frac{2h}{m_0C}$ 3) $\frac{h}{m_0C}$ 4) $\frac{h\sqrt{2}}{m_0C}$

23. **A radioactive nucleus can decay by two different processes. The half life for the first process is 2t and that for the second process is t. The effective disintegration constant of nucleus is**

1) $\frac{3}{2t\ln 2}$ 2) $\frac{3\ln 2}{2t}$ 3) $\frac{\ln 2}{3t}$ 4) $\frac{3\ln 2}{t}$

24. A proton with kinetic eargy, K strikes another proton at rest. If the cislon is head - on, find the correct graph between K and the distance of closest approach, r.

1) 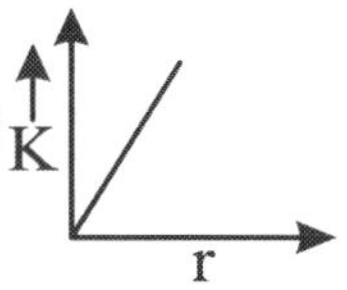

2) 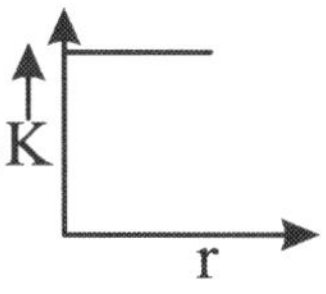

3) 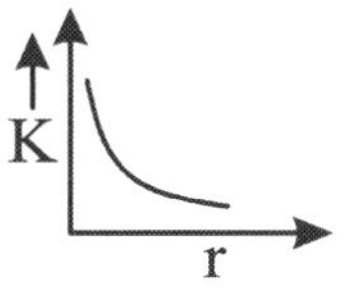

4) 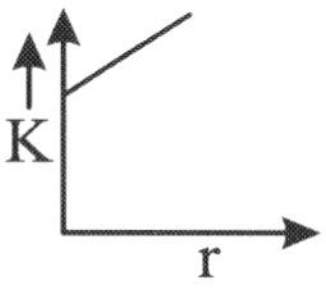

25. The fraction of a radiactive sample wll decay during half of its half-life period is

1) $\frac{1}{\sqrt{2}}$ 2) $\frac{1}{\sqrt{2} < 1}$ 3) $\frac{\sqrt{2} < 1}{\sqrt{2}}$ 4) $\frac{1}{2}$

26. A small quantity of a solution containing N^4 radio -nuclide of half - life T and activity R_0 is injected into blood of a person. $1\,cm^3$ **of sample of blood taken from the blood of the person shows activity R_1. If the total volume of the blood in the body of theerson is V, find the timer after which sample is taken.**

1) $\frac{T}{\ln(2)} < \measuredangle n \frac{R_0}{VR_1} <$ 2) $\frac{T}{\ln(2)} < \measuredangle n \frac{VR_0}{R_1} <$

3) $\frac{T}{\ln(2)} < \measuredangle n \frac{VR_1}{R_0} <$ 4) $\frac{T}{\ln(2)} < \measuredangle n \frac{R_1}{VR_0} <$

27.A nucleus with mass number 220 initially at rest emits an < - particle. If the Q value of the reaction is 5.5 MeV, calculate the kinetic energy of< -particle.

1)4.4 MeV 2)5.4 MeV 3)5.6MeV 4)6.5 MeV

28. Some amo unt of radioactive s ubstance (half-life=10 days) is spread inside a room and consequently the level of radiationeomes 50times the prmissible level for normalcupancy of the room. The room be safe for ocupation after

1)20days 2)34.8 days 3)56.4 days 4)62.9 days

29. In the options given below, let E denote the rest mas energy of a nucleus and n aneutron. The correct option is

1) $E<{}^{236}_{92}U< < E<{}^{137}_{53}I< < E<{}^{97}_{39}Y< < 2E<n<$ 2) $E<{}^{236}_{92}U< < E<{}^{137}_{53}I< < E<{}^{97}_{39}Y< < 2E<n<$

3) $E<{}^{236}_{92}U< < E<{}^{140}_{56}Ba< < E<{}^{94}_{36}Kr< < 2E<n<$ 4) $E<{}^{236}_{92}U< < E<{}^{140}_{56}Ba< < E<{}^{94}_{36}Kr< < 2E<n<$

30. Four different radioactive elements are kept in separate containers. In the begining the container A has 200 g-atom with half-life of 2 days , B has 20 g-atom with half-life of 20 days, C has 2g-atom with half-life 200 days and D has 100g-atoms with half-life of 10 day. In the begining the maximum activity exhibited by the contair is

1) A 2)B 3)C 4)D

31. Binding energy pernucleon versas mass number curve fornuclei is shown in the fig. W,X,Y and Z are fournuclei indicated on the curve. The precess that would release energy is

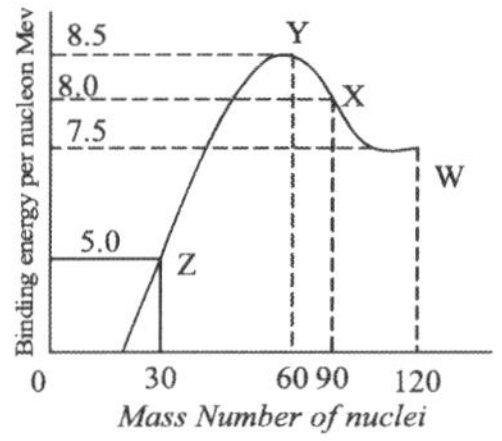

1) $Y < 2Z$ 2) $W < X < Z$ 3) $X < Y < Z$ 4) $W < 2Y$

32. When $_3Li^7$ nuclei are bombarded by protons, and the resultantnuclei are $_4Be^8$, the emitted particles will be.

1)alpha particles 2)beta particles 3)gamma photons 4)neutrons

33. A sample of uranium is a mixture of three isotopes $_{92}U^{234}$, $_{92}U^{235}$ and $_{92}U^{238}$ present in the ratio 0.006%, 0.71% and 99.284% respectively. The half lives of then isotopes are e $2.5<10^5$ years, $7.1<10^8$ years and $4.5<10^9$ years respectively. The contribution to activity (in %) of each isotope in the sample respectively

1) 51.41%,2.13%, 46.46% 2) 51.41%,46.46%,2.13%

3) 2.13%,51.41%,46.46% 4) 46.46%,2.13%, 51.41%

34. The table that follows shows some measurements of the decay rate of a sample of ^{128}I, a radio nudide often used medically as a tracer to measure the rate at which iodine is absorbed by the thyroid gland.

Time(min)	A(counts/s)	Time(min)	A (counts)/s
4	392.2	132	10.9
36	161.4	164	4.56
68	65.5	196	1.86
10	26.8	218	1.00

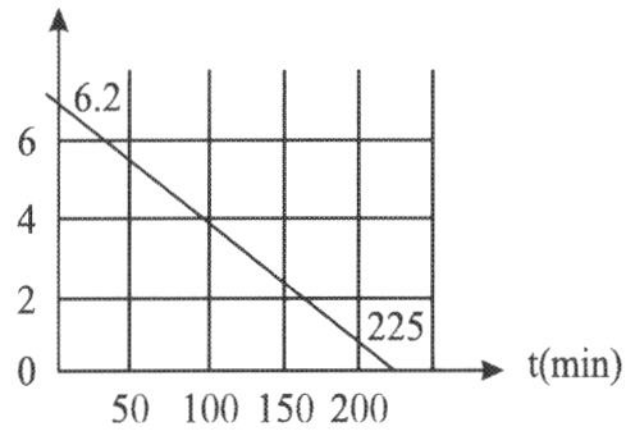

The half life $t_{1/2}$ for this radio nuclide.

1)25 min 2)50 min 3)2.5 min 4)5 min

35. The fraction f of radioactive element decayed change with respect to time (t). The curve representing the correct variation is

1) 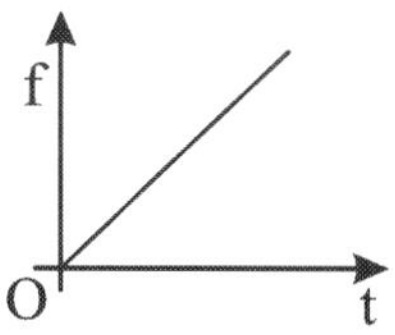

2) 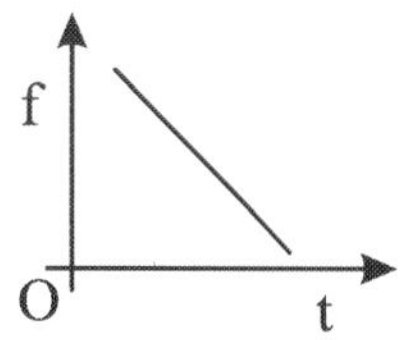

3) 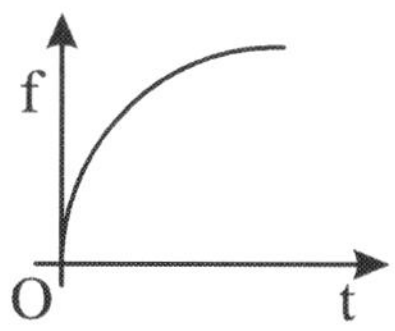

4) 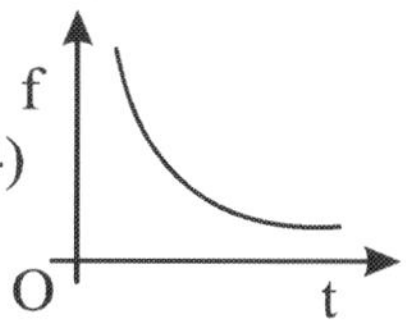

36. The rate of decay (R) of nuclei in a radiocative sample is plotted against time (t). Which of the following best represents the resulting curve?

1) 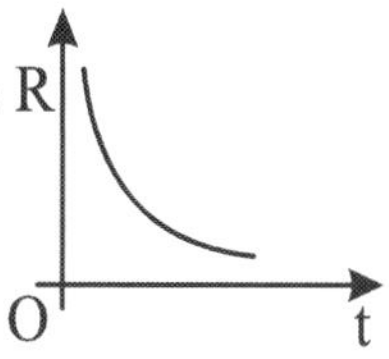

2) 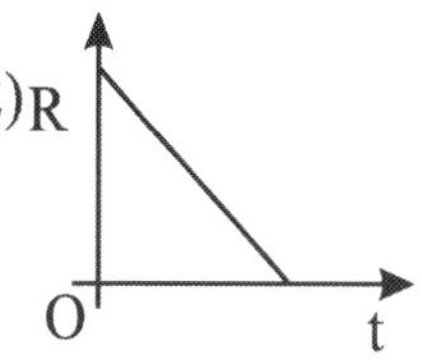

3) 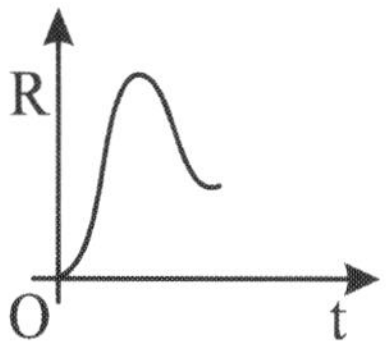

4) 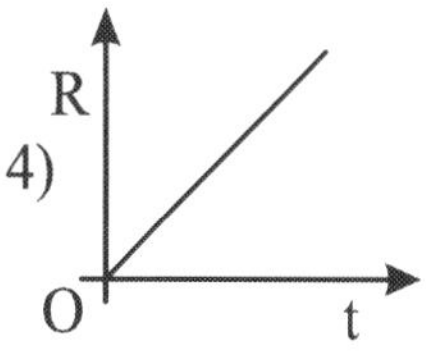

37. The probability of survival of a radioactive nucleus for one mean life is

1) $\frac{1}{e}$ 2) $1<\frac{1}{e}$ 3) $\frac{\ln 2}{e}$ 4) $1<\frac{\ln 2}{e}$

38. A radioactive isotope is being produced at a constant rate A. The isotope has a half-life T initially there are no nuclei, after a time t>>T, the number of nuclei becomes constant. The value of this constant is

1) AT 2) $\frac{A}{T}\ln(2)$ 3) $AT\ln(2)$ 4) $\frac{AT}{\ln(2)}$

LEVEL - III - KEY

1)2	2)2	3)2	4)1	5)1	6)2	7)1	8)4	9)2	10)3
11)2	12) 3	13) 1	14) 2	15) 2	16) 1	17) 2	18) 3	19) 2	20) 3
21)1	22) 3	23) 2	24) 3	25) 3	26) 1	27) 2	28) 3	29) 1	30) 1
31)4	32) 3	33) 1	34) 1	35) 3	36) 1	37) 1	38) 4		

LEVEL -III - HINTS

1. $W < \frac{W_0}{2^n}$ $n < \frac{t}{t_{\frac{1}{2}}}$; $t_{1/2} < 5hrs$

2. $W< \frac{W_0}{2^n}$;no fo half lives $n<\frac{t}{t_{1/2}}$

3. $W< \frac{W_0}{2^n}$; $n < \frac{t}{t_{1/2}}$. initially value of Helium 22.4lits no of halflives =2

4. $W< \frac{W_0}{2^n}$, $n < \frac{t}{t_{\frac{1}{2}}} < \frac{12}{3} < 4$

5. $W < \frac{W_0}{2^n}$; $n < 3$

7. $W< \frac{W_0}{2^n}$. Intial amount W_0, after one hour amount of remaining sample =80% W_0 ; $W_1 < \frac{80}{100} W_0$

8. $N < N_0 e^{<<t}$

9. $\frac{dN}{dt} < <N$; $<< <_1 << _2$

$<< \frac{1}{T}$

10. n = 4 half lifes
11. t = 1 half life
12. t = 2 half lifes

13. Given $N< \frac{N_o}{2}$; $< \frac{N}{N_o} < \frac{1}{2}$

that means t = 1 half life

14. $t< \frac{t_{1/2} \log \frac{N_0}{N}}{\log 2}$

15. At t_1 $\frac{2}{3} < \frac{1}{2^{t_1/20}}$ At t_2 $\frac{1}{3} < \frac{1}{2^{t_2/20}}$

t_2-t_1= 20 mins.

16. $\frac{U_c}{A} <$ constant; $\frac{\frac{1}{2} CV_o^2}{<N_o} < \frac{\frac{1}{2} CV^2}{<N}$

Where $V < Ve^{<t/CR}$, $N < Ne^{<<t}$ and $<< \frac{1}{t}$

17. t = 2 half lifes; $t < nT < 2<5<<10^9$ *year*

$<10^{10}$ *year*

18. Activity = $<N$ where $N < \frac{N_A m}{M}$

19. activity = $<N_o e^{<<t}$ where $N_o < \frac{N_A m}{M}$

20. Let the number of radioactive Potassium atoms present initially (t=0) is N_0 and the number of stable argon atoms at t=0 is zero. After time t the number of stable argon atoms is m and the radioactive potassium atoms is $N_0 < m$ given that

$$\frac{N_0 < m}{m} < \frac{1}{7}, m < \frac{7}{8} N_0 \; and N_0 < m < \frac{1}{8} N_0$$

since after one half life N_0 reduces to $N_0/2$ after 2 half lives $N_0/4$ and after 3 half lives it reduces to $N_0/8$

Thus, $t < nT < 3 < 2.5 < 10^9$ *years*
$< 7.5 < 10^9$ *year*

21. Activity $R < <N$
so that

$$R_1 < <N_1 \; and R_2 < <N_2$$

$$R_1 < R_2 < <<N_1 < N_2 << \frac{0.6931}{T} <N_1 < N_2 <$$

$$N_1 < N_2 < \frac{<R_1 < R_2 <T}{0.6931}$$

$$N_1 < N_2 < <R_1 < R_2 <T ; \quad < t_2 < t_1 << R_1 < R_2 <T$$

22. $$\frac{hc}{<} < \frac{hc}{<} < m_f C^2 < m_f C^2$$

23. $<<< _1 << _2$

24. At closest approach, both protons will be moving with same velocity.

$$<KE < \frac{1}{2} m v_0^2 < 2.\frac{1}{2} m \left(\frac{<v_0}{<2}\right)^2 < \frac{e^2}{4<<_0 r}$$

$$\frac{1}{4} m v_0^2 < \frac{e^2}{4<<_0 r} < \frac{K}{2}$$

$$Kr < \frac{e^2}{2<<_o} < \text{constant}, \quad K < \frac{1}{r}$$

25. $$\frac{N}{N_0} < \left(\frac{<1}{<2}\right)^n < \left(\frac{<1}{<2}\right)^t \quad Here, t < \frac{T}{2} \; or \frac{t}{T} < \frac{1}{2}$$

$$\frac{N}{N_0} < \left(\frac{<1}{<2}\right)^{\frac{1}{2}} < \frac{1}{\sqrt{2}}; \quad \frac{N_0 < N}{N_0} < 1 < \frac{1}{\sqrt{2}} < \frac{\sqrt{2} < 1}{\sqrt{2}}$$

26. Total volume of blood,

$$V < \frac{\text{Total activity R}}{\text{Activity per cm}^3 <R_1<}$$

$$< \frac{R_0 e^{<<}}{R_1}, \qquad \text{(or)} \; \frac{VR_1}{R_0} < e^{<<}$$

$$<\ln\frac{<VR_1<}{<R_0<} << <t < \frac{\ln(2)t}{T};\; t < \frac{T}{\ln(2)}\ln\frac{<R_0<}{<VR_1<}$$

27. $K_1 < K_2 < 5.5\,MeV$

From conservation oflinear momentum

$<2K_1(216m)<^{1/2} < <2K_2(4m)<^{1/2}$; $K_2 < 54\,K_1$

From the above $K_2 < 5.4\,MeV$

28. Since the intial activity is 50 timesactivity for safe ccupancy, therefore $R_0 < 50R$ where $R < <N$ since,

$R < N$

$\frac{R}{R_0} < \frac{N}{N_0} < \left<\frac{1}{2}\right>^{n} < \left<\frac{1}{2}\right>^{t/T}$ or

$\left<\frac{1}{2}\right>^{t/10} < \frac{1}{50}$

29. (a): Rest mass energy of U will be greater than theat mass energy of nuclei into which it breaks. The consistituent nuclei and neutrons will have kinetic energy also, as a result of conservation of linear momentum.

30. The activity of radioactive element is given by $R < <N < \frac{0.6931N}{T}$ $Thus, R < \frac{N}{T}$. The value of $\frac{N}{T}$ is largest for the element in the container A.

31. Energy of products must be more then the reactants to release energy.

32. $_3Li^7 <_1 H^1 < {}_4Be^{8} <$

33: Let m is the total mass of the uranium mixture. The masses of the isotopes $_{92}U^{234}$, $_{92}U^{235}$ and $_{92}U^{238}$ in the mixture are $m_1 < \frac{0.006}{100}m$; $m_2 < \frac{0.71}{100}m$, and $m_3 < \frac{99.284}{100}m$.

If N_A is the Avogadro number, then number of atoms of three isotopes are; $N_1 < \frac{m_1 N_A}{M_1}$, $N_2 < \frac{m_2 N_A}{M_2}$,

and $N_3 < \frac{m_3 N_A}{M_3}$

Activity of radioactive sample $A < \lambda N$.

As $\lambda < \frac{0.693}{t_{1/2}}$, < $\quad A < \frac{0.693}{t_{1/2}}N$

If t_1, t_2 and t_3 be the half lives, then

$$A_1 : A_2 : A_3 < \frac{N_1}{t_1} : \frac{N_2}{t_2} : \frac{N_3}{t_3}$$

or $$A_1 : A_2 : A_3 < \frac{m_1}{Mt_1} : \frac{m_2}{M_2 t_2} : \frac{m_3}{Mt_3}$$

$$< \frac{0.006}{234 < 2.5 < 10^5 <} : \frac{0.71}{235 < 7.5 < 10^8 <} : \frac{99.284}{238 < 4.5 < 10^9 <}$$

$< 51.41\% : 2.13\% : 46.46\%$

34. $A < Ae^{<\lambda t}$ $\quad < \ln A < \ln A_0 < \lambda t$

$$< \lambda < \frac{\ln A_0 < \ln A}{t} < \frac{6.2 < 0}{225} < 0.0275 \text{min}^{<1}$$

$$< T_{1/2} < \frac{0.693}{\lambda} < 25 \text{min}$$

37. Probability of survival for any nucleus at time t is $P < \frac{N}{N_0} < \frac{Ne^{<\lambda t}}{N_0} < e^{<\lambda t}$

So, in one mean life, required probability is $e^{<\lambda < \frac{1}{\lambda}} < \frac{1}{e}$;

38. $A < N< ; < N < \frac{A}{<} < \frac{AT}{\text{In}2}$

LEVEL - IV

Comprehension-1:

The count rate meter is used to measure the activity of a given amo unt of a radio active element. At one instant, the meter shows 475 counts/minute exactly 5 minutes later, is shown 270 counts/minute then

1. **The decay constant is**

1)0.82/ minute 2)0.113/ minute 3)0.166/ minute 4)0.1 82/ minute

2. **Mean life of the sample is**

1)6.35 minutes 2) 7.45 minutes 3)8.85 minutes 4)9.95 minutes

3. **Half life of the sample is**

1)6. 13 minutes 2)8.42 minutes 3)8.85 minutes 4)9.92 minutes

Comprehension-2:

Atomic nucleus is central core of every atom in which the whole of positive charge and almost entire mass of atom is concentrated. It is a tiny sphere of radius R is given by $R < R_o A^{1/3}$, where $R_o < 1.4 < 10^{<15} m$, a constant and A is the mass number of the nucleus

4. A graph between '$\log\left<\frac{R}{R_o}\right<$, and 'log A''

1)Is a parabola
2)Is a straightline passing through origin
3)Is a straight line have an intercept
4)Is an ellipse

5. On increasing the value of 'A' the density of the nucleus

1)Increases
2)Decreases
3)Remains constant
4)None

6. The radius of the nucleus of mass number 125 is

1) $175<10^{<15}m$
2) $35<10^{<15}m$
3) $70<10^{<15}m$
4) $7<10^{<15}m$

Comprehension-3:

The questions given below are based on the following paragraph

A nucleus of mass M + $<m$ is at rest and decays into two daughter nuclei of equal mass $\frac{M}{2}$ each. speed of light is c.

7. The binding energy per nucleon for the parent nucleus is E_1 and that for the daughter nuclei is E_2 then

1) $E_2 = 2E_1$
2) $E_1 > E_2$
3) $E_2 > E_1$
4) $E_1 = 2E_2$

8. The speed of daughter nuclei is

1) $c\frac{<m}{M<<m}$
2) $c\sqrt{\frac{2<m}{M}}$
3) $c\sqrt{\frac{<m}{M}}$
4) $c\sqrt{\frac{<m}{M<<m}}$

9. Choose correct statement from the following.

A)Large mass number nuclei undergo fission
B)Low mass number nuclei undergo fusion
C)For heavy nuclei the decrease in binding energy per nucleon shows the contribution of the increasing coulomb repulsion.

1) A, B are correct
2) A, B, C are correct
3) B , C are correct
4) A , C are correct

10. Which of the following are not fundamental particles

i)Electron
ii) Photon
iii) a - Particle
iv)Deutron

1)Only i & ii are true
2)Only ii & iii are true
3)Only i & iii are true
4)Only iii & iv are true

11. When a nucleus with atomic number Z and mass number A undergoes a radioactive decay process

a)Both Z and A will decrease, if the process is $<$ decay

b)Z will decrease but A will not change, if the process is $<^{+}$ decay

c)Z will increase but A will not change, if the process is $<^{-}$ decay

d)Z and A will remain unchanged, if the process is $<$ decay

1)a & b are true
2)b & d are true
3)a, b & c are true
4)a, b, c, d are true

12. **A nuclide A undergoes a decay and another nuclide B undergoes b^- decay**

a) All the< –particle emitted by A will have almost the same speed

b)The <–particle emitted by A may have widely different speeds

c) All the< –particle emitted by B will have almost the same speed

d)The <–particle emitted by B may have widely different speeds

1)a, b are true 2)b, c are true

3)b, d are true 4)a, d are true

13. **In the fission of U^{235}**

i)Slow neutron is absorbed by U^{235}

ii) The products in the process are not same, their atomic number varies from 34 to 58

iii) About 200 Mev energy is released per fission

iv)The product are always Ba and Kr

1)Only i, ii & iii are true 2)Only ii & iii are true

3)All are true 4)Only i , ii & iv are true

14. **Which of the following statements are correct**

i)Positron is predicted by Dirac and discovered by Anderson

ii) Liquid drop model of nucleus is developed by Bohr and Wheeler

iii) Carbon cycle was proposed by Bethe

iv)Fission reaction is first observed by Otto Hahn and Strassman

1) All are true 2)Only i, ii & iv are true

3)Only i, iii & iv are true 4)Only iii & iv are true

15. **Consider the following two statement A and B and identify the correct answer given below :**
A)Nuclear density is same for all nuclei

B)Radius of the nucleus (R) and its mass number (A) are related as $\sqrt{A} < R^{1/6}$

1) A and B are true 2) A and B are false

3) A is true but B is false 4) A is false but B is true

16. **Consider the following statements A, B and identify the correct choice in the given answers**

A: Density of a nucleus is independent of its mass number

B: Beryllium is used as moderator in nuclear reactors

1) A and B are correct 2) A and B are wrong

3) A is correct, B is wrong 4) A is wrong, B is correct

17. **Consider the following statements (A) and (B) and identify the correct answer given below.**

Statement (A) : Positive values of packing fraction implies a large value of binding energy.

Statement (B) : The difference between the mass of the nucleus and the mass number of the nucleus is called packing fraction

1)(A) and (B) are correct 2)(A) and (B) are false

3)(A) is true (B) is false 4) A is false, B is true

LEVEL - IV - KEY

1)2	2)3	3)1	4)2	5)3	6)4	7)3	8)2	9)2	10)4
11)4	12) 4	13) 1	14) 1	15) 3	16) 1	17) 2			

LEVEL - IV - HINTS

1. $A < A_0 e^{<<t}$; $270 < 475e^{<<t}$

   On solving we get $<< 0.113$/min

2. $<< \frac{1}{<}$

3. T=0.693<

6. $R < R_0 A^{1/3}$

7. binding energy pernucleon of daughter nuclei is more than that of the parent nuclei.

8. According to conservation of momentum

   $0 < \frac{M}{2}V_1 < \frac{M}{2}V_2; V_1 < V_2$; $<mc^2 < \frac{1}{2}.\frac{M}{2}V_1^2 < \frac{1}{2}.\frac{M}{2}V_2^2$

** * * *

Chapter - 8
(Part - 1)
SEMI CONDUCTOR DEVICES

SEMI CONDUCTOR DEVICES

SYNOPSIS

Classification of solids :

- Solids can be classified as metals, insulators and semiconductors basing on conductivity (or) band theory
- Classification according to conductivity is as follows
- Metals have high conductivity $\left(10^8 to 10^2\ Sm^{-1}\right)$ (or) low resistivity
- Insulators have low conductivity $\left(10^{-19} to 10^{-11} Sm^{-1}\right)$ (or) high resistivity $\left(10^{+11} to\ 10^{+19}\Omega m\right)$
- Semiconductors have conductivity $\left(10^5\ to\ 10^{-6} Sm^{-1}\right)$ (or) resistivity $\left(10^{-5} to 10^6 \Omega m\right)$, intermediate to metals and insulators

Energy Band theory in solids:

- An isolated atom has well defined energy levels and energy of an electron depends on its orbit (Principal quantum number)
- But in solids atoms are so close such that outer orbits are very close (or) overlaped to form energy band.
- Inside the crystal each electron has a unique position and no two electrons see exactly same pattern of surrounding charges and each electron has different energy level.
- Different energy level with continuous energy variation form energy bands (According to Pauli's principle)
- The energy band formed by a series of energy bands containing valance electrons is **valance band.**
- At 0 K, electrons start filling energy level in valance band starting from the lowest one.
- The highest energy level, occupied by an electron in the valance band at 0K is called **Fermi level.**
- The lowest unfilled energy band formed just above valance band is called **conduction band.**
- Depending on the forbidden energy gap between valance band and conduction band, the solids are classified as conductors, insulators and semiconductors.

1) Conductors : The energy band structure in conductors have two possibilities

- The valance band may be completely filled and the conduction band partially filled with an extremely small energy gap between them $E_g = 0$

  Ex: Sodium (Na)
- The valence band is completely filled and conduction band is empty but the two overlap each other. Eg: Zn

1) 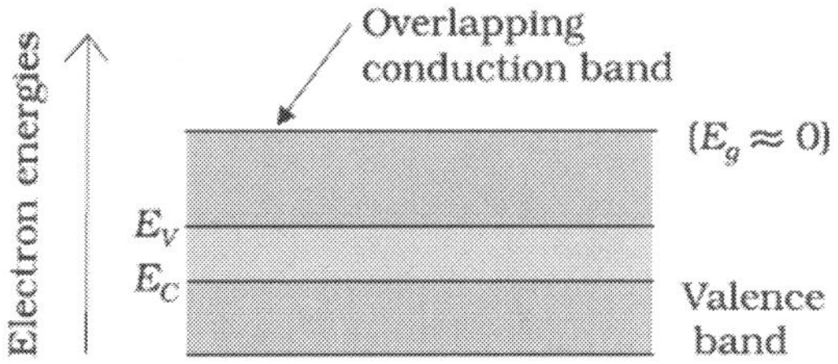

conductors

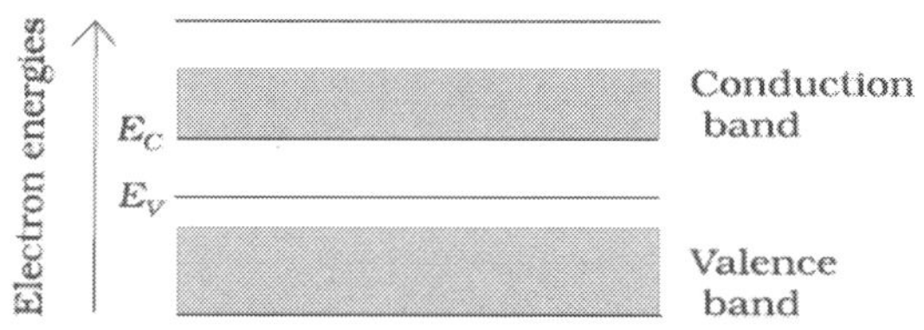

conductors

2) **Insulators :**

- In insulators forbidden energy gap is quite large. $E_g > 3ev$
- Ex. Energy gap for diamond is 5.5 eV.

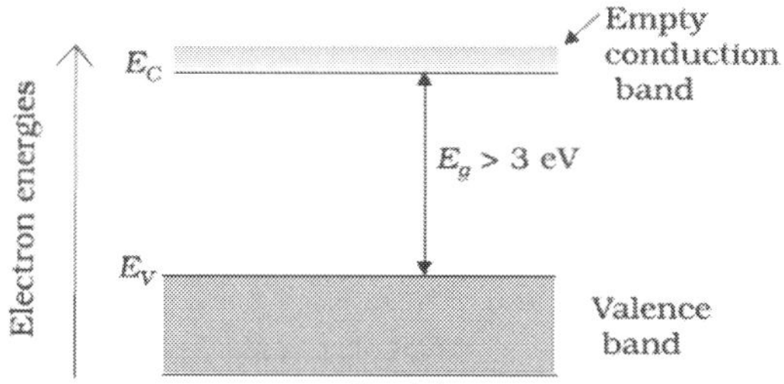

Insulators

3) **Semi conductors :**

- Semi conductors are the basic materials used in the present solid state devices like diode, transistor, Ic's.
- The energy band structure of the semiconductors is similar to insulators but in their case, the size of forbidden energy gap is much smaller than that of insulators. $E_g = 0.2eV \text{ to } 3eV$

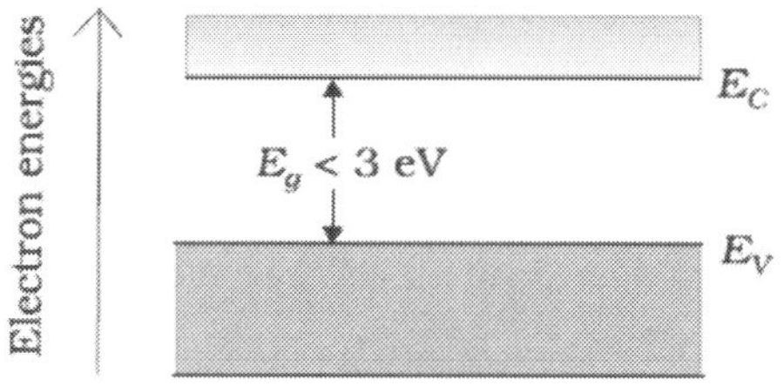

Semiconductors

- Eg: Forbidden energy gap for Ge is 0.67eV, for Si is 1.1 eV and for GaAs is 1.41eV
- The vacancy in the covalent band is called hole.
- The electrons and holes move in opposite directions.
- The difference in the resistivity of C, Si and Ge depends upon the energy gap between their conduction and valence bands. For C (diamond), Si and Ge, the energy gaps are 5.4 eV, 1.1 eV and 0.7 eV, respectively. Sn also is a group IV element but it is a metal because the energy gap in its case is 0 eV.

Intrinsic semi conductor :

- Semi conductors in their purest form having equal number of electrons in CB and holes in VB is called as intrinsic semi conductors.
- These electrons (free from covalent bonds) and holes are called intrinsic carriers and such a semiconductor is called intrinsic semiconductor.

Ex:- Germanium (Ge) and Silicon (Si) are the purest form of intrinsic semi conductors.

- In an intrinsic semiconductor if n_e denotes the electron number density in conduction band, n_h the hole number density in valence band and n_i the number density (or) concentration of intrinsic carriers, then

$n_e = n_h = n_i$

Doping:

- The process of adding impurity to a pure semi conductor (in 1 : 10^6 ratio also called 1ppm) so as to improve its conductivity is called doping.
- Pentavalent impurity atoms are called donor impurity atoms.
- Trivalent impurity atoms are called acceptor impurity atoms.

Extrinsic semi conductor :

- The semi conductor formed by adding a small quantity of impurity (tri or penta valent) to pure semi conductor is called as extrinsic semi conductor.
- The extrinsic semi conductor formed by adding pentanalent impurity atoms is called **n-type semi conductor.**
- The extrinsic semi conductor formed by adding trivalent impurity atoms is called **p-type semi conductor.**
- Fermi level in n-type semi conductors (also known as donor energy level) lies in forbidden energy gap and is very close to conduction band (≈ 0.01eV below conduction band)
- Fermi level in p-type semi conductor (also known as acceptor energy level) lies in forbidden energy gap and is very close to valence band (≈ 0.01 to 0.05 eV above valence band)
- When ever a covalent bond breaks an electron hole pair is formed.
- When valency electron combines with a hole, a covalent bond formed and at the starting point of the electron a hole is formed.
- If a conduction electron combines with a hole a covalent bond is formed without the formation of a hole any where.
- In a doped (or) extrinsic semi conductor number density of electrons in conduction band (n_e), number density of holes in the valence band (n_h) and number density of electrons in conduction band (or) holes in valence band in a pure semi conductor (n_i) then they are related as $n_e n_h = n_i^2$.
- The intrinsic concentration n_i varies with T as

$n_i^2 = A_o T^3 e^{-Eg/KT}$ (A_o is constant)

- The fraction of electrons of valence band present in conduction band is given by

$f \alpha e^{-\frac{Eg}{KT}}$

Where K is Boltzman's constant, E_g is forbidden energy gap and T is temperature

The Energy Gap : Experimentally it has been found that the forbidden energy region E_g depends on temperature.

For silicon $E_g(T) = 1.21 - 3.60 \times 10^{-4} T$

For germanium $E_g(T) = 0.785 - 2.23 \times 10^{-4} T$

At room temperature $(300K)$ for silicon $E_g = 1.1ev$ and for germanium $E_g = 0.72ev$

- Adding equal concentration of donor and acceptor or atoms to P type and N type semiconductor respectively results in an intrinsic semiconductor.
- When concentration of donor atoms exceeds the acceptor concentration in P type semiconductor, it changes from P type to N type semiconductor.
- In semi conductors the total current I is the sum of electron current I_e and holes current I_h

  $I = I_e + I_h$
- Electrical conductivity $\sigma = \sigma_e + \sigma_h$

  $\sigma = n_e \mu_e e + n_h \mu_h e$

  For intrinsic semi conductor $n_e = n_h = n_i$ so that $\sigma = n_i e(\mu_e + \mu_h)$

ELECTRICAL CONDUCTIVITY OF SEMI CONDUCTORS:(FORMULAE DERIVATION)

- Consider a block of semiconductor of length l, area of cross-section A and having number density of electrons and holes as n_e and n_h respectively.
- By that on applying a potential difference, say V, a current I flows through it as shown in figure.
- The electron current (I_e) and the hole current (I_h) constitute the total current I flowing through the semiconductor i.e., $I = I_e + I_h$............(1)

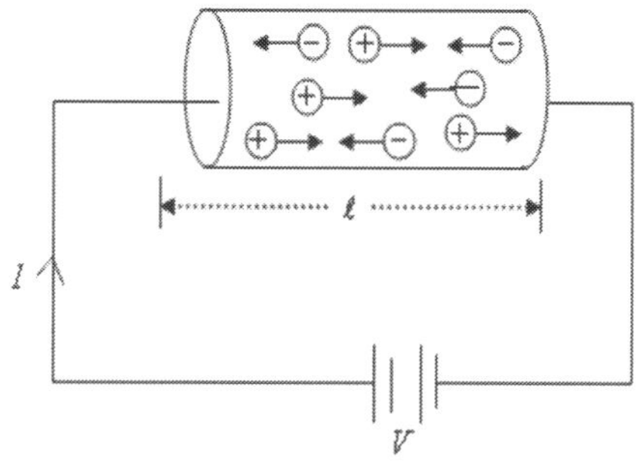

- If n_e is the density of conduction band electrons in the semiconductor and V_e is the drift velocity of electrons, then electron current is given by $I_e = en_e Av_e$........(2)

  Also, the hole current $I_h = en_h Av_h$......(3)

  Using the equation (2) and (3), the equation (1) becomes

  $I = en_e Av_e + en_h Av_h$

  or $I = eA(n_e v_e + n_h v_h)$........(4)

  If ρ is the resistivity of the material of the semiconductor, then the resistance offered by the semiconductor to the flow of current is given by $R = \rho \frac{\ell}{A}$.........(5)

  Since V=RI, from the equations (4) and (5),

  we have $V = RI = \rho \frac{\ell}{A} \times eA(n_e v_e + n_h v_h)$

or $V = \rho \ell e\left(n_e v_e + n_h v_h\right)$.............(6)

If E is the electric field set up across the semiconductor, then $E = \frac{V}{\ell}$........(7)

From the equations (6) and (7) , we have

$E = \rho e\left(n_e v_e + n_h v_h\right)$

or $\frac{1}{\rho} = e\left(n_e \frac{v_e}{E} + n_h \frac{v_h}{E}\right)$.........(8)

On applying electric field, the drift velocity acquired by the electrons (or holes) per unit strength of electric field is called mobility of electrons (or holes)

$\mu_e = \frac{v_e}{E}$ and $\mu_h = \frac{v_h}{E}$

Therefore, the equation (8) becomes

$\frac{1}{\rho} = e\left(n_e \mu_e + n_h \mu_h\right)$.........(9)

Also, $\sigma = \frac{1}{\rho}$ is called the conductivity of the material of semiconductor.

$\therefore \sigma = e\left(n_e \mu_e + n_h \mu_h\right)$.......(10)

- Electrons mobility is greater than the hole mobility .
- Mobility is a property of the semiconductor itself. It does not depends on the doping concentration.
- The mobility of an electron or hole generally decreases with increase temperature.
- Resistance of semi conductors decrease with the increase in temperature so semi conductors are insulators at low temperature but becomes slightly conducting at room tempeture.
- P- type (or) n- type semi conductor material is elctrically neutral.

W.E-1: The number of silicon atoms per m^3 is 5×10^{28}. This is doped simultaneosuly with 5×10^{22} atoms per m^3 of Arsenic and 5×10^{20} per m^3 atoms of Indium. Calculate the number of electrons and holes. Given that $n_i = 1.5\times10^{16}\,m^{-3}$ .Is the material - n type or p-type?

Sol: Arsenic is donor impurity

No. of donor atoms added, $N_D = 5\times10^{22}\,m^{-3}$,

Indium is acceptor impurity, no.of accepter atoms added

$N_A = 5\times10^{20}\,m^{-3}$

Therefore, no.of free electrons created

$n_e = N_D = 5\times10^{22}$

Now, $n_e > n_h$, therefore, net no.of free electrons created,

$n_e^1 = n_e - n_h = 5\times10^{22} - 5\times10^{20} = 4.95\times10^{22} m^{-3}$

Also net no. of holes created $n_h^1 = \frac{n_1^2}{n_e^1} = \frac{(1.5\times10^{16})^2}{4.95\times10^{22}} = 4.55\times10^9 m^{-3}$

AS $n_e^1 > n_h^1$, the resulting material is n-type semiconductor.

W.E-2: A semiconductor has an electron concentration of $0.45\times10^{12} m^{-3}$ and a hole concentration of $5.0\times10^{20} m^{-3}$. Calculate its conductivity . Given electron mobility $= 0.135 m^2V^{-1}s^{-1}$; hole mobility $= 0.048 m^2V^{-1}s^{-1}$,

Solution : The conductivity of a semiconductor is the sum of the conductivities due to electrons and holes and is given by $\sigma = \sigma_e + \sigma_h = n_e e\mu_e + n_h e\mu_h = e(n_e\mu_e + n_h\mu_h)$

As per given date, n_e is negligible as compared to n_h, so that we can write

$\sigma = en_h\mu_h = (1.6\times10^{-19}C)(5.0\times10^{20}m^{-3})(0.048m^2V^{-1}s^{-1}) = 3.84\Omega^{-1}m^{-1} = 3.84Sm^{-1}$

W.E-3: An N-type silicon sample of width $4\times10^{-3}m$ thickness and length $6\times10^{-2}m$ carries a current of 4.8mA when the voltage is applied across the length of the sample. What is the current density ?

If the free electron density is $10^{22}m^{-3}$, then find how much time it takes for the electrons to travel the full length of the sample .

Soution : The current density J is given by

$$J = \frac{I}{A} = \frac{4.8\times10^{-3}}{(4\times10^{-3})(25\times10^{-5})} = \frac{4.8\times10^{-3}}{10^{-6}}$$

The drift velocity v_d given by $v_d = \frac{J}{ne} = \frac{4800}{10^{22}\times1.6\times10^{-19}} = 3m/s$

The time taken 't' is given by $t = \frac{L}{v_d} = \frac{6\times10^{-2}}{3} = 0.02\,\text{sec}$

W.E-4: The energy gap of pure Si is 1.1 eV. The mobilities of electrons and holes are respectively $0.135m^2V^{-1}s^{-1}$ and $0.048m^2V^{-1}s^{-1}$ and can be taken as independent of temperature. The intrinsic carrier concentration is given by $n_i = n_0e^{-Eg/2kT}$. Where n_0 is a constant, E_g The gap width and k The Boltzmann's constant whose value is $1.38\times10^{-23}JK^{-1}$. The ratio of the electrical conductivities of Si at 600K and 300K is.

Solution: The total electrical conductivity of a semiconductor is given by

$$\sigma = e(n_e\mu_e + n_h\mu_h)$$

For an intrinsic semiconductor,

$$n_e = n_h = n_i$$

We can thus write for the conductivity

$$\sigma = e(\mu_e + \mu_h)n_i$$

or $\sigma = e(\mu_e + \mu_h) n_0 e^{E_g/2kT}$

As the mobilities μ_e, μ_h are independent of temperature, they can be regarded as constant. The ratio of the conductivities at 600 K and 300 K is then,

$$\frac{\sigma_{600}}{\sigma_{300}} = \frac{e(\mu_e + \mu_h) n_0 e^{-E_g/2k\times600}}{e(\mu_e + \mu_h) n_0 e^{-E_g/2k\times300}} = e^{-E_g/1200k}$$

As per given data $E_g = 1.1eV$

$$k = 1.38\times10^{-23} JK^{-1} or \left(\frac{1.38\times10^{-23}}{1.6\times10^{-19}}\right) eVK^{-1}$$

$\therefore k = 8.625\times10^{-5} eVK^{-1}$

Solving we get the ratio of electrical conductivities is 4×10^4

W.E-5: In a p-n junction diode, the current I can expressed as $I = I_0 \exp\left(\frac{eV}{2k_BT} - 1\right)$ where I_0 is called the reverse saturation current, V is the voltage across the diode and is positive for forward bias and negative for reverse bias, and I is the current through the diode, K_B is the Boltzmann constant $(8.6\times10^{-5} eV/K)$ and T is the absolute temperature. If for a given diode $I_o = 5\times10^{-12}$ A and T= 300K, then

(a) What will be the forward current at a forward voltage of 0.6V?

(b) What will be the increase in the current if the voltage across the diode is increased to 0.7V?

(c) What is the dynamic resistance?

(d) What will be current if reverse bias voltage changes from 1V to 2V?

Solution:

$I_o = 5\times10^{-12} A, k = 8.6\times10^{-5} eVk^{-1}$

$= 8.6\times10^{-5}\times1.6\times10^{-19} Jk^{-1}$

a) $I = I_0\left(e^{ev/2kT} - 1\right)$,

For V=0.6V,

$$I = 5\times10^{-12}\left(e^{\frac{1.6\times10^{-19}\times0.6}{2\times8.6\times10^{-5}\times1.6\times10^{-19}\times300}} - 1\right) = 5\times10^{-12}\left(e^{23.52} - 1\right)$$

$= 5\times10^{-12}\left(1.256\times10^{10} - 1\right) = 0.0628.A$

b) For v=0.7v, we have

$$I = 5\times10^{-12}\left(e^{\frac{1.6\times10^{-19}\times0.7}{2\times8.6\times10^{-5}\times1.6\times10^{-19}\times300}} - 1\right)$$

$I = 5\times10^{-12}\left(e \quad -1\right)$

$= 5\times10^{-12}\left(6.054\times10^{11}-1\right) = 3.0271A$

$\therefore \Delta I = 3.271 - 0.0628 = 2.9643A$

c) $\Delta I = 2.9643, \Delta v = 0.7 - 0.6 = 0.1V$ dynamic resistance

$= \dfrac{\Delta v}{\Delta I} = \dfrac{0.1}{2.9643} = 0.0337\Omega$

d) For change in voltage from 1 to 2v, the current will remain equal to $I_0 = 5\times10^{-12}A$. It shows that the diode possesses practically infinite resistance in reverse biasing

W.E-6: The energy of a photon of sodium light $(\lambda = 589nm)$ equal to the band gap of a semiconducting material (a) Find the minimum energy E required to create a hole-electron pair (b) Find the value of E/kT at a temperature of 300K

Solution :

(a) The energy of the photon in $eV = \dfrac{12400}{\lambda} = \dfrac{12400}{5890} = 2.1eV$

(Wavelength of photon =589nm=5890 A^0)

Thus the band gap is 2.1ev. This is also the minimum energy E required to push an electron from the valence band into the conduction band. Hence the minimum energy required to create a hole-electron pair is 2.1eV

(b) At T=300K,kT=$\left(8.62\times10^{-5}eV/K\right)\left(300K\right)$

$= 25.86\times10^{-3}eV$ Thus, $\dfrac{E}{kT} = \dfrac{2.1eV}{25.86\times10^{-3}eV} = 81$

The availale thermal energy is nearly 81 times less than that of the required energy to create electron hole pair. So it is difficult for the thermal energy to create the hole-electron pair but a photon of light can do it easily

P-n junction :

- In a single Ge (or) Si crystal by doping one part with trivalent impurity and the other part with pentavalent impurity , a p - n junction is formed
- At p-n junction migration of majority charges i.e. holes form p side to n side and electrons from n side to p - side due to concentration difference is called diffussion
- At p-n junction , the donor impurity atoms become +ve ions and the acceptor impurity atoms become -ve ions, so that the ions act as two electrodes and the junction is called p-n junction diode.
- The ions at junction develop a potential difference called potential barrier and this barrier prevents further diffusion of charges across junction
- Around junction electron combine with holes to form a small region, free of charge carriers is called depletion region.
- The large electric field of intensity (E) is directed from n-type semi conductor to p-type at the junction.
- If potential barrier is V_b, width of depletion layer is d then $E = \dfrac{V_b}{d}$.
- Only inside depletion layer the +ve and -ve immobile ions exist whereas outside this layer the material is neutral

- Potential barrier depends on nature of semi conductor crystal, temperature and amount of doping
- The symbol of p-n junction diode is given below

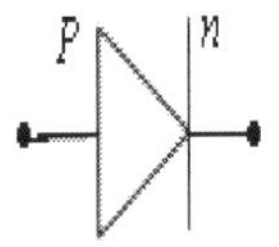

Semi conductor diode :

1) **Forward bias** :

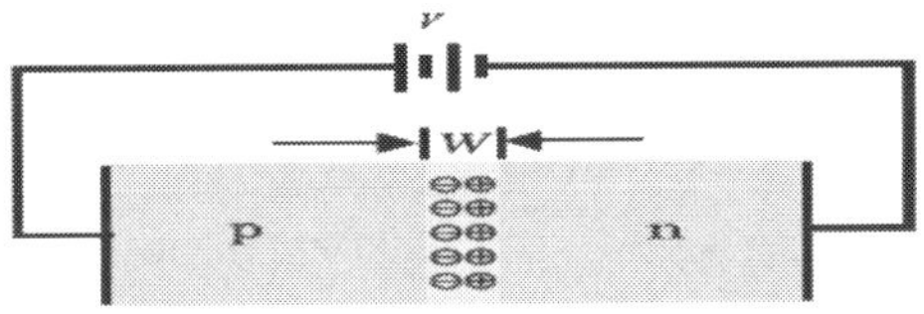

- When an external voltage V is applied across a semiconductor diode such that p-type is connected to +ve terminal and n-type to -ve terminal of battery (in general p - type to high voltage and n-type to low voltage), the diode is said to be forward biased.
- External voltage V is greater and opposite to barrier potential V_b. So width of depletion layer and resistance decrease.
- Effective barrier voltage under forward bias is V_b-V.

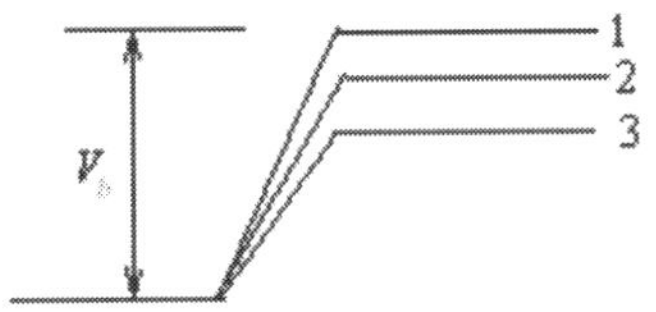

Fig. p-n junction diode under forward bias. Barrier potential (1) without battery, (2) low battery voltage, and (3) high voltage battery.

- Resistance of ideal diode in forward bias is zero
- If external voltage (V) is greater than barrier voltage then majority charge carriers diffuse across the junction and constitute diffusion current ($I = I_e + I_h$)
- Direction of diffusion current is from p- type to n-type
- The current flows through the diode in the below circuit is

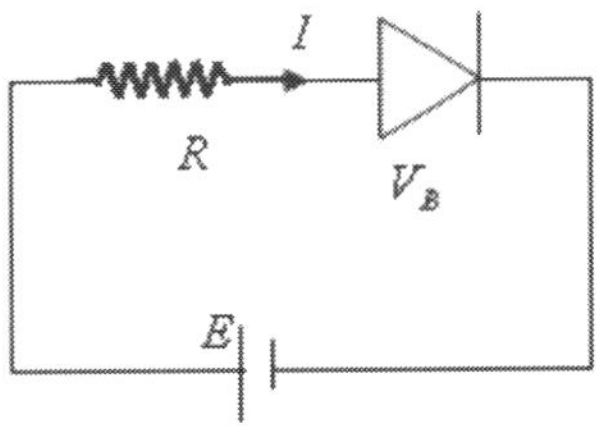

$$I = \frac{E - V_B}{R + r_f}$$

Where R= external resistance,

r_f = resistance of diode in forward bias,

V_B =barrier potential.

i) Power developed across the diode $= V_B I$

ii) Power developed across the resistor $= (E - V_B)I$,

- The external voltage beyond which diode current start increasing rapidly is called knee voltage (V_{knee})
- The below diagram show forward bias of junction diode.

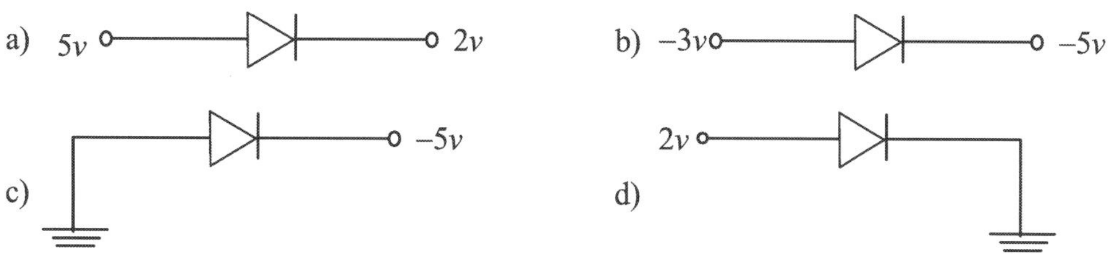

2) Reverse bias :

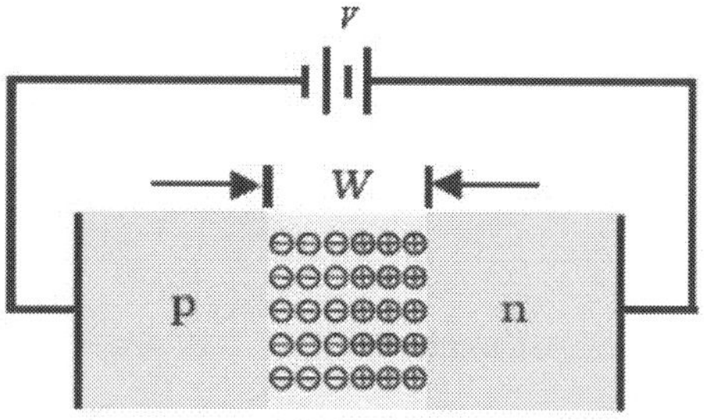

- When an external voltage V is applied across a semiconductor diode such that p-type is connected to -ve terminal and n-type to +ve terminal of battery (in general p - type to low voltage and n-type to high voltage), the diode is said to be reverse biased.
- External voltage V is greater and in same direction to barrier potential V_b, so width of depletion layer and resistance increase.
- Effective barrier voltage under reverse bias is $V_b + V$.

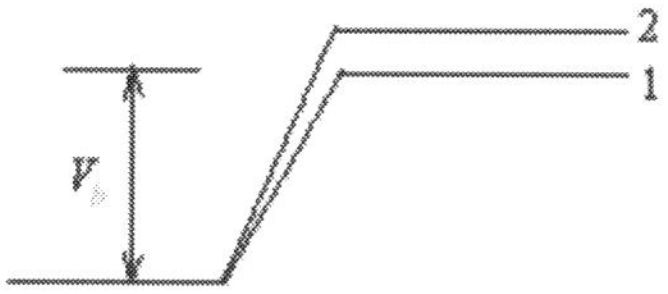

Fig : Diode under reverse bias, Barrier potential under reverse bias.

- Resistance of ideal diode in reverse bias is infinity upto a large external voltage (V).
- At lower external voltage few covalent bonds are broken to liberate electrons and holes and these constitute reverse saturation current
- At very high reverse bias voltage all the covalent bonds are broken to liberate large number of electrons called as breakdown voltage.
- The below diagram show reverse bias of p n junction diode.

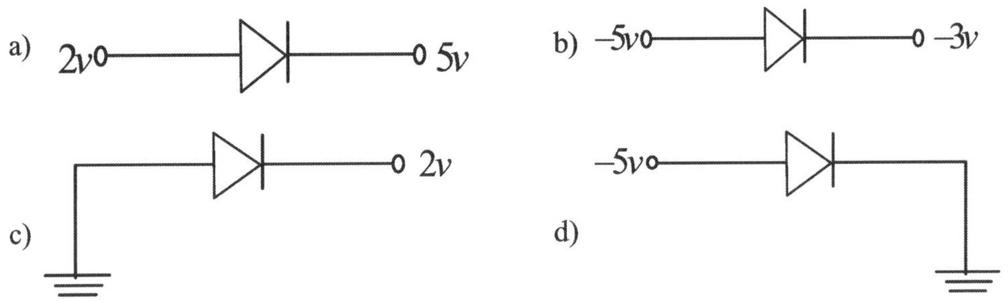

- Thus there is an unexpected release (avalanche) of large number of electrons and holes there by sharp increase in current takes place at a voltage called avalanche breakdown voltage.

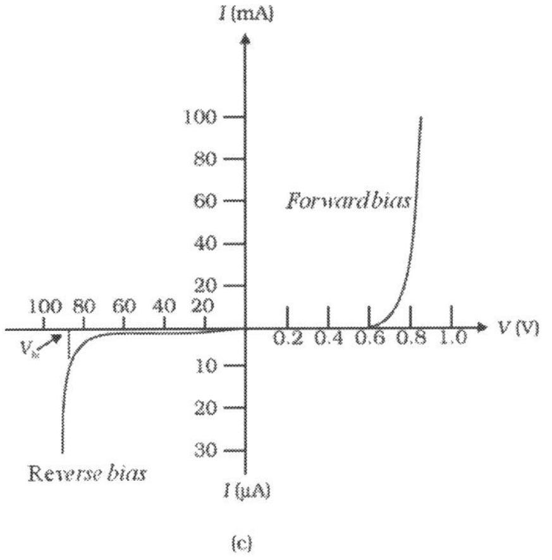

Fig : Avalanche breakdown

NOTE:

The potential barrier existing across an unbiased p-n junction is V_B volt

i) The minimum kinetic energy required by a hole to diffuse from the p-side to the n-side is 'eV_B'

ii) If the junction is forward biased at V volt, then the minimum kinetic energy required by a hole to diffuse from the p-side to the n-side is $e\left(V_B - V\right)$

iii) If the junction is reverse biased at V volt, then the minimum kinetic energy required by a hole to diffuse from the p-side to the n-side is $e\left(V_B + V\right)$

W.E-7:The V-I characteristic of a silicon diode is shown in the Fig. Calculate the resistance of the diode at

a) $I_D = 15mA$ and $(b)V_D = -10V$

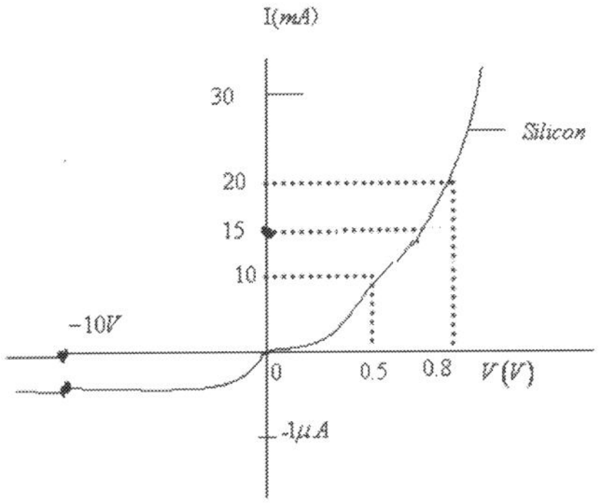

Solution: Considering the diode characteristics as a straight line between I=10mA to I=20mA passing through the origin, we can calculate the resistance using Ohm's law

a) From the curve at I=20mA,V=0.8V

I=10mA,V=0.7V

$r_{fb} = \Delta V / \Delta I = 0.1V / 10mA = 10\Omega$

b) From the curve at $V = -10V$, $I = -1\mu A$

Therefore, $r_{rb} = 10V / 1\mu A = 1.0 \times 10^7 \Omega$

W.E-8: Two junction diodes, one of germanium (Ge) and other of silicon (Si) are connected as shown in fig to a battery of 12V and a load resistance $10k\Omega$. The germanium diode conducts at 0.3V and silicon diode at 0.7V. When current flows in the circuit, the potential of terminal Y will be

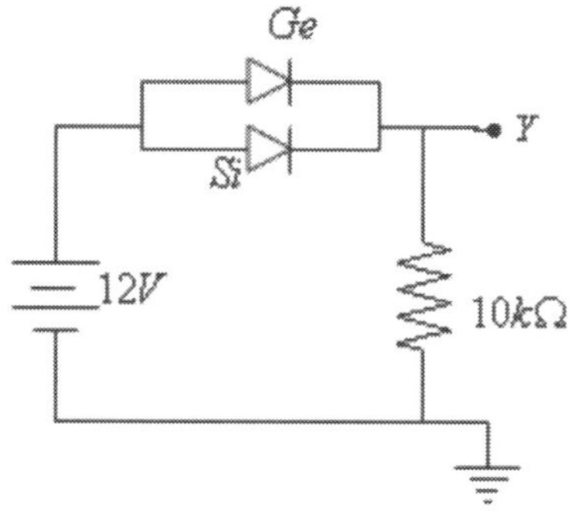

Solution : The Ge diode conducts for a p.d of 0.3V, therefore the current passes through it and the Si diode do not conduct. Hence the potential of terminal $Y = 12 - 0.3 = 11.7V$

W.E-9: Find maximum voltage across AB in the circuit shown in Fig. Assume that diode is ideal

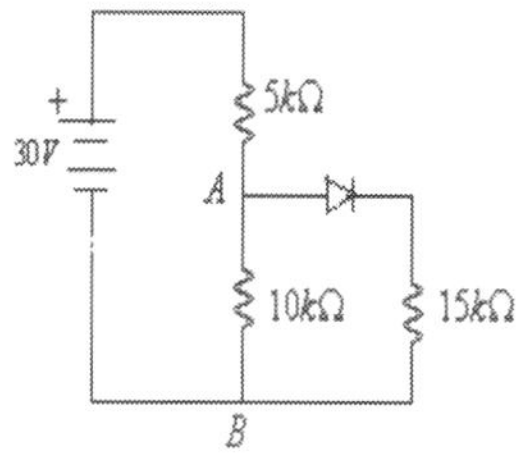

Solution: As the diode is treated ideal, its forward resistance $R_f = zero$. It acts as short circuit. So $10\,k\Omega$ is in parallel with $15\,k\Omega$ and the effective resistance across AB is

$$R_{AB} = \frac{10 \times 15}{10 + 15} = \frac{10 \times 15}{25} = 6k\Omega$$

$6k\Omega$ is in series with $5k\Omega$

$\therefore$ total resistance $= R_T = 6k\Omega + 5k\Omega = 11k\Omega$,

V=30V. Current drawn from the battery is $I = \frac{V}{R_T} = \frac{30V}{11k\Omega} = 2.72mA$

$V_{AB} = IR_{AB} = 2.72mA \times 6k\Omega = 16.32V$

S **W.E-10:Find the voltage V_A in the circuit shown in figure . The potential barrier for Ge is 0.3 V and for Si is 0.7 V**

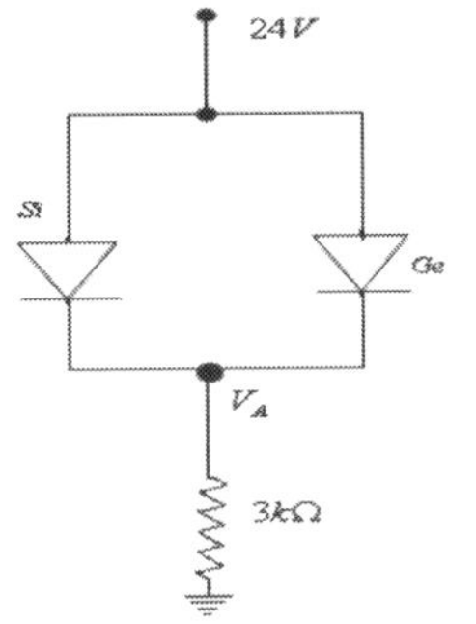

Solution: In the situation given, germanium diode will turn on first because potential barrier for germanium is smaller. The silicon diode will not get the opportunity to flow the current and so remains in open circuit. The equivalent circuit is as in figure

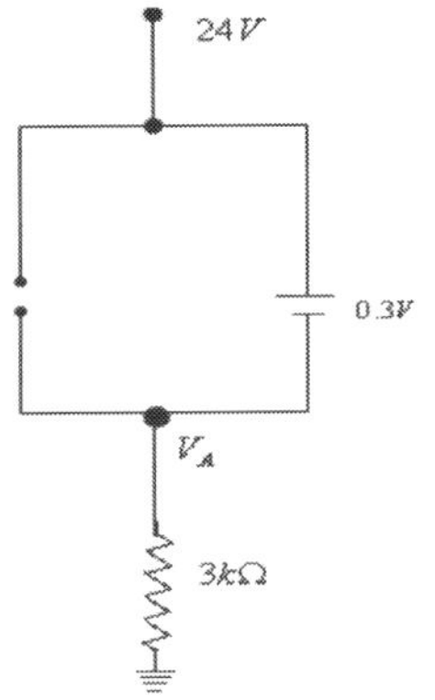

$V_A = 24 - 0.3 = 23.7V$

W.E-11:Considering the circuit and data given in the diagram, calculate the currents flowing in the diodes D_1 and D_2 Forward resistance of D_1 and D_2 is 20Ω

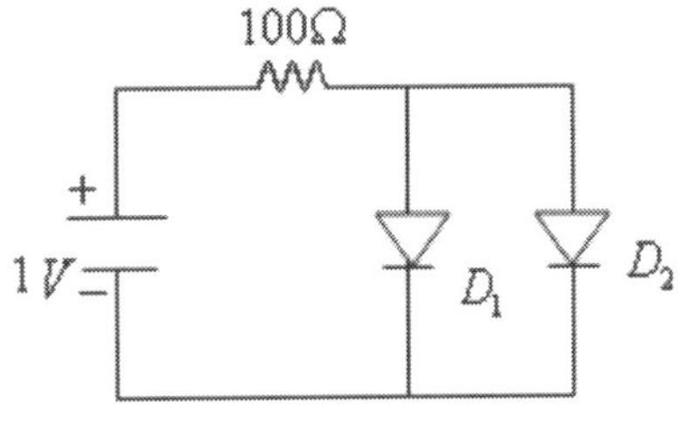

Solution: Since the positive terminal of battery is connected to P-type of both diodes D_1 and D_2, they are forward biased. These diodes are replaced by with their forward resistance as shwon is Fig:

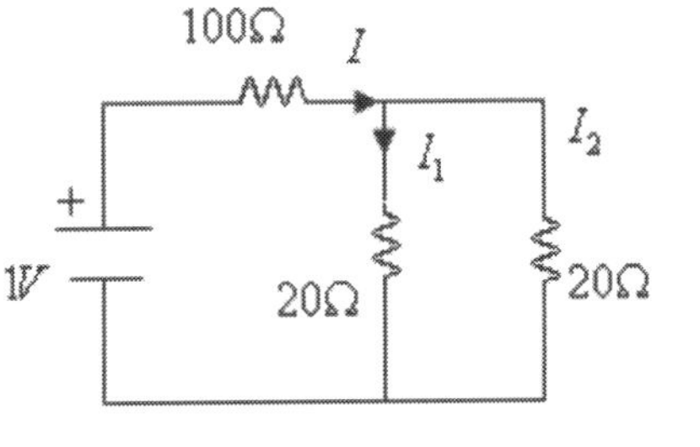

he resistance of 20Ω and 20 Ωin

parallel, $\frac{1}{R}=\frac{1}{20}+\frac{1}{20}(or)R=\frac{20}{2}=10\Omega$

Therefore, total current I in the circuit

$I=\frac{1}{100+10}=\frac{1}{110}$ amp and $I_1=I_2=\frac{1}{2}\times\frac{1}{110}=\frac{1}{220}$ amp

W.E-12:In the circuit shown, the potential drop across each capacitor is (assuming the two diodes are ideal)

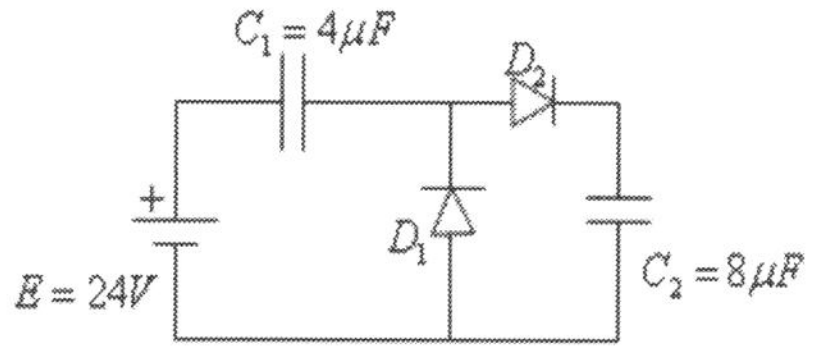

Solution: The diode D_1 is reverse biased (open circuit), but the diode D_2 is forward (short circuit).

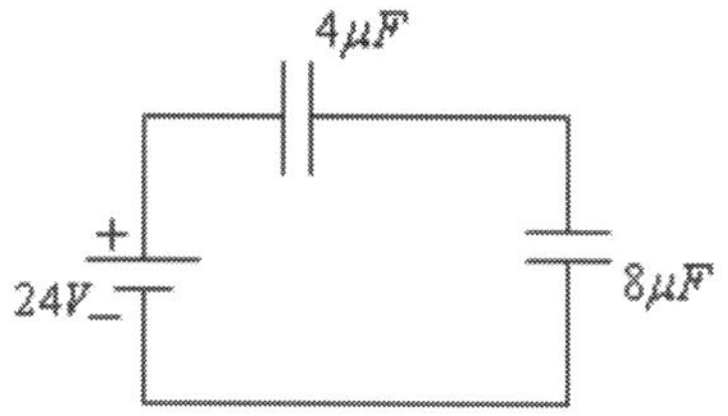

$\therefore$ the potential of the battery divides across the two capacitors in the inverse ratio of their capacities.

i.e, $\frac{V_1}{V_2}=\frac{C_2}{C_1}=\frac{8}{4}=\frac{2}{1}$

$V_1=\frac{2}{3}E=\frac{2}{3}\times 24=16V$ $\quad V_2=\frac{1}{3}E=\frac{1}{3}\times 24=8V$

W.E-13:A potential barrier of 0.50V exists across a p-n junction

a) **If the depletion region is $5.0\times10^{-7}m$ wide,what is the intensity of the electric field in this region?**

b) **An electron with a speed of $5.0\times10^{5}m/s$ approaches the p-n junction from the n-side. With what speed will it enter the p-side ?**

Solution :

a) The electric field is E=V/d $=\frac{0.50V}{5.0\times10^{-7}m}=1.0\times10^{6}V/m$

b) 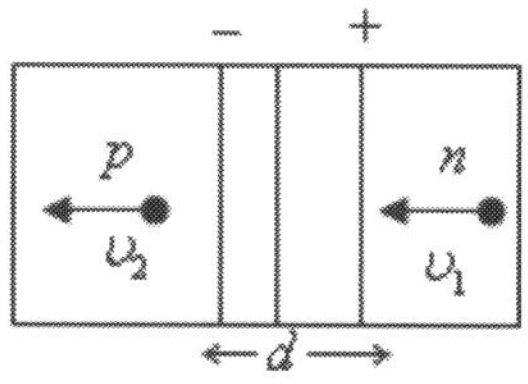

Let the electron has a speed v_1 when

it enters the depletion layer and v_2 when it comes out of it. As the potential energy

increases by $e \times 0.50V$. From the principle of conservation of energy $\frac{1}{2}m\upsilon_1^2 = e \times 0.50 + \frac{1}{2}m\upsilon_2^2$

$$\frac{1}{2}\left(9.1\times10^{-31}\right)\left(5\times10^5\right)^2 = 1.6\times10^{-19}\times0.5+\frac{1}{2}\times9.1\times10^{-31}\times\upsilon_2^2$$

Solving this. $\upsilon_2 = 2.7\times10^5 m/s$

- **Application of junction diode as a RECTIFIER**
- The process of conversion of ac to dc is called rectification, tha arrangement is called rectfier. They are
- **Half wave rectifier**

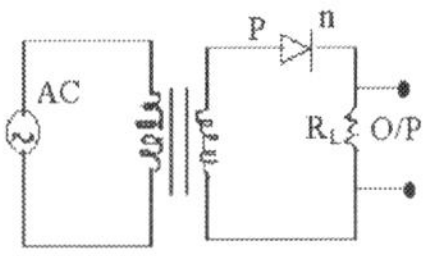

Half Wave Rectifier

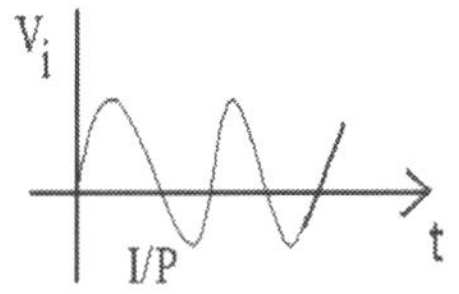

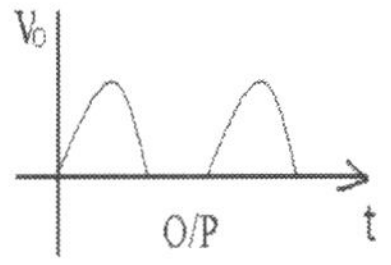

- In halfwave rectification we need atleast one semicondcutor diode.
- In half wave rectifier the efficiency is $\eta = 0.406\frac{R_L}{r_f + R_L}$ (where $r_f \to$ forward resistance of diode and $R_L \to$ load resistance.)

  It's maximum value is 40.6%.
- Maximum current $I_m = \frac{V_m}{R_L + r_f}$ (V_m *is* max *imum voltage*)
- Mean dc current $I_{dc} = \frac{I_m}{\pi}$
- $I_{rms} = \frac{I_m}{2}$
- Value of ripple factor $= 1.21$
- The ripple frequency is equal to the frequency of applied emf.

- The value of dc component in out put voltage is less than the ac.
- Input ac power $= I_{rms}^2 \left(R_L + r_f\right)$
- Out put dc power $= I_{dc}^2 R_L$

Efficiency of a half-wave rectifier
(Formula derivation)

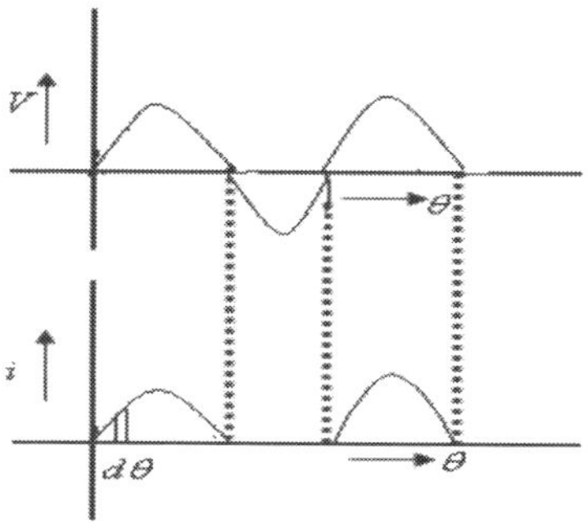

From fig

$$I_{average} = \frac{Area\ under\ the curve\ for\ a\ cycle}{base}$$

$$= \frac{\int_0^{\pi} i d\theta}{2\pi}$$

$$= \frac{1}{2\pi} \int_0^{\pi} \frac{V_m \sin\theta}{r_f + R_L} d\theta$$

$$= \frac{V_m}{2\pi\left(r_1 + R_L\right)} \left[-\cos\theta\right]_0^{\pi}$$

$$= \frac{V_m}{r_1 + R_L} \times \frac{1}{\pi} = \frac{I_m}{\pi}$$

$$\left[where\, I_m = \frac{V_m}{r_f + R_L}\right]$$

Where r_f and R_L denote diode resistance and load resistance respectively

Hence d.c. power

$$= I_{d.c}^2 \times R_L = \left(\frac{I_m}{\pi}\right)^2 \times R_L$$

a.c. power input= $I_{r.m.s}^2 \left(r_f + R_L\right)$

For a half-wave rectifier

$$I_{r.m.s}=\frac{I_m}{2}$$

Hence $P_{a.c}=\frac{I_m^2}{2}\times\left(r_f+R_L\right)$

$$\therefore \text{ Rectifier efficiency } =\frac{P_{dc}}{P_{ac}}=\frac{\left(I_m/\pi\right)^2\times R_L}{\left(I_m/2\right)^2\left(r_f+R_L\right)}=\frac{0.406R_L}{r_f+R_L}$$

$$\eta=\frac{0.406}{1+\frac{r_f}{R_L}}$$

The efficiency is maximum when is neglible.

- **Full wave rectifier:**
  η =

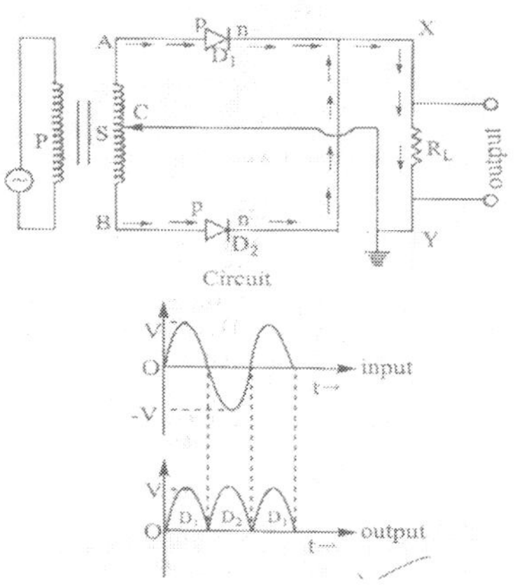

- In fullfwave rectification we need at least two semicondcutor diodes.
- In full wave rectifier the efficiency is $\eta=0.812\frac{R_L}{r_f+R_L}$. Its maximum value is 81.2%

(where $r_f \rightarrow$ forward resistance of diode and $R_L \rightarrow$ load resistance.)

- Maximum current

$$I_m=\frac{V_m}{R_L+r_f}\left(V_m \text{ is maximum voltage}\right)$$

- Mean dc current $I_{dc}=\frac{2I_m}{\pi}$

- $I_{rms} = \frac{I_m}{\sqrt{2}}$
- Value of ripple factor $= 0.482$
- The ripple frequency is twice to the frequency of applied emf.
- The value of dc component in out put voltage is more than the ac.
- Input ac power $= I_{rms}^2 \left(R_L + r_f\right)$
- Out put dc power $= I_{dc}^2 R_L$
- **Efficiency of full-wave rectifier:**

  (Formula derivation)

  Efficiency of full-wave rectifier is given by

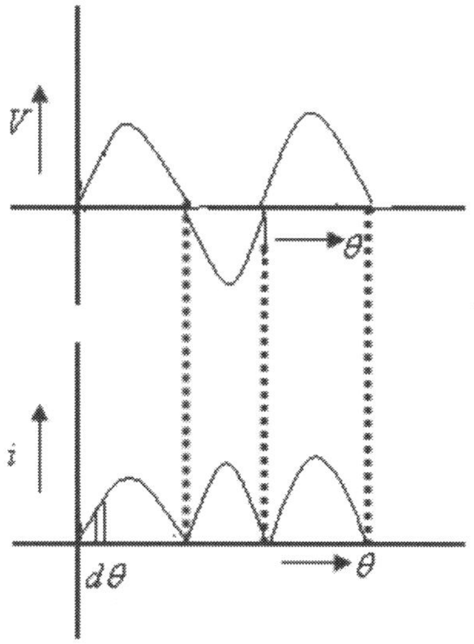

$$\eta = \frac{P_{d.c}}{P_{a.c}}$$

Instantaneous current is $i = \frac{\upsilon}{r_f + R_L}$

Average d.c. current is $I_{d.c} = \frac{2I_m}{\pi}$

$\therefore$ d.c . power output $= I^2{}_{d.c} \times R_L = \left(\frac{2I_m}{\pi}\right)^2 \times R_L$

a.c. input power is given by $P_{d.c} = I_{r.m.s}^2 \left(r_f + R_L\right)$

For a full-wave rectified cycle, $I_{r.m.s} = I_m / \sqrt{2}$

$$P_{a.c} = \left(\frac{I_m}{\sqrt{2}}\right)^2 \left(r_f + R_L\right)$$

$\therefore$ Full-wave rectification efficiency,

$$\eta = \frac{P_{d.c.}}{P_{a.c.}} = \frac{(2I_m/\pi)^2 R_L}{\left(\frac{I_m}{\sqrt{2}}\right)^2 (r_f + R_L)}$$

$$= \frac{0.812 R_L}{r_f + R_L} = \frac{0.812}{1 + \frac{r_f}{R_L}}$$

The efficiency will be maximum if r_f is zero

$\therefore$ maximum efficiency $= 81.2\%$

The efficiency of a full-wave rectifier is double the efficiency of a half-wave rectifier

- **Full-wave bridge rectifier**: In a centre tap full-wave rectifier, it is difficult to locate the centre tap on the secondary winding, which can be overcome in bridge rectifier. The circuit is shown in Figure Four diodes D_1, D_2, D_3 and D_4 are used in the circuit.

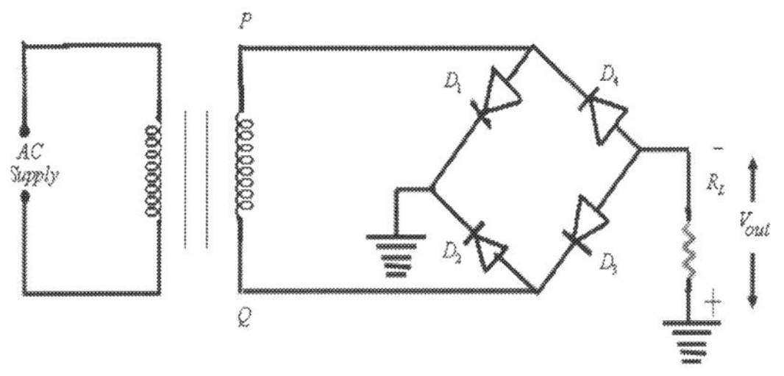

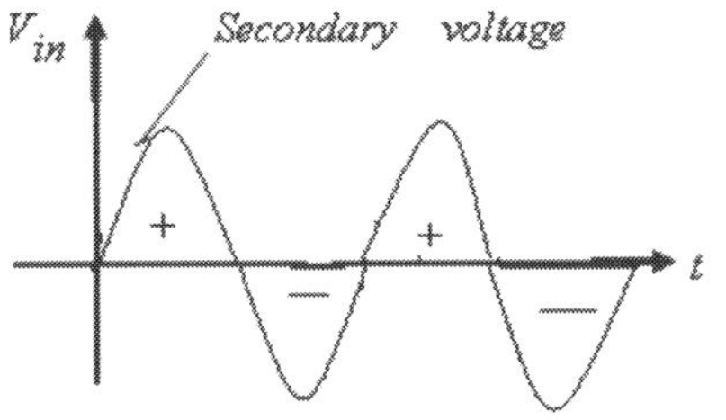

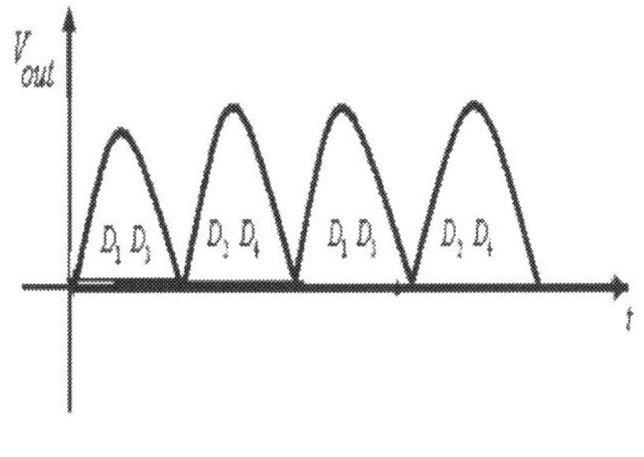

- **Ripple factor**: The ratio of r.m.s value of a.c component to d.c. component in the rectifier output is called ripple factor.

$$\text{Ripple factor} = \frac{I_{a.c.}}{I_{d.c.}} = \sqrt{\left(\frac{I_{r.m.s}}{I_{d.c.}}\right)^2 - 1}$$

Ripple factor decides the effectiveness of a rectifier. The smaller value of ripple factor shows lesser a.c. component; hence more effectiveness of rectifier.

i) For half-wave rectification, $I_{r.m.s.} = \frac{I_m}{2}$; $I_{d.c.} = \frac{I_m}{\pi}$

$$\text{Ripple factor} = \sqrt{\left(\frac{I_m/2}{I_m/\pi}\right)^2 - 1} = 1.21$$

ii) For full-wave rectification

In full-wave rectification,

$$I_{r.m.s.} = \frac{I_m}{\sqrt{2}}; \qquad I_{d.c.} = \frac{2I_m}{\pi}$$

$$\text{Ripple factor} = \sqrt{\left(\frac{I_m/\sqrt{2}}{2I_m/\pi}\right)^2 - 1} = 0.48$$

- **Form factor**: It is ratio of $I_{r.m.s.}$ and $I_{d.c.}$

i) For half-wave rectification:

$$F = \frac{I_{r.m.s}}{I_{d.c.}} = \frac{\pi}{2} = 1.57$$

$$\left[as\ I_{r.m.s} = \frac{I_m}{2}\ and\ I_{d.c} = \frac{I_m}{\pi}\right]$$

ii) For Full-wave rectification:

$$F = \frac{I_{r.m.s.}}{I_{d.c.}} = \frac{\pi}{2\sqrt{2}}$$

$$\left[as\ I_{r.m.s} = \frac{I_m}{\sqrt{2}}\ and\ I_{d.c} = \frac{2I_m}{\pi}\right]$$

The role of capacitor in filtering:

- When the voltage across the capacitor is rising, it gets charged. If there is no external load, it remains charged to the peak voltage of the rectified output.
- When there is a load, it gets discharged through the load and the voltage across it begins to fall. In the next half-cycle of rectified output it again gets charged to the peak value in figure.
- The rate of fall of the voltage across the capacitor depends upon the inverse product of capacitor C and the effective resistance R_L used in the circuit and is called the time cosntant.
- To make the time constant large value of C should be large. So capacitor input filters use large capacitors. The output voltage obtained by using capacitor input filter is nearer to the peak volatge of the rectified voltage. This type of filter is most widely used in power supplies.

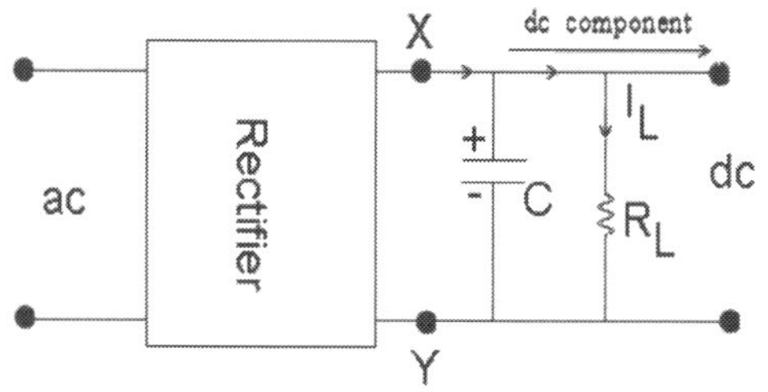

(a)

a)Full wave rectifier with capacitor filter

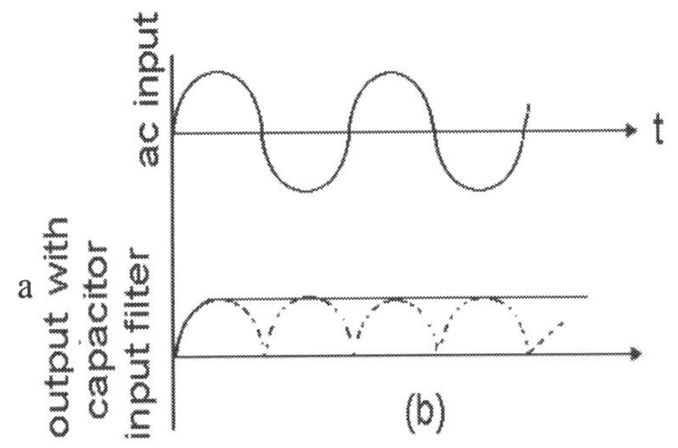

b) Input and out put voltage of rectifier in (a)

WE:14:A p-n diode is used in a half wave rectifier with a load resistance of 1000Ω. If the forward resistance (r_f) of diode is 10Ω, calculate the efficiency of this halfwave rectifier.

Solution: Load resistance $R_L = 1000\Omega$

Forward resistance of the diode $= r_f = 10\Omega$

Efficiency of half wave rectifier $\left[\dfrac{0.406R_L}{r_f + R_L}\right] = \dfrac{0.406\times1000}{1010} = 0.4019$

The percentage efficiency of the half wave rectifier $\eta = 40.19\%$

WE:15:Afull wave rectifier uses two diodes with a load resistance of 100Ω. Each diode is having negligible forward resis tance . Find the efficiency of this full wave rectifier.

Solution: Forward resistance of the diode $r_f = 0$,

Load resistance, $R_L = 100\Omega; \eta = ?$

efficiency of full wave rectifier $= \dfrac{0.812\times100}{100} = 0.812$

The percentage efficiency of the full wave rectifier $= 81.2\%$

WE:16:If a p-n junction diode, a square input signal of 10V is applied as shwon

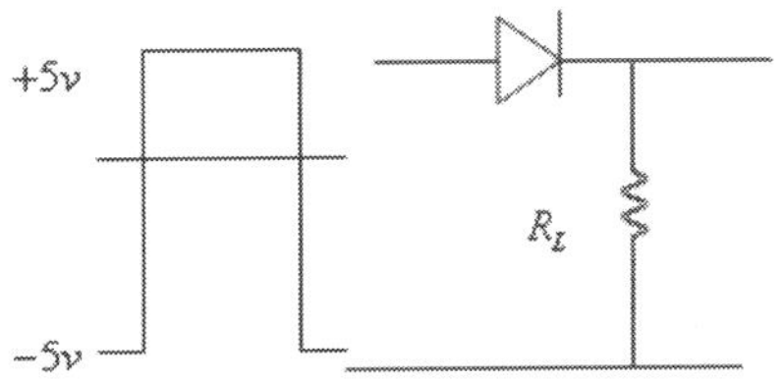

Then the out put signal across R_L will be

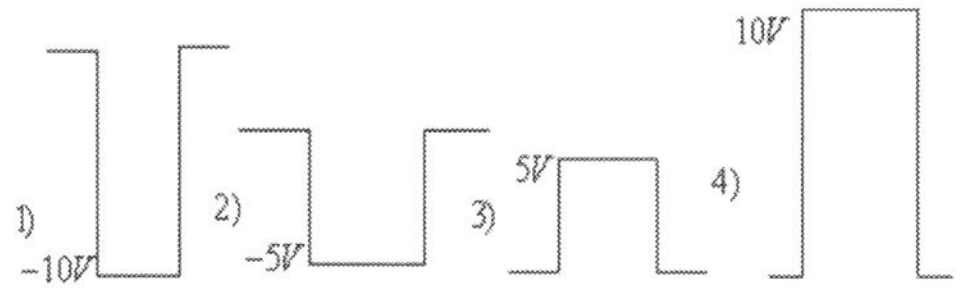

Solution : The junction diode will conduct when it is forward biased. Therefore, the output voltage will be obtainded during positive half cycle only . So option is (3)

- **p-n junction diodes are specially used as :**

  1) Zener diode 2) Opto electronic junction devices

1) Zener diode :

- A heavily doped p-n junction diode used to operate in reverse bias is called zener diode.
- In Zener diode even at low reverse voltage direct breakage of covalent bonds due to very strong [illegible] hole pairs. This voltage where current increases sharply

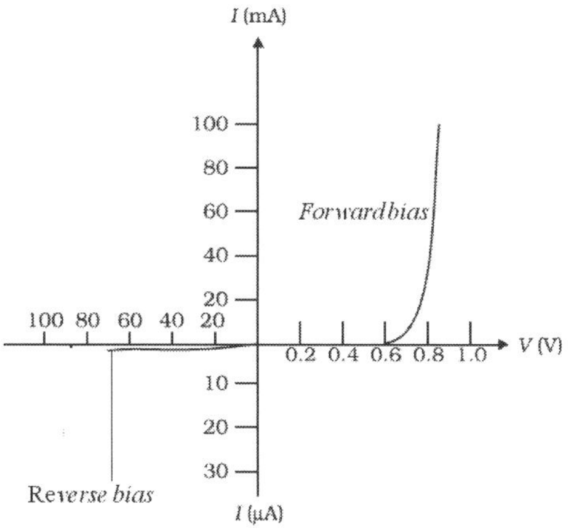

(c)

- Symbol of zener diode is 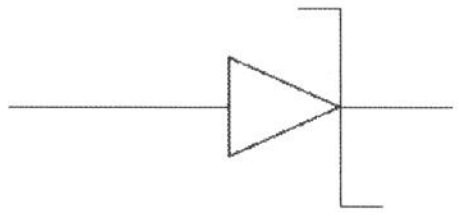

- Zener diode is used as voltage regulator its circuit diagram is

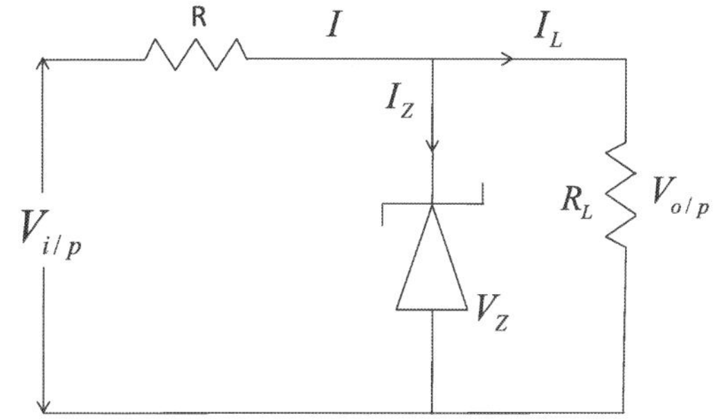

1) $I = I_Z + I_L$ 2) $V_{O/P} = V_Z = I_L R_L$ 3) $V_{I/P} = IR + V_Z$

$V_{I/P} = IR + V_Z \Rightarrow V_{O/P} = V_{I/P} - (I_Z + I_L)R$

As $V_{I/P}$ changes current through zener diode change so that $V_{O/P}$ (or) V_Z remain constant.

WE:17:For the circuit shown in figure, Find

1)the output voltage; 2)the voltage drop across series resistance;

3)the current through Zener diode.

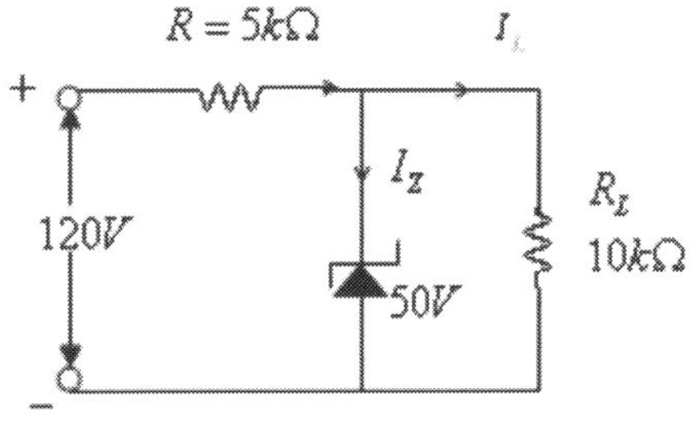

Solution: From the figure $R = 5k\Omega = 5\times10^3\Omega$;

input voltage $V_{in} = 120V$;

zener voltage, $V_z = 50V$

1) Output voltage $V_z = 50V$

2) Voltage drop across series tesistance R= $V_{in} - V_z = 120 - 50 = 70V$

3) Load current $I_L = \dfrac{V_z}{R_L} = \dfrac{50}{10\times10^3} = 5\times10^{-3}A$

Current through R= $i = \dfrac{V_{in} - V_z}{R} = \dfrac{70}{5\times10^3} = 14\times10^{-3}A$

According to Kirchoff's first law $I = I_L + I_z$

$\therefore$ Zener current

$I_Z = I - I_L = 14\times10^{-3} - 5\times10^{-3}$

$= 9\times10^{-3} = 9mA$

2) Opto electronic junction devices

PHOTODIODE

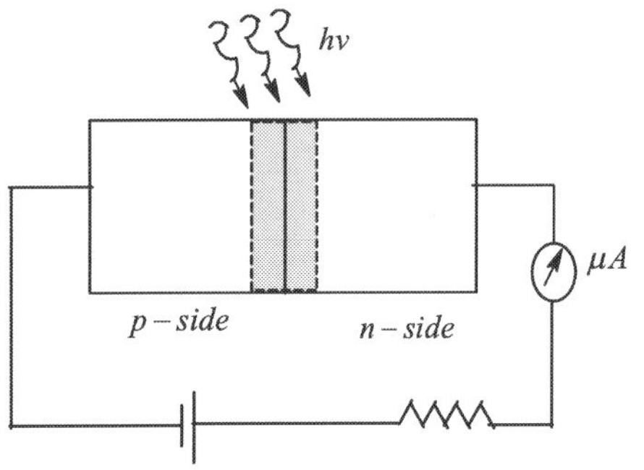

Fig : An illuminated photodiode under reverse bias

- Photocurrent is proportional to incident light intensity.
- Photodiode can be used as a photodetector to detect optical signals.

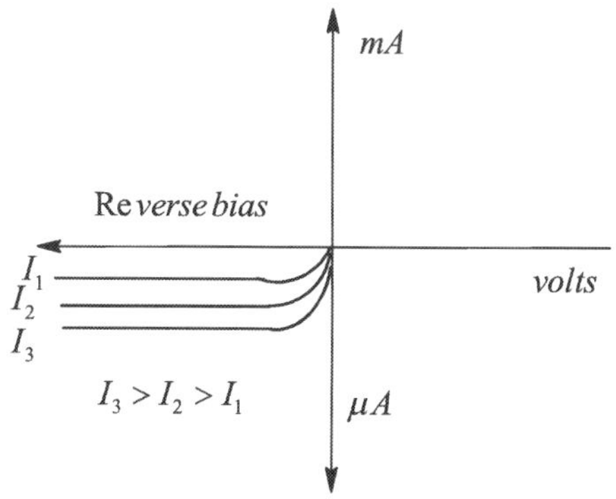

Fig : I-V characteristics of a photodiode for different illumination intensity $I_3 > I_2 > I_1$.

WE:18 The current in the forward bias is known to be more (mA) than the current in the reverse bias (μA). What is the reason then to operate the photodiodes in reverse bias ?

Solution: Consider the case of an n-type semiconductor. The majority carrier density (n) is considerably larger than the minority hole density p(n>>p) On illumination, let the excess electrons and holes generated be Δn *and* Δp, respectively

$n^1 = n + \Delta n$

$p^1 = p + \Delta p$

Here n^1 and p^1 are the electron and hole concentrations at any particular illumination and n and p are carriers concentration when there is no illumination. Remember $\Delta n = \Delta p$ *and* $n >> p$. Hence, the fractional change in the majority carriers (i.e., $\Delta n / n$)would be much less than that in the minority carrier dominated reverse bias current is more easily measurable than the fractional change in the forward bias current. Hence,photodiodes are preferably used in the reverse bias conditionfor measuring light intensity

- **LIGHT EMITTING DIODE (LED)**

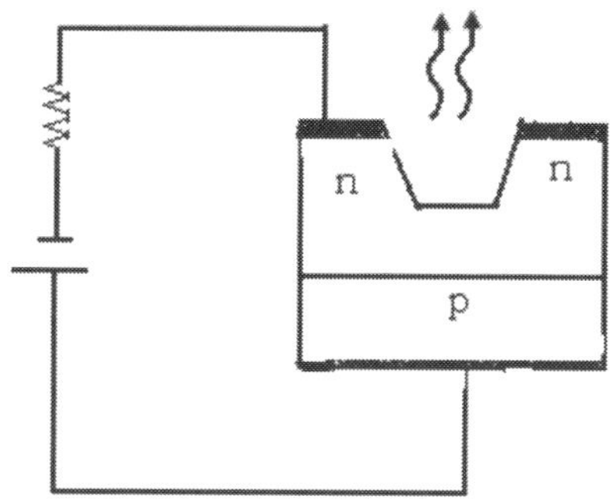

- Light-emitting diode (LED) is a forward-biased p-n junction diode which emits visible light when energised.
- The energy of radiation emitted by LED is equal to or less than the band gap of the semiconductor.
- The band width of emitted light is $100 A^0$ to $500 \overset{0}{A}$ or in other words it is nearly (but not exactly) monochromatic.

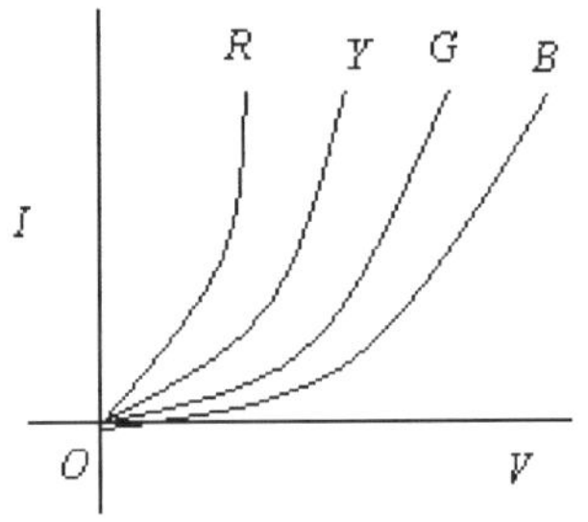

The I-V characterstics of L.E.D:

- **SOLAR CELL :**

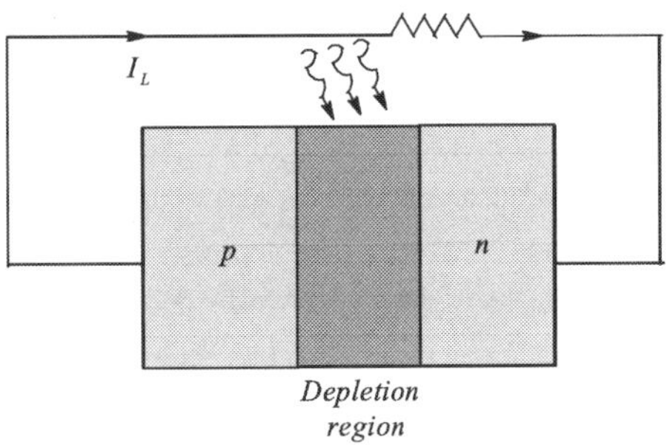

Fig : A typical illuminated p-n junction solar cell

- Unlike a photodiode, a solar cell is not given any biasing. It supplies emf like an ordinary cell.
- Sunlight is not always required for a solar cell.
- Semiconductors with band gap close to 1.5ev. are ideal materials for solar cell fabrication
- Si and GaAs are preferred material for solar cells.

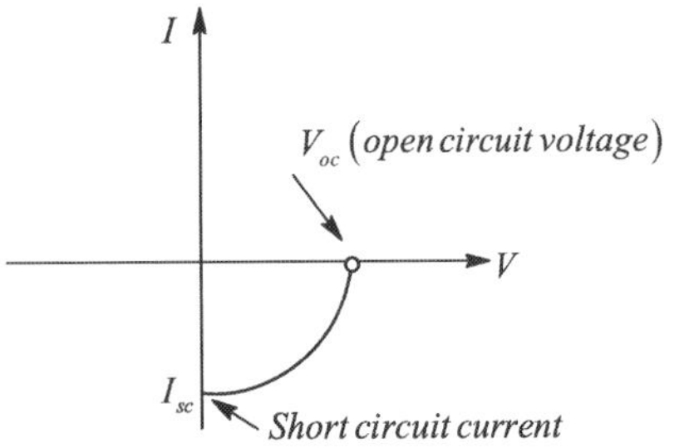

Fig : I-V characteristics of a solar cell

- **SELECTION OF SOLAR MATERIAL:**
- i) band gap (1.0 to 1.8eV) ii) high optical absorption ($10^4 cm^{-1}$)
  iii) good electrical conductivity iv) availability of the raw material and (v) cost.

TRANSISTORS:

- Transfer +resistor= Transistor
- Transistors are current operated solid-state devices.
- Silicon is the element from which most of the transistors and other semiconductor components are made today.
- Transistor has three regions known as the emitter (E), base (B) and collector (C)
  1) Emitter is heavily doped and medium in size.
  2) Base of a transistor is lightly doped and very thin.
  3) Collector of a transistor is moderately doped and large in size.

- Transistor has two junctions

  1) emitter-base junction 2) Collector-base junction
- In a transistor the emitter base junction is forward biased and the collector base junction is reverse biased.
- **In an n-p-n Transistor:**

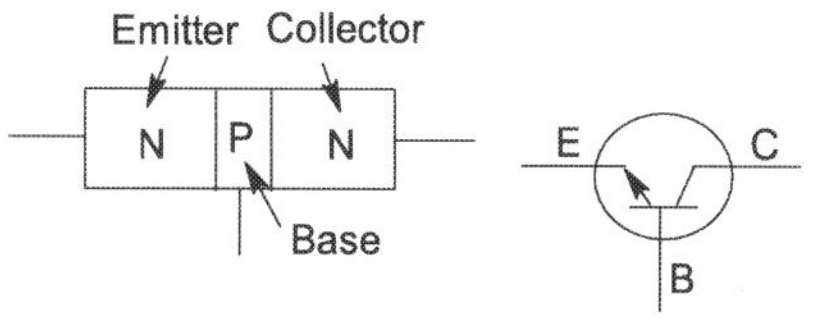

- The current is due to electrons inside and outside the n-p-n transistor and they are the majority charge carriers
- The conventional current flows from base to emitter.
- The emitter junction is forward biased and the collector junction is reverse biased.
- The emitter current (I_E) is the sum of base current (I_B) and collector current (I_C), i.e.,$I_E = I_B + I_C$
- I_C is 97 to 98% of I_E and I_B is 2 to 3% of I_E.
- **In a p-n-p Transistor:**

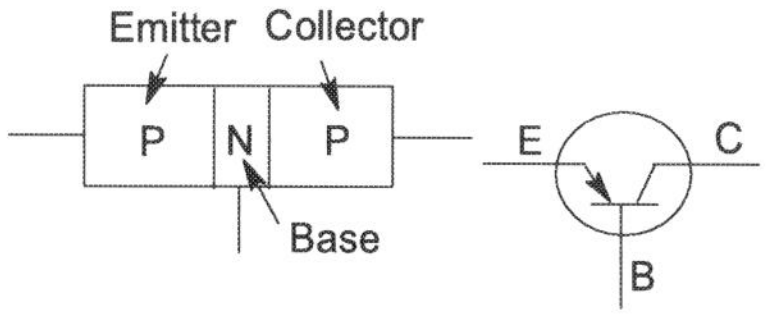

- The current is due to holes inside as they are the majority charge carriers and due to electons outside the p-n-p transistor.
- The conventional current is from emitter to base.
- The emitter junction is forward biased and the collector junction is reverse biased.
- Here also $I_E = I_C + I_B$
- I_C is 97 to 98% of I_E and I_B is 2 to 3% of I_E.

  Thus the collector current is less than the emitter current ($I_C < I_E$)
- **TRANSISTOR CONFIGURATIONS**

  In electronic circuits transistors are connected in three ways. They are

  1) Common base configuration ,

  2) Common emitter configuration,

  3) Common collector configuration
- **COMMON BASE CONFIGURATION**
- In this configuration base is common to both input and output.
- Base terminal is earthed and input is given across base - emitter and output is taken across base - collector as shown in figure
- This mode is called grounded base configuration.

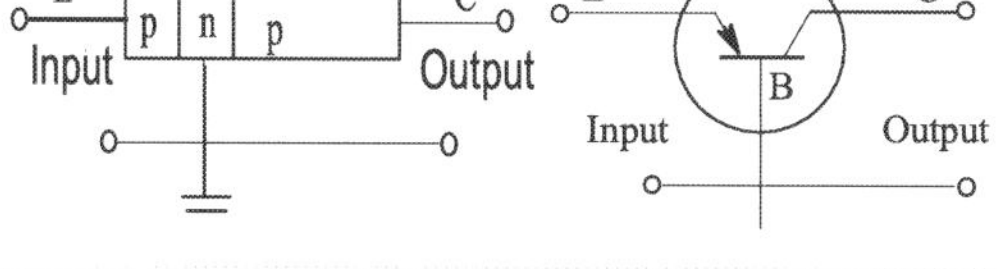

- Current amplification factor of common base configuration for ac $\alpha = \left(\frac{\Delta I_C}{\Delta I_E}\right)_{\text{constant } V_{CB}}$

- Current amplification factor of common base configuration for dc $\alpha = \frac{I_c}{I_E}$

- Values of α range from 0.95 to 0.99.

- Input resistance of transistor in CB configuration is $R_{in} = \left(\frac{\Delta V_{BE}}{\Delta I_B}\right)_{V_{CB}}$

- Output resistance of transistor in CB configuration is $R_{out} = \left(\frac{\Delta V_{CB}}{\Delta I_C}\right)_{I_E}$

- Voltage gain = current gain x resistance gain. $A_v = \alpha \times \frac{R_{out}}{R_{in}}$

- Power gain = Voltage gain x current gain. $A_p = A_v \times \alpha = \alpha^2 \times \frac{R_{out}}{R_{in}}$

- **COMMON EMITTER CONFIGURATION**
- In this configuration emitter is common to both input and output.
- The emitter is earthed and input is given across base - emitter and output is taken across collecter - emitter as shown in fig
- This mode is called grounded emitter configuration.

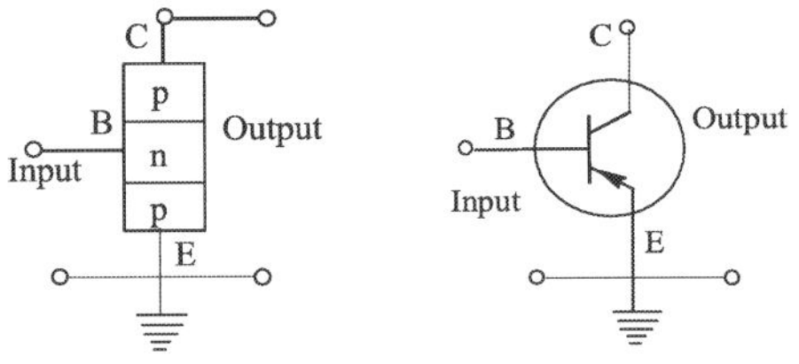

- Current amplification factor of common emitter configuration for ac $\beta = \left(\frac{\Delta I_C}{\Delta I_B}\right)_{\text{constant } V_{CE}}$

- Current amplification factor of common emitter configuration for dc $\beta = \frac{I_c}{I_B}$

- Values of β range from 20 to 500.

- Input resistance of transistor in CE configuration is $R_{in} = \left(\frac{\Delta V_{BE}}{\Delta I_B}\right)_{V_{CE}}$

- Output resistance of transistor in CE configuration is $R_{out} = \left(\frac{\Delta V_{CE}}{\Delta I_C}\right)_{I_B}$

- Voltage gain = current gain x resistance gain.

$$A_v = \beta \times \frac{R_{out}}{R_{in}}$$

- Power gain = Voltage gain x current gain. $A_p = A_v \times \beta = \beta^2 \times \frac{R_{out}}{R_{in}}$

- **COMMON COLLECTOR CONFIGURATION**
- In this configuration collector is common to both input and output.
- The collector is earthed and input is given across base - collector and output is taken across emitter -collector as shown in figure
- This mode is called grounded collector configuration.

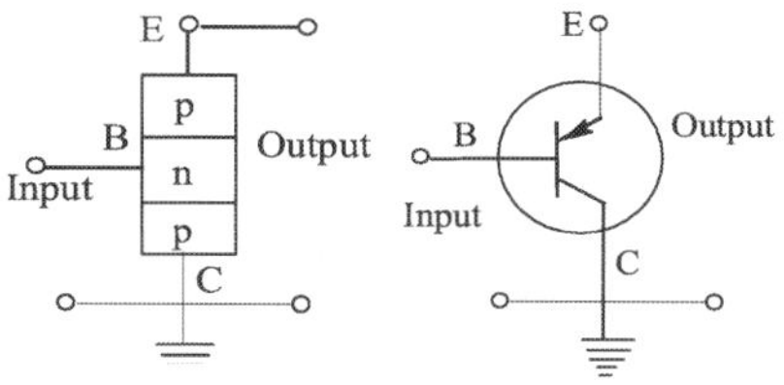

- Current amplification factor of common collector configuration for ac $\gamma = \left(\frac{\Delta I_E}{\Delta I_B}\right)_{\text{constant } V_{CE}}$

- Current amplification factor of common collector configuration for dc $\gamma = \frac{I_E}{I_B}$

- Input resistance of transistor in CC configuration is $R_{in} = \left(\frac{\Delta V_{BE}}{\Delta I_B}\right)_{V_{CE}}$

- Output resistance of transistor in CCconfiguration is $R_{out} = \left(\frac{\Delta V_{CE}}{\Delta I_E}\right)_{I_B}$

- **Relation between α & β :-**

ac current gain in C.B , $\alpha = \frac{\Delta I_C}{\Delta I_E}$

ac current gain in C.E , $\beta = \frac{\Delta I_C}{\Delta I_B}$

$$\beta = \frac{\Delta I_C}{\Delta I_B} = \frac{\Delta I_C}{\Delta I_E - \Delta I_C} = \frac{\frac{\Delta I_C}{\Delta I_E}}{1 - \frac{\Delta I_C}{\Delta I_E}} = \frac{\alpha}{1-\alpha} \Rightarrow \beta = \frac{\alpha}{1-\alpha} \text{ (or) } \alpha = \frac{\beta}{1+\beta}$$

- **Relation between α, β and γ :-** $\gamma = \frac{\Delta I_E}{\Delta I_B} = \frac{1}{1-\alpha} = \frac{\beta}{\alpha} = \beta + 1$

- **Comparative study of CB, CE, CC configurations**

	parameter	CB configuration	CE configuration	CC configuration
1	Input resistance	Minimum (50-20 Ω)	Medium (1-2K Ω)	Maximum (150-180K Ω)
2	Output resistance	Maximum (1-2M Ω)	More (≈50K Ω)	Minimum (≈1K Ω)
3	Current gain	Minimum $\alpha = 0.95 - 0.99$	More $\beta = 20 - 500$	Maximum $\gamma = 20 - 500$
4	Voltage gain	Medium	maximum	minimum
5	Power gain	Medium (20-30)	Maximum (30-40)	Minimum (≈10)
6	Useful in application of	Current	power	Impedance matching
7	Phase difference	0	π rad (or) 180°	0

- **Characterestics of a transistor:-**
  n–p–n Transistor (C.E)

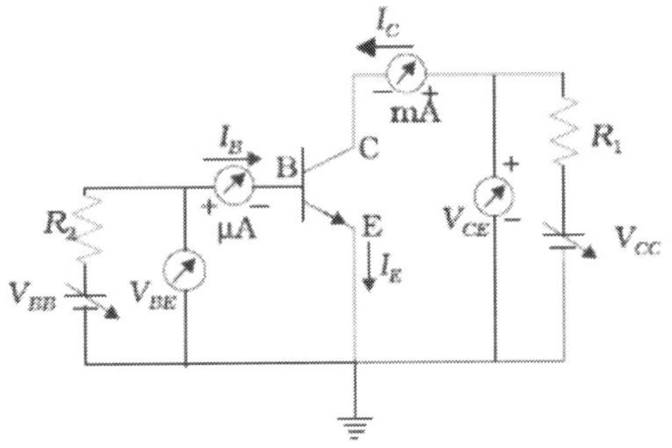

1) **Input characterestics**

Input characterstics are drawn between V_{BE} verses I_B at constant V_{CE}. The output voltage V_{ce} is fixed (say at zero volts). The input voltage V_{be} is changed in steps (say 0.1 V) upto 1 volt and the corresponding base current I_b is noted down. This process is repeated for different values (say 10V, 20V etc.) of V_{ce}.

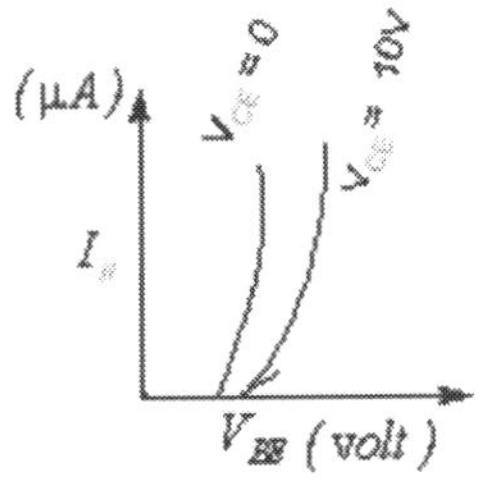

2) **Output characterestics**

Out put characterstics are drawn between V_{CE} verses I_C at constant I_B

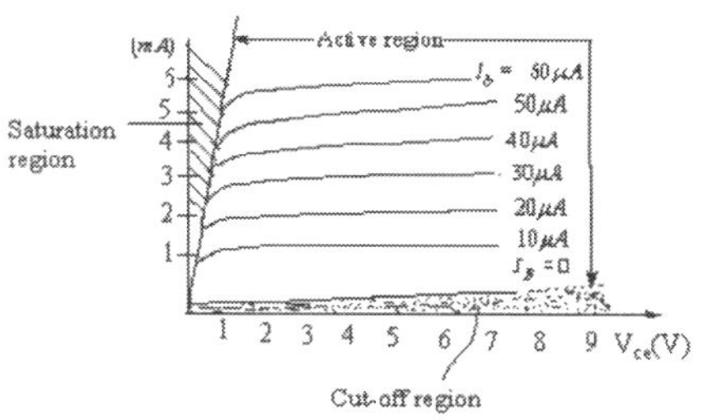

- **Saturation region :-** In this region the collector current becomes almost independent of the base current. This happens when both junctions are forward biased.
- **Cut off region:-** In this region the collector curent is almost zero. This happens when both the junctions are reverse biased.
- **Active (or) Linear region:-** In this region collector current (I_c) is many times greater than base current (I_b). A small change in input current (ΔI_b) produces a large change in the output current (ΔI_c). This happens when emitter junction is forward biased and collector junctions is reverse biased.
- The transistor works as an amplifier when operated in the active region.
- When the transistor is used in the cut off (or) saturation state it acts as a switch

- **Transistor as a switch**

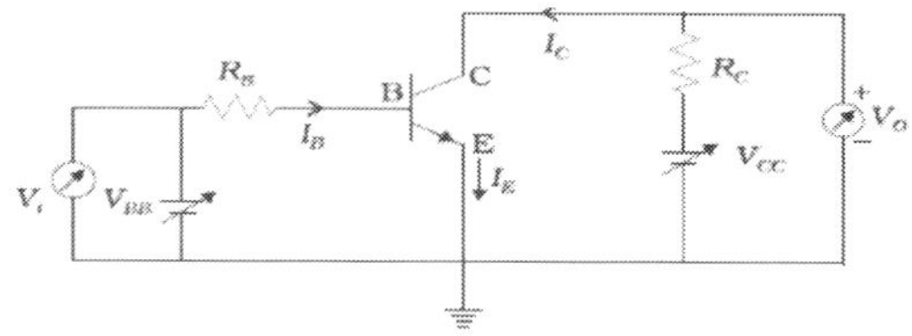

- Appling KVL to the input and output sides of the circuit, We get

  $V_i = I_B R_B + V_{BE}$ and $V_O = V_{CC} - I_C R_C$
- In the case of Si transistor, as long as input V_i is less than 0.6 V, the transistor will be in cut off state and current I_C will be zero. Hence $V_o = V_{CC}$ When V_i becomes greater than 0.6 V the transistor is in active state ,so V_o decreases linearly till its value becomes less than about 1.0 V.
- **graph:**

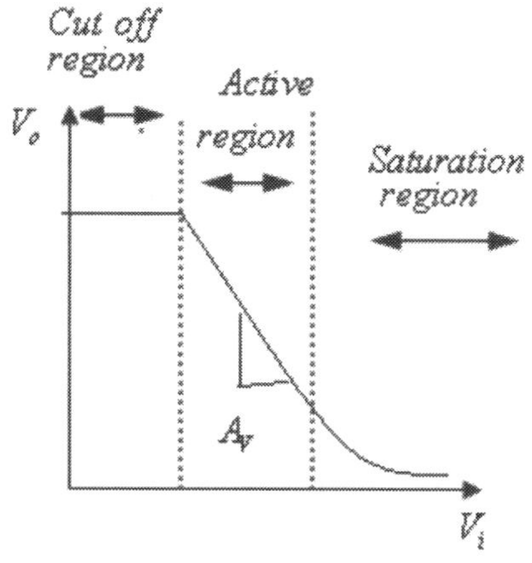

- If we plot the V_0 vs V_i curve [also called the transfer characterstics of the base biased transistor in figure,we see that between cut off state and active state and also between active state and saturation state there are regions of non-linearty showing that the transition from cut off state to active state and from active state to saturation state are not sharply defined.
- **TRANSISTOR AS AN AMPLIFIER (CE CONFIGURATION)**

The process of raising the strength of a weak input signal to a strong output signal is called 'amplification'.

Amplifier : It is a device which increases the weak input signal into strong output signal

Amplifier has wide applications in industries, T.V, radio and communication systems.

Amplifiers are of two types

1) Power amplifiers 2) Voltage amplifiers

1) **Power amplifier:** Amplifier which is used to raise the power level is known as "Power amplifier."

2) **Voltage amplifier:** The amplifier which is used to raise voltage level is known as voltage amplifier.

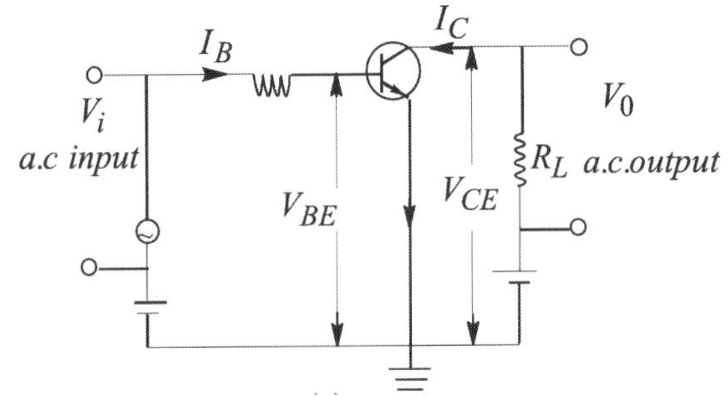

Fig:N-P-N TRANSISTOR AS AMPLIFIER

i) **Voltage gain :**

It is defined as the ratio of change in output voltage to the change in input voltage.

$$A_v = \frac{\Delta V_{CE}}{\Delta V_{BE}} = -\frac{R_L(\Delta I_c)}{R_i(\Delta I_b)} = -\beta\frac{R_L}{R_i}$$

Negative sign indicates input and output voltages are in opposite phase.

ii) **Power gain :**

It is defined as the ratio of output power to the input power.

$$\text{power gain} = \frac{output\ \ power}{input\ \ power} = \frac{I_{out}V_{out}}{I_{in}V_{in}}$$

Power gain= current gain × voltage gain $\Rightarrow A_p = \beta \times A_v = \beta^2 \times \frac{R_{out}}{R_{in}}$

NOTE:1 In common base amplifier, the phase difference between the input and output signals is zero

NOTE:2 In common emitter amplifier, the phase difference between input and output signals is π

WE:19:Current amplification factor of a common base configuration is 0.88. Find the value of base current when the emitter current is 1 m A.

Solution: In a common -base arragement, the current amplification factor

$$\alpha = \left(\frac{\Delta I_C}{\Delta I_E}\right)_{V_{CB}} = \frac{I_C}{I_E}$$

Given $\alpha = 0.88, I_E = 1mA$

$\therefore$ Collector current

$I_C = \alpha I_E = 0.88 \times 1 = 0.88mA$

Now since $I_E = I_B + I_C$

$\therefore$ Base current $\quad I_B = I_E - I_C = 1 - 0.88 = 0.12mA$

WE:20:In a transistor, the emitter circuit resistance is 100Ω and the collector resistance is 100Ω. The power gain, if th emitter and collector currents are as sumed to be equal, will be

Solution:

If $I_C \approx I_B \Rightarrow \beta \approx 1$

$$\therefore\ A_P = \beta^2\left(\frac{R_L}{R_i}\right) = \left(\frac{R_L}{R_i}\right)$$

$$= \frac{100 \times 10^3}{100} = 1000$$

WE:21:From the output characteristics shown in Fig .Calculate the values of β_{ac} *and* β_{dc} of the transistor when V_{CE} is 10V and $I_c = 4.0mA$

Solution:

Consider any two characteristics for two values of I_B which are above and below the given value of I_C, Here $I_C = 4.0mA.$

(Choose characteristics for $I_B = 30$ and $20\mu A$) At $V_{CE} = 10V$ we read the two values of I_C from the graph Then

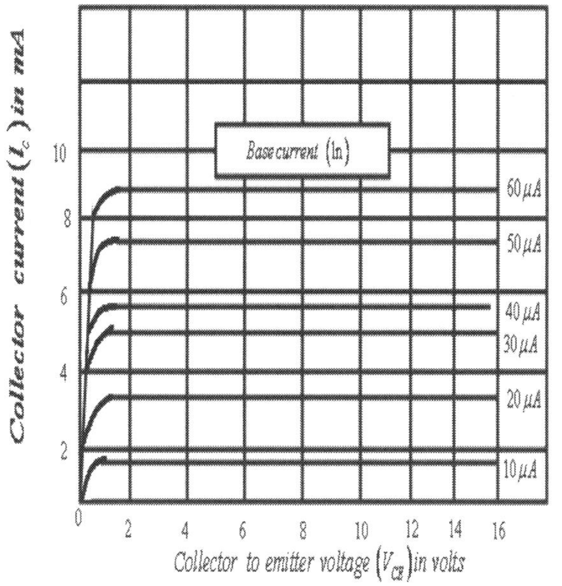

$\Delta I_B = (30 - 20)\,\mu A$

$\Delta I_C = (4.5 - 3.0)\,mA = 1.5mA$

Therefore

$\beta_{ac} = \left(\frac{\Delta i_c}{\Delta I_B}\right)_{V_{CE}} = 1.5mA/10\mu A = 150$

For determining β_{dc} calculate the two values of β_{dc} for the two characteristics chosen and find their mean . Therefore for $I_C = 4.5mA$ *and* $I_B = 30\mu A$ $\beta_{dc} = \frac{I_C}{I_B} = 4.5mA\ /30\mu A = 150$ and for

$I_C = 3.0mA$ *and* $I_B = 20\mu A$

$\beta_{dc} = 3.0mA/20\mu A = 150$

Hence $\beta_{dc} = (150+150)/2 = 150$

WE:22:In Figure the V_{BB} supply can be varie form 0V to 5.0V. The Si transistor has $\beta_{dc} = 250$ and $R_c = 1K\Omega, V_{CC} = 5.0V$. Assume that when the transistor is saturated, $V_{CE} = 0V$ and $V_{BE} = 0.8V$. Calculate (a) the minumum base current, for which the transistor will reach saturation. Hence, (b) determine V_1 when the transistor is switched on(c) find the ranges of V_1 for which the transistor is switched off and switched on.

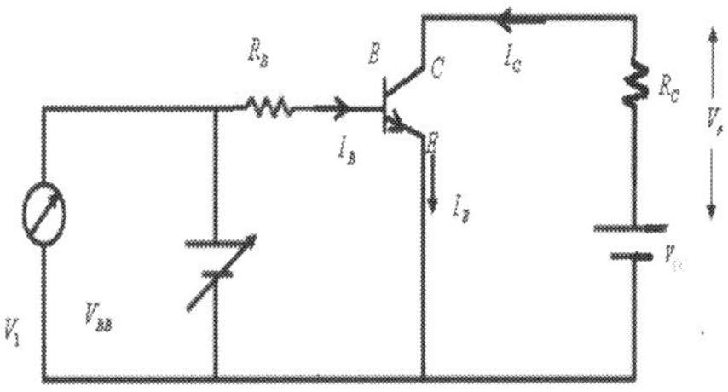

Solution:

Given at saturation $V_{CE} = 0V, V_{BE} = 0.8V$ $V_{CE} = V_{CC} - I_C R_C \Rightarrow I_C = V_{CC}/R_C$
$= 5.0V/1.0K\Omega = 5.0mA$

Therefore $I_B = I_C/\beta = 5.0mA/250 = 20\mu A$

The input voltage at which the transistor will go into saturation is given by

$V_{IH} = V_{BB} = I_B R_B + V_{BC} = 20\mu A \times 100K\Omega + 0.8V = 2.8V$

The value of input voltage below which the transistor remains cutoff is given by $V_{IL} = 0.6V, V_{IH} = 2.8V$. Between 0.0V and 0.6V, the transistor will be in the 'switched off state. Between 2.8V and 5.0V,it will be in 'switched on' state Note that the transistor is in active state when I_B varies from 0.0mA to 0.20 mA.

In this range, $I_C = \beta I_B$ is valid. In the

In this range, $I_C = \beta I_B$ is valid. In the

saturation range, $I_C \le \beta I_B$

WE:23:Two amplifiers are connected one after the other in series (cascaded). The first amplifier has a voltage gain of 10 and the second has a voltage gain of 20. If the input signal is 0.01 volt, calculate the out put ac signal.

Solution: When the amplifiers are Connected in series, the net voltage gain is equal to the product of the gains of the individual amplifiers.

$\therefore Av = Av^1 \times Av^{11}$,hear $Av^1 = 10$,and $Av^{11} = 20$

also $Av = \dfrac{V_{output}}{V_{input}}$

$\therefore$ we can write $\dfrac{V_{output}}{V_{input}} = Av^1 \times Av^{11} : V_{in} = 0.01V$

$V_{out} = V_{in} \times Av^1 \times Av^{11} = 0.01 \times 10 \times 20 = 2V$

WE:24:In a single state transistor amplifier, when the signal changes by 0.02V, the base current by $10\mu A$ and collector current by 1mA. If collector load $R_C = 2k\Omega$ and $R_L = 10k\Omega$, Calculate:(i)

Current Gain (ii)Input impedance,(iii) Effective AC load,(iv) Voltage gain and (v) Power gain.

Solution :

i) Current Gain $\beta = \dfrac{\Delta i_c}{\Delta i_b} = \dfrac{1mA}{10\mu A} = 100$

ii) Input impedance $R_i = \dfrac{\Delta V_{BE}}{\Delta i_b} = \dfrac{0.02}{10\mu A} = 2000\Omega = 2k\Omega$

iii) Effective (a,c) load $R_{AC} = R_C \,||\, R_L \;\therefore R_{AC} = \dfrac{2 \times 10}{2+10} = 1.66k\Omega$

iv) Voltage gain $A_V = \beta \times \dfrac{R_{AC}}{R_{in}} = \dfrac{100 \times 1.66}{2} = 83$

v) Power gain, A_p = Current gain$\times$ Voltage gain $= 100 \times 83 = 8300$

WE:25:An n-p-n transistor in a common emitter mode is used as a simple voltage amplifier with a collector current of 4mA. The terminal of a 8V battery is connected to the collector through a load resistance R_L and to the base through a resistance R_B. The collector-emitter voltage $V_{CE} = 4V$, base-emitter voltage $V_{BE} = 0.6V$ and base current amplification factor $\beta_{d.c} = 100$.Calculate the values of R_L and R_B

Solution :

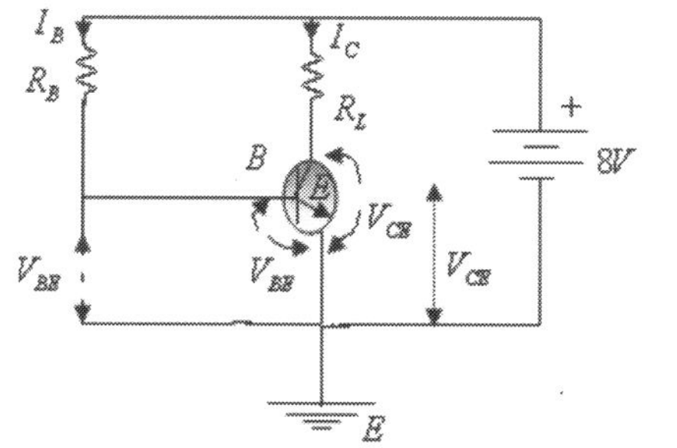

Potential difference across R_L

$= 8V - V_{CE} = 8V - 4V = 4V$

Now $I_C R_L = 4V$

$$R_L = \frac{4}{4\times10^{-3}} = 10^3\Omega = 1k\Omega$$

Further for base emitter equation

$V_{CC} = I_B R_B + V_{BE}$

or $I_B R_B$ = Potential difference across $R_B = V_{CC} - V_{BE} = 8. - 0.6 = 7.4V$

$$\text{Again, } I_B = \frac{I_C}{\beta} = \frac{4\times10^{-3}}{100} = 4\times10^{-5}A$$

$$\therefore R_B = \frac{7.4}{4\times10^{-5}} = 1.85\times10^5\Omega = 185k\Omega$$

- **CONCEPT OF FEEDBACK**
- When a part of the output voltage (or current) of an amplifier is injected back into the input circuit, feedback is said to exist.
- If the voltage feedback is in phase with the applied voltage, the feedback is said to be positive or regenerative;
- If the voltage feedback is in opposite phase to the incoming signal, the feedback is said to be negative or degenerative.
- Fig. illustrates the principles of feedback.

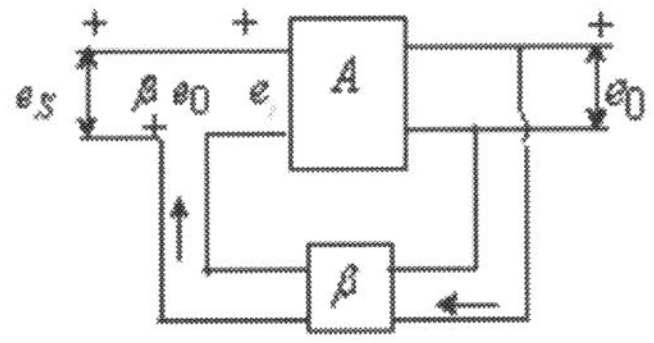

- The gain of the amplifier without feedback is A. If a signal e_S is applied at the input terminals of the amplifier, then let the output voltage be e_0.
- If a fraction β ($\beta = \frac{e_f}{e_0}$) of this output voltage is fedback into the input in phase with the applied signal, then the actual input voltage of the amplifier,

  $e_i = e_s + e_f$

  $e_i = e_s + \beta e_0$ (for positive feed back)
- This total input voltage multiplied by the gain A of the amplifier must be equal to the output voltage i.e. $e_0 = Ae_i \Rightarrow e_0 = A(e_s + \beta e_0) = Ae_s + A\beta e_0$

$\Rightarrow e_0 - A\beta e_0 = Ae_s \Rightarrow e_0(1-\beta A) = Ae_s$ from which the gain A_f with feedback is, $A_f = \frac{e_0}{e_s} = \frac{A}{1-\beta A}$

- The positive feedback thus, increases the gain of the amplifier. If too much positive feedback is applied so that $1-\beta A=0$, the gain of the amplifier becomes infinite.
- For stable oscillation $\beta A=1$ (Barkhausen's criteria)
- In this case the amplifier becomes unstable and the output can be obtained with no external input signal, i.e., the amplifier becomes an oscillator.
- In the case of negative feedback $e_i = e_s - e_f$
- In the case of negative feedback the voltage feedback βe_0 is in opposite phase to the applied voltage e_S, so that gain with negative feedback becomes, $A_f = \dfrac{e_0}{e_s} = \dfrac{A}{1+\beta A}$
- The negative feed back also reduces noise & distortion in an amplifier.

NOTE:

i) When $|1+A\beta|>1, |A_f|<|A|$, feed back is negative

ii) When $|1+A\beta|<1, |A_f|>|A|$, feed back is positive

- **Transistor as an oscillator:**
- The simplest eletrical oscillating system consists of an inductance L and capacitor C connected in parallel.

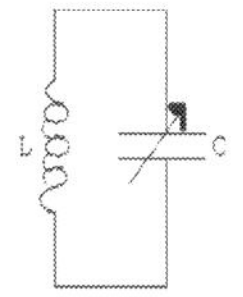

- Once an eletrical energy is given to the circuit, this energy oscillates between capacitance (in the form of eletrical energy) and inductance (in the form of magnetic energy) with a frequency

$$v = \frac{1}{2\pi\sqrt{LC}}$$

- The amplitude of oscillations is damped due the presence of inherent resistance in the circuit
- In order to obtain oscillations of constant amplitide, an arrangement of regenerative or positive feedback from an output circuit to the input circuit is made so that the circuit losses may be compensated.

WE:26:In a negative feedback amplifier, the gain without feedback in 100, feed back ratio is 1/25 and input voltage is 50m V. Calculate (i)gain with feedback (ii) feedback factor(iii) output voltage (iv) feedback voltage (v) new input voltage so that output voltage with feedback equals the output voltage without feedback

Solution :

i) Gain with feedback

$$A_f = \frac{A}{1+\beta A} = \frac{100}{1+(1/25)\times 100} = 20$$

ii) $\beta = \dfrac{1}{25}$

iii) Out put voltage

$V_0' = A_f V_i = 20 \times 50mV = 1volt$

iv) Feedback voltage $\beta V_0' = \frac{1}{25} \times 1 = 0.04\ volt$

v) New increased input voltage

$$V_i^1 = V_i(1+\beta A) = 50\left(1+\frac{1}{25}\times 100\right) = 250mV$$

Digital Eletronics:

1) **Binary number system:**

- It is a two-valued system developed by George Boole. Only two digits 0 and 1, called bits, are used in binary system. But binary addition (in this 1+1=10) is different from addition in Boolean algebra (in this 1+1=1 and remaining will be same i.e., 0+0=0, 1+0=1, 0+1=1).
- A number in a decimal system can be converted into binary by the successive division of 2 until the quotient is zero. The remainders obtained in the successive divisions, taken in the reverse order (from bottom to top shown below) give the binary representation of that number.

WE:27:Convert the decimal number 23 into binary number.

Solution:

2|23
2|11−1
2| 5−1
2| 2−1
2| 1−0
2| 0−1

Arranging the remainders in the reverse order(from bottom to top shown below) the binary equivalent of 23 is found to be 10111. i.e., $23_{(10)} = 10111_{(2)}$

- In binary representation of any number, the first bit is the 'most significant bit' (MSB) and the last bit is the 'least significant bit' (LSB).
- Binary number can be converted into decimal number by multiplying each bit with 2^n and adding them as mention below. where n is positon of the bit from right side (LSB) to leftside (MSB).

WE:28: **Convert the binary number 10111 into decimal number.**

Solution:

4 3 2 1 0

$10111_{(2)} = 1\times 2^4 + 0\times 2^3 + 1\times 2^2 + 1\times 2^1 + 1\times 2^0$

$= 16 + 0 + 4 + 2 + 1 = 23_{(10)}$

$10111_{(2)} = 23_{(10)}$

2) Boolean Algebra:

- Only two states or values of a varible are allowed in Boolean algebra. In logic these two states correspond to 'on' and 'off' and 'saturation' of electronic devices.
- The two allowed states in Boolean alegbra are represented by the digits 0 and 1.

  0 can also represent Off, low, false, No, 0 V

  1 can also represent ON, high, true, Yes, 5 V
- The variables of Boolean algebra are subjected to three operations.
- The OR addition indicated by a plus (+) sign.
- The AND multiplication indicated by a cross (X) or a dot (.) sign.
- The NOT operation indicated by a bar over a variable.
- **OR Addition:** The + sign in Boolean algebra represents OR addition. The equation Y = A+B is read as 'Y equals A OR B'. The
- **AND multiplication:** In Boolean algebra the equation Y=AXB or Y=A.B or Y=AB is read as 'Y equals A AND B'
- **NOT operation:** The NOT operation on a variable A is represented by $\overline{A}$.

  The equation $Y = \overline{A}$ is read as 'Y equals NOT A'.
- Rule for OR, AND and NOT functions in Boolean algebra

OR	AND	NOT
0+0=0	0.0=0	$\overline{0} = 1$
0+1=1	0.1=0	$\overline{1} = 0$
1+0=1	1.0=0	
1+1=1	1.1=1	

- Some useful laws of Boolean algebra
- **Commulative laws:**

  A+B=B+A

  A.B=B.A
- **Associative laws:**

  A+(B+C) = (A+B) +C

  A.(B.C) = (A.B).C
- **Distributive laws:**

  A.(B+C) = A.B +A.C
- **Basic OR and AND Relations:**

OR	**AND**
A+0=A	A.0=0
A+1=1	A.1=A
A+A=A	A.A=A
$A+\overline{A}=1$	$A.\overline{A}=0$

- **Double complement function:**

  $\overline{\overline{A}} = A$

- **De Morgan's theorems:**

  DeMorgan's theorems (or rules) are very useful in simplifying complicated logical expressions and can be stated as under :

  Theorem 1 : The complement of the sum of two (or more) variables is equal to the product of complements of the variables i.e. $\overline{A+B} = \overline{A} \bullet \overline{B} \Rightarrow \overline{A+B} = \overline{A}.\overline{B}$

  Theorem 2 : The complement of the product of two (or more) variables is equal to the sum of complements of the variables i.e

  $\overline{A \bullet B} = \overline{A} + \overline{B}. \Rightarrow \overline{A.B} = \overline{A} + \overline{B}$

- **LOGIC GATES :-**
- A digital circuit having a certain logical relationship between the input and the output voltages is called a logic gate.
- There are three basic logic gates

  (1) OR gate (2) AND gate (3) NOT gate

- **The OR gate**

  1) It is a logic gate which has two or more inputs and one output.

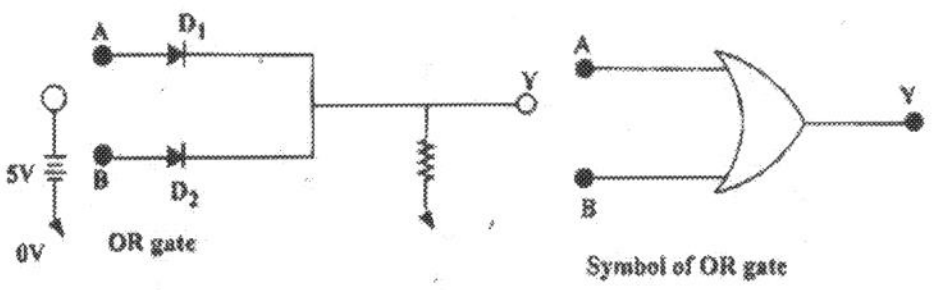

OR gate   Symbol of OR gate

- If its one or more inputs are high, then iis output will be high. Therefore it has a logic of OR.
- **Boolean expression for OR gate**

  A+B=Y

  which reads as A OR B is equal to Y.

- Truth table for OR gate -

A	B	Y=A+B
0	0	0
0	1	1
1	0	1
1	1	1

- **The AND gate**
- It is logic gate which has two or more inputs and one output

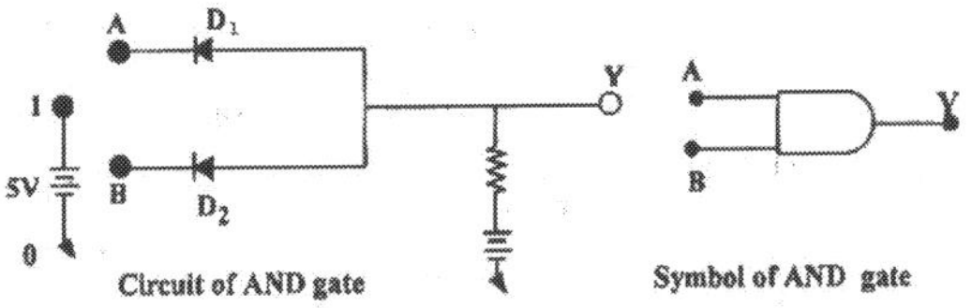

Circuit of AND gate   Symbol of AND gate

- If its all inputs are high, then it is output will be high (1). Therefore it has a logic of AND.
- **Boolean expression for AND gate -**

  A. B = Y reads as A AND B is equal to Y.
- Truth table for AND gate

A	B	Y = A . B
0	0	0
0	1	0
1	0	0
1	1	1

- **The NOT gate -**
- It is the logic gate which has one input and one output

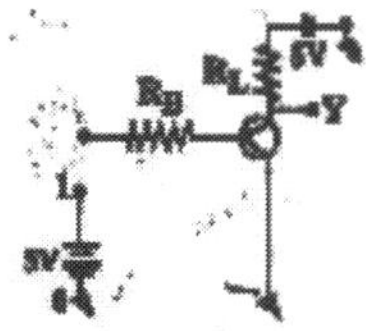

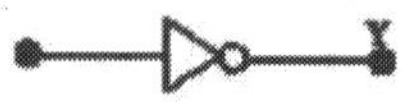

Circuit of NOT gate   Symbol of NOT gate

- If its input is high (1), then its output will be low(0). Therefore it has a logic of NOT.
- **Boolean expression for NOT gate** $\overline{A} = Y$ reads as 'A NOT is equal to Y'.
- Truth table for NOT gate

A	$Y = \overline{A}$
0	1
1	0

- **Combination of gates -**

  In complicated digital circuits used in calculators, computers etc. the different types of combination of three basic logic gates are used.
- **NOR gate**
- This logic gate is the combination of OR gate and NOT gate.

  OR + NOT = NOR
- In this logic gate the output of OR gate is given to the input of NOT gate as shown in the below figure.

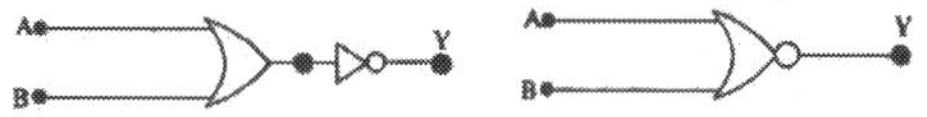

OR   NOT   NOR gate

- **Boolean expression for NOR gate**

  $Y = \overline{A + B}$

  Which reads as A OR B negated.

- Truth table for NOR gate

A	B	A+ B	$Y = \overline{A+B}$
0	0	0	1
0	1	1	0
1	0	1	0
1	1	1	0

- **NAND gate -**
- This logic gate is the combination of AND gate and NOT gate
  AND + NOT $\rightarrow$ NAND
- In this logic gate the output of AND gate is given to the input of NOT gate as shown below

AND NOT NAND gate

- Boolean expression for NAND gate $Y = \overline{A.B}$
- Truth table for NAND gate

A	B	A.B	$\overline{A.B}$
0	0	0	1
0	1	0	1
1	0	0	1
1	1	1	0

- **Uses of NOR gate and NAND gate -**
- The NAND gate and NOR gates are the building blocks of digital circuits. All the basic gates (OR, AND and NOT) can be obtained by the repeated use of NAND or NOR gates.
- **NOT gate from NAND gate -**

1) Diagram

2) **Truth table**

A	B	$Y = \overline{A.A} = \overline{A}$
0	0	1
1	1	0

- **AND gate from NAND gate**

1) Diagram -

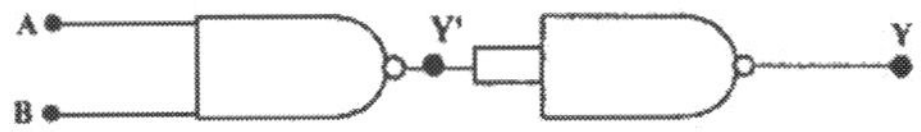

2) Truth table

A	B	Y
0	0	0
0	1	0
1	0	0
1	1	1

- **OR gate from NAND gate -**

  1) Diagram -

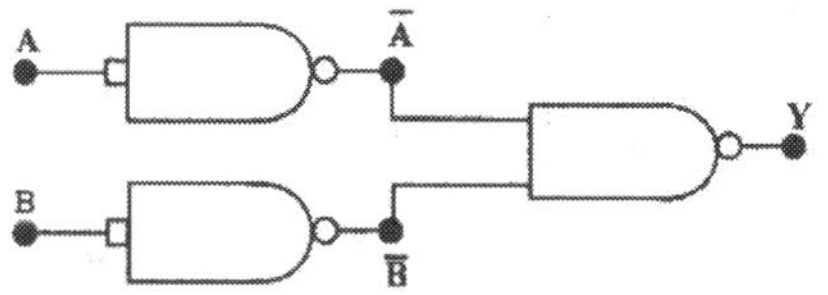

2) Truth table

A	B	$\overline{A}$	$\overline{B}$	$Y = (A+B)$
0	0	1	1	0
1	0	0	1	1
0	1	1	0	1
1	1	0	0	1

- **NOT gate from NOR gate -**

  1) Diagram

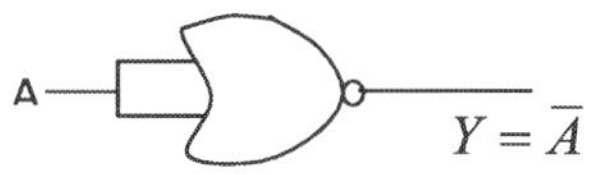

2) Truth table

A	B	$Y = \overline{A.A} = \overline{A}$
0	0	1
1	1	0

- **AND gate from NOR gate**

  1) Diagram -

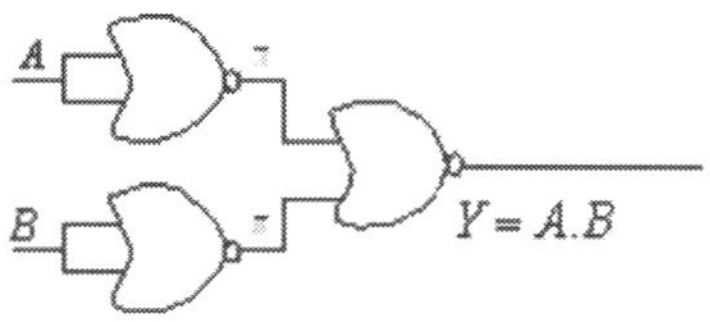

2) Truth table

A	B	Y
0	0	0
0	1	0
1	0	0
1	1	1

- **OR gate from NOR gate -**

  1) Diagram -

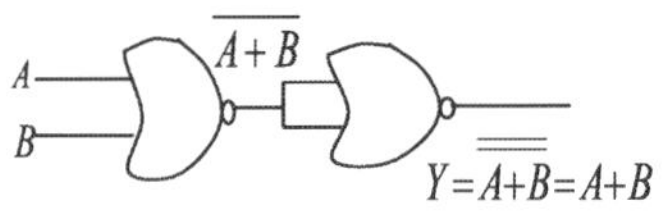

  2) Truth table

A	B	Y
0	0	0
0	1	1
1	0	1
1	1	1

- **XOR GATE**
- XOR gate is obtained by using OR, AND and NOT gates. It is also called exlusive OR gate.
- The output of a two input XOR gate is 1 only when the two inputs are different.
- The Boolean equation is $Y = A.\overline{B} + B.\overline{A}$

  1) two input XOR gate

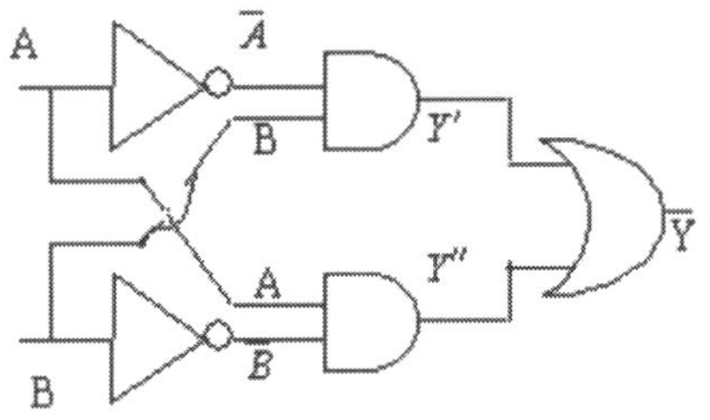

  2) circuit symbol

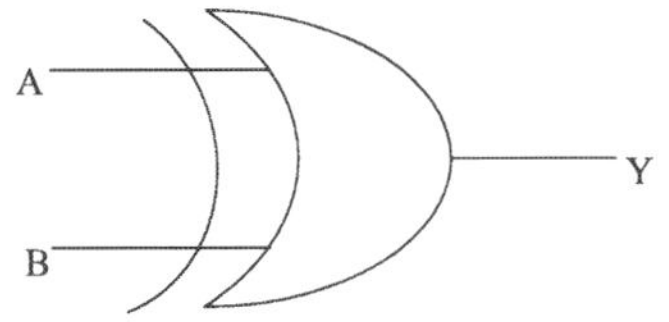

  3) truth table

A	B	Y
0	0	0
0	1	1
1	0	1
1	1	0

XNOR GATE

- XNOR gate is obtained by using OR, AND and NOT gates.
- It is also called exlusive NOR gate.
- The output of a two input XNOR gate is 1 only when both the inputs are same.
- The Boolean equation is $Y = A.B + \overline{A}.\overline{B}$

XNOR gate is inverse of XOR gate.

1) Diagram -

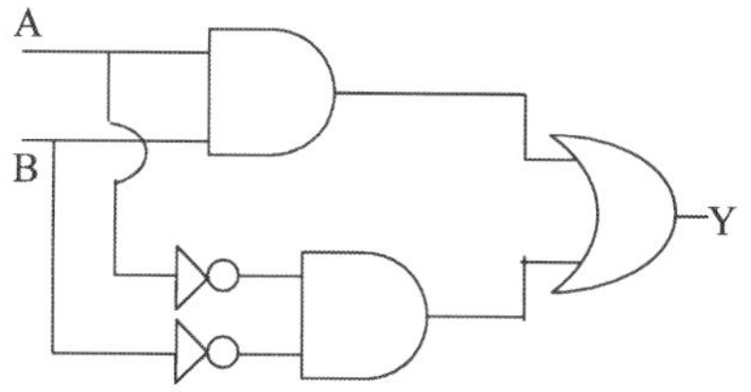

2) circuit symbol

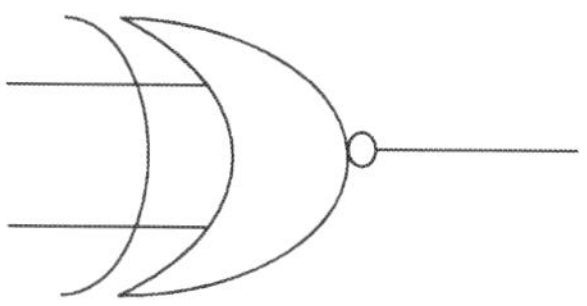

3) truth table

A	B	Y
0	0	1
0	1	0
1	0	0
1	1	1

WE:29:The Boolean expression of the output Y of the inputs A and B for the circuit shown in the fig:

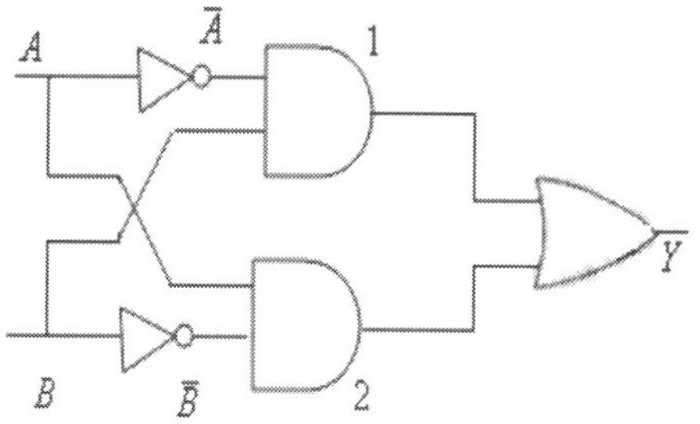

Solution : The output of AND gate 1 is $\overline{A}B$

The output of AND gate 2 is $A\overline{B}$

$\therefore$ the output of OR gate is Y= $= \overline{A}B + A\overline{B}$

WE:30: The diagram of a logic circuit is given below. The output of the circuit is represented by

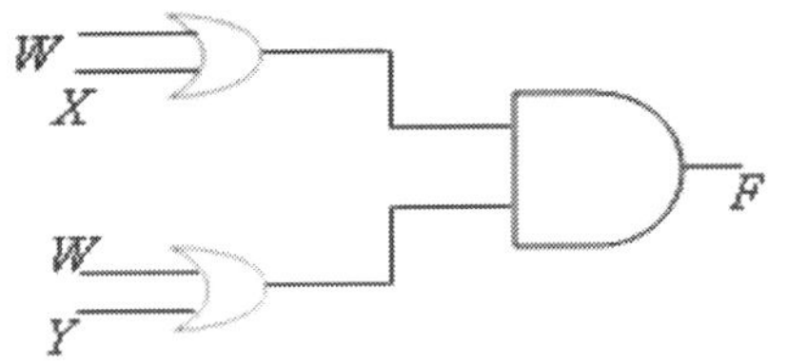

Solution :

$$F=(W+X).(W+Y)\ =W.W+W.Y+X.W+X.Y$$

$$=W+W.Y+X.W+X.Y\ =W(1+Y)+X.W+X.Y$$

$$=W+XW+X.Y\ =W(1+X)+X.Y=W+X.Y$$

WE:31: The logic circuit and its truth table are given, what is the gate X in the diagram

A	B	Y
1	1	1
1	0	1
0	1	1
0	0	0

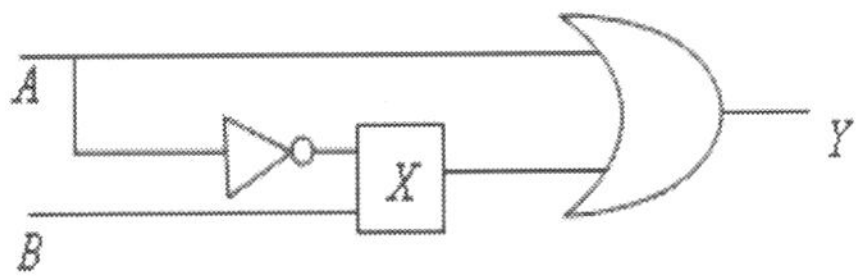

Solution: From the truth table we note that $Y=A+B$

i.e., it is for OR gate (or)

$$A+X=A+B=A+B.(A+\bar{A})(\because A+\bar{A}=1)$$

then, $A+X=A+B.A+B.\bar{A}\ =A.(1+B)+\bar{A}.B=A+\bar{A}.B$

So $X=A+\bar{A}.B$**, which is AND gate with inputs as** $\bar{A}$ **and B**

WE:32: You are given two circuits as shown in Figure which consist of NAND gates. Identify the logic operation carried out by the two circuits.

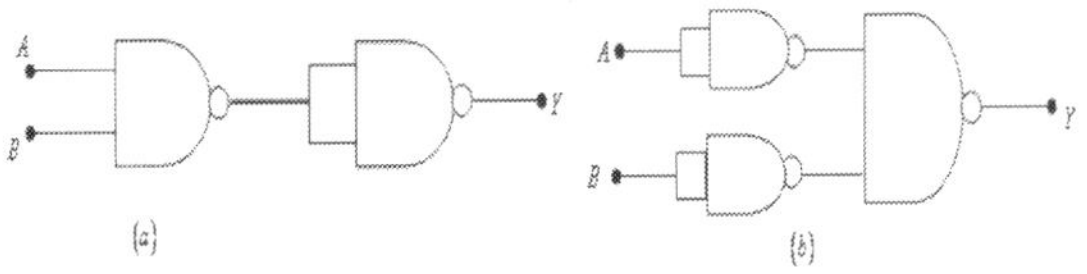

Solution: From fig(a). The output of NAND gate is connected to NOT gate (obtained from NAND gate) Let Y^1 be the output of NAND gate and the final output of the combination of two gates is Y. The output of a NAND gate is O only when both the inputs are zero, while in NOT gate, the input gets inverted.

Truth table for the arrangement

A	B	Y^1	Y
0	0	1	0
1	0	1	0
0	1	1	0
1	1	0	1

=

A	B	Y
0	0	0
1	0	0
0	1	0
1	1	1

It is the truth tables of AND gate. Therefore the given circuit acts as AND gate.

b) The output of two NOT gates are Connected to NAND gate Let Y_1 and Y_2 be the outputs of the two NOT gates and the final output of the combination of three gates be Y. In a NOT gate. the input gets inverted, while the output of a NAND gate is 'O' only when both the inputs are zero. Truth table for given arrangement.

A	B	Y
0	0	0
1	0	1
0	1	1
1	1	1

WE:33:Justify the output waveform (Y) of the OR gate for the following inputs A and B given is Fig

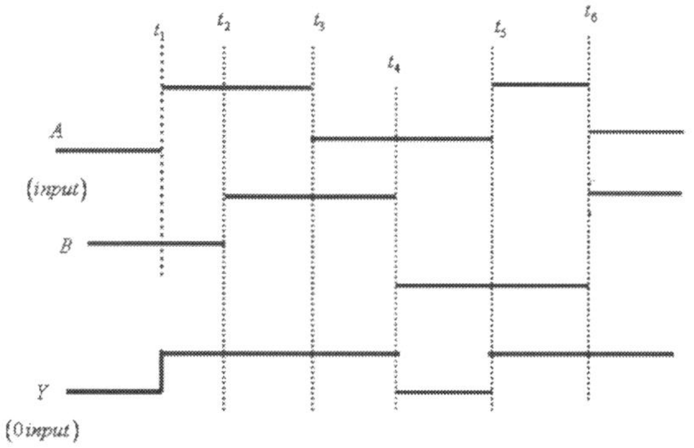

Solution: Note the following:

. At $t < t_1$; $A = 0, B = 0$; $Hence Y = 0$. For t_1 to t_2; $A = 1, B = 0$; $Hence Y = 1$

. For t_2 to t_3; $A = 1, B = 1$; $Hence Y = 1$. For t_3 to t_4; $A = 0, B = 1$; $Hence Y = 1$

. For t_4 to t_5; $A = 0, B = 0$; $Hence Y = 0$. For t_5 to t_6; $A = 1, B = 0$; $Hence Y = 1$

. For $t > t_6$; $A = 0, B = 1$; $Hence Y = 1$

Therefore the wave form Y will be as shown in the Fig

WE:34:Draw logic diagrams for the Boolean expressions given below.

i) $A \bullet \bar{B} + \bar{A} \bullet B = Y$ **ii)** $(A + \bar{B}) \bullet (\bar{A} + B) = Y$

Solution: (i) The required logic diagram for the given Boolean expression is given in figure. Here the input B before applying to first AND gate and input A before applying to second AND gate have been inverted. The output of these gates are, therefore, $A\bar{B}$ and $\bar{A}B$ respectively. These outputs are fed to

OR gate which gives $Y = A\overline{B} + \overline{A}B$ as shown in figure.

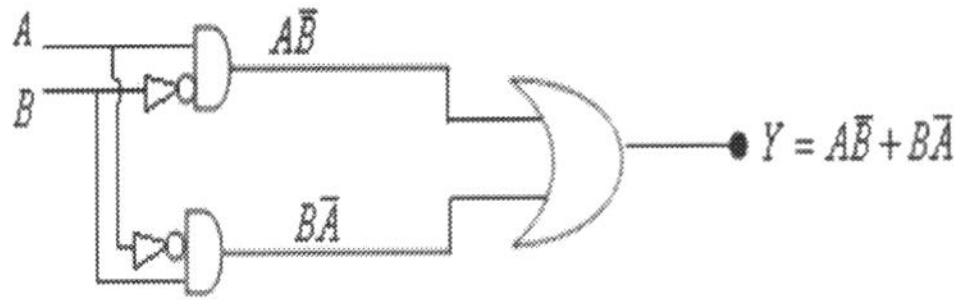

ii) The required logic diagram for the given Boolean expression is shown in figure. The input B to first OR gate and input A to second OR gate have been inverted. The output of these gates are, therefore, $A+\overline{B}$ and $\overline{A}+B$ as shown. These inputes when applied to AND gate give the required output.

$Y = (A+\overline{B}).(\overline{A}+B)$

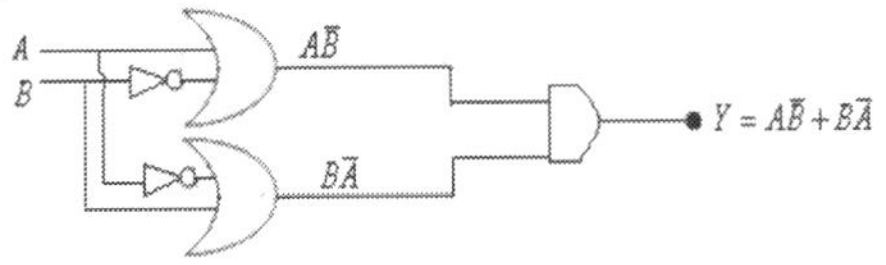

INTEGRATED CIRCUITS:

An entire circuit (consisting of many passive components like R and C active devices like diode and transistor) on a small single block (or chip) of a semiconductor is known as Integrated Circuit (IC). The most widely used technology is the Monolithic Integrated Circuit.

- The chip dimensions are as small as $1mm \times 1mm$ or it could even be smaller.
- Depending on nature of input signals, IC's can be grouped in two categories:
- (a) linear or analogue IC's and
- (b) digital IC's. The linear IC's process analogue signals varies linearly with the input.
- One of the most useful linear IC's is the operational amplifier.
- The digital IC's process signals that have only two value.
- They contain circuits such as logic gates. Depending upon the level of integration (i.e,. the number of circuit components or logic gates), the IC's are termed as

  i) Small Scale Integration, SSI(logic gates ≤ 10)

  ii) Medium Scale Integration, MSI (logic gates < 100)

  iii) Large Scale Integration, LSI (logic gates <1000)

  iv) very large scale integration VLSI (logic gates > 1000).

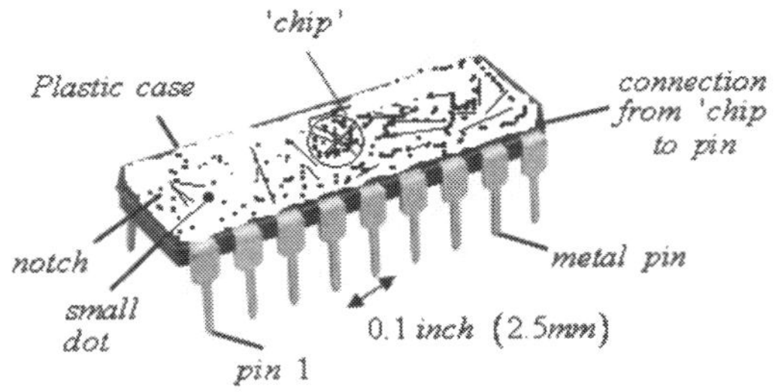

The casing and connection of a 'chip'

Chapter - 9
(Part - 2)
SEMI CONDUCTOR DEVICES

CONCEPTUAL QUESTIONS

INTRINSIC AND EXTRINSIC SEMICONDUCTORS

1. **Which property makes the crystalline solids to have sharp melting point?**
1) Equal strength of all the interatomic bonds 2) anisotropic
3) long range order 4) short range order

2. **Which of the following is not a property of crystalline substance**
1) sharp melting point 2) bounded by flat surface 3) isotropic 4) long range order

3. **A strip of copper and another of germanium are cooled from room temperature to 80K. The resistance of:**
1) each of these decreases
2) copper strip decreases and that of germanium decreases
3) copper strip decreases and that of germanium increases
4) each of these increases

4. **The difference in the variation of resistance with temperature in a metal and a semiconductor arises essentially due to the difference in the :**
1) crystal structure
2) variation of the number of charge carriers with temperature
3) type of bonding
4) variation of scattering mechanism with temperature

5. **The manifestation of band structure in solids is due to :**
1) Heisenberg's uncertainty principle 2) Pauli's exclusive principle
3) Bohr's correspondence principle 4) Boltzman's law

6. **A solid which is not transparent to visible light and whose conductivity increases with temperature is formed by:**
1) ionic bonding 2) covalent bonding 3) vander Wall's bonding 4) metallic bonding

7. **To a germanium sample, traces of gallium are added as an impurity. The resultant sample would behave like :**
1) A conductor 2) A p-type semiconductor
3) A n-type semiconductor 4) An insulator

8. **In semiconductors at a room temperature**
1) The valence band is partially empty and the conduction band is partially filled.
2) The valence band is completely filled and the conduction band is partially filled
3) The valence band is completely filled 4) The conduction band is completely empty

9. **Identify the property which is not characteristic for a semiconductor?**
1) at a very low temperatures, it behaves like an insulator
2) at higher temperatures two types of charge carriers will cause conductivity
3) the charge carriers are electrons and holes in the valence band at higher temperatures
4) the semiconductor is electrically neutral.

10. **Carbon, silicon, and germanium have four valence electrons each. At room temperature, which one of the following statements is most appropriate?**
1) The number of free conduction electrons is negligibly small in all the three.
2) The number of free electrons for conduction is significant in all the three.
3) The number of free electrons for conduction is significant only in Si and Ge but small in C.
4) The number of free conduction electrons is significant in C but small in Si and Ge

11. **The relation between number of free electrons (n) in a semiconductor and temperature (T) is given by**

1) $n \propto T$ 2) $n \propto T^2$ 3) $n \propto \sqrt{T}$ 4) $n \propto T^{3/2}$

12. **An electrically neutral semi conductor has**

1) equal amounts of negative and positive charges 2) no minority charge carriers
3) no majority charge carriers 4) no free charges

13. **There is no hole current in conductors because they have**

1) high conductivity 2) high electron density 3) no valence band
4) overlapping of valence and conduction bands.

14. **In the insulators**

1) the valence band is partially filled with electrons
2) the conduction band is partially filled with electrons
3) the conduction band is partially filled with electrons and valence band is empty
4) the conduction band is empty and the valence band is filled with electrons.

15. **In semiconductors the forbidden energy gap between V.B. and C.B. is of the order of**

1) 1eV 2) 5eV 3) 1 keV 4) 1MeV

16. **The level formed due to impurity atom, in the forbidden energy gap, very near to the valence band in a p-type semicoductor is called**

1) acceptor level 2) donar level 3) conduction level 4) forbidden level

17. **The bond in semiconductors is**

1) covalent 2) ionic 3) metallic 4) hydrogen

18. **On increasing temperature, the conductivity of pure semiconductors**

1) decreases 2) increases 3) remains unchanged 4) becomes zero

19. **The mobility of free electrons is greater than that of free holes because**

1) they carry negative charge 2) they are light
3) their mutual collisions are less 4) they require low energy to continue their motion

20. **A semiconductor at 0 K behaves as**

1) conductor 2) insulator
3) super conductor 4) extrinsic semiconductor

21. **Carbon, Silicon, and Germanium have four valence electrons each. These are characterised by valence and conduction bands separated by energy band gap respectively equal to $(E_g)_C$, $(E_g)_{Si}$ and $(E_g)_{Ge}$. Which of the following statement is true?**

1) $(E_g)_{Si} < (E_g)_{Ge} < (E_g)_C$ 2) $(E_g)_C < (E_g)_{Ge} > (E_g)_{Si}$

3) $(E_g)_C > (E_g)_{Si} > (E_g)_{Ge}$ 4) $(E_g)_C = (E_g)_{Si} = (E_g)_{Ge}$

22. **The valency of impurity element for making p-type semiconductor is**

1) 5 2) 4 3) 3 4) 7

23. **In n-type semiconductors the electron concentration is equal to**

1) density of donor atoms
2) density of acceptor atoms
3) density of both type of atoms
4) neither density of acceptor atoms nor density of donor atoms

24. **Which of the following statements is not true?**

1) the resistance of intrinsic semiconductors decreases with increase of temperature.
2) doping pure Si with trivalent impurities give p-type semiconductors
3) the majority charge carries in n-type semiconductors are holes
4) a p-n junction can act as a semiconductor diode

25. **p-type semi conductor is**

1) negatively charged
2) positively charged
3) neutral
4) may be positive or negative

26. **n-type semi conductor is**

1) negatively charged
2) positively charged
3) neutral
4) may be positive or negative

27. **An electric field is applied across a semi conductor. Let n be the number of charge carriers. As temperature increases, n will**

1) increase
2) decrease
3) does not change
4) may increase or decrease

28. **In a n - type semiconductor , the fermi energy level lies**

1) in the forbidden energy gap nearer to the conduction band.
2) in the forbidden energy gap nearer to the valence band.
3) in the middle of forbidden energy gap
4) outside the forbidden energy gap

29. **An n-type and p-type silicon can be obtained by doping pure silicon with**

1) Arsenic and phosphrous
2) Indium and aluminium
3) Phosphorous and indium
4) aluminium and boron

30. **The width of forbidden gap in silicon crystal is 1.1 ev. When the crystal is converted into a n-type semiconductor the distance of fermi level from conduction band is**

1) Greater than 0.55eV
2) Equal to 0.55 eV
3) lesser than 0.55eV
4) Equal to 1.1 eV

31. **In extrinsic semiconductors**

1) the conduction band and valence band overlap
2) the gap between conduction band and valence band is near about 16eV
3) the gap between conduction band and valence band is near about 1eV
4) The gap between conduction band and valence band will be 100eV and more

32. **The element that can be used as acceptor impurity to dope silicon is**

1) Antimony
2) Arsenic
3) Boron
4) phosphorous

33. **Among the following, the wrong statement in the case of semiconductor is**

1) Resistivity is in between that of a conductor and insulator
2) Temperature coefficient of resistance is negative
3) Doping increases conductivity
4) At absolute zero temperature it behaves like a condutor

34. **The value indicated by fermi energy level in an intrinsic semiconductor is**
1) the average energy of electrons and holes 2) the energy of electrons in conduction band
3) the energy of holes in valence band 4) the energy of forbidden region

35. **The conduction band and valency band of a good conductors are**
1) well separated 2) just touch 3) very close 4) overlap

36. **Two pieces one of germinium and the other of aluminium are cooled from T_1K to T_2K. The resistance of**
1) aluminium increases and that of germanium decrease
2) each of them decreases
3) aluminium decreases and that of germanium increases
4) each of them increases

37. **In intrinsic semiconductor at room temperature the no. of electrons and holes are**
1) equal 2) zero 3) unequal 4) infinite

38. **Band gap in insulator is of the order**
1) 6 eV 2) 0.60 eV 3) –6 eV 4) 0eV

39. **In p-type semiconductor conduction is due to**
1) greater number of holes and less number of electrons
2) only electrons 3) only holes
4) greater number of electrons and less number of holes

40. **In an intrinsic semiconductor, the fermi nergy level is**
1) nearer to valence band than conduction band
2) equidistant from conduction band and valence band
3) nearer to conduction band than valence band 4) bisecting the conduction band

41. **With increase in temperature in an intrinsic semiconducter the ratio of conduction electrons and holes is**
1) 1:1 2) 1: 2 3) 2 :1 4) 1 : 3

42. **To obtain n-type extrinsic semiconductor, the impurity element to be added to germanium should be of valency**
1) 2 2) 5 3) 4 4) 3

43. **The majority carriers in a p-type semi-conductor are........**
1) Electrons 2) Holes 3) Both 4) Impurities

44. **The objective of adding impurities in the extrinsic semiconductor is**
1) to increase the conductivity of the semiconductor
2) to increase the density of total current carries
3) to increase the density of either holes or electrons but not both
4) to eliminate the electron -hole pairs produced in intrinsic semiconductor.

45. **In intrinsic semiconductor conductivity is**
1) low 2) average 3) high 4) very low

46. **In intrinsic semiconductor conductivity is due to**
1) doping 2) breaking of covalent bonds 3) free electrons 4) holes

47. **When the conductivity of a semi conductor is only due to breaking of covalent bonds, the semi conductor is called**
1) n-type 2) p-type 3) intrinsic 4) extrinsic

48. **Assertion (A) : C,Si and Ge have 4 Valancy each but C is an insulator where as Si and Ge are semi conductors**

Reason (R) : Energy gap is least for Ge, less for Si compared to C, So that free electrons for conduction in Ge and Si are significant but negligible small for C.

1) A, R are true and R explains A correctly 2) A, R are true and R do not explain A correctly

3) A is true, but R is false 4) R is true, but A is false

JUNCTION DIODE

49. **The potential barrier at PN junction is due to**

1) fixed acceptor and donor ions on either side of the junction

2) minority carriers on either side of the junction 3) majority carriers on either side of the junction

4) both majority and minority carriers on either side of junction

50. **A PN junction diode cannot be used**

1) as rectifier 2) for converting light energy to electric energy

3) for getting light radiation 4) for increasing the amplitude of an ac signal

51. **A full wave rectifier along with the output is shown in fig. the contributions from the diode (2)are**

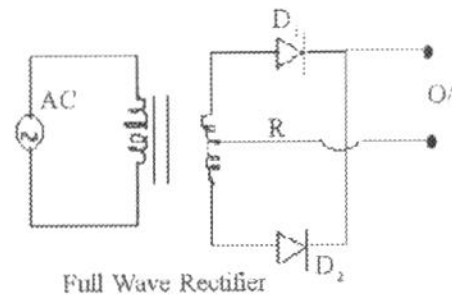

Full Wave Rectifier

1) C 2) A,C 3) B, D 4) A,B,C,D

52. **A full-wave rectifier is used to convert 'n' Hz a.c into d.c,then the number of pulses per second present in the rectified voltage is.**

1) n 2) n/2 3)2n 4)4n

53. **If the input frequency of half-wave rectifier is n Hz ac, then its output is**

1) a constant dc 2) n/2 Hz pulsating dc

3) n Hz pulsating dc 4) 2n Hz pulsating dc

54. **p-n junction diode acts as**

1) ohmic resistance 2) non-ohmic resistance 3) both 1 and 2 4) amplifier

55. **The process of converting alternating current into direct current is known as**

1) modulation 2) amplification 3) detection 4) rectification

56. **On increasing reverse voltage in a p-n junction diode the value of reverse current will**

1) gradually increases

2) first remains constant and then suddenly increase

3) remains constant 4) gradually decrease

57. **In forward bias the depletion layer behaves like**

1) an insulator 2) a conductor 3) a semiconductor 4) capacitor

58. p-n junction in reverse bias behaves like

1) an inductor 2) a condenser 3) amplifier 4) an off switch

59. The main cause of avalenche breakdown is

1) collision by ionisation 2) high doping

3) recombination of electrons and holes 4) low doping

60. The main cause of Zener breakdown is

1) the base semiconductor being germanium

2) production of electron - hole pairs due to thermal exitation

3) low doping 4) high doping

61. When p-n junction is foward biased, the current across the junction is mainly due to

1) diffusion of charges 2) drifting of charges

3) both diffusion and drifting of charges 4) holes only

62. The current through any p-n junction is due to

a) drift of charge carriers **b) diffusion of charge carriers**

c) different concentrations of same type of charge carriers in different regions

d) same concentrations of same type of charge carriers in different regions

1) a,b and c 2) a and b only 3) only d 4) a,b,c,d

63. The thickness of depletion layer is approximately

1) 1 μ m 2) 1mm 3) 1cm 4) 1m

64. The depletion region is

1) region of opposite charges 2) neutral region

3) region of infinite energy 4) region of free current carriers

65. The diffusion current in a p-n junction is greater than the drift current when the junction is

1) forward biased 2) reverse biased 3) un biased

4) both forward and reverse biased

66. Germanium diode

1) may be used as rectifier because it offers a relatively low resistance for forward bias and very high resistance for reverse bias.

2) may be used as a rectifier because it offers a relatively high resistance for forward bias and very low resistance for reverse bias.

3) cannot be used as a rectifier 4) may be used as an amplifier

67. The output of the given circuit in Fig.

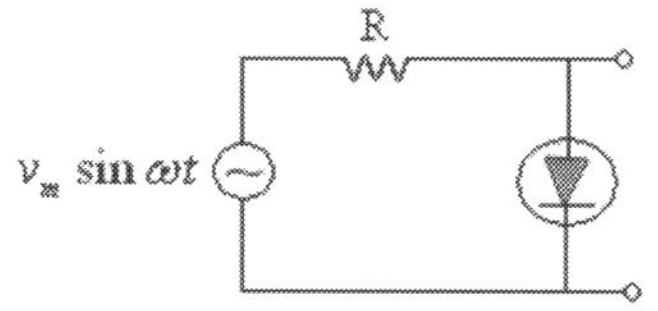

1) would be zero at all times

2) would be like a half-wave rectifier with positive cycles in output

3) would be like a half-wave rectifier with negative cycles in output

4) would be like that of a full-wave rectifier

68. **Diode is forward biased and the applied voltage is greater than the potential barrier then**
I) resistance of the junction in the forward bias decreases
II) potential barrier remains same **III) width barrier remains decreases**
IV)p-type is at higher potential than the n-type
1) all are true 2)all are false 3) I,III,IV are true 4)I,II, III are true

69. **When a junction diode is reverse biased, the current called drift current is due to**
1) majority charge carriers of both n&p sides 2) minority charge carriers of both n&p sides
3) holes of both n &p sides 4) conduction band electrons of n-side only

70. **Among the following one statement is not correct when a junction diode is in forward bias**
1) the width of depletion region decreases
2) free electron on n- side will move towards the junction
3) holes on p -side move towards the junction
4) electron on n- side and holes on p-side will move away from junction

71. **Consider a.p-n junction as a capacitor, formed with p and n material acting as thin metal electrodes and depletion layer width acting as seperation between them. Basing on this assume that a n-p-n transistor is working as an amplifier in CE configuration. If C_1 and C_2 are base-emiter and collector emitter junction capacitances, then :**
1) $C_1 > C_2$ 2) $C_1 < C_2$ 3) $C_1 = C_2$ 4) $C_1 = C_2 = 0$

72. **Consider the following statements A and B and identify the correct answer**
1) : Germanium is preferred over silicon in the construction of zener diode.
2) : Germanium has high thermal stability than silicon in the construction of Zener diode
1) Both (1) and (2) are true 2) Both (1) and (2) are false
3) (1) is true but (2) is false 4) (1) is false but (2) is true

73. **A Zener diode when used as a voltage regulator is connected**
(a) in forward bias **(b) in reverse bias**
(c) in parallel to the load **(d) in series to the load**
1) (a) and (b) are correct 2) (b) and (c) are correct
3) (a) only is correct 4) (d) only is correct

74. **Consider the following statements A and B and identify the correct answer**
(A) A Zener diode is always connected in reverse bias to use it as voltage regulator
(B) The potential barrier of a p - n junction lies between 0.1 to 0.3V, approximately
1) A and B are correct 2) A and B are wrong
3) A is correct but B is wrong 4) A is wrong but B is correct

75. **Consider the following statements A and B identify the correct of the give answer.**
A) The width of the depletion layer in a p-n junction diode increases in forward bias.
B) In an intrinsic semiconductor the fermi energy level is exactly in the middle of the forbidden gap
1) A is true and B is false 2) Both A and B are false
3) A is false and B is true 4) Both A and B are true

76. **The potential in the depletion layer due to**
1) Electrons 2) Holes 3)Ions 4) Forbidden band

77. Pickout the incorrect statement regarding reverse saturation current in the p-n junction diode

1) this current doubles for every 100^0C rise of temperature

2) this current is due to minority carriers

3) the current carriers are produced by thermal agitation

4) reverse saturation current is also known as leakage current

78. When the p-n junction diode is reverse biased, the thickness of the depletion layer

1) increases 2) decreases 3) becomes zero 4) remains constant

79. p-n junction diode can be used as

1) amplifer 2) detector 3) oscillator 4) capacitor

80. A p-n junction diode is reverse biased. Then

1) more current flows 2) the barrier potential decreases

3) the barrier potential increases 4) resistance offered is low

81. In the middle of the depletion layer of reverse biased p-n junction, the :

1) electric field is zero 2) potential is maximum

3) electric field is maximum 4) potential is zero

82. When p-n junction diode is forward biased then

1) the depletion region is reduced and barrier height is increased

2) the depletion region is widened and barrier height is reduced

3) both the depletion region and barrier height are reduced

4) both the depletion region and barrier height are increased

83. In a p-n junction photo cell, the value of the photo electromotive force produced by monochromatic light is proportional to

1) The barrier voltage at the p-n junction. 2) The intensity of the light falling on the cell

3) The frequency of the light falling on the cell. 4) The voltage applied at the p-n junction

84. In Fig. V_0 is the potential barrier across a p-n junction, when no battery is connected across the junction:

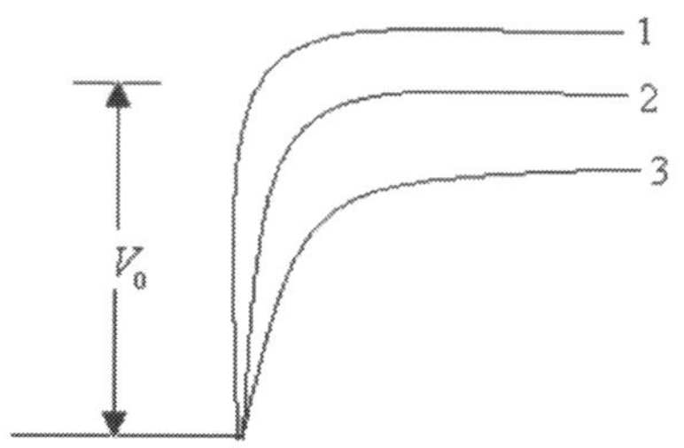

1) 1 and 3 both correspond to forward bias of junction

2) 3 corresponds to forward bias of junction and 1 corresponds to reverse bias of junction

3) 1 corresponds to forward bias and 3 corresponds to reverse bias of junction

4) 3 and 1 both correspond to reverse bias and 3 corresponds to reverse bias of junction

85. There is a sudden increase in current in zener diode is

1) Due to rupture of bonds 2) Resistance of deplection layer becomes less

3) Due to high doping 4) Due to less doping

86. Application of a forward bias to a p-n junction:

1) increases the number of donors on the n-side

2) increases the electric field in the depletion zone

3) increases the potential difference across the depletion zone 4) widens the depletion zone

87. In a p-n junction diode, not connected to any circuit

1) the potential is the same everywhere
2) the p-type side is at a higher potential than the n-type side.
3) there is an electric field at the junction directed from the n-type side to the p-type side.
4) there is an electric field at the junction directed from the p-type side to the n-type side.

88. Select the correct statement from the following:

1) A diode can be used as a rectifier
2) A triode cannot be used as a rectifier
3) The current in a diode is always proportional to the applied voltage
4) The linear portion of the I -V characteristic of a triode is used for amplification without distortion

89. When a p-n junction diode is forward-biased, energy is released at the junction due to the recombination of electrons and holes. This energy is in

1) Visible region 2) Infrared region 3) UV region 4) X-ray region

90. In Fig. assuming the diodes to be ideal:

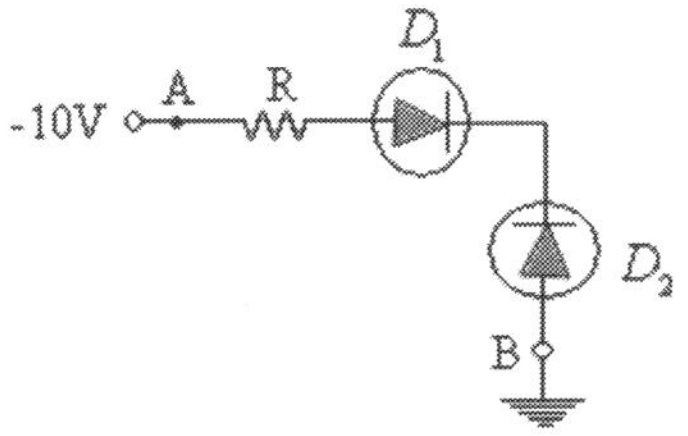

1) D_1 is forward biased and D_2 is reverse biased and hence current flows from A to B
2) D_2 is forward biased and D_1 is reverse biased and hence no current flows from B to A and vice versa
3) D_1 and D_2 are both forward biased and hence current flows from A to B
4) D_1 and D_2 are both reverse biased and hence no current flows from A to B and vice-versa

91. **A Si and a Ge diode has identical physical dimensions.The band gap in Si is larger than that in Ge.An identical reverse bias is applied across the diodes.**

1)The reverse current in Ge is larger than that in Si
2)The reverse current in Si is larger than that in Ge
3)The reverse current is identical in the two diodes
4)The relative magnitude of the reverse currents cannot be determined from the given data only

92. **The correct curve between potential(V) and distance(d) near p-n junction is**

1) 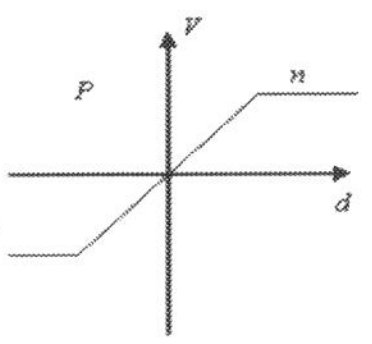

2) 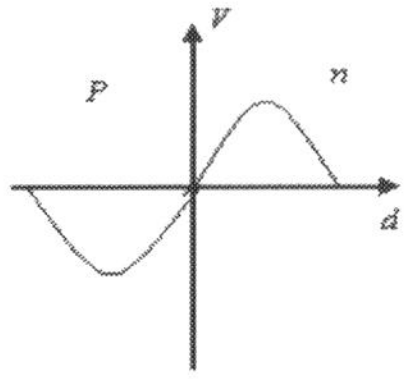

3) 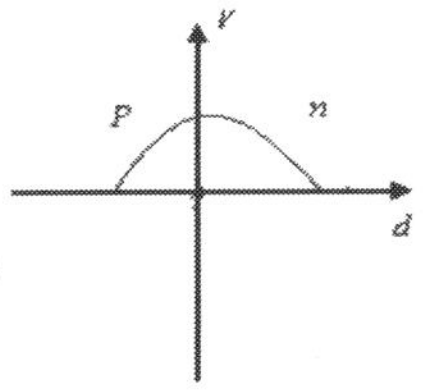

4) 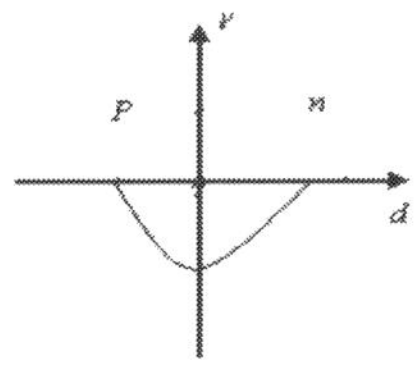

93. **The graph given below represents the I-V characterstics of a zener diode.Which part of the characterstics curve is most relavant for its operation as a voltage regulator.**

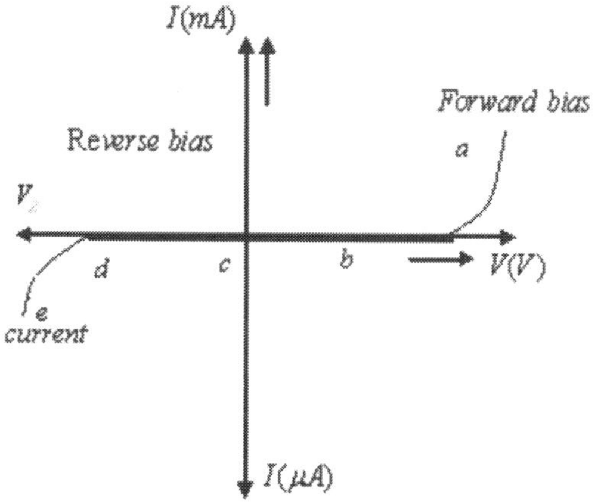

1)ab 2)bc 3)cd 4)de

94. **Assertion (A) : Si and GaAs are preferred materials for solar cells**
Reason (R) : Energy gap of Si is 1.1eV and that of GaAs is 1.53eV which give maximum irradiance where as other materials like CdS or CdSe $\left(E_g = 2.4eV\right)$ and PbS $\left(E_g = 0.4eV\right)$ give minimum irradiance.
1) A, R are true and R explains A correctly 2) A, R are true and R do not explain A correctly
3) A is true, but R is false 4) R is true, but A is false

95. **Assertion (A) : Putting p-type semiconductor slabs directly in physical contact can not form p-n junction**
Reason (R) : The roughness at contact will be much more than inter atomic crystal spacing $\left(=2\overset{0}{A}\,to\,3\overset{0}{A}\right)$ and continuous flow of charge carriers is not possible
1) A, R are true and R explains A correctly 2) A, R are true and R do not explain A correctly
3) A is true, but R is false 4) R is true, but A is false

96. **Assertion (A) : Eventhough forward bias is known to be more (mA) than the current in the reverse bias (μA), photodiodes are still used in reverse bias.**
Reason (R) : The fractional change in majority charge carriers would be much less than that in minority charge carriers due to photo-effects so that minority carrier reverse bias current is more easily to measure with photo diodes than majority carrier forward current.
1) A, R are true and R explains A correctly 2) A, R are true and R do not explain A correctly
3) A is true, but R is false 4) R is true, but A is false

97. **Match List-I with List-2**

LIST-1	**LIST-2**
A) Intrinsic	**E) Donor level semiconductor**
B) N-type	**F) Acceptor level semiconductor**
C) p-type	**G) Immobile ions semiconductor**
D) Depletion layer centre of forbidden	**H) Fermi energy levels is at the energy gap**
	I) Fermi energy level is near the conduction band

	A	B	C	D		A	B	C	D
1)	F	E,G	H	I	2)	H	E,I	F	G
3)	E	F	G	H,I	4)	I	H	G	E,F

98. Match the following

LIST - I	**LIST - II**
a. Antimony	**e. P - type impurity**
b. Indium	**f. N - type impurity**
c. Carbon	**g. not semiconductor**
d. Silicon	**h. semiconductor**

1)a-f, b-e, c-g, d-h　2)a- e, b- g, c- h, d- f　3) a-g, b-h, c-f, d-e　4) a-h, b-f, c-e, d- g

TRANSISTORS

99. The correct relation between current gains α and β is

1) $\beta=\frac{\alpha}{1-\alpha}$　2) $\beta=\frac{\alpha}{1+\alpha}$　3) $\beta=\alpha(1-\alpha)$　4) $\beta=\frac{1-\alpha}{\alpha}$

100. Transistors are made of

1) insulators　2) conductors

3) alloys　4) doped semi-condcutors

101. In n-p-n transistor the arrow head on emitter represents that the conventional current flows from

1) base to emitter　2) emitter to base　3) emitter to collector　4) base to collector

102. In a junction transistor the emitter, base and collector are made of

1) extrinsic semi conductors　2) intrinsic semi conductors

3) both 1 and 2　4) metal

103. In a transistor

1) both emitter and the collector are equally doped

2) base is more heavily doped than collector

3) collector is more heavily doped than the emitter

4) the base is made very thin and is lightly doped

104. In a transistor

1) length of emitter is greater than that of collector　2) length of collector is greater than that of emitter

3) both emitter and collector have same length

4) any one of emitter and collector can have greater length

105. In transistor the emitter current is

1) slightly more than the collector current　2) slightly less than the collector current

3) equal to the collector current　4) equal to the base current

106. In the use of transistor as an amplifier

1) the emitter - base junction is reverse biased and the collector base junction is also reverse biased

2) the emitter - base junction is forward biased and the collector - base junction is reverse biased

3) both the junctions are forward biased

4) any of the two junctions may be forward biased.

107. One way in which the operation of an npn transistor differ from that of a pnp transistor is that

1) the emitter junction is reverse biased in npn

2) the emitter junction injects minority carriers into the base region of the pnp.

3) the emitter injects holes into the base of the pnp and electrons into the base region of npn

4) the emitter injects holes into the base of npn

108. npn transistors are preferred to pnp transistors because they have

1) low cost

2) low dissipation energy

3) capable of handling large power

4) electrons have high mobility than holes and hence high mobility of energy

109. A CE transistor amplifies a weak current signal because collector current is

1) β times I_b 2) β times I_C 3) α times I_b 4) α times I_C

110. When a positive voltage signal is applied to the base of a common emitter npn amplifier

1) The emitter current decreases

2) The collector voltage becomes more positive

3) The collector voltage becomes less positive

4) The collector current decreases

111. In case of common emitter p-n-p transistor input characteristic is a graph drawn

1) With I_C on y-axis and V_{CE} on x-axis keeping I_B constant

2) With I_B on y-axis and V_{BE} on x-axis keeping V_{CE} constant

3) With I_C on y-axis and I_B on x-axis keeping V_{CE} constant

4) With V_{BE} on y-axis and V_{CE} x-axis keeping I_B constant

112. **The output characterstics of an n-p-n transistor represent,[I_C =Collector current, V_{CE} =potential difference between collector and emitter, I_B =Base current, V_{BB} =voltage given base; V_{BE} =the potential difference between base and emitter]**

1)change in I_C *as* I_B and V_{BB} are changed

2)Changes in I_C with changes in $V_{CE}(I_B = \text{cosntant})$

3)changes in I_B with changes in V_{CE}

4)Change in I_C as V_{BE} is changed

113. In a transistor the base is made very thin and is lightly doped with an impurity, because

1) to enable the collector to collect about 95% of the holes or electrons coming from the emitter side

2) to enable the emitter to emit small number of holes or electrons

3) to save the transistors from high current effects

4) to enable the base to collect about 95% of holes or electrons coming from the emitter side

114. A p-n-p transistor is said to be in active region of operation, When

1) Both emitter junction and collector junction are forward biased

2) Both emitter junction and collector junction are reverse biased

3) Emitter junction is forward biased and collector junction is reverse biased

4) Emitter junction is reverse biased and collector junction is forward biased

115. An n-p-n transistor power amplifier in C-E configuration gives

1) Voltage amplification only

2) Current amplification only

3) Both current and voltage amplification

4) Only power gain of unity

116. When n-p-n transistor is used as an amplifier:

1) electrons move from base to collector

2) holes moves from emitter to base

3) holes move from collector to base

4) holes move from base to emitter

117. The part of a trasistor which is heavily doped to produce large number of majority carriers is :

1) emitter 2) base 3) collector
4) can be any of the above three

118. When n-p-n transistor is used as an amplifier :

1) electrons move from collector to base 2) holes move from collector to base
3) electrons move from base to emitter 4) holes move from base to emitter

119. A n-p-n transistor conducts when

1) both collector and emitter are positive with respect to the base
2) collector is positive and emitter is negative with respect to the base
3) collector is positive and emitter is at same potential as the base
4) both collector and emitter are negative with respect to the base

120. In a common -base amplifier, the phase difference between the input signal voltage and output voltage is :

1) $\frac{\pi}{4}$ 2) π 3) zero 4) $\frac{\pi}{2}$

121. In a common emitter amplifier, the phase difference between the input signal voltage and output voltage is

1) $\frac{\pi}{2}$ 2) 0 3) π 4) $\frac{\pi}{4}$

122. When n-P-n transistor is used as an amplifer

1) Electrons move from emitter to collector 2) Holes move from emitter to base
3) Electrons move form collector to base 4) Holes move from base to collector

123. In a PNP transistor the base is the N-region. Its width relative to the P-region is

1) smaller 2) larger 3) same 4) not related

124. A three terminal device with one terminal common to both the output and input is called

1) rectifier 2) transistor 3) diode 4) triode

125. Input and output signal of an amplifier in CE configuration are always

1) Equal 2) Inphase 3) Having a phase difference 4) Out of phase

126. Ge transistor can be operated at a temperature

1) upto 90^0C 2) upto 40^0C 3) upto 100^0C 4) upto 20^0C

127. Transistor amplifier circuit with a feed back circuit is called

1) oscillator 2) detector 3) modulator 4) all

128. A pulsating voltage is a mixture of an a.c componet and a d.c component. the circuit used to seperate a.c and d.c component is called

1) an oscillator 2) an amplifier 3) a rectifier 4) a filter

129. The α and β of a transistor are always

1) $\alpha > 1$, $\beta < 1$ 2) $\alpha < 1$, $\beta > 1$ 3) $\alpha = \beta$ 4) $\alpha\beta = 1$

130. In case of NPN transistor, emitter current is always greater than collector current, because:

1) Collector side is revese biased and emitter side is forward biased

2) Collector being reverse biased attracts more electrons

3) Some electrons are lost in base

4) Collector side is forward biased and emitter side is reverse biased.

131. When a transistor amplifier having current gain of 75 is given an input signal, $V_I = 2\sin(157t + \pi/2)$, the output signal is found to be $V_O = 200\sin(157t + 3\pi/2)$. The transistor is connected as:

1) A common collector amplifier
2) A common base amplifier
3) A common emitter amplifier
4) An oscillator

132. An oscillator is an amplifier with

1) A large gain
2) Negative feedback
3) Positive feedback
4) No feedback

133. In which of the transistor configurations, the voltage gain is highest?

1) Common-base
2) Common-emitter
3) Common-collector
4) Same in all threee

134. A working transistor with its three legs marked P, Q, and R is tested using a multimeter to R and the other (positive) terminal to P or Q. Some resistance is seen on the multimeter. Which of the following is true for the transistor?

1) It is a pnp transistor with R as emitter
2) It is an npn transistor with R as collector
3) It is an npn transistor with R as base
4) It is pnp transistor with R as collector

135. Match the following Column I with Column II

Column I	Column II
i n-p-n transistor	A 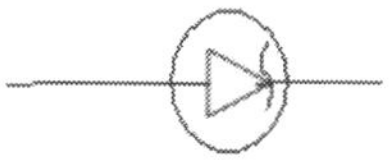
ii p-n-p transistor	B 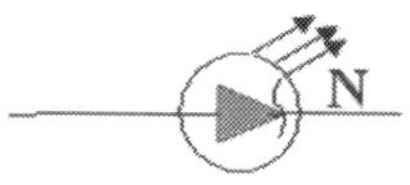
iii Light emmiting diode	C 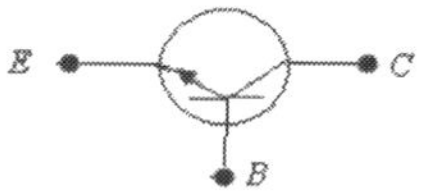
iv Zener diode	D 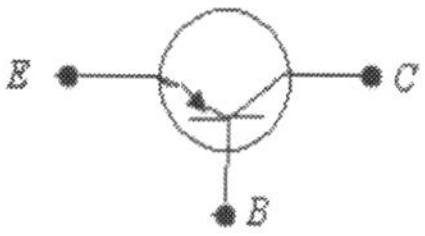

1) I-A,II-B,III-C, IV-D
2) I-D, II-A, III-B, IV-C
3) I-C, II-D, III-B, IV-A
4) I-B, II-A, III-C, IV-D

136. Match List-I with List-2

List-1	**List-2**
A) Emitter	E) Transistor
B) Base	F) Moderately doped
C) Collector	G) lightly doped
D) Transfer of	H) Heavily doped resistance
	I) Largest physical size

	A	B	C	D		A	B	C	D
1)	F	E	H,I	G	2)	G	F,I	E	H
3)	H	G	F,I	E	4)	E	H,I	G	F

137. Match the following for C E transistor amplifier

LIST - I	**LIST - II**
a. Current gain	**e.** $\frac{\Delta I_c}{\Delta I_b}$
b. Voltage gain	**f.** $\frac{\Delta I_c}{\Delta I_b}\frac{R_L}{R_i}$
c. Power gain	**g.** $\left(\frac{\Delta I_c}{\Delta I_b}\right)^2 \frac{R_L}{R_i}$
d. Resistance gain	**h.** $\frac{R_L}{R_i}$

1) a-e, b-f, c-g, d-h
2) a- f, b- g, c- h, d-e
3) a-g, b-h, c-e, d-f
4) a- h, b- e, c- f, d- g

138. Assertion (A) :- Common emiter mode of a transistor is widely used

Reason (R) :-Current gain, voltage gain, and power gain are maximum in C.E mode of a transistors.

1) If both A and R are correct and R is the correct explanation of A
2) If both A and R are correct and R is not the correct explanation of A
3) If 'A' is correct and 'R'is incorrect
4) If 'A' is incorrect and 'R'is correct

139. ***Assertion (A)*** **:- Transistor in C.E. mode can be used as amplifier**

Reason (R) **:- A small change in base current produces a relatively large change in collector current**

1) If both A and R are correct and R is the correct explanation of A
2) If both A and R are correct and R is not the correct explanation of A
3) If 'A' is correct and 'R'is incorrect
4) If 'A' is incorrect and 'R'is correct

LOGIC GATES

140. In the Binary number system the number 100 represents :

1) one
2) three
3) four
4) hundred

141. Among the following is not the function of NOT gate is

1) stop a signal
2) invert an input signal
3) complement a signal
4) change the logic in a digital circuit

142. Digital circuit can be made by repetitive use of

1) OR gates 2) AND gates 3) NOT gates 4) NAND gates

143. Among the following one gives output 1 in the AND gate

1) A = 0, B = 0 2) A = 1, B = 1 3) A = 1, B = 0 4) A = 0, B = 1.

144. Person who use Boolean algebra for describing the operation of logic gates first was

1) Boole 2) Shannon 3) Schottky 4) Zener

145. NAND and NOR gates are called universal gates because they

1) are universally available

2) can be combined to produce OR, AND and NOT gates

3) are widely used in the Integrated circuits 4) can be easily manufactured.

146. In positive logic, the logic state 1 corresponds to

1) positive voltage 2) zero voltage

3) lower voltage level 4) higher voltage level.

147. In Boolean algebra A + B = Y implies that :

1) sum of A and B is Y

2) Y exists when A exists or B exists or both A and B exist

3) Y exists only when A and B both exist

4) Y exists when A or B exist but not when both A and B exist.

148. In the Boolean algebra, the following one is wrong

1) 1+0 = 1 2) 0+1=1 3) 1+1=1 4) 0+0 =1

149. In Boolean algebra A.B = Y implies that:

1) product of A and B is Y 2) Y exists when A exists or B exists

3) Y exists when both A and B exist but not when only A or B exists

4) Y exists when A or B exists but not both A and B exist.

150. In the Boolean algebra, the following one is wrong

1) 1.0 = 0 2) 0.1 = 0 3) 1.1 = 0 4) 1.1 = 1

151. The following truth table is for

A	B	Y
1	1	0
1	0	1
0	1	1
0	**0**	**1**

1) NAND gate 2) AND gate 3) XOR gate 4) NOT gate

152. The output of a 2-input OR gate is zero only when its

1) both inputs are 0 2) either input is 1 3) both inputs are 1 4) either input is zero.

153. Boolean algebra is essentially based on

1) symbols 2) logic 3) truth 4) numbers

154. The value of $\overline{A}+A$ in the Boolean algebra is

1) A 2) $\overline{A}$ 3) 0 4) 1

155. The value of $A.\overline{A}$ in Boolean algebra is

1) 0 2) 1 3) A 4) $\overline{A}$

156. The following is NOT equal to 0 in the Boolean algebra is

1) $\overline{A.0}$ 2) $A.\overline{A}$ 3) A.0 4) $\overline{A+\overline{A}}$

157. An AND gate is followed by a NOT gate in series. With two inputs A & B, the Boolean expression for the out put Y will be :

1) A.B 2) A + B 3) $\overline{A+B}$ 4) $\overline{A.B}$

158. NOR gate is the series combination of

1) NOT gate followed by OR gate
2) OR gate followed by NOT gate
3) AND gate followed by OR gate
4) OR gate followed by AND gate

159. The gate that has only one input terminal

1) NOT 2) NOR 3) NAND 4) XOR

160. AND gate:

1) It has no equivalence to switching circuit.
2) It is equivalent to series switching circuit.
3) It is equivalent to parallel switching circuit
4) It is a mixture of series and parallel switching circuit

161. The gate that can act as a building block for the digital circuits is

1) OR 2) NOT 3) AND 4) NAND

162. Statement -1: NAND or NOR gates are called digital building blocks.

Statement-2: The repeated use of NAND (or NOR) gates can produce all the basis or complicated gates.

1) Statement -1 is false, statement -2 is true
2) statement 1-is true statement -2 is true statement -2 is correct explanation of statement-1.
3) statement 1-is true statement -2 is true statement -2 is not correct explanation of statement-1.
4) statement 1-is true statement -2 is false

163. Statement-1: NOT gate is also called invertor circuit

Statement-2: NOT gate inverts the input order.

1) Statement -1 is false, statement -2 is true
2) statement 1-is true statement -2 is true statement -2 is correct explanation of statement-1.
3) statement 1-is true statement -2 is true statement -2 is not correct explanation of statement-1.
4) statement 1-is true statement -2 is false

164. Match the quantities of column I with the column II when A=1 and B=0

Column - I	**Column - II**
a) $(A.B)+\overline{A}$	(p) A
b) $(A.\overline{B})+A$	(q) B
c) $(\overline{A}+B).A$	(r) A + B
d) $(A+\overline{B}).B$	(s) A.B

1) a-q,s b-p,r, c-q,s, d-q,s
2) a-p,s, b-p,s, c-q,r, d-r,q
3) a-q,r, b-p,s, c-q,r, d-s,p
4) a-r,s, b-p,r, c-q,s, d-p,s

KEY - CONCEPTUAL QUESTIONS

01)1	**02)3**	**03)3**	**04)2**	**05)2**	06)2	07)2	08)1	09)3	10)3
11)4	12)1	13)4	14)4	15)1	16)1	17)1	18)2	19)4	20)2
21)3	22)3	23)1	24)3	25)3	26)3	27)1	28)1	29)3	30)3
31)3	32)3	33)4	34)1	35)4	36)3	37)1	38)1	39)1	40)2
41)1	42)2	43)2	44)3	45)1	46)2	47)3	48)1	49)1	50)4
51)3	52)3	53)3	54)2	55)4	56)2	57)2	58)4	59)1	60)4
61)1	62)1	63)1	64)4	65)1	66)1	67)3	68)3	69)2	70)4
71)1	72)2	73)2	74)3	75)3	76)3	77)1	78)1	79)2	80)3
81)3	82)3	83)2	84)2	85)1	86)1	87)3	88)1	89)2	90)2
91)3	92)1	93)4	94)1	95)1	96)1	97)2	98)1	99)1	100)4
101)1	102)1	103)4	104)2	105)1	106)2	107)3	108)4	109)1	110)3
111)2	112)2	113)1	114)3	115)3	116)1	117)1	118)4	119)2	120)3
121)3	122)1	123)1	124)2	125)4	126)1	127)1	128)4	129)2	130)3
131)3	132)3	133)2	134)3	135)3	136)3	137)1	138)1	139)1	140)3
141)1	142)4	143)2	144)2	145)2	146)4	147)2	148)4	149)3	150)3
151)1	152)1	153)2	154)4	155)1	156)1	157)4	158)2	159)1	160)2
161)4	162)2	163)2	164)1						

LEVEL - I -1

INTRINSIC, EXTRINSIC SEMICONDUCTORS AND DIODES

1. **The electrical conductivity of a semiconductor increases when electomagnetic radiation of wavelength shorter than 2480nm is incident on it. The band gap in (eV) for the semiconductor is**

   1) $0.7eV$ 2) $0.5eV$ 3) $2.5eV$ 4) $1.2eV$

2. **Pure Si at 300 K has equal electron (n_i) concentrations of $1.5\times10^{16}m^{-3}$. Doping by indium increases n_h $4.5\times10^{22}m^{-3}$. n_e in the doped Si is**

   1) 5×10^9 2) 7×10^9 3) 9×10^9 4) 8×10^9

3. **In a p-n junction the depletion region is 400nm wide and electric field of $5\times10^5Vm^{-1}$ exists in it. The minimum energy of a conduction electron, which can diffuse from n-side to the p-side is**

   1) 4eV 2) 5eV 3) 0.4eV 4) 0.2eV

4. **The reverse bias in a junction diode is changed from 5V to 15V then the value of current changes from 38 μ A to 88 μ A. The resistance of junction diode will be**

   1) $4 \times 10^5\Omega$ 2) $3 \times 10^5\Omega$ 3) $2 \times 10^5\Omega$ 4) $10^6\Omega$

5. **A diode made of silicon has a barrier potential of 0.7V and a currrent of 20mA passes through the diode when a battery of emf 3V and a resistor is connected to it. The wattage of the resistor and diode are**

   1) 0.046W, 0.014W 2) 4.6W, 0.14W 3) 0.46W, 0.14W 4) 46W, 14W

6. In a half wave rectifier output is taken across a 90 ohm load resistor. If the resistance of diode in forward biased condition is 10ohm, the efficiency of rectification of ac power into dc power is

1) 40.6% 2) 81.2% 3) 73.08 % 4) 36.54%

7. In a full wave rectifier output is taken across a load resistor of 800 ohm. If the resistance of diode in forward biased condition is 200 ohm, the efficiency of rectification of ac power into dc power is

1) 64.96% 2) 40.6% 3) 81.2% 4) 80%

TRANSISTORS

8. In a P-N-P transistor, the collector current is 10 mA. If 90% of the holes reach the collector, then emitter current will be:

1) 13 mA 2) 12 mA 3) 11 mA 4) 10 mA

9. A transistor has a base current of 1mA and emitter current 100mA. The current transfer ratio will be

1) 0.9 2) 0.99 3) 1.1 4) 10.1

10. When the base -emitter voltage of a transistor connected in the common -emitter mode is changed by 20 mV the collector current is changed by 25 mA. Find the transconductance.

1) $1.25\Omega^{-1}$ 2) $2.50\Omega^{-1}$ 3) $0.5\Omega^{-1}$ 4) $5.5\Omega^{-1}$

11. In a transistor circuit the base current changes from 30 μA to 90 μA. If the current gain of the transistor is 30, the change in the collector current is

1) 4 mA 2) 2 mA 3) 3.6 mA 4) 1.8 mA

12. The current gain of transistor in a common emitter circuit is 40. The ratio of emitter current to base current is

1) 40 2) 41 3) 42 4) 43

13. In a common base configuration the emitter current changes by 5mA when emitter voltage is changed by 200 mV at a fixed collector to base voltage. The input resistance is

1) $40\,\Omega$ 2) $1000\,\Omega$ 3) $2.5\,\Omega$ 4) $4\,\Omega$

14. For a common base amplifier, the values of resistance gain and voltage gain are 3000 and 2800 respectively. The current gain will be

1) 0.93 2) 0.83 3) 0.73 4) 0.63

15. In a transistor amplifier β = 62, $R_L = 5000\,\Omega$ and internal resistance of the transistor is $500\,\Omega$. Its power amplification will be

1) 25580 2) 33760 3) 38440 4)55280

LOGIC GATES

16. Decimal number 15 is equivalent to the binary number:

1) 110001 2) 000101 3) 101101 4) 001111

17. Binary number 1001001 is equivalent to the decimal number:

1) 37 2) 73 3) 41 4) 32

18. In the Binary number system 1+1=

1)2 2)1 3)10 4)100

19. If A = B = 1, then in terms of Boolean algebra the value of A.B + A is not equal to

1) B.A+B 2) B+A 3) B 4) $\overline{A}$.B

20. In the Boolean algebra, the following one which is not equal to A is

1) A.A 2) A + A 3) $\overline{A}.A$ 4) $\overline{\overline{A}+\overline{A}}$

21. The logic expression which is NOT true in Boolean algebra is

1) $\left[\bar{1}+\bar{1}\right].1=0$ 2) $\left[\bar{1}+0\right].1=0$ 3) $\left[\bar{1}+0\right].\bar{1}=0$ 4) $\left[1+1\right].1=0$

KEY LEVEL - I -1

1)2	2)1	3)4	4)3	5)1	6)4	7)1	8)3	9)2	10)1
11)4	12)2	13)1	14)1	15)3	16)4	17)2	18)3	19)4	20)3
21)4									

HINTS LEVEL - I -1

1. $$E_g=\frac{hc}{\lambda}=\frac{6.6\times10^{-34}\times3\times10^{8}}{2480\times10^{-9}\times\left(1.6\times10^{-19}\right)}=0.5eV$$

2. We know that ofr a doped semiconductor in thermal equilibrium, we have

$n_e n_h=n_i^2$

As per given data, $n_i=1.5\times10^{16}m^{-3}$

$n_h=4.5\times10^{22}m^{-3}$

Thus $n_e=\frac{n_i^2}{n_h}=\frac{\left(1.5\times10^{16}\right)^2 m^{-6}}{4.5\times10^{22}m^{-3}}=5.0\times10^{9}m^{-3}$

3. $W=Vq$ =Edq

$=5\times10^{5}\times400\times10^{-9}\times1.6\times10^{-19}J$

$=\frac{5\times10^{5}\times400\times10^{-9}\times1.6\times10^{-19}J}{1.6\times10^{-19}}eV$

$= 2000 \times 10^{-4}$ eV = 0.2 eV

4. $$R=\frac{\Delta V}{\Delta I}=\frac{15-5}{\left(88-38\right)\times10^{-6}}=\frac{10}{50\times10^{-6}}=\frac{10^{6}}{5}=\frac{10}{5}\times10^{5}=2\times10^{5}\Omega$$

5. P = VI (for resistor)

$=\left(3-0.7\right)20\times10^{-3}=2.3\times20\times10^{-3}=46\times10^{-3}=0.046W$

for diode, $P=0.7\times20\times10^{-3}=0.014W$

6. $$\eta=\frac{0.406R_L}{r_f+R_L}$$

7. $$\eta=\frac{0.812R_L}{r_f+R_L}$$

8. $I_C = \frac{90}{100} I_E \Rightarrow I_E = \frac{10}{9} I_C$

9. $\alpha = \frac{I_c}{I_e}, I_e = I_b + I_c$

10. $g_m = \frac{\Delta I_c}{\Delta V_{BE}} = \frac{25\times10^{-3}}{20\times10^{-3}} = 1.25\Omega^{-1}$

11. $\Delta I_B = \Delta I_E - \Delta I_C$ and $\beta = \frac{\Delta I_C}{\Delta I_B}$

12. $\beta = \frac{I_C}{I_B}$ and $\frac{I_E}{I_B} = \frac{I_C + I_B}{I_B} = \frac{I_C}{I_B} + 1 = \beta + 1$

13. $V_{BE} = I_e R_i$

14. $A_v = A_R \beta$

15. $A_P = \beta^2 \frac{R_L}{R_i}$

16.
$$\begin{array}{r|l} 2 & 15 \\ \hline 2 & 7-1 \\ \hline 2 & 3-1 \\ \hline 2 & 1-1 \\ \hline & 0-1 \end{array}$$

$\therefore 15_{(10)} = 1111_{(2)}$

17. $\overset{6}{1}\overset{5}{0}\overset{4}{0}\overset{3}{1}\overset{2}{0}\overset{1}{0}\overset{0}{1}_{(2)} = 1\times2^6 + 0\times2^5 + 0\times2^4 + 1\times2^3$

$+ 0\times2^2 + 0\times2^1 + 1\times2^0 = 73$

18. 1+1=10 in binary number system

19. If A=B=1; then A.B+A=1.1+1=1

then $\bar{A}.B = \bar{1}.1 = 0.1 = 0$

20. $\bar{A} \bullet A = 0$

21. logic, OR and AND operation

LEVEL - I - 2

INTRINSIC, EXTRINSIC SEMICONDUCTORS AND DIODES

1. **The electrical conductivity of a semi conductor increases when electromagnetic radiation of wavelength shorter than 1240 nm is incident on it. The forbidden band energy for the semi conductor is (in eV)**

1) 0.5 2) 0.97 3) 0.7 4) 1.1

2. **A semiconductor is known to have an electron concentration of $5\times10^{13}/cm^3$ and hole concentration of $8\times10^{12}/cm^3$. The semiconductor is**

1) n-type 2) p-type 3) intrinsic 4) insulator

3. **A potential barrier of 0.5V exists across a p-n junction. If the width of depletion layer is 10^{-6}m, then intensity of electric field in this region will be**

1) 1×10^6 V/m 2) 5×10^5 V/m 3) 4×10^4 V/m 4) 2×10^6 V/m

4. **A p - n junction diode has breakdown voltage of 28V. If applied external voltage in reverse bias is 40V the current through it is**

1) Zero 2) infinite 3) 10 A 4) 15A

5. **The value of current in the following diagrams is (diode assumed to be ideal one)**

1) 0.1 amp 2) 0.01amp 3) 1 amp 4) zero

6. **A half -wave rectifier is used to convert 50Hz A.C. to D.C. voltage. The number of pulses per second in the rectified voltage are**

1) 50 2) 25 3) 100 4) 75

7. **If a full wave rectifier circuit is operating from 50Hz mains, the fundamental frequency in the ripple will be**

1) 25 Hz 2) 50 Hz 3) 70.7 Hz 4) 100 Hz

TRANSISTORS

8. **In an npn transistor the base and collector currents are 100 μ A and 9mA respectively. Then the emitter current will be**

1) 9.1mA 2) 18.2mA 3) 3.91 μ A 4) 18.2 μ A

9. **A change of 8mA in the emitter current brings a change of 7.9mA in the collector current. The change in base current required to have the same change in the collector current is**

1) 0.01mA 2) 1A 3) 10mA 4) 0.1mA

10. For a p-n-p transistor in CB configuration, the emitter current I_E **is 1mA and α = 0.95. The base current and collector current are**

1) 0.95 mA, 0.05mA 2) 0.05mA, 0.95mA 3) 9.5mA, 0.5mA 4) 0.5mA, 9.5mA

11. **If a change of 100 μ A in the base current of an n-p-n transistor in CE causes a change of 10mA in the collector current, the ac current gain of the transistor is**

1) 10 2) 100 3) 1000 4) 10000

12. **For a common emitter amplifier, current gain is 70. If the emitter current is 8.4mA, then the base current is**

1) 0.236mA 2) 0.118mA 3) 0.59mA 4) 8.3mA

13. **The base current of a transistor is 105 μ A and the collector current is 2.05mA. Then β of the transistor is**

1) 1.952 2) 19.52 3) 195.2 4) 1952

14. **For a transistor the value of α is 0.9. Its β value is**

1) 9 2) 0.9 3) 0.09 4) 90

15. **For a transistor the current amplification factor is 0.8. The transistor is connected in common emitter configuration, the change in collector current when the base current changes by 6mA is**

1) 6mA 2) 4.8mA 3) 24mA 4) 8mA

16. **A change of 400mV in base-emitter voltage causes a change of 200μ A in the base current. The input resistance of the transistor is**

1) $1K\Omega$ 2) $6K\Omega$ 3) $2K\Omega$ 4) $8K\Omega$

17. **In a common base circuit, if the collector base voltage is changed by 0.6V, collector current changes by 0.02mA. The output resistance will be**

1) $10^4\Omega$ 2) $2 \times 10^4\Omega$ 3) $3 \times 10^4\Omega$ 4) $4 \times 10^4\Omega$

18. **A common emitter transistor amplifier has a current gain of 50. If the load resistance is 4k Ω, and input resistance is 500 Ω, the voltage gain of amplifier is**

1) 100 2) 200 3) 300 4) 400

LOGIC GATES

19. **Equivalent of decimal number 8 in the binary number is**

1)10 2)101 3)1000 4)1011

20. **The equivalent of 110in the decimal number is**

1)2 2)4 3)8 4)6

21. **If A = 1, B = 0 then the value of $\overline{A}+B$ in terms of Boolean algebra is**

1) A 2) B 3) B + A 4) $A.\overline{B}$

22. **In the Boolean algebra : A + B =**

1) $\overline{A}+\overline{B}$ 2) A.B 3) $\overline{\overline{A}}+\overline{\overline{B}}$ 4) $\overline{\overline{A}}+\overline{B}$

23. **The following one represents logic addition is**

1) 1 + 1 = 2 2) 1 + 1 = 10 3) 1 + 1 = 1 4) 1 + 1 = 11

24. **In the Boolean algebra : $\overline{A}.\overline{B}$ = ...**

1) $\overline{A+B}$ 2) A.B 3) $\overline{\overline{A}+\overline{B}}$ 4) A + B

25. **In the Boolean algebra, which gate is expressed as $Y = \overline{A+B}$**

1) OR 2) NAND 3) AND 4) NOR

26. **The truth table for NOT gate is**

1) $\begin{bmatrix} 1 & 1 \\ 0 & 0 \end{bmatrix}$ 2) $\begin{bmatrix} 1 & 0 \\ 0 & 0 \end{bmatrix}$ 3) $\begin{bmatrix} 1 & 0 \\ 0 & 1 \end{bmatrix}$ 4) $\begin{bmatrix} 0 & 1 \\ 1 & 1 \end{bmatrix}$

KEY LEVEL - I - 2

1) 2	2)1	3)2	4)2	5)4	6)1	7)4	8)1	9)4	10)2
11)2	12)2	13)2	14)1	15)3	16)3	17)3	18)4	19)3	20)4
21)2	22)3	23)3	24)1	25)4	26)3				

HINTS LEVEL - I -2

1. $E = \frac{hc}{\lambda} = \frac{12000}{12400 A^0} (e_v)$

2. $n_e > n_h$

3. $E = \frac{V}{d}$

$\frac{10^{24}}{10^{18}} = 10^6$

4. Since $V_a > V_b$

5. diode does not conduct in reverse bias as $V_p < V_n$

6. n Hz

7. 2n Hz

8. $I_e = I_b + I_c$

9. $\Delta I_B = \Delta I_E - \Delta I_C$

10. $\alpha = \frac{I_c}{I_e}, I_e = I_b + I_c$

11. $\beta = \frac{\Delta I_c}{\Delta I_b}$

12. $\beta = \frac{I_c}{I_b}$ and $I_e = I_b + I_c$

13. $\beta = \frac{I_c}{I_b}$

14. $\beta = \frac{\alpha}{1-\alpha}$

15. $\beta = \frac{\alpha}{1-\alpha}$ and $\beta = \frac{\Delta I_c}{\Delta I_b}$

16. $\Delta V_{BE} = \Delta I_e \times R_i$

17. $V_{CB} = I_C \times R_{out}$

18. $A_V = \beta \frac{R_L}{R_i}$

19.
$$\begin{array}{r|l} 2 & 8 \\ \hline 2 & 4-0 \\ \hline 2 & 2-0 \\ \hline 2 & 1-0 \\ \hline & 0-1 \end{array} \quad \therefore 8_{(10)} = 1000_{(2)}$$

20. $\overset{2\,1\,0}{110}_{(2)} = 1\times 2^2 + 1\times 2^1 + 0\times 2^0 = 6$

21. If A=1, B=0 then $\overline{A}+B = \overline{1}+0 = 0+0 = 0 = B$

22. $\overline{\overline{A}} = A$ and $\overline{\overline{B}} = B$

23. acording to Boolean algebra, 1+1=1

24. According to Demorgon's theorem $\overline{A}\bullet\overline{B} = \overline{A+B}$

25. $Y = \overline{A+B}$ is the output of NOR gate

26. conceptual

<u>LEVEL - II -1</u>

INTRINSIC,EXTRINSIC SEMI CONDUCTORS AND DIODES

1. Mobilities of electrons and holes in a sample of intrinsic germanium at room temperature are $0.36m^2/Vs$ and $0.17m^2/Vs$. The electron and hole densities are each equal to 2.5 $\times 10^{19}m^{-3}$. The electrical conductivity of germanium is

1) $0.47\,S/m$ 2) $5.18\,S/m$ 3) $2.12\,S/m$ 4) $1.09\,S/m$

2. In a p-n junction diode the thickness of deplection layer is $2\times10^{-6}m$ and barrier potential is 0.3V. The intensity of the electrical field at the junction is

1) $0.6\times10^{-6}Vm^{-1}$ from n to p side 2) $0.6\times10^{-6}Vm^{-1}$ from p to n side

3) $1.5\times10^{5}Vm^{-1}$ from n to p side 4) $1.5\times10^{5}Vm^{-1}$ from p to n side

3. A potential barrier V volts exists across a P-N junction. The thickness of the depletion region is 'd'. An electron with velocity 'v' approches P-N junction from N-side. The velocity of the electron acrossing the junction is

1) $\sqrt{v^2+\frac{2Ve}{m}}$ 2) $\sqrt{v^2-\frac{2Ve}{m}}$ 3) v 4) $\sqrt{\frac{2Ve}{m}}$

4. In the following , reverse biased diode is

1) 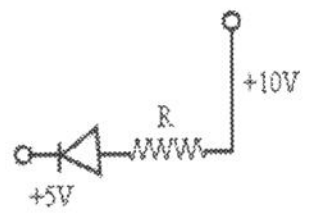

2) 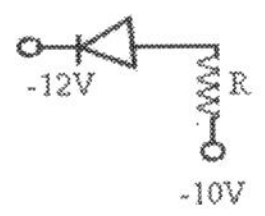

3) 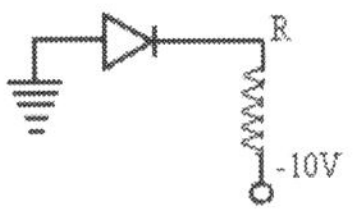

4) 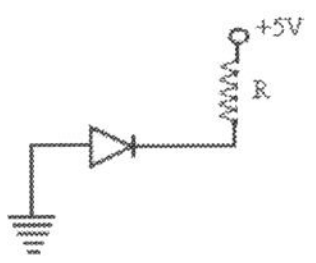

5. Two similar p-n junctions can be connected in three different ways as shown in the figures. The two connections across which the potential difference is same are

a) 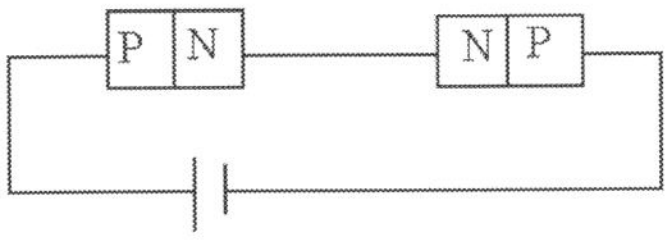

b) 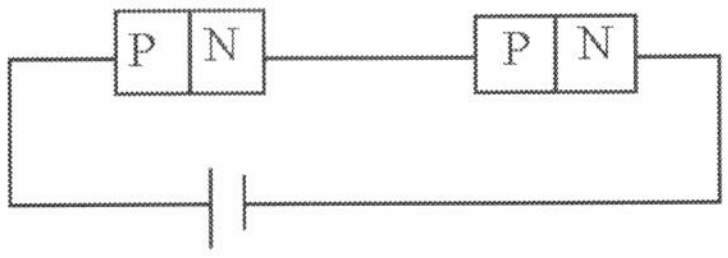

c) 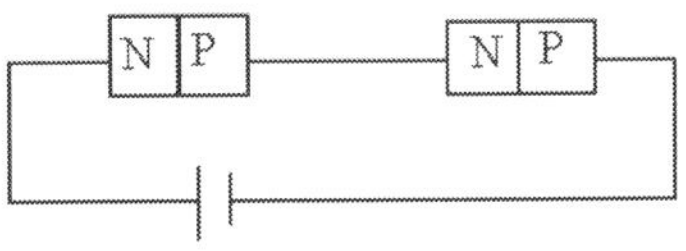

1) circuits a and b 2) circuits b and c 3) circuits a and c 4) all the circuits

6. **If $V_1 > V_2$, r is resistance offered by diode in forward bias then current through the diode is**

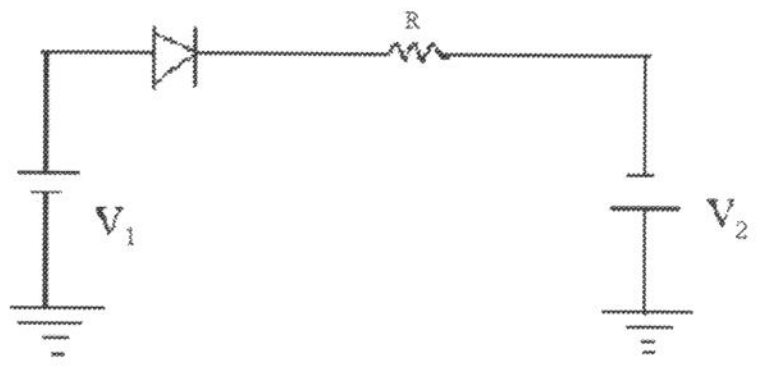

1) 0

2) $\frac{V_1 + V_2}{R + r}$

3) $\frac{V_1 - V_2}{R + r}$

4) None

7. **A PN junction diode when forward biased has a drop of 0.5V which is assumed to be independent of current. The current in excess of 10mA through the diode produces large joule heating which damages the diode. If we want to use a 1.5V battery to forward bias the diode, the resistor used in series with the diode so that the maximum curent does not exceed 5mA is**

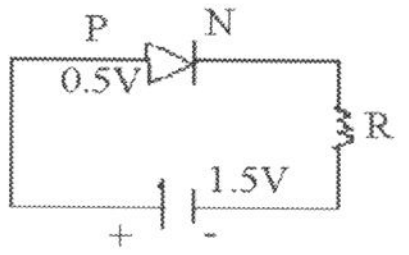

1) $2 \times 10^2 \Omega$

2) $2 \times 10^5 \Omega$

3) $2 \times 10^3 \Omega$

4) $2 \times 10^4 \Omega$

8. **The circuit shown in figure (1) Contains two diodes each with a forward resistance of 50 ohm and with infinite reverse resistance. If the battery voltage is 6 V, the current through the 100 ohm resistance is.**

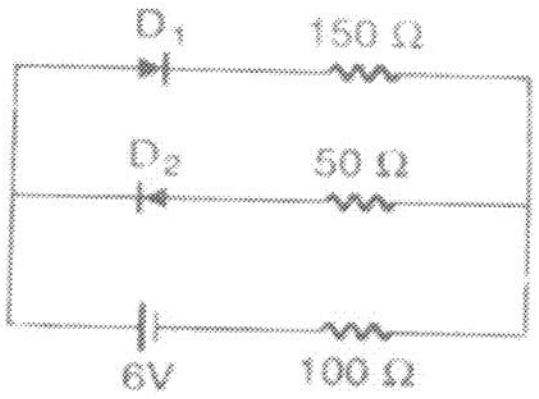

1) 0.01A

2) 0.02A

3) 0.03A

4) 0.04A

9. **4 ideal diodes are connected as shown in the circuit the current through 50Ω is**

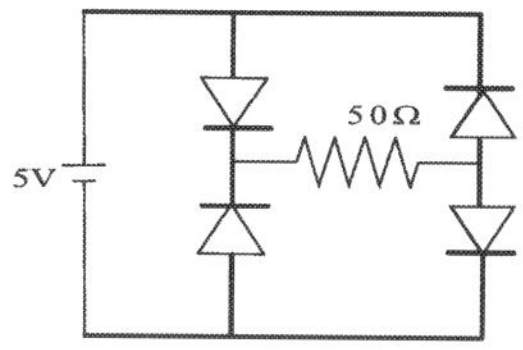

1) 0.1 A

2) 0.5 A

3) 0.6 A

4) 1 A

10.. Find the effective resistance between A & B

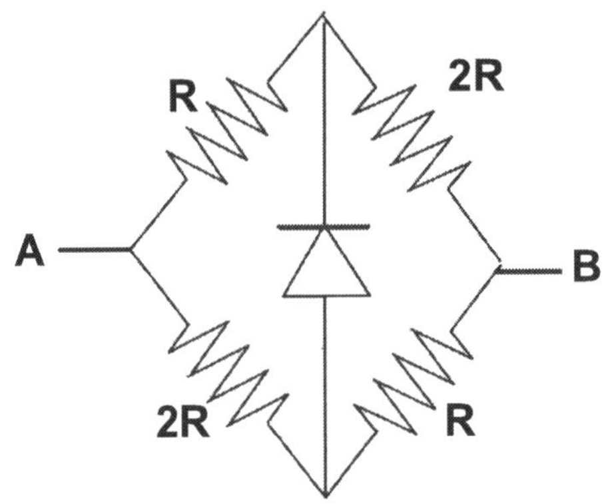

1) $\frac{2}{3R}$ 2) $\frac{3R}{2}$ 3) $\frac{2R}{3}$ 4) R

11 The equivalent resistance of the circuit across AB is given by

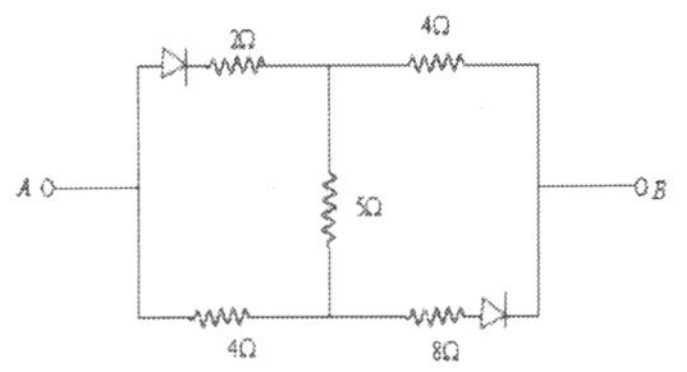

1) 4Ω 2) 13Ω 3) 4Ω *or* 13Ω 4) 4Ω zero

12. The equivalent resistance between A and B is

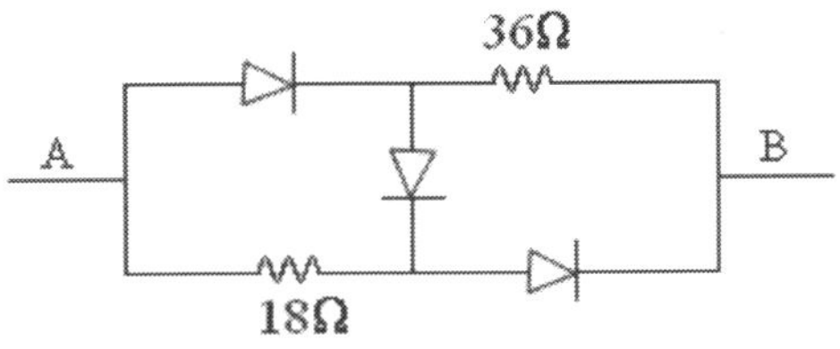

1) $36\Omega\, if\, V_A > V_B$ 2) $18\Omega\ if\ V_A < V_B$

3) Zero if $V_A < V_B$ and $54\Omega\, if\, V_A > V_B$ 4) Zero if $V_A > V_B$ and $54\Omega\, if\, V_A < V_B$

13. A junction diode is connected to a 10 V source and $10^3\,\Omega$ rheostat. The slope of load line on the characteristic curve of diode by (in A/V)

1) 10^{-1} 2) 10^{-2} 3) 10^{-3} 4) 10^{-4}

14. A p–n junction (D) shown in the figure can act as a rectifier. An alternating curent source (V) is connected in the circuit

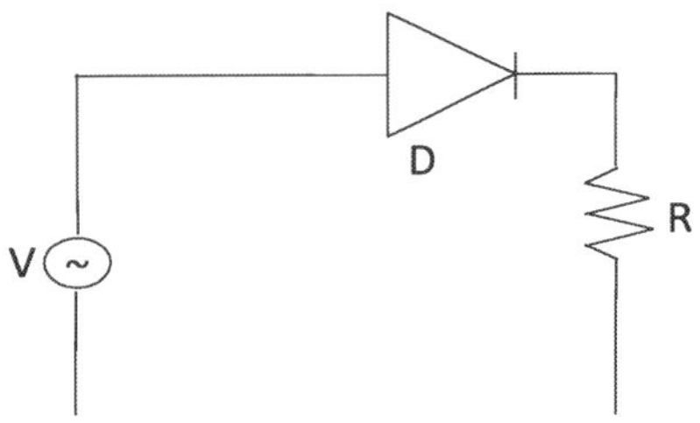

1) 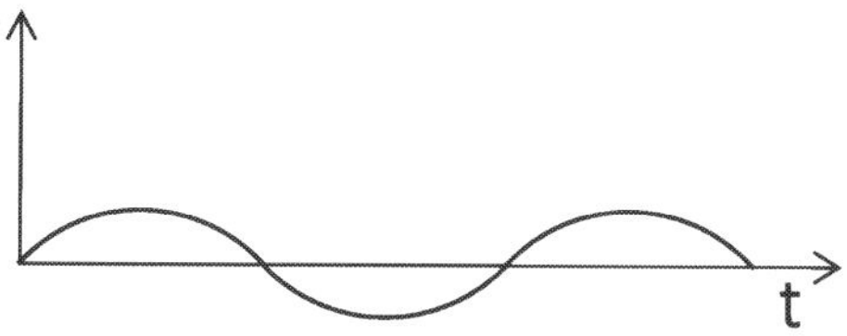

2) 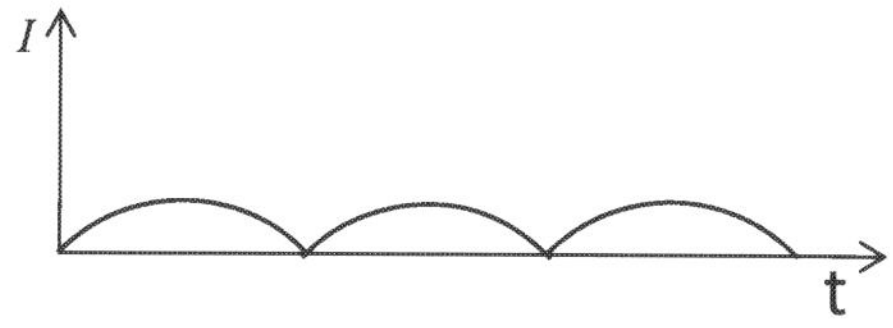

3) 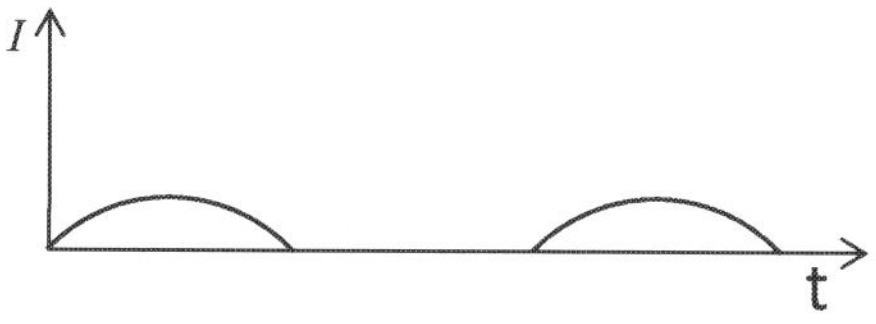

4) 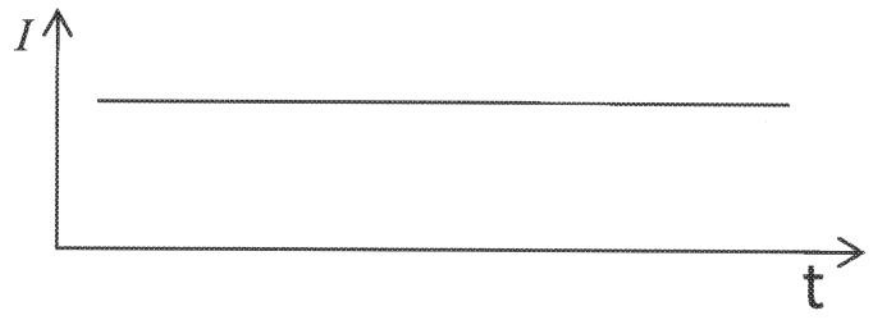

15. In the figure, the input is across terminals A and C and the output is across B and D. Then the output is

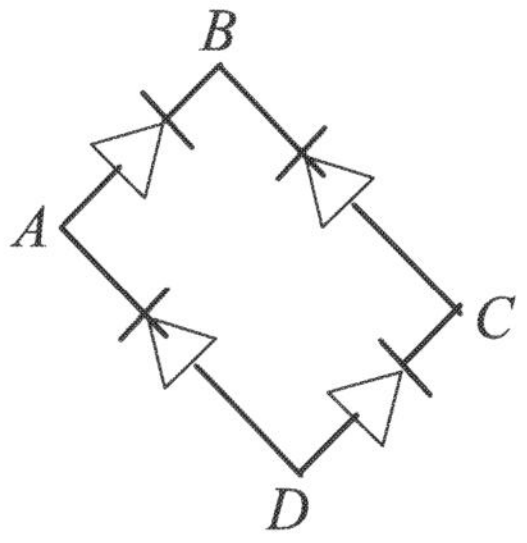

1) Zero
2) Same as the input
3) Full wave rectified
4) Half-wave rectified

16. The I-V characteristic of an LED is

1) 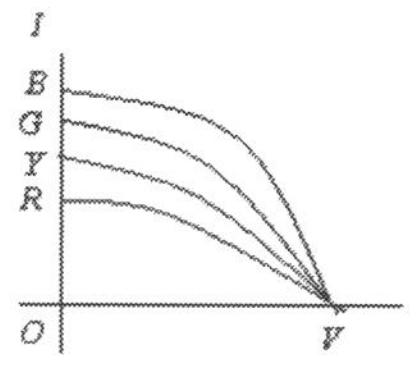

2) 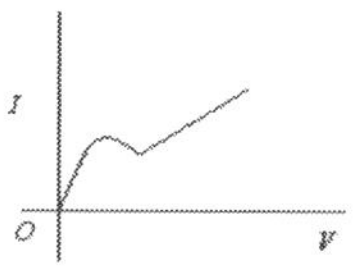

3) 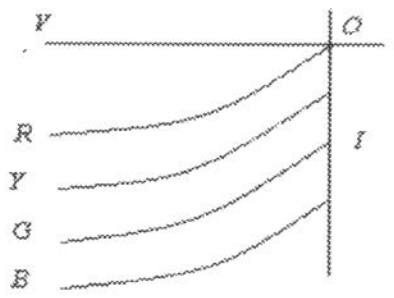

4) 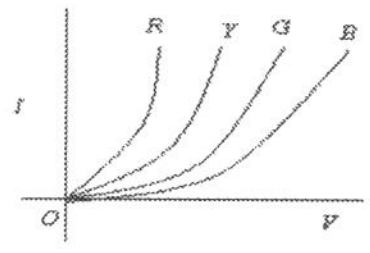

17. In the figure shown, the potential drop across each capacitor is (assume diodes to be ideal)

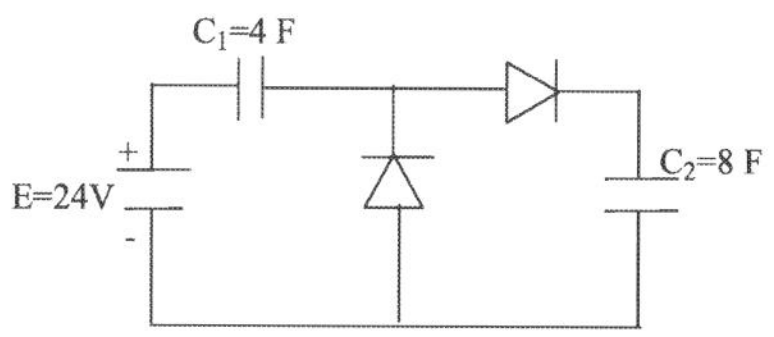

1) 12V, 12V
2) 16V, 8V
3) zero, 24V
4) 8V, zero

18. **From the circuit shown below, the maximum and minimum values of zener diode current are**

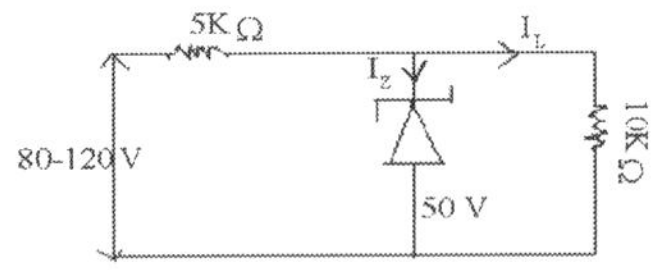

1) 6mA, 5mA 2) 14mA, 5mA 3) 9 mA, 1mA 4) 3mA, 2mA

TRANSISTORS

19. **In an n-p-n transistor circuit, the collector current is 10mA. If 90% of the electrons emitted reach the collector.**

1) the emitter current , the will be 9mA 2) the base current will be 1 mA
3) the emitter current will be 11mA 4) both 2 & 3

20. **The constant α of a transistor is 0.9. What would be the change in the collector current corresponding to a change of 4 mA in the base current in a common emitter arrangement ?**

1) 30mA 2) 63mA 3)36 mA. 4) 3.6 mA

21. **A voltage amplifier operated from a 12 volt battery has a collector load $6k\Omega$. Calculate the maximum collector current in the circuit.**

1) 0.5mA 2) 1 mA 3) 3 mA 4) 2mA

22. **In a transistor circuit shown here the base curent is $35\mu A$ The value of the resistor R_b is**

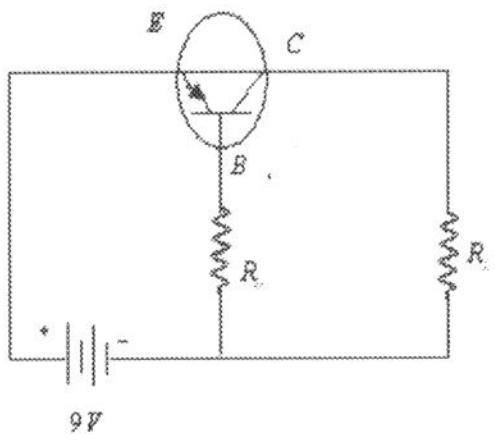

1) $124k\Omega$ 2) $257k\Omega$ 3) $352k\Omega$ 4) None of these

23. **In a single state transistor amplifier, when the signal changes by 0.02 V, the base current change by 10μA and collector current by 1mA. If collector load $R_C = 2k\Omega$ and $R_L = 10k\Omega$, Calculate Current Gain, Input impedance, Effective a.c.load, Voltage gain and Power gain.**

1)50, 2kΩ,1.66kΩ, 83, 8300 2) 100, 1kΩ,1.66kΩ, 83, 8300
3) 100, 2kΩ,1.66kΩ, 83, 830 4) 100, 2kΩ,1.66kΩ, 83, 8300

LOGIC GATES

24. **On subtracting 010101 from 101010, we get:**

1) 001011 2) 001100 3) 010101 4) 011111

25. **The minimum number of gates required to realise this expression** $Z = DABC + D\overline{ABC}$ **is**

1) One 2) Two 3) Eight 4) Five

26. **In the circuit below, A and B represent two inputs and C represents the output, the circuit reperesents**

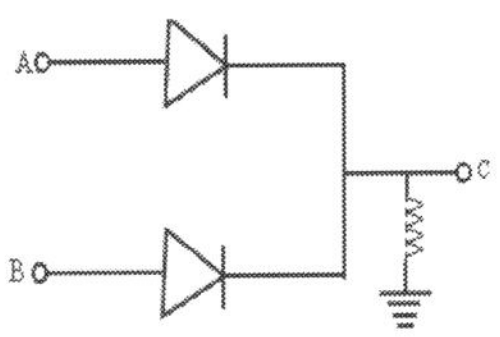

1) NOR gate 2) AND gate 3) gate 4) OR gate

27. **In the following circuit the output Y becomes zero for the inputs**

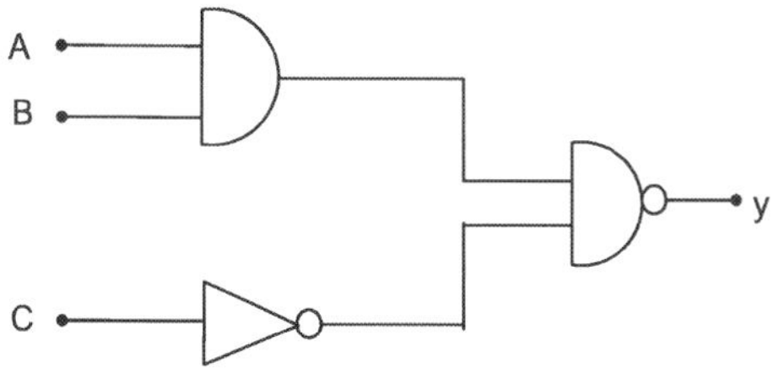

1) $A=1,B=0,C=0$ 2) $A=0,B=1,C=1$

3) $A=0,B=0,C=0$ 4) $A=1,B=1,C=0$

28. **The combinatin of gates shown below yields**

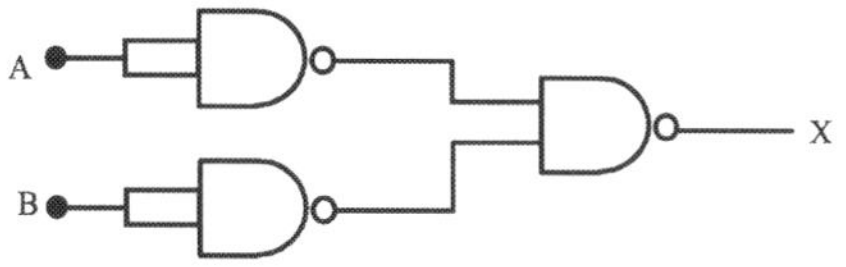

1) NAND gate 2) OR gate 3) NOT gate 4) XOR gate

29. **The logic gate having an output of 1 is**

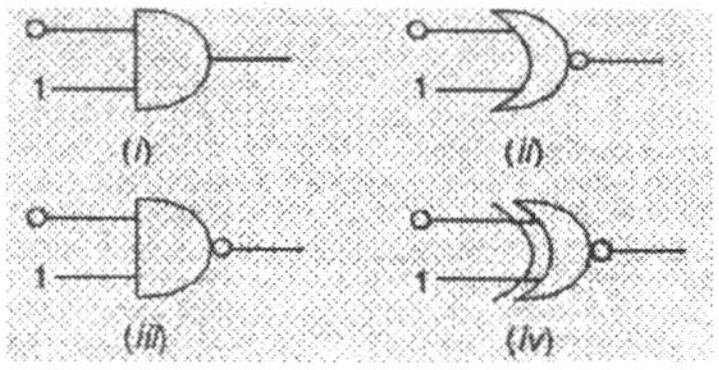

1)(iv) 2)(i) 3)(ii) 4)(iii)

30. **In a given circuit as shown the two input wave forms A and B are applied simultaneously**

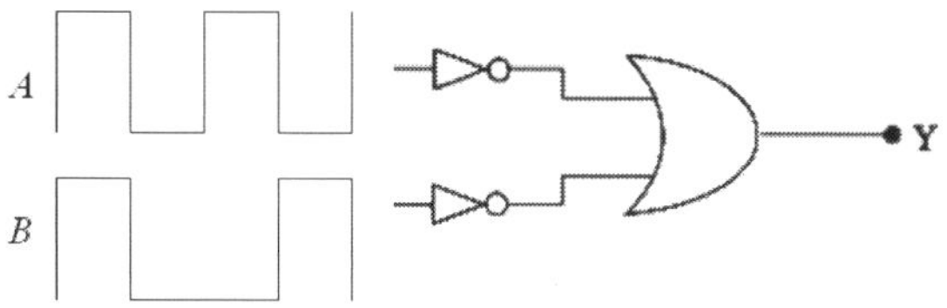

The resultant wave form at Y is

1) 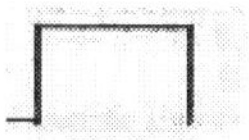 2) 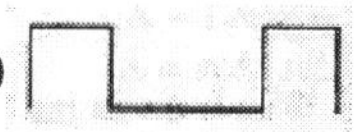 3) 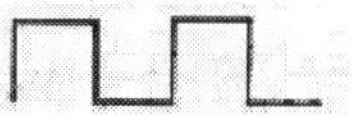 4) 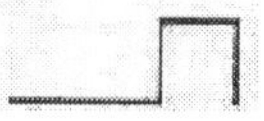

31. **The output of the combination of the gates shown in the figure below is**

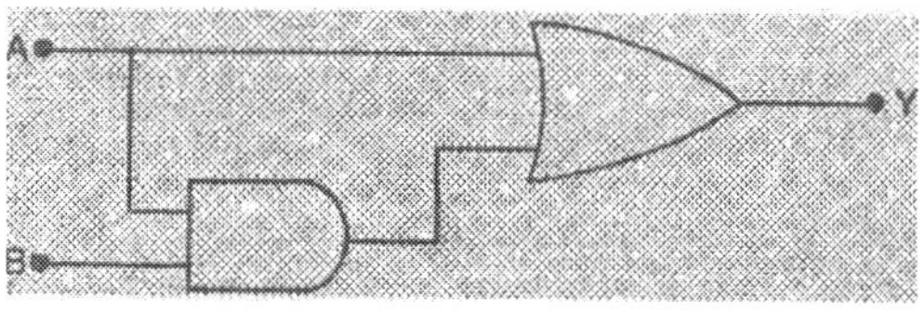

1) A + A. B 2) $(A+B)A+\overline{B}$ 3) $(A.B)+(\overline{A}.\overline{B})$ 4) $(A+B)(\overline{A.B})$

32. The expression of Y in following circuit is

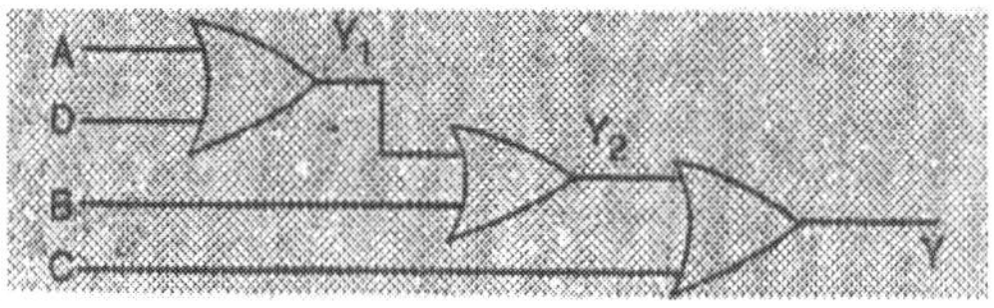

1) ABCD 2) A+BCD 3) A+B+C+D 4) AB+CD

33. How many NAND gates are used in an OR gate ?

1) four 2) two 3) three 4) Five

KEY LEVEL - II

1)3	2)3	3)2	4)4	5)2	6)2	7)1	8)2	9)1	10)2
11)3	12)4	13)3	14)3	15)3	16)4	17)2	18)3	19)4	20)3
21)4	22)2	23)4	24)3	25)1	26)4	27)4	28)2	29)4	30)1
31)1	32)3	33)3							

HINTS LEVEL - II (CW)

1. $\sigma = \frac{1}{\rho} = e(\mu_e n_e + \mu_h n_h)$

2.. $E = \frac{v}{d} = \frac{0.3}{2\times10^{-6}} = 1.5\times10^5 Vm^{-1}$

Its direction from n to p side.

3. $v^2 - u^2 = 2as$

4. For reverse biasing of an ideal diode, the potential of n-side should be higher then potential of p-side. Only option (4) is satisfying the criterion for reverse biasing.

5. In (1) one is forward biased and the other one is reverse biased.In (2) both are forward biased In (3) both are reverse biased

6. $i = \frac{total\ P.D}{total\ resistence}$

7. $R = \frac{E - V_b}{i} = \frac{1.5 - 0.5}{5\times10^{-3}} = \frac{1}{5\times10^{-3}}$

$=0.2' 10^3 = 2' 10^2 W$

8. As per given circuit, diode D_1 is forward biased and offers a resistance of 50 ohm. Diode D_2 is reverse biased and as its corresponding resistance is infinite, no current flows through it. Thus the equivalent circuit is as shown in Figure(2). As all the three resistances are in series, the current through then is

$I = \frac{6V}{(50+150+100)\Omega} = \frac{6}{300} A = 0.02A$

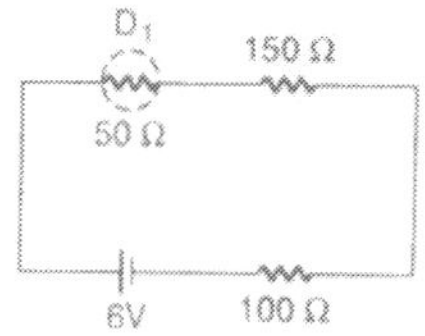

9. $i = \frac{V}{R} = \frac{5}{50}$

10. current does not flow through the diode.

11 If $V_A > V_B$ both the junction diodes are forward biased and the given circuit diagram becomes a balanced Wheatestone bridge The equivalent resistance in this case becomes 4Ω ,if $V_A < V_B$ the diodes are reveres biased in that case $4\Omega, 5\Omega$ and 4Ω are in series

12. During FB, diode conducts and during RB, diode does not conduct.

13. Slope $= \frac{\Delta I}{\Delta V}$

14 Given figure is halfwave rectifier

15. The given question is based on Bridge Rectifier. This is the most widely used full-wave rectifier. It makes use of four diodes D_1, D_2, D_3, D_4 connected in the four arms of a bridge. Bridge rectifier does not require a centre-tapped transformer.

16. Conceptual

17. The diode connected parallel to the battery is reverse bias. so current will not pass through it. so total emf divided among C_1 and C_2 and in the inverse ratio of their capacities

$$V = \frac{q}{C}; \frac{V_1}{V_2} = \frac{C_2}{C_1}$$

$$V_1 = \frac{VC_2}{C_1 + C_2}, V_2 = \frac{VC_1}{C_1 + C_2}$$

18. $V = iR$

19. $I_C = \frac{90}{100} I_E \Rightarrow I_E = \frac{10}{9} I_C$ and $I_e = I_b + I_c$

20. $\beta = \frac{\alpha}{1-\alpha}$ But $\beta = \left(\frac{\delta I_C}{\delta I_B}\right)_{V_{CE}}$

$$\Rightarrow \delta I_C = \beta . \delta I_B = 9 \times 4 \text{mA}$$

21. $I_c = \frac{V_{cc}}{R_L}$

22 $V = i_b R_b$

$$\therefore \; R_B = \frac{V}{i_b} = \frac{9}{35 \times 10^{-6}} = 257 k\Omega$$

23. (i) Current Gain $\beta = \frac{\delta i_c}{\delta i_b}$

(ii) Input impedance $R_i = \frac{\delta V_{BE}}{\delta i_b}$

(iii) Effective (ac) load $R_{AC} = \frac{R_C R_L}{R_C + R_L}$

(iv) Voltage gain $A_v = \beta \times \frac{R_{AC}}{R_{in}}$

(v) Power gain, $A_p = A_v \times \beta$

24. $$\begin{array}{r} {}^{1}\ \ \ {}^{1}\ \ \ {}^{1} \\ 1'\,{}^{1}01'\,{}^{1}01'\,{}^{1}0 \\ -0\ 1\,0\ 10\ 1 \\ \hline 0\ 1\,0\ 10\ 1 \end{array}$$

25. $Z = DABC + DA\overline{BC}$

$= DA\left(BC + \overline{BC}\right) = DA(1)$

$= DA$ i.e One AND gate.

26. It is diode-diode logic ckt of OR gate

27. Here $Y = \overline{(A.B).\overline{C}} = \left(\overline{A.B}\right) + C = \overline{A} + \overline{B} + C$

$\therefore Y = 0$ if $A = 1$, $B = 1$ and $C = 0$

28. $X = \overline{\overline{A}.\overline{B}} = \overline{\overline{A}} + \overline{\overline{B}} = A + B$. It is OR gate

29. $Y = \overline{A.B} = \overline{1.0} = 1 = NAND$

30. Logic of A is 1010, Logic of B is 1001

Logic of Y is 0111

31. The input to OR gate is A and A.B. Hence Y = A + BA

32. $Y_1 = A + D, Y_2 = A + D + B$,

$Y = A + D + B + C$

33. For this purpose we use NAND gate in manner as shown. The first two NAND gates are operated as NOT gates adn their output are fed to the third The resulting circuit is OR gate. Here, there gates are used

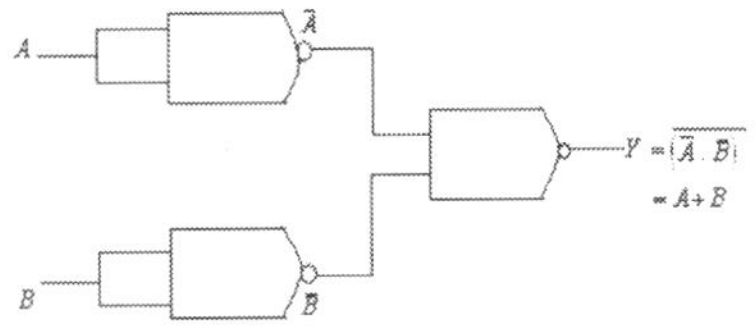

LEVEL - II -2

INTRINSIC,EXTRINSIC SEMI CONDUCTORS AND DIODES

1. If the lattice constant of this semiconductor is decreased, then which of the following is correct?

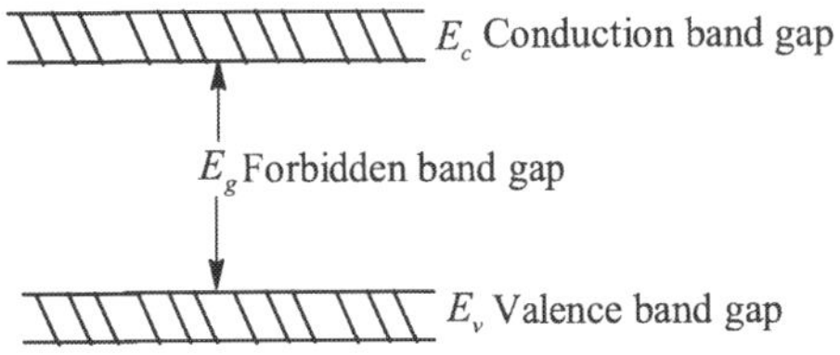

1) All E_c, E_g & E_v decrease

2) All E_c, E_g & E_v increase

3) E_c and E_v increase, but E_g decrease

4) E_c and E_v decrease but E_g increase

2. **A Ge specimen is doped with Al. The concentration of acceptor atoms is $10^{21}\ atoms/m^3$. Given that the intrinsic concentration of electron -hole pairs is $10^{19}\ /m^3$, the concentration of electrons in the specimen is**

1) $10^{17}/m^3$ 2) $10^{15}/m^3$ 3) $10^4/m^3$ 4) $10^2/m^3$

3. **The following data are for intrinsic germanium at 300 K.** $n_i = 2.4\times10^{19}/m^3$,

$\mu_e = 0.39 m^2/Vs, \mu_h = 0.19 m^2/Vs$. **Calculate the conductivity of intrinsic germanium.**

1) $4.3\,Sm^{-1}$ 2) $1.21\,Sm^{-1}$ 3) $2.22\,Sm^{-1}$ 4) $4.22\,Sm^{-1}$

4. **The diagram correctly represent the direction of flow of charge carriers in the forward bias of p-n junction is**

1) 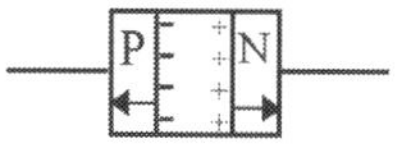

2) 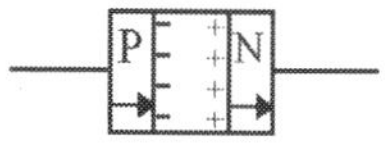

3) 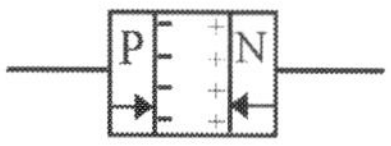

4) 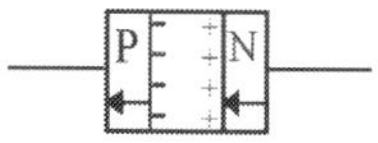

5. **In the figurs shown below**

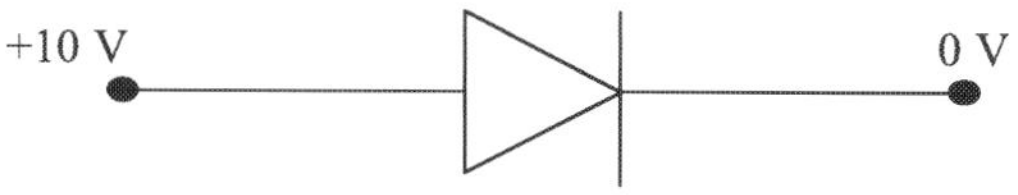

Fig a

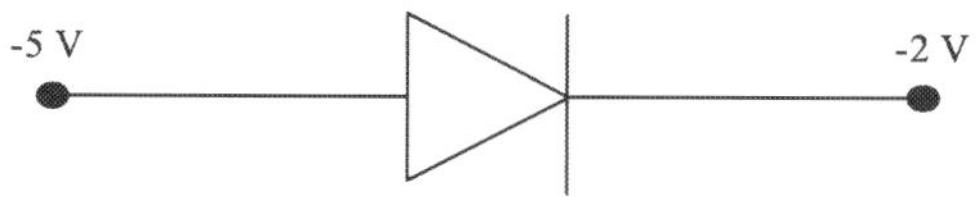

Fig b

1) In both Fig a and Fig b the diodes are forward biased

2) In both Fig, a and Fig b the diodes are reverse biased

3) In Fig a the diode is foward biased and in Fig b, the diode is reverse biased

4) In Fig a the diode is reverse biased and in fig b, it is forward biased

6. **A P-N junction diode can withstand currents up to 10 mA. Under forward bias, The diode has a potential drop of 0.5 V across it which is assumed to be independent of current.The maximum voltage of the battery used to forward bias the diode when a resistance of 200Ω is connected in series with it is**

1) 2.5V 2) 2.6V 3) 2.7V 4) 2.8V

7. **A cell of emf 4.5V is connected to a junction diode whose barrier potential is 0.7V. If the external resistance in the circuit is 190 Ω, the current in the circuit is**

1) 20 mA 2) 2mA 3) 23mA 4) 200mA

8. V_A *and* V_B **denote potentials of A and B, then the equivalent resistance between A and B in the adjoining circuit is(ideal diode)**

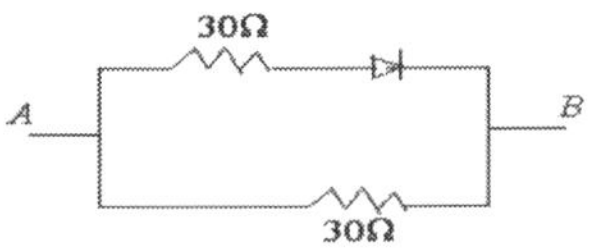

1) 15Ω *if* $V_A > V_B$ 2) 30Ω *if* $V_A < V_B$ 3) Both 1 and 2 4) neither 1 nor 2

9. **Two ideal junction diodes D_1, D_2 are connected as shown in the figure. A 3V battery is connected between A and B. The current supplied by the battery if its positive terminal is connected to A is**

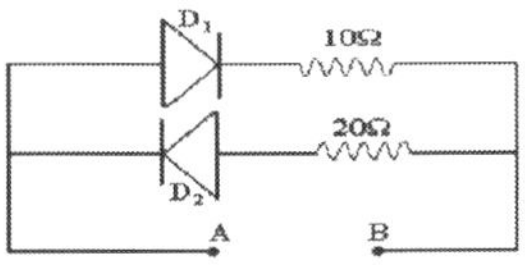

1) 0.1A 2) 0.3 A 3) 0.9 A 4) 90 A

10. **Find the effective resistance between A and B**

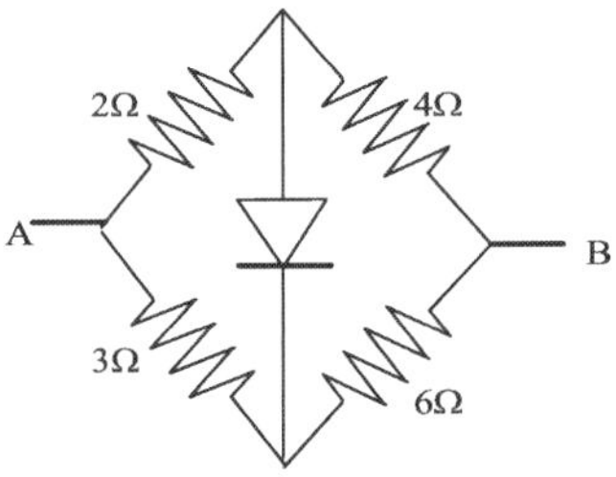

1) 5/18 Ω 2) 9/5 Ω 3) 18/5 Ω 4) 5/9 Ω

11. **The peak voltage in the output of a half-wave diode rectifier fed with a sinusoidal signal without filter is 10V. The d.c. component of the output voltage is**

1) $10/\sqrt{2}\ V$ 2) $10/\pi\ V$ 3) 10 V 4) $20/\pi\ V$

12. **In the figure shown the potential drop across the series resistor is**

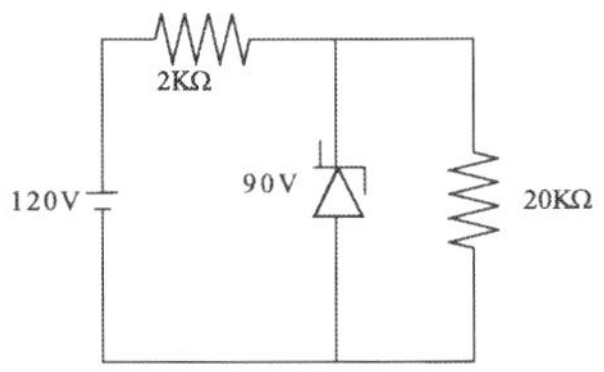

1) 30 V 2) 60 V 3) 90 V 4) 120 V

13. **A 220V ac supply is connected between points A and B (Fig). What will be the potential difference V across the capacitor?**

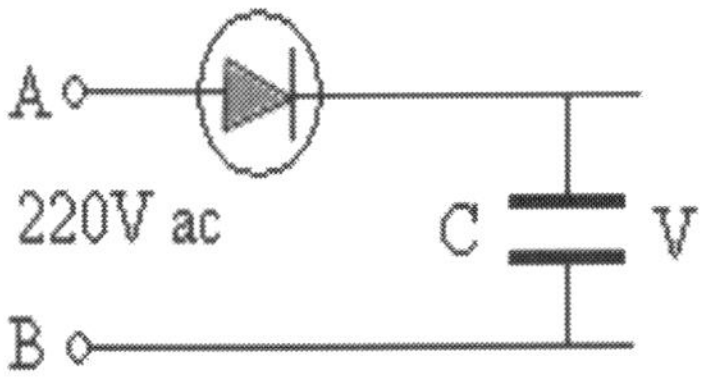

1) 220V 2) 110V 3) 0V 4) $220\sqrt{2}V$

14 In the circuit shown (Fig). if the diode forward voltage drop is 0.3V, the voltage difference between A and B is :

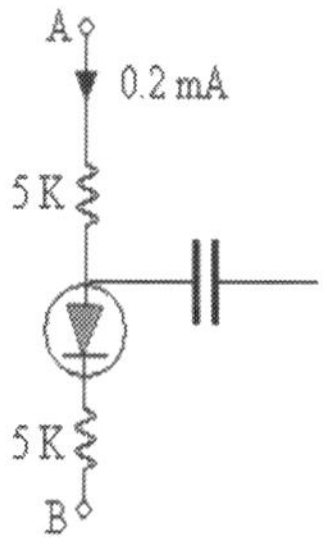

1) 1.3V 2) 2.3V 3) 0 4) 0.5V

TRANSISTORS

15. In a n-p-n transistor 10^{10} electron enter the emitter in 10^{-6}s. If 2% of the electrons are lost in the base, the current transfer ratio and the current amplification factor are

1) 0.98, 49 2) 0.49, 49 3) 0.98, 98 4) 0.49, 98

16. In a common base mode of transistor, collector current is 5.488 mA for an emitter current of 5.60mA. The value of the base current amplification factor (β) will be:

1) 48 2) 49 3) 50 4) 51

17. Current amplification factor of a common base configuration is 0.88. Find the value of base current when the emitter current is 1 mA.

1) 0.12 mA. 2)0.1 mA 3) 0.5 mA 4) 0.2 mA

18. For a transistor $\beta = 40$ and $I_B = 25\mu A$. Find the value of I_E.

1) 1mA 2) 1.025mA 3) 2mA 4) 1.2 mA

19. In a transistor if $\frac{I_C}{I_E} = \alpha$ and $\frac{I_C}{I_B} = \beta$, If α varies between $\frac{20}{21}$ and $\frac{100}{101}$, then the value of β lied between

1)1-10 2)0.95-0.99 3)20-100 4)200-300

20. For a transistor $x = \frac{1}{\alpha}$ & $y = \frac{1}{\beta}$ where α & β are current gains in common base and common emitter configuration. Then

1) $x + y = 1$ 2) $x - y = 1$ 3) $2x = 1 - y$ 4) $x + y = 0$

21. A voltage amplifier operated from a 12 volt battery has a collector load $6k\Omega$. Calculate the maximum collector current in the circuit.

1) 0.5mA 2) 1 mA 3) 3 mA 4) 2mA

22. A CE amplifier is designed with a transistor having $\alpha = 0.99$. Input impedance is 1 $k\Omega$ and load is 10 $k\Omega$. Voltage gain will be:

1) 9900 2) 99000 3) 99 4) 990

23. In a common emitter amplifier the load resistance of the output circuit is 792 times the resistance of the input circuit. If α = 0.99, the voltage gain is

1) 79200 2) 39600 3) 7920 4) 3960

24. In a transistor amplifier β = 62, $R_L = 5000\,\Omega$ and internal resistance of the transistor is 500 Ω. Its power amplification will be

1) 25580 2) 33760 3) 38440 4) 55760

25. The tuned collector oscillator circuit used in the local oscillator of a radio receiver makes use of a tuned circuit with L = 60 μH and C = 400 pF. Calculate the frequency of oscillations.

1) 1.03 KHz 2) 1.03 Hz 3) 1.03 GHz 4) 1.03 MHz

LOGIC GATES

26. When we add binary numbers 111 and 111 we get the binary number:

1) 222 2) 1000 3) 1110 4) 000

27. If A= B=C=1 and $X = A\overline{BC} + B\overline{CA} + C\overline{AB}$, then X=

1) 0 2) 1 3) 100 4) 110

28. The input of A and B for the Boolean expression $\left(\overline{A+B}\right).\left(\overline{A.B}\right) = 1$ is

1) 0,0 2) 0, 1 3) 1,0 4) 1, 1

29. Write the name of the follwoing gate that the circuit shown in figure.

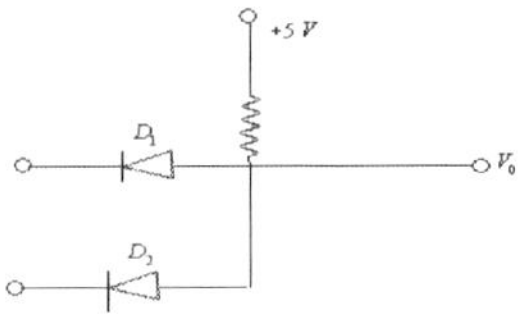

1) AND gate 2) OR gate 3) NOR gate 4) XOR gate

30. Consider a two-input AND gate of figure below. Out of the four entries for the Truth Table given here, the correct ones are

	Input		Output
	A	B	Y
1	0	1	0
2	1	0	0
3	1	1	1
4	0	0	1

1) All 2) 1 and 2 only 3) 1, 2 and 3 only 4) 1, 3 and 4 only

31. A Truth table is given below. The logic gate having following truth table is

A	B	Y
0	0	1
1	0	0
0	1	0
1	1	0

1) NAND gate 2) NOR gate 3) AND gate 4) OR gate

32. For a logic 0101 the waveform is

1) 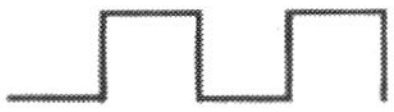

2) 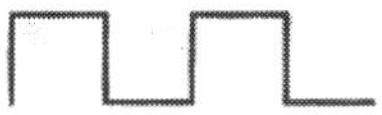

3) 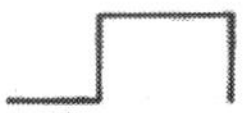

4) 

33. For the given combination of gates, if the logic states of inputs A, B, C are as follows A=B=C=0 and A=B=1, C=0 then the logic states of output D are

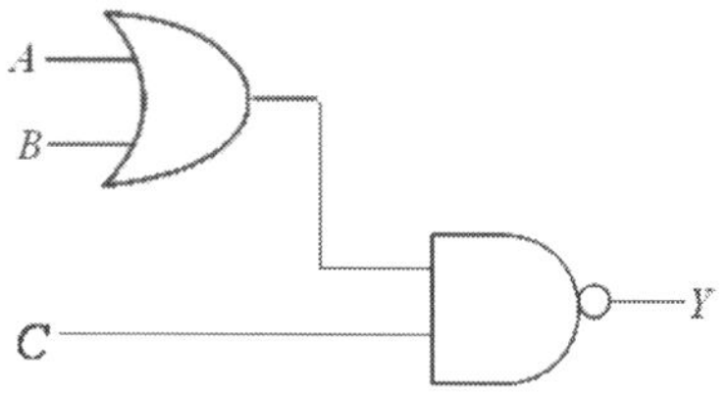

1) 0,0 2) 0,1 3) 1, 0 4) 1,1

34. Identify the gate represented by the block diagram as shown in fig.

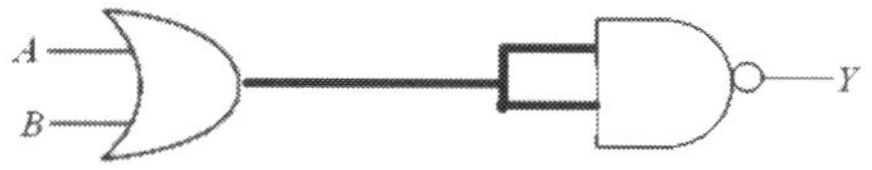

1) AND gate 2) NOT gate 3) NAND gate 4) NOR gate

35. The Boolean expression for the gate circuit shown below is

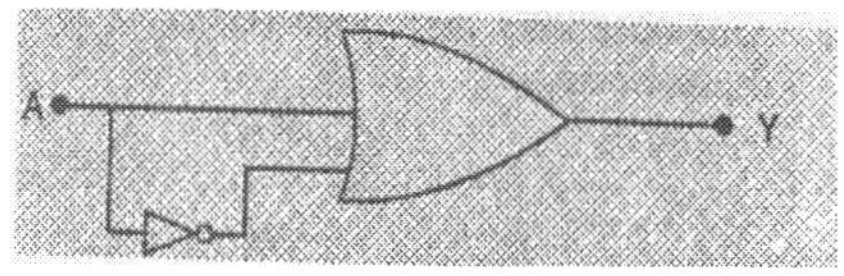

1) $A+\overline{A}=1$ 2) $A+1=1$ 3) $A+A=A$ 4) $A+0=A$

36. The output Y of the gate circuit shown in the figure below is

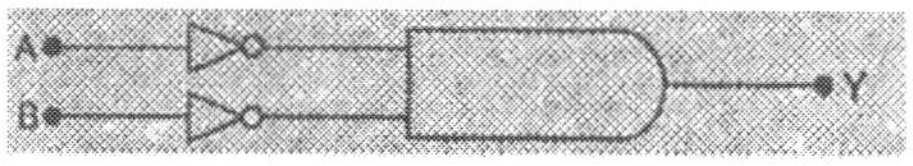

1) $\overline{A.B}$ 2) $\overline{A}.\overline{B}$ 3) $\overline{\overline{A.B}}$ 4) $\overline{A}+\overline{B}$

KEY LEVEL - II -2

1) 4	2)1	3)3	4)3	5)3	6)1	7)1	8)3	9)2	10)3
11)2	12)1	13)4	14)2	15)1	16)2	17)1	18)2	19)3	20)2
21)4	22)4	23)1	24)3	25)4	26)3	27)1	28)1	29)1	30)3
31)2	32)1	33)4	34)4	35)1	36)2				

HINTS - LEVEL - II -2

1. Conceptual
2. When Ge specimen is doped with Al, then concentration of acceptor atoms is also called concentration of holes. Using formula, $n_i^2 = n_0 p_0$ where

n_1 = concentration of electron hole pair $= 10^{19} m^{-3}$

n_0 = concentration of electron

p_0 = concentration of holes $= 10^{21} atom\ m^{-3}$

$\Rightarrow (10^{19})^2 = 10^{21} \times n_0 \qquad \Rightarrow n_0 = 10^{17} m^{-3}$

3. $\sigma_i = n_i e(\mu_e + \mu_h)$
4. In forward bias holes move from P to n, side electrons move from n to P
5. for forward bias, $V_p > V_n$, and for reverse bias $V_p < V_n$
6. $i = \frac{E - V_b}{R}$
7. V=4.5- 0.7=3.8V ; $R = 190\Omega$, $i = \frac{V}{R}$
8. $R = \infty$ in the case of reversebias
9. If positive terminal is connected to A, D_1 is in forward bias and D_2 is in reverse bias. So 20W is ineffective.

   $\therefore i = \frac{V}{R} = \frac{3}{10} = 0.3A$
10. Use Wheat stones bridge principle
11. $V_{dc} = \frac{V_m}{\pi} = \frac{10}{\pi} volt$
12. $V_{in} = V + V_Z$
13. Capacitor once gets charged upto maximum potential .After that for any other lesser value of p.d across A and B diode is reverse biased and it does not allow charge to flow in opposite direction.

   $From\, V_{rms} = \frac{V_{max}}{\sqrt{2}}$

   $\Rightarrow V_{max} = V_{rms}\sqrt{2}$

   $= 220\sqrt{2}\ volts$
14. $V_A - 0.2\times10^{-3}\times5\times10^3 - 0.3$

   $-0.2\times10^{-3}\times5\times10^3 - V_B = 0$

   $\Rightarrow V_A - V_B = 2.3\ volt$
15. $\alpha = \frac{I_c}{I_e}, I_c = \frac{98}{100} I_e, \beta = \frac{\alpha}{1-\alpha}$
16. $I_e = I_b + I_c$ and $\beta = \frac{I_C}{I_B}$
17. In a common-base arrangement, the current amplifica-tion factor.

   $\alpha = \frac{I_C}{I_E} \Rightarrow I_C = \alpha I_E = 0.88 \times 1 = 0.88mA.$ and $I_B = I_E - I_C$
18. $\beta = \frac{I_C}{I_B} \Rightarrow I_C = \beta I_B = 40 \times 25 \times 10^{-6}$ and $I_E = I_B + I_C$
19. $\beta = \frac{\alpha}{1-\alpha}$

20. $\alpha = \frac{\Delta I_c}{\Delta I_e}, \beta = \frac{\Delta I_c}{\Delta I_b}$

21. $I = \frac{\text{Battery voltage}}{\text{Collector load}} = \frac{12\,\text{volt}}{6k\Omega} =$

22. $\beta = \frac{\alpha}{1-\alpha}$ and $A_v = \beta \times \frac{R_{AC}}{R_{in}}$

23. $\beta = \frac{\alpha}{1-\alpha}, A_V = \beta \frac{R_L}{R_i}$

24. $A_P = \beta^2 \frac{R_L}{R_i}$

25. $v = \frac{1}{2\pi\sqrt{LC}}$

26.
$$\begin{array}{r} 111 \\ +111 \\ \hline 1110 \end{array}$$
(carries: 1 1 1)

27. Check the value of x by substituing A=B=C=1

28. $\overline{A+B} = 1$, $\overline{A.B} = 1$
$A+B = 0$, $A.B = 0$

29. conceptual

30. Check the truth table

31. $Y = \overline{A+B}$

32. concpetual

33.

A	B	Y
1	1	1
1	0	0
0	1	0
0	0	0

34. The boolean expression of this gate is
$Y = \overline{\overline{(A+B)}.\overline{(A+B)}} = \overline{(A+B)} + \overline{(A+B)} = \overline{(A+B)}$
which is for NOR gate.

35. The input of OR gate is A & $\overline{A}$
$Y = A + \overline{A} = 1$

36. The input of AND gate is $\overline{A}$ and $\overline{B} \Rightarrow Y = \overline{A}.\overline{B}$

LEVEL-III

1. **If the ratio of the concentration of electrons to that of holes in a semi-conductors is 7/5 and the ratio of currents is 7/4, then the ratio of drift velocities is**

   1) 5/8 2) 4/5 3) 5/4 4) 4/7

2. **If the resistivity of copper is $1.7\times10^{-6}\Omega$ cm, then the mobility of electrons in copper, if each atom of copper contributes one free electron for conduction, is [The atomic weight of copper is 63.54 and its density is 8.96 g/cc]:**

   1) $23.36 cm^2/Vs$ 2) $503.03 cm^2/Vs$ 3) $43.25 cm^2/Vs$ 4) $88.0 cm^2/Vs$

3. **A pure silicon crystal of length $l(0.1m)$ and area $A(10^{-4}m^2)$ has the mobility of electron (μ_e) and holes (μ_h) as $0.135m^2/Vs$ and $0.048m^2/Vs$, respectively, If the voltage applied across it is 2V and the intrinsic charge concen-tration is $n_i = 1.5\times10^6 m^{-3}$, then the total current flowing through the crystal is.**

   1) $8.78\times10^{-17}A$ 2) $6.25\times10^{-17}A$ 3) $7.89\times10^{-17}A$ 4) $2.456\times10^{-17}A$

4. **Find the current produced at room temperature in a pure germanium plate of area $2\times10^{-4}m^2$ and of thickness 1.2×10^{-3} m when a potential of 5 V is applied across the faces. Concentration of carriers in germanium at room temperature is 1.6×10^6 per cubic metre. The mobilities of electrons and holes are $0.4m^2V^{-1}s^{-1}$ and $0.2m^2V^{-1}s^{-1}$ respectively. The heat energy generated in the plate in 100 second is**

   1) $2.4\times10^{-11}J$ 2) $3.4\times10^{-11}J$ 3) $5.4\times10^{-11}J$ 4) $6.4\times10^{-11}J$

5. **An n-type semiconductor has impurity level 20meV below the conduction band. In a thermal collision, transferble energy is KT. The value of T for which electrons start to jump in conduction band is :**

   1) 232 K 2) 348 K 3) 400 K 4) 600 K

6. **Assume that the number of hole-electron pair in an intrinsic semiconductor is proportional to $e^{-\Delta E/2KT}$. Here ΔE = energy gap and $k = 8.62\times10^{-5}eV/kelvin$. The energy gap for silicon is $1.1eV$. The ratio of electron hole pairs at 300K and 400 K is :**

   1) $e^{-5.31}$ 2) e^{-5} 3) e 4) e^2

7. **In the circuit shown in figure (1), the V_0, I_1, I_{D_1}, and I_{D_3} are respectively**

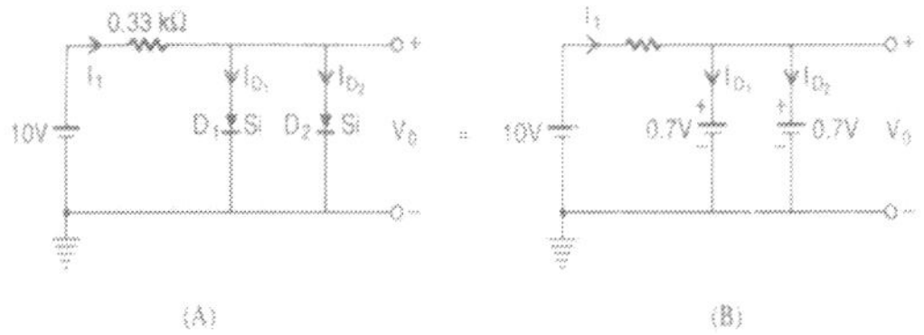

   1) 0.5V, 25 mA, 15 mA 2) 0.7V, 28.18 mA, 14.09 mA

   3) 0.4V, 15 mA, 20 mA 4) 0.3V, 15.06 mA, 20.18 mA

8. **For a junction diode, the ratio of forward current (I_f) and reverse current is**

   [I_e = electronic charge, V = voltage applied across junction,

   k = Boltzmann constant T = temperature in kelvin]

   1) $e^{-v/kT}$ 2) $e^{V/kT}$ 3) $\left(e^{eV/kT}-1\right)$ 4) $\left(e^{V/kT}-1\right)$

9. **In the diagram D an ideal diode and an alternating voltage of peak value 10V is connected as input V_1. Which of the following diagram represents the correct waveform of output voltage V_θ ?**

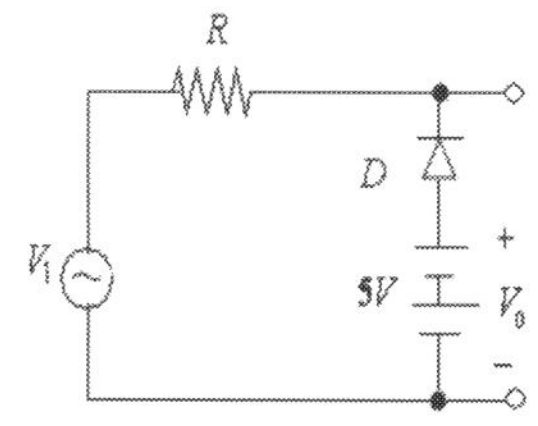

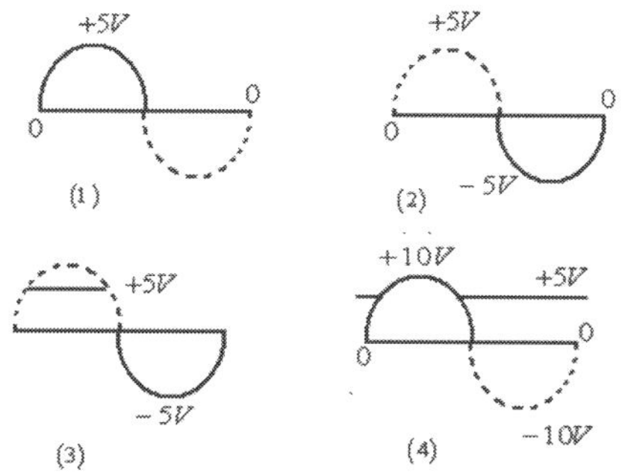

10. **For a CE-transistor amplifier, the audio signal voltage across the collector resistance of $2k\Omega$ is 2V. Suppose the current amplification factor of the transistor is 100. Find the input signal voltage and base current, if the base resistance is $1k\Omega$.**

1) 0.02V 2) 0.01V 3) 0.03V 4) 0.04V

11. **In the circuit, transistor has a current**

$(\beta) = 100$. **What should be the base resistor R_B neglect V_{BE}, so that $V_{CE} = 5V$:**

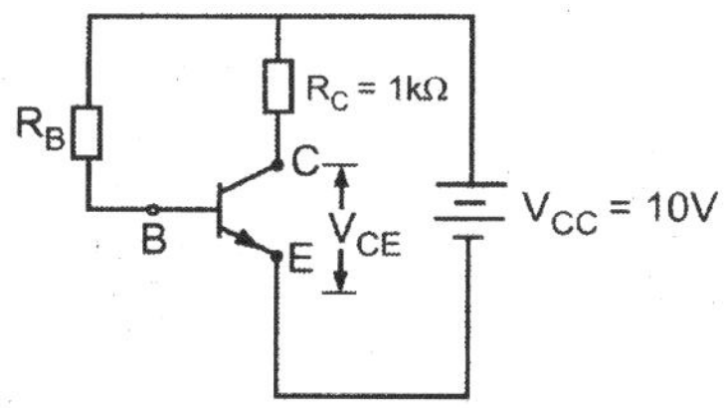

1) 200 kΩ 2) 1 kΩ 3) 500 kΩ 4) 2 kΩ

12. **An $N-P-N$ transistor is connected in common - emitter configuration in which collector supply is 8V and the voltage drop across the load resistance of 800 Ω connected in the collector circuit is 0.8V. If current amplification factor is 25/26 (If the internal resistance of the transistor is 200 Ω), the collector-emitter voltage, voltage gain and power gain are respectively.**

1) 5.2V, 1.86, 3 2) 6.2V, 186, 5.5 3) 7.2V, 3.86, 3.698 4) 8.2V, 4.91, 3.15

13 For a CE transistor amplifier, the audio signal voltage across the collector resistance of $2k\Omega$ is 2V. Suppose the current amplification factor of the transistor is 100. The value of R_B in series with V_{BB} supply of 2V, if the DC base current has to be10 times the signal current is

1) $4k\Omega$ 2) $14k\Omega$ 3) $28k\Omega$ 4) $54k\Omega$

14 Figure shows the transfer characteristics of a base biased CE transistor. Which of the following statements are true ?

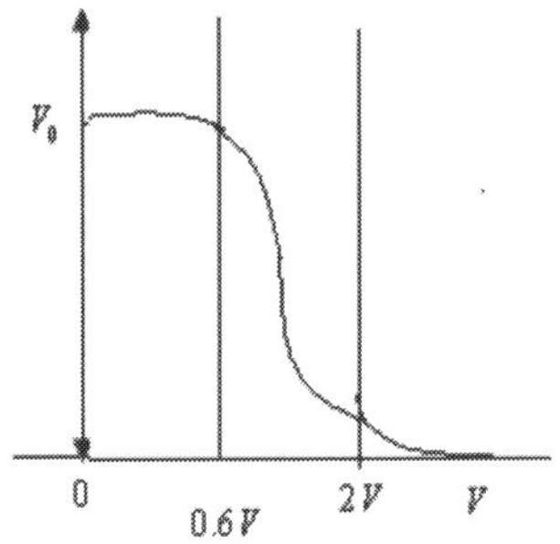

A) At $V_1 = 0.4\ V$ transistor is in active state

B) At $V_1 = 1\ V$ it can be used as an amplifier

C) At $V_1 = 0.5\ V$, It can be used as a switch turned off

D) At $V_1 = 2.5\ V$, it can be used as a switch turned on

1)A,B,C 2)B,C,D 3)A,C,D 4)A,B,D

15. The following configuration of gates is equivalent to

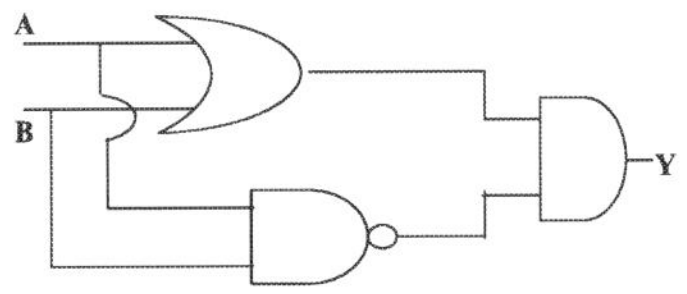

1) NAND gate 2) XOR gate 3) OR gate 4) NOR gate

16. The combination of the gates shown below produces

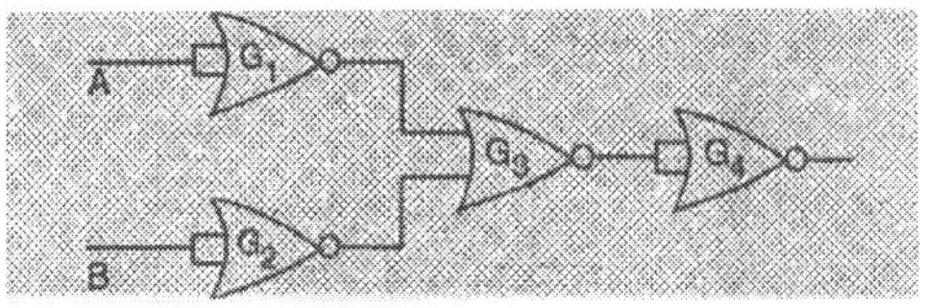

1) AND gate 2) XOR gate 3) NOR gate 4) NAND gate

17. **Which of the following truth tables is true?**

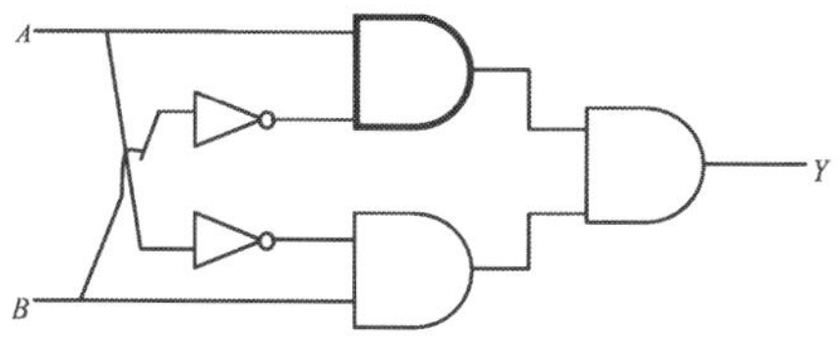

1)

A	B	Y
0	0	0
1	0	0
0	1	0
1	1	0

2)

A	B	Y
0	0	0
1	0	0
0	1	1
1	1	1

3)

A	B	Y
0	0	0
1	0	1
0	1	1
1	1	1

4)

A	B	Y
0	0	0
1	0	1
0	1	1
1	1	0

18. **Truth table for system of four NAND gates as shown in figure is :**

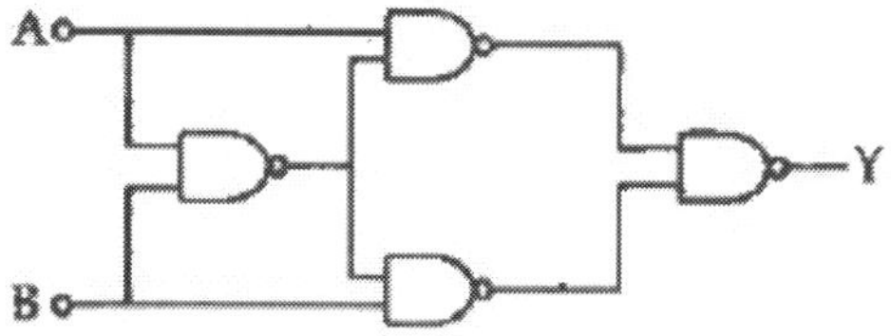

1)

A	B	Y
0	0	0
0	1	1
1	0	1
1	1	0

2)

A	B	Y
0	0	0
0	1	0
1	0	1
1	1	1

3)

A	B	Y
0	0	1
0	1	1
1	0	0
1	1	0

4)

A	B	Y
0	0	1
0	1	0
1	0	0
1	1	1

19. **The logic circuit shown below has the input waveforms 'A' and 'B' as shown. Pick out the correct output waveform**

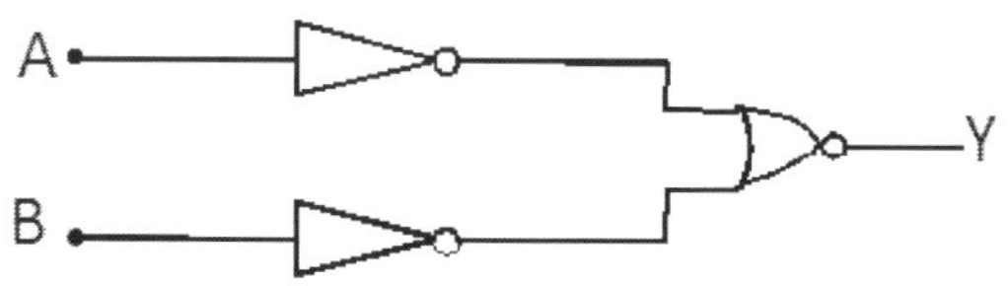

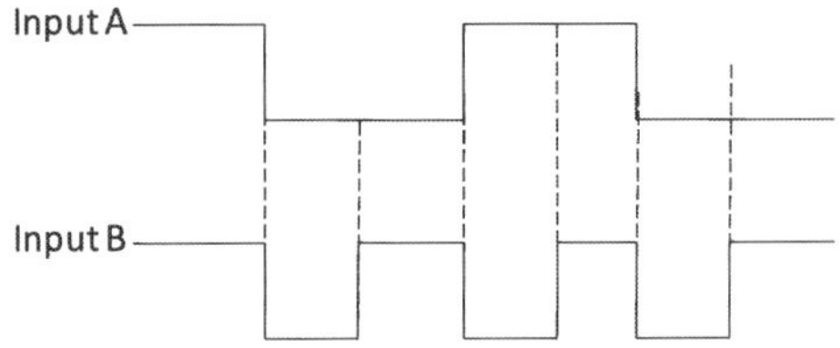

1) 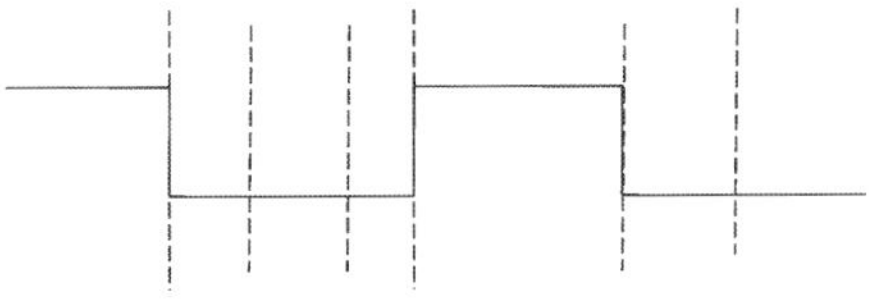

2) 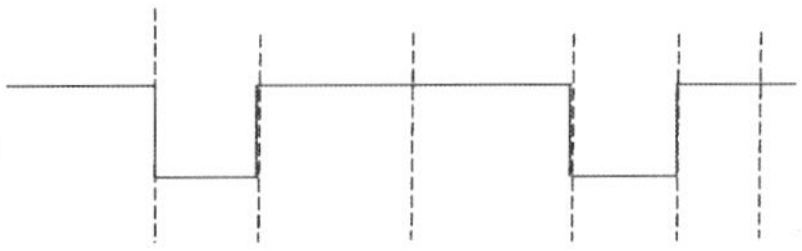

3) 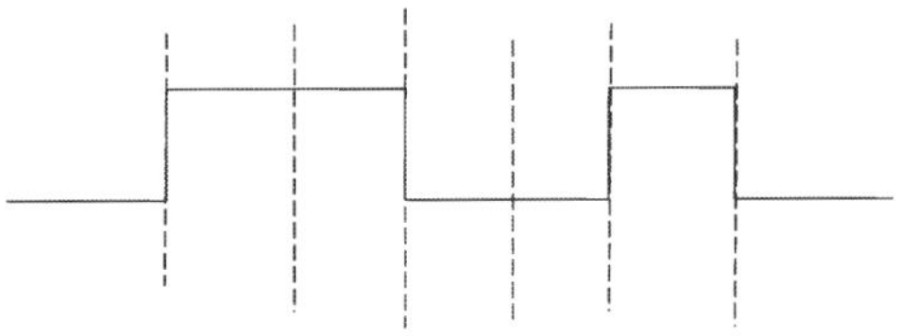

4) 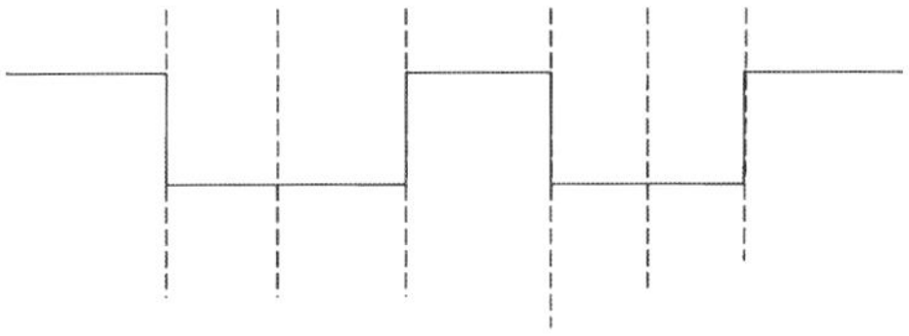

20. Logic gates X and Y have the truth tables shwon below

P	Q	R	P	R
0	0	0	0	1
1	0	0	1	0
0	1	0		
1	1	1		

When the output of X is connected to the input of Y, the resulting combination is equivalent to a single .

1) NOT gate 2) OR gate 3) NOR gate 4) NAND gate

KEY LEVEL - III

1) 3	2)3	3)1	4)4	5)1	6) 1	7)2	8)3	9)4	10)2
11)1	12)3	13)2	14)2	15)2	16)4	17)1	18)1	19)1	20)4

HINTS LEVEL - III

1. $I = n\ e\ Av_d$

$$\frac{I_e}{I_h} = \frac{n_e \times (v_d)_e}{n_h \times (v_d)_h}$$

Here $\frac{n_e}{n} = \frac{7}{5}, \frac{I_e}{I} = \frac{7}{4}$

$$\therefore \frac{7}{4} = \frac{7}{5} \times \frac{(v_d)_e}{(v_d)_h} \quad \Rightarrow \frac{(v_d)_e}{(v_d)_h} = \frac{5}{7} \times \frac{7}{4} = \frac{5}{4}$$

2. Mobility of electron $\mu = \frac{\sigma}{ne}$ - (1)

also resistivity $\rho = \frac{1}{\sigma}$

From Eqs. (i) and (ii), we get $\mu = \frac{1}{ne\rho}$

Here, n = number of free electrons per unit volume

$$n = \frac{N_0 \times d}{atomicweight}$$

$$n = \frac{6.023 \times 10^{23} \times 8.96}{63.54} = 8.5 \times 10^{22}$$

From Eqs. (iii) and (iv), we get

$$\mu = \frac{1}{8.5 \times 10^{22} \times 1.6 \times 10^{-19} \times 1.7 \times 10^{-6}}$$

$$\therefore \mu = 43.25 cm^2 / Vs$$

3. $I = e(n_e \mu_e + n_h \mu_h) \frac{AV}{l}$

Putting the values, we get

$I = 8.78 \times 10^{-17} A$

4. $n_e = n_h = n_i$

$\therefore \sigma = n_i e(\mu_e + \mu_h) = 1.6 \times 1.6 \times 0.6 \times 10^{-13}$

Current produced in germanium plate.

$$I = JA = \sigma E.A = \sigma \left(\frac{V}{d}\right) A$$

Heat generated in the plate,

$H = V \times I \times t$

$= 5 \times 1.28 \times 10^{-13} \times 100 = 6.4 \times 10^{-11}$ joule

5. $KT = 20 \times 10^{-3} eV$

$$\therefore T = \frac{20 \times 10^{-3}}{k} eV = \frac{20 \times 10^{-3} eV}{8.62 \times 10^{-5} eV / K} = 232 \text{ K}$$

6. $\frac{N_1}{N_2} = \frac{e^{-\Delta E / 2KT_1}}{e^{-\Delta E / 2KT_2}}$

$$= e^{\frac{\Delta E}{2K}\left(\frac{1}{T_2} - \frac{1}{T_1}\right)} = \frac{1.1}{e^{2 \times 8.62 \times 10^{-5}}} \left(\frac{1}{400} - \frac{1}{300}\right)$$

7. Since applied voltage is greater than 0.7 V and direction of current is same as direction of arrow head, both the diodes are in "on" state (see above figure(2)). Now V_0 is parallel to D_1 and D_2.

Hence V_0=0.7 V

$$\text{Current } I_1 = \frac{V_R}{R} = \frac{E - V_D}{R} = \frac{10V - 0.7V}{0.33k\Omega} = 28.18 \text{ mA}$$

Bothe the silecon diodes have similar characteristics.

$$\text{Hence } I_{D_1} = I_{D_2} = \frac{I_1}{2} = 14.09mA$$

8. Here; $v_0 = 2V, \beta = 100,$

R_C (collector resistance) $= 2k\Omega, R_B = 1k\Omega$

$$\text{As } A_v = \frac{v_0}{v_1} = \beta\left(\frac{R_C}{R_B}\right), v_i = v_0\left(\frac{R_B}{R_C}\right)\left(\frac{1}{\beta}\right)$$

$$=(2V)\left(\frac{1k\Omega}{2k\Omega}\right)\left(\frac{1}{100}\right)=0.01V$$

9. Current in junction diode, $I_f = I_0\left(e^{eV/kT} - 1\right)$ In forward baising, V is positive and in reverse bias V is negative Then,

$$I_r = I_0$$

$$\frac{I_f}{I_r} = \frac{I_0(e^{eV/KT} - 1)}{I_0} =$$

$$=\left(e^{eV/kT} - 1\right)$$

10. For $V_1 < 5V$, the diode is forward-biased, output will be fixed at 5V. For $V_1 < 5V$, the diode is reverse biased. The output will follow V

11. Applying KVL to QCETSRQ, we get

$V_{CC} = I_C R_C + V_{CE}$ and calculate I_C

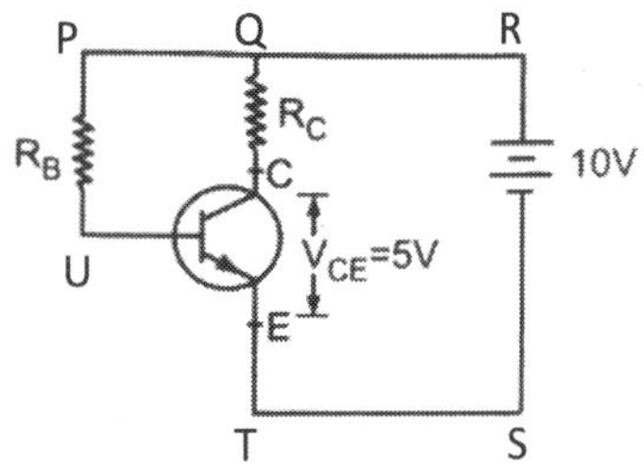

$$\beta = \frac{I_C}{I_B} \Rightarrow I_B = \frac{I_C}{\beta}$$

Applying KVL to PUTSRQP, we get

$V_{CC} = I_B R_B + V_{BE}$ here V_{BE} is negligible

$$\therefore V_{CC} = I_B R_B \Rightarrow I_B = \frac{V_{CC}}{R_B}$$

12. The circuit arrangement is shown in figure

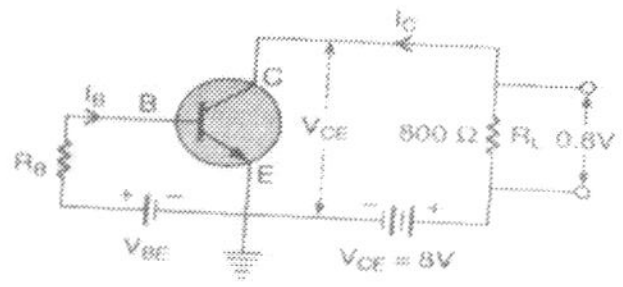

collector current $I_C = \frac{\text{Voltage drop across } R_L}{R_L}$

$= \frac{0.8}{800} = 10^{-3}\ amp.$

Now $V_{CE} = 8 - 0.8 = 7.2\ volt$

Current gain $\beta = \frac{I_C}{I_B}$ (or) $\frac{25}{26} = \frac{10^{-3}}{I_B}$

$\therefore\ I_B = \frac{26}{25} \times 10^{-3} = 1.04 \times 10^{-3}$ amp.

Voltage gain $= \beta \frac{R_L}{R_C} = \frac{25}{26} \times \frac{800}{200} = \frac{100}{26} = 3.846$

Power gain $= \beta^2 \frac{R_L}{R_C} = \left(\frac{25}{26}\right)^2 \times \frac{800}{200} = \left(\frac{25}{26}\right)^2 \times 4 = 3.698$

13 The output AC voltage is 2V. So the AC collector current $i_c = \frac{2}{2000} = 1mA$

The signal current through the base is therefore , given by $i_B = \frac{i_c}{\beta}$

The D.C base current has to be $10 \times i_B$

$\therefore\ R_B = \frac{(V_{BB} - V_{BE})}{I_B}, V_{BE} = 0.6V$

$\therefore\ R_B = \frac{(2 - 0.6)}{0.10} = 14\,k\Omega$

14. From the given transfer characteristics of a base biased common emitter transsistor, we note that

ii) When $V_1 = 1V$ (which is in between 0.6V to 2.V), the transistor circuit is in active state and it used as an amplifier

iii) When $V_1 = 0.5V$ there is no collector current , The transistor is in cut-off state. The transistor circuit can be used as a switch turned off

iv) When $V_1 = 2.5V$ the collector current becomes maximum and transistor is in saturation and can be used as switch turned on state

15. $Y = (A+B).(\overline{A.B}) = (A+B).(\overline{A}+\overline{B})$

$= A.\overline{A} + A.\overline{B} + \overline{A}.B + B.\overline{B}$ $Y = (A.\overline{B}) + (\overline{A}.B)$. it is XOR gate

16. The output of G_1 is $\overline{A}$ G_2 is $\overline{B}$. Hence output of G_3 is AB $\therefore$ $y=\overline{AB}$ i.e., NAND gate.

17. $Y=\left(A.\overline{B}\right).\left(\overline{A}.B\right)$

18. $Y=\overline{\left(\overline{A.\overline{A.B}}\right).\left(\overline{\overline{A.B}.B}\right)}=\left(\overline{\overline{A.\overline{A.B}}}\right)+\left(\overline{\overline{\overline{A.B}.B}}\right)$

$Y=A.\left(\overline{A}+\overline{B}\right)+\left(\overline{A}+\overline{B}\right).B$

$=A.\overline{A}+A.\overline{B}+\overline{A}.B+B.\overline{B}$ $Y=\left(A.\overline{B}\right)+\left(\overline{A}.B\right)$.

19. Truth table:

A	B	Y
1	1	1
1	0	0
0	1	0
0	0	0

20. Verify truth table

LEVEL-IV - COMPREHENSION TYPE

PASSAGE - I

A block of pure silicon at 300K has a length of 10cm and an area of 1.0cm^2. A battery of emf 2V is connected across it. The mobility of electron is $0.14\,m^2v^{-1}S^{-1}$ and their number density is 1.5$\times$10^{16} m^{-3}. The mobility of holes is $0.05\,m^2v^{-1}S^{-1}$

1. The electron current is

1) $6.72\times10^{-4}A$ 2) $6.72\times10^{-5}A$ 3) $6.72\times10^{-6}A$ 4) $6.72\times10^{-7}A$

2. The hole current is

1) $2.0\times10^{-7}A$ 2) $2.2\times10^{-7}A$ 3) $2.4\times10^{-7}A$ 4) 2.6

3. The total current in the block is

1) $2.4\times10^{-7}A$ 2) $6.72\times10^{-7}A$ 3) $4.32\times10^{-7}A$ 4) $9.12\times10^{-7}A$

PASSAGE:2

The input and output resistances in a common base amplifier circuits are 400Ω and 400KΩ respectively. The emitter current is 2mA and current gain is 0.98.

4. The collector current is

1) 1.84mA 2)1.96mA 3)1.2mA 4)2.04mA

5. The base current is

1) 0.012mA 2) 0.022mA 3)0.032mA 4) 0.042mA

6. Voltage gain of transistor is

1)960 2)970 3)980 4)990

7. Power gain of transistor is

1) 950 2) 960 3)970 4)980

8. The logic circuits are given in column I and the Boolean expressions in column II.

Column - I	Column - II
A) 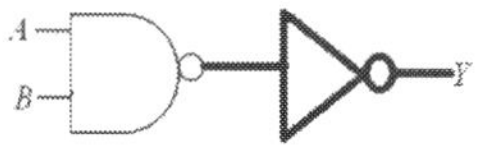	(P) $y=\overline{A}+\overline{B}$
B) 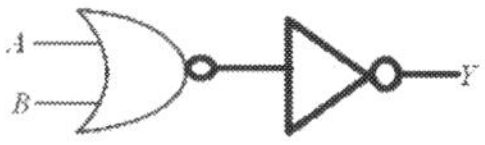	(Q) $y=\overline{A}.\overline{B}$
C) 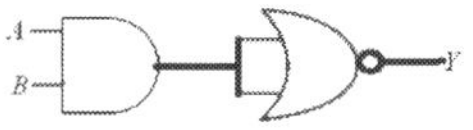	(R) $y=A.B$
D) 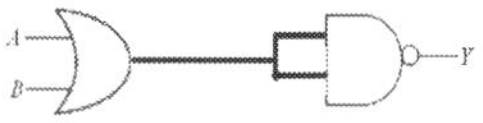	(S) $y=A+B$

1) $A\to R, B\to S, C\to p, D\to Q$
2) $A\to S, B\to R, C\to P, D\to Q$
3) $A\to P, B\to Q, C\to R, D\to S$
4) $A\to S, B\to P, C\to Q, D\to R$

KEY LEVEL - IV

1) 4 2)3 3)4 4)2 5)4 6)3 7)2 8)1

HINTS LEVEL - IV

1. $E=\frac{V}{l}=\frac{2}{0.1}=20V/m$

$A=1.0cm^2, v_e=\mu_e E=0.14\times 20=2.8ms^{-1}$

$I_e=n_e AeV_e=6.72\times 10^{-7}A$

2. In a pure semi conductor,

$n_e=n_h=1.5\times 10^{16}m^{-3}$

$V_h=\mu_h E=0.05\times 20=1.0m/s$

$I_h=n_h Aev_h=2.4\times 10^{-7}A$

3. Total current

$I=I_e+I_h$

$=9.12\times 10^{-7}A$

4. $I_c=\alpha I_e=0.98\times 2=1.96mA$

5. $I_b=I_e-I_c=2-1.96=0.04mA$

6. $A_v=\alpha\frac{R_0}{R_i}=0.98\times\frac{400\times 10^3}{400}=980$

7. Power amplification, $A_P=\alpha A_v$
$=0.98\times 980=960$

Chapter - 10

COMMUNICATION SYSTEMS

COMMUNICATION SYSTEMS

SYNOPSIS

Introduction: Communication is an act of exchange of information between the sender and the receiver. Over decades, methods have been evolved to develop languages, codes, signals etc to make communication effective. Communication through electrical signals has made things much simpler because they can be transmitted over extremely large distances in extremely short time as their speed is $3\times10^{8} m/s$.

Modern communicaton has its roots in the 19th and 20th century in the work of scientists like J.C. Bose, F.B. Morse, G.Macroni and Alexander Graham Bell. The pace of development seems to have increased dramatically after the first half of the 20th century. We can hope to see many more accomplishments in the coming decades. The aim of this chapter is to introduce the concepts of communication, namely the mode of communication, the need for modulation, production and deduction of amplitude modulation.

➢ **Communication is basically of two types:**

a) Point to point :- This takes place between a transmitter and a receiver. Telephonic conversion between two persons is a good example of it.

b) Broad cast mode :- Here, a large number of receivers receive the information from a single transmitter. Radio and television are good examples of broadcast mode.

Elements of Communication System

Basic units of a communication system.

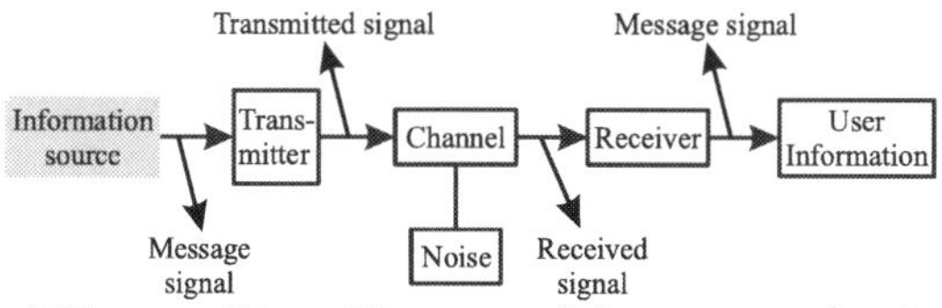

a) Transmitter: The part of the communication system, which sends out the information is called transmitter.

b) Transmission channel: The medium or the link, which transfers message signal from the transmitter to the receiver of a communication system is called channel.

c) Receiver: The part of the communication system, which picks up the information sent out by the transmitter is called **receiver.** The receiver consists of

➢ **Basic Terminology Used in Electronic Communication System**

Some important terms needed to understand the basic elements of communication

a) Information : It is nothing but, the message to be conveyed. The message may be a symbol, code, group of words etc. Amount of information in message is measured in "bits"

b) Communication Channel : Physical medium through which signals propagate between transmitting and receiving stations is called communication channel.

➢ **Transmitter:** Essential components of transmitter are as follows.

a)Transducer **:** Converts sound signals into electric signal. The device which converts a physical quantity (information) into electrical signal is known as ***transducer.***

b)Modulator **:** Mixing of audio electric signal with high frequency radio wave.

c)Amplifier: Boosting the power of modulated signal.

d)Antenna **:** Signal is radiated in the space with the aid of an antenna.

Receiver: Basic componenets of receiver.

a) ***Pickup antenna:*** To pick the signal

b) ***Demodulator:*** To separate out the audio signal from the modulated signal

c) ***Amplifier:*** To boost up the weak audio signal

d) ***Transducer:*** To convert back audio signal in the form of electrical pulses into sound waves.

Message Signal: Information converted in electrical form and suitable for transmission is called **signal.**

➢ A signal is defined as a single-valued function of time (that conveys the information) and which, at every instant of time has a unique value.

- ***Types of message signals***

  a) ***Analog signal:*** A signal, which is a continuous function of time (usually a sinusoidal function) is called *analog signal.*

  b) ***Digital signal:***A discrete signal (discontinuous function of time) which has only two levels is called digital signal.

- **Noise :** This refers to undesired signals which disturb the transmission and processing of signals.
- **Attenuation :** It is the loss of strength of a signal during propagation in a medium.
- **Amplification :** It is the process of increasing the strength of the signal (amplitude) using an amplifier.
- **Range :**It is the maximum distance from a source upto which the signal is received with sufficient strength.
- **Repeaters :** These are the devices used to increase the range of communication system
- **Band width of signals (speech, T.V and digital data) :** Band width is the frequency range over which an equipment operates.

(or)

It is the portion of the spectrum occupied by the signal.

Bandwidth of signals :

In a communication system, the message signal may be voice, music, picture or data etc. Each of these signals has a spread of different range of frequencies. Hence, the type of communication system needed depends upon the band of frequencies involved. Speech signal requires the band width of 2800 Hz (3100 Hz to 300 Hz). For music, a bandwidth of about 20KHz is required (due to high frequency produced by musical instruments). The audible range of frequencies extends from 20Hz to 20KHz. Video signals require band width of 4.2 MHz for picture transmission. However, a band width of 6MHz is needed for T.V signals. (as it contains both voice and picture)

Digital signals are in the form of rectangular waves as shown in Fig. One can show that this rectangular wave can be decomposed into a superposition of sinuosidal waves of frequencies $f_0 .\ 2f_0, 3f_0, 4f_0 ....nf_0$ where n is an integer extending to infinity and $f_0 = 1/T_0$. The fundamental (f_0), fundamental $(f_0)+$ second harmonic $(2f_0)$ and fundamental $(f_0)+$ second harmonic $(2f_0)+$ third harmonic $(3f_0)$, are shown in the same figure to illustrate this fact. It is clear that to reproduce the rectangular wave shape exactly we need to superimpose all the harmonics $f_0, 2f_0, 3f_0, 4f_0.....$ which implies an infinite bandwidth. However, for practical purposes, the contribution from higher harmonics can be neglected this limiting the bandwidth. As a result, received waves are a distorted version of the transmitted one. If the bandwidth is large enough to accommodate a few harmonics, the information is not lost and the rectangular signal is more or less recovered. This is so because the higher the harmonic. Less is its contribution to the wave form.

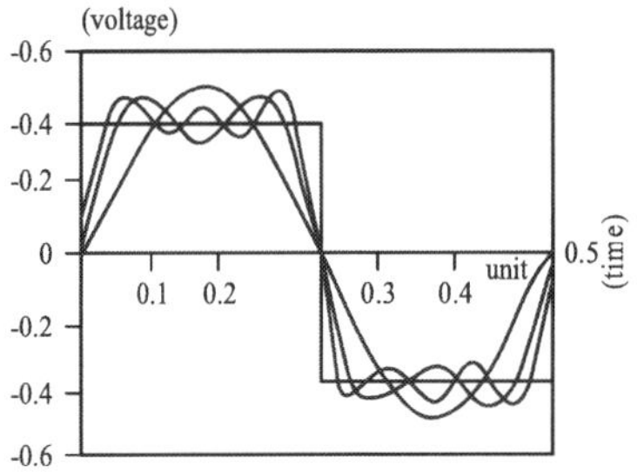

a) Rectangular wave

b) Fundamental (f_0)

c) Fundamental $(f_0) +$ second harmonic $(2f_0)$

d) Fundamental (f_0) Second harmonic $(2f_0)+$ third harmonic $(3f_0)$

➢ **Bandwidth of transmission medium :**

The most used transmission media are wire, free space, and fibre optic cable. Different transmission media offer different band width. Coaxial cable offers a band width of about 750 MHz. Radio wave communication through free space takes place over a wide range of frequencies from 100kHz-GHz.

Service	Frequency bands	Comments
Standard AM broadcast	540-1600 kHz	
FM broadcast	88-108 MHz	
Television	54-72 MHz	VHF(Very high frequencies)
	76-88 MHz	TV
	174-216 MHz	UHF(ultra high frequencies)
	420-890 MHz	TV
Cellular Mobile	896-901 MHz	Mobile to base station
	840-935 MHz	Base station to mobile
Satellite	5.925-6.425 GHz	Uplink
Communication	3.7 - 4.2 GHz	Downlink

➢ Optical communication using fibres is performed in the frequency range of 1 THz to 1000 THz (microwaves to ultraviolet).

➢ An optical fibre can offer a transmission band width in excess of 100 GHz.

➢ **COMMUNICATION CHANNELS:**

The medium or the link, which transfers message signal from the transmitter to the receiver of a communication system is called ***Communication channel***.

(i) Space communication

(i) Ground wave propagation

(ii) Space wave propagation. (Tropospheric wave propagation. Surface wave propagation.)

(iii) Sky wave propagation: A new dimension recently added to space communication is satellite communication.

2) Line communication

i) Two wire transmission line

ii) Coaxial cable

iii) Optical fibre cable

(1)SPACE COMMUNICATION

➢ **Propagation of EM waves in the atmosphere**

The communication process utilizing the physical space around the earth is termed as space communication. Electromagnetic waves which are used in Radio, Television and other communication system are radio waves and microwaves.

The velocity of electromagnetic waves of different frequency in a medium is different. It is more for red light and less for violet light. Electromagnetic waves are of transverse nautre.

➢ **Earth's atmosphere**

(i) Earth's atmosphere is a gaseous envelope which surrounds the earth.

(ii) Earth atmosphere mainly consists of nitrogen 78% , oxygen 21% along with a little protion of argon, carbon dioxide, water vapour, hydrocarbons, sulphur compounds and dust particles.

(iii) The density of the atmospheric air goes on decreasing as we go up.

(iv) The electrical conductivity of the atmospheric air increases as we go up.

(v) The various regions of earth's atmosphere are:

- **Troposphere.** It extends upto a height of 12km
- **Stratosphere.** It extends from 12km to 50km. There is an ozone layer in stratosphere which mostly absorbs high energy radiations like ultraviolet radiations. etc. coming from outer space.
- **Mesosphere.** It extends from 50km to 80km.
- **Ionosphere.** (i) It extends from 80 km to 400km.

(ii) In this region, the temperature rises to some extent with height, hence it is called Thermosphere.

(iii) The ionosphere which is composed of ionised matter (i.e. electrons and positive ions) plays an important role in space communication.

(iv) The ionosphere is subdivided into four main layers as D, E, F_1 and F_2.

(v) D- layer is at a virtual height of 80km from surface of earth and having electron density $\approx 10^9 m^{-3}$.

(vi) The extent of ionisation of D layer depends upon the altitude of sun. This layer disappears at night. It reflects very low frequency (VLF) and low frequency (LF) electromagnetic waves, but absorbs medium frequency (MF) and high frquency (HF) electromagnetic waves to a certain degree,

(vii) E-layer is at a virtual height of 110km, from the surface of earth, having electron density $\approx 10^{11} m^{-3}$. The critical frequency* of this layer is about 4MHz. This layer helps to MF surface-wave propagation a little but reflects some high frequency waves in day time. It exists in day as well as in night time.

(viii) $F_1 -$ layer is at a virtual height of 180km from the surface of earth, having electron density $\approx 5 \times 10^{11} m^{-3}$. The critical frequency for this layer is 5 MHz. It reflects some of the high frequency waves but most of the high frequencywaves pass through it and they get reflected from layer F_2 at night time.

(ix) F_2 layer is at a virtual height of about 300km in day time and about 350km in night time. The electron density of this layer is $\approx 8 \times 10^{11} m^{-3}$. The cirtical frequency of this layer is 8MHz in day time and 6MHz in night time. It reflects back the electromagnetic waves of frequency upto 30 MHz but cannot reflect back the electromagnetic waves of frequency 40MHz or more. It exists in day as well as night time

3. The electromagnetic waves of frequency ranging from a few kilo hertz to a few hundered mega hertz are called radio waves.

The various frequency ranges used in radio waves or micro wave communication system are as follows:

(a) Medium frequency band (M.F) 300 to 3000kHz.

(b) High frequency band (H.F) 3 to 30 MHz

(c) very high frequency band (V.H.F) 30 to 300 MHz.

(d) Ultra high frequency band (U.H.F) 300 to 3000MHz

(e) Super high frequency band (S.H.F) 3000 to 30,000 MHz.

(f) Extra high frequency band (E.H.F) 30 to 300 GHz

The radio waves emitted from a transmitter antenna can reach the receiver antenna by the following mode of operation.

- **KENNELY HEAVISIDE LAYER :**

i) At 110 km above the surface of earth the concentration of electrons is very large. This layer is called Kennely Heaviside layer.

ii) The thickness of this layers is about of few km.

iii) Beyond this layer the electron concentration decreases upto 250 km

iv) From 250 km to 400 km, a layer of large concentration of electrons called Apple ton layer exists.

v) Above appleton layer, ie above Ionosphere the temperature is $927.6^0 C$.

	Ground Wave Propagation	Sky Wave Propagation	Space Wave Propagation
Channel	Ground	Layers of atmosphere	Line of sight communication
Method	Wave glides over the surface of earth diffraction effect	Due to reflection of radio waves from the layers having higher electron density	The radio waves travel from transmitting antenna to receiving antenna along a straight line
Frequency	Depends on power and frequency Less than 2 MHz	3 MHz to 30 MHz	Greater than 40 MHz
Uses	In medium wave Broad casting	Short wave broadcasting	FM broadcasting and microwave links
Range	Depends on height of the Antenna and Curvature of earth	Depends on the angle of incidence on the ionosphere. 150 km to 3000 km	Due to curvature of the earth the waves are blocked at a point
Attenuation	Attenuation increases with frequency		

Different layers of atmosphere and their interaction with the propagating electromagnetic waves :

Name of the layer	Approximate height	Exists during	Frequencies most affected
Troposphere	10 km	Day and night	VHF (up to several GHz)
Parts of Ionosphere : D (part of Stratosphere)	65-75 km	Day only	Reflects LF, absorbs MF and HF to some degree
E (Part of Stratosphere)	100 km	Day only	Helps surface waves reflects HF
FI (Parts of Mesophere)	170-190 km	Day time, merges with F2 at night	Partially absorbs HF waves yet allowing them to reach F2
F2 (Thermosphere)	300 km at night 250-400 km during day time	Day and night	Efficiently reflects HF waves, particularly at night

- **Ground wave propagation :** In this method, the radio waves are guided along the surface. The wave induces charges on the earth. These charges travel with the wave and this forms a current. Now the earth behaves like a leaky capacitor in carrying the induced current. The wave loses some energy, as energy is spent due to flow of charge through the earth's resistance. The wave also looses energy due to diffraction as it glides along the ground. The loss of energy increases as the frequency increases. Thus ground propagation is suitable upto 2MHz. As they loose energy they cannot go to long distances on the ground. Maximum range of the ground wave can be increased by increasing the power of the transmitter.
- **Sky wave propagation :** Above 2MHz and upto 30MHz, long distance communication takes place through ionosphere. The ionosphere reflects the radio waves back to the earth. This method is called sky wave propagation. It is used for shortwave broad casting services. Ionosphere is a thick blanket of 65 km to 400 km above the earth's surface. UV rays and other higher energy radiation coming from space results in the ionization of air molecules. The ionosphere is further divided into serval layers as shown in table below. It should be understood that degree of ionization changes with height. This is because the density of atmosphere decreases with height. At great heights, the radiation is intense, but the molecules available are few. On the other hand, near the earth's surface the molecular concentration is high but the intensity of radiation is low and thus again the ionization is low. Logically, the peak of ionization density occurs at some intermediate heights. The ionosphere acts as a mirror (reflector) for

frequencies of 3-30 MHz. Electromagnetic waves of frequences greater than 30 MHz pass through the atmosphere and skip.

The process of bending of EM waves is similar to total internal reflection in optics. The bending of waves can be easily explained on the basis of variation of refractive index of the ionosphere with change in electron density. Suppose that a radio wave enters the ionosphere from the underlying unionized medium. Since the refractive index of ionosphere decreases from D layer to F_2 layer, consequently, the incident ray will move away from the normal drawn at the point of incidence following the ordinary laws of refraction

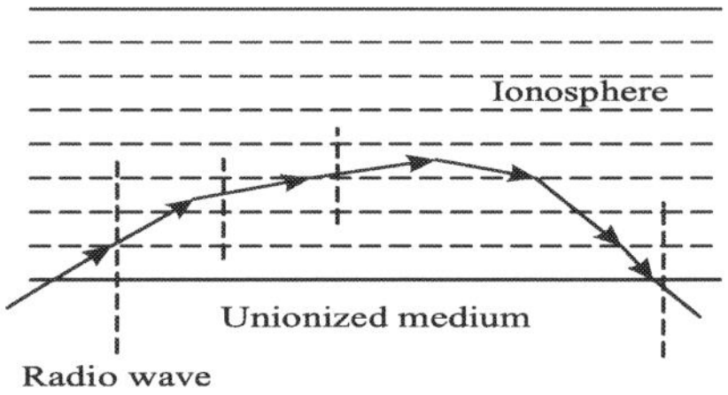

During the propagation in ionosphere the angle of refraction gradually increases and the ray goes on bending more and more till at some point, the angle of refraction becomes 90^0 and the wave travels parallel to the earth surface. This point is called point of reflection. Then the ray tends to move in the down ward direction and comes back to earth because of symmetry. Super high frequency (SHF) waves propagate as sky waves taking reflection at satellite.

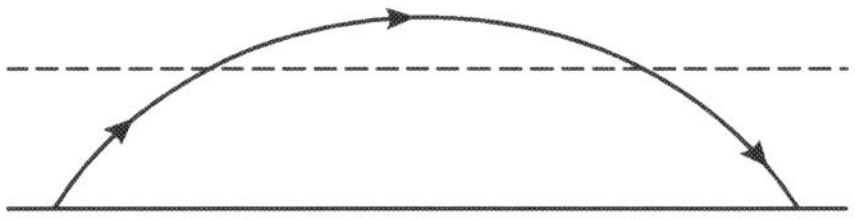

The sky wave propagation can cover a very long distance and so round the globe communication is possible.

(c) The sky waves being electromagnetic in nature, changes the dielectric constant and refractive index of the ionosphere. The effective refractive index of the ionosphere is

$$n_{eff} = n_0\left[1 - \frac{Ne^2}{\varepsilon_0 m\omega^2}\right]^{1/2} = n_0\left[1 - \frac{80.5N}{f^2}\right]^{1/2}$$

Where n_0 = refractive index of free space, N = electron density of ionosphere, ε_0 = permittivity of free space, e = charge on electron, m = mass of electron w = angular frequency of EM wave.

(d) As we go deep into the ionosphere, N increases so n_{eff} decreases. The refractions or bending of the beam will continue and finally it reflects back.

(e) The highest frequency of radio wave, which gets reflected to earth by the ionosphere after having been sent straight to it is

➢ ***Critical frequency (f_c)***

If maximum electron density of the ionosphere is N_{max} per m^3, then $f_c \approx 9(N_{max})^{1/2}$. Above f_c, a wave will penetrate the ionosphere and is not reflected by it.

(f): The highest frequency of radio waves which when sent at some angle of incidence, towards the ionosphere, get reflected and return to the earth is ***Maximum usable***

- ***frequency (MUF)*** $MUF = \frac{f_c}{\cos\theta}$

(g) The smallest distance from a transmitter along the earth's surface at which a sky wave of a fixed frequency but more than f_c is sent back to the earth is ***Skip distance.***

(h) The fluctuation in the strength of a signal at a receiver due to interference of two waves is ***fading***. Fading is more at high frequencies. It results into errors in data transmission and retrieval.

- **Wave propagation :** This method is used for line-of-sight [LOS] communication and also for satellite communication. At frequencies above 40MHz, communication is mainly by LOS method. At such frequencies, relatively smaller antenna can be erected above the ground. Because of LOS propagation, the direct waves get blocked, at some point due to the curvature of the earth as shown in the figure.

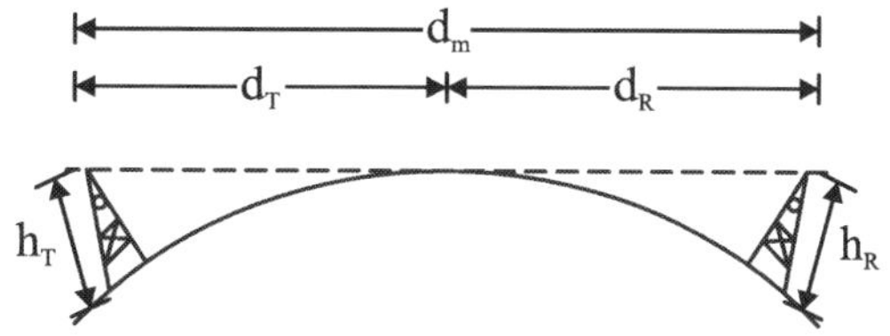

For the signal to be received beyond the horizon, the receiving antenna must be high enough to intercept the LOS waves. If the transmitting antenna is at a height h_T then it can be shown that the distance to the horizon d_T is given by $d_T = \sqrt{2Rh_T}$ where 'R' is the radius of earth. Similarly if the receiving antenna is at a height h_R, the distance to the horizon d_R is $d_R = \sqrt{2Rh_R}$

$\therefore$ The maximum distance d_M between the two antennas is $d_M = \sqrt{2Rh_T} + \sqrt{2RH_R}$ where

R = Radius of the earth.

h_T = height of the transmitting antenna and

h_R = height of the receiving antenna.

If the Population density around the tower is given, the number of persons covered by the transmitting tower

= (Area covered by the tower) $\times$ Population density.

$\therefore$ No. of persons $= \pi d^2 \times$ covered population density (Here d = radius of the area covered by single transmitting tower of height h_T)

Television broadcast, microwave and satellite communications are a few examples of communication systems that use space wave propagation. The figure below illustrates the various modes of wave propagation.

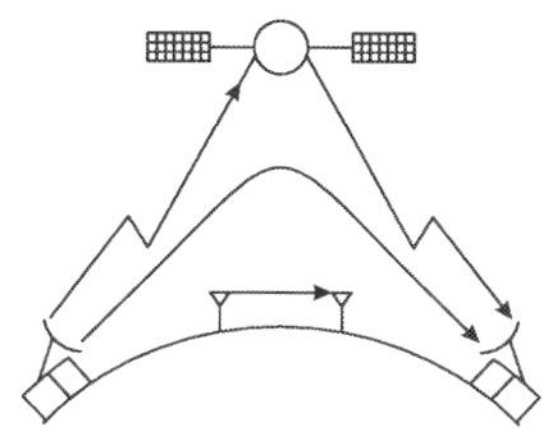

- **Range of TV transmission :**

As the frequency range of TV signals is 100-200 MHz, such signal transmission via ground waves is not possible. In such situations, we use line of sight transmission.

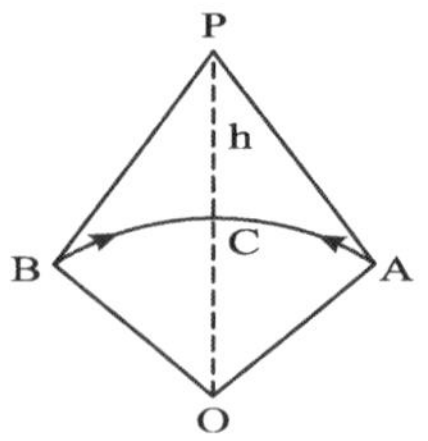

Let CP be the TV tower on the earth's surface. It's antenna is at P. Let PC = h. When TV broadcast is made, the signal can reach the earth upto A to B. There will be no reception of the signal beyond A and B. Arc length CA and CB is the range of TV transmission. If O is the centre of the earth, OA = OB = R is the radius of the earth, from right angled triangle OAP

$OP^2 = OA^2 + PA^2$

$(h+R)^2 = R^2 + PA^2$

$PA = PB = d$

$(h+R)^2 = R^2 + d^2$

$h^2 + R^2 + 2Rh = R^2 + d^2$

As $h << R$ we can ignore h^2

$d^2 = 2Rh$ and $d = \sqrt{2Rh}$

Range of TV transmission depends upon the height of the transmission antenna. Broadcasts are made from tall transmitting antenna.

- A **repeater** is a combination of a receiver and a transmitter. A repeater, picks up the signal from the transmitter, amplifies and retransmits it to the receiver sometimes with a change in carrier frequency. Repeaters are used to extend the range of a communication system as shown in figure. A communication satellite is essentially a repeater station in space. Use of repeater station to increase the range of communication

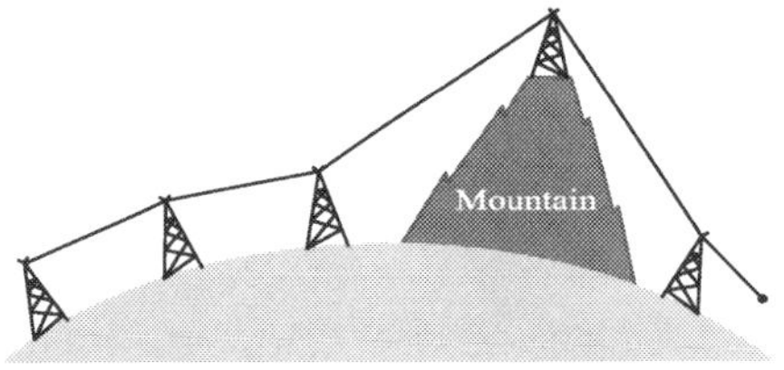

These problems are solved by using geostationary satellite as a communication satellite.

- **Satellite Communication:** Long distance communication beyond 10 to 20 MHz was not possible before 1960 because all the three modes of communication discussed above failed (ground waves due to conduction losses, space wave due to limited line of sight and sky wave due to the penetration of the ionosphere by the high frequencies beyond f_c).

Ionosphere behaves as a rarer medium by which carrier wave is reflected back if its frequency $f\,(\le f_c)$

where f_c is called a "critical frequency" and is given by $f_c = f_0\,(N_{max})^{\frac{1}{2}}$

[N_{max} = maximum electron density.]

Satellite communication made this possible. The basic principle of satellite communication is shown in figure. A communication satellite is a spacecraft placed in an orbit around the earth. The frequencies used in satellite communication lie in UHF/ microwave regions. These waves can cross the ionosphere and reach the satellite.

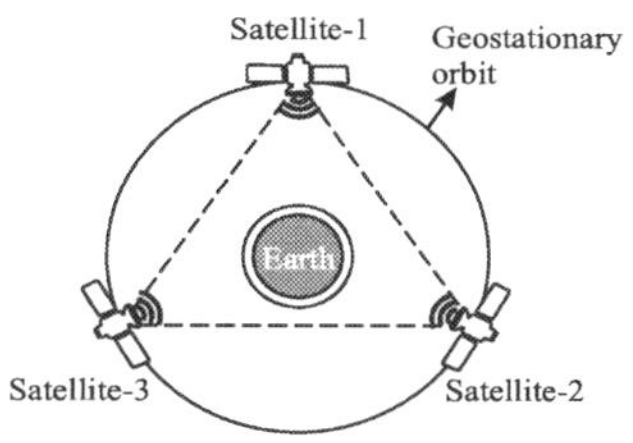

- A geostationary satellite has the same time period of revolution of earth. It locates at the height of 36000 *km* above the earth's surface (well above the ionosphere).
- A communication satellite is a spacecraft placed in an orbit around the earth which carriers a transmitting and receiving equipment called radio transponder. It amplifies the microwave signals emitted by the transmitter from the surface of earth and send to the receiving station on earth.
- The transmitted signal is UP-LINKED and received by the satellite station which DOWN-LINKS it with the ground station through its transmitter.

The up-link and down-link frequencies are kept different (both frequencies being in the regions of UHF/ microwave).

- At least three geo-stationary satellites are required which are 120° apart from each other to have the communication link over the entire globe of earth.
- Satellite technology is very useful in collecting information about various factors of the atmosphere which governs the weather and climatic conditions.

The satellite communication can be used for establishing mobile communication with great use the communication satellites are now being used in Global Positioning System (GPS). The ordinary users can find their positions within accuracy of $100m$.

There are two types of satellites used for long distance transmission.

(i) Passive satellite: It acts as reflector only for the signals transmitted from earth. Moon the natural satellite of earth is a passive satellite.

(ii) Active satellite: It carries all the equipment used for receiving signals sent from the earth, processing them and then re-transmitting them to the earth. Now a days active satellites are in use.

- The Indian communication satellites INSAT-2B and INSAT-2C are positioned in such away in the outer space that they are accessible from any place in India.

REMOTE SENSING AND APPLICATION OF SATELLITE COMMUNICATION.

Remote sensing is the technique to obtain information about an object by observing it from a distance and without coming to actual contact with it.

- There are two types of remote sensing instruments: active and passive. Active instruments provide their own energy to illuminate the object of interest, as radar does. They send an energy pulse to the object and then receive and process the pulse reflected from the object. Passive instruments sense only radiations emitted by the object or solar radiation reflected from the object.
- The remote sensing is done through a satellite. The satellite used in remote sensing should move in an orbit around the earth in such a way that it always passes over the particular area of the earth at the same local time.

The orbit of such a satellite is known as sun-synchronous orbit. A remote sensing orbit can be circular polar orbit or in highly inclined elliptical orbit.

- A remote sensing satellite takes, photographs of a particular region which nearly the same illumination every time it passes through that region.
- The most useful remote sensing technology is that it makes possible the repetitive surveys of vast areas in a very short time, even if the areas are inaccessible.
- Space based remote sensing is a new technology. It has high potential for applications in nearly all aspects of resource management.
- The Indian remote sensing satellites are

  IRS-1A, IRS-1B, and IRS-1C.
- Remote sensing is applied in (i) Meteorology (ii) Climatology (iii) Oceanography

  (iv) Archaeology, geological surveys. (v) Water resource surveys, (vi) Urban land use surveys. (vii) Agriculture and forestry and natural disaster.

  (viii) To detect movements of enemy army. (ix)To locate the place where underground nuclear explosion has carried out.
- **2.LINE COMMUNICATION**

  Line communication means interconnection of two points with the help of wires for exchange of information. There are three principal types

  (i)Two Wire Transmission Line

  (ii) Coaxial wire lines (coaxial cables)

  (iii) optical fibers
- **Two Wire Transmission Line**

  The most commonly used two wire lines are: Parallel wire, twisted pair wires and co-axial cable.

(1) ***Parallel wire line:*** In a two wire transmission line, two metallic wires (may be hard or flexible) are arranged parallel to each other inside a protective insulation coating . Commonly used to connect an antenna with TV receiver. Such wires can suffer from interferences and losses.

(2) ***Twisted pair wire:*** It consists of two insulated copper wires twisted around each other at regular intervals to minimize electrical interference (to connect telephone systems). Used to connect telephone systems. It works well up to small distances. They cannot transmit signals over very large distances. They transmit both, the analog and digital signals. They are easy to install and cost effective.

- **Coaxial wire lines:** It consists of a central copper wire (which transmits surrounded by a PVC insulation over which a sleeve of copper mesh (outer conductor) is placed. The outer conductor is normally connected to ground and thus it provides an electrical shield to the signals carried by the central conductor. The outer conductor is externally covered with a polymer jacket for protection.

(a) Co-axial line wires can be used for microwaves and ultra-high frequency waves.

(b) The communication through co-axial lines is more efficient than through a twisted pair wire lines.

(c) Co-axial cables can be gas filled also. To reduce flash over between the conductor handling high power, N_2-gas is used in the cable.

- ***Impedance of Line***

  (1)Each portion of the transmission line can be considered as a small inductor, resistor and capacitor. As a result each length of transmission line has characteristic impedance.

(2) In case of co-axial cable, the dielectric can be represented by a shunt resistance G.

(3) When co-axial cable is used to transmit a radio frequency signal, X_L and X_C are large as compared to R and G respectively. Hence R and G can be neglected.

(4) In co-axial cable, R is zero, so no loss of energy and hence no attenuation of frequency signal occurs when transmitted along it. That's why co-axial cables are specially used in cable TV network.

- ***Characteristic impedance (Z_0)* :** It is defined as the impedance measured at the input of a line of infinite length.

a)For parallel line $Z_0 = \frac{276}{\sqrt{k}} \log \frac{2s}{d}$

d = Diameter of each wire

s = Separation between the two wires

k = Dielectric constant of the insulating medium

b) For co-axial line wire $Z_0 = \frac{138}{\sqrt{k}} \log \frac{D}{d}$

d = Diameter of inner conductor

D = Diameter of outer conductor

c) At radio frequency $Z_0 = \sqrt{\frac{L}{C}}$

The usual range of characteristic impedance for parallel wire lines is 150W to 600 W and for co-axial wire it is 40 W to 150 W.

- **Velocity factor of a line (*v. f.*) :** It is the ratio of reduction of speed of light in the dielectric of the cable

$$v.f. = \frac{v}{c} = \frac{\text{Speed of light in medium}}{\text{Speed of light in vacuum}} = \frac{1}{\sqrt{K}}$$

For a line *v.f.* is generally of the order of 0.6 to 0.9.

- **TELEPHONE LINKS**

(1) A telephone (the most common means of communication) link can be established with the help of co-axial cables, ground waves, sky waves, microwaves or optical fiber cables.

(2) Simultaneous transmission of a number of messages over a single channel without their interfering with one another is called multiplexing.

(3) Twisted pair wire lines provide a band width of 2 *MHz*, while co-axial cable provides a band width of 20 *MHz*. For further increase in band width, we use (i) microwave link (ii) communication satellite link.

- **OPTICAL COMMUNICATION**

The use of optical carrier waves for transmission of information from one place to another is called optical communication.

The information carrying capacity $\propto$ bandwidth $\propto$ frequency of carrier wave. Because of high frequency (10^{12} *Hz* to 10^{16} *Hz*) optical communication is better than others. (radio and microwave frequencies, 10^6 *Hz* – 10^{11} *Hz*).

Basic optical communication link is a point to point link having transmitter at one end, receiver at the other end and consists of three components namely

(1) Optical source and modulator

(2) Optical signal detector or photodetector (3) Optical fibre cable through which optical signal is transmitted.

- **Optical sources for communication links**

Light emitting diodes (LED) and diode lasers are preferred for optical source. LEDs are used for small distance transmission while diode laser is used for very large distance transmission.

For optical communication, light is to be **modulated** with the information signal. The frequency and intensity of light is sensitive to temparature changes,which is to be avoided. So suitable arrangement isrequired to obtain **thermal stability.**

- **Optical signal detector or photodetector:**

The optical signal reaching the receiving end has to be detected by a detector which converts light intoelectrical signals, So that the transmitted information may be decoded.

The optical detector should have

(i) size compatible with the fibre

(ii) High sensitivity at the desired optical wavelength

(iii) High response for fast speed data transmission/reception.

Semiconductor based photo-electors are used because they fullfill the above criteria

- **MODULATION AND ITS NECESSITY :**

Message signals are also called base band signals. Which essentially designate the band of frequencies representing the original signal, as delivered by the source of information. No signal, is a single frequency sinusoid, but it spreads over a range of frequencies called the signal bandwidth to transmit an electric signal frequency less than 20 kHz) over a long distance directly. It is clear that low frequency waves, can not travel long distances. Hence, to transmit low frequency wave over long distance, we take the help of high frequency waves called carrier wave. The low frequency wave is superposed over high frequency carrier wave. This process is called the modulation. The low frequency wave is called the modulating wave and the high frequency wave is called the carrier wave, and the resultant wave is called modulated wave. In this section we will discuss in detail about modulation. What is it? What is the need of modulation or how modulation is done etc.

No signal in general, is a single frequency but it spreads over a range of frequencies called the signal bandwidth. Suppose we with to transmit an electronic signal in the audio-frequency (20Hz-20kHz) range over a long distance. Can we do it? No it cannot because of the following problems.

- **Size of antenna :** For transmitting a signal we need an antenna. This antenna should have a size comparable to the wavelength of the signal. For an electromagnetic wave of frequency 20kHz, wave length is 15km. Obviously such a long antenna is not possible and hence direct transmission of such signal is not practical.

The linear size of the antenna must be the order of the wave length and for effective transmission its length must be $h = \frac{\lambda}{4}$

so that antenna properly senses the time variation of the signal.

Example1: For an electromagnetic wave of

$f = 20$ kHz, $\lambda = 15$ km Obviously, such a long antenna is not possible to construct and operate.

Hence direct transmission of such baseband signals is not practical.

Example2: If $f = 1$ MHz, then $\lambda = 300$ m

$h = 75$ m

Therefore, there is a need of translating the information contained in our original low frequency baseband signal into high or radio frequencies before transmission.

- ***Additional Information:***

(a) The distance between transmitting antenna and the horizon, $D_t = \sqrt{2Rh_t}$.

Where h_t = height of transmitting antenna R = Radius of the earth

(b) The distance between receiving antenna and the horizon, $D_r = \sqrt{2Rh_r}$.

Where h_r = height of receiving antenna

(c) The maximum distance between the transmitting antenna and receiving antenna D_m.

$D_m = D_r + D_t$

$D_m = \sqrt{2Rh_r} + \sqrt{2Rh_t}$.

Where R is the radius of earth.

$h_r > h_t$ so then the receiving antenna intercepts the line of sight waves.

➤ **Single antenna**

(d) The radius "d" of the area covered by a single transmitting tower of height h is given by $d = \sqrt{2R_e h}$. Where R_e is the radius of the Earth.

(e) If the Population density around the tower is given, the number of persons covered by the tower is = (Area covered by the tower) x Population density No. of persons covered = πd^2 x Population-density.

➤ **Effective power radiated by an antenna**

Power radiated by an antenna is proportional to $\left(\frac{l}{\lambda^2}\right)$. Where l is length of the antenna.

For a good transmissmission,high powers are required, hence low wavelength i.e high frequency transmissions are needed.

➤ **Mixing up of signals from different transmitters**

Suppose many people are talking at the same time or many transmitters are transmitting baseband information signals simultaneously. All these signals will get mixed up and there is no simple way to distinguish between them. This points out towards a possible solution by using communication at high frequencies and alloting a band of frequencies to each mesage signal for its transmission.

In doing so, we take the help of a high frequency signal, known as the carrier wave, and a process known as **modulation** which attaches information to it.

➤ **Modulation:** The process of superimposing information contained in the low frequency, message signal on a high frequency carrier wave, near transmitter is known as **modulation**.

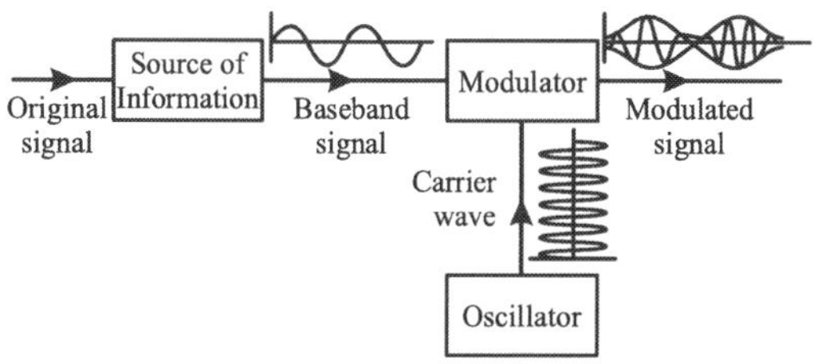

➤ **TYPES OF MODULATION:**

The carrier wave may be continuous (sinusoidal) or in the form of pulses as shown in figure(2).

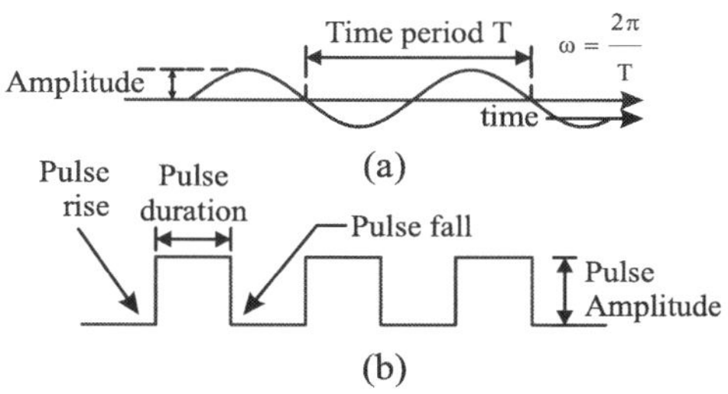

Therefore depending upon the specific characteristicof carrier wave which is being varied in accordance with the message signal, modulation can basically be differentiated as

(i) continuous wave modulation; and

(ii) pulse wave modulation.

- **According to the type of modulation**

For sinusodial continuous carrier waves

i) Amplitude Modulation (AM)

ii) Frequency Modulation (FM)

iii) Phase Modulation

- **For pulsed carrier waves**

i) Pulse Amplitude Modulation (PAM)

ii) Pulse Time Modulation (PTM)

a) Pulse Position Modulation (PPM)

b) Pulse Width Modulation (PWM) or Pulse Duration Modulation (PDM)

iii) Pulse Code Modulation (PCM)

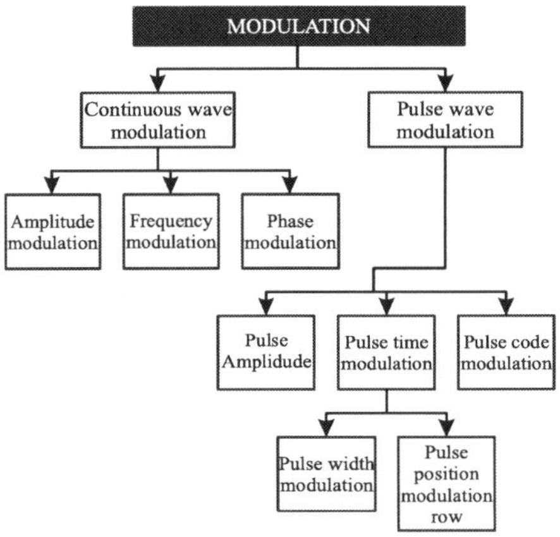

I) Continuous Wave Modulation

Eqution representing sinusoidal carrier wave can

$c(t) = A_c \sin(\omega_c t + \phi)$ ----(1)

where c(t) is the signal strength (voltage or current),

A_c is the amplitude

$(\omega_c t + \phi)$ is called argument ofPhase angle of the carrier wave

$\omega_c (= 2\pi f_c)$ is the angular frequency

ϕ is the initial phase of the carrier wave. During the process of modulation, any of the two parameters, viz amplitude or phase angle, of the carrier wave can be controlled by the message or information signal. This results in two **types of modulations:**

i) Amplitude modulation (AM)

ii) Angle modulation

Angle modulation again can be of two types. They are

i) Frequency modulation (FM)

ii) Phase modulation (PM)

As shown in figure.

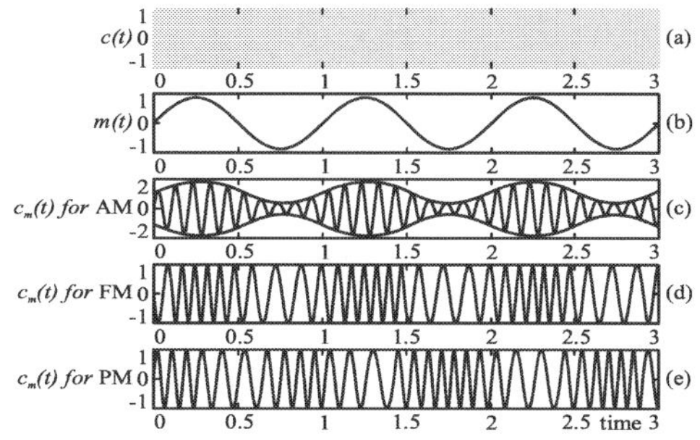

II) Pulse Wave Modulation.

The significant characteristics of a pulse are: i) Pulse amplitude

ii) Pulse duration or pulse width

iii) pulse position (denoting the time of rise or fall of the pulse amplitude) as shown in figure (3).

Types of pulse modulation:

a) pulse amplitude modulation (PAM),

b) pulse duration modulation (PDM) or pulse width modulation (PWM)

c) pulse position modulation (PPM).

I Continuous Wave Modulation:

1) Amplitude Modulation:

The method in which the amplitude of carrier is varied in accordance with the modulating signal keeping the frequency and phase of carrier wave constant is called amplitude modulation (AM).

Here we explain amplitude modulation process using a sinusoidal signal as the modulating signal.

Let $m(t) = E_m \sin \omega_m t$ represent the message or the modulating or base band signal. Here $\omega_m = 2\pi f_m$ is the angular frequency of the message signal.

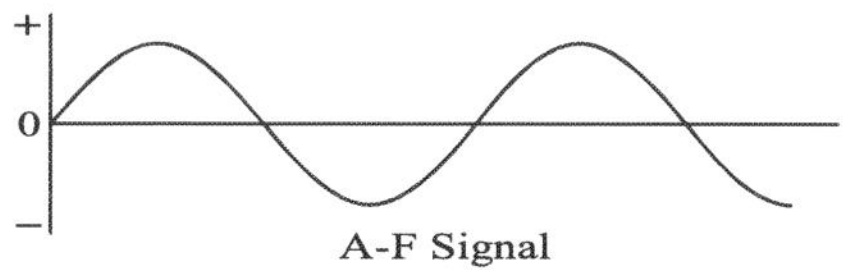

A-F Signal

$c(t) = E_c \sin \omega_c t$ represent carrier wave. Here $\omega_c = 2\pi f_c$ is the angular frequncy of the carrier signal

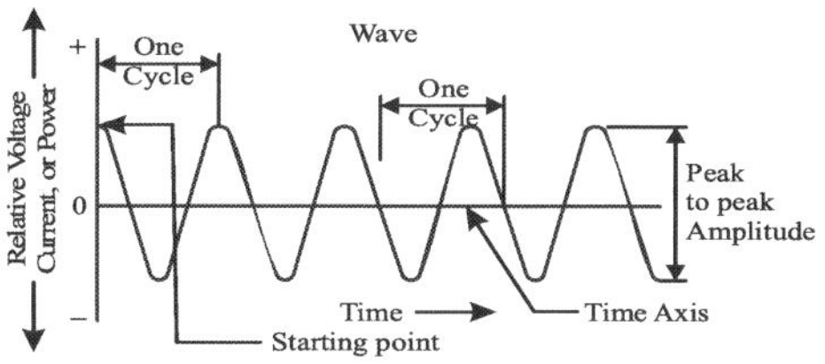

The modulated signal $c_m(t)$ can be written as

$$c_m(t) = (E_c + E_m \sin\omega_m t)\sin\omega_c t \text{ ----(1)}$$

$$c_m(t) = A_c\left(1 + \frac{E_m}{E_c}\sin\omega_m t\right)\sin\omega_c t \text{ -----(2)}$$

$$c_m(t) = E_c \sin\omega_c t + \mu E_c \sin\omega_m t \sin\omega_c t \text{ --(3)}$$

Using the trignometric relation

$\sin A \sin B = \frac{1}{2}\left(\cos(A-B) - \cos(A+B)\right)$, we can write $c_m(t)$ of equation (3) as

$$c_m(t) = A_c \sin\omega_c t + \frac{\mu A_c}{2}\cos(\omega_c - \omega_m)t - \frac{\mu A_c}{2}\cos(\omega_c + \omega_m)t \text{ -----(4)}$$ Here

$\omega_c - \omega_m = 2\pi(f_c - f_m)$ = Lower side band frequency (LSB)

$\omega_c + \omega_m = 2\pi(f_c + f_m)$ =Upper side band frequency (USB)

Here (m or $\mu = E_m/E_c$) is the modulation index; (or) modulating factor.

In practice, μ is kept ≤ 1 to avoid distorition.

Depth of modulation $= \frac{A_m}{A_C} \times 100 = \mu \times 100$

Depth of modulation interms of E_{max} and E_{min}

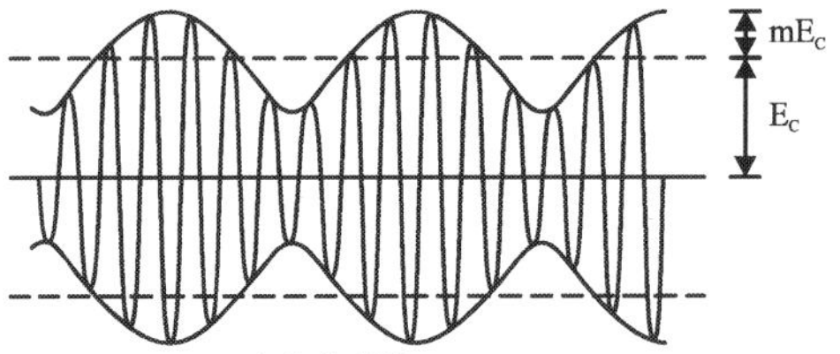

A.M. Wave

$$E_{max} = E_c + E_m = E_c(1+m)$$

$$E_{min} = E_c - E_m = E_c(1-m)$$

$$\frac{E_{max}}{E_{min}} = \frac{E_c + E_m}{E_c - E_m} = \frac{E_c(1+m)}{E_c(1-m)}$$

$$\boxed{\frac{E_{max} - E_{min}}{E_{max} + E_{min}} = m_a}$$

The Band width of AM wave is " $2f_m$ "

The modulated signal now consists of the carrier wave of frequency ω_c plus two sinusoidal waves each with a frequency slightly different from ω_c, known as side bands. The frequency spectrum of the amplitude modulated signal is shown in

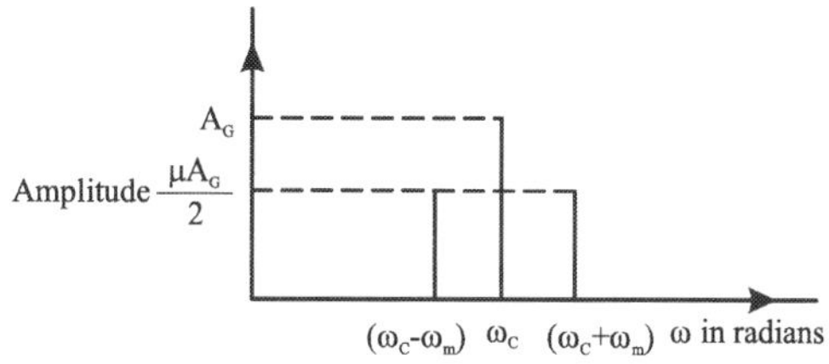

figure (5) A plot or amplitude versus ω for an amplitude modulated signal

As long as the broadcast frequencies (carrier waves) are sufficiently spaced out so that sidebands do not overlap, different stations can operate without interfering with each other.

Special cases of Amplitude modulation:

CaseI: In the absence of signal.

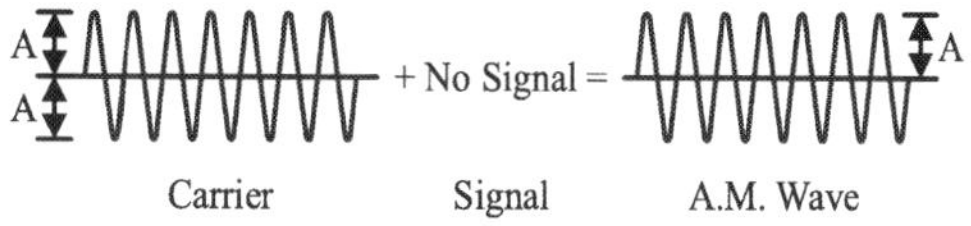

Modulation factor $m_a = \frac{O}{A} \times 100 = 0\%$

Case-II: When the signal amplitude is equal to CW wave.

Amplitude varies from 2A to zero.

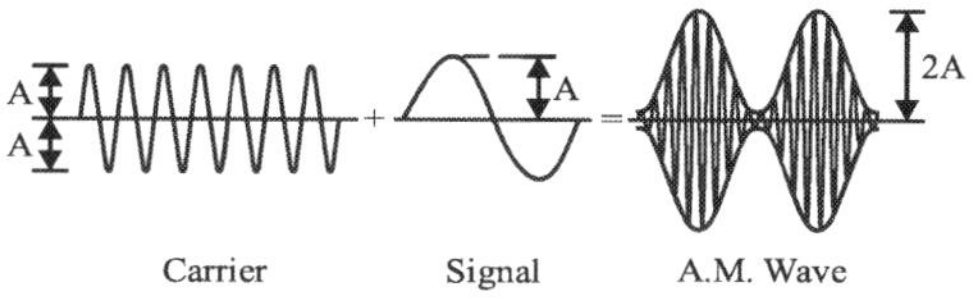

$$\frac{\text{Amplitude change in carrier wave}}{\text{Amplitude of CW}} = \frac{2A - A}{A} = 100\%$$

Case-III: When the amplitude of the signal is half of that of CW.

Amplitude of CW changes from A to $\left(A + \frac{A}{2}\right) = 1.5\,A$

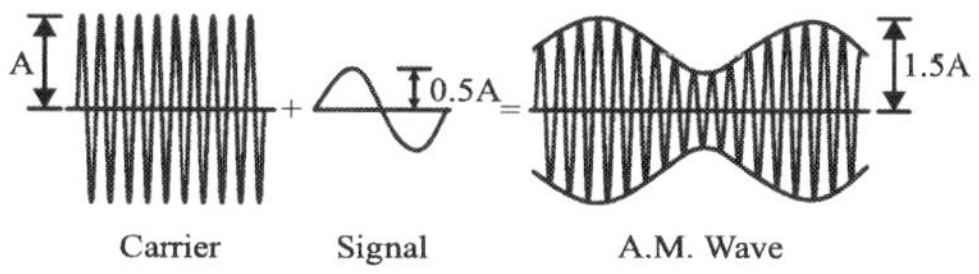

$$\text{Modulation factor} = \frac{0.5\,A}{A} = 0.5$$

$$= 50\,\%$$

Case-IV: When the amplitude of signal is 1.5 times that of the CW.

Amplitude of the modulated wave changes from 2.5 A to A

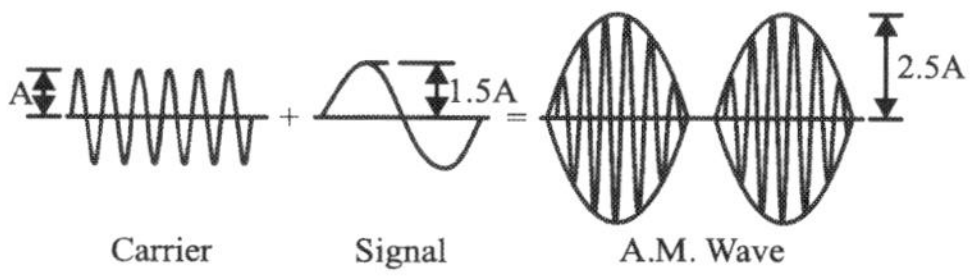

Modulation factor $m_a = \frac{2.5A - A}{A} = 1.5 = 150\,\%$.

In this case the quality of signal is lost

Note: A carrier wave is modulated by a number of sine waves with modulation indices m_1, m_2 and m_3. The total modulation index of the wave is

$$m = \sqrt{m_1^2 + m_2^2 + m_3^2}$$

- **Power out put in AM wave**

$$P_t = P_c + P_s$$

where P_t is power transmitted

P_c is power of carrier wave

P_s is total power of side bands

The equation of a carrier wave $Y_c = A_c \sin(w_c t + \phi)$

Power of carrier wave $P_c = \frac{[A_{rms}]^2}{R} = \frac{\left(\frac{A_c}{\sqrt{2}}\right)^2}{R} = \frac{A_c^2}{2R}$

The power of side bands = The power of lower side band + the power of upper side band

$$P_s = \frac{\left[\frac{(\mu A_c)}{2}/\sqrt{2}\right]^2}{R} + \frac{\left(\mu\frac{A_c}{2}/\sqrt{2}\right)^2}{R}$$

$$= \frac{\mu^2 A_c^2}{4R} = \frac{\mu^2}{2} P_c$$

$$P_t = P_c + P_s = P_c + \frac{\mu^2}{2} P_c$$

$$= P_c\left(1 + \frac{\mu^2}{2}\right)$$

$$\therefore P_t = P_c\left(1 + \frac{\mu^2}{2}\right) ----(1)$$

$$\Rightarrow \frac{P_t}{P_c} = 1 + \frac{\mu^2}{2} ----(2)$$

$$\Rightarrow \left(\frac{i_t}{i_c}\right)^2 = 1 + \frac{\mu^2}{2} ----(3)\left[\because P \alpha i^2\right]$$

Example: If the modulation factor is 1 ie 100 % modulation then the useful power is 1/3 of the total power radiated. The remaining 2/3 power is contained by carrier wave

$$\boxed{\frac{P_S}{P_T} = \frac{m_a^2}{2 + m_a^2} = \frac{1}{3}} \text{ and } \boxed{\frac{P_c}{P_T} = \frac{2}{2+m_a^2} = \frac{2}{3}}$$

Transmission Efficiency $\eta = \dfrac{m^2}{2+m^2}$

PRODUCTION OF AMPLITUDE MODULATED WAVE

Amplitude modulation can be produced by a variety of methods. A conceptually simple method is shown in the block diagram of Fig.

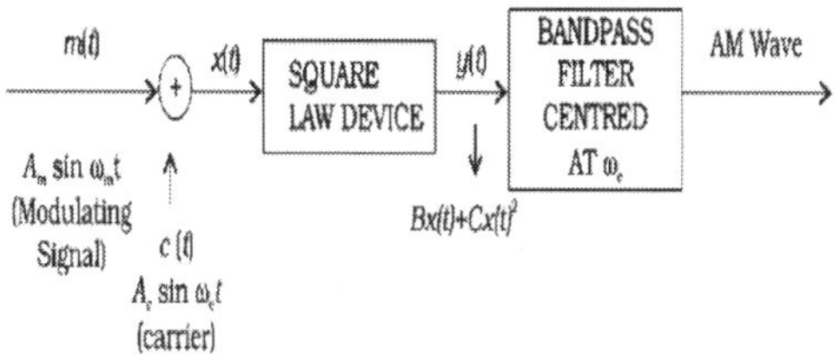

Here the modulating signal $A_m \sin \omega_m t$ is added to the carrier signal $A_c \sin \omega_c t$ to produce signal

$x(t) = A_m \sin \omega_m t + A_c \sin \omega_c t$

This signal is passed through a square law device which is a non linear device that can give tha output

$y(t) = Bx(t) + Cx^2(t)$ where B and C are constants

Thus,

$$y(t) = BA_m \sin \omega_m t + BA_c \sin \omega_c t$$

$$+C\left[A_m^2 \sin^2 \omega_m t + A_c^2 \sin^2 \omega_c t + 2A_m A_c \sin \omega_m t \sin \omega_c t\right] \text{-(6)}$$

$$= BA_m \sin \omega_m t + BA_c \sin \omega_c t + \frac{CA_m^2}{2} + A_c^2 - \frac{CA_m^2}{2}\cos 2\omega_m t - \frac{CA_c^2}{2}\cos 2\omega_c t$$

$$+CA_m A_c \cos(\omega_c - \omega_m)t - CA_m A_c \cos(\omega_c + \omega_m)t \text{ -(7)}$$

Where the trignometric relations $\sin^2 A = (1 - \cos 2A)/2$ and the relation for sinA sinB mentioned earilier are used.

In equation, there is a dc term $c/2\left(A_m^2 + A_c^2\right)$ and sinusoids of frequencies

$\omega_m, 2\omega_m, \omega_c, 2\omega_c, \omega_c - \omega_m$ and $\omega_c + \omega_m$. As shown in figure this signal is passed through a band pass filter which rejects dc and the sinusoids of frequencies $\omega_m, 2\omega_m$ and $2\omega_c$ and retains the frequencies $\omega_c, \omega_c - \omega_m$ and $\omega_c + \omega_m$. The output of the band pass filter therefore is of the same form as equation(4) and is therefore an AM wave.

It is to be mentioned that the modulated signal cannot be transmitted as such. The modulator is to be followed by a power amplifier which provides the necessary power and then the modulated signal is fed to an antenna of appropriate size for radiation as shown in figure(7).

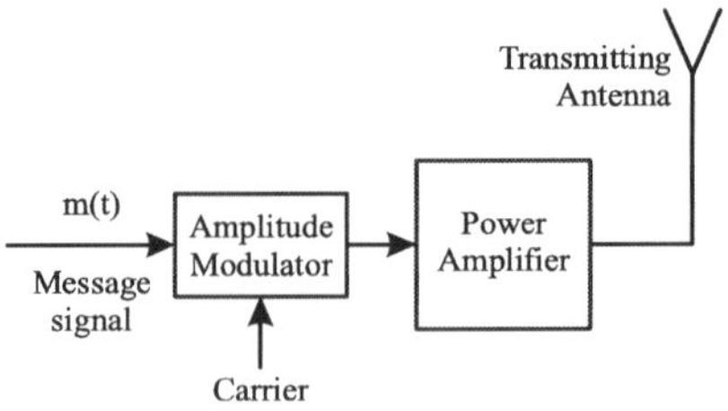

Figure(7)

- **A-M Transmitter.**

  In the block diagram of the AM transmitter the r-f section consists of an oscillator feeding a buffer, which in turn feeds a system of frequency multipliers and/or intermediate power amplifiers. If frequency multiplication is unneccessary, the buffer feeds directly into the intermediate power amplifiers which, in turn, drive the final power amplifier. The input to the antenna is taken from the final power amplifier.

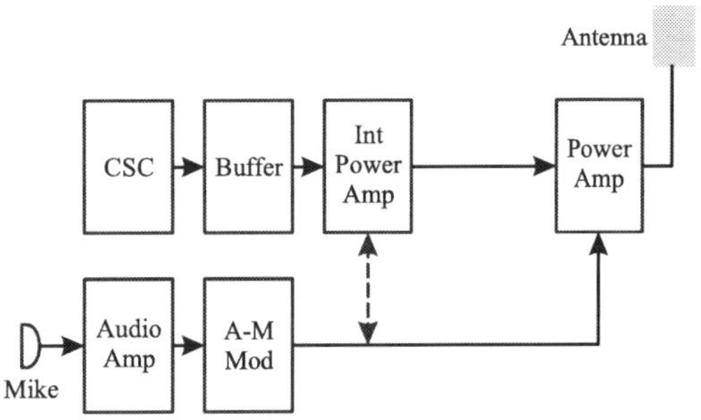

Amplitude-Modulated Transmitter

- **Detection of amplitude modulated wave**

  The transmitted message gets attenuated in propagating through the channel. The receiving antenna is therefore to be followed by an amplifier and a detector. In addition, to facilitate further processing, the carrier frequency is usually changed to a lower frequency by what is called an intermediate frequency (IF) stage preceding the detection. The detected signal may not be strong enough to be made use of and hence is required to be amplified. A block diagram of a typical receiver is shown in figure.

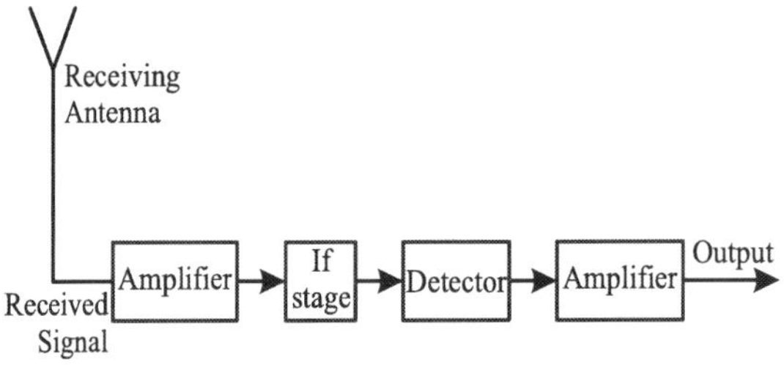

- **Simple demodulator circuit :**

  1) A diode can be used to detect or demodulate an amplitude modulated (AM) wave.

  2) A diode basically acts as a rectifier *i.e.* it reduces the modulated carrier wave into positive envelope only.

  3) The AM wave input is shown in figure. It appears at the output of the diode across PQ as a rectified wave (since a diode conducts only in the positive half cycle). This rectified wave after passing through the RC network does not contain the radio frequency carrier component. Instead, it has only the envelope of the modulated wave.

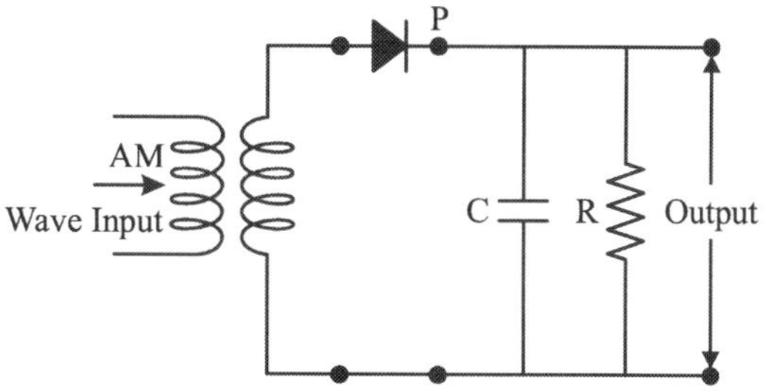

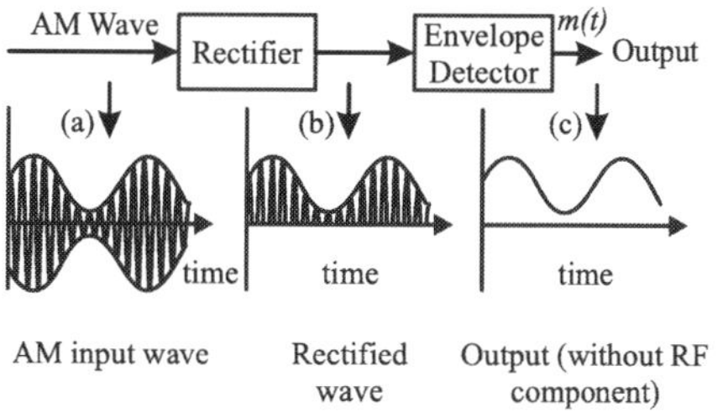

The capacitor connected in parallel with resistance R provides very low impedance at the carrier frequency and a much higher impedance at the modulating frequency. As a result capacitor effectively shorts or filters out the carrier, there by leaving the original modulating signal

In the actual circuit the value of RC (The time constant, t=RC) is chosen such that $\frac{1}{f_c} << RC$; where f_C = frequency of carrier signal.

Distortion in diode detectors: There are two types of distortions in diode detectors. Namely

a) Negative peak clipping

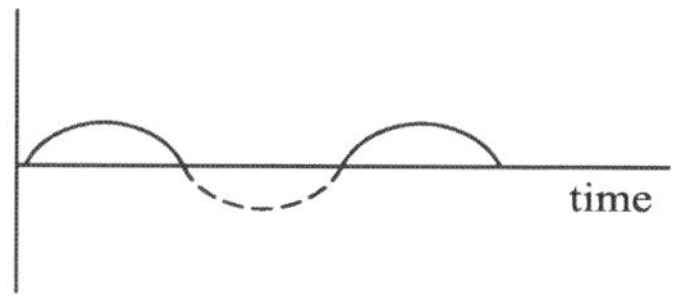

Figure shows the negative peak missing in the output message We know that

Modulation Index(m)= $\frac{E_m}{E_C} = \frac{I_m}{I_c}$ where $I_m = \frac{E_m}{Z_m}$ and $I_c = \frac{E_c}{R_c}$

Here z_m is audio load resistance of diode and R_c is dc diode resistance while $z_m < R_c$

hence $I_m > I_c$. This makes the modulation index in the demodulated wave higher than it was in modulated wave applied at the detector. In turn there is an increase in the chance of over modulation for modulation index nearer to 100% Due to this over modulation there is a negative clipping of the detector wave.

b) Diagonal clipping:

1. Diode ac load may no longer be purely resistive, it can contain reactive component.
2. At high modulation depths current will be changing so fast that the time constant of the load may be too slow to follow the changes. As a result current will decay exponentially. Hence output voltage follows the discharge law of the CR circuit

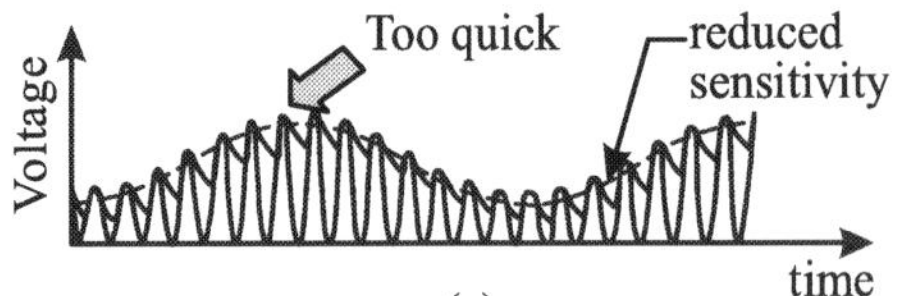

(a)

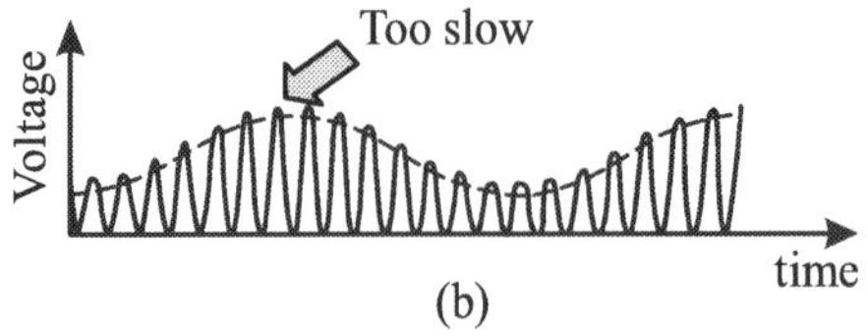

(b)

3. Condition necessary for avoiding distortion of this type is as follows.

$$I_m = \frac{E_m}{Z_m} = \frac{mE_c}{Z_m} \text{ and } I_c = \frac{E_c}{R_c}$$

$$m_d = \frac{I_m}{I_c} = \frac{mE_c/Z_m}{E_c/R_c} = \frac{mR_c}{Z_m}$$

Maximum value of $m_{dMaximum} = 1$

So maximum permissible transmitted modulation index will be $M_{\max imum} = m_d \frac{Z_m}{R_c} = 1 \times \frac{Z_m}{R_c}$

W.E-1: ***The load on an Am diode detector consists of a resistance of 50 KΩ in parallel with a capacitor of 0.001 μF. Determine the maximum modulation index that the detector can handle without distortion when modulation frequency is (i) 1 kHz (ii) 5 kHz.***

Sol: $$Z_m = R_c \| C = \frac{1}{\sqrt{\frac{1}{(R_c)^2} + \frac{1}{(X_c)^2}}}$$

$$M_{\max} = \frac{Z_m}{R_c}$$

$$= \frac{1}{R_c\sqrt{\frac{1}{(R_c)^2} + \frac{1}{(X_c)^2}}}$$

$$= \frac{1}{\sqrt{1 + (2\pi fCR_c)^2}}$$

For f=1 kHz

$$M_{\max} = \frac{1}{\sqrt{1 + 0.098696}} = 0.945$$

for f= 5 kHz

$$M_{max} = \frac{1}{\sqrt{1+2.4674}} = 0.537$$

- **Limitation of amplitude modulation**

  (i) Noisy reception

  (ii) Low efficiency

  (iii) Small operating range

  (iv) Poor audio quality

2) Angle modulation:

The angle of the carrier wave is varied according to the base band signal while the amplitude is maintained constant. This method provides better discrimination against NOISE and INTERFERENCE then the Amplitude modulation

There are two ways of varying the angle of the carrier.

$\theta_i(t)$ is the angle of moduilated sinusoidal carrier assumed to be a function of the message signal.

Equation of the angle modulated wave is $S(t) = A_c \cos(\theta_i(t))$

where A_c is amplitude of the carrier.

Instantaneous frequency of the angle modulated wave is $f_i(t) = \frac{1}{2\pi}\frac{d\theta_i(t)}{dt}$

here $\frac{d\theta_i(t)}{dt}$ is the angular velocity of the PHASOR of length $\mathbf{A_c}$

a) By varying the frequency ω_c, Frequency Modulation.

b) By varying the phase ϕ_c, Phase Modulation.

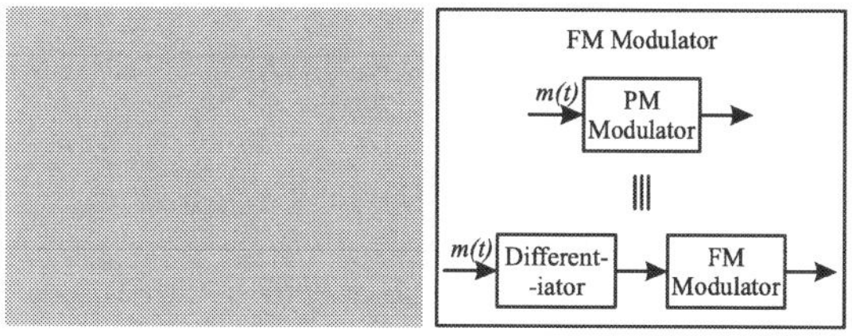

A comparison of FM and PM modulators.

i)Frequency modulation (FM)

The method in which the frequency of carrier is varied in accordance to the modulating signal, keeping the amplitude and phase of the carrier the same is called **Frequency modulation** (FM)

a) Audio quality of AM transmission is poor. There is need to eliminate amplitude sensitive noise. This is possible if we **eliminate amplitude variation.**

b) In FM the overall **amplitude** of FM wave remains constant at all times.

c) In FM, the total transmitted **power** remains constant.

d) ***Frequency deviation*:** The maximum change in frequency from mean value (f_c) is known as frequency deviation. This is also the change or shift either above or below the frequency f_c and is called as frequency deviation.

$$\delta = (f_{max} - f_c) = f_c - f_{min} = k_f . \frac{E_m}{2\pi}$$

k_f = Constant of proportionality.

It determines the maximum variation in frequency of the modulated wave for a given modulating signal.

e) Carrier swing (CS): The total variation in frequency from the lowest to the highest is called the carrier swing *i.e.* $CS = 2\ \delta f$

f) Frequency modulation index (m_f-) :

The ratio of maximum frequency deviation to the modulating frequency is called modulation index.

$$m_f = \frac{\delta}{f_m} = \frac{f_{max} - f_c}{f_m} = \frac{f_c - f_{min}}{f_m} = \frac{k_f E_m}{f_m}$$

g) Frequency spectrum: FM side band modulated signal consist of infinite number of side bands whose frequencies are

$$(f_c \pm f_m), (f_c \pm 2f_m), (f_c \pm 3f_m) \ldots\ldots$$

The number of side bands depends on the modulation index m_f

Band width $= 2n \times f_m$; where n = number of significant side band pairs

h) **Deviation ratio:** The ratio of maximum permitted frequency deviation to the maximum permitted audio frequency is known as deviation ratio. $(= \frac{(\Delta f)_{max}}{(f_m)_{max}})$

i) Percent modulation: The ratio of actual frequency deviation to the maximum followed frequency deviation

$$m = \frac{(\Delta f)_{actual}}{(\Delta f)_{max}}$$

j) Frequency spectra of fm waves under various conditions.

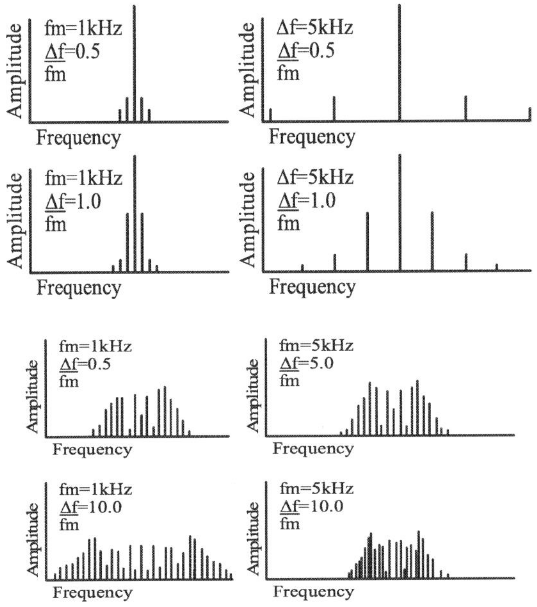

(A) Constant Modulating Frequency

(B) Constant Frequency Deviation

Δf Frequency Deviation

fm Modulation Frequency

$\frac{\Delta f}{fm}$ Modulation Index

W.E-2: ***The carrier frequency is 500 kHz. The modulating frequency is 15 kilohertz and the deviation frequency is 75 kilohertz. Find***

a) modulation index b) Number of side bands c) Band width

Sol: $MI = \frac{\Delta f}{f_m} = 5$

We can have that there are 16 significant sidebands for a modulation index of 5.

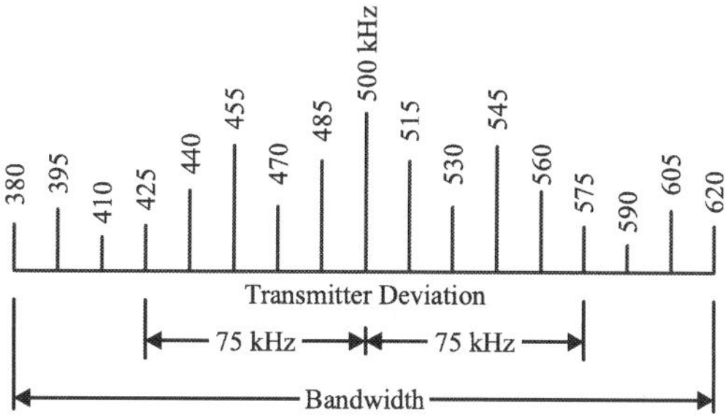

To determine total bandwidth for this case,

we use: $bW = f_m \times (\text{number of sidebands})$

bW = 15 x 16=240 kHz

ii) Phase Modulation (PM), phase of carrier is varied in accordance with modulating signal keeping namplitude and frequency constant. We use the term phase shift to characterize such changes. If phase changes after cycle k, the next sinusoidal wave will start slightly later than the time at which cycle k completes.

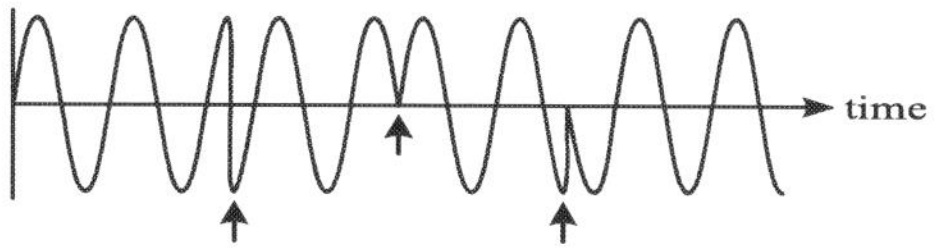

II) pulse wave modulation.

Here the carrier wave is in the form of pulses. Pulse modulation is an analog process as the modulating signal is analog.

The common pulse modulating systems are:

i) **Pulse amplitude modulation(PAM):**

The amplitude of the pulse varies in accordance with the modulating signal.

ii)Pulse time modulation (PTM):

Or

Pulse duration modulation(PDM)

a) Pulse width modulation (PWM):

The pulse duration varies in accordance with the modulating signal, or the width of the unmodulated signal is constant.

b) Pulse position modulation (PPM):

In PPM, the position of the pulses of the carrier wave train is varied in accordance with the instantaneous value of the modulating signal.

iii) Pulse code modulation (PCM):

The pulse amplitude, pulse width and pulse position modulations are not completely digital. A completely digital modulation is obtained by pulse code modulation (PCM) by following three operations.

(a) Sampling: It is the process of generating pulses of zero width and of amplitude equal to the instantaneous amplitude of the analog signal. The number of samples taken per second is called sampling rate.

(b) Quantization: The process of dividing the maximum amplitude of the analog voltage signal into a fixed number of levels is called quantization.

(c) **Coding:** The process of digitizing the quantised pulses according to some code is called coding.

➢ **Digital Communication And Quantisation Of Message Signal**

(Data Transmission and Retrieval)

Data means facts, concepts or instructions suitable for communication, interpretation or processing by human beings or by automatic means. Data in most cases consists of pulse type of signals.

In digital communication the modulating signals are discrete and are coded as represen tation of message signals to be transimitted. There are a number of *encoding steps in* digital communication, which makes its circutary complicated. Digital communication is error free and noise free.

The source encoder converts the informa tion into binery code. Encoder first digitise the analog waveform. Some times an additional encoding called *channel encoding* is carried out. In the final step, before transmission,the channel codes modulate a continuous wave form.

The pulse code modulated (PCM) signal is a series of 1's and 0's. The following three modulation techniques are used to transmit a PCM signal.

(1) Amplitude shift keying (ASK):

Two different amplitudes of the carrier represent the two binary values of the PCM signal. This method is also known as on-off keying (OOK)

1: Presence of carrier of same constant amplitude.

0: Carrier of zero amplitude.

(2) Frequency shift keying (FSK): The binary values of the PCM signal are represented by two frequencies.

1: Increase in frequency

0: Frequency unaffected

(3) Phase shift keying (PSK): The phase of the carrier wave is changed in accordance with modulating data signal.

1: Phase changed by p

0: Phase remains unchanged.

The analog signal is sampled by the sampler. The sampled pulses are then quantized. The encoder codes the quantized pulses according to the binary codes. After modulating the PCM signal (by ASK, FSK or PSK method)the modulated signal is, then transmitted into free space in the form of bits.

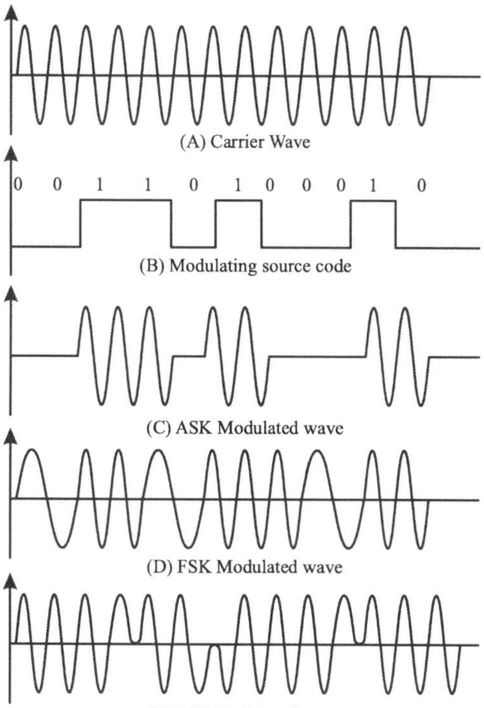

(A) Carrier Wave
(B) Modulating source code
(C) ASK Modulated wave
(D) FSK Modulated wave
(E) PSK Modulated wave

- **MODEM AND FAX**

(1) **Modem:** Modems are used to interface two digital sources/receivers.

i) Word modem implies MODulator and DEModulator. Both the functions (modulation and demodulation) are included in a signal unit.

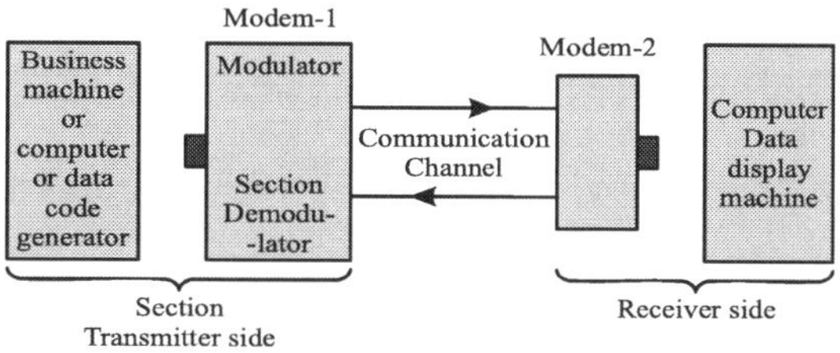

ii) Modems are placed at both ends of the communication circuit.

iii) The modem at the transmitting station changes the digital output from a computer to an analog signal, which can be easily sent via communication channel. While the receiving modem reverses the process.

iv)There are three modes of operation of a modem.

(a) Simplex mode: Data is transmitted in only one direction.

(b) Half duplex: Data is transmitted between the transmitter and the receiver in both direction, but only in one direction at a time.

(c) Full duplex: Data are transmitted between the transmitter and receiver in both directions at the same time.

(2) Fax (Facsimile transmission): The electronic reproduction of a document at a distance place is known as facsimile transmission (FAX).

The original written document is converted into transmittable codes and is converted in to electrical signals, which are then modulated and transmitted on to the receiving end.

- **The Internet :** It is a system with billions of users worldwide. It permits communication and sharing of all types of information between any two or more computers connected through a large and complex network. It was started in 1960's and opened for public use in 1990's. With the passage of time it has witnessed tremendous growth and it is still expanding its reach. Its applications include.
- **Email:** It permits exchange of text/ graphic material using email software. We can write a letter and send it to the recipient through ISP's (internet Service Providers) who work like the dispatching and receiving post offices.
- **File transfer:** A FTP (File Transfer Programmes) allows transfer of files/software from one computer to another connected to the internet.
- **World Wide Web (WWW):** Computers that store specific information for sharing with others provide websites either directly or through web service providers. Government departments, companies, NGO's (Non-Government Organizations) and individuals can post information about their activities for resticted or free use on their websites. This information becomes accessible to the users. Several search engines like Google, Yahoo!etc. help us in finding information by listing the related websites. Hypertext is a powerful feature of the web that automatically links relevant information from one page on the web to another using HTML (hypertext markup language)
- **E-commerce:** Use of the internet to promote business using electronic means such as using credit cards is called E-commerce. Customers view images and receive all the information about various products or services of companies through their websites. They can do on-line shopping from home/office. Goods are dispatched or services are provided by the company through mail/courier.
- **Chat:** Real time conversation among people with common interests through typed messages is called chat. Everyone belonging to the *chat group* gets the message instantaneously and can respond rapidly.
- **D.Mobile telephony :** The concept of mobile telephony was developed first in 1970's and it was fully implemented in the following decade. The central concept of this sytem is to divide the service area into a suitable number of *cells* centered on an office called MTSO (Mobile Telephone Switching Office). Each cell contains a low-power transmitter called a base station and caters to a large number of mobile receivers (popularly called cell phones). Each cell could have a service area of a few square kilometers or even less depending upon the number of customers. When a mobile receiver crosses the coverage area of one base station, it is necessary for the mobile user to be transferred to another base staion. This process is called *handover or handoff*. This process is carried out very rapidly, to the extent that the consumer does not even notice it. Mobile telephones operate typically in the UHF range of frequencies (about 800-950 MHz)

EXAMPLE PROBLEMS

W.E-3: ***How many AM broadcast stations can be accommodated in a 100 kHZ bandwidth if the highest modulating frequency of carrier is 5 kHZ?***

Sol. Any station being modulated by a 5 kHz singal will produce an upper side frequency 5 kHz above its carrier and a lower side frequency 5 kHz below its carrier, thereby requiring a bandwidth of 10 kHz. Thus, Number of stations accommodated

$$\frac{\text{Total bandwidth}}{\text{Bandwidth per station}}=\frac{100}{10}=10$$

W.E-4: ***How many 500 kHz waves can be on a 10km transmission line simultaneously?***

Sol: Let λ be the wavelength of 500 kHz signal. Then, $\lambda=\frac{c}{f}=\frac{3.0\times10^8}{5.0\times10^5}m=600m$

The number of waves on the line can be found from,

$$n=\frac{d}{\lambda}=\frac{10\times10^3}{600}=16.67$$

W.E-5: ***A two wire transmission line has a capacitance of 20 pF/m and a characteristic impedance of*** 50Ω

a) What is the inductance per metre of this cable?

b) Determine the impedance of an infinitely long section of such cable.

Sol: a) The characteristic impedance. $Z = \sqrt{L/C}$ $L=(Z^2)(C)=(50)^2(20\times10^{-12})H=0.05H/m$

b) The characteristic impedance of a transmission line is the impedance that an infinite length of line would present to a power supply at the input end of the line. Thus, $Z_\infty = Z_0 = 50\Omega$

W.E-6: ***T.V. transmission tower at a particular station has a height of 160m.***

a) What is the coverage range?

b) How much population is covered by transmission, if the average population density around the tower is 1200 per km^2 ***?***

c) What should be the height of tower to double the coverage range

Sol: a) Coverage range $d = \sqrt{2Rh}$

$= \sqrt{2\times6400\times10^3\times160m}$

$= 45.254km$

b) Population covered

$= (population\ density)\times(area\ \text{covered})$

$= (1200)\times(\pi d^2)$

$= (2400\pi Rh) = 2400\times3.14\times6.4\times10^3\times0.16$

$= 77.17\,lac$

c) Coverage range $\propto \sqrt{h}$

Therefore coverage range can be doubled by making height of the tower four times to 640m. So, height of the tower should be increased by 480 m.

W.E-7: ***An audio signal given by*** $e_s = 15\sin 2\pi(200t)$ ***modulates a carrier wave given by*** $e_s = 60\sin 2\pi(100,000t)$. ***If calculate***

a) Percent modulation

b) Frequency spectrum of the modulated wave.

Sol: a) Signal Amplitude, $B = 15$

Carrier amplitude, $A = 60$

$m = \frac{B}{A} = \frac{15}{60} = 0.25$

$\therefore$ Percentage modulation $= 0.25\times100 = 25\%$

b) By comparing the given equations of signal and carrier with their standard form

$e_s = E_s \sin\omega_s t = E_s \sin 2\pi f_s t$ and

$e_c = E_c \sin\omega_c t = E_c \sin 2\pi f_c t$

we have signal frequency $f_s = 2000Hz$ and carrier frequency $f_c = 100,000Hz$

The frequencies present in modulated wave

i) $f_c = 100,000Hz = 100kHz$ ii) $f_c - f_s = 100,000 - 2000 = 98kHz$

iii) $f_c + f_s = 100kHz + 2kHz = 102kHz$

Therefore, frequency spectrum of modulated wave extends from 98kHz to 102 kHz is called band width.

W.E-8: ***The antenna current of an AM transmitter is 8A when only the carrier is sent but it increases to 8.93A when the carrier is modulated. Find percent modulation.***

Sol. The modulated or total power carried by AM wave $P_T = P_C\left(1+\frac{m^2}{2}\right)$. If R is load resistance. I_m is the current when carrier is modulated and I_c the current when unmodulated, then

$$\frac{P_T}{P_C} = \frac{I_m^2 R}{I_c^2 R}; \quad \therefore I + \frac{m^2}{2} = \frac{I_m^2 R}{I_c^2 R}$$

Given $I_m = 8.93A, I_c = 8A$

$$\therefore m^2 = 2\left[\left(\frac{8.93}{8.0}\right)^2 - 1\right] \therefore m = 0.7$$

Therefore, percentage modulation = 70%

W.E-9: ***A sinusoidal carrier voltage of 80 volts amplitude and 1 MHz frequency is amplitude modulated by a sinusoidal voltage of frequency 5kHz producing 50% modulation. Calculate the amplitude and frequency of lower and upper side bands.***

Sol: Amplitude of both LSB and USB are equal and given by

$$= \frac{mE_c}{2} = \frac{0.5 \times 80}{2} = 20 volts$$

Now frequency of LSB = $f_c - f_s = (1000 - 5)kHz = 995kHz$

Frequency of USB = $f_c + f_s = (1000 + 5)kHz = 1005kHz$

W.E-10: ***The load current in the transmitting antenna of an unmodulated AM transmitter is 6 Amp. What will be the antenna current when modulation is 60%.***

Sol: Total power carried by AM wave

$$P_T = P_C\left(1+\frac{m^2}{2}\right) \text{......}(1)$$

Where P_c is the power of carrier component and m is the modulation factor. If R is the resistance, I_m the antenna load current when modulation is 60% and I_c is the antenna load current when un modulated, then

$$\frac{P_T}{P_C}=\frac{I_m^2R}{I_c^2R},\therefore 1+\frac{m^2}{2}=\frac{I_m^2}{I_c^2}$$ using (1)

or $I_m=I_c\sqrt{\left\{\left(1+\frac{m^2}{2}\right)\right\}}$

Given $I_c=6Amp, m=0.6$

$$I_m=6\left[1+\frac{(0.6)^2}{2}\right]^{1/2}=6[1.086]=6.52Amp$$

W.E-11: ***A carrier wave of 1000 W is subjected to 100% modulation. Calculate (i) Power of modulated wave, (ii) power is USB, (iii) power is LSB***

Sol: i) Total power of modulated wave

$$P_T=P_C\left(1+\frac{m^2}{2}\right);=1000\left(1+\frac{1^2}{2}\right)=1500watt$$

ii) Power in USB $=\frac{1}{2}P_{SB}$

Where power carried by side bands is given by amplitude modulation and detection

$$P_{SB}=P_C\left(\frac{m^2}{2}\right);=1000\left(\frac{1^2}{2}\right)=500watt$$

$$P_{USB}=\frac{1}{2}P_{SB}=\frac{1}{2}\times 500=250watt$$

iii) Since power in LSB = Power in USB

$$P_{LSB}=P_{USB}=250watt$$

W.E-12: ***A transmitting antenna at the top of a tower has a height 32m and the height of the receiving antenna is 50m. What is the maximum distance between them for satisfactory communication in LOS mode? Given radius of earth*** $6.4\times10^6 m$.

Sol: $d_m=\sqrt{2\times64\times10^5\times32}+\sqrt{2\times64\times10^5\times50m}$

$=64\times10^2\times\sqrt{10}+8\times10^3\times\sqrt{10}m$

$=144\times10^2\times\sqrt{10}m=45.5km$

W.E-13: ***A message signal of frequency 10 kHz and peak voltage of 10 volts is used to modulate a carrier of frequency 1 MHz and peak voltage of 20 volts. Determine***

a) modulation index

b) the side bands produced.

Sol: a) modulation index = 10/20 = 0.5

b) The side bands are at

$(1000+10)kHz=1010\ kHz\ \&(1000-10kHz)=990kHz$. are frequency.

CONCEPTUAL

1. **A transducer used at the transmitting end, serves the purpose of converting**
1) electrical signal to sound form 2) sound signal to electrical form
3) electrical signal to magnetic form 4) sound signal to magnetic form

2. **In a communication system, noise is most likely to affect the signal**
1) at the transmitter 2) in the medium of transmission
3) information source signal 4) at the destination

3. **Device that converts one form of energy into another is called**
1) transmitter 2) transducer 3) receiver 4) channel

4. **The part of communication system that extracts the signal at the output of the channel is**
1) transducer 2) transmitter 3) receiver 4) receiver or transmitter

5. **The attenuation of a signal is compensated by**
1) rectifier 2) oscillator 3) modulator 4) amplifier

BAND WIDTH OF SIGNALS & BAND WIDTH OF RANSMISSION

6. **Modern communication systems use**
1) analog circuits 2) digital circuits
3) combination of analog & digital circuits 4) radio circuits

7. **The audio signal**
1) can be sent directly over the air for large distance
2) can not be sent directly over the air for large distance
3) possesses very high frequency 4) possesses very low frequency

8. **A digital signal possess**
1) continuously varying values 2) only two discrete values
3) only four discrete values 4) constant values

9. **Digital signals**
1) provide continuous set of values 2) represent values as randomly
3) Utilise Decimal code system 4) Utilise binary code system

10. **Digital signals**
i) do not provide a continuous set of values. **ii) represents values as descrete steps.**
iii) can utilize binary system
iv) can utilize decimal as well as binary system.
The true option is.
1) (i) & (ii) only 2) (ii) & (iii) only 3) (i), (ii) & (iii) only 4) (i),(ii),(iii)& (iv)

11 **A digital signal**
1) is less reliable than analog signal 2) is more reliable than analog signal
3) is equally reliable as the analog signal 4) Not at all reliable

12. **The band width required for transmiting video signal is**
1) 50 KHz 2) 1 MHz 3) 4.2 MHz 4) 6 MHz

13. **Band width of an optical fiber is**
1) more than 100 GHz 2) few kHz 3) less than 1MHz 4) less than 1GHz

14. **The short wave Radio broadcasting band is**
1) 7 MHz to 22 MHz 2) 88 MHz to 108 MHz
3) 30 KHz to 300 KHz 4) 3 GHz to 30 GHz

15. **The FM Radio broad casting band is**
1) 5 MHz to 30 MHz
2) 88 MHz to 108 MHz
3) 30 KHz to 300 KHz
4) 3 GHz to 30 GHz

16. **The TV broad casting bands are**
1) MF and HF bands
2) VHF and UHF bands
3) UHF and SHF bands
4) SHF and EHF band

17. **A: Satellite communication uses different frequency bands for uplink and downlink**
B: Bandwidth of video signals is 4.2 MHz
1) A is true but B is false
2) A is false but B is true
3) A and B are false
4) A and B are true

18. **A: The frequency band of VHF is greater than UHF of TV transmission**
B: Optical fiber transmission has frequency band of 1 THz to 1000 THz
1) A is true but B is false
2) A is false but B is true
3) A and B are false
4) A and B are true

19. **The higher frequency TV broad casting bands range is**
1) 54 - 72 MHz and 76 to 88 MHz
2) 174 - 216 MHz and 420 to 890 MHz
3) 896 to 901 MHz and 840 to 935 MHz
4) 5.925 to 6.425 GHz and 3.7 to 4.2 GHz

20. **Frequency ranges for micro waves are :**
1) 3×10^9 to 3×10^4 Hz
2) 3×10^{13} to 3×10^9 Hz
3) 3×10^{14} to 3×10^9 Hz
4) 3×10^{11} to 3×10^9 Hz

21. **The frequency band used for radar relay systems & T.V is**
1) UHF 2) VLF 3) VHF 4) EHF

22. **For TV transmission the frequency range employed**
1) 30 - 300 MHz 2) 30 - 300 GHz 3) 30 - 300 KHz 4) 30 - 300 Hz

23. **The frequency which is not part of AM broadcast**
1) 100 kHz 2) 700 kHz 3) 600 kHz 4) 1500 kHz

24. **Cellular Mobile works in the frequency range of**
1) 840 to 935 MHz 2) 3.7 to 4.2 GHz 3) 420 to 890 MHz 4) 30 to 300 GHz

25. **Frequency range used in down linking in satellite communication is**
1) 0.896 to 0.901 GHz 2) 0.420 to 0.890 GHz 3) 5.925 to 6.425 GHz 4) 3.7 to 4.2 GHz

26. **In the satellite communication, the uplinking frequency range is**
1) 0.896 to 0.901 GHz
2) 0.420 to 0.890 GHz
3) 5.925 to 6.425 GHz
4) 3.7 to 4.2 GHz

27. **In a communication system, noise is most likely to affect the signal**
1) At the transmitter
2) In the channel or in the transmission line
3) In the information source
4) At the receiver

28. **The frequency of a FM transmitter without signal input is called**
1) Lower side band frequency
2) Upper side band frequency
3) Resting frequency
4) None of these

29. **Indicate which one of the following system is digital**
1) Pulse position modulation
2) Pulse code modulation
3) Pulse width modulation
4) Pulse amplitude modulation

30. Television signals on earth cannot be received at distances greater than 100 ***km*** **from the transmission station. The reason behind this is that**

1) The receiver antenna is unable to detect the signal at a distance greater than 100 *km*
2) The TV programme consists of both audio and video signals
3) The TV signals are less powerful than radio signals
4) The surface of earth is curved like a sphere

31. Audio signal cannot be transmitted because

1) The signal has more noise
2) The signal cannot be amplified for distance communication
3) The transmitting antenna length is very small to design
4) The transmitting antenna length is very large and impracticable

PROPAGATION OF EM WAVES IN THE ATMOSPHERE

32. A signal emitted by an antenna from a certain point can be received at another point of the surface in the form of

1) sky wave 2) ground wave 3) sea wave 4) both 1 and 2

33. An antenna is a device

1) That converts electromagnetic energy into radio frequency signal
2) That converts radio frequency signal into electromagnetic energy
3) That converts guided electromagnetic waves into free space electromagnetic waves and vice-versa
4) None of these

34. An antenna

1) Converts AF wave to RF wave 2) RF signal into electromagnetic energy
3) Converts the guided EM waves into free space EM waves and vice versa
4) Super imposes AF wave on RF wave

35. An antenna behaves as a resonant circuit only when the length is

1) equal to $\lambda/4$ 2) equal to $\lambda/2$
3) equal to the integral multiples of $\lambda/2$ 4) equal to $3\lambda/4$

36. If audio signal is transmitted directly into space, the length of the transmitting antenna required will be

1) extremely small 2)extremely large 3) infinitely large 4) moderate

37. The height of the antenna

a) limits the population covered by the transmission
b) limits the ground wave propagation
c) effectively used in line of sight communication

1) a & b are true 2) b & c are true 3) c & a are true 4) a, b,c are true

38. Statement A: If the antenna is vertical the vertically polarised EM wave is radiated
Statement B: The vertically polarised EM wave has electrical variations in the vertical plane

1) A is true but B is false 2) A is false but B is true
3) A and B are false 4) A and B are true

39. A: It is necessary for transmitting antenna must be at same height as that of receiving antenna for line of sight communication.
B: EM waves of frequency beyond 40 MHz, propagate as space waves.

1) both A and B are correct 2) both A and B are wrong
3) only A is correct 4) only B is correct

40. Broadcasting antennas are generally

1) Omnidirectional type
2) Vertical type
3) Horizontal type
4) None of these

MODULATION

41. The process of superimposing signal frequency (*i.e.* audio wave) on the carrier wave is known as

1)Transmission 2) Reception 3)Modulation 4) Detection

42. In frequency modulation

1) The amplitude of modulated wave varies as frequency of carrier wave
2) The frequency of modulated wave varies as amplitude of modulating wave
3) The amplitude of modulated wave varies as amplitude of carrier wave
4) The frequency of modulated wave varies as frequency of modulating wave

43. For transmitting audio signal properly

1) it is first superimposed on high frequency carrier wave
2) it is first superimposed on low frequency carrier wave
3) It is sent directly without superimposing on any wave
4) it is superposed with carrier wave of high velocity

44. The process of recovering the audio signal from the modulated wave is known as

1) amplification 2) rectification 3) modulation 4) demodulation

45. The most commonly employed analog modulation technique in satellite communication is the

1) amplitude modulation
2) frequency modulation
3) phase modulation
4) amplitude & phase modulation

46. The need for doing modulation is

1) to increase the intensity of audio signal
2) to decrease the intensity of audio signal
3) to transmit audio signal to large distances
4) to increase the frequency of the audio signal

47. The type of modulation is employed in India for radio transmission is

1) pulse modulation
2) frequency modulation
3) amplitude modulation
4) phase modulation

48. Modulation is used to

1) reduce the bandwidth
2) to seperate the transmission of different users
3) to ensure that intelligence may be transmitted to long distances
4) to allow the uses of practical antenna

49. The process of translating the information contained by the low base band signal to high frequencies is called

1) Detection 2) Modulation 3) Amplification 4) Demodulation

50. During the process of modulation the RF wave is called

1) Modulating wave 2) Modulated wave 3) Carrier wave 4) Audio wave

51. Modulation is required to

a) distinguish different transmissions

b) ensure that the information may be trans mitted over long distances

c) allow the information accessible for different people

1) a & b are true 2) b & c are true 3) c & a are true 4) a, b & c are true

52. The physical quantities of the wave used for modulation

1) Amplitude only
2) Amplitude and frequency
3) Amplitude, frequency and phase
4) Only frequency

53. In Amplitude modulation

1) The amplitude of the carrier wave varies in accordance with the amplitude of the modulating signal
2) The amplitude of carrier wave remains constant, frequency changes in accordance with the modulating signal
3) The amplitude of carrier wave varies in accordance with the frequency of the modulating signal
4) The amplitude changes in accordance with the wave length of the modulating signal

54. Amplitude modulation is used for broad casting because

1) it is more noise immune
2) it requires less transmitting power
3) it has simple circuit
4) it has high fidelity (faithful reproduction)

55. In amplitude modulation, carrier wave frequencies are .... than that compared to those in frequency modulation

1) lower
2) higher
3) same
4) lower or higher

56. AM is used for broad casting because,

1) it is more noise immune than other modulating systems
2) it requires less transmitting power compared with other systems
3) its use avoids receiver complexity
4) no other modulation system can provide the necessary bandwidth, faithful transmission.

57. Draw backs of Amplitude modulation

1) During transmission extreneous noise creeps in.
2) Most of the transmitting power is wasted, as it does not contain useful information.
3) The reception is not clear in the case of weak signals due to noise
4) The receiver set is complex

58. In amplitude modulation

1) only amplitude is changed but frequency remains same
2) both amplitude & frequency changes equally
3) both amplitude & frequency changes unequally
4) only frequency changes but amplitude remains constant.

59. The limitation of amplitude modulation is

1) clear reception
2) high efficiency
3) small operating range
4) good audio quality

60. In frequency modulation

1) Frequency of CW remains constant but amplitude changes in accordance with modulating wave frequency
2) Frequency of CW changes in accordance with the modulating wave frequency but the amplitude also changes.
3) Frequency of CW changes in accordance with the frequency of modulating wave frequency but the amplitude remains constant.
4) Frequency of CW changes in accordance with the amplitude of modulating wave amplitude

61. In T.V. broadcasting both picture and sound are transmitted simultaneously. In this

1) audio signal is frequency modulated and video signal is amplitude modulated
2) both audio and video signals are frequency modulated
3) audio signal is amplitude modulated and video signal is frequency modulated
4) both audio and video signals are amplitude modulated

62. **Effective power radiated by an antenna is**

1) Proportional to the square at the length of the antenna
2) inversely proportional to the wavelength
3) inversely proportional to the square of the wavelength
4) proportional to the wavelength

63. **The concepts of communication are**

a) mode of communication
b) need for modulation
c) types of modulation
d) detection of modulated wave

1) a, b, c are true 2) b, c, d are true 3) c, d, a are true 4) a, b, c & d are true

64. **The difference between phase and frequency modulation**

1) practically they are same but theoretically they differ
2) lies in the poorer audio response of phase modulation
3) lies in the poorer audio response of frequency modulation
4) lies in the definitions of modulation and their modulation index

65. **Basically, the product modulator is**

1) An amplifier 2) A mixer
3) A frequency separator 4) A phase separator

66. **Which of the following is the disadvantage of FM over AM**

1) Larger band width requirement 2) Larger noise
3) Higher modulation power 4) Low efficiency

67. **Audio signal cannot be transmitted as such because**

1) the signal has more noise
2) the signal cannot be amplified for distance communication
3) the transmitting antenna length is very small to design
4) the transmiting antenna length is very large and impracticable

68. **The examples of broadcast are**

A) radio **B) television** **C) telephony** **D) internet**

1) A & B 2) A, B & D 3) A,B & C 4) B & D

69. **The waves relavent to telecommunications are**

1) visible light 2) infrared 3) ultraviolet 4) microwave

70. **While tuning in a certain broad cast station with a receiver, we are actually**

1) varying the local oscillator
2) varying the resonant frequency of the circuit for the radio signal to be picked up
3) tuning the antenna 4) varying the current of receiver set

71. **Long distance short-wave radio broadcasting uses**

1) Ground wave 2) Ionospheric wave 3) Direct wave 4) Sky wave

72. **Radio waves of constant amplitude can be generated with**

1) filter 2) rectifier 3) FET 4) oscillator

73. **Advantage of HF transmission is**

A) that the length of antenna is small
B) that the antenna can be mounted at larger heights
C) that the power radiated is more for a given length of antenna

1)a & b 2) b& c 3) a & c 4) a,b & c

74. **The intensity of the ground waves decrease with increase of distance due to**
1) Interference 2) Diffraction 3) Polarization 4) Due to unknown reason

75. **High frequency waves are**
1) absorbed by F layer 2) reflected by the E layer
3) capable of use for long distance transmission 4) affected by the solar cycle

76. **As the e.m. waves travel in free space**
1) absorption takes place 2) attennuation takes place
3) refraction takes place 4) reflection takes place

77. **The electromagnetic waves of frequency 80 MHz and 200 MHz**
1) can be reflected by troposphere 2) can be reflected by ionosphere
3) can be reflected by mesosphere 4) cannot be reflected by any layer of earth's atmosphere

78. **The better propagation mode to propagate television frequency and radar signals is**
1) satellite communication 2) ground propagation
3) polarized communication 4) skywave communication

79. **Micro wave link repeaters are typically 50 km apart**
1) because of atmospheric attenuation 2) because of the earths curvature
3) to ensure that signal voltage may not harm the repeater
4) to reduce the interference of microwaves

80. **Attenuation of ground waves is due to**
1) Diffraction effect
2) Radio waves induce currents in the ground because of the polarisation
1) a & b are true 2) Only a is true 3) Only b is true 4) Both a & b false.

81. **The ground wave eventually disappears, as one moves away from the transmitter, because of**
1) interference from the sky wave 2) loss of line of signal condition
3) maximum single - hop distance limitation 4) diffraction effect causing tilting of the wave

82. **The range of ground wave transmission can be increased by**
1) increasing the power of transmitter with the use of HF
2) increasing the power of transmitter with the use of VLF
3) decreasing the power and increasing the frequency of radio waves
4) decreasing both power and frequency of radio waves

83. **Space wave propagation is used in**
a) microwave communication **b) satellite communication**
c) TV transmission
1) Only a 2) Both a & b 3) Both b & c 4) a ,b & c

84. **Frequencies in the UHF range normally propagate by means of:**
1) Ground waves 2) Sky waves. 3) Surface waves 4) Space waves.

85. **When a sky wave is reflected onto the ground**
1) frequency of the reflected wave is different to that of incident wave
2) there is a phase difference introduced to the reflected wave
3) the reflected wave is out of phase with incident wave and reach the receving antenna along with the direct wave from transmitting antenna causing interference.
4) the waves are not reflected by the ground.

86. **The electromagnetic waves of frequency 2 MHz to 30 MHz are**

(1)In ground wave propagation (2)In sky wave propagation

(3)In microwave propagation (4)In satellite communication

87. **Among the following frequencies one will be suitable for beyond-the horizon communication using sky waves is**

1) 10 kHz 2) 10 MHz 3) 1GHz 4)1000GHz

88. **Among the following, the waves which can penetrate the ionosphere are**

1)10GHz 2)10MHz 3)20MHz 4) 25 MHz

89. **A: At great heights from surface of earth and close to earth ionisation of air molecules is low.**
B: EM waves of frequencies beyond 30 MHz penetrate ionosphere and escape.

1) both A and B are correct 2) both A and B are wrong

3) only A is correct 4) only B is correct

90. **Through which mode of propagation, the radio waves can be sent from one place to another**

1) Ground wave propagation 2) Sky wave propagation

3) Space wave propagation 4) All of them

91. **The frequency at which communication will not be reliable is**

1) 100 KHz 2)1 MHz 3) 10 GHz 4)100 GHz

92. **The frequency above which radiation of electrical energy is practical is**

1) 0.2 kHz 2) 2 kHz 3) 20 kHz 4) 2Hz

93. **The radio waves of frequency 300 MHz to 3000 MHz belong to**

1) High frequency band 2) Very high frequency band

3) Ultra high frequency band 4) Super high frequency band

94. **Coaxial cable is an example of**

1) Optical fibre 2) Free space 3) Wire medium 4) Sea medium

95. **Optical fibre communication is generally preferred over general communication system because**

1) it is more efficient 2) of signal security 3) both (1) & (2) 4) it is easily available

96. **The attenuation in optical fibre is mainly due to**

1)Absorption 2)Scattering 3)Neither absorption nor scattering 4)Both 1 and 2

97. **Consider telecommunication through optical fibres. Which of the following statements is not true**

(1)Optical fibres may have homogeneous core with a suitable cladding

(2)Optical fibres can be of graded refractive index

(3)Optical fibres are subject to electromagnetic interference from outside

(4)Optical fibres have extremely low transmission loss

98. **The phenomenon by which light travels in an optical fibres is**

1)Reflection 2)Refraction 3)Total internal reflection 4) Transmission

99. **A laser is a coherent source because it contains**

1) Many wavelengths 2) Uncoordinated wave of a particular wavelength

3) Coordinated wave of many wavelengths 4) Coordinated waves of a particular wavelength

100. **In which of the following remote sensing technique is not used**

1)Forest density 2)Pollution 3)Wetland mapping 4)Medical treatment

CONCEPTUAL - KEY

1) 2	2) 2	3) 2	4) 3	5) 4	6) 2	7) 2	8) 2	9) 4	10) 3
11) 2	12) 3	13) 1	14) 1	15) 2	16) 2	17) 4	18) 2	19) 2	20) 2
21) 1	22) 1	23) 1	24) 1	25) 4	26) 3	27) 2	28) 3	29) 2	30) 4
31) 4	32) 4	33) 3	34) 3	35) 1	36) 2	37) 4	38) 4	39) 4	40) 2
41) 3	42) 2	43) 1	44) 4	45) 2	46) 3	47) 3	48) 1	49) 2	50) 3
51) 4	52) 3	53) 1	54) 3	55) 1	56) 3	57) 2	58) 1	59) 3	60) 3
61) 1	62) 3	63) 4	64) 1	65) 2	66) 1	67) 4	68) 2	69) 4	70) 3
71) 3	72) 4	73) 4	74) 2	75) 2	76) 2	77) 4	78) 1	79) 2	80) 1
81) 4	82) 2	83) 4	84) 4	85) 3	86) 2	87) 2	88) 1	89) 1	90) 4
91) 1	92) 3	93) 3	94) 3	95) 3	96) 4	97) 3	98) 3	99) 4	100) 4

LEVEL - I -1

1. **An optical communication system is operating at a wavelength of 800 nm, it's optical source frequency is**

1) 3.8×10^{14} Hz 2) 3.8×10^{12} Hz 3) 3.8×10^{10} Hz 4) 3.8×10^{9} Hz

2. **A carrier wave of 1000 kHz is used to carry the signal, the length of the transmitting antenna will be equal to**

1) 3 m 2) 30 m 3) 300 m 4) 3000 m

3. **A transmitting antenna is at a height of 40 m and the receiving antenna is at a height of 60 m. The maximum distance between them for satisfactory communication is nearly**

1) 22.5 km 2) 27.5 km 3) 50 km 4) 25 km

4. **A T V transmitting antenna is 80 m tall. If the receiving antenna is on the ground the service area is**

1) 12 sq km 2) 3215 sq km 3) 144sq km 4) 32 sq km

5. **The maximum distance upto which TV transmission from a TV tower of height 'h' can be received is proportional to**

1) $h^{1/2}$ 2) h 3) $h^{3/2}$ 4) h^2

6. **In short wave communication waves of which of the following frequencies will be reflected back by the ionospheric layer, having electron density 10^{11} per m^3**

1) 2 *MHz* 2) 10 *MHz* 3) 12 *MHz* 4) 18 *MHz*

7. **In an amplitude modulated wave for audio frequency of 500 *cycle/second*, the appropriate carrier frequency will be**

1) 50 *cycles/sec* 2) 100 *cycles/sec* 3) 500 *cycles/sec* 4) 50,000 *cycles/sec*

8. **The modulation index of an FM carrier having a carrier swing of 200 *kHz* and a modulating signal 10 *kHz* is**

1) 5 2) 10 3) 20 4) 25

9. **If a number of sine waves with modulation indices n_1, n_2, n_3 .... Modulate a carrier wave, then total modulation index (n) of the wave is**

1) $n_1 + n_2 .... + 2(n_1 + n_2 .....)$ 2) $\sqrt{n_1 - n_2 + n_3 ....}$

3) $\sqrt{n_1^2 + n_2^2 + n_3^2 ....}$ 4) $n_1 + n_2 ....$

10. A sky wave with a frequency 55 *MHz* is incident on *D*-region of earth's atmosphere at 45°. The angle of refraction is (electron density for *D*-region is 400 *electron/cm*³)

1)60° 2)45° 3) 30° 4)15°

11. What should be the maximum acceptance angle at the aircore interface of an optical fibre if n_1 and n_2 are the refractive indices of the core and the cladding, respectively

1) $\sin^{-1}(n_2/n_1)$ 2) $\sin^{-1}\sqrt{n_1^2-n_2^2}$ 3) $\left[\tan^{-1}\frac{n_2}{n_1}\right]$ 4) $\left[\tan^{-1}\frac{n_1}{n_2}\right]$

12. The characteristic impedance of a coaxial cable is of the order of

1)50 Ω 2)200 Ω 3)270 Ω 4)None of these

13. The velocity factor of a transmission line *x*. If dielectric constant of the medium is 2.6, the value of *x* is

1)0.26 2)0.62 3)2.6 4)6.2

14. For television broadcasting, the frequency employed is normally

1)30-300 *MHz* 2)30-300 *GHz* 3)30-300 *KHz* 4)30-300 *Hz*

15. Because of tilting which waves finally disappear

1) Microwaves 2)Surface waves 3)Sky waves 4)Space waves

16. A radar has a power of 1kW and is operating at a frequency of 10 GHz. It is located on a mountain top of height 500m. The maximum distance upto which it can detect object located on the surface of the earth (Radius of earth $6.4\times10^6 m$) is

1) 80 km 2) 16 km 3) 40 km 4) 64km

LEVEL - I -1- KEY

1) 1	2) 3	3) 3	4) 2	5) 1	6) 1	7) 4	8) 2	9) 3	10) 2
11) 2	12) 3	13) 2	14) 1	15) 2	16) 1				

LEVEL - I -1- HINTS

1. $c = n\lambda \Rightarrow n = c/\lambda$
2. length of antenna $\approx \lambda$

   Here $l = \lambda = \frac{C}{v} = \frac{3\times10^8}{1000\times10^3} = 300m$
3. distance between transmitter and receiver is $s = \sqrt{2Rh_T} + \sqrt{2Rh_R}$
4. area covered, $A = \pi d^2$ where $d = \sqrt{2Rh}$
5. $d = \sqrt{2Rh} \Rightarrow d\,\alpha\sqrt{h}$
6. By using $f_c \approx 9(N_{max})^{1/2} \Rightarrow f_c \approx 2\,MHz$
7. Carrier frequency > audio frequency
8. $CS = 2\,'\Delta f$ or $\Delta f = CS/2$

   $\Delta f = \frac{200}{2} = 100\,kHz$; Now $m_f = \frac{\Delta f}{f_m} = \frac{100}{10} = 10$

10. $n_{eff} = n_0\sqrt{1-\left(\frac{80.5N}{\nu^2}\right)} = 1\sqrt{1-\frac{80.5\times(400\times10^6)}{(55\times10^6)^2}} \approx 1$

Also $n_{eff} = \frac{\sin i}{\sin r} \Rightarrow \sin r = \sin i \Rightarrow r = i = 45°$

11. Core of acceptance angle $\theta = \sin^{-1}\sqrt{n_1^2 - n_2^2}$

12. conceptual

13. $v.f. = \frac{1}{\sqrt{k}} = \frac{1}{\sqrt{2.6}} = 0.62$

14. VHF (Very High Frequency) band having frequency range 30 MHz to 300 MHz is typically used for TV and radar transmission.

15. conceptual

16. Range of radar on earth surface (optical distance, for space wave, i.e., line of view)

$$Range = \sqrt{(R+h)^2 - R^2} = \sqrt{2Rh + h^2} = \sqrt{2Rh} = \sqrt{2\times6400\times\frac{1}{2}km} = 80km$$

LEVEL - I -2

1. **An optical communication system is operating at a wavelength of 600 nm, it's optical source frequency is**

1) 3.8×10^{14} Hz 2) 3.8×10^{12} Hz 3) 3.8×10^{10} Hz 4) 5×10^{14} Hz

2. **A carrier wave of 2000 kHz is used to carry the signal, the length of the transmitting antenna will be equal to**

1) 3 m 2) 30 m 3) 150 m 4) 3000 m

3. **A transmitting antenna is at a height of 25 m and the receiving antenna is at a height of 64 m. The maximum distance between them for satisfactory communication is nearly**

1) 22.5 km 2) 46.5 km 3) 50 km 4) 25 km

4. **A T V transmitting antenna is 100 m tall. If the receiving antenna is on the ground the service area is**

1) 12 sq km 2) 4000sq km 3) 144sq km 4) 32 sq km

5. **The maximum distance upto which TV transmission from a TV tower of height 'h' can be received is proportional to (R earth radius)**

1) $(Rh)^{\frac{1}{2}}$ 2) h 3) $h^{3/2}$ 4) h^2

6. **In short wave communication waves of which of the following frequencies will be reflected back by the ionospheric layer, having electron density 10^{12} per m^3**

1) 2 MHz 2) 9 MHz 3) 12 MHz 4) 18 MHz

7. **In an amplitude modulated wave for audio frequency of 1000 *cycle/second*, the appropriate carrier frequency will be**

1) 50 *cycles/sec* 2) 100 *cycles/sec* 3) 500 *cycles/sec* 4) 40,000 *cycles/sec*

8. **The modulation index of an FM carrier having a carrier swing of 300 *kHz* and a modulating signal 10 *kHz* is**

1) 15 2) 10 3) 20 4) 25

9. **A sky wave with a frequency 64.4 *MHz* is incident on *D*-region of earth's atmosphere at 45°. The angle of refraction is (electron density for *D*-region is 400 *electron/cm*^{3})**

1) 60° 2) 90° 3) 45° 4) 15°

10. **The velocity factor of a transmission line *x*. If dielectric constant of the medium is 1.8, the value of *x* is**

1) 0.26 2) 0.62 3) 0.74 4) 6.2

11. **A radar has a power of 1kW and is operating at a frequency of 10 GHz. It is located on a mountain top of height 700m. The maximum distance upto which it can detect object located on the surface of the earth (Radius of earth $6.4\times10^6 m$) is**

1) 80 km 2) 16 km 3) 40 km 4) 92 km

LEVEL - II -2 - KEY

1) 4 2) 3 3) 2 4) 2 5) 1 6) 2 7) 4 8) 1 9) 3 10) 3
11) 4

LEVEL - II -2-HINTS

1. $C = n\lambda$; $n = \dfrac{C}{\lambda}$

2. Length of antenna $\approx \lambda, v = 2000KHz$

here $l = \lambda = \dfrac{C}{v} = \dfrac{3\times10^8}{2000\times10^3} = 150m$

3. Distance between transmitter and receiver is $d = \sqrt{2Rh_T} + \sqrt{2Rh_R}$

$= \sqrt{2\times6400\times10^3\times25} + \sqrt{2\times6400\times10^3\times64}$ $= 46.5$ km

4. Area covered $A = \pi d^2$; where $d = \sqrt{2Rh}$

5. $d = \sqrt{2Rh}$; $d \propto \sqrt{Rh}$; $d \propto (Rh)^{\frac{1}{2}}$

6. By using $f_c \approx 9(N_{\max})^{\frac{1}{2}}$; $f_c \approx 9(10^{12})^{\frac{1}{2}}$

$f_c \approx 9MHz$

7. Carrier frequency > audio frequency

8. $CS = 2\Delta f$

$\Delta f = \dfrac{CS}{2} = \dfrac{300KHz}{2} = 150KHz$

$m_f = \dfrac{\Delta f}{f_m} = \dfrac{150KHz}{10KHz} = 15$

9. $n_{eff} = n_0\sqrt{1-\frac{80.5N}{v^2}}$

$= 1\sqrt{1-\frac{80.5\times400\times10^6}{\left(64.4\times10^6\right)^2}}$; ≈ 1

Also $n_{eff} = \frac{\sin i}{\sin r} \Rightarrow \sin i = \sin r$; $\Rightarrow i = r = 45^0$

10. $V.f = \frac{1}{\sqrt{k}}$; $= \frac{1}{\sqrt{1.8}} = 0.74$

11. Range of Radar on earth surface (optical distance, for space wave, i.e line of view)

Range $= \sqrt{(R+h)^2 - R^2} = \sqrt{2Rh+h^2}$

$= \sqrt{2Rh} = \sqrt{2\times6400\times0.7}$

$[700m = 0.7km] =$ 92 km

LEVEL - II -3

1. **The maximum distance between the transmitting and receving TV towers is 72 km . If the ratio of the heights of the TV transmitting tower to receiving tower is 16 : 25, the heights of the transmitting and receiving towers are**

1) 51.2 m ; 80 m 2) 40 m ; 80 m 3) 80 m ; 125 m 4) 25 m ; 75 m

2. **The maximum distance between the transmitting and receiving TV towers is D. If the heights of both transmitting and receiving towers are doubled then the maximum distance between them becomes**

1) 2D 2) $\sqrt{2}$ D 3) 4D 4) D/2

3. **A T.V. tower has a height of 10 m is in a region of average population density 100 π /km2. The number of people that can recieve the transmission is nearly**

1) 1,28,000 2) 64,000 3) 2,56,000 4) 32,000

4. **At certain distance from a transmitting tower a receiver tower of height 20 m is used to receive direct signal. Another tower is installed beyond the first along the same line of sight to receive the signals from the same transmitter. Its height is 44% more than the first receiving tower. Then the separation between the two receiving towers is**

1) 6.4 km 2) 3.2 km 3) 1.6 km 4) 0.8 km

5. **The TV signals have a band width of 6MHz. The number of TV channels that can be accommodated in a band width 12 GHz is**

1) 2 2) 20 3) 200 4) 2000

6. **A carrier wave of peak voltage 12 V is used to transmit a message signal. The peak voltage of the modulating signal in order to have a modulation index of 75% is**

1) 6 V 2) 9 V 3) 4 V 4) 15 V

7. A modulating signal is a square wave as shown in the figure

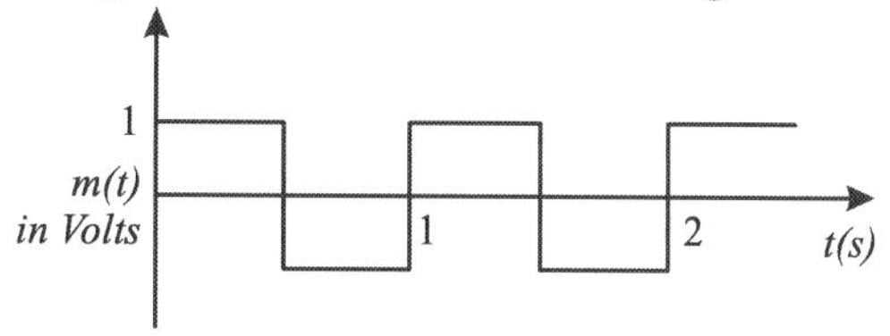

The carrier wave is given by $c(t) = 2\sin(8\pi t)$ *volts*.

The modulation index is

1) 2 2) 0.75 3) 0.5 4) 1.5

8. A audio signal $15\sin 2\pi(1500t)$ amplitude modulates $60\sin 2\pi(10^5)t$. The depth of modulation is

1) 25% 2) 20% 3) 50% 4) 40%

9. An AM wave is given by $V = 1500[1+0.5\sin 12560t]\sin(5.26\times10^5 t)$.

The modulating frequency is

1) 2.0 kHz 2) 1.0 kHz 3) 12.5 kHz 4) 50 kHz

10. The amplitude modulated current is given by $i = 125[1+0.6\sin 2900t]\sin(5.50\times10^5 t)$.

The RMS value of carrier current will be

1) $\frac{125}{\sqrt{2}}A$ 2) $\frac{100}{\sqrt{2}}A$ 3) $\frac{75}{\sqrt{2}}A$ 4) $\frac{50}{\sqrt{2}}A$

11. An audio signal $25\sin 2\pi(1400t)$ amplitude modulates $80\sin 2\pi(10^5 t)$. The two side band frequencies are

1) 98.6 kHz, 101.4 kHz
2) 92.5 kHz, 105.5 kHz
3) 94 kHz, 102.5 kHz
4) 96 kHz, 106 kHz

12. If f_c and f_m are the frequencies of carrier wave and signal, then the band width is

1) f_m 2) $2f_m$ 3) f_c 4) $2f_c$

13. The number of AM broadcast stations that can be accommodated in a 300kHz band width for the highest modulating frequency 15 kHz will be

1) 10 2) 5 3) 7 4) 12

14. A 1000 kHz carrier is simultaneously modulated with 300Hz, 800 Hz and 2 kHz audio waves. The frequencies present in the output are

1) 999.7 kHz, 1000.3kHz, 999.2kHz
2) 1000.8kHz, 998 kHz, 1002kHz
3) 1002.8 kHz, 996kHz, 1106 kHz
4) Both (1) and (2)

15. Depth of modulation in terms of E_{max} and E_{min} is

1) $m_a = \frac{E_{max}+E_{min}}{E_{min}}$
2) $m_a = \frac{E_{max}-E_{min}}{E_{max}}$
3) $m_a = \frac{E_{max}-E_{min}}{E_{max}+E_{min}}$
4) $m_a = \frac{E_{max}+E_{min}}{E_{max}-E_{min}}$

16. **If V_c is amplitude of carrier wave in AM transmitter where modulation factor is m then amplitude of side bands can be**

1) $\frac{V_c}{2m}$ 2) $\frac{m}{2}V_c$ 3) mV_c 4) $\frac{V_c}{2}$

17. **If a carrier wave of amplitude 10mV is modulated by an audio signal of amplitude 2mV, then amplitude of LSF or USF is given by**

1) 0.2 mV 2) 0.5 mV 3) 1 mV 4) 2 mV

18. **The amplitude modulated current is given by $i = 125[1+0.6\sin 2900t]\sin 10^6 t$. The depth of modulation is**

1) 60% 2) 6% 3) 36% 4) 66%

19. **If minimum voltage in an AM wave was found to be 2V and maximum voltage 10V. The % modulation index**

1) 80% 2) 66.67% 3) 64.25% 4) 76.25%

20. **Sinusoidal carrier voltage of frequency 1.5 *MHz* and amplitude 50 *V* is amplitude modulated by sinusoidal voltage of frequency 10 *kHz* producing 50% modulation. The lower and upper side-band frequencies in *kHz* are**

1) 1490, 1510 2) 1510, 1490 3) $\frac{1}{1490}, \frac{1}{1510}$ 4) $\frac{1}{1510}, \frac{1}{1490}$

21. **A transmitter supplies 9 *kW* to the aerial when unmodulated. The power radiated when modulated to 40% is**

1) 5 *kW* 2) 9.72 *kW* 3) 10 *kW* 4) 12 *kW*

22. **The total power content of an AM wave is 1500 *W*. For 100% modulation, the power transmitted by the carrier is**

1) 500 *W* 2) 700 *W* 3) 750 *W* 4) 1000 *W*

23. **The bit rate for a signal, which has a sampling rate of 8 *kHz* and where 16 quantization levels have been used, is**

1) 32000 *bits/sec* 2) 16000 *bits/sec* 3) 64000 *bits/sec* 4) 72000 *bits/sec*

24. **The antenna current of an AM transmitter is 8 *A* when only carrier is sent but increases to 8.96 *A* when the carrier is modulated sinusoidally. The percentage modulation is**

1)50% 2)60% 3)65% 4)71%

25. **In a diode AM-detector, the output circuit consist of $R = 1k$W and $C = 10\,pF$. A carrier signal of 100 *kHz* is to be detected. Is it good**

1) Yes 2) No
3) Information is not sufficient 4) None of these

26. **In an FM system a 7 *kHz* signal modulates 108*MHz* carrier so that frequency deviation is 50 *kHz*. The carrier swing is**

1) 7.143 2) 8 3) 0.71 4) 350

27. **Maximum useable frequency (MUF) in *F*-region layer is *x*, when the critical frequency is 60 *MHz* and the angle of incidence is 70°. Then *x* is**

(1)150 *MHz* 2)170 *MHz* (3)175 *MHz* 4)190 *MHz*

28. **Suppose that the modulating signal is** $m(t)=2\cos(2\pi f_m t)$ **and the carrier signal is** $x_c(t)=A_c\cos(2\pi f_c t)$ **which one of the following is a conventional AM signal without over modulation**

1) $x(t)=A_c m(t)\cos(2\pi f_c t)$
2) $x(t)=A_c(1+m(t))\cos(2\pi f_c t)$
3) $x(t)=A_c\cos(2\pi f_c t)+\frac{A_c}{4}m(t)\cos(2\pi f_c t)$
4) $x_c(t)=A_c\cos(2\pi f_m t)+\cos(2\pi f_c t)+A_c\sin(2\pi f_m t)\sin(2\pi f_c t)$

29. $C(t)$ **and** $m(t)$ **are used to generate an AM signal. The modulation indes of generated AM signal is 0.5. Then the quantity** $\frac{P_{Total\ SB}}{P_{Carrier}}=$

1) 1/8
2) 1/4
3) 2/3
4) 9/8

30. **Which of one of the following modulated signals are recovered up to a scaling factor using envelope detedtor**

1) $20\cos 200\pi t+30m(t)\cos 200\pi t$
2) $20\cos 200\pi t+16m(t)\cos 200\pi t$
3) $10m(t)\cos 400\pi t$
4) None

31. **A message signal** $m(t)=4\cos 2000\pi t$ **modulates the carrier** $C(t)=\cos 2\pi f_c t$ **where** $f_c=1MHz$ **to produce an AM signal. For demodulation using envelope detector the time constant RC should satisfy**

1) $0.5ms<RC<1ms$
2) $1\mu s\ll RC\ll 0.5ms$
3) $RC\gg 1\mu s$
4) $RC\gg 1\mu s$

32. **A 1 MHz carrier signal is modulated by a symmetrical sinusoidal wave for period of 100** μs **in a nonlinear (square law device). Which of the following frequencies will not be present in the modulated signal**

1) 990 kHz
2) 1010 kHz
3) 1020 kHz
4) 1030 kHz

33. **A given AM transmitter develops an unmodulated power out put of 1kW across 50**Ω **resistance. When a message signal of amplitude 5 V is applied on it then the side bands carry 40% of power of carrier. Amplitude of the carrier signal used is**

1) 505.952 V
2) 126.488
3) 252.976
4) 316.22

34. **Let** $m(t)=\cos(4\pi\times10^3 t)$ **and** $C(t)=5\cos(2\pi\times10^{6}t)$ **are the message and carrier signals modulation index is 0.5. What is the efficiency achieved?**

1) 8.33%
2) 11.11%
3) 20 %
4) 25%

LEVEL– II -3- KEY

1) 3	2) 2	3) 1	4) 2	5) 4	6) 2	7) 3	8) 1	9) 1	10) 1
11) 1	12) 2	13) 1	14) 4	15) 3	16) 2	17) 3	18) 1	19) 2	20) 1
21) 2	22) 4	23) 1	24) 4	25) 2	26) 1	27) 3	28) 3	29) 1	30) 2
31) 2	32) 3	33) 4	34) 2						

LEVEL - II -3-HINTS

1. $\frac{h_t}{h_r} = \frac{16}{25}$; $d = \sqrt{2Rh_t} + \sqrt{2Rh_r}$

$72 \times 10^3 = \sqrt{2R\frac{16}{25}h_r} + \sqrt{2Rh_r} \Rightarrow h_r = 125m \;\&\; h_t = 80m$

2. $s = \sqrt{2Rh_T} + \sqrt{2Rh_R}$ when both h_T and h_R are doubled then s becomes $\sqrt{2}$ times.

3. Number of people that can receive
$=$ area $\times$ population density
$= \pi d^2 \times$ population density

4. distance between transmitter and first receiver is x and between transmitter and second receiver is y.

$x = \sqrt{2Rh_T} + \sqrt{2Rh_1}$

$y = \sqrt{2Rh_T} + \sqrt{2R(1.44h_1)}$

find y -x

5. Number of channels $= \frac{12 \times 10^9}{6 \times 10^6} = 2000$

6. $\mu = \frac{A_m}{A_c} \Rightarrow \frac{3}{4} = \frac{A_m}{12}$

7. $\mu = \frac{A_m}{A_c} = \frac{1}{2} = 0.5$

8. $\mu = \frac{A_m}{A_c} \times 100 = \frac{15}{60} \times 100 = 25\%$

9. $C_m(t) = A_c(1 + \mu \sin \omega_m t)\sin \omega_c t$

$f_m = \frac{\omega_m}{2\pi} = \frac{12560}{2\pi} = 2khz$

10. $I_{rms} = \frac{I_{peak}}{\sqrt{2}} = \frac{125}{\sqrt{2}}$

11. $LSB = f_c - f_m = 10^5 - 1400 = 98.6khz$

$USB = f_c + f_m = 10^5 + 1400 = 101.4khz$

12. band width $= (f_c + f_m) - (f_c - f_m) = 2f_m$

13. $n = \frac{\text{band width}}{\text{signal band width}} = \frac{300}{2 \times 15} = 10$

14. frequencies in output are $f_c - f_m \;\&\; f_c + f_m$

15. $E_{max} = A_c + A_m$

$E_{min} = A_c - A_m \Rightarrow \mu = \frac{A_m}{A_c} = \frac{E_{max} - E_{min}}{E_{max} + E_{min}}$

16. Amplitude of LSB = USB $= \mu \frac{A_c}{2}$

17. Amplitude of LSB = USB $= \mu \frac{A_c}{2} = \frac{A_m}{A_c} \times \frac{A_c}{2} = \frac{A_m}{2}$

18. $\mu = 0.6 \times 100 = 60\%$

19. $\mu = \frac{10-2}{10+2} \times 100$

20. (1) Here, $f_c = 1.5\ MHz = 1500\ kHz$, $f_m = 10\ kHz$

\ Low side band frequency

$= f_c - f_m = 1500\ kHz - 10\ kHz = 1490\ kHz$

Upper side band frequency

$= f_c + f_m = 1500\ kHz + 10\ kHz = 1510\ kHz$

21. (2) $P_t = P_c\left[1 + \frac{m^2}{2}\right] = 9\left[1 + \frac{(0.4)^2}{2}\right]$

$= 9\left[1 + \frac{0.16}{2}\right]$ (Q $m = 40\% = 0.4$)

$= 9\ (1.08) = 9.72\ kW$

22. (4) $\frac{P_t}{P_c} = 1 + \frac{m^2}{2}$ or $P_c = P_t\left[\frac{2}{2+m^2}\right]$

\ $P_c = 1500\left[\frac{2}{2+1}\right]$ $\therefore\ m = 100\% = 1 = 1000\ W$

23. (1) If n is the number of bits per sample, then number of quantisation level $= 2^n$ Since the number of quantisation level is 16

$\Rightarrow 2^n = 16$ Þ $n = 4$

bit rate = sampling rate × no. of bits per sample = 8000 × 4 = 32,000 *bits/sec.*

24. (4) We know that $\left(\frac{I_t}{I_c}\right)^2 = 1 + \frac{m^2}{2}$

Here, $I_t = 8.96A$ and $I_c = 8A$

$\left(\frac{8.96}{8}\right)^2 = 1 + \frac{m^2}{2}$ or $1.254 = 1 + \frac{m^2}{2}$

or $\frac{m^2}{2} = 0.254$ or $m^2 = 0.508$ or $m = 0.71 = 71\%$

25. (2)For demodulation $\frac{1}{f_c} << RC$

$\frac{1}{f_c} = \frac{1}{100 \times 10^3} = 10^{-5}\ s$

$RC = 10^3 \times 10 \times 10^{-12}\ s = 10^{-8}\ s$

We see that $\frac{1}{f_c}$ here is not less than RC as required by the above condition. Hence, this is not good.

26. (1) Carrier swing

$$=\frac{\text{Frequency deviation}}{\text{Modulating frequency}}=\frac{50}{7}=7.143$$

27. (3) $MUF=\frac{f_c}{\cos\theta}=\frac{60}{\cos 70°}=175\ MHz$

28. Convention AM signal withour over modulation means m<1

$$x(t)=A_c(1+m(t))\cos(2\pi f_c t)$$

For option C $x(t)=A_c\left(1+\frac{1}{4}m(t)\right)\cos(2\pi f_c t);\ \therefore m=\frac{1}{4}$

29. $\frac{P_{Total\ SB}}{P_{Carrier}}=\frac{P_c\left(m^2/2\right)}{P_c}=\frac{1}{8}$

30. For option 1 $m=\frac{30}{20}=1.5$ m>1

For option 2 $m=\frac{16}{20}=0.8$ m<1

For option 3 signal contains only side bands so message cannot be detected.

31. $f_c=1\ MHz$ and $f_m=2000Hz$

$$\frac{1}{f_c}\ll RC\ll\frac{1}{f_m}$$

32. The frequency components of modulation signal are from nonlinear device are

$f_c,\ f_c\pm f_m,\ f_c\pm 3f_m....$

So the frequencies present in the AM are 100 kHz, 990 kHz, 1010 kHz, 970 kHz, 1030 kHz

33. $\frac{A_C^2}{2R}=1kW\Rightarrow\frac{A_c^2}{2\times 50}=10^3$

34. $\eta=\frac{m^2}{2+m^2}$

LEVEL - II -4

1. **The maximum distance between the transmitting and receving TV towers is 64 km . If the ratio of the heights of the TV transmitting tower to receiving tower is $4:9$, the heights of the transmitting and receiving towers are**

1) 51.2 m ; 80 m 2) 1280 m ; 2880 m 3) 80 m ; 125 m 4) 25 m ; 75 m

2. **The maximum distance between the transmitting and receiving TV towers is D. If the heights of both transmitting and receiving towers are halved then the maximum distance between them becomes**

1) 2D 2) $D/\sqrt{2}$ 3) 4D 4) D/2

3. **A T.V. tower has a height of 5 m is in a region of average population density 100π /km2. The number of people that can recieve the transmission is nearly**

1) 1,28,000 2) 64000 3) 2,56,000 4) 32,000

4. At certain distance from a transmitting tower a receiver tower of height 180 m is used to receive direct signal. Another tower is installed beyond the first along the same line of sight to receive the signals from the same transmitter. Its height is 21% more than the first receiving tower. Then the separation between the two receiving towers is

1) 6.4 km 2) 3.2 km 3) 1.6 km 4) 4.8 km

5. The TV signals have a band width of 4MHz. The number of TV channels that can be accommodated in a band width 16 GHz is

1) 2 2) 20 3) 200 4) 4000

6. A carrier wave of peak voltage 10 V is used to transmit a message signal. The peak voltage of the modulating signal in order to have a modulation index of 80% is

1) 6 V 2) 9 V 3) 8 V 4) 15 V

7. A modulating signal is a square wave as shown in the figure

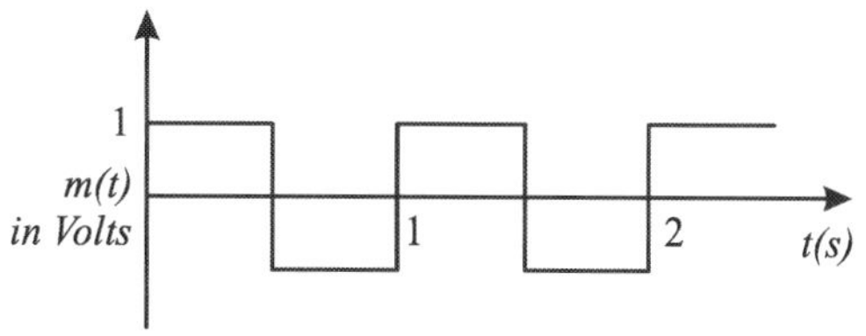

The carrier wave is given by . $c(t) = 1.33\sin(8\pi t)$ *volts* The modulation index is

1) 2 2) 0.75 3) 0.5 4) 1.5

8. A audio signal $20\sin 2\pi(1500t)$ amplitude modulates $40\sin 2\pi(10^5)t$. The depth of modulation is

1) 25% 2) 20% 3) 50% 4) 40%

9. An AM wave is given by $V = 1500[1+0.5\sin 62800t]\sin(5.26\times10^5 t)$.

The modulating frequency is

1) 20 kHz 2) 10 kHz 3) 12.5 kHz 4) 50 kHz

10. The amplitude modulated current is given by $i = 75[1+0.3\sin 2003t]\sin(7\times10^5 t)$.

The RMS value of carrier current will be

1) $\frac{125}{\sqrt{2}}A$ 2) $\frac{100}{\sqrt{2}}A$ 3) $\frac{75}{\sqrt{2}}A$ 4) $\frac{50}{\sqrt{2}}A$

11. An audio signal $5\sin 2\pi(1600t)$ amplitude modulates $20\sin 2\pi(2\times10^5 t)$. The two side band frequencies are

1) 98.4 kHz, 101.6 kHz 2) 92.5 kHz, 105.5 kHz

3) 32.1 kHz 31.59 kHz 4) 96 kHz, 106 kHz

12. If $2f_c$ and $3f_m$ are the frequencies of carrier wave and signal, then the band width is

1) f_m 2) $6f_m$ 3) f_c 4) $2f_c$

13. The number of AM broadcast stations that can be accommodated in a 140kHz band width for the highest modulating frequency 10 kHz will be

1) 10 2) 5 3) 7 4) 12

14. A 2000 kHz carrier is simultaneously modulated with 600Hz, 800 Hz and 4 kHz audio waves. The frequencies present in the output are

1) 2000.6 kHz,2000.8 kHz,1996 kHz 2) 1999.4 kHz, 2004 kHz, 1999.2kHz
3) 1002.8 kHz, 996kHz, 1106 kHz 4) Both (1) and (2)

15. If a carrier wave of amplitude 20mV is modulated by an audio signal of amplitude 4mV, then amplitude of LSF or USF is given by

1) 0.2 mV 2) 0.5 mV 3) 1 mV 4) 2 mV

16. The amplitude modulated current is given by $i = 70[1 + 0.06\sin 2900t]\sin 10^6 t$. The depth of modulation is

1) 60% 2) 6% 3) 36% 4) 66%

17. If minimum voltage in an AM wave was found to be 1.1 V and maximum voltage 10V. The % modulation index

1) 80.2% 2) 66.67% 3) 64.25% 4) 76.25%

18. Sinusoidal carrier voltage of frequency 3 *MHz* and amplitude 50 *V* is amplitude modulated by sinusoidal voltage of frequency 10 *kHz* producing 50% modulation. The lower and upper side-band frequencies in *kHz* are

1) 1490, 1510 2) 3010, 2990 3) $\frac{1}{1490}, \frac{1}{1510}$ 4) $\frac{1}{1510}, \frac{1}{1490}$

19. A transmitter supplies 6 *kW* to the aerial when unmodulated. The power radiated when modulated to 60% is

1) 5 *kW* 2) 9.72 *kW* 3) 7.08 *kW* 4) 12 *kW*

20. The total power content of an AM wave is 6000 *W*. For 100% modulation, the power transmitted by the carrier is

1) 500 *W* 2) 700 *W* 3) 750 *W* 4) 4000 *W*

21. The bit rate for a signal, which has a sampling rate of 8 *kHz* and where 32 quantization levels have been used, is

1) 32000 *bits/sec* 2) 40000 *bits/sec* 3) 64000 *bits/sec* 4) 72000 *bits/sec*

22. The antenna current of an AM transmitter is 10 *A* when only carrier is sent but increases to 10.1 *A* when the carrier is modulated sinusoidally. The percentage modulation is

1)50% 2)60% 3) 200% 4)71%

23. In a diode AM-detector, the output circuit consists of $R = 1k\Omega$ and $C = 10\,pF$. A carrier signal of 1000 *MHz* is to be detected. Is it good

1) Yes 2) No 3) Information is not sufficient 4) None of these

24. In an FM system a 9 *kHz* signal modulates 108*MHz* carrier so that frequency deviation is 60 *kHz*. The carrier swing is

1) 7.143 2) 8 3) 6.67 4) 350

25. Maximum useable frequency (MUF) in *F*-region layer is *x*, when the critical frequency is 60 *MHz* and the angle of incidence is 60°. Then *x* is

1)150 *MHz* 2)170 *MHz* 3)120 *MHz* 4)190 *MHz*

LEVEL -II -4- KEY

1) 2	2) 2	3) 2	4) 4	5) 4	6) 3	7) 2	8) 3	9) 2	10) 3
11) 1	12) 2	13) 3	14) 4	15) 4	16) 2	17) 1	18) 2	19) 3	20) 4
21) 2	22) 3	23) 1	24) 3	25) 3					

LEVEL - II -4-HINTS

1. $d=\sqrt{2Rh_T}+\sqrt{2Rh_R}$; $\frac{h_T}{h_R}=\frac{4}{9};d=64km$

2. $D=\sqrt{2Rh_T}+\sqrt{2Rh_R}$

3. $A=\pi D^2=\pi\left[2Rh_2\right]$

   $n=\left(Population\,density\right)A$

4. $d=\sqrt{2Rh_2}-\sqrt{2Rh_1}$; $h_2=h_1-21\%$ of h_1

5. $n=\frac{16GHz}{4MHz}$

6. Modulation index

   $=\frac{\text{Amplitude of modulating signal}}{\text{Amplitude of carrier wave}}$

7. $m=\frac{a_m}{a_c}$

8. depth $=\frac{a_c-a_m}{a_c}\times 100$

9. $f_m=\frac{w_m}{2\pi}$

10. $i_{rms}=\frac{i_0}{\sqrt{2}}$

11. $LSB=f_c-f_m=10^5-1600=98.4KHz$

    $USB=f_c+f_m=10^5+1600=101.6KHz$

12. Frequencies in out put are $2f_c-3f_m$ and $2f_c+3f_m$

13. $n=\frac{band\ width}{signal\ band\ with}=\frac{140}{2\times 10}=7$

14. Frequencies in out put are f_c-f_m and f_c+f_m

15. Amplitude of $L.S.B=U.S.B=\mu\frac{A_c}{2}=\frac{A_m}{A_c}.\frac{A_c}{2}=\frac{A_m}{2}$

16. Depth of modulation $=0.06\times 100=6\%$

17. $\mu=\frac{10-1.1}{10+1.1}\times 100=80.2\%$

18. f_c-f_m and f_c+f_m

19. $P_t = P_c\left[1+\frac{m^2}{2}\right] = (6KW)\left[1+\frac{(0.6)^2}{2}\right]$

20. $P_c = P_t\left[\frac{2}{2+m^2}\right]$

21. No. of quantization levels = 2^n

bit rate = sampling power x no. of bits per sample

22. $\left(\frac{I_t}{I_c}\right)^2 = 1+\frac{m^2}{2}$

23. For demodulation $\frac{1}{f_c} << RC$

24. Carrier swing $= \frac{frequency\ deviation}{Modulating\ frequency}$

25. $MUF = \frac{f_c}{\cos\theta}$

LEVEL - III

1. **The tuned circuit of an oscillator in a simple AM transimitter employs a 250 micro henry coil and 1nF condenser. If the oscillator output is modulated by audio frequency upto 10KHz, the frequency range occupied by the side bands in KHz is**

1) 210 to 230 2) 258 to 278 3) 308 to 328 4) 118 to 128

2. **A TV tower has a height of 70m. If the average population density around the tower is $1000 km^{-2}$, the population covered by the TV tower**

1) 2.816×10^6 2) 2.86×10^9 3) 2.816×10^3 4) 2.816×10^{12}

3. **A carrier wave is modulated by a number of sine waves with modulation indices 0.1, 0.2, 0.3. The total modulation index (m) of the wave is**

1) 0.6 2) 0.2 3) $\sqrt{0.14}$ 4) $\sqrt{0.07}$

4. **The maximum peak-to-peak voltage of an AM wave is 16 mV and the minimum peak-to-peak voltage is 4mV. The modulation factor is equal to**

1) 0.6 2) 0.3 3) 0.8 4) 0.25

5. **An AM wave is exressed as $e = 10(1+0.6\cos 2000\pi t)\cos 2\times10^8 \pi t$ volts, the minimum and maximum values of modulated carrier wave are**

1) 10 V, 20 V 2) 4 V, 8 V 3) 16 V, 4 V 4) 8 V, 20 V

6. **The audio signal voltage is given by $V_m = 2\sin 12\pi\times10^3 t$. The band width and LSB if carrier wave has a frequency** $3.14\times10^6\ rad/s$

1) 12 KHz; 494 KHz 2) 6 KHz; 313 KHz

3) 6 KHz; 494 KHz 4) 18 KHz; 494 KHz

7. **A TV transmission tower at a particular station has a height of 160m. Radius of earth is 6400km**

   i. The range it covers is 45255 m

   ii. The population that it covers is77.42 lakhs. When population density is $1200 km^{-2}$

   iii. The height of antenna should be increased by 480 m to double the coverage range

   1) i and ii are true 2) ii and iii are true 3) i and iii are true 4) i, ii and iii are true

8. **The tuned circuit of the oscillator in a simple AM transmitter employs a 40 μH coil and 12 nanofarad(nF) capacitor. If the oscillator output is modulated by audio frequency of 5 kHz,Which of the following frequencies doesn't appear in the out put AM?**

   1) $f_{USB} = 225$ KHz 2) $f_{USB} = 235$ kHz 3) f_c=230 kHz 4) $f_c = 235$ kHz

9. **A 400 watt carrier is modulated to a depth of 80%. Caluculate the total power in the modulated wave.**

   1) 528 W 2) 128 W 3) 256 W 4) 400 W

10. **Caluculate modulation index if carrier waves is modulated by three signals with modulation indices as 0.6, 0.3 and 0.4**

    1) 1.0 2) 0.70 3) 0.78 4) 1.3

11. **A 1000 KHz carrier is simoultaneously modulated with f_{m1}= 300 Hz, f_{m2}= 800 Hz and f_{m3}= 1 KHz audio sine waves. What will be the frquencies present in the output?**

    a) $f_{LSB1} = 999.7 KHz$, $f_{USB1} = 1000.3 KHz$

    b) $f_{LSB2} = 999.2 KHz$, $f_{USB2} = 1000.8 KHz$

    c) $f_{LSB3} = 999 KHz$, $f_{USB3} = 1001 KHz$

    d) $f_{LSB3} = 990 KHz$, $f_{USB3} = 1010 KHz$

    1) a only 2) b, c and d 3) a, b and c 4) a, c only

LEVEL-III - KEY

1) 3 2) 1 3) 3 4) 1 5) 3 6) 1 7) 4 8) 2 9) 1 10) 3
11) 3

LEVEL-III - HINTS

1. $f_c = \frac{1}{2\pi\sqrt{LC}} = 318 khz,\ f_m = 10 khz$

   $LSB = f_c - f_m = 308$

   $USB = f_c + f_m = 328$

2. Population = Population density × Area

   = Population density × $\pi \times 2Rh$

3. $m = \sqrt{m_1^2 + m_2^2 + m_3^2}$

4. $\mu = \frac{(16-4)/2}{(16+4)/2}$

5. $E_{max} = (1+\mu)E_c$

   $E_{min} = (1-\mu)E_c$

6. band width $= 2f_m$

LSB $= f_c - f_m$

7. $d = \sqrt{2Rh}$

Population = Population density$\times \pi d^2$

$d \alpha \sqrt{h}$

9. $P_{Total} = P_C\left(1+\dfrac{m^2}{2}\right) = 528$ watts

10. $m_t = \sqrt{m_1^2 + m_2^2 + m_3^2} = 0.78$

LEVEL - IV

1. **Consider telecommunication through optical fibres. Among the following statements one is not true**

1)optical fibers can be of graded refractive index

2)optical fibers are subjected to electromagnetic interference from out side

3)optical fibers have extremely low transmission loss

4)optical fibers have homogeneous core with a suitable cladding

2. **A 600 W carrier is modulated to a depth of 75% by a 400Hz sine wave. The total antenna power is**

1) 769 W   2) 796 W   3) 679 W   4)637.5 W

3. **The antenna current of an AM transmitter is 8A when only the carrier is sent, but it increases to 8.93A when the carrier is modulated by a single wave. The percentage modulation is**

1) 60.1%   2) 70.1%   3) 80.1%   4) 50.1%

4. **The antenna current is 8A only when only carrier wave is transmitted and 9.6A when AM wave is transmitted. If carrier power is 10kw, the modulating power is**

1) 14.4 kW   2) 1.2 kW   3) 2 kW   4) 4.4 kW

Passage - I

A message signal of frequency 10kHz and peak voltage 10V is used to modulate a carrier wave of frequency 1MHz and peak voltage 20volts.

5. **The modulation index is**

1) 0.2   2) 0.3   3) 0.5   4) 0.8

6. **Among the following are USB and LSB are**

1) 1010kHz, 990 kHz   2) 990 kHz, 1010kHz   3) 950 kHz, 1050kHz   4) 1050kHz, 990 kHz

7. **The bandwidth of signal is**

1) 10kHz   2) 20 kHz   3) 30 kHz   4) 40 kHz

Passage - II

8. **When only carrier is transmitted antenna current observed is 8A when it is modulated with 500 Hz sine wave antenna current becomes 9.6A. The percentage of modulation is**

1) 80%   2) 20%   3) 94.26%   4) 83.76%

9. **1% of 10^{12} Hz of a satellite link was used for telephony. The number of channels or subscribers if each channel is of 8 kHz are**

1) 2.5×10^7 2) 1.25×10^6 3) 2.5×10^8 4) 1.25×10^8

10. **A carrier is modulated simultaneously by 3 sine waves of modulation indices 0.3, 0.4 and 0.45 respectively. The net modulation index is**

1) 1.15 2) 0.67 3) 0.57 4) 0.84

11. **A 1 kW carrier is modulated to a depth of 80%. The total power in the modulated wave is**

1) 1.32 2) 1.56 3) 1.84 4) 1.96

Passage - III

The expression for an AM voltage is $e = 5(1+0.6\cos 1000\pi t)\cos 5\times10^6\pi t$

12. **The peak value of carrier wave is**

1) 30 V 2) 6 V 3) 3 V 4) 5 V

13. **The percentage of modulation is**

1) 0.6% 2) 6% 3) 60% 4) 66%

14. **Frequency of audio signal is**

1) 1000 Hz 2) 5000 Hz 3) 500Hz 4) 250 Hz

15. **Frequency of carrier signal is**

1) $5\times10^6\,Hz$ 2) 1000Hz 3) $2.50\times10^3\,Hz$ 4) $2.50\times10^6\,Hz$

16. **Frequencies of sideband is**

1) 2.5005×10^6 Hz, 2.4995×10^6 Hz 2) 2.505×10^6 Hz, 2.495×10^6 Hz

3) 2.505×10^6 kHz, 2.495×10^6 kHz 4) 2.505 MHz, 2.495kHz

Passage - IV

An amplitude modulated voltage is expressed as $e = 10(1+0.8\cos 2000\pi t)\cos 3\times10^6\pi t\ volt$

17. **The peak value of carrier wave is**

1) 5 V 2) 8 V 3) 10 V 4) 100 V

18. **The depth of modulation is**

1) 8 2) 0.8 3) 800 4) 80

19. **The frequency of carrier wave is**

1) $3\times10^6\,Hz$ 2) $1.5\times10^3\,Hz$ 3) $3\times10^3\,Hz$ 4) $1.5\times10^6\,Hz$

20. **The minimum and maximum value of modulated carrier wave is**

1) 2V, 18V 2) 4V, 16V 3) 2V, 8V 4) 4V, 12V

21. **The frequency of modulating signal is**

1) 1000 Hz 2) 100Hz 3) 10MHz 4) 100 MHz

22. **The bandwidth of modulated signal is**

1) 1000 Hz 2) 2000 Hz 3) 1500 Hz 4) 2500 Hz

ASSERTION AND REASON

Instructions for the following questions:

In each of the following questions a statement of Assertion (A) is given followed by corresponding statement of Reason (R) just below it. Of the statements mark the correct answers.

1) A and R are true and R is the correct explanation of A

2) A and R are true but R is not correct explanation of A

3) A is true but R is false **4) A is false but R is true.**

23. **A:** It is not necessary for a transmitting antenna to be at the same height as that of receiving antenna for line-of sight communication

R: If the signal is to be received beyond the horizon then the receiving antenna must be high enough to intercept the line-of sight waves

24. **A:** Low frequency audio signals cannot be transmitted directly over long distances

R: To transmit low frequency audio signals over long distances, the audio signals are superimposed on a high frequency carrier signal

25. **A:** The microphone acts as a transducer.

R: In radio communication, microphone is used to convert sound signal into electrical signal.

26. **A:** A FM signal is less susceptible to noise than an AM signal.

R: In FM transmission, the message signal is in the form of frequency variations of carrier waves. During modulation process, the noise gets amplitude modulated.

27. **A:** Artifical satellites are not necessary for long distance TV transmission.

R: The TV signals are not reflected by the ionosphere. They are reflected back to the earth from an earth station by making use of artifical satellites.

28. **A:** Greater the height of a TV transmitting antenna, greater is its coverage.

R: The height of transmitting antenna is directly proportional to the square of the distance of TV coverage

29. **A:** Long distance radio broadcasts use short wave bands.

R: The short waves are absorbed by the earth but reflected by F layer in ionosphere. After reflection, they reach the surface of earth back only at a large distance from the transmitter.

30. **A:** TV signals cannot be transmitted via sky wave propagation.

R: Ionosphere is unable to reflect radio waves of frequencies greater than 40MHz.

MATRIX MATCHING

31. **Match the frequency band with the type of use**

Frequency Band	**Type of use**
a) LF	e) Radio Broad casting
b) HF	f) Marine and navigational aid
c) VHF	g) Satellite communication
d) SHF	h) TV Broad casting

1) a - e; b - f; c - h; d - g 2) a - f; b - e; c- g; d- h

3) a - f; b - e; c - h; d - g 4) a - e; b - g; c - f; d - h

32. **Match layer of ionosphere and height of the layer from surface of earth for atmosphere**

a) D	e) 250 - 400 km
b) E	f) 100 km

c) F g) 65 - 75 km

d) F_2 h) 170 - 190 km

1) a-g; b-f; c-h; d-e 2) a-h; b-g; c-e; d-f 3) a-g; b-h; c-f; d-e 4) a-g; b-h; c-e; d-f

33. Match layer of ionosphere and frequencies most affected

a) D e) helps surface waves and reflects HF

b) E f) efficiently reflects HF waves

c) F g) partially absorbs HF

d) F_2 h) reflects LF and absorbs MF

1) a-g; b-f; c-h; d-e 2) a-h; b-e; c-g; d-f

3) a-h; b-e; c-f;d -g 4) a-e; b-h; c-g; d-f

34. Match the frequency band with the type of use service frequency band

a) AM broadcast e) 88 - 108 MHz

b) satellite communication f) 896 - 935 MHz

c) FM broadcast g) 540 - 1600 kHz

d) Cellular mobile h) 3.7 - 6.5 GHz

1) a-g; b-h; c-f; d-e 2) a-h; b-g; c-e; d-f

3) a-f; b-h; c-e; d-g 4) a-g; b-h; c-e; d-f

35. Match List-1 with List-2

List-1	**List-2**
Communication mode	Example
a) Point to point communication	e) RADAR
b) broadcast communication	f) AM Radio
c) Line of sight communication	g) FM Radio
d) Satellite communication	h) Traditional telephony
	i) Mobile telephony
	j) TV

1) a-h; b-f,g,j; c-e;d-i,j 2) a-e; b-i,j; c-h;d-g,f

3) a-f ; b-h,i,j; c-g;d-e 4) a-g; b-f,g,j; c-e;d-i,j

36. Match List-1 with List-2

List-1	**List-2**
Name of device	**use**
a) Antenna	e) sends out information
b) Transmitter	f) picksup information
c) Receiver	g) converts energy in one form to another form
d) Transducer	h) radiates signal
	i) recieves signal

1) a-e; b-g,h;c-f;d-h 2) a-h,i;b-e;c-f;d-g 3) a-f; b-e;c-f,i;d-h 4) a-h;b-g,i;c-f;d-e

37. Match List-1 with List-2

List-1	List-2
Name	Frequencies most affected
a) Part of stratosphere(D)	e) VHF(upto several GHz)
b) Part of mesophere (F_1)	f) reflects LF
c) Part of thermosphere(F_2)	g) Partially absorbs HF
d) Troposphere	h) Efficiently reflects HF

1) a-e;b-f,g,h;c-i;d-h 2) a-f;b-f,g,h;c-f;d-e 3) a-f,g,h;b-g;c-i;d-e 4) a-g;b-f,g,h;c-f;d-e

38. Match List-1 with List-2

List-1	**List-2**
Type of propagation	Frequencies
a) sky waves	e) 1.5 MHz
b) space wave	f) 20 MHz
c) ground wave	g) 30MHz
d) micro wave	h) 50MHz
	i) 3 GHz

1) a-e,i;b-e;c-h;d-i 2) a-f,g;b-h;c-e;d-i 3) a-i,g;b-e;c-h;d-f 4) a-g;f-h;c-i,b;d-e

LEVEL - IV - KEY

1) 2	2) 1	3) 2	4) 4	5) 3	6) 1	7) 2	8) 3	9) 2	10) 2
11) 1	12) 4	13) 3	14) 3	15) 4	16) 1	17) 3	18) 2	19) 4	20) 1
21) 1	22) 2	23) 1	24) 1	25) 1	26) 1	27) 4	28)1	29) 1	30) 1
31) 3	32) 1	33) 2	34) 4	35) 1	36) 2	37) 3	38) 2		

LEVEL - IV - HINTS

2. $P_t = P_c\left(1+\frac{\mu^2}{2}\right) = 600\left[1+\frac{(0.75)^2}{2}\right] = 768.5W$

3. $\left(\frac{i_t}{i_c}\right)^2 = 1+\frac{\mu^2}{2} \Rightarrow \left(\frac{8.93}{8}\right)^2 = 1+\frac{\mu^2}{2}$

$\Rightarrow \mu = 0.701 = 70.1\%$

4. $\frac{P_t}{P_c} = \left(\frac{i_t}{i_c}\right)^2 \Rightarrow \frac{P_t}{10} = \left(\frac{9.6}{8}\right)^2 = 14.4kw$

$\therefore P_m = 14.4-10 = 4.4\ kW$

5. $\mu = \frac{10}{20}$

6. $USB = f_m + f_c$

$LSB = f_m - f_c$

7. bandwidth of signal $= 2f_m$

8. $\frac{i_t}{i_c} = \sqrt{1+\frac{\mu^2}{2}}$

9. 1% of 10^{12} Hz $= 10^{10}$ Hz

Number of channels $= \frac{10^{10}}{8\times10^3} = 1.25\times10^6$

10. $m = \sqrt{m_1^2 + m_2^2 + m_3^2}$

11. $P_t = P_c\left[1+\frac{\mu^2}{2}\right]$

12. eqn of AM voltage is given by

$e = A_c(1+\mu\sin\omega_m t)\sin\omega_c t$

$\therefore A_c = 5\text{ v}$

13. $\mu = 0.6\times100 = 60\%$

14. $f_m = \frac{\omega_m}{2\pi} = \frac{1000\pi}{2\pi} = 500$

15. $f_c = \frac{\omega_c}{2\pi} = \frac{5\times10^6\pi}{2\pi}$

16. $USB = f_c + f_m$

$LSB = f_c - f_m$

17. eqn of AM voltage is given by

$e = A_c(1+\mu\sin\omega_m t)\sin\omega_c t$

$\therefore A_c = 10\text{ V}$

18. $\mu = 0.8$

19. $f_c = \frac{\omega_c}{2\pi} = \frac{3\times10^6\pi}{2\pi}$

20. $E_{min} = (1-\mu)A_c$, $E_{max} = (1+\mu)A_c$

21. $f_m = \frac{\omega_m}{2\pi} = \frac{2000\pi}{2\pi}$

22. band width $= 2f_m$

Made in United States
Orlando, FL
06 November 2026

90301604R00304